The Mycota

Edited by
K. Esser

The Mycota

I *Growth, Differentiation and Sexuality*
1st edition ed. by J.G.H. Wessels and F. Meinhardt
2nd edition ed. by U. Kües and R. Fischer
3rd edition ed. by J. Wendland

II *Genetics and Biotechnology*
1st and 2nd edition ed. by U. Kück
3rd edition ed. by J.P. Benz and K. Schipper

III *Biochemistry and Molecular Biology*
1st and 2nd edition ed. by R. Brambl and G. Marzluf
3rd edition ed. by D. Hoffmeister

IV *Environmental and Microbial Relationships*
1st edition ed. by D. Wicklow and B. Söderström
2nd edition ed. by C.P. Kubicek and I.S. Druzhinina
3rd edition ed. by I.S. Druzhinina and C.P. Kubicek

V *Plant Relationships*
1st edition ed. by G. Carroll and P. Tudzynski
2nd edition ed. by H.B. Deising

VI *Human and Animal Relationships*
1st edition ed. by D.H. Howard and J.D. Miller
2nd edition ed. by A.A. Brakhage and P.F. Zipfel

VII *Systematics and Evolution*
1st edition ed. by D.J. McLaughlin, E.G. McLaughlin, and P.A. Lemke†
2nd edition ed. by D.J. McLaughlin and J.W. Spatafora

VIII *Biology of the Fungal Cell*
1st and 2nd edition ed. by R.J. Howard and N.A.R. Gow
3rd edition ed. by D. Hoffmeister and M. Gressler

IX *Fungal Associations*
1st and 2nd edition ed. by B. Hock

X *Industrial Applications*
1st edition ed. by H.D. Osiewacz
2nd edition ed. by M. Hofrichter

XI *Agricultural Applications*
1st and 2nd edition ed. by F. Kempken

XII *Human Fungal Pathogens*
1st edition ed. by J.E. Domer and G.S. Kobayashi
2nd edition ed. by O. Kurzai

XIII *Fungal Genomics*
1st edition ed. by A.J.P. Brown
2nd edition ed. by M. Nowrousian

XIV *Evolution of Fungi and Fungal-Like Organisms*
Ed. by S. Pöggeler and J. Wöstemeyer

XV *Physiology and Genetics: Selected Basic and Applied Aspects*
1st edition ed. by T. Anke and D. Weber
2nd edition ed. by T. Anke and A. Schüffler

The Mycota

A Comprehensive Treatise on Fungi as Experimental Systems for Basic and Applied Research

Edited by K. Esser

II

Genetics and Biotechnology
3rd Edition

Volume Editors:
J. Philipp Benz and Kerstin Schipper

Springer

Editors

J. Philipp Benz
TUM School of Life Sciences Weihenstephan
Holzforschung München - Fungal Biotechnology in Wood Science
Technische Universität München
Freising, Germany

Kerstin Schipper
Institut für Mikrobiologie
Department Biologie
Heinrich-Heine-Universität Düsseldorf
Düsseldorf, Germany

ISBN 978-3-030-49923-5 ISBN 978-3-030-49924-2 (eBook)
https://doi.org/10.1007/978-3-030-49924-2

This Springer imprint is published by the registered company Springer Nature Switzerland AG.
The registered company address is: Gewerbestrasse 11, 6330 Cham, Switzerland

Series Preface

Mycology, the study of fungi, originated as a sub discipline of botany and was a descriptive discipline, largely neglected as an experimental science until the early years of this century. A seminal paper by Blakeslee in 1904 provided evidence for self-incompatibility, termed "heterothallism," and stimulated interest in studies related to the control of sexual reproduction in fungi by mating-type specificities. Soon to follow was the demonstration that sexually reproducing fungi exhibit Mendelian inheritance and that it was possible to conduct formal genetic analysis with fungi. The names Burgeff, Kniep and Lindegren are all associated with this early period of fungal genetics research.

These studies and the discovery of penicillin by Fleming, who shared a Nobel Prize in 1945, provided further impetus for experimental research with fungi. Thus, began a period of interest in mutation induction and analysis of mutants for biochemical traits. Such fundamental research, conducted largely with Neurospora crassa, led to the one gene:one enzyme hypothesis and to a second Nobel Prize for fungal research awarded to Beadle and Tatum in 1958. Fundamental research in biochemical genetics was extended to other fungi, especially to *Saccharomyces cerevisiae*, and by the mid-1960s fungal systems were much favored for studies in eukaryotic molecular biology and were soon able to compete with bacterial systems in the molecular arena.

The experimental achievements in research on the genetics and molecular biology of fungi have benefited more generally studies in the related fields of fungal biochemistry, plant pathology, medical mycology, and systematics. Today, there is much interest in the genetic manipulation of fungi for applied research. This current interest in biotechnical genetics has been augmented by the development of DNA-mediated transformation systems in fungi and by an understanding of gene expression and regulation at the molecular level. Applied research initiatives involving fungi extend broadly to areas of interest not only to industry but to agricultural and environmental sciences as well.

It is this burgeoning interest in fungi as experimental systems for applied as well as basic research that has prompted publication of this series of books under the title The Mycota. This title knowingly relegates fungi into a separate realm, distinct from that of either plants, animals, or protozoa. For consistency throughout this series of volumes, the names adopted for major groups of fungi (representative genera in parentheses) are as follows:

Pseudomycota

Division:	Oomycota (*Achlya, Phytophthora, Pythium*)
Division:	Hyphochytriomycota

Eumycota

Division:	Chytridiomycota (*Allomyces*)
Division:	Zygomycota (*Mucor, Phycomyces, Blakeslea*)
Division:	Dikaryomycota
Subdivision:	Ascomycotina
Class:	Saccharomycetes (*Saccharomyces, Schizosaccharomyces*)
Class:	Ascomycetes (*Neurospora, Podospora, Aspergillus*)
Subdivision:	Basidiomycotina
Class:	Heterobasidiomycetes (*Ustilago, Tremella*)
Class:	Homobasidiomycetes (*Schizophyllum, Coprinus*)

We have made the decision to exclude from The Mycota the slime molds which, although they have traditional and strong ties to mycology, truly represent nonfungal forms insofar as they ingest nutrients by phagocytosis, lack a cell wall during the assimilative phase, and clearly show affinities with certain protozoan taxa.

The series throughout will address three basic questions: what are the fungi, what do they do, and what is their relevance to human affairs? Such a focused and comprehensive treatment of the fungi is long overdue in the opinion of the editors.

A volume devoted to systematics would ordinarily have been the first to appear in this series. However, the scope of such a volume, coupled with the need to give serious and sustained consideration to any reclassification of major fungal groups, has delayed early publication. We wish, however, to provide a preamble on the nature of fungi, to acquaint readers who are unfamiliar with fungi with certain characteristics that are representative of these organisms and which make them attractive subjects for experimentation.

The fungi represent a heterogeneous assemblage of eukaryotic microorganisms. Fungal metabolism is characteristically heterotrophic or assimilative for organic carbon and some nonelemental source of nitrogen. Fungal cells characteristically imbibe or absorb, rather than ingest, nutrients and they have rigid cell walls. The vast majority of fungi are haploid organisms reproducing either sexually or asexually through spores. The spore forms and details on their method of production have been used to delineate most fungal taxa. Although there is a multitude of spore forms, fungal spores are basically only of two types: (1) asexual spores are formed following mitosis (mitospores) and culminate vegetative growth, and (2) sexual spores are formed following meiosis (meiospores) and are borne in or upon specialized generative structures, the latter frequently clustered in a fruit body. The vegetative forms of fungi are either unicellular, yeasts are an example, or hyphal; the latter may be branched to form an extensive mycelium.

Regardless of these details, it is the accessibility of spores, especially the direct recovery of meiospores coupled with extended vegetative haploidy, that have made fungi especially attractive as objects for experimental research.

The ability of fungi, especially the saprobic fungi, to absorb and grow on rather simple and defined substrates and to convert these substances, not only into essential metabolites but into important secondary metabolites, is also noteworthy. The metabolic capacities of fungi have attracted much interest in natural products chemistry and in the production of antibiotics and other bioactive compounds. Fungi, especially yeasts, are important in fermentation processes. Other fungi are important in the production of enzymes, citric acid, and other organic compounds as well as in the fermentation of foods.

Fungi have invaded every conceivable ecological niche. Saprobic forms abound, especially in the decay of organic debris. Pathogenic forms exist with both plant and animal hosts. Fungi even grow on other fungi. They are found in aquatic as well as soil environments, and their spores may pollute the air. Some are edible; others are poisonous. Many are variously associated with plants as copartners in the formation of lichens and mycorrhizae, as symbiotic endophytes or as overt pathogens. Association with animal systems varies; examples include the predaceous fungi that trap nematodes, the microfungi that grow in the anaerobic environment of the rumen, the many insect associated fungi, and the medically important pathogens afflicting humans. Yes, fungi are ubiquitous and important. There are many fungi, conservative estimates are in the order of 100,000 species, and there are many ways to study them, from descriptive accounts of organisms found in nature to laboratory experimentation at the cellular and molecular level. All such studies expand our knowledge of fungi and of fungal processes and improve our ability to utilize and to control fungi for the benefit of humankind.

We have invited leading research specialists in the field of mycology to contribute to this series. We are especially indebted and grateful for the initiative and leadership shown by the Volume Editors in selecting topics and assembling the experts. We have all been a bit ambitious in producing these volumes on a timely basis and therein lies the possibility of mistakes and oversights in this first edition. We encourage the readership to draw our attention to any error, omission, or inconsistency in this series in order that improvements can be made in any subsequent edition.

Finally, we wish to acknowledge the willingness of Springer-Verlag to host this project, which is envisioned to require more than 5 years of effort and the publication of at least nine volumes.

Bochum, Germany
Auburn, AL, USA
April 1994

KARL ESSER
PAUL A. LEMKE
Series Editors

Volume Preface to the Third Edition

It is now more than 15 years since the release of the second edition of *The Mycota* Volume II—*Genetics and Biotechnology*—in 2004. Since then, great methodological progress, particularly in the field of molecular biology, has led to substantial developments in these fields. On the one hand, novel techniques based on next generation sequencing have sped up the analysis of genomes and transcriptomes. On the other hand, tremendous advances have been made in the fields of cloning and genome editing, which have led to many novel discoveries and a refined understanding of the genetics and molecular mechanisms underlying cellular processes. At the same time, the relevance of fungi for research and development is unbroken and has only increased in the past, both due to unsolved threats to human health and welfare by fungal pests and pathogens and by the many benefits that fungal biotechnology can offer to the diverse emerging markets and processes that form the basis of the modern bioeconomy. In this respect, we felt that the compilation of a new third edition was timely and highly warranted.

The great amount of novelty in the field is well reflected by the updated nature of this third edition: Out of the 17 chapters, 15 are new additions to the book, while the other two chapters have been completely refurbished. In line with the title of the series, the book is separated into two parts: Part I combines nine contributions to basic research in molecular genetics and Part II eight articles on different topics of fungal biotechnology. Our goal was to compile a broad range of excellent articles on up-to-date topics. Major fungal reference systems, including both yeasts and filamentous fungi like *Neurospora crassa*, *Aspergillus niger*, *Saccharomyces cerevisiae*, *Pichia pastoris*, and *Ustilago maydis*, were considered, but also novel fungi from so far underexplored environments, such as the sea or animal guts, as well as mycorrhizal and wood-degrading fungi.

In Part I, basic genetic processes including chromosome organization and inheritance (chapters "Chromatin Structure and Function in *Neurospora crassa*" and "Origin, Function, and Transmission of Accessory Chromosomes"), the unfolded protein response (chapter "Genetics of the Unfolded Protein Response in Fungi"), the circadian clock (chapter "From Genetics to Molecular Oscillations: The Circadian Clock in *Neurospora crassa*"), and small RNAs (chapter "Small RNAs in Fungi") are discussed. Furthermore, an overview of the latest discoveries regarding self/non-self recognition in fungi (chapter "NLR Function in Fungi as Revealed by the Study of Self/Non-self Recognition Systems") is provided as well as insights into the fascinating world of fungal symbioses with the example of arbuscular mycorrhiza (chapter "Genetics and Genomics Decipher Partner Biology in Arbuscular Mycorrhizas"). Two chapters on the border of basic and applied science conclude this section: A comprehensive summary of the genetics behind secondary metabolism and fungal development (chapter "Coordination of Fungal Secondary Metabolism and Development") and a chapter on the most recent advances in fungal genomics (chapter "Fungal Genomics"), which we have decided to include due to its timeliness and its broad impact on all fields of fungal genetics and biotechnology. For interested readers who want to learn more, we refer to *The Mycota* Volume XIII—*Fungal Genomics*, in which an excellent overview about the many facets of this topic can be found.

In Part II, chapters "Filamentous Fungi as Hosts for Heterologous Production of Proteins and Secondary Metabolites in the Post-Genomic Era", "New Avenues Toward Drug Discovery in Fungi", "Exploiting Fungal Photobiology as a Source of Novel Bio-blocks for Optogenetic Systems", "Yeast Cell Factories", and "Engineering *Saccharomyces cerevisiae* for Production of Fatty Acids and Their Derivatives" highlight recent progress and future prospects in the development of molecular tools and synthetic biology for the engineering and optimization of microbial cell factories. This includes both yeasts and filamentous fungi used for the production of recombinant proteins, fatty acids, and secondary metabolites. The chapter "Fungi Involved in the Biodeterioration and Bioconversion of Lignocellulose Substrates" contains a comprehensive review on the current state of our understanding regarding wood-degrading fungi, while the remaining chapters cover the applied aspects and future potentials of less well-studied fungi, such as from the marine environment or the anaerobic digestive tracts of ruminants (chapters "Biotechnology of Marine Fungi: New Workhorses and Applications" and "The Biotechnological Potential of Anaerobic Gut Fungi").

We sincerely thank all authors for their great contributions. We are well aware that the writing of a book chapter requires an enormous amount of work that is often perceived as non-rewarding without the benefits of an open-access publication. While hoping that open access will become the standard also for book chapters in the near future, this was not possible for this volume, which we regret as editors. However, we are therefore even more grateful for the personal engagement and cooperativity that we experienced from the authors, which allowed us to compile this exciting new edition nevertheless.

We would furthermore like to thank Louise Glass (UC Berkeley) for her assistance during the editing process as well as all the reviewers who helped us to give constructive feedback to the chapters and thereby improve the book.

Unfortunately, this volume will be the last to be published by Karl Esser as the series editor. He passed away just shortly before all manuscripts could be finalized. However, he still managed to personally revise the vast majority of the chapters in this edition, and we are thankful for his motivation and support all along the process—this book has benefited tremendously from his experience and his passion for fungal biology.

Freising, Germany
Düsseldorf, Germany
April 2020

J. Philipp Benz
Kerstin Schipper
Volume Editors

Obituary: Karl Esser 1924–2019

Karl Esser, Emeritus Professor of General Botany and Director of the Botanical Garden at the Ruhr University Bochum, died on December 3, 2019, at the age of 95. Karl Esser was an internationally respected and recognized researcher in fungal biology and genetics. In 1994, he and Paul Lemke (Auburn, Alabama) became the founding editors of the MYCOTA, which was meant to highlight developments in both basic and applied research and to present an overview of fungal systematics and cell biology. Even a few weeks before his death, Karl Esser managed the present volume of this series. He published 4 textbooks and more than 220 original publications in scientific journals. Further, he was editor or member of the editorial board of 14 scientific journals.

Karl Esser graduated from Cologne University with a PhD in biology and biochemistry in 1952. Later on, he was research fellow at the Genetic Institute of Paris and of the Botanical Institute in Cologne. After his habilitation (second PhD) in 1958, he was a research fellow at the Department of Microbiology, Yale University, New Haven, Connecticut, USA. After his return to Germany, he took the chair for General Botany at the Ruhr University and was member of the Founding Committee to build up the newly established Ruhr University. In 1967, he became the first director of the Botanical Garden in Bochum, one of the largest at a German university. On April 1, 1989, he retired from his position as a professor of botany. He was active in many international organizations. To name only a few, he was chairman of the German National Committee for the International Union of Biological Sciences (IUBS), he was member of the Executive Committee of IUBS, he was secretary of the Division of Botany and Microbiology, and finally, he was president of the 14th International Botanical Congress 1987 in Berlin. He received awards from several societies. For example, he got the ONR Award of the Society of Industrial Microbiology USA, and he was a fellow of the American Association for the Advancement of Science (AAAS), the American Academy of Microbiology, and the Society of Industrial Microbiology (SIM, USA). Further, he became honorary member of the Polish and the German Botanical Society, and he received the Docteur honoris causa of the Université d'Orleans, the Université de Toulouse, and the Université de Lille, all in France.

His scientific work focused in general on the genetics of plants and fungi. He was trained as a plant geneticist, but recognized early in his career that fungi are ideal organisms to conduct genetics. Already in 1967, he published a textbook entitled *Genetics of Fungi*, which was for long main literature for his students. In Germany, he was a pioneer of fungal genetics, and he was very highly influential for this discipline. Using filamentous fungi as his main model organisms, he investigated two basic phenomena of biology. In 1958, he set up a system to study fruit body development in fungi using the ascomycete *Sordaria macrospora*. In the late 1970s and the early 1980s, he focused his research work on senescence, using the ascomycete *Podospora anserina*. In 1985, this research was honored through the SANDOZ Prize for Gerontological Research. He was one of the first who detected plasmids in higher fungi, and his group published one of the first papers on the DNA-mediated transformation of filamentous fungi. As an academic teacher, he wrote a very much recognized textbook, where he compared the lifecycle, morphology, and genetics of algae and fungi. The German book entitled *Kryptogamen* was later translated into English and was published by Cambridge University Press in 1982. Karl Esser has a long and distinguished academic career in mycology and was always interested in combining basic with applied research. From this, it follows that his group conducted a lot of projects with filamentous fungi that are important for industrial microbiology.

The research of Karl Esser had a great impact on the scientific careers of many of his students. Thus, many of the internationally recognized projects led by his students have their origin in the laboratory of Karl Esser. To name a few, these are genetics of industrial yeasts (Stahl, Berlin), killer plasmids in yeasts (Meinhardt, Münster), genetics of *Claviceps purpurea*, the producer of the ergot alkaloids (Tudzynski, Münster), senescence in *Podospora anserina* (Osiewacz, Frankfurt), genetics of mitochondria and transposons in fungi (Kempken, Kiel), fungal genomics (Nowrousian, Bochum), autophagy (Pöggeler, Göttingen), fungal cell biology and RNA editing (Teichert, Bochum), and sexual development and secondary metabolism (Kück, Bochum).

Ulrich Kück, Frank Kempken, Friedhelm Meinhardt, Minou Nowrousian, Heinz-Dieter Osiewacz, Stefanie Pöggeler, Ines Teichert, and Paul Tudzynski

Contents

Molecular Genetics

Biotechnology

Editors and Contributors

About the Editors

J. Philipp Benz (born 1977) studied biology at the TU Braunschweig (2003). The research work for his thesis was done in the laboratory of Nigel Crawford at UC San Diego (CA, USA). He subsequently performed his dissertation at the LMU Munich, working in the research group of Jürgen Soll (Department of Biochemistry and Physiology of Plants) (2009). He then returned to the USA to become a postdoctoral fellow in the laboratory of Chris Somerville at the Energy Biosciences Institute (UC Berkeley, CA, USA) where he started to work on filamentous fungi. In 2014, he was appointed assistant professor for Wood Bioprocesses at the Technical University of Munich (TUM School of Life Sciences Weihenstephan, Holzforschung München). In 2020, he was tenured at TUM and advanced to become associate professor for Fungal Biotechnology in Wood Science. His research group is studying the processes used by filamentous fungi to degrade lignocellulosic biomass. Central questions are how fungi perceive the composition of the plant biomass on a molecular level and how they adapt their metabolism to effectively degrade the substrate.

Kerstin Schipper (born 1980) completed her studies in biology at Osnabrück University in 2005. For the research work of her master's thesis, she moved to the laboratory of Heinrich Jung at LMU Munich. She then performed her dissertation and a short postdoctoral time under the supervision of Regine Kahmann at the Max Planck Institute for Terrestrial Microbiology in Marburg in the field of plant–pathogen interactions. Since 2011, she is a group leader at the Institute for Microbiology at Heinrich Heine University Düsseldorf. Her work focuses on fungal biotechnology using the basidiomycete *Ustilago maydis*. Her major interest is a novel unconventional protein secretion pathway and its exploitation for production of heterologous proteins.

Contributors

MARTIN ALTVATER
Department of Biotechnology, University of Natural Resources and Life Sciences Vienna, Vienna, Austria
Austrian Centre of Industrial Biotechnology (ACIB GmbH), Vienna, Austria

HUGO AMOEDO
Institut für Mikrobiologie und Genetik, Abteilung Molekulare Mikrobiologie und Genetik, Göttinger Zentrum für Molekulare Biowissenschaften (GZMB), Georg-August-Universität Göttingen, Göttingen, Germany

LEONIE BAUMANN
Institute of Molecular Biosciences, Faculty of Biological Sciences, Goethe-University Frankfurt, Frankfurt, Germany

PAOLA BONFANTE
Dipartimento di Scienze della Vita e Biologia dei Sistemi, Università di Torino, Torino, Italy

SANDRA BORN
Institute of Molecular Biosciences, Faculty of Biological Sciences, Goethe-University Frankfurt, Frankfurt, Germany

AXEL A. BRAKHAGE
Department of Molecular and Applied Microbiology, Leibniz Institute for Natural Product Research and Infection Biology (HKI), Jena, Germany
Institute of Microbiology, Friedrich Schiller University, Jena, Germany

GERHARD H. BRAUS
Institut für Mikrobiologie und Genetik, Abteilung Molekulare Mikrobiologie und Genetik, Göttinger Zentrum für Molekulare Biowissenschaften (GZMB), Georg-August-Universität Göttingen, Göttingen, Germany

JOSÉ T. CÁNOVAS-MÁRQUEZ
Departamento de Genética y Microbiología, Facultad de Biología, Universidad de Murcia, Murcia, Spain

GENNARO CAROTENUTO
Dipartimento di Scienze della Vita e Biologia dei Sistemi, Università di Torino, Torino, Italy

ZACHARY A. CHASE
Department of Biological Sciences, Rensselaer Polytechnic Institute, Troy, NY, USA

YANFEN CHENG
Laboratory of Gastrointestinal Microbiology, Nanjing Agricultural University, Nanjing, China

ABIGAIL J. COURTNEY
Department of Microbiology, University of Georgia, Athens, GA, USA

Asen Daskalov
The Plant and Microbial Biology Department, University of California, Berkeley, Berkeley, CA, USA
Institut Européen de Chimie et Biologie (IECB), Pessac, France

Verónica Delgado
Millennium Institute for Integrative Biology (iBio), Santiago, Chile
Departamento de Genética Molecular y Microbiología, Facultad de Ciencias Biológicas, Pontificia Universidad Católica de Chile, Santiago, Chile

Veronika Flad
Bavarian State Research Center for Agriculture (LfL), Central Department for Quality Assurance and Analytics, Freising, Germany

Witold Dyrka
Politechnika Wrocławska, Wydział Podstawowych Problemów Techniki, Katedra Inżynierii Biomedycznej, Wrocław, Poland

Mostafa Elshahed
Department of Microbiology and Molecular Genetics, Oklahoma State University, Stillwater, OK, USA

Aileen R. Ferraro
Department of Microbiology, University of Georgia, Athens, GA, USA

Michal Flak
Department of Molecular and Applied Microbiology, Leibniz Institute for Natural Product Research and Infection Biology (HKI), Jena, Germany
Institute of Microbiology, Friedrich Schiller University, Jena, Germany

Kateřina Fliegerová
Institute of Animal Physiology and Genetics, v.v.i, Academy of Sciences of the Czech Republic, Prague 4, Czech Republic

Malgorzata E. Futyma
Technical University of Denmark, Kgs. Lyngby, Denmark

Victoriano Garre
Departamento de Genética y Microbiología, Facultad de Biología, Universidad de Murcia, Murcia, Spain

Brigitte Gasser
Department of Biotechnology, University of Natural Resources and Life Sciences Vienna, Vienna, Austria
Austrian Centre of Industrial Biotechnology (ACIB GmbH), Vienna, Austria
CD Laboratory for Growth-decoupled Protein Production in Yeast, Vienna, Austria

Andrea Genre
Dipartimento di Scienze della Vita e Biologia dei Sistemi, Università di Torino, Torino, Italy

Jennifer Gerke
Institut für Mikrobiologie und Genetik, Abteilung Molekulare Mikrobiologie und Genetik, Göttinger Zentrum für Molekulare Biowissenschaften (GZMB), Georg-August-Universität Göttingen, Göttingen, Germany

BARRY GOODELL
University of Massachusetts, Amherst, MA, USA

GARETH W. GRIFFITH
Institute of Biological, Environmental and Rural Sciences, Aberystwyth University, Penglais, Aberystwyth, Ceredigion, Wales, UK

LEONARDO GUZMAN-ZAMORA
Millennium Institute for Integrative Biology (iBio), Santiago, Chile
Departamento de Genética Molecular y Microbiología, Facultad de Ciencias Biológicas, Pontificia Universidad Católica de Chile, Santiago, Chile

MICHAEL HABIG
Environmental Genomics, Christian-Albrechts University of Kiel and Max Planck Institute for Evolutionary Biology, Plön, Germany

REBEKKA HARTING
Institut für Mikrobiologie und Genetik, Abteilung Molekulare Mikrobiologie und Genetik, Göttinger Zentrum für Molekulare Biowissenschaften (GZMB), Georg-August-Universität Göttingen, Göttingen, Germany

KAI HEIMEL
Institut für Mikrobiologie und Genetik, Abteilung Molekulare Mikrobiologie und Genetik, Göttinger Zentrum für Molekulare Biowissenschaften (GZMB), Georg-August-Universität Göttingen, Göttingen, Germany

CASEY HOOKER
Agriculture and Biological Engineering, Purdue University, West Lafayette, IN, USA

JENNIFER M. HURLEY
Department of Biological Sciences, Rensselaer Polytechnic Institute, Troy, NY, USA
Center for Biotechnology and Interdisciplinary Sciences, Rensselaer Polytechnic Institute, Troy, NY, USA

MEAGHAN S. JANKOWSKI
Department of Biological Sciences, Rensselaer Polytechnic Institute, Troy, NY, USA

ZOFIA D. JARCZYNSKA
Technical University of Denmark, Kgs. Lyngby, Denmark

ANDREW D. KLOCKO
Department of Chemistry & Biochemistry, University of Colorado Colorado Springs, Colorado Springs, CO, USA

ANNA M. KÖHLER
Institut für Mikrobiologie und Genetik, Abteilung Molekulare Mikrobiologie und Genetik, Göttinger Zentrum für Molekulare Biowissenschaften (GZMB), Georg-August-Universität Göttingen, Göttingen, Germany

ANNEMARIE KRAMER
Flensburg University of Applied Science, Flensburg, Germany

Mario K. C. Krespach
Department of Molecular and Applied Microbiology, Leibniz Institute for Natural Product Research and Infection Biology (HKI), Jena, Germany
Institute of Microbiology, Friedrich Schiller University, Jena, Germany

Antje Labes
Flensburg University of Applied Science, Flensburg, Germany

Luisa Lanfranco
Dipartimento di Scienze della Vita e Biologia dei Sistemi, Università di Torino, Torino, Italy

Luis F. Larrondo
Millennium Institute for Integrative Biology (iBio), Santiago, Chile
Departamento de Genética Molecular y Microbiología, Facultad de Ciencias Biológicas, Pontificia Universidad Católica de Chile, Santiago, Chile

Zachary A. Lewis
Department of Microbiology, University of Georgia, Athens, GA, USA

Yuanfei Li
Laboratory of Gastrointestinal Microbiology, Nanjing Agricultural University, Nanjing, China

Diethard Mattanovich
Department of Biotechnology, University of Natural Resources and Life Sciences Vienna, Vienna, Austria
Austrian Centre of Industrial Biotechnology (ACIB GmbH), Vienna, Austria

Cindy Meister
Institut für Mikrobiologie und Genetik, Abteilung Molekulare Mikrobiologie und Genetik, Göttinger Zentrum für Molekulare Biowissenschaften (GZMB), Georg-August-Universität Göttingen, Göttingen, Germany

Uffe H. Mortensen
Technical University of Denmark, Kgs. Lyngby, Denmark

Laura Murcia
Departamento de Genética y Microbiología, Facultad de Biología, Universidad de Murcia, Murcia, Spain

Magdalena Nagler
Institute of Microbiology, Universität Innsbruck, Innsbruck, Austria

Eusebio Navarro
Departamento de Genética y Microbiología, Facultad de Biología, Universidad de Murcia, Murcia, Spain

Francisco E. Nicolás
Departamento de Genética y Microbiología, Facultad de Biología, Universidad de Murcia, Murcia, Spain

Robin A. Ohm
Microbiology, Department of Biology, Utrecht University, Utrecht, The Netherlands

Michelle A. O'Malley
Department of Chemical Engineering, University of California, Santa Barbara, Santa Barbara, CA, USA

Mislav Oreb
Institute of Molecular Biosciences, Faculty of Biological Sciences, Goethe-University Frankfurt, Frankfurt, Germany

Sabine M. Podmirseg
Institute of Microbiology, Universität Innsbruck, Innsbruck, Austria

Annica J. Pschibul
Department of Molecular and Applied Microbiology, Leibniz Institute for Natural Product Research and Infection Biology (HKI), Jena, Germany
Institute of Microbiology, Friedrich Schiller University, Jena, Germany

Matthew Reilly
Department of Biology, University of York, York, UK

Jakob K. H. Rendsvig
Technical University of Denmark, Kgs. Lyngby, Denmark

Vicente Rojas
Millennium Institute for Integrative Biology (iBio), Santiago, Chile
Departamento de Genética Molecular y Microbiología, Facultad de Ciencias Biológicas, Pontificia Universidad Católica de Chile, Santiago, Chile

Andrés Romero
Millennium Institute for Integrative Biology (iBio), Santiago, Chile
Departamento de Genética Molecular y Microbiología, Facultad de Ciencias Biológicas, Pontificia Universidad Católica de Chile, Santiago, Chile

Francisco Salinas
Millennium Institute for Integrative Biology (iBio), Santiago, Chile
Instituto de Bioquímica y Microbiología, Facultad de Ciencias, Universidad Austral de Chile (UACH), Valdivia, Chile

Michael Sauer
Department of Biotechnology, University of Natural Resources and Life Sciences Vienna, Vienna, Austria
Austrian Centre of Industrial Biotechnology (ACIB GmbH), Vienna, Austria
CD Laboratory for Biotechnology of Glycerol, Vienna, Austria

Sven J. Saupe
Non-self Recognition in Fungi, UMR5095 Institut de Biochimie et Génétique Cellulaires, CNRS - Université de Bordeaux, Bordeaux Cedex, France

Bernhard Schmelzer
Department of Biotechnology, University of Natural Resources and Life Sciences Vienna, Vienna, Austria
Austrian Centre of Industrial Biotechnology (ACIB GmbH), Vienna, Austria

Volker Schroeckh
Department of Molecular and Applied Microbiology, Leibniz Institute for Natural Product Research and Infection Biology (HKI), Jena, Germany

Susanna Seppälä
Department of Chemical Engineering, University of California, Santa Barbara, Santa Barbara, CA, USA

Kevin V. Solomon
Agriculture and Biological Engineering, Purdue University, West Lafayette, IN, USA

Eva H. Stukenbrock
Environmental Genomics, Christian-Albrechts University of Kiel and Max Planck Institute for Evolutionary Biology, Plön, Germany

Michael K. Theodorou
Agriculture Centre for Sustainable Energy Systems, Harper Adams University, Newport, Shropshire, UK

Karl G. Thieme
Institut für Mikrobiologie und Genetik, Abteilung Molekulare Mikrobiologie und Genetik, Göttinger Zentrum für Molekulare Biowissenschaften (GZMB), Georg-August-Universität Göttingen, Göttingen, Germany

Florian Wernig
Institute of Molecular Biosciences, Faculty of Biological Sciences, Goethe-University Frankfurt, Frankfurt, Germany

Diana Young
Bavarian State Research Center for Agriculture (LfL), Central Department for Quality Assurance and Analytics, Freising, Germany

Noha Youssef
Department of Microbiology and Molecular Genetics, Oklahoma State University, Stillwater, OK, USA

Molecular Genetics

1 Chromatin Structure and Function in *Neurospora crassa*

Abigail J. Courtney[1], Aileen R. Ferraro[1], Andrew D. Klocko[2], Zachary A. Lewis[1]

CONTENTS

I. Introduction

In eukaryotic cells, the relevant substrate for most DNA-based processes is **chromatin**, a complex of DNA and proteins that functions to organize the genome and regulate DNA-templated processes inside the nucleus. The most basic unit of chromatin is the nucleosome core particle, comprised of a histone octamer wrapped ~1.5 times by ~146 base pairs (bp) of DNA (Luger et al. 1997). Repeating nucleosome units are separated by a short region of linker DNA, giving rise to the well-known "beads on a string" structure of chromatin (Olins and Olins 1974; Kornberg 1974; Oudet et al. 1975). Inside the cell, the chromatin fiber is extremely dynamic and heterogeneous. A number of diverse mechanisms alter local chromatin structure, providing an important layer of regulation for many genome functions. For example, DNA-templated processes are profoundly impacted by (1) differential covalent modification of histone proteins and DNA, (2) binding of nonhistone proteins to the chromatin fiber, (3) replacement of core histones with histone variants, (4) restructuring and remodeling of nucleosomes on DNA by ATP-dependent chromatin remodeling factors, and (5) assembly of higher-order chromatin structures that compact the chromatin fiber and organize the genome within the three-dimensional space inside the nucleus (Henriksen 2007; Clapier and Cairns 2009; Henikoff and Smith 2015; Schmitt et al. 2016; Allis and Jenuwein 2016; Clapier et al. 2017).

Historically, chromatin has been simply classified into two different types, euchromatin and heterochromatin, which were originally defined based on their distinct cytological appearance (Heitz 1928; Zacharias 1995). DNA associated with euchromatin is relatively accessible, gene-rich, and transcriptionally active, whereas DNA associated with heterochromatin is typically highly condensed, transcriptionally repressed, and closely associated with the nuclear periphery (Pueschel et al. 2016; Solovei et al. 2016). Molecular analyses have now described a large number of distinct chromatin environments based on differential enrichment of chromatin modifications and binding of nonhistone proteins. Indeed, the idea that eukaryotic chromosomes are partitioned into many structurally and functionally distinct domains is now widely appreciated, and local chromatin structure is known to play important

[1] Department of Microbiology, University of Georgia, Athens, GA, Georgia; e-mail: abigail.courtney@uga.edu; aileen.ferraro25@uga.edu; zlewis@uga.edu
[2] Department of Chemistry & Biochemistry, University of Colorado Colorado Springs, Colorado Springs, CO, USA; e-mail: aklocko@uccs.edu

Genetics and Biotechnology, 3rd Edition
The Mycota II
J.P. Benz, K. Schipper (Eds.)

regulatory roles in nearly all nuclear DNA-based processes, including transcription, DNA repair, DNA replication, and chromosome segregation (Allis and Jenuwein 2016). For the purposes of this chapter, we will use the terms **euchromatin** and **heterochromatin** to describe active and silent chromatin compartments, respectively, but it should be clear that a still undetermined number of chromatin states exist within each of these nuclear compartments.

For over three decades, the fungus *Neurospora crassa* has been an important gene and concept discovery platform for many aspects of chromatin structure and function. Work with *N. crassa* continues to yield discoveries in this field with broad relevance to fungi and to eukaryotes in general. Identifying the structural and functional relationships within chromatin domains remains an active area of research in *N. crassa* with many open questions that warrant further study.

II. Euchromatin

Most genes in *N. crassa* are arranged closely together on the chromosome, with adjacent genes typically separated by only 500–2000 base pairs of intergenic DNA sequences (Galagan et al. 2003). The vast majority of genes in *N. crassa* can be classified as "euchromatic" based on the absence of conserved heterochromatin-associated modifications (see Sect. III, "Heterochromatin") (Lewis et al. 2009; Jamieson et al. 2013; Bicocca et al. 2018). Chromatin structure within the euchromatin compartment of the *N. crassa* genome has not been as extensively studied as heterochromatin, but important structural and regulatory information has been generated from genetic and genomic studies by a number of research laboratories. For example, genome-wide approaches have been applied to examine chromatin accessibility, nucleosome occupancy, nucleosome positioning, the presence of the conserved histone variant H2A.Z, and several conserved chromatin modifications. These analyses revealed that gene promoters and gene coding sequences reside in strikingly different chromatin environments. Promoters and regulatory regions are highly accessible and often depleted for nucleosomes. On the other hand, the coding sequences and most intergenic sequences are characterized by low DNA accessibility and high histone occupancy. General features of chromatin structure in regulatory regions and gene bodies are discussed separately below.

A. Promoters and Regulatory Regions

A key structural feature of euchromatin is the presence of "open" or "accessible" regions corresponding to gene promoters and/or distal gene regulatory regions (Fig. 1a). These open regions have been identified by mapping nucleosome occupancy using an approach called MNase-seqMicrococcal nuclease (MNase) (Lai and Pugh 2017). MNase-seq has been carried out with *N. crassa* by several groups (Sancar et al. 2015; Seymour et al. 2016; Liu et al. 2017; Klocko et al. 2019). This method first uses micrococcal nuclease (MNase) to partially digest chromatin to yield mono-nucleosomes (a histone octamer and approximately 146 bp of DNA). The DNA bound by histones is protected from MNase digestion, whereas "free" linker DNA is degraded. Protein is then removed and the remaining DNA fragments are analyzed by high-throughput sequencing. Subsequent bioinformatic analysis is performed to determine nucleosome occupancy across the entire genome. Inaccessible regions with high nucleosome occupancy are identified by relatively high sequence coverage, whereas regions that are depleted for nucleosomes (i.e., accessible) can be identified based on low sequence coverage. In addition to occupancy, this approach can provide information about the positioning of nucleosomes at specific loci. Nucleosomes that have a stable position within a population of nuclei give rise to a clear alternating pattern of high and low sequence coverage corresponding to bound nucleosomes and linker DNA, respectively (Lai and Pugh 2017). Inspection of MNase-seq data from *N. crassa* reveals a characteristic Nucleosome *F*ree *R*egion (NFR) near the transcriptional start site of many genes (Sancar et al. 2015; Seymour

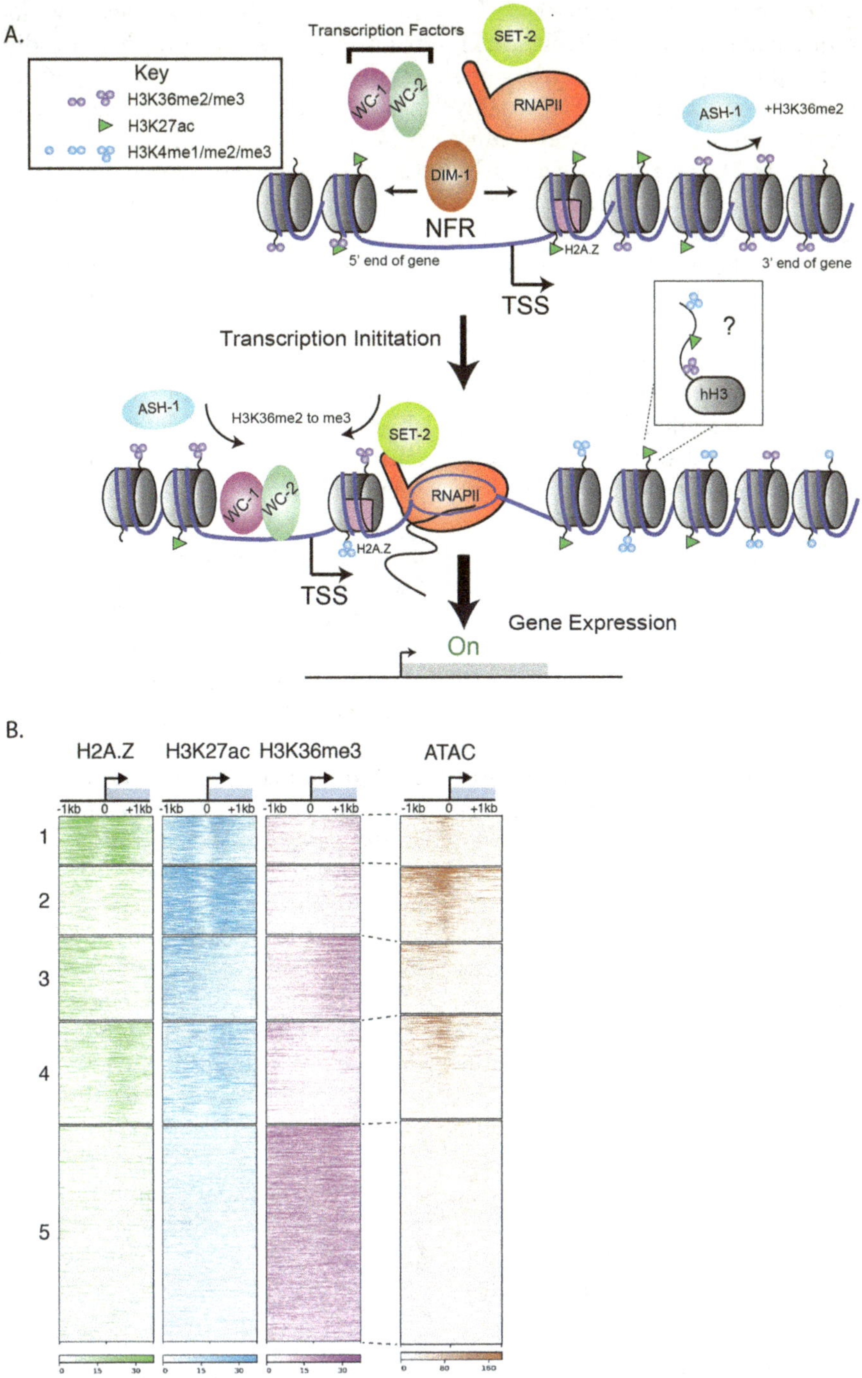

Fig. 1 The structure of genes in the euchromatin of *Neurospora crassa*. (**a**) Top: Structure of a typical euchromatic gene in Neurospora prior to transcription initiation. The nucleosome-free region (NFR) is cleared by ATP-dependent chromatin remodelers, allowing transcription factors and RNA polymerase II (RNAPII) to bind to the accessible DNA; nucleosomes near the transcription start site (TSS) are enriched for acetylation and the histone variant H2A.Z. Bottom: Upon recruitment of RNAPII, transcription is initiated, and the repressive H3K36me2 is converted to H3K36me3 by RNAPII-associated SET-2/KMT3 or the RNAPII-

et al. 2016; Klocko et al. 2019), as observed for many other organisms (Lai and Pugh 2017). Additional accessible regions within intergenic sequences can also be identified, presumably corresponding to important distal regulatory sequences that control transcription of a nearby promoter. On the other hand, well-positioned nucleosomes can be observed adjacent to the NFR extending into the gene coding sequence (see Sect. II.B).

Several mechanisms operate to create or maintain open chromatin within gene promoters. In other eukaryotes, promoter regions adjacent to the transcriptional start site are enriched for sequences that are energetically unfavorable for nucleosome binding, such as poly(dA:dT) tracts (Iyer and Struhl 1995; Segal and Widom 2009; Krietenstein et al. 2016). It is not known if this also occurs in *N. crassa*, but given the similarity in nucleosome occupancy pattern surrounding transcriptional start sites (i.e., an NFR is often observed just upstream of the transcriptional start site), it is reasonable that the NFRs of *N. crassa* may be enriched for similar sequences. Other mechanisms that promote chromatin accessibility involve regulatory proteins or protein complexes. ATP-dependent **chromatin remodeling enzymes** can function to enhance chromatin accessibility, for example. The major class of ATP-dependent remodeling enzymes is the SWI/SNF (*swi*tching defective/*s*ucrose *n*on*f*ermenting) superfamily, named after the phenotypes of yeast mutants that led to the discovery of the founding family member (Stern et al. 1984; Neigeborn and Carlson 1984). SWI/SNF family remodeling enzymes harness the energy from ATP hydrolysis to alter local chromatin structure by changing the absolute position or composition of the histone octamer on DNA through a variety of mechanisms including sliding nucleosomes along the DNA, ejecting nucleosomes from the DNA, ejecting histone dimers from the histone octamer, and, in the case of the multi-subunit SWR1 complex, depositing the histone variant H2A.Z in place of the canonical H2A (Clapier and Cairns 2009; Clapier et al. 2017). These activities can increase or decrease chromatin accessibility. A number of *N. crassa* remodelers have been implicated in gene regulation, primarily linked to the regulation of circadian clock genes. For example, the SNF2 family enzyme CLOCKSWITCH (NCU09106) and the multi-subunit SWI/SNF complex were shown to remodel chromatin in the promoter and upstream regulatory sequences controlling *frequency* (*frq*; NCU02265), a key component of circadian oscillator (Belden et al. 2007; Wang et al. 2014). Another chromatin remodeler, the Clock-ATPase (CATP; NCU06484), is necessary to displace histones from the C-box, a critical *cis*-regulatory element required for activation of *frq* transcription. Disruption of the *catp* gene caused increased histone H3 density at the C-box, reduced *frq* transcription, and an aberrant circadian rhythm (Cha et al. 2013). Interestingly, the *catp* gene was also identified as the *defective in methylation* mutant *dim-1* locus (Sect. IIIA) (Klocko et al. 2019), suggesting this chromatin remodeler acts across the genome at both euchromatin and heterochromatin.

Another major mechanism for creating "open" chromatin is the binding of transcription factors. Nucleosome occupancy and transcription factor binding are typically mutually exclusive, such that transcription factors can compete with nucleosomes for access to the underlying DNA at promoters. In *N. crassa*, the DNA-binding photoreceptor White Collar Complex (WCC), comprised of WC1 [NCU02356] and WC2 [NCU00902], is required

Fig. 1 (continued) independent ASH1, thereby activating gene expression. Methylation of H3K4 occurs as well, with trimethylation enrichment at the TSS and monomethylation enrichment at the 3′ end of genes. Inset: Each histone H3 N-terminal tail can be enriched for multiple posttranslational modifications, although it is unclear what proteins bind these marks. (**b**) Integration of ChIP-seq and ATAC-seq data reveals at least five distinct promoter structures in *N. crassa* (Lewis lab, unpublished). Divergent genes with shared promoters were removed, and k-means clustering of genes with unambiguous promoter assignments was performed based on patterns of H2A.Z, H3K27 acetylation, and H3K36me3 ChIP-seq enrichment. The heat map shows relative enrichment of each mark in clusters 1–5. ATAC-seq enrichment for chromatin accessibility is also shown

to drive open chromatin at certain genome sites. A study of WCC and another transcription factor necessary for induction of a subset of light-inducible genes, SUB-1, revealed that WCC was needed to create open chromatin in response to light (Sancar et al. 2015). On the other hand, open chromatin at SUB-1 (NCU01154) binding sites did not depend on SUB-1 (Sancar et al. 2015). Thus, some transcription factors play critical roles in opening chromatin, which could in turn allow for additional transcription factors to subsequently bind promoters at an adjacent sequence motif, thereby enabling combinatorial regulation of gene expression by multiple transcription factors. Interestingly, interactions of WCC with SWI/SNF are required for circadian activation of *frq* expression in the dark, but not for its light-induced activation (Wang et al. 2014). Together, these results indicate that WCC may promote chromatin accessibility through multiple mechanisms. More broadly, these data are consistent with a model in which a subset of *N. crassa* transcription factors, such as the WCC, function as master or "pioneer" transcription factors.

In addition to their elevated accessibility compared to the rest of the genome, promoters can be distinguished from coding sequences and heterochromatin by additional structural features. *N. crassa* histones can be acetylated at a number of residues (Xiong et al. 2010), and many promoters are highly enriched for histones with acetylated lysines (Anderson et al. 2010; Smith et al. 2010; Bicocca et al. 2018). Mutation of the *ngf* (NCU10847) gene encoding a homolog of the highly conserved GCN5 lysine acetyltransferase or a key lysine residue in histone H3 (NCU01635) causes defects in WCC-dependent activation of the light-induced gene *albino-3* (NCU01427) (Grimaldi et al. 2006; Brenna et al. 2012), demonstrating that histone lysine acetylation positively regulates transcription in *N. crassa* similar to other organisms. Unmodified lysines in the histone tails are proposed to create a relatively inaccessible chromatin structure that is refractory to transcription due to strong interactions between positively charged lysines in the histone tail with adjacent nucleosomes and the negatively charged DNA backbone (Verdone et al. 2005). Thus, histone lysine acetylation may promote transcription in two ways: (1) Acetylation neutralizes the positive charge of lysines in the histone tail, which is proposed to weaken histone-DNA interactions, increase the accessibility of the underlying DNA, and lower the energetic barrier to transcription. (2) Acetylated lysines can be recognized and bound by specific chromatin-binding proteins (e.g., see Gong et al. 2016), facilitating transcription factor recruitment to promoters for enhancing transcription. Additional work is needed to elucidate the complex role(s) of acetylation in *N. crassa* promoters, but hyperacetylated lysines of promoter histones are strongly correlated with active transcription (Smith et al. 2010; Bicocca et al. 2018).

Finally, a common feature of *N. crassa* promoters is the histone variant, H2A.Z (NCU05347). This histone variant is highly conserved from yeast to humans and is commonly enriched in promoter nucleosomes (Talbert and Henikoff 2010). Despite extensive study in multiple organisms, however, the function of H2A.Z has remained mysterious. Genetic studies implicate H2A.Z in transcriptional activation, transcription repression, and DNA repair, yet how this histone variant performs different functions in different genomic contexts is poorly understood (Talbert and Henikoff 2010). Two published studies of *N. crassa* H2A.Z suggest that it may repress transcription (Liu et al. 2017; Dong et al. 2018), but RNA-seq analysis suggests that H2A.Z is required for repression of some genes and activation of others (A.J. Courtney and Z.A. Lewis, unpublished). The paradoxical functions of H2A.Z likely depend on posttranslational modifications of H2A.Z as well as other promoter features that co-localize with H2A.Z. Hierarchical clustering of transcriptional start sites in *N. crassa* was performed based on multiple chromatin features including H2A.Z. This approach resolved at least five distinct promoter structures present in this fungus (A.J. Courtney and Z.A. Lewis, unpublished) (Fig. 1b). The recruitment of unidentified H2A.Z binding proteins may also impact these transcriptional effects.

More recently, *A*ssay for *T*ransposase-*A*ccessible *C*hromatin coupled to high-throughput sequencing (ATAC-seq) (Buenrostro et al. 2013) has provided even higher resolution analysis of open chromatin within *N. crassa* promoters and regulatory regions (A.R. Ferraro and Z.A. Lewis, unpublished). Curiously, a subset of *N. crassa* genes has promoters that are unusually large (>2000 bp of upstream intergenic sequence) and rich in accessible regions. Based on these chromatin accessibility profiles, these rare genes are likely bound by multiple transcription factors. It is also interesting to note that this subset of genes is enriched for regulatory proteins such as transcription factors or signal transduction proteins. One hypothesis is that these relatively large, hyper-accessible regions seem to resemble evolutionary precursors to metazoan enhancers.

B. Coding Sequences

In contrast to promoters, gene coding sequences are largely inaccessible. Nucleosome mapping by MNase-seq shows high occupancy of well-positioned nucleosomes within gene bodies, especially for those nucleosomes immediately adjacent to the transcriptional start site and the NFR (Sancar et al. 2015; Seymour et al. 2016; Klocko et al. 2019), although the periodicity of nucleosome positioning degrades further from the 5′ end of genes.

Coding sequences themselves have a unique pattern of epigenetic marks. Trimethylation of lysine 36 on histone H3 (H3K36me3), which is generally thought of as a "euchromatic" chromatin modification, has a critical role in gene bodies and some promoters to limit DNA accessibility. In *S. cerevisiae*, this modification is deposited co-transcriptionally by the enzyme SET-2, also referred to as KMT3, using the recommended histone methyltransferase nomenclature (Allis et al. 2007).

The SET-2 protein contains a canonical SET [*S*u(var)3-9, *e*nhancer of zeste, *t*rithorax] catalytic domain typical of all histone methyltransferases, including SET-1/KMT2, DIM-5/KMT1, and SET-7/KMT6 enzymes discussed below. SET2/KMT3 enzymes directly interact with the C-terminal repeats of the RNA polymerase II (RNAPII) subunit RPB1 (Kizer et al. 2005).

Co-transcriptional deposition of H3K36me3 in yeast gives rise to a characteristic pattern of H3K36me3 in gene bodies of expressed genes, with highest levels observed in the 3′ end of gene bodies (Krogan et al. 2003). It is thought that this H3K36me3 prevents reinitiation of the RNAPII following transcription, as deletion of *set2* in yeast causes aberrant transcription from cryptic promoters within gene bodies (Carrozza et al. 2005; Keogh et al. 2005; Joshi and Struhl 2005), consistent with a role for this histone mark in limiting DNA accessibility. Further, H3K36me3 recruits a histone deacetylase (Keogh et al. 2005; Joshi and Struhl 2005) to further compact chromatin, thereby limiting accessibility of underlying DNA and preventing transcription initiation outside of promoter regions.

These roles for methylation of H3K36 may hold true in *N. crassa* as well. The *N. crassa set-2* gene (NCU00269) is required for H3K36me3, like its yeast homolog, and SET-2-catalyzed H3K36me3 is also found within expressed *N. crassa* genes, suggesting that SET-2 also functions to deposit H3K36me3 during gene transcription in *N. crassa* (Adhvaryu et al. 2005; Bicocca et al. 2018). In contrast to yeast, however, virtually all gene bodies in *N. crassa* contain high levels of H3K36me3 regardless of their expression level (Bicocca et al. 2018). The differences in H3K36me3 localization patterns can be explained by the fact that *N. crassa* encodes a second H3K36 methyltransferase, ASH1 (NCU01932), that dimethylates K36 in histone H3 within the gene bodies and promoters of repressed genes (Bicocca et al. 2018). H3K36me2 deposited by ASH1 can apparently be converted to H3K36me3 by SET-2 through a transcription-independent mechanism to elicit the full H3K36me3 across all gene bodies; these data suggest the *N. crassa* SET-2/KMT3 enzymes have a "roaming activity" to convert dimethyl H3K36 to trimethyl H3K36, even at repressed genes. Dimethylation of H3K36 by ASH1, and the subsequent conversion to

H3K36me3 by SET-2/KMT3, is associated with facultative heterochromatin (see Sect. C2) (Bicocca et al. 2018), in contrast to the transcription-coupled H3K36 methylation by SET-2/KMT3 at active genes. Widespread initiation from cryptic promoters has not been reported for the *N. crassa set-2* deletion strain, but SET-2 is implicated in transcriptional repression of the clock gene *frequency* (*frq*). The large and accessible promoter region of *frq* is highly regulated by multiple mechanisms, including by a natural antisense RNA (Hurley et al. 2015; Cha et al. 2015). Antisense transcription through the *frq* promoter leads to transcription-dependent H3K36 methylation by the SET-2/KMT3 histone methyltransferase and subsequent gene repression (Xue et al. 2014). Other factors that serve to suppress unregulated *frq* transcription include the transcriptional repressor RCO-1 (NCU06205) and the chromatin remodeler CHD-1 (NCU03060) (Zhou et al. 2013; Sun et al. 2016), the latter suggesting that nucleosome positioning plays a critical role in *frq* repression.

Euchromatin compartments are also highly enriched for H3 that is dimethylated or trimethylated on lysine 4 (H3K4me2/3). Similar to H3K36me3, H3K4 methylation is also cotranscriptionally deposited. In *S. cerevisiae*, additionally, states of H3K4 methylation exhibit a gradient of enrichment from the 5′ to 3′ end of genes, with trimethylated lysine 4 (H3K4me3) on histone H3 exhibiting highest enrichment near the 5′ end of genes, H3K4me2 enrichment shifted towards the middle of the gene bodies, and peaks of monomethyl H3K4 enrichment shifted further towards the 3′ end of genes (Liu et al. 2005; Pokholok et al. 2005). The enrichment of H3K4 methylation occurs in most genes, yet the absolute levels of enrichment are tightly correlated with transcriptional levels: high H3K4me corresponds to a high presence of mRNA, although it is unclear if the H3K4me causes elevated transcription or is simply a result of transcriptional output. While the distribution of the different H3K4 methylation states has not been well studied in *N. crassa*, available data suggest that H3K4 methylation has a complex regulatory network for deposition: for example, many expressed genes have no H3K4 methylation. The function of different H3K4 methylated states (H3K4me3/2/1) in *N. crassa* is also understudied and poorly understood. Like H3K36me3, methylation of K4 on H3 is linked to regulation of *frq* expression and repression of gene expression (Raduwan et al. 2013; Zhu et al. 2019), though additional work is needed to understand how H3K4 methylation contributes to genome function. Additional work is also needed to determine the mechanisms that control H3K4 methylation states and levels at specific genes.

III. Heterochromatin

Emil Heitz coined the term "heterochromatin" to describe structurally distinct segments of chromosomes based on their cytological staining pattern (Heitz 1928; Zacharias 1995). Subsequent genetic and molecular studies revealed that (1) heterochromatin compartments are transcriptionally and recombinationally suppressed, (2) heterochromatin is characterized by molecular features that are conserved in many organisms, and (3) heterochromatin is important for critical cell functions such as genome defense, chromosome segregation, gene regulation, and genome organization (Grewal and Jia 2007; Janssen et al. 2018). Heterochromatin domains can be subclassified into two types, **constitutive** and **facultative** heterochromatin, which are functionally distinct and characterized by different molecular components including a unique set of posttranslational modifications on histones deposited and bound by heterochromatin-specific proteins. These types of heterochromatin are discussed separately below. Notably, Neurospora has served as an important model system for understanding heterochromatin because it shares many critical molecular features of heterochromatin with higher eukaryotes, including plants and animals. The shared hallmarks of heterochromatin include cytosine methylation on DNA and histone H3 methylated on lysine 9 (H3K9me) or lysine 27 (H3K27me). Notably, some or all of these features are absent from major yeast models such as *Saccharomyces cerevisiae* and *Schizosaccharomyces pombe*. Application of powerful genetic approaches in *N. crassa* has thus led to several important discoveries regarding heterochromatin.

A. Constitutive Heterochromatin (Repeat-Associated Heterochromatin)

Constitutive heterochromatin in *N. crassa* shares many features with constitutive heterochromatin found in multicellular eukaryotes. Like plants and animals, *N. crassa* heterochromatin domains are rich in repeated DNA sequences, but repeats in *N. crassa* are degenerate due to the action of *r*epeat-*i*nduced *p*oint mutation (RIP), highlighting a key difference between *N. crassa* and other systems (Selker et al. 1987; Selker 2002; Galagan and Selker 2004). RIP is a genome defense system that introduces C to T mutations in duplicated DNA sequences during the sexual phase of the *N. crassa* life cycle (Selker et al. 1987; Selker 2002; reviewed in Gladyshev 2017) and requires a DNA methyltransferase homolog RID (RIP-defective, NCU02034) (Freitag et al. 2002). Approximately 20% of the *N. crassa* genome exhibits sequence hallmarks of RIP, and most of these DNA sequences appear to be nonfunctional transposon sequences (Galagan et al. 2003). These regions make up the constitutive heterochromatin component of the *N. crassa* genome and are easily identified by their high content of A/T nucleotides relative to gene-rich euchromatin (Lewis et al. 2009). In addition, these genomic regions are enriched for chromatin features that are associated with constitutive heterochromatin domains in plants and animals including 5-methylcytosine ($5^{m}C$), histone H3 trimethylated at lysine 9 (H3K9me3), and heterochromatin protein 1 (HP1; NCU04017), a conserved chromatin-binding protein that interacts specifically and directly with H3K9me3 (Nielsen et al. 2002; Freitag et al. 2004a; Lewis et al. 2009). Within the *N. crassa* genome, heterochromatin domains can be relatively large, ranging from ~300 bp to over 300 kilobases (kb) in the centromeres. In the case of Linkage Group VII, the heterochromatic centromere domain is enriched for both H3K9me3 and the centromere-specific H3 variant CenH3 and makes up to approximately 10% of the entire chromosome (Lewis et al. 2009; Smith et al. 2011).

1. Assembly of Constitutive Heterochromatin Domains

A great deal of information regarding assembly of constitutive heterochromatin in *N. crassa* has come from genetic studies designed to elucidate control of cytosine methylation in DNA, which was first identified in *N. crassa* within duplicated sequences that were products of RIP (Selker et al. 1993; Margolin et al. 1998). Isolation of mutants with abolished or reduced levels of $5^{m}C$, the so-called *defective in methylation (dim)* mutants, uncovered key genes whose protein products are needed to assemble and regulate constitutive heterochromatin in *N. crassa*. To date, nine *dim* mutants have been identified, although some of these have yet to be described. Additional information obtained from proteomics and molecular studies has led to a detailed working model for assembly of constitutive heterochromatin domains.

A/T-rich DNA can recruit the H3K9 methyltransferase complex, DCDC (*D*IM-5,7,9, *C*ULLIN4, *D*NA damage-binding protein 1 *C*omplex, named after its confirmed components, gene IDs NCU04402, NCU04152, NCU01656, NCU00272, and NCU06605) (Tamaru and Selker 2001; Tamaru et al. 2003; Zhao et al. 2010; Lewis et al. 2010b, a; Xu et al. 2010). Once recruited, DCDC catalyzes the covalent addition of three methyl groups to lysine 9 of H3 (Fig. 2a) (Tamaru et al. 2003).

The catalytic subunit of the complex is DIM-5, a SET domain methyltransferase that belongs to the KMT1 family of enzymes. DIM-5 is homologous with KMT1 family enzymes including *Drosophila* SuVar3-9 (*su*ppressor of *var*iegation *3–9*) and its mammalian homologs SUV39H1 and SUV39H2, as well as the plant enzyme KRYPTONITE. All of these homologs methylate H3K9 in constitutive heterochromatin domains (Rea et al. 2000; Jackson et al. 2002; Allis et al. 2007). DIM-7 is required for proper recruitment of DCDC to

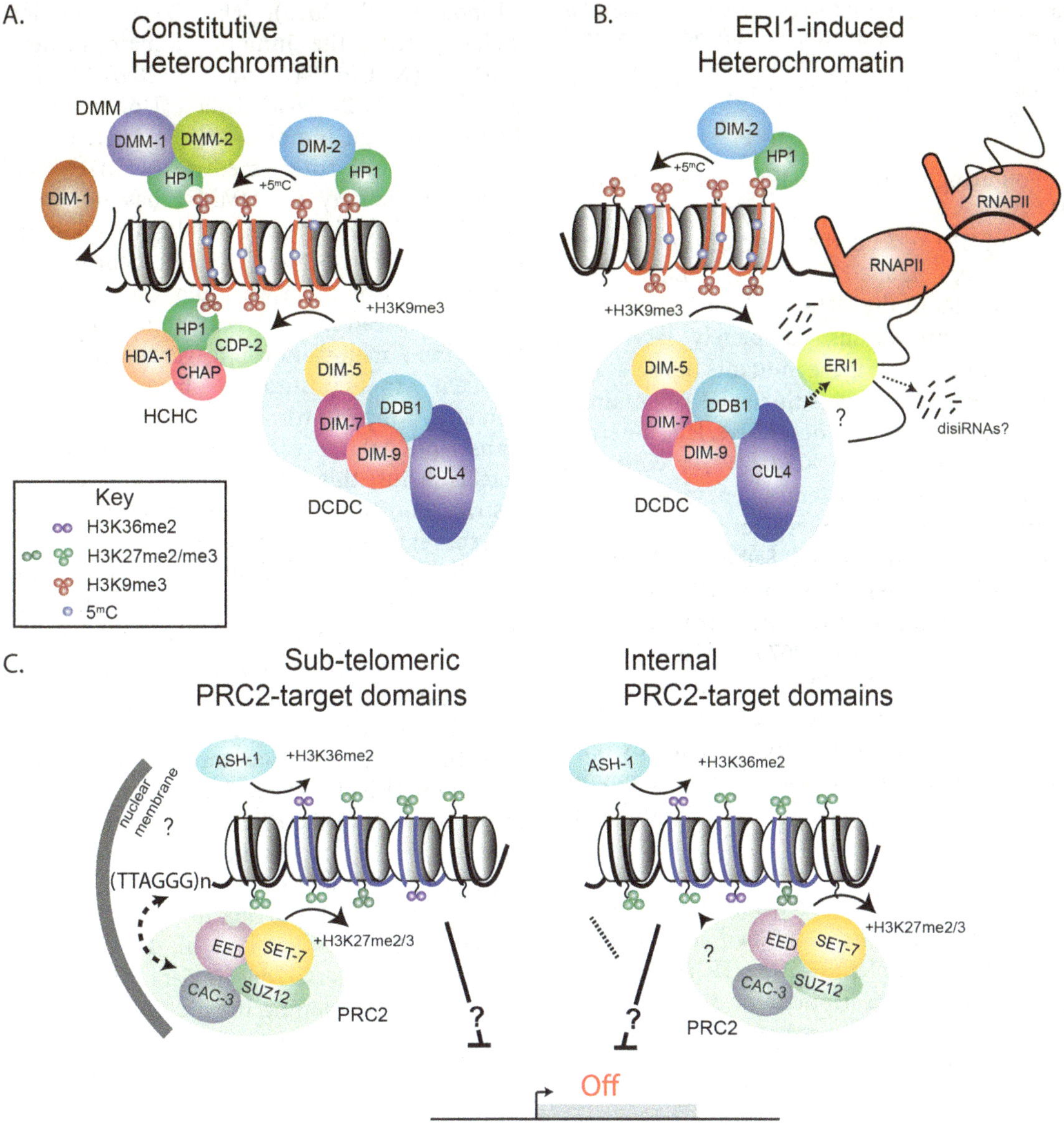

Fig. 2 Current models for the formation of heterochromatin in *Neurospora crassa.* (**a**) H3K9me3-dependent constitutive heterochromatin catalyzed by the DCDC formed at genomic loci containing A/T-rich, repetitive DNA and intergenic loci defined by convergent genes and (**b**) directed by binding of ERI1 to nascent RNA associated with stalled RNA polymerase. (**c**) H3K27me2/3- and H3K36me2-demarcated facultative heterochromatin catalyzed by the PRC2 and ASH1, respectively, formed at subtelomeric loci (left) and internal, position-independent genomic loci (right)

RIP'd DNA regions (Lewis et al. 2010a), whereas DIM-9, DDB1, and CUL4 are required for DIM-5 catalytic activity in vitro (Lewis et al. 2010a). However, the latter three complex members, which also function in DNA repair, appear to have a structural role for DCDC assembly, as the C-terminal domain of CUL4, which contains the critical lysine required for the posttranslational addition of a NEDD peptide for DNA repair, is dispensable for heterochromatin formation (Adhvaryu et al. 2015). Further, the Neurospora protein DIM-3 (NCU01249), whose homologs in yeast (importin α) and humans (karyopherin α) transport proteins into the nucleus, may be required for DCDC localization to heterochromatin following nuclear shuttling (Klocko et al. 2015). Interestingly, DIM-5 does not require these accessory factors to methylate the H3 K9 residue in vitro (Zhang et al. 2002, 2003; Tamaru et al. 2003; Collins et al. 2005; Adhvaryu et al. 2011). In addition to regulation by interacting proteins, DIM-5 activity can be impacted by modification of nearby residues providing an additional layer of regulation governing constitutive heterochromatin assembly (Adhvaryu and Selker 2008; Adhvaryu et al. 2011).

Following DIM-5 action, the resulting H3K9me3 mark is bound by the chromodomain of HP1 (encoded by *heterochromatin protein one*, *hpo*; NCU04017). Like other components in the DNA methylation pathway, HP1-deficient strains completely lack DNA methylation (i.e., these mutants have a Dim phenotype). To date, three HP1-containing complexes have been described. HP1 interacts directly with DIM-2 (NCU02247), a conserved cytosine methyltransferase enzyme that is required for all $5^{m}C$ in vegetative tissues of *N. crassa* (Kouzminova and Selker 2001; Honda and Selker 2008). A second HP1 complex, HCHC, is comprised of HP1, *c*hromo-domain protein-2 (*C*DP-2; NCU00738), *h*istone deacetylase-1 (*H*DA-1; NCU01525), and CDP-2- and HDA1-associating protein (*C*HAP; NCU01796) (Honda et al. 2016). The HCHC complex regulates heterochromatin assembly by removing acetyl groups from lysine residues within histone tails (Honda et al. 2012, 2016); acetyl marks are considered active (euchromatic) marks that loosen the DNA-histone interaction (see Sect. IIA). The combined action of the HP1-DIM2 complex and the HCHC complex generates a transcriptionally repressed and compacted chromatin structure that is characterized at the molecular level by low levels of histone acetylation and high levels of $5^{m}C$. Interestingly, by using an in vivo tethering system to ectopically localize individual components of the heterochromatin machinery, it was found that tethered HP1 can actually induce H3K9me3 in an HCHC/HDA-1-dependent manner, suggesting deacetylated histone tails recruit DCDC-dependent catalytic activity and, more broadly, that some positive feedback between HCHC and DCDC occurs (Gessaman and Selker 2017).

A third HP1 complex, called the DNA methylation modulator complex (DMM), functions to prevent uncontrolled spreading of heterochromatin into adjacent gene sequences (Honda et al. 2010). The DMM complex includes HP1, the Jumonji domain protein DMM-1 (NCU01554) and the DNA-binding protein DMM-2 (NCU08289) (Honda et al. 2016). Although Jumonji domain proteins are predicted to be lysine demethylases (KDMs), an enzymatic activity for DMM-1 has not been described. Genetic studies highlight the important role of DMM-1 at heterochromatin/euchromatin boundaries. Mutation of *dmm-1* leads to spreading of DNA methylation and H3 lysine 9 methylation into genes that reside adjacent to constitutive heterochromatin domains, which in turn leads to their aberrant transcriptional repression and severe growth defects. Thus, this HP1 complex functions to restrict constitutive heterochromatin to the correct regions of the genome.

Characterization of another *dim* mutant, *dim-1* (NCU06484), whose phenotype includes the hypomethylation of constitutive heterochromatin and hypermethylation of intergenic euchromatic regions, revealed that positioning of nucleosomes within heterochromatic regions is distinct from euchromatin and functionally important (Klocko et al. 2019). The causative mutations within the *dim-1* strain were identified in the Neurospora homolog of an ATP-dependent chromatin remodeler conserved from yeast (Yta7) (Lombardi et al. 2011) to humans (ATAD2) (Zou et al. 2007). Using MNase-seq as described in Sect. IIA above, nucleosome positioning in wild-type and *dim-1* strains was examined. Interestingly, heterochromatic nucleosomes of a wild-type Neurospora strain were found to be less well positioned than their euchromatic counterparts (Klocko et al. 2019). Peaks of MNase-protected DNA were broad and short, a classic phenotype for moving, less well-positioned nucleosomes (Lai and Pugh 2017), in contrast to the well-defined, high-amplitude nucleosomal peaks surrounding the transcriptional start sites of euchromatic genes.

It can be speculated that the underlying repetitive, AT-rich DNA in heterochromatin may not be conducive for proper positioning of histone octamers. Nucleosomes just outside heterochromatic regions were often well-positioned and stable, perhaps reflecting a local binding site for complexes that manage heterochromatin borders. Interestingly, the periodicity of nucleosomes in heterochromatin was antiparallel to cytosine methylation enrichment (Klocko et al. 2019), suggesting regions of DNA not protected by nucleosomes (i.e., linker DNA) were more apt to contain DNA methylation. In support, a deletion strain lacking histone H1, which is thought to bind linker DNA, exhibited increased methylation at typical constitutive heterochromatin domains (Seymour et al. 2016). The loss of cytosine methylation within a *dim-1* strain could be explained by the nucleosome disorder that was observed. Heterochromatic nucleosome positioning was exacerbated upon inactivation of the DIM-1 chromatin remodeler—any positioning of nucleosomes was essentially abolished and linker DNA availability for methylation was thus reduced (Klocko et al. 2019). This same nucleosome disorder was observed at intergenic euchromatic regions gaining cytosine methylation in a Δ*dim-1* strain, leading to speculation that disordered nucleosomes may be a signal for heterochromatin formation.

Heterochromatin domains in *N. crassa* are also enriched for a phosphorylated form of H2A, referred to as ɣH2A (Sasaki et al. 2014). In both *Saccharomyces cerevisiae* and humans, phosphorylation of a C-terminal serine on H2A or H2A.X, respectively, to yield ɣH2A occurs typically at sites of DNA damage, where it is contributes to proper DNA repair (Chambers and Downs 2007; Dickey et al. 2009). Thus, one can speculate the role of ɣH2A in Neurospora heterochromatin is to maintain the integrity of the underlying A/T-rich DNA. In *S. cerevisiae*, ɣH2A is deposited at sites of DNA damage by the Tel1 and Mec1 kinases (homologs of human ATM and ATR) (Downs et al. 2000), but the enzyme that phosphorylates H2A to yield ɣH2A in *N. crassa* heterochromatin domains is unknown.

Despite several decades of work, the mechanisms that control assembly of constitutive heterochromatin remain poorly understood. In the fission yeast *Schizosaccharomyces pombe*, small interfering RNAs are important for directing H3K9 methylation to the appropriate regions of the genome (Martienssen and Moazed 2015). In *N. crassa*, however, the RNA interference pathway is not required (Freitag et al. 2004b). In contrast, the ectopic introduction of A/T-rich DNA is sufficient to direct $5^{m}C$ and presumably heterochromatin formation (Miao et al. 2000; Tamaru and Selker 2003), but how this occurs is unknown. In addition, repeated DNA (i.e., paired repeated sequences) may play an important role in directing assembly of constitutive heterochromatin. A tandem array of the euchromatic *albino-1* gene generated by transformation was associated with H3K9me3, and a synthetic sequence duplication was subjected to DIM-2-dependent RIP during the sexual cycle (Chicas et al. 2005; Gladyshev and Kleckner 2017). In both cases, the DNA sequences could recruit components of the constitutive heterochromatin machinery even though they were not A/T-rich. Together, these observations support the idea that multiple, redundant, or partially redundant mechanisms regulate assembly of heterochromatin domains. Future studies are needed to uncover these regulatory mechanisms.

2. Functions of Constitutive Heterochromatin

In *N. crassa*, heterochromatin domains are devoid of genes, yet it is clear that proper assembly of heterochromatin is important for genome function. Indeed, mutants that lack key components of the constitutive heterochromatin pathway exhibit severe growth defects (e.g., *dim-5*, *hpo*, and others). The most obvious function for constitutive heterochromatin is to defend the genome against transposable elements. The presence of cytosine methylation in DNA was shown to inhibit transcription elongation by RNA polymerase II in vivo leading to transcriptional repression of the methylated sequence (Rountree and Selker 1997). Due to the action of RIP, *N. crassa* lab strains contain only a single active DNA transposon typically present as a single copy (Wang et al. 2015). Functional retrotransposons are present in some wild isolates, however, and the DNA methyltransferase DIM-2 is required to suppress mobilization of active Tad elements (Zhou et al. 2001).

Proper assembly of constitutive heterochromatin domains is also needed for genome stability. Deletion mutants that lack HP1 or the H3K9 methyltransferase DIM-5 displays defects in centromere structure (Smith et al. 2011), which leads to defects in chromosome segregation, as DCDC mutants have increased segregation defects including chromosome bridges (Lewis et al. 2010a). These heterochromatin-defective mutants display other signs of genome instability as well, including elevated levels of γH2A and high sensitivity to DNA-damaging agents (Sasaki et al. 2014; Basenko et al. 2015). It is not entirely clear why *hpo* and *dim-5* strains display increased sensitivity to DNA-damaging agents, but this is at least partially caused by mis-localization of the repressive chromatin modifications dimethyl and trimethyl H3K27 (H3K27me2/3). In the absence of DIM-5 or HP1, H3K27me2/3 aberrantly accumulates in genomic regions that are typically assembled into constitutive heterochromatin (see Sect. IIIC). Why mis-localization of H3K27me2/3 leads to genotoxic stress is unknown, but one reasonable hypothesis is that heterochromatin structure is improperly regulated during S-phase. It may also be related to misregulation of *catalase-3* expression (Wang et al. 2016).

Recent work suggests that constitutive heterochromatin may play a key role in genome organization within the three-dimensional space of the nucleus. The Neurospora genome is non-stochastically organized within the nucleus; this organization is similar to other eukaryotes but has properties specific to the unique composition of Neurospora DNA. The first evidence of the 3D architecture of Neurospora was gleaned from electron microscopy of Neurospora nuclei, which showed contrasting light and dark stained material, which we now realize comprise active euchromatic DNA and silent heterochromatic DNA, respectively (Shatkin and Tatum 1959). Much of the heterochromatin was observed to associate with the nuclear membrane. Further cytological experiments using fluorescently tagged centromeric and telomeric proteins showed clustering of these genomic structures. GFP fusions to a centromeric protein showed that all seven centromeres cluster into a single focus (Smith et al. 2011), while GFP fusions with proteins that form the telomere protection complex shelterin showed that the 14 telomeres cluster into two to four foci (Galazka et al. 2016). Interestingly, both centromeric and telomeric foci are in close proximity to the nuclear membrane, which speaks to their heterochromatic nature. This chromosome organization is consistent with what had been observed by Rabl (Rabl 1885) and what has now become known as the Rabl chromosome configuration: clusters of centromeres and telomeres at the nuclear membrane are essential for organizing eukaryotic nuclear DNA. Despite the characterization of chromosomal structure, little information has been gleaned about the interactions of individual loci from cytological studies.

Recent technological advances have enabled the genome-wide analysis of chromosomal contacts across the Neurospora genome, revealing important aspects of three-dimensional (3D) chromosome structure within the nucleus. The extremely powerful protocol chromosome conformation capture (3C) monitors the interactions between genomic loci: in vivo chromatin is nonspecifically cross-linked, digested with a restriction enzyme, and ligated, thus covalently linking interacting genomic loci into a single DNA molecule (Dekker et al. 2002). When 3C is coupled with **high-throughput sequencing** (Hi-C) (Lieberman-Aiden et al. 2009), it became possible to monitor the long-range contacts of every locus in the genome, with the only limitations being the density of restriction enzyme sites and sequencing depth. Two noteworthy publications detailed the genome organization of wild-type *N. crassa* and initially characterized the epigenetic factors that contribute to proper 3D genome architecture (Galazka et al. 2016; Klocko et al. 2016). Importantly, due to the action of RIP, the degenerate nature of heterochromatin-associated repeats in *N. crassa* allows for contacts across the whole genome to be analyzed, including sequences found in silent heterochromatin. This is not possible in other organisms.

The seven chromosomes of the Neurospora genome form both strong intrachromosomal and interchromosomal contacts: the chromosomal territories of near-exclusive intrachromosomal interactions that dominate the human Hi-C heatmaps are less prevalent in Neurospora (Galazka et al. 2016). Strong long-range interchromosomal contacts between the seven centromeres or between the 14 telomeres are readily observed, confirming microscopic observations of centromere or telomere clustering and the Rabl organization pattern of Neurospora chromosomes (Galazka et al. 2016). Interestingly, strong interactions between heterochromatin dominate the organizational landscape. Silent DNA marked by H3K9me3, or to a lesser extent H3K27me2/3, strongly interacts to form a "heterochromatic bundle" of intrachromosomal and interchromosomal interactions; few paired interactions between heterochromatin and euchromatin are observed (Galazka et al. 2016). On a single chromosome, interactions between interspersed heterochromatic regions marked by H3K9me3 often form the base of euchromatin loops (Galazka et al. 2016). These loops of active DNA, which average < 100 kilobases in size, may play a structural role for genome compaction or may be essential for optimal transcription induction. Surprisingly, deletion of the genes encoding the H3K9 methyltransferase DIM-5 or its cognate binding partner HP1 has little impact on the genome organization, as the heterochromatin bundle still forms in their absence. Hi-C datasets of Δ*dim-5* and Δ*hpo* strains only have decreased centromere-flank interactions and reduced packaging of interspersed heterochromatic regions (Galazka et al. 2016). Only a *dim-3* strain encoding a neomorphic importin-α allele that compromises, but does not abolish, H3K9me3 and cytosine methylation levels (Klocko et al. 2015) shows any drastic differences in genome organization. As the average volume of nuclei is significantly larger with defective importin α, sub-telomeres have a greater tendency to dissociate from the nuclear membrane, compromising the normal folding of chromosomes, as Hi-C of a *dim-3* strain shows increased centromere—euchromatin interactions and decreased intra- and inter-telomeric contacts (Galazka et al. 2016). Thus, noncanonical proteins can have a profound yet potentially indirect impact on the Neurospora genome architecture.

B. ERI1-Directed Facultative Heterochromatin

In contrast to constitutive heterochromatin, facultative heterochromatin assembles on gene sequences under certain conditions and may contribute to gene regulation (Trojer and Reinberg 2007). Based on genome-wide analysis of DNA methylation and H3K9me3 by microarray and high-throughput sequencing-based approaches, it was initially concluded that both modifications were restricted to repeat-rich constitutive heterochromatin domains and therefore were not associated with facultative heterochromatin (Lewis et al. 2009). However, subsequent studies have identified low levels of DNA methylation and H3K9me3 at certain euchromatic sequences in the *N. crassa* genome. This was first observed at the *frq* locus of *N. crassa*, when DNA methylation was detected in the promoter by Southern blotting (Belden et al. 2011). Interestingly, inspection of the data revealed that most copies of *frq* are unmethylated, while a fraction of nuclei exhibit heavy DNA methylation in which most Cs are methylated. High-throughput sequencing of small RNAs revealed a second unusual feature of the *frq* locus. High levels of small RNAs were produced from the *frq* promoter region on both sense and antisense strands and were processed via a Dicer-independent mechanism (Lee et al. 2010). Together, these observations raised the intriguing possibility that small RNAs could direct DNA methylation, H3K9 methylation, or both modifications to these regions. This idea was particularly attractive given the well-established links between small RNAs and H3K9 methylation in *Schizosaccharomyces pombe* and the existence of the extensively studied RNA-directed DNA methylation pathway in *Arabidopsis thaliana* (Martienssen and Moazed 2015; Wendte and Pikaard 2017). In addition to *frq*, production of Dicer-independent **small-interfering RNAs** (disiRNAs) from other loci is reportedly correlated with low levels of DNA methylation (Dang et al.

2013). Interestingly, the conserved RNA exonuclease enhanced RNAi-1 (ERI-1; NCU06684) is necessary and sufficient for DNA methylation at these sites (Fig. 2b)(Dang et al. 2016). Based on these studies, a model for heterochromatin formation at disiRNA-producing loci has been proposed. In this model, convergent transcription is thought to stall RNA polymerase, allowing ERI1 binding to the nascent RNA and resulting in subsequent recruitment of the heterochromatin machinery. It is important to note that this process must only occur in a small subset of nuclei. The observation that *frq* is unmethylated in most nuclei but contains dense cytosine methylation in a small fraction of nuclei is consistent with the idea that ERI1-induced deposition of H3K9me3 and $5^{m}C$ is a rare event, which may explain the inability to detect high levels of these chromatin modifications in genome-wide studies. Additional work is needed to fully explain the role of ERI1-induced heterochromatin in *N. crassa*. One idea is that certain loci may be subject to occasional epigenetic silencing within a population (i.e., multiple epialleles exist) as a form of epigenetic bet-hedging. In other words, epialleles could create additional phenotypic diversity that would allow certain individuals in a genetically identical population to thrive under different selective pressures (Grimbergen et al. 2015).

C. The Polycomb System and Facultative Heterochromatin

One powerful advantage of *N. crassa* as a model organism for epigenetic studies is the presence of a conserved Polycomb repression system, which in metazoans is important for assembling facultative, or temporary, heterochromatin which maintains cell type-specific or environmentally specific gene repression (Simon and Kingston 2013; Lewis 2017). In animals, plants, and some fungi, facultative heterochromatin is demarcated by the presence of histone H3 containing trimethylated lysine 27 (H3K27me3). In *N. crassa*, both dimethylated and trimethylated H3K27 are found together covering ~7% of the Neurospora genome. These regions typically have high gene content, in contrast to constitutive heterochromatin, and are located near the telomeres of all the seven chromosomes (Smith et al. 2008; Jamieson et al. 2013). Subtelomeric H3K27me2/3-associated domains are highly enriched for genes that are present only in fungi and are less conserved with other eukaryotes, though a few of these repressed and lowly expressed genes have known functions (Jamieson et al. 2013).

1. Assembly of Facultative Heterochromatin by Polycomb Repressive Complex-2

Recent efforts have focused on elucidating the mechanisms that control methylation of H3 lysine 27 and assembly of facultative domains. H3K27me2/3 is catalyzed by the *P*olycomb *R*epressive *C*omplex 2 (PRC2), a four-member complex comprised of the histone methyltransferase SET-7/KMT6 (NCU07496) and three additional members: EED (NCU05300), SUZ12 (NCU05460), and P55/CAC-3 (Neurospora p55/chromatin assembly complex-3; NCU06679) (Borkovich et al. 2004; Jamieson et al. 2013). Neurospora PRC2 members are conserved with related fungal species and with higher eukaryotes (Jamieson et al. 2013); however, genes encoding members of the H3K27me2/3 reader complex PRC1, found in *Drosophila*, mammals, and plants, are not encoded in Neurospora (Jamieson et al. 2013; Lewis 2017), leading to the question of how the facultative heterochromatin mark is read and interpreted. Loss of PRC2 in Neurospora does not confer any noticeable growth defect (Jamieson et al. 2013), in contrast to other closely related fungal species (Connolly et al. 2013; Studt et al. 2016).

Depending on the target locus, placement of the H3K27me2/3 mark within chromosomal domains can occur in a position-dependent or position-independent manner (Fig. 2c) (Jamieson et al. 2018). Analysis of H3K27me2/3 in translocation strains revealed that new telomere creation as a result of chromosome rearrangement leads to establishment of a new facultative heterochromatin domain adjacent to the new chromosome end (Jamieson et al.

2018), suggesting proximity to the chromosome end is sufficient to drive assembly of a facultative heterochromatin domain. In fact, telomere repeats (TTAGGG) themselves can induce H3K27me2/3 placement, as ectopically placing 8 or 17 telomere repeats drives formation of novel facultative heterochromatin domains (Jamieson et al. 2018). However, position-independent domains of facultative heterochromatin exist as well; these domains are typically internal on the chromosome and are maintained upon chromosome translocation. No underlying signal for H3K27me2/3 has been identified in position-independent domains, as ectopic placement of underlying DNA from one such domain failed to induce H3K27me2/3 (Jamieson et al. 2018).

An interesting dynamic exists between H3K27me2/3-marked facultative heterochromatin and constitutive heterochromatin marked by H3K9me3: the placement of H3K9me3 at some genomic loci actually represses the establishment of facultative heterochromatin (Fig. 2c). Deletion or mutation of components of DCDC or HP1 results in relocalization of the H3K27me2/3 mark to former sites of constitutive heterochromatin (Basenko et al. 2015; Jamieson et al. 2016). The reverse relationship does not hold true: loss of PRC2 function does not cause H3K9me3 movement (Basenko et al. 2015; Jamieson et al. 2016). The inhibition of PRC2 components actually rescues two phenotypes of constitutive heterochromatin mutants: defective growth rates and sensitivity to DNA-damaging agents (Basenko et al. 2015; Jamieson et al. 2016). Interestingly, the movement of facultative heterochromatin to formerly constitutive heterochromatin domains is also observed in a Δ*hpo* strain, in which the H3K9me3 binding HP1 is removed, but H3K9me3 is still catalyzed on histone tails. Mass spectrometry of Δ*hpo* histones showed tails with substantial levels of both H3K9me3 and the dimethylated posttranslational modification of H3K27 (Jamieson et al. 2016). Thus, the loss of the canonical H3K9me3-binding protein HP1 must signal, directly or indirectly, for PRC2 recruitment and specific dimethylation on lysine 27.

It is currently unknown how H3K27me2/3-marked heterochromatin repression is established and maintained, given the lack of PRC1 homologs in the Neurospora genome—clearly H3K27me2/3 must be bound by and interpreted as a repressive mark. All four members of the Neurospora PRC2Polycomb Repressive Complex 2 (PRC2) complex are required for wild-type levels of H3K27me2/3, but deletion of *npf/cac-3* causes loss of only the most subtelomeric H3K27me2/3, whereas facultative heterochromatin domains located internally on the chromosome are maintained (Jamieson et al. 2013; Klocko et al. 2016). Other fungal species encode a PRC2 complex that itself, through the EED protein, allosterically binds the H3K27me2/3 mark for positive induction of catalytic activity (Jiao and Liu 2015), thereby providing a convenient mechanism for propagating the dimethylation or trimethylation to nearby histones. Moreover, one of the most common mechanisms to bind a methylated histone tail is mediated by chromodomains, such as that found in HP1 for H3K9me3.

Within the chromodomain, three aromatic residues—most often tyrosines and tryptophans—form an aromatic cage enclosing the methylated, and now nonpolar, lysine side chain; further amino acid determinants of the chromodomain interact with the surrounding residues within the N-terminal histone tail to provide specificity to that specific histone residue (Nielsen et al. 2002).

Currently, it is unclear whether chromodomain-containing proteins have a role in binding H3K27me2/3. Further, the first H3K27me2/3 reader protein, Early Bolting in Short Day (EBS), which contains a bromo-adjacent homology (BAH) domain that also forms an aromatic cage for binding H3K27me2/3, was recently characterized in *Arabidopsis* (Yang et al. 2018), thus providing a new research direction with a different domain. Neurospora appears to have a weak homolog of EBS, but it is unclear what role, if any, this homolog may play in H3K27me2/3 recognition. Lastly, many eukaryotic model systems rely on noncoding RNAs for the establishment of H3K27me2/3, including on the repressive environment of the inactive X-chromosome (e.g., see Colognori et al. 2019), yet a possible role of noncoding RNAs for fungal facultative heterochromatin has yet to be explored.

2. Methylation of H3K36 by ASH1 at Facultative Heterochromatin

Recently, it was determined that another epigenetic mark that is traditionally thought to demarcate active genes, methylation of lysine 36 on histone H3 (H3K36me), is essential for eukaryotic gene repression and establishment of facultative heterochromatin. In Neurospora, methylation of H3K36 is catalyzed by two different SET domain-containing proteins: the RNA polymerase II (RNAPII)-associated SET-2, which monomethylates, dimethylates, and trimethylates H3K36 on actively transcribed genes and ASH1, which acts independently of RNAPII to dimethylate and occasionally trimethylate unmarked H3K36; the action of both proteins provides the full complement of H3K36me2 and H3K36me3 in wild-type Neurospora cells (Bicocca et al. 2018). In an elegant series of experiments, Bicocca et al. (2018) were able to separate the levels of H3K36me2 and H3K36me3 that were exclusively catalyzed by ASH1 and found that ASH1-catalyzed H3K36me2/3 demarcates lowly expressed genes, in contrast to the enrichment of SET-2-catalyzed H3K36me2/3 across the gene bodies of highly transcribed genes (Fig. 2c). Thus, ASH1-catalyzed H3K36me2 and H3K36me3 act as a repressive mark. Indeed, peaks of ASH1-catalyzed H3K36me2 co-localize with H3K27me2/3 enrichment at facultative heterochromatin (Bicocca et al. 2018), suggesting ASH1 catalytic activity is also critical for gene repression, although other ASH1-dependent H3K36me2 peaks exist independent of PRC2 activity. Interestingly, in mutants that exhibit alternative patterns of facultative heterochromatin relative to a wild-type strain, the newly placed H3K27me2/3 co-localizes with ASH1-catalyzed H3K36me2, and ~ 30% of H3K27me2/3 domains first require the deposition of H3K36me2 by ASH-1, although ASH1-catalyzed H3K36me2 can also inhibit H3K27me2/3 deposition (Bicocca et al. 2018). Together, these data suggest that for some position-independent facultative heterochromatin domains, the initial dimethylation of H3K36 by ASH1 may help recruit PRC2 for dimethylation and trimethylation of H3K27, although a direct interaction between these complexes has not been reported to date and it is unknown how ASH1 is recruited to these sites. Thus, the power of the model organism *Neurospora crassa* has helped elucidating a complex dynamic between two histone marks important for gene repression.

3. Functions of Facultative Heterochromatin

Facultative heterochromatin marked by PRC2-catalyzed H3K27me2/3 and ASH1-catalyzed H3K36me2 mainly functions to repress gene expression (Jamieson et al. 2013; Basenko et al. 2015; Klocko et al. 2016). As previously mentioned, genes marked by H3K27me2/3 in a wild-type strain are typically silent or expressed at very low levels. Paradoxically, removal of the H3K27me2/3 mark in a Δ*set-7* mutant only derepressed a subset of the genes in facultative heterochromatin at sub-telomeres, although the overall low level of the messenger RNA from these genes may confound any RNA-sequencing analysis. It is probable that the remaining association of the sub-telomeres at the nuclear membrane (below) may continue to enforce a repressive environment on these genes (Klocko et al. 2016), or the presence of ASH-1-catalyzed H3K36me2, which often co-localizes with the H3K27me2/3 mark inhibits any remaining transcription initiation at these promoters (Bicocca et al. 2018).

Facultative heterochromatin is an important determinant in organizing the Neurospora genome. As detailed above, interactions between H3K27me2/3-marked intrachromosomal and interchromosomal loci are some of the strongest interactions within the wild-type Neurospora genome and facilitate the formation of the "heterochromatin bundle" in Neurospora nuclei (Galazka et al. 2016). However, the loss of H3K27me2/3 following the deletion of *set-7* (*kmt-6*) did impact the 3D architecture of Neurospora DNA, as monitored by Hi-C. Namely, while the heterochromatin bundle was still able to form in a Δ*set-7* strain, the genome architecture of a Δ*set-7* strain showed a decrease in contacts between intrachromosomal and interchromosomal subtelomeric regions, as well as a moderate increase in contacts between centromeres and euchromatic arms (Klocko et al. 2016). These changes could be explained by the H3K27me2/3 subtelomeres becoming more detached from the nuclear membrane. Indeed, microscopy of fluorescently labeled telomeres showed a greater propensity of the telomere foci to be found in the central part of the nucleus and less associated with the nuclear membrane (Klocko et al. 2016). In support, Hi-C of a strain lacking *npf* (Neurospora P55), which causes the loss of only subtelomeric H3K27me2/3, produced an almost identical genome interaction map that supported sub-telomere detachment upon loss of sub-telomeric facultative heterochromatin (Klocko et al. 2016). Neurospora sub-telomeres are unique in that these are the only genomic loci in which both H3K9me3 and H3K27me2/3 co-localize in a wild-type strain; however, removal of both histone marks in a Δ*dim-5*;Δ *set-7* mutant did not exacerbate the telomere detachment phenotype. Only a decrease in centromere flank interactions was evident in this strain relative to a single Δ*set-7* mutant (Klocko et al. 2016); this double mutant also refutes the hypothesis that H3K27me2/3 relocalized to centromeres in a Δ*dim-5* strain helps to maintain the heterochromatin bundle with compromised constitutive heterochromatin.

VI. Conclusions

Neurospora crassa has served as a powerful model system for understanding chromatin structure and function in fungi and in other eukaryotes. Work in *N. crassa* has generated important information about how chromatin structure contributes to gene regulation within euchromatin through studies of promoter structure, DNA accessibility, activating epigenetic marks, and histone variants. Similarly, *N. crassa* has played an important role in advancing our understanding of silent chromatin and the factors necessary for its assembly and maintenance, from the permanent, or constitutive, heterochromatin necessary to inactivate repetitive DNA to facultative heterochromatin that silences a multitude of evolutionarily novel genes. It is now, more than ever, clear that chromatin—both activating and silencing—impacts every genome function in the nucleus from nucleotide base pairing to three-dimensional organization of chromosomes. In many ways, chromatin regulation in *N. crassa* appears more complex than in the budding yeast *S. cerevisiae*, primarily because a number of key chromatin regulatory pathways are absent from the Saccharomycotina. For this reason, *N. crassa* is a more relevant model system for elucidating general principles and mechanisms of chromatin structure in fungi than the more extensively studied budding yeast. Indeed, key discoveries in *N. crassa* have motivated work in a number of other fungal genera. Namely, work with species of *Aspergillus*, *Fusarium*, *Leptosphaeria*, *Magnaporthe*, *Zymoseptoria*, and others has uncovered important roles for chromatin regulators in secondary metabolite production and regulation of pathogenicity genes, highlighting the need for future work [e.g., see Shwab et al. 2007; Soukup et al. 2012; Palmer et al. 2013; Connolly et al. 2013; Soyer et al. 2014; Gacek-Matthews et al. 2015, 2016; Niehaus et al. 2016; Gu et al. 2017; Maeda et al. 2017; Janevska et al. 2018; He et al. 2018; Pfannenstiel et al. 2018; Möller et al. 2019; Lan et al. 2019; and others].

Despite a great deal of progress, many important questions about chromatin structure and function remain unanswered. Given the importance of chromatin structure to fungal pathogenesis and secondary metabolism, identification of chromatin-based mechanisms that are fungal-specific could provide new avenues to limit the economic costs of fungal disease and crop contamination. A better understanding of chromatin regulatory processes could also enable more efficient engineering of fungi for biotechnology purposes. *N. crassa* is an ideal experimental system to address key open questions and serve as an important model for the fungal kingdom due to its rich history in the field, a diversity of chromatin-based mechanisms shared among many fungi, and the availability of powerful molecular genetic tools to drive future studies. Important goals for future work include developing a more complete understanding of promoter structure and how it contributes to gene regulation, defining unknown mechanisms that control assembly and maintenance of both facultative and constitutive heterochromatin, defining the genes and mechanisms that govern three-dimensional chromosome organization in the nucleus and how organization might change under different conditions, and, lastly, defining the complete complement of histone modifications in *N. crassa* and resolving additional subtypes of chromatin within euchromatin and heterochromatin based on co-occurrence of histone modifications and histone binding proteins.

References

Adhvaryu KK, Berge E, Tamaru H et al (2011) Substitutions in the amino-terminal tail of Neurospora histone H3 have varied effects on DNA methylation. PLoS Genet 7:e1002423

Adhvaryu KK, Gessaman JD, Honda S et al (2015) The cullin-4 complex DCDC does not require E3 ubiquitin ligase elements to control heterochromatin in *Neurospora crassa*. Eukaryot Cell 14:25–28

Adhvaryu KK, Morris SA, Strahl BD, Selker EU (2005) Methylation of histone H3 lysine 36 is required for normal development in *Neurospora crassa*. Eukaryot Cell 4:1455–1464

Adhvaryu KK, Selker EU (2008) Protein phosphatase PP1 is required for normal DNA methylation in Neurospora. Genes Dev 22:3391–3396

Allis CD, Berger SL, Cote J et al (2007) New nomenclature for chromatin-modifying enzymes. Cell 131:633–636

Allis CD, Jenuwein T (2016) The molecular hallmarks of epigenetic control. Nat Rev Genet 17:487–500

Anderson DC, Green GR, Smith K, Selker EU (2010) Extensive and varied modifications in histone H2B of wild-type and histone Deacetylase 1 mutant of *Neurospora crassa*. Biochemistry 49:5244–5257

Basenko EY, Sasaki T, Ji L et al (2015) Genome-wide redistribution of H3K27me3 is linked to genotoxic stress and defective growth. Proc Natl Acad Sci U S A 112:E6339–E6348

Belden WJ, Lewis ZA, Selker EU et al (2011) CHD1 remodels chromatin and influences transient DNA methylation at the clock gene frequency. PLoS Genet 7:e1002166

Belden WJ, Loros JJ, Dunlap JC (2007) Execution of the circadian negative feedback loop in Neurospora requires the ATP-dependent chromatin-remodeling enzyme CLOCKSWITCH. Mol Cell 25:587–600

Bicocca VT, Ormsby T, Adhvaryu KK et al (2018) ASH1-catalyzed H3K36 methylation drives gene repression and marks H3K27me2/3-competent chromatin. elife 7:e41497. https://doi.org/10.7554/eLife.41497

Borkovich KA, Alex LA, Yarden O et al (2004) Lessons from the genome sequence of *Neurospora crassa*: tracing the path from genomic blueprint to multicellular organism. Microbiol Mol Biol Rev 68:1–108

Brenna A, Grimaldi B, Filetici P, Ballario P (2012) Physical association of the WC-1 photoreceptor and the histone acetyltransferase NGF-1 is required for blue light signal transduction in *Neurospora crassa*. Mol Biol Cell 23:3863–3872

Buenrostro JD, Giresi PG, Zaba LC et al (2013) Transposition of native chromatin for fast and sensitive epigenomic profiling of open chromatin, DNA-binding proteins and nucleosome position. Nat Methods 10:1213–1218

Carrozza MJ, Li B, Florens L et al (2005) Histone H3 methylation by Set2 directs deacetylation of coding regions by Rpd3S to suppress spurious intragenic transcription. Cell 123:581–592

Cha J, Zhou M, Liu Y (2013) CATP is a critical component of the Neurospora circadian clock by regulating the nucleosome occupancy rhythm at the frequency locus. EMBO Rep 14:923–930

Cha J, Zhou M, Liu Y (2015) Mechanism of the Neurospora circadian clock, a FREQUENCY-centric view. Biochemistry 54:150–156

Chambers AL, Downs JA (2007) The contribution of the budding yeast histone H2A C-terminal tail to DNA-damage responses. Biochem Soc Trans 35:1519–1524

Chicas A, Forrest EC, Sepich S et al (2005) Small interfering RNAs that trigger posttranscriptional gene silencing are not required for the histone H3 Lys9 methylation necessary for transgenic tandem repeat stabilization in *Neurospora crassa*. Mol Cell Biol 25:3793–3801

Clapier CR, Cairns BR (2009) The biology of chromatin remodeling complexes. Annu Rev Biochem 78:273–304

Clapier CR, Iwasa J, Cairns BR, Peterson CL (2017) Mechanisms of action and regulation of ATP-dependent chromatin-remodelling complexes. Nat Rev Mol Cell Biol 18:407–422

Collins RE, Tachibana M, Tamaru H et al (2005) In vitro and in vivo analyses of a Phe/Tyr switch controlling product specificity of histone lysine methyltransferases. J Biol Chem 280:5563–5570

Colognori D, Sunwoo H, Kriz AJ et al (2019) Xist Deletional analysis reveals an interdependency between Xist RNA and Polycomb complexes for spreading along the inactive X. Mol Cell 74:101–117.e10

Connolly LR, Smith KM, Freitag M (2013) The *Fusarium graminearum* histone H3 K27 methyltransferase KMT6 regulates development and expression of secondary metabolite gene clusters. PLoS Genet 9:e1003916

Dang Y, Cheng J, Sun X et al (2016) Antisense transcription licenses nascent transcripts to mediate transcriptional gene silencing. Genes Dev 30:2417–2432

Dang Y, Li L, Guo W et al (2013) Convergent transcription induces dynamic DNA methylation at disiRNA loci. PLoS Genet 9:e1003761

Dekker J, Rippe K, Dekker M, Kleckner N (2002) Capturing chromosome conformation. Science 295:1306–1311

Dickey JS, Redon CE, Nakamura AJ et al (2009) H2AX: functional roles and potential applications. Chromosoma 118:683–692

Dong Q, Wang Y, Qi S et al (2018) Histone variant H2A. Z antagonizes the positive effect of the transcriptional activator CPC1 to regulate catalase-3 expression under normal and oxidative stress conditions. Free Radic Biol Med 121:136–148

Downs JA, Lowndes NF, Jackson SP (2000) A role for *Saccharomyces cerevisiae* histone H2A in DNA repair. Nature 408:1001–1004

Freitag M, Hickey PC, Khlafallah TK et al (2004a) HP1 is essential for DNA methylation in Neurospora. Mol Cell 13:427–434

Freitag M, Lee DW, Kothe GO et al (2004b) DNA methylation is independent of RNA interference in Neurospora. Science 304:1939

Freitag M, Williams RL, Kothe GO, Selker EU (2002) A cytosine methyltransferase homologue is essential for repeat-induced point mutation in *Neurospora crassa*. Proc Natl Acad Sci U S A 99:8802–8807

Gacek-Matthews A, Berger H, Sasaki T et al (2016) KdmB, a Jumonji histone H3 Demethylase, regulates genome-wide H3K4 trimethylation and is required for normal induction of secondary metabolism in *Aspergillus nidulans*. PLoS Genet 12:e1006222

Gacek-Matthews A, Noble LM, Gruber C et al (2015) KdmA, a histone H3 demethylase with bipartite function, differentially regulates primary and secondary metabolism in *Aspergillus nidulans*. Mol Microbiol 96:839–860

Galagan JE, Calvo SE, Borkovich KA et al (2003) The genome sequence of the filamentous fungus *Neurospora crassa*. Nature 422:859–868

Galagan JE, Selker EU (2004) RIP: the evolutionary cost of genome defense. Trends Genet 20:417–423

Galazka JM, Klocko AD, Uesaka M et al (2016) Neurospora chromosomes are organized by blocks of importin alpha-dependent heterochromatin that are largely independent of H3K9me3. Genome Res 26:1069–1080

Gessaman JD, Selker EU (2017) Induction of H3K9me3 and DNA methylation by tethered heterochromatin factors in. Proc Natl Acad Sci U S A 114:E9598–E9607

Gladyshev E (2017) Repeat-induced point mutation and other genome defense mechanisms in fungi. Microbiol Spectr 5:FUNK-0042-2017. https://doi.org/10.1128/microbiolspec.FUNK-0042-2017

Gladyshev E, Kleckner N (2017) DNA sequence homology induces cytosine-to-thymine mutation by a heterochromatin-related pathway in Neurospora. Nat Genet 49:887–894

Gong F, Chiu L-Y, Miller KM (2016) Acetylation reader proteins: linking acetylation signaling to genome maintenance and cancer. PLoS Genet 12:e1006272

Grewal SIS, Jia S (2007) Heterochromatin revisited. Nat Rev Genet 8:35–46

Grimaldi B, Coiro P, Filetici P et al (2006) The *Neurospora crassa* White Collar-1 dependent blue light response requires acetylation of histone H3 lysine 14 by NGF-1. Mol Biol Cell 17:4576–4583

Grimbergen AJ, Siebring J, Solopova A, Kuipers OP (2015) Microbial bet-hedging: the power of being different. Curr Opin Microbiol 25:67–72

Gu Q, Ji T, Sun X et al (2017) Histone H3 lysine 9 methyltransferase FvDim5 regulates fungal development, pathogenicity and osmotic stress responses in *Fusarium verticillioides*. FEMS Microbiol Lett 364. https://doi.org/10.1093/femsle/fnx184

He M, Xu Y, Chen J et al (2018) MoSnt2-dependent deacetylation of histone H3 mediates MoTor-dependent autophagy and plant infection by the rice blast fungus *Magnaporthe oryzae*. Autophagy 14:1543–1561

Heitz E (1928) Das Heterochromatin der Moose

Henikoff S, Smith MM (2015) Histone variants and epigenetics. Cold Spring Harb Perspect Biol 7: a019364

Henriksen MA (2007) Epigenetics. In: Allis CD, Jenuwein T, Reinberg D, Caparros M-L (eds) The quarterly review of biology, vol 82. Cold Spring Harbor Laboratory Press, New York, pp 416–416. $150.00. x 502 p; ill.; index. ISBN: 0-87969-724-5. 2007

Honda S, Bicocca VT, Gessaman JD et al (2016) Dual chromatin recognition by the histone deacetylase complex HCHC is required for proper DNA methylation in *Neurospora crassa*. Proc Natl Acad Sci U S A 113:E6135–E6144

Honda S, Lewis ZA, Huarte M et al (2010) The DMM complex prevents spreading of DNA methylation from transposons to nearby genes in *Neurospora crassa*. Genes Dev 24:443–454

Honda S, Lewis ZA, Shimada K et al (2012) Heterochromatin protein 1 forms distinct complexes to direct histone deacetylation and DNA methylation. Nat Struct Mol Biol 19:471–477. S1

Honda S, Selker EU (2008) Direct interaction between DNA methyltransferase DIM-2 and HP1 is

required for DNA methylation in *Neurospora crassa*. Mol Cell Biol 28:6044–6055
Hurley J, Loros JJ, Dunlap JC (2015) Dissecting the mechanisms of the clock in Neurospora. Methods Enzymol 551:29–52
Iyer V, Struhl K (1995) Poly(dA:dT), a ubiquitous promoter element that stimulates transcription via its intrinsic DNA structure. EMBO J 14:2570–2579
Jackson JP, Lindroth AM, Cao X, Jacobsen SE (2002) Control of CpNpG DNA methylation by the KRYPTONITE histone H3 methyltransferase. Nature 416:556–560
Jamieson K, McNaught KJ, Ormsby T et al (2018) Telomere repeats induce domains of H3K27 methylation in Neurospora. elife 7:e31216. https://doi.org/10.7554/eLife.31216
Jamieson K, Rountree MR, Lewis ZA et al (2013) Regional control of histone H3 lysine 27 methylation in Neurospora. Proc Natl Acad Sci U S A 110:6027–6032
Jamieson K, Wiles ET, McNaught KJ et al (2016) Loss of HP1 causes depletion of H3K27me3 from facultative heterochromatin and gain of H3K27me2 at constitutive heterochromatin. Genome Res 26:97–107
Janevska S, Baumann L, Sieber CMK et al (2018) Elucidation of the two H3K36me3 histone methyltransferases Set2 and Ash1 in *Fusarium fujikuroi* unravels their different chromosomal targets and a major impact of Ash1 on genome stability. Genetics 208:153–171
Janssen A, Colmenares SU, Karpen GH (2018) Heterochromatin: guardian of the genome. Annu Rev Cell Dev Biol 34:265–288
Jiao L, Liu X (2015) Structural basis of histone H3K27 trimethylation by an active polycomb repressive complex 2. Science 350:aac4383
Joshi AA, Struhl K (2005) Eaf3 chromodomain interaction with methylated H3-K36 links histone deacetylation to pol II elongation. Mol Cell 20:971–978
Keogh M-C, Kurdistani SK, Morris SA et al (2005) Cotranscriptional set2 methylation of histone H3 lysine 36 recruits a repressive Rpd3 complex. Cell 123:593–605
Kizer KO, Phatnani HP, Shibata Y et al (2005) A novel domain in Set2 mediates RNA polymerase II interaction and couples histone H3 K36 methylation with transcript elongation. Mol Cell Biol 25:3305–3316
Klocko AD, Ormsby T, Galazka JM et al (2016) Normal chromosome conformation depends on subtelomeric facultative heterochromatin in *Neurospora crassa*. Proc Natl Acad Sci U S A 113:15048–15053
Klocko AD, Rountree MR, Grisafi PL et al (2015) Neurospora importin α is required for normal heterochromatic formation and DNA methylation. PLoS Genet 11:e1005083
Klocko AD, Uesaka M, Ormsby T et al (2019) Nucleosome positioning by an evolutionarily conserved chromatin remodeler prevents aberrant DNA methylation in. Genetics 211:563–578
Kornberg RD (1974) Chromatin structure: a repeating unit of histones and DNA. Science 184:868–871
Kouzminova E, Selker EU (2001) Dim-2 encodes a DNA methyltransferase responsible for all known cytosine methylation in Neurospora. EMBO J 20:4309–4323
Krietenstein N, Wal M, Watanabe S et al (2016) Genomic nucleosome organization reconstituted with pure proteins. Cell 167:709–721.e12
Krogan NJ, Kim M, Tong A et al (2003) Methylation of histone H3 by Set2 in *Saccharomyces cerevisiae* is linked to transcriptional elongation by RNA polymerase II. Mol Cell Biol 23:4207–4218
Lai WKM, Pugh BF (2017) Understanding nucleosome dynamics and their links to gene expression and DNA replication. Nat Rev Mol Cell Biol 18:548–562
Lan H, Wu L, Sun R et al (2019) The HosA histone deacetylase regulates alatoxin biosynthesis through direct regulation of aflatoxin cluster genes. Mol Plant-Microbe Interact 32(9):1210–1228. MPMI01190033R
Lee H-C, Li L, Gu W et al (2010) Diverse pathways generate microRNA-like RNAs and dicer-independent small interfering RNAs in fungi. Mol Cell 38:803–814
Lewis ZA (2017) Polycomb Group Systems in Fungi: new models for understanding Polycomb repressive complex 2. Trends Genet 33:220–231
Lewis ZA, Adhvaryu KK, Honda S et al (2010a) DNA methylation and normal chromosome behavior in Neurospora depend on five components of a histone methyltransferase complex, DCDC. PLoS Genet 6:e1001196
Lewis ZA, Adhvaryu KK, Honda S et al (2010b) Identification of DIM-7, a protein required to target the DIM-5 H3 methyltransferase to chromatin. Proc Natl Acad Sci U S A 107:8310–8315
Lewis ZA, Honda S, Khlafallah TK et al (2009) Relics of repeat-induced point mutation direct heterochromatin formation in *Neurospora crassa*. Genome Res 19:427–437
Lieberman-Aiden E, van Berkum NL, Williams L et al (2009) Comprehensive mapping of long-range interactions reveals folding principles of the human genome. Science 326:289–293
Liu X, Dang Y, Matsu-Ura T et al (2017) DNA replication is required for circadian clock function by regulating rhythmic nucleosome composition. Mol Cell 67:203–213.e4
Liu CL, Kaplan T, Kim M et al (2005) Single-nucleosome mapping of histone modifications in *S. cerevisiae*. PLoS Biol 3:e328
Lombardi LM, Ellahi A, Rine J (2011) Direct regulation of nucleosome density by the conserved AAA-ATPase Yta7. Proc Natl Acad Sci U S A 108:E1302–E1311
Luger K, Mäder AW, Richmond RK et al (1997) Crystal structure of the nucleosome core particle at 2.8 a resolution. Nature 389:251–260
Maeda K, Izawa M, Nakajima Y et al (2017) Increased metabolite production by deletion of an HDA1-

type histone deacetylase in the phytopathogenic fungi, *Magnaporthe oryzae* (*Pyricularia oryzae*) and Fusarium asiaticum. Lett Appl Microbiol 65:446–452

Margolin BS, Garrett-Engele PW, Stevens JN et al (1998) A methylated Neurospora 5S rRNA pseudogene contains a transposable element inactivated by repeat-induced point mutation. Genetics 149:1787–1797

Martienssen R, Moazed D (2015) RNAi and heterochromatin assembly. Cold Spring Harb Perspect Biol 7: a019323

Miao VP, Freitag M, Selker EU (2000) Short TpA-rich segments of the zeta-eta region induce DNA methylation in *Neurospora crassa*. J Mol Biol 300:249–273

Möller M, Schotanus K, Soyer JL et al (2019) Destabilization of chromosome structure by histone H3 lysine 27 methylation. PLoS Genet 15:e1008093

Neigeborn L, Carlson M (1984) Genes affecting the regulation of SUC2 gene expression by glucose repression in *Saccharomyces cerevisiae*. Genetics 108:845–858

Niehaus E-M, Studt L, von Bargen KW et al (2016) Sound of silence: the beauvericin cluster in *Fusarium fujikuroi* is controlled by cluster-specific and global regulators mediated by H3K27 modification. Environ Microbiol 18:4282–4302

Nielsen PR, Nietlispach D, Mott HR et al (2002) Structure of the HP1 chromodomain bound to histone H3 methylated at lysine 9. Nature 416:103–107

Olins AL, Olins DE (1974) Spheroid chromatin units (v bodies). Science 183:330–332

Oudet P, Gross-Bellard M, Chambon P (1975) Electron microscopic and biochemical evidence that chromatin structure is a repeating unit. Cell 4:281–300

Palmer JM, Bok JW, Lee S et al (2013) Loss of CclA, required for histone 3 lysine 4 methylation, decreases growth but increases secondary metabolite production in *Aspergillus fumigatus*. PeerJ 1: e4

Pfannenstiel BT, Greco C, Sukowaty AT, Keller NP (2018) The epigenetic reader SntB regulates secondary metabolism, development and global histone modifications in *Aspergillus flavus*. Fungal Genet Biol 120:9–18

Pokholok DK, Harbison CT, Levine S et al (2005) Genome-wide map of nucleosome acetylation and methylation in yeast. Cell 122:517–527

Pueschel R, Coraggio F, Meister P (2016) From single genes to entire genomes: the search for a function of nuclear organization. Development 143:910–923

Rabl C (1885) Über Zelltheilung. Morphologisches Jahrbuch 10

Raduwan H, Isola AL, Belden WJ (2013) Methylation of histone H3 on lysine 4 by the lysine methyltransferase SET1 protein is needed for normal clock gene expression. J Biol Chem 288:8380–8390

Rea S, Eisenhaber F, O'Carroll D et al (2000) Regulation of chromatin structure by site-specific histone H3 methyltransferases. Nature 406:593–599

Rountree MR, Selker EU (1997) DNA methylation inhibits elongation but not initiation of transcription in *Neurospora crassa*. Genes Dev 11:2383–2395

Sancar C, Ha N, Yilmaz R et al (2015) Combinatorial control of light induced chromatin remodeling and gene activation in Neurospora. PLoS Genet 11:e1005105

Sasaki T, Lynch KL, Mueller CV et al (2014) Heterochromatin controls γH2A localization in *Neurospora crassa*. Eukaryot Cell 13:990–1000

Schmitt AD, Hu M, Ren B (2016) Genome-wide mapping and analysis of chromosome architecture. Nat Rev Mol Cell Biol 17:743–755

Segal E, Widom J (2009) Poly(dA:dT) tracts: major determinants of nucleosome organization. Curr Opin Struct Biol 19:65–71

Selker EU (2002) Repeat-induced gene silencing in fungi. Adv Genet 46:439–450

Selker EU, Cambareri EB, Jensen BC, Haack KR (1987) Rearrangement of duplicated DNA in specialized cells of Neurospora. Cell 51:741–752

Selker EU, Fritz DY, Singer MJ (1993) Dense nonsymmetrical DNA methylation resulting from repeat-induced point mutation in Neurospora. Science 262:1724–1728

Seymour M, Ji L, Santos AM et al (2016) Histone H1 limits DNA methylation in *Neurospora crassa*. G3 6:1879–1889

Shatkin AJ, Tatum EL (1959) Electron microscopy of *Neurospora crassa* mycelia. J Biophys Biochem Cytol 6:423–426

Shwab EK, Bok JW, Tribus M et al (2007) Histone deacetylase activity regulates chemical diversity in Aspergillus. Eukaryot Cell 6:1656–1664

Simon JA, Kingston RE (2013) Occupying chromatin: polycomb mechanisms for getting to genomic targets, stopping transcriptional traffic, and staying put. Mol Cell 49:808–824

Smith KM, Dobosy JR, Reifsnyder JE et al (2010) H2B- and H3-specific histone deacetylases are required for DNA methylation in *Neurospora crassa*. Genetics 186:1207–1216

Smith KM, Kothe GO, Matsen CB et al (2008) The fungus *Neurospora crassa* displays telomeric silencing mediated by multiple sirtuins and by methylation of histone H3 lysine 9. Epigenetics Chromatin 1:1–20

Smith KM, Phatale PA, Sullivan CM et al (2011) Heterochromatin is required for normal distribution of *Neurospora crassa* CenH3. Mol Cell Biol 31:2528–2542

Solovei I, Thanisch K, Feodorova Y (2016) How to rule the nucleus: divide et impera. Curr Opin Cell Biol 40:47–59

Soukup AA, Chiang Y-M, Bok JW et al (2012) Overexpression of the *Aspergillus nidulans* histone 4 acetyltransferase EsaA increases activation of secondary metabolite production. Mol Microbiol 86:314–330

Soyer JL, El Ghalid M, Glaser N et al (2014) Epigenetic control of effector gene expression in the plant pathogenic fungus *Leptosphaeria maculans*. PLoS Genet 10:e1004227

Stern M, Jensen R, Herskowitz I (1984) Five SWI genes are required for expression of the HO gene in yeast. J Mol Biol 178:853–868

Studt L, Rösler SM, Burkhardt I et al (2016) Knock-down of the methyltransferase Kmt6 relieves H3K27me3 and results in induction of cryptic and otherwise silent secondary metabolite gene clusters in *Fusarium fujikuroi*. Environ Microbiol 18:4037–4054

Sun G, Zhou Z, Liu X et al (2016) Suppression of WHITE COLLAR-independent frequency transcription by histone H3 lysine 36 methyltransferase SET-2 is necessary for clock function in Neurospora. J Biol Chem 291:11055–11063

Talbert PB, Henikoff S (2010) Histone variants—ancient wrap artists of the epigenome. Nat Rev Mol Cell Biol 11:264–275

Tamaru H, Selker EU (2001) A histone H3 methyltransferase controls DNA methylation in *Neurospora crassa*. Nature 414:277–283

Tamaru H, Selker EU (2003) Synthesis of signals for de novo DNA methylation in *Neurospora crassa*. Mol Cell Biol 23:2379–2394

Tamaru H, Zhang X, McMillen D et al (2003) Trimethylated lysine 9 of histone H3 is a mark for DNA methylation in *Neurospora crassa*. Nat Genet 34:75–79

Trojer P, Reinberg D (2007) Facultative heterochromatin: is there a distinctive molecular signature? Mol Cell 28:1–13

Verdone L, Caserta M, Di Mauro E (2005) Role of histone acetylation in the control of gene expression. Biochem Cell Biol 83:344–353

Wang Y, Dong Q, Ding Z et al (2016) Regulation of Neurospora catalase-3 by global heterochromatin formation and its proximal heterochromatin region. Free Radic Biol Med 99:139–152

Wang B, Kettenbach AN, Gerber SA et al (2014) Neurospora WC-1 recruits SWI/SNF to remodel frequency and initiate a circadian cycle. PLoS Genet 10:e1004599

Wang Y, Smith KM, Taylor JW, et al (2015) Endogenous small RNA mediates meiotic silencing of a novel DNA transposon. G3 5:1949–1960

Wendte JM, Pikaard CS (2017) The RNAs of RNA-directed DNA methylation. Biochim Biophys Acta Gene Regul Mech 1860:140–148

Xiong L, Adhvaryu KK, Selker EU, Wang Y (2010) Mapping of lysine methylation and acetylation in core histones of *Neurospora crassa*. Biochemistry 49:5236–5243

Xu H, Wang J, Hu Q et al (2010) DCAF26, an adaptor protein of Cul4-based E3, is essential for DNA methylation in *Neurospora crassa*. PLoS Genet 6: e1001132

Xue Z, Ye Q, Anson SR et al (2014) Transcriptional interference by antisense RNA is required for circadian clock function. Nature 514:650–653

Yang Z, Qian S, Scheid RN et al (2018) EBS is a bivalent histone reader that regulates floral phase transition in Arabidopsis. Nat Genet 50:1247–1253

Zacharias H (1995) Emil Heitz (1892-1965): chloroplasts, heterochromatin, and polytene chromosomes. Genetics 141:7–14

Zhang X, Tamaru H, Khan SI et al (2002) Structure of the Neurospora SET domain protein DIM-5, a histone H3 lysine methyltransferase. Cell 111:117–127

Zhang X, Yang Z, Khan SI et al (2003) Structural basis for the product specificity of histone lysine methyltransferases. Mol Cell 12:177–185

Zhao Y, Shen Y, Yang S et al (2010) Ubiquitin ligase components Cullin4 and DDB1 are essential for DNA methylation in *Neurospora crassa*. J Biol Chem 285:4355–4365

Zhou Y, Cambareri EB, Kinsey JA (2001) DNA methylation inhibits expression and transposition of the Neurospora tad retrotransposon. Mol Gen Genomics 265:748–754

Zhou Z, Liu X, Hu Q et al (2013) Suppression of WC-independent frequency transcription by RCO-1 is essential for Neurospora circadian clock. Proc Natl Acad Sci U S A 110:E4867–E4874

Zhu Q, Ramakrishnan M, Park J, Belden WJ (2019) Histone H3 lysine 4 methyltransferase is required for facultative heterochromatin at specific loci. BMC Genomics 20:350

Zou JX, Revenko AS, Li LB et al (2007) ANCCA, an estrogen-regulated AAA+ ATPase coactivator for ERalpha, is required for coregulator occupancy and chromatin modification. Proc Natl Acad Sci U S A 104:18067–18072

2 Origin, Function, and Transmission of Accessory Chromosomes

MICHAEL HABIG[1], EVA H. STUKENBROCK[1]

CONTENTS

I. Introduction

Accessory chromosomes are found in some but not all individuals of a population, in contrast to essential chromosomes that are always present. This presence/absence polymorphism of entire chromosomes was first observed in 1907 in a dipteran species using classical karyotyping (Wilson 1907). Since then accessory chromosomes, also known as B, supernumerary, lineage-specific, or (conditionally) dispensable chromosomes, have been described in a large variety of plant, animal, and fungal species, including approximately 14% of karyotyped orthopteran insect species (Jones 1995) as well as 8% of monocot and 3% of eudicot species (Levin et al. 2005). Across this wide range of taxa, accessory chromosomes share the following defining characteristics: (i) they are not essential for growth and development of the organism, (ii) they do not recombine with the essential chromosomes, and (iii) they often do not follow Mendelian inheritance (Jones 1995). Historically, the amenability of some accessory chromosomes in plants and animals (called B chromosomes in these species) to light microscopy allowed for earlier detection and analysis in these kingdoms. In fungi however, the relatively small size of accessory chromosomes hindered their detection using conventional light microscopy (Mehrabi et al. 2017). The development of pulsed-field electrophoresis in 1984 (Schwartz and Cantor 1984) enabled the separation of fungal chromosomes, and with this technique the comparison of karyotypes between isolates led to the first discovery of fungal accessory chromosomes in *Nectria haematococca* in 1991 (Miao et al. 1991). The advent of next-generation sequencing and availability of whole genome sequences have since then fueled detailed analyses including population genomic analyses of accessory chromosomes (Plissonneau et al. 2018).

What is the origin of accessory chromosomes? Two non-exclusive models have been proposed to explain their presence. They may originate from essential core chromosomes and diverge or degenerate over time (Galazka and Freitag 2014), or they may be acquired by horizontal chromosome transfer (Mehrabi et al. 2011; Mehrabi et al. 2017). Both models are supported by current data: in some species,

This chapter partly overlaps with a chapter in a thesis which MH submitted to the Christian-Albrechts University of Kiel, Germany, in order to obtain a PhD.

[1] Environmental Genomics, Christian-Albrechts University of Kiel and Max Planck Institute for Evolutionary Biology, Plön, Germany; e-mail: mhabig@bot.uni-kiel.de; estukenbrock@bot.uni-kiel.de

Genetics and Biotechnology, 3rd Edition
The Mycota II
J.P. Benz, K. Schipper (Eds.)

high similarity in sequence composition between core and accessory chromosomes suggests an origin within the same genome, while in other species significant differences in sequence composition may indicate horizontal transfer (Akagi et al. 2009a; Ma et al. 2010). Phylogenetic analyses of fungal plant pathogens, for example, indicate that the accessory chromosomes of *Fusarium oxysporum* f. sp. *lycopersici* causing *Fusarium* wilt of tomato have a different origin than the core chromosome (Ma et al. 2010). Similarly, the accessory chromosomes in asexual *Alternaria alternata* strains appear to have been horizontally transferred between lineages (Akagi et al. 2009a). The presence of mechanisms that allow horizontal transfer of entire chromosomes between distinct lineages has furthermore been experimentally demonstrated (Akagi et al. 2009a, b; He et al. 1998; Ma et al. 2010; Masel 1996). It is therefore possible that mechanisms of both models are involved in the origin of new accessory chromosomes in different taxa, although their relative importance appears to vary between species.

In plants and animals, B chromosomes are widespread, and no unifying pattern of taxonomic distribution is apparent. In contrast, **accessory chromosomes in fungi are restricted almost exclusively to plant symbionts with a large fraction of these being plant pathogens** (Bertazzoni et al. 2018; Mehrabi et al. 2017; Soyer et al. 2018). Why are fungal accessory chromosomes mainly found in plant pathogens? Fungal plant pathogens represent a very intensively studied group of organisms, and notably genetic and genomic variation has been a focus of research (Möller and Stukenbrock 2017). It is therefore possible that the excess of examples in this group of organisms simply reflects the research coverage. J. Taylor and colleagues have moreover suggested that the enrichment of accessory chromosomes in plant-associated fungi may reflect the need to rapidly adapt to changes in plant defense (Taylor et al. 2017). Host specificity determinants of several plant-associated fungi locate on accessory chromosomes (Ahn and Walton 1996; Coleman et al. 2009; Condon et al. 2013; Johnson et al. 2001; Ma et al. 2010). Possibly, this particular genomic location provides an advantage for virulence-related traits as they can be modified readily without affecting core processes. In this way, accessory chromosomes may represent specific genome compartments that allow these fungi to rapidly respond to changes in host plant defenses (Taylor et al. 2017).

The function of accessory chromosomes appears to be highly variable. B chromosomes in plants and animals often have no known function, and the fitness effect is assumed to be neutral or even negative (Houben et al. 2014). However, a few examples demonstrate that B chromosomes in plants and animals can also confer a fitness advantage (Jones 1995; Pereira et al. 2017; Yoshida et al. 2011). **In several pathogenic fungi, accessory chromosomes directly influence pathogenicity** as they are required for the infection of specific hosts and hence provide a benefit (Ahn and Walton 1996; Coleman et al. 2009; Condon et al. 2013; Johnson et al. 2001; Ma et al. 2010). In these cases, the respective chromosomes encode traits that confer a fitness advantage and natural selection should favor individuals carrying these chromosomes. By contrast, the accessory chromosomes of the wheat pathogen *Zymoseptoria tritici* were shown to cause a fitness disadvantage (Habig et al. 2017), highlighting the variety of fitness effects associated with accessory chromosomes. However, for many fungal accessory chromosomes, function and fitness effects have yet to be determined, and the influence of selection on the presence of these chromosomes still remains unclear (Soyer et al. 2018).

How are accessory chromosomes transmitted during cell division? Based on cytological studies, much more is known about the transmission of accessory chromosomes during mitosis and meiosis in plants and animals than in fungi. Many B chromosomes in plant and animals appear to offset their negative fitness effect by increasing their relative frequency during cell divisions compared to the essential A chromosomes. This increase in frequency has been described as **chromosome drive** which results from **segregation distortion before, during, or after meiosis** (Houben 2017).

Therefore, many B chromosomes of plants and animals are considered to be selfish genetic elements propagating themselves at a cost for the organism (Houben 2017). Similarly, in fungi, the accessory chromosomes of *Botrytis cinerea*, *Cochliobolus heterostrophus*, *Gibberella fujikuroi* mating population A, Leptosphaeria maculans, *Magnaporthe oryzae*, *N. haematococca* mating population VI, and *Z. tritici* show non-Mendelian inheritance during meiosis (Table 1). This non-Mendelian inheritance involves either frequent losses or transmission advantages, i.e., transmission of an accessory chromosome to more progeny than predicted by Mendelian segregation (Coleman et al. 2009; Croll et al. 2013; He et al. 1998; Mehrabi et al. 2017; Orbach et al. 1996; Soyer et al. 2018; Wittenberg et al. 2009; Xu and Leslie 1996). In this respect, it is likely that accessory chromosomes are also propagated by a drive mechanism in fungi. In support of this, we recently demonstrated a meiotic chromosome drive of accessory chromosomes in *Z. tritici* (Habig et al. 2018).

For fungi without a sexual cycle, only mitotic transmission of accessory chromosomes is relevant. In these cases, **fungal accessory chromosomes** can again **show distinct mitotic transmission patterns** with extremely high frequencies of chromosome losses and chromosome rearrangements (Möller et al. 2018; Vlaardingerbroek et al. 2016a). We argue that the distinct transmission patterns of accessory chromosomes during mitosis and meiosis determine the distribution and frequency of accessory chromosomes in fungal populations. A plausible explanation for this chromosome variation is the ability of fungi to rapidly generate new phenotypes in changing environments. However, there is no direct evidence for this hypothesis in natural populations of fungi.

We here address the origin, function, and transmission of accessory chromosomes in fungi. A main focus of our review is the **mitotic and meiotic transmission of accessory chromosomes**, which in many fungi clearly differs from the transmission of core chromosomes. Since the accessory chromosomes of fungi, plants, and animals share characteristics, we first briefly discuss the B chromosomes of plants and animals where previous findings can complement the knowledge available on the accessory chromosomes of fungi. We first give an overview on the occurrence and characteristics of fungal accessory chromosomes with a focus on their function and possible origin. We use the term accessory chromosome as a general term and only mention the specific names used in distinct species when referring to examples in detail. Similarly, we will use the term core chromosomes to specify the chromosomes that are shared by all members of a population.

II. Accessory Chromosomes Are Widespread and Diverse but Share Specific Characteristics

In plants, accessory chromosomes are often described as B chromosomes contrasting to the essential A chromosomes shared by all individuals of a species. One of the best-studied examples is the rye (*Secale cereale*) B chromosome of which up to eight copies can be found within a single cell. In animals, B chromosomes have been described, among others, in fish, amphibian, and insects, including a recent example from the model system *Drosophila melanogaster* (Bauerly et al. 2014). These plant and animal B chromosomes share some characteristics: they usually comprise highly repetitive DNA sequences which show sequence similarities to either A chromosomes within the same genome or A chromosomes in closely related species (Pansonato-Alves et al. 2014). These sequences can be derived from mobile genetic elements, satellite DNA, or ribosomal DNA (Bougourd and Jones 1997; Dhar et al. 2002; Houben et al. 2001; Houben 2017; Lamb et al. 2007; Pansonato-Alves et al. 2014) and are usually found to be heterochromatic (Cheng 2010; Östergren 1947) and gene-poor (Banaei-Moghaddam et al. 2015).

The first fungal accessory chromosomes were described for two plant pathogenic species: *Nectria haematococca* (Miao et al. 1991) and *Cochliobolus heterostrophus* (Tzeng et al.

Table 1 Summary of properties of fungal accessory chromosomes

Species	Number of accessory chromosomes	Size range of accessory chromosomes (Mb)	Genes on accessory chromosomes influencing pathogenicity	Phenotype when present	Mitotic transmission	Meiotic transmission	References
Alternaria alternata	1	< 2	*Tox* genes	Host-specific pathogenicity	Nd (spontaneous loss observed in apple pathotype)	Na	(Akagi et al. 2009b; Akamatsu et al. 1997; Akamatsu et al. 1999; Hatta et al. 2002; Hu et al. 2012; Johnson et al. 2001, 2001; Kohmoto 1993; Nakashima et al. 1985; Nakatsuka et al. 1986; Salamiah et al. 2001b; Salamiah et al. 2001a)
Botrytis cinerea	3	0.22–0.58	None identified	None identified	Nd	Non-Mendelian	(Kan et al. 1993; Van Kan et al. 2017)
Cochliobolus carbonum	2	0.75–3.5	*TOX2*	High virulence on certain genotypes of maize	Nd	Mendelian	(Ahn and Walton 1996; Condon et al. 2013)
Cochliobolus heterostrophus	1	0.75–1.2	None identified	None identified	Nd	Non-Mendelian	(Condon et al. 2013; Tzeng et al. 1992)
Colletotrichum gloeosporioides	2	1.2–2	None identified	None identified	Nd	Na	(Masel et al. 1993; Masel 1996)
Fusarium oxysporum f. sp. *lycopersici*	4	< 3.5	Secreted in xylem (*Six*) genes	Pathogenicity on tomato	Loss in 1 of 35,000 spores	Na	(Ma et al. 2010; Rep et al. 2004; Vlaardingerbroek et al. 2016a)
Gibberella fujikuroi MP A	1	0.7	None identified	None identified	Nd	Non-Mendelian	(Xu 1995; Xu and Leslie 1996)
Leptosphaeria maculans	2	0.7–0.95	*AvrLm11*,Five effector candidates	Avirulence on hosts containing *Rlm11*	Nd	Non-Mendelian	(Balesdent et al. 2013; Rouxel et al. 2011; Rouxel and Balesdent 2017)
Magnaporthe oryzae	2	0.5–2	*AVR-Pita*	Unknown	Nd	Non-Mendelian	(Chuma et al. 2003; Chuma et al. 2011; Orbach et al. 1996; Yoshida et al. 2009)
Nectria haematococca MP VI	3	0.53–1.6	Pea pathogenicity (PEP) cluster	Pathogenicity on pea	Nd	Non-Mendelian	(Coleman et al. 2009; Miao et al. 1991)
Zymoseptoria tritici	8	0.409–0.773	None identified	Decreased virulence	Loss at high frequency (up to 1 in 2 spores)	Non-Mendelian	(Croll et al. 2013; Goodwin et al. 2011; Habig et al. 2017; Möller et al. 2018; Wittenberg et al. 2009)

[a]Abbreviations: *nd* not determined, *na* not applicable, *MP* mating population

1992). Today we know that the genomes of a **wide range of fungal species, in particular plant pathogenic fungi, comprise accessory chromosomes** (Bertazzoni et al. 2018; Mehrabi et al. 2017; Soyer et al. 2018). Here, we will focus on a few cases that more generally exemplify properties of accessory chromosomes in fungi.

N. haematococca can be found in diverse habitats: as a soil saprophyte, as a commensal organism in the rhizosphere, as a pathogen of many different plant species, and as an opportunistic pathogen in humans (Vanetten 1978; Zhang et al. 2006). The ability to thrive in these diverse habitats is partly determined by the presence of accessory chromosomes which have been called conditionally dispensable chromosomes in this fungus (Han et al. 2001; Miao et al. 1991). Three of the 17 chromosomes found in *N. haematococca* mating population VI (MPVI), namely, chromosome (chr.) 14, chr. 15, and chr. 17, were shown to be dispensable (Coleman et al. 2009). Similar to the B chromosomes in plants and animals, the conditionally dispensable chromosomes of *N. haematococca* MPVI contain an increased abundance of repetitive sequences and show a lower GC content in comparison to the core chromosomes (Coleman et al. 2009). They harbor more unique genes (i.e., genes without homologs), and these may reflect adaptation to different environmental niches of individual isolates (Coleman et al. 2009).

The accessory chromosomes of the economically important wheat pathogen *Zymoseptoria tritici* comprise 12% of the genome in the reference isolate IPO323 with eight distinct accessory chromosomes that range in size from 409 to 773 kb (Goodwin et al. 2011). This represents one of the largest complements of accessory chromosomes described to date. In comparison to the core chromosomes, the accessory chromosomes of *Z. tritici* are relatively **gene-poor** (Goodwin et al. 2011; Grandaubert et al. 2015). The genes located on these chromosomes show a **distinct codon usage** which differs from the codon usage of the genes located on the core chromosomes (Goodwin et al. 2011). Interestingly, only few genes on the accessory chromosomes encode proteins of known function, and most of the predicted genes do not comprise any annotated functional domains (Grandaubert et al. 2015). In contrast to accessory chromosomes of other fungal pathogens, the accessory chromosomes of *Z. tritici* contain a significantly lower number of genes encoding secreted proteins and putative virulence determinants (Grandaubert et al. 2015; Rep et al. 2004; Vlaardingerbroek et al. 2016a). Only few genes belong to gene families that include members on both the core and accessory chromosomes, and a detailed survey of gene duplicates showed a low number of paralogous sequences on the accessory chromosomes (Kellner et al. 2014). The accessory chromosomes of *Z. tritici* are **enriched in repetitive DNA and transposable elements** (Grandaubert et al. 2015) and, similar to the B chromosomes in plants, show **histone modifications associated with heterochromatin** (Schotanus et al. 2015). In comparison to the core chromosomes, they are highly enriched in post-translational modification of the histone H3 by trimethylation of lysine 27 (H3K27me3) (Schotanus et al. 2015). Sequencing of the IPO323 genome led to the hypothesis that the accessory chromosomes in this organism originated from horizontal chromosome transfer from another organism (Goodwin et al. 2011). However, the fact that closely related species of *Z. tritici* like *Zymoseptoria pseudotritici*, *Zymoseptoria ardabiliae*, and *Zymoseptoria brevis* also harbor accessory chromosomes including regions syntenic to regions of the accessory chromosomes of *Z. tritici* suggest that these chromosomes represent an ancient trait in the genus (Feurtey, Lorrain et al., in prep).

The characteristic structural features that distinguish accessory from core chromosomes are also present in *Leptosphaeria maculans*, a pathogen that causes stem canker (blackleg) of oilseed rape (*Brassica napus*) and related crucifers (Rouxel and Balesdent 2005; West et al. 2001). *L. maculans* contains a dispensable mini-chromosome of 700–950 kb in size (Leclair et al. 1996) which mostly comprises AT-enriched isochores (>60% AT). Besides being AT-rich, this mini-chromosome is gene-poor (Balesdent et al. 2013; Rouxel et al. 2011) and heterochromatic (Soyer et al. 2014). Interestingly, the AT isochores are enriched in transposable elements which are affected by repeat-induced point mutation (RIP), a fungal-specific genome defense mechanism against repetitive sequences (Selker 1990; Soyer et al. 2014). RIP can affect neighboring sequences (Gladyshev and Kleckner 2017), and therefore genes located within the AT isochores are subject to increased mutation rates (Rouxel et al. 2011). Among the genes located within the AT isochores, many encode effectors—small secreted proteins that manipulate the host defense (Jones and Dangl 2006; Lo Presti et al. 2015)—and their location within these AT isochores could promote rapid sequence diversification of these genes.

Genome comparison between *Fusarium oxysporum* f. sp. *lycopersici* strain *Fol4287* and *Fusarium verticillioides* showed that *Fol4287* contains chromosomes 3, 6, 14, and 15 and parts of the chromosomes 1 and 2 which do not have syntenic chromosomes/regions in *F. verticillioides* (Ma et al. 2010). These accessory chromosomes and regions of chromosomes are called lineage-specific in this tomato pathogen. Again, the lineage-specific regions are **rich in transposons**; show a **lower gene density**, a **higher proportion of unique genes**, and **a different codon usage;** and **may have a distinct phylogenetic history** compared the essential chromosomes (Ma et al. 2010). In several additional fungal species, accessory chromosomes have been described. These include the plant pathogens *Magnaporthe oryzae* (Chuma et al. 2011; Dean et al. 2005), *Cochliobolus carbonum* (Ahn and Walton 1996), *Colletotrichum gloeosporioides* (He et al. 1998; Masel et al. 1993), *Gibberella fujikuroi*, and several host-specific lineages of *Alternaria alternata* (Akamatsu et al. 1999; Hatta et al. 2002; Hu et al. 2012; Johnson et al. 2001). Recently, an accessory chromosome has also been described in *Botrytis cinerea* (Van Kan et al. 2017). It is striking that accessory chromosomes have been found mostly in plant pathogenic fungi. One exception is the insect pathogenic fungus *Metarhizium anisopliae*, which harbors a single accessory chromosome (Wang et al. 2003) and is used as a biocontrol agent (Schrank and Vainstein 2010).

In conclusion, the accessory chromosomes in fungi show characteristics similar to accessory chromosomes found in other organisms as they are enriched in repetitive elements, are mainly heterochromatic, and often show a different GC content and codon usage compared to the core chromosomes. However, these characteristics are not exclusive and thus do not allow unequivocal identification of accessory chromosomes because sequences with similar characteristics can also be found within essential core chromosomes (Möller and Stukenbrock 2017).

III. Fungal Accessory Chromosomes Are Generally Associated with Function

The dispensability of accessory chromosomes for growth and development raises questions on their functional role. B chromosomes in plants and animals are mostly heterochromatic, and therefore gene expression has been assumed to be absent or repressed (Banaei-Moghaddam et al. 2013). Recent studies, however, report active transcription of coding sequences located on a number of plant and animal B chromosomes, indicating a functional role of genes encoded on these chromosomes (Banaei-Moghaddam et al. 2013; Carchilan et al. 2007; Graphodatsky et al. 2005; Ma et al. 2017; Pereira et al. 2017; Ramos et al. 2017; Trifonov et al. 2013; Yoshida et al. 2011). However, transcription is in general found to be lower compared to genes located on the core chromosomes (Banaei-Moghaddam et al. 2015). To date, a fitness benefit of individuals carrying B chromosomes has only been reported for a small number of plant and animal species (Jones 1995; Yoshida et al. 2011). One of these few cases is the exemplary rye B chromosome, which protects meiocytes against heat stress-induced damage (Pereira et al. 2017).

In contrast, **genes located on fungal accessory chromosomes are often associated with a function** (see Table 1 and references therein).

Some genes directly influence pathogenicity and host range. One of the conditionally dispensable chromosomes of *Nectria haematococca*, the PDA1-CD chromosome, carries, for example, a cluster of genes required for pea pathogenicity (PEP) (Han et al. 2001). *N. haematococca* isolates with the PDA1-CD chromosome are highly virulent on pea plants due to the presence of the PDA1 gene which is part of the PEP cluster and codes for a cytochrome P450 enzyme that detoxifies the pea phytoalexin pisatin (Maloney and VanEtten 1994). By contrast, *N. haematococca* isolates lacking the PDA1-CD chromosome are strongly attenuated in the ability to cause lesions on pea (Han et al. 2001). The dispensable chromosome of *Leptosphaeria maculans* also contains genes that affect its host range. These include the avirulence gene *AvrLm11* conferring avirulence on *Brassica rapa* (Balesdent et al. 2013) located on one of the dispensable chromosomes of *L. maculans* as well as five additional putative effector genes (Rouxel et al. 2011). The loss of the dispensable chromosome of *L. maculans* leads to susceptibility of *B. rapa* carrying the resistance gene *Rlm11* (Rouxel and Balesdent 2017). However, the reason for the continued maintenance of the accessory chromosome encoding the *AvrLm11* in *L. maculans* populations remains unclear to date. Individuals with this accessory chromosome are avirulent on *Rlm11* carrying hosts and therefore would have a substantial fitness disadvantage. We speculate that a counteracting selection, possibly a transmission advantage during meiotic or mitotic reproduction, maintains this chromosome in the population.

In *Zymoseptoria tritici*, the functional role of the accessory chromosomes is less clear. Based on quantitative trait loci mapping, small but significant fitness benefits for certain accessory chromosomes were detected (Stewart et al. 2016). However, isogenic *Z. tritici* strains with deletions of whole accessory chromosomes could produce more pycnidia than the wild type, and therefore the **accessory chromosomes 14, 16, 18, 19, and 21 rather confer a fitness cost** (Habig et al. 2017). Interestingly, this fitness cost was dependent on the host wheat cultivar suggesting an interaction of host genotype-specific traits and traits encoded by the accessory chromosomes. *Z. tritici* switches to necrotrophic growth after a prolonged period of symptomless, biotrophic growth. Several accessory chromosomes of *Z. tritici* influence the timing of this lifestyle switch (Habig et al. 2017), but yet it is unclear how this affects the fitness of the pathogen in the field.

Host specificity is also influenced by the accessory chromosomes in lineages of the genus *Alternaria*. These plant-pathogenic fungi infect a remarkably broad range of hosts (Rotem 1994). Accessory chromosomes, which are called conditionally dispensable chromosomes in *Alternaria*, are carried by several *Alternaria* lineages (Akamatsu et al. 1999; Hatta et al. 2002; Johnson et al. 2001). These are generally smaller than 2.0 MB in size (Akagi et al. 2009b; Salamiah et al. 2001a, b) and represent a prominent example of how host specificity is determined by traits encoded on accessory chromosomes. Several **genes that are encoding host-specific toxins (HSTs)** or are involved in the biosynthesis of toxins are located within gene clusters **on the conditionally dispensable chromosomes** (Nakatsuka et al. 1986; Nakashima et al. 1985; Kohmoto 1993; Johnson et al. 2001; Akamatsu et al. 1997). Thereby, these genes confer a strong selection advantage during host-pathogen interaction, and this selection appears to maintain these chromosomes in the *Alternaria* populations.

Fusarium oxysporum reproduces asexually and consists of many pathogenic clonal lineages which are grouped into host-specific *formae speciales* (Baayen et al. 2000; Lievens et al. 2008; O'Donnell et al. 1998; Recorbet et al. 2003). The ability of *F. oxysporum* to infect and avoid recognition in its host plant depends on small Secreted-in-xylem (SIX) proteins that are secreted into the host xylem during the infection (Rep et al. 2004). In the tomato-pathogen *F. oxysporum* f. sp. *lycopersici*, the lineage-specific chromosome 14 harbors six SIX genes, and the lineage-specific chromosomes in general are enriched in genes coding for secreted effectors and virulence factors (Ma et al. 2010; Schmidt et al. 2013). In the pea-pathogen *F. oxysporum* f. sp. *pisi*, homologs of the pisatin-detoxifying PDA and PEP genes of *N. haematococca* are located on chromosomes considered to be dispensable (Coleman et al. 2011; Milani et al. 2012; Williams et al. 2016). Furthermore, in *F. oxysporum* f. sp. *medicaginis* and *F. oxysporum* f. sp. *ciceris* genes with possible pathogenicity-related functions are located on dispensable sequences (Williams et al. 2016). Although a direct link between pathogenicity and host specificity is lacking for *Magnaporthe oryzae*, it is remarkable that the accessory sequences in this fungus harbor 316 candidate effector genes, defined as genes encoding secreted proteins (Yoshida et al. 2009). In addition, the avirulence gene *AVR-Pita* was found to be located at multiple genomic locations including loci on the accessory chromosome (Chuma et al. 2011; Yoshida et al. 2009). Hence, the presence/absence of the accessory chromosome may influence the ability to infect different host genotypes. Polymorphic localization of *AVR-Pita* (including on the accessory chromosomes) was proposed to be an adaptive strategy by allowing loss and rapid recovery of this avirulence gene, consistent with the idea that the accessory chromosomes might serve as a particular genomic environment allowing the rapid evolution of new adaptive traits (Chuma et al. 2011).

In conclusion, **genes located on accessory chromosomes of fungi have often been associated with a fitness benefit**—in many cases by encoding traits that directly influence pathogenicity and host specificity. In contrast to the B chromosomes in plants and animals, fungal accessory chromosomes are rarely associated with a fitness cost. Nevertheless, they can confer a fitness cost under some conditions as demonstrated in *Z. tritici* and the potential effects of avirulence factors present on the accessory chromosomes in *L. maculans* and *M. oryzae*.

IV. The Origin of Fungal Accessory Chromosomes

How did accessory chromosomes originate in fungal genomes? Genome data has shed light on the possible origin of accessory chromosomes, and experimental studies have revealed mechanisms of de novo chromosome acquisition. Thereby evidence for two distinct scenarios have accumulated: (i) accessory chromosomes can be derived from essential core chromosomes, and (ii) they can be acquired by horizontal chromosome transfer from closely or distantly related taxa.

Initially, plant and animal B chromosomes were thought to be non-homologous to the essential A chromosomes. This view later changed due to their mosaic-like composition of sequences derived from essential A chromosomes and organelles, and therefore, they are now mainly considered to originate from A chromosomes (Banaei-Moghaddam et al. 2015; Houben et al. 2014; Martis et al. 2012). The origin of B chromosomes has been addressed using genomic data in rye, barley, dogs, and several fish species (Banaei-Moghaddam et al. 2015; Martis et al. 2012; Mayer et al. 2011; Pansonato-Alves et al. 2014; Valente et al. 2014; Valente et al. 2017). These analyses suggested a multi-step model for the origin of accessory chromosomes from several core chromosomes where initial genome duplication (either partial or comprising the whole genome) has been followed by reductive chromosome translocations decreasing the size of the B chromosome (Banaei-Moghaddam et al. 2015; Houben et al. 2014). Chromosomes that have evolved in this way diverged over time by the accumulation of repetitive sequences and further structural variation that impaired homologous recombination with the original core chromosome (Galazka and Freitag 2014; Houben et al. 2014). Alternatively, small accessory chromosomes could be generated by mis-segregation or Robertsonian translocations at or near the centromeres of two acrocentric A chromosomes. This could result in the duplication of a short region of A chromosomes that, if it contains a functional centromere and origin of replication, could be considered a very small B chromosome and later may increase in size by acquiring additional sequences (Galazka and Freitag 2014).

In some fungal species, there is evidence that accessory chromosomes have originated from core chromosomes. In other species, it is conceivable that fusion of hyphae has mediated **horizontal transfer of chromosomes between lineages or even species**. So far, little is known about the exact mechanism that transforms core chromosome copies into accessory chromosomes. In *Zymoseptoria tritici*, experimental mating and karyotyping of progeny strains revealed the evolution of a new accessory chromosome by a process called "breakage-fusion-bridge" (BFB) cycles that involves chromosomal fusions followed by degenerative breakage (Croll et al. 2013; McClintock 1941). BFB cycles can be initiated by non-allelic homologous recombination between repeats, whereby dicentric and acentric chromosomes can be generated (Croll et al. 2013; McClintock 1941). Acentric chromosomes are lost due to the absence of centromeres, but dicentric chromosomes can form a bridge at anaphase and undergo BFB cycles (Fig. 1) (Croll et al. 2013; Gisselsson et al. 2000). In *Z. tritici* many repeat families are shared between core and accessory chromosomes, possibly facilitating an origin of the accessory chromosomes within the genome (Grandaubert et al. 2015). Size variation due to non-allelic sister chromatid recombination was also observed in *Magnaporthe oryzae* (Chuma et al. 2003). The high content of repetitive DNA on many accessory chromosomes may indeed favor the occurrence of non-allelic homologous recombination.

In *Fusarium oxysporum* f. sp. *lycopersici*, the lineage-specific chromosomes appear to have a different origin than the core chromosomes. These chromosomes have a high proportion of unique genes and vary in codon usage as well as GC content from the core chromosomes and are therefore believed to have been acquired from another *Fusarium* species (Ma et al. 2010). Interestingly, under experimental conditions, **transfer of the lineage-specific chromosome 14 of *F. oxysporum* can occur between asexual lineages of the pathogen by vegetative hyphal fusion** (Ma et al. 2010; Vlaardingerbroek et al. 2016b). For the analysis, strains were used that carried two selection markers (neomycin and hygromycin), located on the chromosome 14 of the donating strain (pathogenic on tomato) and on a core chromosome of the receiving strain (non-pathogenic on tomato), respectively. Co-inoculating these two strains on agar plates allowed the isolation of colonies resistant to both antibiotics. These double resistant strains were pathogenic on tomato and contained large portions of the lineage-specific chromosome 14. This experimentally validates the possibility of chromosomal transfers by vegetative fusion (in this case between different strains of the same species) as a possible mechanism for the acquisi-

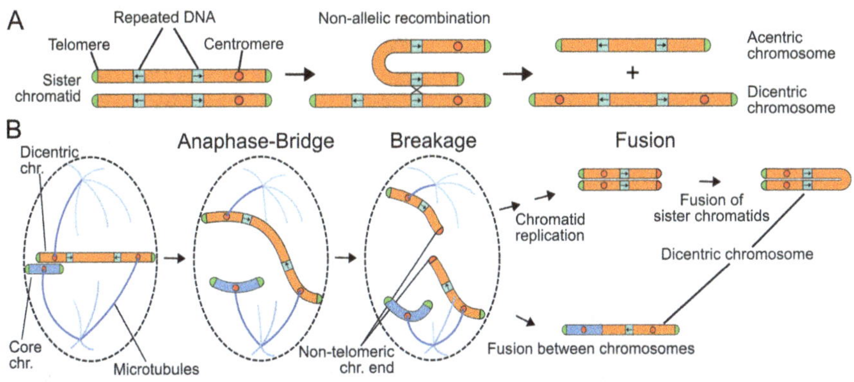

Fig. 1 Non-allelic recombination between sister chromatids of an accessory chromosome initiates a breakage-fusion-bridge (BFB) cycle. The generation of dicentric chromosome followed by BFB cycles may play a role in the plasticity of the accessory chromosomes and the bidirectional transfer of sequences between accessory and core chromosomes. (**a**) Non-allelic recombination between repeated sequences on an accessory chromosome results in an acentric chromosome which is lost during subsequent cell divisions and a dicentric chromosome (adapted from (Croll et al. 2013)). (**b**) The dicentric chromosome (chr.) is subject to a BFB cycle when the two centromeres are pulled to opposite poles of the dividing cell during anaphase. Thereby, the chromatid breaks and two non-telomeric chromosome ends are generated. These ends can either fuse between the sister chromatids after sister chromatid replication or fuse to other chromosomes creating a translocation on, in this example, a core chromosome. In both cases, dicentric chromosomes are generated which might again be subjected to a BFB cycle. Note: for the sake of clarity, sister chromatids were omitted when possible

tion of accessory chromosomes (Ma et al. 2010). Further examples of chromosome transfers include the asexual plant pathogen *Colletotrichum gloeosporioides* that infects a wide variety of crops. The supernumerary chromosome of *C. gloeosporioides* is considered to have been transferred by vegetative fusion, which can also be observed under laboratory conditions (He et al. 1998; Masel 1996). Similarly, chromosome transfer occurs in *Alternaria* spp. Here, pathogenic strains cause leaf spots and blights on different host plants (Rotem 1994). Phylogenetic analysis of toxin-encoding, conditionally dispensable chromosomes supports chromosome transfer between different strains (Akagi et al. 2009a). This hypothesis could be confirmed under laboratory conditions (Akagi et al. 2009b). In addition, transfer of accessory chromosomes was suggested by comparative analyses of transposon or repeat content and codon usage of the accessory and the core chromosomes of *Alternaria arborescens* (Hu et al. 2012). Although the origin of the conditionally dispensable chromosomes in *Alternaria* spp. is still unknown, these potential **horizontal transfers would be an example for a possible mechanism by which pathogens could acquire novel determinants of host specificity** (Mehrabi et al. 2011).

It is, however, important to note that the observed differences in sequence composition between core and accessory chromosomes do not provide a proof of a horizontal chromosome transfer. Different recombination and mutation rates, the differences in transposable element content, and higher frequencies of pseudogenes could likewise explain some of the observed differences in sequence composition (Galazka and Freitag 2014). This is exemplified by the wheat pathogen *Fusarium poae* which carries several accessory chromosomes (Fekete and Hornok 1997). Here, the accessory genome was shown to be lacking repeat-induced point mutations (RIP). RIP is a genome defense mechanism initially described in *Neurospora crassa* that acts to mutate repetitive elements to prevent their further propagation in the genome (Selker 1990). Interestingly, in *F. poae* RIP occurs on the core chromosomes (Vanheule et al. 2016). The sole absence of RIP on the accessory may confer differences in sequence composition between core and accessory chromosomes. Similarly, sequences on the accessory chromosomes of *Z. tritici* show signatures of accelerated evolution, which could also account for the observed sequence differences between the core and the accessory chromosomes (Stukenbrock et al. 2010; Stukenbrock et al. 2011).

Although accessory chromosomes are thought to be non-recombining with the core chromosomes, **transfer of genetic material from accessory to core chromosomes appears to happen frequently.** In several cases an accessory chromosome became part of a core chromosome (Ma et al. 2010; Schotanus et al. 2015; Vanheule et al. 2016). In *Z. tritici*, the distal 0.865 Mb segment of the long right arm of the core chromosomes 7 shares characteristics like histone modifications, gene content, and gene organization with the accessory chromosomes (Schotanus et al. 2015). This chromosome segment is therefore considered to be either a translocated fragment or an entire accessory chromosome fused to a core chromosome (Schotanus et al. 2015). This type of chromosomal fusion could be the result of BFB cycles as already observed in the genome of *Z. tritici* (Croll et al. 2013) (Fig. 1b). Similarly, the right arms of the core chromosomes 1 and 2 in *F. oxysporum* f. sp. *lycopersici* share sequence characteristics with the lineage-specific chromosomes, and hence these chromosomal regions are also considered to be lineage-specific (Ma et al. 2010). In addition, the core chromosome of *F. poae* contains large blocks (>200 kb) of sequences that were recently translocated from the accessory genome (Vanheule et al. 2016). Hence, sequence exchange between accessory and core chromosomes occurs, but the extent of this exchange and the functional and evolutionary consequences remain to be elucidated.

V. Accessory Chromosomes Are Mitotically Instable

Accessory chromosomes often show distinct patterns of mitotic transmission compared to the core chromosomes. These transmission differences can also affect their meiotic transmission if cells of the germline are affected (e.g., non-disjunction during first pollen mitosis of rye B chromosomes) (Houben 2017). Further examples of non-equal transmission during mitosis in plants and animals were described for *Aegilops speltoides* and grasshoppers in which the number of accessory chromosome differs between organs of the organism (Mendelson and Zohary 1972; Remis and Vilardi 2004).

In fungi, the differentiation between somatic cells and germline cannot be defined. Therefore, any differences in transmission of accessory chromosomes affect their overall propagation. Indeed, asexual fungi rely solely on mitotic transmission. **Accessory chromosomes in several fungi can be lost during mitotic transmission.** In *Fusarium oxysporum* f. sp. *lycopersici*, the two smallest lineage-specific chromosomes (chr. 1, chr. 14) are lost at a frequency of 1 in 35,000 spores. Interestingly, the core chromosome 12 could also be lost but at a much lower frequency than the accessory chromosomes (Vlaardingerbroek et al. 2016a). Similarly, in *Alternaria alternata* loss of the conditionally dispensable chromosome occurred spontaneously in vitro (Johnson et al. 2001) and rendered the strain non-pathogenic. In *Zymoseptoria tritici* and its closely related sister species *Zymoseptoria ardabiliae*, spontaneous loss of accessory chromosomes was observed at a very high frequency of up to 1 in 50 spores (Möller et al. 2018). Surprisingly, temperature had a dramatic effect on the loss rate of accessory chromosomes in *Z. tritici*. A rise in temperature from 18 °C to 28 °C increased the losses of accessory chromosomes to 1 in 2 spores (Möller et al. 2018). Importantly, during mitotic growth in planta, which results in the production of asexual pycnidiospores, losses of accessory chromosomes were also observed at a similar rate to those in vitro at 18 °C (Möller et al. 2018). This demonstrates that accessory chromosome losses occur throughout the asexual part of the life cycle of *Z. tritici*. Interestingly, accessory chromosome 18, which was frequently lost during in planta experiments, is also often absent in field isolates of *Z. tritici* (Croll et al. 2013; McDonald et al. 2016). Hence, the presence/absence polymorphism for the accessory chromosomes appears to be—at least partially—generated by frequent losses during mitotic divisions. Interestingly, in addition to frequent chromosome losses, the same study reported frequent chromosome fusions, chromosome breakages, and accessory chromosome duplications at elevated temperatures (Möller et al. 2018). Chromosomal fusions involved core as well as accessory chromosomes. This further highlights the putative plasticity of the *Z. tritici* genome and the potential for bidirectional sequence exchange between core and accessory chromosomes.

Why do the accessory chromosomes show a different transmission during mitosis? Interestingly, the **accessory chromosomes** in *Z. tritici*, *F. oxysporum*, *Fusarium asiaticum*, and *Gibberella fujikuroi* **are enriched in the histone modification H3K27me3** (Galazka and Freitag 2014; Schotanus et al. 2015; Fokkens et al. 2018). On core chromosomes this modification is restricted to subtelomeric regions and involved in the localization of these subtelomeric regions to the nuclear envelope (Erlendson et al. 2017; Harr et al. 2015, 2016). Therefore, entire accessory chromosomes might be localized at the nuclear envelope. This different localization would also be supported by the fact that in *Z. tritici* centromeres do not co-localize in one focus, as observed in many fungi, plants, and animals (Dong and Jiang 1998; Ross et al. 2013; Smith et al. 2012) but to several distinct foci in the nucleus (Schotanus et al. 2015). We hypothesize that the different **epigenetic marks** on the accessory chromosomes **affect their location in the nucleus and thereby influencing their DNA replication and transmission.** Recently, we could show that removal of the H3K27me3 histone modification leads to an increased transmission fidelity of the accessory chromosomes, supporting the pivotal role for histone modifications in distinguishing core and accessory chromosomes (Möller et al. 2019) (Habig et al., in preparation).

VI. Accessory Chromosomes Are Frequently Transmitted in a Non-Mendelian Way During Meiosis

By definition, accessory chromosomes are present in some but not all members of a population. During sexual mating this presence/

absence polymorphism can result in unpaired (i.e., univalent) accessory chromosomes during meiosis if the two parental strains have different chromosome complements (Fig. 2). How are univalent accessory chromosomes transmitted during meiotic divisions? In animals and plants, the widespread occurrence of accessory chromosomes is generally attributed to their **selfish mode of transmission** (Houben 2017; Jones 2012), which increases their fre-

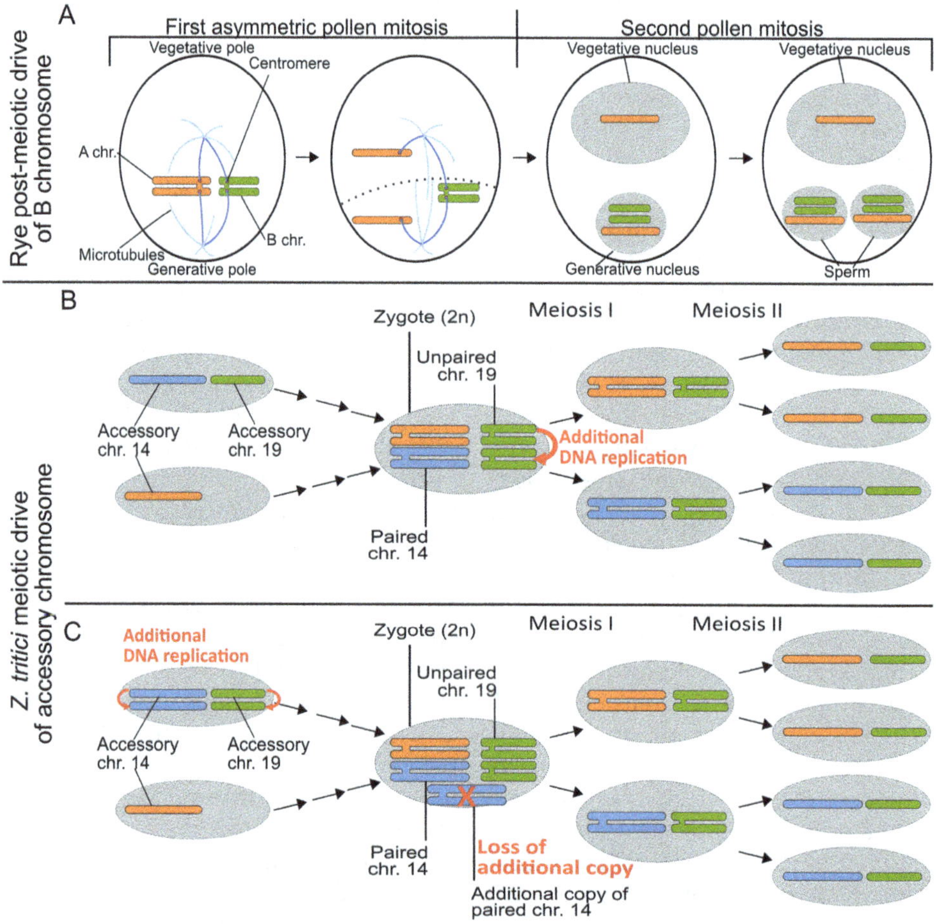

Fig. 2 Potential drive mechanisms for accessory chromosomes in plants and fungi. (**a**) Suggested mechanism for a post-meiotic drive affecting the B chromosome (chr.) in rye. The centromeres of the B chromosome are not separating during the first pollen mitosis. Due to the non-symmetric cell division the B chromosome sister-chromatids are more likely to become part of the generative nucleus of which the two sperm cells are generated in the second pollen mitosis. This mechanism leads to an increase in the B chromosome frequency in the progeny without additional DNA replication. (**b** & **c**) Two possible mechanisms for a meiotic chromosome drive in *Z. tritici* involving additional replication of accessory chromosomes by the example of a paired (i.e., homologs in both parental strains) accessory chromosome 14 and an unpaired chromosome 19. (**b**) Unpaired accessory chromosomes undergo an additional round of DNA replication initiated after the pairing of homologs within the zygote suggesting an additional feedback mechanism between pairing of homologs and DNA replication. (**c**) Alternatively, additional DNA replication of unpaired chromosomes takes place prior to meiosis affecting all accessory chromosomes. Since no additional copies of the paired accessory chromosomes are found, these must be lost during the zygote stage where pairing occurs. Please note that for the sake of clarity, the subsequent mitotic cell division resulting in eight ascospores is not displayed, and recombination events were omitted. b & c adapted from (Habig et al. 2018)

quency by preferential segregation during cell divisions including meiosis. However, little is known about the meiotic transmission of fungal accessory chromosomes. For several fungal accessory chromosomes, a non-Mendelian mode of inheritance has been reported, which appears to be similar to the mode in plants and animals. We will therefore first describe some of the well-understood examples of B chromosome transmission in plants and animals, which may direct our research of accessory chromosome inheritance in fungi.

B chromosome accumulation mechanisms, also termed chromosome drive, involve non-Mendelian modes of transmission and have been described in approximately 60% of the plant species that carry B chromosomes (Jones 2012). Chromosome drive can either occur at the pre- or post-meiotic stages or during meiosis (Akera et al. 2017; Gregory 2011; Houben 2017; Mroczek et al. 2006). The B chromosome of rye is a well-characterized example for a chromosome drive mechanism acting post-meiotically (Banaei-Moghaddam et al. 2012; Endo et al. 2008; Hasegawa 1934; Houben 2017). Following meiosis the pollen development involves two mitotic divisions resulting in tri-cellular pollen. One pollen consists of two sperm cells and a vegetative nucleus, which does not provide genetic material to the offspring (Twell 2011). During the first pollen mitosis, the two sister-chromatids of B chromosomes show non-disjunction (Hasegawa 1934) and will in most cases become part of the generative nucleus. Thereby the resulting sperm cells will contain two B chromosomes instead of the expected one (Fig. 2a) (Hasegawa 1934; Houben 2017). What are the underlying mechanisms that lead to the exploitation of asymmetric cell divisions by B chromosomes? The different segregation of A versus B chromosomes during mitotic or meiotic cell divisions predicts functional differences between their centromeres (or the loci that mediate attachment of the chromosomes to the spindle). Indeed, the centromeres of A and B chromosomes in rye differ in their repeat composition, with the B chromosome centromeres including additional classes of repeats (Banaei-Moghaddam et al. 2012). Here, a non-disjunction control region (NCR) acting *in trans* controls chromosome segregation of the B chromosome, possibly by long noncoding RNAs which affect centromere organization (Carchilan et al. 2007; Endo et al. 2008; Houben 2017).

In general, centromeric sequences and the centromere-associated histone CenH3 were shown to evolve rapidly (reviewed in (Rosin and Mellone 2017)), inspiring the **centromere drive** hypothesis (Henikoff and Malik 2002; Kursel and Malik 2018). Accordingly, centromeric DNAs act as selfish genetic elements promoting their own transmission. Originally, this hypothesis considered this preferential transmission to only occur in asymmetrical female meiosis. However, every asymmetrical cell division that eventually results in germline cells should be susceptible to a non-random segregation of chromosomes as seen in the example of the rye B chromosome. B chromosomes are often considered selfish genetic elements. Other selfish genetic elements have also been shown to increase their relative frequency by killing or disabling gametes or embryos that lack the selfish element (Hurst and Werren 2001). To date, no example of such a killing mechanism has been described for an accessory chromosome. Yet, it is worth noting that drive of entire chromosomes has been described for the sex chromosomes in several species within the order Rodentia and Diptera. Here these drives are indeed caused either by asymmetrical segregation during mitotic or meiotic cell divisions or by the selective killing of gametes that do not carry the sex chromosomes (reviewed in (Helleu et al. 2014; Hurst and Werren 2001; Jaenike 2008)).

In fungi, little is known about the meiotic transmission of accessory chromosomes. Those few cases for which the inheritance of the accessory chromosomes was analyzed have shown (i) loss of accessory chromosomes during meiotic and mitotic cell divisions and (ii) non-Mendelian segregation with an increase of chromosome frequencies (Balesdent et al. 2013; Camacho et al. 2011; Chuma et al. 2003; Coleman et al. 2009; Fouché et al. 2018; Goodwin et al. 2011; Miao et al. 1991; Orbach et al. 1996; Tzeng et al. 1992; Wittenberg et al. 2009). Loss of accessory chromosomes during meiosis was observed for many fungal accessory chromosomes and may, in part, explain the observed presence/absence polymorphism. Examples are the accessory mini-chromosome of *Magnaporthe oryzae* that fails to segregate during crosses (Orbach et al. 1996) and the dispensable chromosome of Leptosphaeria *maculans*, which is lost in approximately 5% of the progeny following meiosis (Balesdent et al. 2013; Leclair et al. 1996). In these cases,

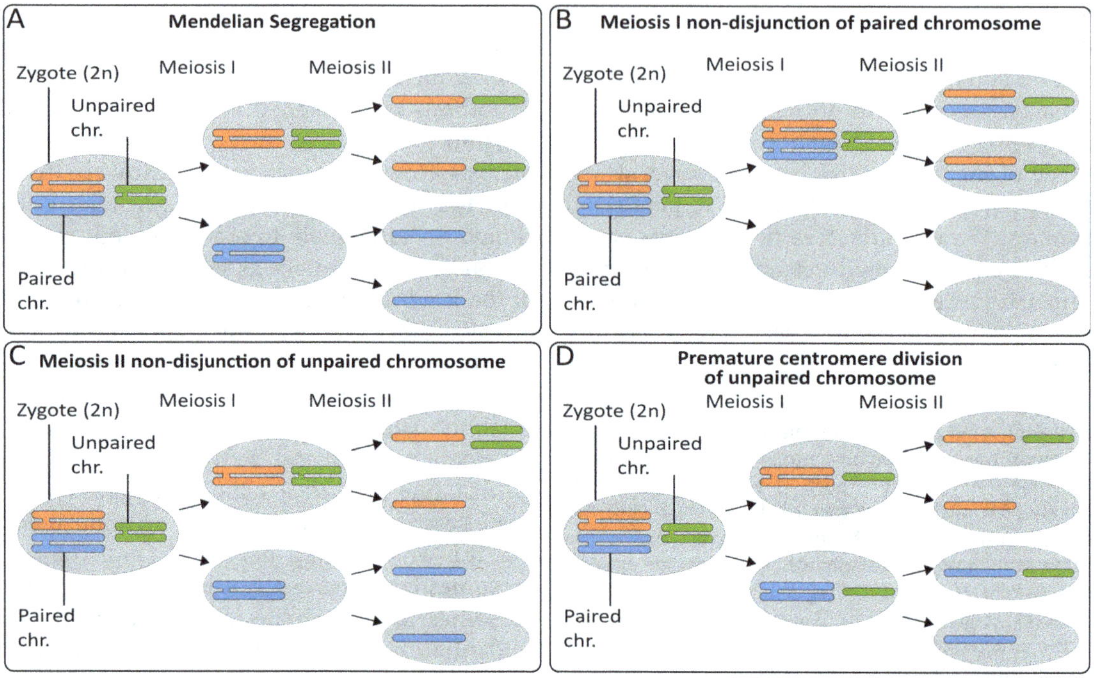

Fig. 3 Modes of accessory chromosome transmission during the cell divisions of meiosis. None of these mechanism leads to a change in the absolute frequency of the accessory chromosome within the progeny. Three accessory chromosomes are illustrated, two of which are homologs to each other (orange/blue) and one accessory chromosome (green) lacking a homolog. Recombination events were omitted to increase clarity. (**a**) During Mendelian segregation, the homologous chromosomes are segregating during meiosis I and sister-chromatids during meiosis II. (**b**) Non-disjunction of a paired chromosome during meiosis I leads to two disomic meiotic progeny and the loss of the paired accessory chromosome in the remaining two meiotic products. (**c**) Non-disjunction of sister-chromatids of an unpaired chromosome during meiosis II leads to one meiotic product disomic for the unpaired chromosome. (**d**) Premature centromere division of an unpaired chromosome during meiosis I generates single chromatid chromosomes that are segregated during meiosis II

the frequent losses and disomies found in progenies indicate non-disjunction during the meiotic divisions I or II (Fig. 3). In *Zymoseptoria tritici*, accessory chromosomes are missing in up to 20% of the progeny of meiotic crosses in which both parental strains contain the respective accessory chromosome (Fouché et al. 2018; Wittenberg et al. 2009). Importantly, meiotic progeny disomic of accessory chromosomes are frequent, which also indicates non-disjunction during one of the meiotic divisions (Fouché et al. 2018). While non-disjunction does not change the absolute frequency of the accessory chromosomes but leads to a re-distribution of the copies among the progeny, losses will cause a reduction in the frequency. Currently, the relative importance of these two processes is unknown. However, by using tetrad analysis, we could recently show that both losses and non-disjunction of accessory chromosomes do occur in *Z. tritici* (Habig et al. 2018).

If chromosome losses reduce the number of accessory chromosomes, a counteracting mechanism must be in place, which increases the frequency of the accessory chromosomes to avoid their complete loss. Maintenance of chromosomes could occur if natural selection would favor individuals carrying a particular accessory chromosome under specific conditions. As described above, many fungal accessory chromosomes confer a fitness advantage on certain hosts. However, a few examples demonstrate that chromosome maintenance also can be conferred by a meiotic transmission advantage (Balesdent et al. 2013; Fouché et al. 2018; Goodwin et al. 2011; Tzeng et al. 1992; Wittenberg et al. 2009). In *Cochliobolus heterostrophus* two-thirds of randomly selected ascospores contained the accessory chromosome 16, although only one of the parental strains carried the chromosome (Tzeng et al. 1992). Similarly, in *L. maculans* 83% of the progeny received an accessory chromosome that was unpaired during meiosis (Balesdent et al. 2013). However, the mechanistic basis for these transmission advantages has not been described. Transmission rates higher than 50% for unpaired accessory chromosomes have also been determined for *Z. tritici* with frequent observation of meiotically produced ascospores containing disomic accessory chromosomes (Fouché et al. 2018; Goodwin et al. 2011; Wittenberg et al. 2009). Tetrad analysis, in which all meiotic products of a single meiosis were isolated and analyzed, revealed that the *Z. tritici* **accessory chromosomes are subject to a meiotic drive mechanism** (Habig et al. 2018). Interestingly, the meiotic drive appears to be restricted to those chromosomes that are inherited from the female parent (considered to be the parent that also provides the mitochondria to the offspring). Inheritance of the accessory chromosome from the male parent does not lead to a chromosome drive. Moreover, the meiotic drive only affected those accessory chromosomes without a homolog resulting in transmission of the unpaired chromosomes to all meiotic products when inherited from the female. If the chromosomes were inherited from the male or had a homolog, they showed Mendelian inheritance. This unique transmission pattern of unpaired accessory chromosomes might be explained by a feedback mechanism during meiosis that initiates an additional round of DNA replication for unpaired accessory chromosomes inherited from the female parent (Fig. 2b). Alternatively, a pre-meiotic DNA replication is followed by a regulated loss of the redundant copies of paired accessory chromosomes (Fig. 2c). The centromeres of the accessory and the core chromosomes of *Z. tritici* do not show significant differences in size, location, and sequence characteristics (Schotanus et al. 2015). Therefore, it appears that, in contrast to the rye B chromosome, the centromeres are not involved in the chromosome drive in *Z. tritici*.

Although other mechanisms could be involved, this particular transmission pattern may explain some of the observed characteristics of the accessory chromosomes in *Z. tritici*. The meiotic drive may cause the observed lower recombination rate on the accessory chromosomes (Croll et al. 2015; Stukenbrock and Dutheil 2017). The lower recombination rate would in turn account for the accumulation of transposable elements on these chromosomes (Goodwin et al. 2011) and the observed lower efficacy of selection in removing non-adaptive mutations from the coding sequences on the accessory chromosomes (Stukenbrock et al. 2010). Most importantly, their additional amplification could explain the maintenance of accessory chromosomes in this fungus despite their fitness costs (Habig et al. 2017). *Z. tritici* therefore represents the first example for fungal accessory chromosomes that show meiotic chromosome drive similar to the transmission pattern of B chromosomes in plants and animals.

Tetrad analyses to decipher the meiotic transmission of accessory chromosomes have also been performed in *Nectria haematococca* MP VI carrying the accessory chromosome PDA1-CDC (Miao et al. 1991). In contrast to the meiotic drive observed in *Z. tritici*, tetrad analysis of *N. haematococca* MP VI showed Mendelian segregation of unpaired and paired PDA1-CDC. Interestingly, when both parental strains contained one PDA1-CDC, losses were frequent (10–19% of the progeny), but also progenies disomic for PDA1-CDC were highly abundant (Garmaroodi and Taga 2015). Moreover, when a strain with four copies of the accessory chromosome was crossed with a strain lacking PDA1-CDC, only about 50% of the progeny inherited the accessory chromosome (Garmaroodi and Taga 2015). Therefore, even when several copies of an accessory chromosome are present in the zygote, their transmission varies between cases where all copies are inherited from one parent and cases where each parent contributed one copy of the accessory chromosome.

In conclusion, meiotic instability and non-Mendelian inheritance appear to be common among accessory chromosomes in fungi. Losses and disomies are frequent, and for *Z. tritici* there is evidence for a chromosome drive resembling the scenarios in plants and animals. Frequently observed aneuploidy of accessory chromosomes in fungi and the particular meiotic inheritance of these chromosomes could represent interesting starting points to further dissect the mechanism of meiotic transmission of fungal accessory chromosomes. In general, the non-essential fungal accessory chromosomes represent interesting models for studies of meiotic mechanisms. The presence/absence polymorphism allows for the comparative analysis of transmission studying the same chromosomes when paired or unpaired. Hence pairing of homologous chromosomes—which is central in meiosis and other closely related mechanisms like recombination and DNA-repair—can be addressed using fungal accessory chromosomes as model systems.

VII. Concluding Remarks

Accessory chromosomes are a diverse set of genomic entities that are widespread in plants, animals, and fungi. There is neither a single sequence characteristic that is shared among all accessory chromosomes between these kingdoms nor a unifying sequence characteristic that has been identified among the accessory chromosomes in fungi. However, fungal cells appear to be able to differentiate between core and accessory chromosomes. Many fungal accessory chromosomes show differences in their mitotic and meiotic transmission compared to core chromosomes. However, the transmission of accessory chromosomes during meiosis and its effect on the maintenance of these chromosomes are not well understood. The meiotic drive identified for the accessory chromosomes of *Zymoseptoria tritici* appears to be responsible for the maintenance of these chromosomes despite conferring a fitness costs during host interaction. This may be indicative for similar mechanisms in other fungal systems where accessory chromosomes confer fitness costs, like the accessory chromosomes of *Leptosphaeria maculans* and *Magnaporthe oryzae*, which contain avirulence genes. Interestingly, in both organisms the accessory chromosomes also show a non-Mendelian transmission during meiosis (Balesdent et al. 2013; Chuma et al. 2003; Orbach et al. 1996; Rouxel et al. 2011; Rouxel and Balesdent 2017). Thus, a more detailed analysis of their meiotic transmission pattern could be highly relevant.

The underlying mechanisms responsible for the distinctly different meiotic and mitotic transmission of accessory chromosomes are unknown. How does the fungal cell distinguish between accessory and core chromosomes? In the absence of common sequence characteristics of fungal accessory chromosomes, epigenetic marks could be promising targets of future research. The involvement of the histone mark H3K27me3 in the transmission fidelity of the accessory chromosomes of *Z. tritici* could be indicative of a mechanism that connects the transmission pattern to the localization of the accessory chromosomes in the nucleus. To date, the epigenetic landscapes for a very limited number of accessory chromosomes have been determined. In those cases the histone mark H3K27me3 has been identified as being enriched on the accessory chromosomes compared to the core chromosome (Galazka and Freitag 2014; Schotanus et al. 2015; Fokkens et al. 2018; Soyer et al. 2018). However, the functional effect of this H3K27me3 enrichment on the meiotic and mitotic transmission of the accessory chromosomes is unknown for most species.

In conclusion, we propose that fungal accessory chromosomes provide an intriguing opportunity to dissect novel chromosome drive mechanisms and may serve as models to improve our understanding of the underlying mechanisms of meiotic and mitotic transmission of accessory as well as core chromosomes.

References

Ahn JH, Walton JD (1996) Chromosomal organization of TOX2, a complex locus controlling host-selective toxin biosynthesis in *Cochliobolus carbonum*. Plant Cell 8:887–897. https://doi.org/10.1105/tpc.8.5.887

Akagi Y, Akamatsu H, Otani H, Kodama M (2009a) Horizontal chromosome transfer, a mechanism for the evolution and differentiation of a plant-pathogenic fungus. Eukaryot Cell 8:1732–1738. https://doi.org/10.1128/EC.00135-09

Akagi Y, Taga M, Yamamoto M, Tsuge T, Fukumasa-Nakai Y, Otani H, Kodama M (2009b) Chromosome constitution of hybrid strains constructed by protoplast fusion between the tomato and strawberry pathotypes of *Alternaria alternata*. J Gen Plant Pathol 75:101–109. https://doi.org/10.1007/s10327-009-0149-1

Akamatsu H, Itoh Y, Kodama M, Otani H, Kohmoto K (1997) AAL-toxin-deficient mutants of *Alternaria alternata* tomato Pathotype by restriction enzyme-mediated integration. Phytopathology 87:967–972. https://doi.org/10.1094/PHYTO.1997.87.9.967

Akamatsu H, Taga M, Kodama M, Johnson R, Otani H, Kohmoto K (1999) Molecular karyotypes for Alternaria plant pathogens known to produce host-specific toxins. Curr Genet 35:647–656

Akera T, Chmátal L, Trimm E, Yang K, Aonbangkhen C, Chenoweth DM, Janke C, Schultz RM, Lampson MA (2017) Spindle asymmetry drives non-Mendelian chromosome segregation. Science 358:668–672. https://doi.org/10.1126/science.aan0092

Baayen RP, O'Donnell K, Bonants PJ, Cigelnik E, Kroon LP, Roebroeck EJ, Waalwijk C (2000) Gene genealogies and AFLP analyses in the *Fusarium oxysporum* complex identify monophyletic and nonmonophyletic Formae Speciales causing wilt and rot disease. Phytopathology 90:891–900. https://doi.org/10.1094/PHYTO.2000.90.8.891

Balesdent M-H, Fudal I, Ollivier B, Bally P, Grandaubert J, Eber F, Chèvre A-M, Leflon M, Rouxel T (2013) The dispensable chromosome of *Leptosphaeria maculans* shelters an effector gene conferring avirulence towards Brassica rapa. New Phytol 198:887–898. https://doi.org/10.1111/nph.12178

Banaei-Moghaddam AM, Martis MM, Macas J, Gundlach H, Himmelbach A, Altschmied L, Mayer KFX, Houben A (2015) Genes on B chromosomes: old questions revisited with new tools. Biochim Biophys Acta 1849:64–70. https://doi.org/10.1016/j.bbagrm.2014.11.007

Banaei-Moghaddam AM, Meier K, Karimi-Ashtiyani R, Houben A (2013) Formation and expression of pseudogenes on the B chromosome of rye. Plant Cell 25:2536–2544. https://doi.org/10.1105/tpc.113.111856

Banaei-Moghaddam AM, Schubert V, Kumke K, Weiβ O, Klemme S, Nagaki K, Macas J, González-Sánchez M, Heredia V, Gómez-Revilla D, González-García M, Vega JM, Puertas MJ, Houben A (2012) Nondisjunction in favor of a chromosome: the mechanism of rye B chromosome drive during pollen mitosis. Plant Cell 24:4124–4134. https://doi.org/10.1105/tpc.112.105270

Bauerly E, Hughes SE, Vietti DR, Miller DE, McDowell W, Hawley RS (2014) Discovery of supernumerary B chromosomes in *Drosophila melanogaster*. Genetics 196:1007–1016. https://doi.org/10.1534/genetics.113.160556

Bertazzoni S, Williams A, Jones DAB, Syme RA, Tan K-C, Hane JK (2018) Accessories make the outfit: accessory chromosomes and other dispensable DNA regions in plant-pathogenic fungi. Mol Plant-Microbe Interact 8:779–788. https://doi.org/10.1094/MPMI-06-17-0135-FI

Bougourd SM, Jones RN (1997) B chromosomes: a physiological enigma. New Phytol 137:43–54. https://doi.org/10.1046/j.1469-8137.1997.00823.x

Camacho JPM, Schmid M, Cabrero J (2011) B chromosomes and sex in animals. Sex Dev 5:155–166. https://doi.org/10.1159/000324930

Carchilan M, Delgado M, Ribeiro T, Costa-Nunes P, Caperta A, Morais-Cecílio L, Jones RN, Viegas W, Houben A (2007) Transcriptionally active heterochromatin in rye B chromosomes. Plant Cell 19:1738–1749. https://doi.org/10.1105/tpc.106.046946

Cheng Y-M (2010) Evolution of the heterochromatic regions on maize B long arm based on the sequence structure of CL-repeat variants. Chromosom Res 18:605–619. https://doi.org/10.1007/s10577-010-9136-9

Chuma I, Isobe C, Hotta Y, Ibaragi K, Futamata N, Kusaba M, Yoshida K, Terauchi R, Fujita Y, Nakayashiki H, Valent B, Tosa Y (2011) Multiple translocation of the AVR-Pita effector gene among chromosomes of the rice blast fungus *Magnaporthe oryzae* and related species. PLoS Pathog 7:e1002147. https://doi.org/10.1371/journal.ppat.1002147

Chuma I, Tosa Y, Taga M, Nakayashiki H, Mayama S (2003) Meiotic behavior of a supernumerary chromosome in *Magnaporthe oryzae*. Curr Genet 43:191–198. https://doi.org/10.1007/s00294-003-0390-7

Coleman JJ, Rounsley SD, Rodriguez-Carres M, Kuo A, Wasmann CC, Grimwood J, Schmutz J, Taga M, White GJ, Zhou S, Schwartz DC, Freitag M, Ma L-J, Danchin EGJ, Henrissat B, Coutinho PM, Nelson DR, Straney D, Napoli CA, Barker BM, Gribskov M, Rep M, Kroken S, Molnár I, Rensing C, Kennell JC, Zamora J, Farman ML, Selker EU, Salamov A, Shapiro H, Pangilinan J, Lindquist E, Lamers C, Grigoriev IV, Geiser DM, Covert SF, Temporini E, Vanetten HD (2009) The genome of *Nectria haematococca*: contribution of supernumerary chromosomes to gene expansion. PLoS Genet 5:e1000618. https://doi.org/10.1371/journal.pgen.1000618

Coleman JJ, Wasmann CC, Usami T, White GJ, Temporini ED, McCluskey K, Vanetten HD (2011) Characterization of the gene encoding pisatin demethylase (FoPDA1) in *Fusarium oxysporum*.

Mol Plant-Microbe Interact 24:1482–1491. https://doi.org/10.1094/MPMI-05-11-0119
Condon BJ, Leng Y, Wu D, Bushley KE, Ohm RA, Otillar R, Martin J, Schackwitz W, Grimwood J, MohdZainudin N, Xue C, Wang R, Manning VA, Dhillon B, Tu ZJ, Steffenson BJ, Salamov A, Sun H, Lowry S, LaButti K, Han J, Copeland A, Lindquist E, Barry K, Schmutz J, Baker SE, Ciuffetti LM, Grigoriev IV, Zhong S, Turgeon BG (2013) Comparative genome structure, secondary metabolite, and effector coding capacity across Cochliobolus pathogens. PLoS Genet 9:e1003233. https://doi.org/10.1371/journal.pgen.1003233
Croll D, Lendenmann MH, Stewart E, McDonald BA (2015) The impact of recombination hotspots on genome evolution of a fungal plant pathogen. Genetics 201:1213–1228. https://doi.org/10.1534/genetics.115.180968
Croll D, Zala M, McDonald BA, Heitman J (2013) Breakage-fusion-bridge cycles and large insertions contribute to the rapid evolution of accessory chromosomes in a fungal pathogen. PLoS Genet 9:e1003567. https://doi.org/10.1371/journal.pgen.1003567
Dean RA, Talbot NJ, Ebbole DJ, Farman ML, Mitchell TK, Orbach MJ, Thon M, Kulkarni R, Xu J-R, Pan H, Read ND, Lee Y-H, Carbone I, Brown D, Oh YY, Donofrio N, Jeong JS, Soanes DM, Djonovic S, Kolomiets E, Rehmeyer C, Li W, Harding M, Kim S, Lebrun M-H, Bohnert H, Coughlan S, Butler J, Calvo S, Ma L-J, Nicol R, Purcell S, Nusbaum C, Galagan JE, Birren BW (2005) The genome sequence of the rice blast fungus *Magnaporthe grisea*. Nature 434:980–986. https://doi.org/10.1038/nature03449
Dhar MK, Friebe B, Koul AK, Gill BS (2002) Origin of an apparent B chromosome by mutation, chromosome fragmentation and specific DNA sequence amplification. Chromosoma 111:332–340. https://doi.org/10.1007/s00412-002-0214-4
Dong F, Jiang J (1998) Non-Rabl patterns of centromere and telomere distribution in the interphase nuclei of plant cells. Chromosom Res 6:551–558
Endo TR, Nasuda S, Jones N, Dou Q, Akahori A, Wakimoto M, Tanaka H, Niwa K, Tsujimoto H (2008) Dissection of rye B chromosomes, and nondisjunction properties of the dissected segments in a common wheat background. Genes Genet Syst 83:23–30. https://doi.org/10.1266/ggs.83.23
Erlendson AA, Friedman S, Freitag M (2017) A matter of scale and dimensions: chromatin of chromosome landmarks in the fungi. Microbiol Spectr 5. https://doi.org/10.1128/microbiolspec.FUNK-0054-2017
Fekete C, Hornok L (1997) A repetitive DNA sequence associated with karyotype variability in *Fusarium poae*. Acta phytopathologica et entomologica hungarica 32:29–38
Fokkens L, Shahi S, Connolly LR, Stam R, Schmidt SM, Smith KM, Freitag M, Rep M (2018) The multispeed genome of *Fusarium oxysporum* reveals association of histone modifications with sequence divergence and footprints of past horizontal chromosome transfer events. bioRxiv:465070
Fouché S, Plissonneau C, McDonald BA, Croll D (2018) Meiosis leads to pervasive copy-number variation and distorted inheritance of accessory chromosomes of the wheat pathogen *Zymoseptoria tritici*. Genome Biol Evol 10:1416–1429
Galazka JM, Freitag M (2014) Variability of chromosome structure in pathogenic fungi-of 'ends and odds'. Curr Opin Microbiol 20C:19–26. https://doi.org/10.1016/j.mib.2014.04.002
Garmaroodi HS, Taga M (2015) Meiotic inheritance of a fungal supernumerary chromosome and its effect on sexual fertility in *Nectria haematococca*. Fungal Biol 119:929–939. https://doi.org/10.1016/j.funbio.2015.07.004
Gisselsson D, Pettersson L, Höglund M, Heidenblad M, Gorunova L, Wiegant J, Mertens F, Dal Cin P, Mitelman F, Mandahl N (2000) Chromosomal breakage-fusion-bridge events cause genetic intratumor heterogeneity. Proc Natl Acad Sci U S A 97:5357–5362. https://doi.org/10.1073/pnas.090013497
Gladyshev E, Kleckner N (2017) Recombination-independent recognition of DNA homology for repeat-induced point mutation. Curr Genet 63:389–400. https://doi.org/10.1007/s00294-016-0649-4
Goodwin SB, M'Barek SB, Dhillon B, Wittenberg AH, Crane CF, Hane JK, Foster AJ, van der Lee TA, Grimwood J, Aerts A, Antoniw J, Bailey A, Bluhm B, Bowler J, Bristow J, van der Burgt A, Canto-Canché B, Churchill AC, Conde-Ferràez L, Cools HJ, Coutinho PM, CSUKAI M, Dehal P, de Wit P, Donzelli B, van de Geest HC, van Ham RC, Hammond-Kosack KE, Henrissat B, Kilian A, Kobayashi AK, Koopmann E, Kourmpetis Y, Kuzniar A, Lindquist E, Lombard V, Maliepaard C, Martins N, Mehrabi R, Nap JP, Ponomarenko A, Rudd JJ, Salamov A, Schmutz J, Schouten HJ, Shapiro H, Stergiopoulos I, Torriani SF, Tu H, de Vries RP, Waalwijk C, Ware SB, Wiebenga A, Zwiers L-H, Oliver RP, Grigoriev IV, Kema GH, Malik HS (2011) Finished genome of the fungal wheat pathogen *Mycosphaerella graminicola* reveals dispensome structure, chromosome plasticity, and stealth pathogenesis. PLoS Genet 7: e1002070. https://doi.org/10.1371/journal.pgen.1002070
Grandaubert J, Bhattacharyya A, Stukenbrock EH (2015) RNA-seq based gene annotation and comparative genomics of four fungal grass pathogens in the genus Zymoseptoria identify novel orphan genes and species-specific invasions of transposable elements. G3 (Bethesda) g3:115.017731

Graphodatsky AS, Kukekova AV, Yudkin DV, Trifonov VA, Vorobieva NV, Beklemisheva VR, Perelman PL, Graphodatskaya DA, Trut LN, Yang F, Ferguson-Smith MA, Acland GM, Aguirre GD (2005) The proto-oncogene C-KIT maps to canid B-chromosomes. Chromosom Res 13:113–122. https://doi.org/10.1007/s10577-005-7474-9

Gregory TR (2011) The evolution of the genome. Elsevier Science, San Diego

Habig M, Kema G, Holtgrewe Stukenbrock E (2018) Meiotic drive of female-inherited supernumerary chromosomes in a pathogenic fungus. elife 7: e40251. https://doi.org/10.7554/eLife.40251

Habig M, Quade J, Stukenbrock EH (2017) Forward genetics approach reveals host genotype-dependent importance of accessory chromosomes in the fungal wheat pathogen *Zymoseptoria tritici*. MBio 8:e01919–e01917. https://doi.org/10.1128/mBio.01919-17

Han Y, Liu X, Benny U, Kistler HC, VanEtten HD (2001) Genes determining pathogenicity to pea are clustered on a supernumerary chromosome in the fungal plant pathogen *Nectria haematococca*. Plant J 25:305–314

Harr JC, Gonzalez-Sandoval A, Gasser SM (2016) Histones and histone modifications in perinuclear chromatin anchoring: from yeast to man. EMBO Rep 17:139–155. https://doi.org/10.15252/embr.201541809

Harr JC, Luperchio TR, Wong X, Cohen E, Wheelan SJ, Reddy KL (2015) Directed targeting of chromatin to the nuclear lamina is mediated by chromatin state and A-type lamins. J Cell Biol 208:33–52. https://doi.org/10.1083/jcb.201405110

Hasegawa N (1934) A cytological study on 8-chromosome Rye. Cytologia 6:68–77. https://doi.org/10.1508/cytologia.6.68

Hatta R, Ito K, Hosaki Y, Tanaka T, Tanaka A, Yamamoto M, Akimitsu K, Tsuge T (2002) A conditionally dispensable chromosome controls host-specific pathogenicity in the fungal plant pathogen *Alternaria alternata*. Genetics 161:59–70

He C, Rusu AG, Poplawski AM, Irwin JAG, Manners JM (1998) Transfer of a supernumerary chromosome between vegetatively incompatible biotypes of the fungus *Colletotrichum gloeosporioides*. Genetics 150:1459–1466

Helleu Q, Gérard PR, Montchamp-Moreau C (2014) Sex chromosome drive. Cold Spring Harb Perspect Biol 7:a017616. https://doi.org/10.1101/cshperspect.a017616

Henikoff S, Malik HS (2002) Centromeres: selfish drivers. Nature 417:227. https://doi.org/10.1038/417227a

Houben A (2017) B chromosomes–a matter of chromosome drive. Front Plant Sci 8:210. https://doi.org/10.3389/fpls.2017.00210

Houben A, Banaei-Moghaddam AM, Klemme S, Timmis JN (2014) Evolution and biology of supernumerary B chromosomes. Cell Mol Life Sci 71:467–478. https://doi.org/10.1007/s00018-013-1437-7

Houben A, Verlin D, Leach CR, Timmis JN (2001) The genomic complexity of micro B chromosomes of brachycome dichromosomatica. Chromosoma 110:451–459. https://doi.org/10.1007/s00412-001-0173-1

Hu J, Chen C, Peever T, Dang H, Lawrence C, Mitchell T (2012) Genomic characterization of the conditionally dispensable chromosome in *Alternaria arborescens* provides evidence for horizontal gene transfer. BMC Genomics 13:171. https://doi.org/10.1186/1471-2164-13-171

Hurst GDD, Werren JH (2001) The role of selfish genetic elements in eukaryotic evolution. Nat Rev Genet 2:597

Jaenike J (2008) X chromosome drive. Curr Biol 18: R508–R511. https://doi.org/10.1016/j.cub.2008.03.051

Johnson LJ, Johnson RD, Akamatsu H, Salamiah A, Otani H, Kohmoto K, Kodama M (2001) Spontaneous loss of a conditionally dispensable chromosome from the *Alternaria alternata* apple pathotype leads to loss of toxin production and pathogenicity. Curr Genet 40:65–72. https://doi.org/10.1007/s002940100233

Jones RN (1995) Tansley review no. 85. B chromosomes in plants. New Phytol 131:411–434

Jones N (2012) B chromosomes in plants. Plant Biosystems - An International Journal Dealing with all Aspects of Plant Biology 146:727–737. https://doi.org/10.1080/11263504.2012.713406

Jones JDG, Dangl JL (2006) The plant immune system. Nature 444:323–329. https://doi.org/10.1038/nature05286

Kan JAL, Goverse A, Vlugt-Bergmans CJB (1993) Electrophoretic karyotype analysis of *Botrytis cinerea*. Neth J Pl Path 99:119–128. https://doi.org/10.1007/BF03041402

Kellner R, Bhattacharyya A, Poppe S, Hsu TY, Brem RB, Stukenbrock EH (2014) Expression profiling of the wheat pathogen *Zymoseptoria tritici* reveals genomic patterns of transcription and host-specific regulatory programs. Genome Biol Evol 6:1353–1365. https://doi.org/10.1093/gbe/evu101

Kohmoto K (1993) Isolation and biological activities of two host-specific toxins from the tangerine Pathotype of *Alternaria alternata*. Phytopathology 83:495. https://doi.org/10.1094/Phyto-83-495

Kursel LE, Malik HS (2018) The cellular mechanisms and consequences of centromere drive. Curr Opin Cell Biol 52:58–65. https://doi.org/10.1016/j.ceb.2018.01.011

Lamb JC, Riddle NC, Cheng Y-M, Theuri J, Birchler JA (2007) Localization and transcription of a retrotransposon-derived element on the maize B chromosome. Chromosom Res 15:383–398. https://doi.org/10.1007/s10577-007-1135-0

Leclair S, Ansan-Melayah D, Rouxel T, Balesdent M-H (1996) Meiotic behaviour of the minichromosome in the phytopathogenic ascomycete *Leptosphaeria maculans*. Curr Genet 30:541–548

Levin DA, Palestis BG, Jones RN, Trivers R (2005) Phyletic hot spots for B chromosomes in angiosperms. Evolution 59:962–969

Lievens B, Rep M, Thomma BPHJ (2008) Recent developments in the molecular discrimination of formae speciales of *Fusarium oxysporum*. Pest Manag Sci 64:781–788. https://doi.org/10.1002/ps.1564

Lo Presti L, Lanver D, Schweizer G, Tanaka S, Liang L, Tollot M, Zuccaro A, Reissmann S, KAHMANN R (2015) Fungal effectors and plant susceptibility. Annu Rev Plant Biol 66:513–545. https://doi.org/10.1146/annurev-arplant-043014-114623

Ma W, Gabriel TS, Martis MM, Gursinsky T, Schubert V, Vrána J, Doležel J, Grundlach H, Altschmied L, Scholz U, Himmelbach A, Behrens S-E, Banaei-Moghaddam AM, Houben A (2017) Rye B chromosomes encode a functional Argonaute-like protein with in vitro slicer activities similar to its a chromosome paralog. New Phytol 213:916–928. https://doi.org/10.1111/nph.14110

Ma L-J, van der Does HC, Borkovich KA, Coleman JJ, Daboussi M-J, Di Pietro A, Dufresne M, Freitag M, Grabherr M, Henrissat B, Houterman PM, Kang S, Shim W-B, Woloshuk C, Xie X, Xu J-R, Antoniw J, Baker SE, Bluhm BH, Breakspear A, Brown DW, Butchko RAE, Chapman S, Coulson R, Coutinho PM, Danchin EGJ, Diener A, Gale LR, Gardiner DM, Goff S, Hammond-Kosack KE, Hilburn K, Hua-Van A, Jonkers W, Kazan K, Kodira CD, Koehrsen M, Kumar L, Lee Y-H, Li L, Manners JM, Miranda-Saavedra D, Mukherjee M, Park G, Park J, Park S-Y, Proctor RH, Regev A, Ruiz-Roldan MC, Sain D, Sakthikumar S, Sykes S, Schwartz DC, Turgeon BG, Wapinski I, Yoder O, Young S, Zeng Q, Zhou S, Galagan J, Cuomo CA, Kistler HC, Rep M (2010) Comparative genomics reveals mobile pathogenicity chromosomes in Fusarium. Nature 464:367–373. https://doi.org/10.1038/nature08850

Maloney AP, VanEtten HD (1994) A gene from the fungal plant pathogen *Nectria haematococca* that encodes the phytoalexin-detoxifying enzyme pisatin demethylase defines a new cytochrome P450 family. Mol Gen Genet 243:506–514

Martis MM, Klemme S, Banaei-Moghaddam AM, Blattner FR, Macas J, Schmutzer T, Scholz U, Gundlach H, Wicker T, Šimková H, Novák P, Neumann P, Kubaláková M, Bauer E, Haseneyer G, Fuchs J, Doležel J, Stein N, Mayer KFX, Houben A (2012) Selfish supernumerary chromosome reveals its origin as a mosaic of host genome and organellar sequences. Proc Natl Acad Sci U S A 109:13343–13346. https://doi.org/10.1073/pnas.1204237109

Masel AM (1996) Molecular evidence for chromosome transfer between biotypes of *Colletotrichum gloeosporioides*. Mol Plant-Microbe Interact 9:339. https://doi.org/10.1094/MPMI-9-0339

Masel A, Irwin KG, Manners J (1993) DNA addition or deletion is associated with a major karyotype polymorphism in the fungal phytopathogen *Colletotrichum gloeosporioides*. Mol Gen Genet 237(1–2):73–80. https://doi.org/10.1007/BF00282786

Mayer KFX, Martis M, Hedley PE, Simková H, Liu H, Morris JA, Steuernagel B, Taudien S, Roessner S, Gundlach H, Kubaláková M, Suchánková P, Murat F, Felder M, Nussbaumer T, Graner A, Salse J, Endo T, Sakai H, Tanaka T, Itoh T, Sato K, Platzer M, Matsumoto T, Scholz U, Dolezel J, Waugh R, Stein N (2011) Unlocking the barley genome by chromosomal and comparative genomics. Plant Cell 23:1249–1263. https://doi.org/10.1105/tpc.110.082537

McClintock B (1941) The stability of broken ends of chromosomes in *Zea mays*. Genetics 26:234–282

McDonald MC, McGinness L, Hane JK, Williams AH, Milgate A, Solomon PS (2016) Utilizing gene tree variation to identify candidate effector genes in *Zymoseptoria tritici*. G3 (Bethesda) 6:779–791. https://doi.org/10.1534/g3.115.025197

Mehrabi R, Bahkali AH, Abd-Elsalam KA, Moslem M, Ben M'Barek S, Gohari AM, Jashni MK, Stergiopoulos I, Kema GHJ, de Wit PJGM (2011) Horizontal gene and chromosome transfer in plant pathogenic fungi affecting host range. FEMS Microbiol Rev 35:542–554. https://doi.org/10.1111/j.1574-6976.2010.00263.x

Mehrabi R, Mirzadi Gohari A, Kema GH (2017) Karyotype variability in plant-pathogenic fungi. Annu Rev Phytopathol 55:483–503. https://doi.org/10.1146/annurev-phyto-080615-095928

Mendelson D, Zohary D (1972) Behaviour and transmission of supernumerary chromosomes in *Aegilops speltoides*. Heredity (Edinb) 29:329–340

Miao V, Covert S, VanEtten H (1991) A fungal gene for antibiotic resistance on a dispensable ("\B\") chromosome. Science 254:1773–1776. https://doi.org/10.1126/science.1763326

Milani NA, Lawrence DP, Arnold AE, Vanetten HD (2012) Origin of pisatin demethylase (PDA) in the genus Fusarium. Fungal Genet Biol 49:933–942. https://doi.org/10.1016/j.fgb.2012.08.007

Möller M, Habig M, Freitag M, Stukenbrock EH (2018) Extraordinary genome instability and widespread chromosome rearrangements during vegetative growth. Genetics 210:517–529. https://doi.org/10.1534/genetics.118.301050

Möller M, Schotanus K, Soyer JL, Haueisen J, Happ K, Stralucke M, Happel P, Smith KM, Connolly LR, Freitag M, Stukenbrock EH, Madhani HD (2019) Destabilization of chromosome structure by histone H3 lysine 27 methylation. PLoS Genet 15: e1008093. https://doi.org/10.1371/journal.pgen.1008093

Möller M, Stukenbrock EH (2017) Evolution and genome architecture in fungal plant pathogens. Nat Rev Microbiol 15:756

Mroczek RJ, Melo JR, Luce AC, Hiatt EN, Dawe RK (2006) The maize Ab10 meiotic drive system maps to supernumerary sequences in a large complex haplotype. Genetics 174:145–154. https://doi.org/10.1534/genetics.105.048322

Nakashima T, Ueno T, Fukami H, Taga T, Masuda H, Osaki K, Otani H, Kohmoto K, Nishimura S (1985) Isolation and structures of AK-toxin I and II. Host-specific phytotoxic metabolites produced by *Alternaria alternata* Japanese pear pathotype. Agric Biol Chem 49:807–815. https://doi.org/10.1271/bbb1961.49.807

Nakatsuka S-I, Ueda K, Goto T, Yamamoto M, Nishimura S, Kohmoto K (1986) Structure of AF-toxin II, one of the host-specific toxins produced by *Alternaria alternata* strawberry pathotype. Tetrahedron Lett 27:2753–2756. https://doi.org/10.1016/S0040-4039(00)84635-3

O'Donnell K, Cigelnik E, Casper HH (1998) Molecular phylogenetic, morphological, and mycotoxin data support reidentification of the Quorn mycoprotein fungus as *Fusarium venenatum*. Fungal Genet Biol 23:57–67. https://doi.org/10.1006/fgbi.1997.1018

Orbach MJ, Chumley FG, Valent B (1996) Electrophoretic karyotypes of *Magnaporthe grisea* pathogens of diverse grasses. Mol Plant-Microbe Interact 9:261–271

Östergren G (1947) Heterochromatic B-chromosomes in anthoxanthum. Hereditas 33:261–296. https://doi.org/10.1111/j.1601-5223.1947.tb02804.x

Pansonato-Alves JC, Serrano ÉA, Utsunomia R, Camacho JPM, da Costa Silva GJ, Vicari MR, Artoni RF, Oliveira C, Foresti F (2014) Single origin of sex chromosomes and multiple origins of B chromosomes in fish genus Characidium. PLoS One 9: e107169. https://doi.org/10.1371/journal.pone.0107169

Pereira HS, Delgado M, Viegas W, Rato JM, Barão A, Caperta AD (2017) Rye (*Secale cereale*) supernumerary (B) chromosomes associated with heat tolerance during early stages of male sporogenesis. Ann Bot 119:325–337. https://doi.org/10.1093/aob/mcw206

Plissonneau C, Hartmann FE, Croll D (2018) Pangenome analyses of the wheat pathogen *Zymoseptoria tritici* reveal the structural basis of a highly plastic eukaryotic genome. BMC Biol 16:5. https://doi.org/10.1186/s12915-017-0457-4

Ramos É, Cardoso AL, Brown J, Marques DF, Fantinatti BEA, Cabral-de-Mello DC, Oliveira RA, O'Neill RJ, Martins C (2017) The repetitive DNA element BncDNA, enriched in the B chromosome of the cichlid fish *Astatotilapia latifasciata*, transcribes a potentially noncoding RNA. Chromosoma 126:313–323. https://doi.org/10.1007/s00412-016-0601-x

Recorbet G, Steinberg C, Olivain C, Edel V, Trouvelot S, Dumas-Gaudot E, Gianinazzi S, Alabouvette C (2003) Wanted: pathogenesis-related marker molecules for *Fusarium oxysporum*. New Phytol 159:73–92. https://doi.org/10.1046/j.1469-8137.2003.00795.x

Remis MI, Vilardi JC (2004) Mitotically unstable B chromosome polymorphism in the grasshopper *Dichroplus elongatus*. Cytogenet Genome Res 106:359–364. https://doi.org/10.1159/000079313

Rep M, van der Does HC, Meijer M, van Wijk R, Houterman PM, Dekker HL, de Koster CG, Cornelissen BJC (2004) A small, cysteine-rich protein secreted by *Fusarium oxysporum* during colonization of xylem vessels is required for I-3-mediated resistance in tomato. Mol Microbiol 53:1373–1383. https://doi.org/10.1111/j.1365-2958.2004.04177.x

Rosin LF, Mellone BG (2017) Centromeres drive a hard bargain. Trends Genet 33:101–117. https://doi.org/10.1016/j.tig.2016.12.001

Ross BD, Rosin L, Thomae AW, Hiatt MA, Vermaak D, La Cruz AFA, de Imhof A, Mellone BG, Malik HS (2013) Stepwise evolution of essential centromere function in a Drosophila neogene. Science 340:1211–1214. https://doi.org/10.1126/science.1234393

Rotem J (1994) The genus Alternaria: biology, epidemiology, and pathogenicity. American Phytopathological Society

Rouxel T, Balesdent MH (2005) The stem canker (blackleg) fungus, *Leptosphaeria maculans*, enters the genomic era. Mol Plant Pathol 6:225–241

Rouxel T, Balesdent M-H (2017) Life, death and rebirth of avirulence effectors in a fungal pathogen of brassica crops. *Leptosphaeria maculans*. New Phytol 214:526–532

Rouxel T, Grandaubert J, Hane JK, Hoede C, van de Wouw AP, Couloux A, Dominguez V, Anthouard V, Bally P, Bourras S, Cozijnsen AJ, Ciuffetti LM, Degrave A, Dilmaghani A, Duret L, Fudal I, Goodwin SB, Gout L, Glaser N, Linglin J, Kema GH, Lapalu N, Lawrence CB, May K, Meyer M, Ollivier B, Poulain J, Schoch CL, Simon A, Spatafora JW, Stachowiak A, Turgeon BG, Tyler BM, Vincent D, Weissenbach J, Amselem J, Quesneville H, Oliver RP, Wincker P, Balesdent M-H, Howlett BJ (2011) Effector diversification within compartments of the *Leptosphaeria maculans* genome affected by repeat-induced point mutations. Nat Commun 2:202. https://doi.org/10.1038/ncomms1189

Salamiah A, Akamatsu H, Fukumasa-Nakai Y, Otani H, Kodama M (2001b) Construction and genetic analysis of hybrid strains between apple and tomato pathotypes of *Alternaria alternata* by protoplast fusion. J Gen Plant Pathol 67:97–105. https://doi.org/10.1007/PL00013006

Salamiah A, Fukumasa-Nakai Y, Akamatsu H, Otani H, Kohmoto K, Kodama M (2001a) Genetic analysis of pathogenicity and host-specific toxin production of *Alternaria alternata* tomato Pathotype by protoplast fusion. J Gen Plant Pathol 67:7–14. https://doi.org/10.1007/PL00012991

Schmidt SM, Houterman PM, Schreiver I, Ma L, Amyotte S, Chellappan B, Boeren S, Takken FLW, Rep M (2013) MITEs in the promoters of effector genes allow prediction of novel virulence genes in *Fusarium oxysporum*. BMC Genomics 14:119. https://doi.org/10.1186/1471-2164-14-119

Schotanus K, Soyer JL, Connolly LR, Grandaubert J, Happel P, Smith KM, Freitag M, Stukenbrock EH (2015) Histone modifications rather than the novel regional centromeres of *Zymoseptoria tritici* distinguish core and accessory chromosomes. Epigenetics Chromatin 8:41. https://doi.org/10.1186/s13072-015-0033-5

Schrank A, Vainstein MH (2010) *Metarhizium anisopliae* enzymes and toxins. Toxicon 56:1267–1274. https://doi.org/10.1016/j.toxicon.2010.03.008

Schwartz DC, Cantor CR (1984) Separation of yeast chromosome-sized DNAs by pulsed field gradient gel electrophoresis. Cell 37:67–75

Selker EU (1990) Premeiotic instability of repeated sequences in *Neurospora crassa*. Annu Rev Genet 24:579–613

Smith KM, Galazka JM, Phatale PA, Connolly LR, Freitag M (2012) Centromeres of filamentous fungi. Chromosom Res 20:635–656. https://doi.org/10.1007/s10577-012-9290-3

Soyer JL, Balesdent M-H, Rouxel T, Dean RA (2018) To B or not to B: a tale of unorthodox chromosomes. Curr Opin Microbiol 46:50–57. https://doi.org/10.1016/j.mib.2018.01.012

Soyer JL, El Ghalid M, Glaser N, Ollivier B, Linglin J, Grandaubert J, Balesdent M-H, Connolly LR, Freitag M, Rouxel T, Fudal I (2014) Epigenetic control of effector gene expression in the plant pathogenic fungus *Leptosphaeria maculans*. PLoS Genet 10: e1004227. https://doi.org/10.1371/journal.pgen.1004227

Stewart EL, Croll D, Lendenmann MH, Sanchez-Vallet A, Hartmann FE, Palma-Guerrero J, Ma X, McDonald BA (2016) Quantitative trait locus mapping reveals complex genetic architecture of quantitative virulence in the wheat pathogen *Zymoseptoria tritici*. Mol Plant Pathol. https://doi.org/10.1111/mpp.12515

Stukenbrock EH, Bataillon T, Dutheil JY, Hansen TT, Li R, Zala M, McDonald BA, Wang J, Schierup MH (2011) The making of a new pathogen: insights from comparative population genomics of the domesticated wheat pathogen *Mycosphaerella graminicola* and its wild sister species. Genome Res 21:2157–2166. https://doi.org/10.1101/gr.118851.110

Stukenbrock EH, Dutheil JY (2017) Fine-scale recombination maps of fungal plant pathogens reveal dynamic recombination landscapes and intragenic hotspots. Genetics. https://doi.org/10.1534/genetics.117.300502

Stukenbrock EH, Jørgensen FG, Zala M, Hansen TT, McDonald BA, Schierup MH, Malik HS (2010) Whole-genome and chromosome evolution associated with host adaptation and speciation of the wheat pathogen *Mycosphaerella graminicola*. PLoS Genet 6:e1001189. https://doi.org/10.1371/journal.pgen.1001189

Taylor JW, Branco S, Gao C, Hann-Soden C, Montoya L, Sylvain I, Gladieux P (2017) Sources of fungal genetic variation and associating it with phenotypic diversity. Microbiol Spectr 5. https://doi.org/10.1128/microbiolspec.FUNK-0057-2016

Trifonov VA, Dementyeva PV, Larkin DM, O'Brien PCM, Perelman PL, Yang F, Ferguson-Smith MA, Graphodatsky AS (2013) Transcription of a protein-coding gene on B chromosomes of the Siberian roe deer (*Capreolus pygargus*). BMC Biol 11:90. https://doi.org/10.1186/1741-7007-11-90

Twell D (2011) Male gametogenesis and germline specification in flowering plants. Sex Plant Reprod 24:149–160. https://doi.org/10.1007/s00497-010-0157-5

Tzeng TH, Lyngholm LK, Ford CF, Bronson CR (1992) A restriction fragment length polymorphism map and electrophoretic karyotype of the fungal maize pathogen *Cochliobolus heterostrophus*. Genetics 130:81–96

Valente GT, Conte MA, Fantinatti BEA, Cabral-de-Mello DC, Carvalho RF, Vicari MR, Kocher TD, Martins C (2014) Origin and evolution of B chromosomes in the cichlid fish *Astatotilapia latifasciata* based on integrated genomic analyses. Mol Biol Evol 31:2061–2072. https://doi.org/10.1093/molbev/msu148

Valente GT, Nakajima RT, Fantinatti BEA, Marques DF, Almeida RO, Simões RP, Martins C (2017) B chromosomes: from cytogenetics to systems biology. Chromosoma 126:73–81. https://doi.org/10.1007/s00412-016-0613-6

Van Kan JAL, Stassen JHM, Mosbach A, der Lee V, Theo AJ, Faino L, Farmer AD, Papasotiriou DG, Zhou S, Seidl MF, Cottam E, Edel D, Hahn M, Schwartz DC, Dietrich RA, Widdison S, SCALLIET G (2017) A gapless genome sequence of the fungus *Botrytis cinerea*. Mol Plant Pathol 18:75–89. https://doi.org/10.1111/mpp.12384

Vanetten HD (1978) Identification of additional habitats of *Nectria haematococca* mating population VI. Phytopathology 68:1552. https://doi.org/10.1094/Phyto-68-1552

Vanheule A, Audenaert K, Warris S, van de Geest H, Schijlen E, Höfte M, de Saeger S, Haesaert G, Waalwijk C, van der Lee T (2016) Living apart together: crosstalk between the core and supernumerary genomes in a fungal plant pathogen. BMC Genomics 17:670. https://doi.org/10.1186/s12864-016-2941-6

Vlaardingerbroek I, Beerens B, Rose L, Fokkens L, Cornelissen BJC, Rep M (2016b) Exchange of core chromosomes and horizontal transfer of lineage-specific chromosomes in *Fusarium oxysporum*. Environmental Microbiology:n/a-n/a. https://doi.org/10.1111/1462-2920.13281

Vlaardingerbroek I, Beerens B, Schmidt SM, Cornelissen BJC, Rep M (2016a) Dispensable chromosomes in *Fusarium oxysporum* f. sp. lycopersici. Mol. Plant Pathol 17:1455–1466. https://doi.org/10.1111/mpp.12440

Wang C, Skrobek A, Butt TM (2003) Concurrence of losing a chromosome and the ability to produce destruxins in a mutant of *Metarhizium anisopliae*. FEMS Microbiol Lett 226:373–378

West JS, Kharbanda PD, Barbetti MJ, Fitt BDL (2001) Epidemiology and management of *Leptosphaeria maculans* (phoma stem canker) on oilseed rape in Australia, Canada and Europe. Plant Pathol 50:10–27. https://doi.org/10.1046/j.1365-3059.2001.00546.x

Williams AH, Sharma M, Thatcher LF, Azam S, Hane JK, Sperschneider J, Kidd BN, Anderson JP, Ghosh R, Garg G, Lichtenzveig J, Kistler HC, Shea T, Young S, Buck S-AG, Kamphuis LG, Saxena R, Pande S, Ma L-J, Varshney RK, Singh KB (2016) Comparative genomics and prediction of conditionally dispensable sequences in legume-infecting *Fusarium oxysporum* formae speciales facilitates identification of candidate effectors. BMC Genomics 17:191. https://doi.org/10.1186/s12864-016-2486-8

Wilson EB (1907) The supernumerary chromosomes of Hemiptera. Science 26:870–871. https://doi.org/10.1126/science.26.677.870-a

Wittenberg AHJ, van der Lee TAJ, Ben M'Barek S, Ware SB, Goodwin SB, Kilian A, Visser RGF, Kema GHJ, Schouten HJ (2009) Meiosis drives extraordinary genome plasticity in the haploid fungal plant pathogen *Mycosphaerella graminicola*. PLoS One 4:e5863. https://doi.org/10.1371/journal.pone.0005863

Xu J-R (1995) Electrophoretic karyotypes distinguish the biological species of *Gibberella fujikuroi* (Fusarium section Liseola). Mol Plant-Microbe Interact 8:74. https://doi.org/10.1094/MPMI-8-0074

Xu J-R, Leslie JF (1996) A genetic map of *Gibberella fujikuroi* mating population a (*Fusarium moniliforme*). Genetics 143:175–189

Yoshida K, Saitoh H, Fujisawa S, Kanzaki H, Matsumura H, Yoshida K, Tosa Y, Chuma I, Takano Y, Win J, Kamoun S, Terauchi R (2009) Association genetics reveals three novel avirulence genes from the rice blast fungal pathogen *Magnaporthe oryzae*. Plant Cell 21:1573–1591. https://doi.org/10.1105/tpc.109.066324

Yoshida K, Terai Y, Mizoiri S, Aibara M, Nishihara H, Watanabe M, Kuroiwa A, Hirai H, Hirai Y, Matsuda Y, Okada N (2011) B chromosomes have a functional effect on female sex determination in Lake Victoria cichlid fishes. PLoS Genet 7:e1002203. https://doi.org/10.1371/journal.pgen.1002203

Zhang N, O'Donnell K, Sutton DA, Nalim FA, Summerbell RC, Padhye AA, Geiser DM (2006) Members of the *Fusarium solani* species complex that cause infections in both humans and plants are common in the environment. J Clin Microbiol 44:2186–2190. https://doi.org/10.1128/JCM.00120-06

3 Genetics of the Unfolded Protein Response in Fungi

Rebekka Harting[1], Kai Heimel[1]

CONTENTS

I. Introduction

Fungi inhabit almost all ecological niches on earth and exhibit a wide variety of different lifestyles, requiring efficient strategies to adapt the intracellular signaling pathways to the extracellular environment. Secretion of proteinaceous molecules including cell wall components and enzymes for nutrient mobilization and/or for interaction with other organisms is dependent on their correct folding, processing, and transfer within the **secretory pathway**. The unfolded protein response (UPR) is a conserved pathway in eukaryotic organisms. It functions as a central regulatory instance required for homeostasis of the endoplasmic reticulum (ER) by adapting the ER folding capacity to increased folding demands, reflected by the accumulation of un- or misfolded protein aggregates in the ER. After its initial discovery in the 1990s, it has become evident that the UPR features a unique mechanism for signal transduction, referred to as unconventional cytoplasmic splicing, to rapidly adapt the transcriptional program and initiate a comprehensive restructuring of the secretory pathway. The UPR is of central importance for saprotrophic growth of filamentous fungi and for disease development in human and plant pathogenic fungi, as well as in the optimization of protein production within industrial settings of **fungal biotechnology**. Comprehensive cellular and transcriptomic studies uncovered intricate connections between the UPR and other conserved signaling pathways including **autophagy**, the **cell wall integrity (CWI) pathway**, **hypoxia adaptation**, and even the control of **fungal development**. Focusing on the UPR in various model organisms, the basic principles, physiological roles, but also specific functions, interactions, and applications of the UPR will be discussed.

II. General Concept of the UPR

A. Common Principles of UPR Signaling

The unfolded protein response represents a safeguarding mechanism that protects the cell

[1] Institut für Mikrobiologie und Genetik, Abteilung Molekulare Mikrobiologie und Genetik, Göttinger Zentrum für Molekulare Biowissenschaften (GZMB), Georg-August-Universität Göttingen, Göttingen, Germany; e-mail: kheimel@gwdg.de

Genetics and Biotechnology, 3rd Edition
The Mycota II
J.P. Benz, K. Schipper (Eds.)

from imbalances between demand and capacity on the protein folding machinery in the endoplasmic reticulum (ER). A consequence of such imbalances that may arise from different intracellular and extracellular influences on ER-localized or secreted proteins is the accumulation of unfolded or **misfolded proteins** in the ER lumen. Detection of these potentially toxic or harmful proteins (preproteins/protein intermediates) occurs by the ER membrane-localized sensor proteins *PERK* (*P*rotein kinase R (PKR)-like *E*ndoplasmic *R*eticulum *K*inase), *ATF6* (*A*ctivating *T*ranscription *F*actor 6), and **Ire1** (*I*nositol *re*quiring enzyme 1), of which only the latter is conserved from fungi to mammals and best studied in the baker's yeast *Saccharomyces cerevisiae*. Ire1 is composed of an ER luminal sensor-domain that is connected by an ER membrane-spanning transmembrane domain to the cytoplasmic part harboring **kinase** and **RNase** functions (Fig. 1), both of which are required for the induction of ER stress-responsive genes (Kohno et al. 1993; Kozutsumi et al. 1988; Mori et al. 1993; Rose et al. 1989).

In *S. cerevisiae*, sensing of unfolded or misfolded proteins in the ER lumen can occur by (1) direct binding of misfolded proteins to the ER luminal sensor domain or indirectly by (2) unfolded protein-dependent depletion of the major ER chaperone **Kar2/Bip1** (*Kar*yogamy/*B*inding *i*mmunoglobulin *p*rotein) that under unstressed conditions binds to the Ire1 sensor domain and prevents formation of oligomeric and active Ire1 complexes (Zhou et al. 2006). In higher eukaryotes, activation of *ATF6* and *PERK* is similarly triggered by the reduced interaction between Bip1 and *ATF6* or *PERK*. In case of *ATF6* this leads to relocalization of *ATF6* to the Golgi and subsequent S2P (*S*ite *2 P*rotease)-mediated cleavage to liberate the transcriptionally active *ATF6* bZIP protein (Ron and Walter 2007). In case of *PERK*, activation of its kinase domain promotes phosphorylation of eIF2α (*e*ukaryotic *I*nitiation *F*actor *2*α) to globally inhibit protein biosynthesis via attenuated translation initiation. This mode of action is analogous to GCN2 (*G*eneral *C*ontrol *N*onderepressible 2)-dependent phosphorylation of eIF2α under amino acid starvation conditions (Hinnebusch 1994). Interestingly, all three pathways share common principles in their mode of activation and individually activate expression of specific bZIP **transcription factors** that mediate the transcriptional response towards ER stress.

B. Ire1 and Hac1: Key Players of the UPR

1. Activation of Ire1

In *S. cerevisiae*, detection of unfolded proteins in the ER lumen triggers Ire1 oligomerisation, trans-autophosphorylation, ADP binding, and activation of the **RNAse** domain (Gardner et al. 2013) that resides on the cytoplasmic portion of the Ire1 protein. Activity of Ire1 is induced in response to a wide variety of different stimuli. These affect either protein folding in the ER, such as increased demands for protein secretion, or treatment with protein folding inhibitors (dithiothreitol, DTT; tunicamycin, TM) and heat stress. In addition, Ire1 is also activated by ER membrane aberrancies associated with or connected to cell wall stress, inositol depletion (Gardner et al. 2013; Gardner and Walter 2011; Korennykh et al. 2011; Rubio et al. 2011), and iron depletion (Cohen et al. 2017). A multitude of different compounds has been identified that either directly or indirectly activate Ire1 and induce the UPR, for most of which the mode of action has not been fully resolved. Interestingly, quercetin, a plant-derived flavanone, directly binds to an unconventional ligand binding pocket and promotes increased Ire1 activity in vitro (Wiseman et al. 2010). In *Aspergillus nidulans* and *Candida albicans*, the isoprenoid alcohol farnesol and the monoterpene carvacrol trigger UPR activity, respectively (Chaillot et al. 2015; Colabardini et al. 2010). The activity of Ire1 is moreover modulated by various factors. For example, in *S. cerevisiae*, the phosphatase Ptc2 (*P*hospha*t*ase *T*wo *C* 2) mediates Ire1 dephosphorylation and thus contributes to inactivation of the UPR (Welihinda et al. 1998). Interestingly, Ire1 mutants lacking the trans-autophosphorylation activity do not show loss of UPR activity, but prolonged UPR activation, suggesting that

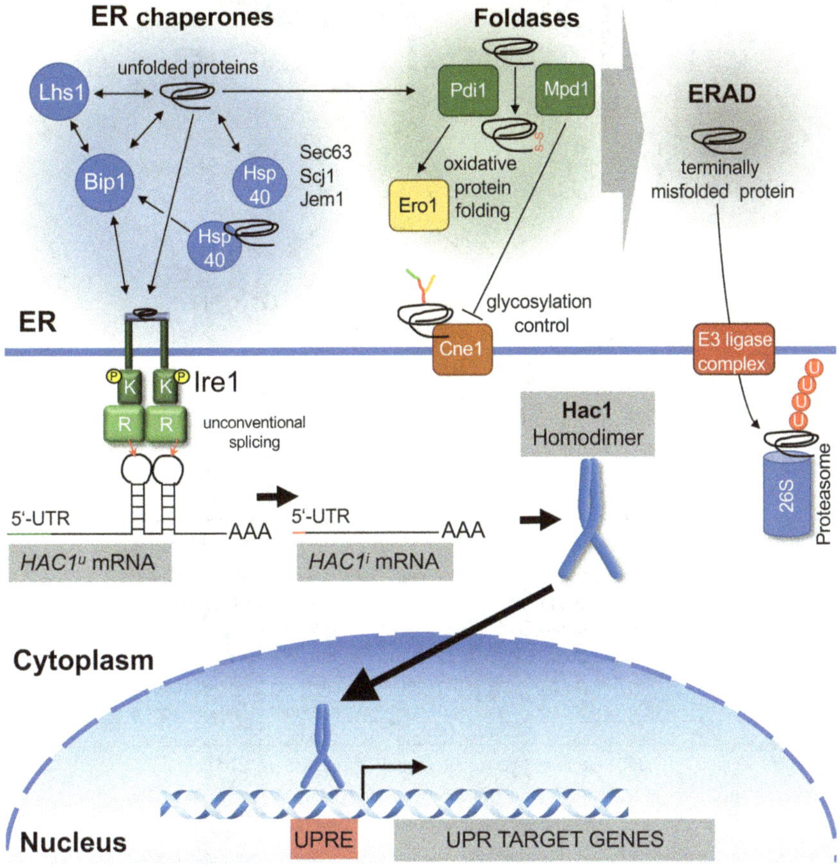

Fig. 1 Overview on the general mechanism, key factors, and their interactions within the fungal UPR. Sensing of unfolded or misfolded proteins by the ER membrane-localized Ire1 kinase/RNase results in Ire1 oligomerisation, trans-autophosphorylation, and activation of the RNase activity required for unconventional cleavage of the native *HAC1*^{u} mRNA. The processed *HAC1*^{i} mRNA is translationally derepressed, promoting Hac1 expression, which in turn activates UPR target genes by binding to unfolded protein response elements (UPREs). Conserved target genes encode the ER chaperones Bip1 and Lhs1 which also contribute to protein translocation into the ER as well as the foldases Pdi1 and Mpd1 and the Pdi1 interacting ER oxidoreductase Ero1. The lectin chaperone calnexin facilitates quality control of N-glycosylated proteins in conjunction with Pdi1 and in regulatory crosstalk with Mpd1. Terminally misfolded and potentially toxic proteins and protein aggregates are targeted for retro-translocation into the cytoplasm and ubiquitylation by ER membrane-localized ERAD complexes for subsequent degradation by the 26S proteasome

phosphorylation is required for de-oligomerisation during recovery from ER stress (Chawla et al. 2011; Rubio et al. 2011).

2. Ire1-Mediated Unconventional Splicing

The overall domain structure and the functional requirement of Ire1 for ER stress resistance is conserved among eukaryotes. In most fungi, and in contrast to higher eukaryotes (see also B. 3), the sole and specific target of Ire1-dependent endonucleolytic cleavage is the *HAC1* mRNA (Kimata et al. 2007; Korennykh et al. 2009; Li et al. 2010; Oikawa et al. 2007; Sidrauski and Walter 1997). The *HAC1* mRNA is constitutively expressed and exhibits a conserved secondary structure in which two hairpin stem-loops are recognized and cleaved by the RNAse domain of Ire1 (Fig. 2). This mechanism is referred to as **unconventional cytoplas-**

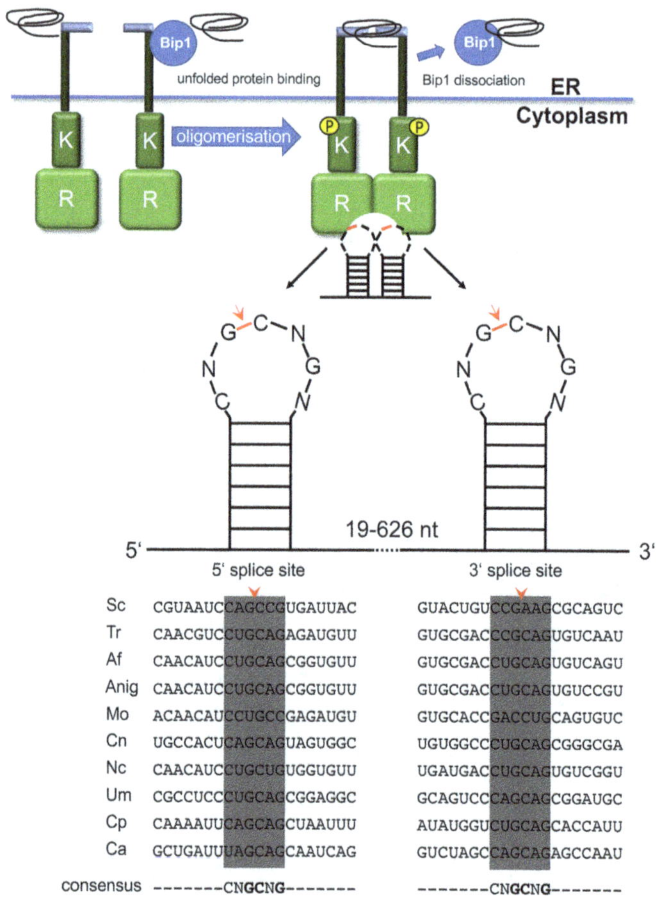

Fig. 2 Unconventional splicing by Ire1. Activation of Ire1 is mediated by direct binding of misfolded proteins to the ER luminal domain of Ire1 and by sequestration of Ire1-interacting Bip1 to misfolded proteins, both of which promote Ire1 oligomerisation and trans-autophosphorylation. Activated Ire1 oligomers recognize the conserved twin stem-loop structures comprising the CNG'CNGN consensus sequence in the loop regions of mRNAs encoding Hac1-like proteins. An overview of experimentally validated intron sequences in selected fungi is provided below. *Sc Saccharomyces cerevisiae*, *Tr Trichoderma reesei*, *Af Aspergillus fumigatus*, *Anig Aspergillus niger*, *Mo Magnaporthe oryzae*, *Cn Cryptococcus neoformans*, *Nc Neurospora crassa*, *Um Ustilago maydis*, *Cp Candida parapsilosis*, *Ca Candida albicans*

mic splicing and represents the singular example of extranuclear mRNA splicing in eukaryotes. Ligation of the two exons is mediated by the tRNA ligase Rlg1/Trl1 (*t*RNA *L*igase 1) that is also important to promote the subsequent translation of the spliced *HAC1* mRNA (Mori et al. 2010; Sidrauski et al. 1996) and to produce the central transcriptional regulator of the UPR termed Hac1 in *S. cerevisiae* (*H*omologous to *A*TF/*C*REB 1) and **XBP1** (*X*-box *B*inding *P*rotein 1) in humans.

In contrast to the conserved Ire1 function, the size of the unconventionally cleaved intron shows increased variation in different fungi. Experimentally validated unconventionally spliced introns range in their size from 20 to 23 nucleotides (nt) in filamentous ascomycetes (Hooks and Griffiths-Jones 2011) and from 56 to 65 nt in the basidiomycetes *Cryptococcus neoformans* and *Ustilago maydis*. In *S. cerevisiae* the unconventional intron encompassing 252 nt mediates translation repression of the unspliced *HAC1* mRNA via base pairing with the 5′ untranslated region (UTR) of the mRNA. Recent studies in *Candida parapsilosis* identified a 626 nt unconventional intron, the putative function of which has not been resolved, yet (Iracane et al. 2018). Interestingly, although heterogeneous in size, the secondary structure of unconventional introns and the consensus cleavage site (CNG'CNGN) (Fig. 2) are conserved among eukaryotes, although endogenous Ire1 and foreign Hac1-encoding mRNAs display limited interspecies compatibility (Zhang et al. 2016b). In filamentous fungi, translation of the unspliced *Hac1* mRNA is controlled by an extended 5' UTR (untranslated region) of the mRNA, which is shortened under ER stress conditions and translationally derepressed (Joubert et al. 2011; Mulder et al. 2004; Saloheimo et al. 2003).

3. Ire1 Functions Independent of *Hac1* mRNA Splicing

While in most fungi the Hac1 mRNA represents the sole Ire1 target, in higher eukaryotes, Ire1 activity is as well connected to the degradation of various ER-bound mRNAs, a process referred to as **regulated Ire1-dependent decay (RIDD)** (Hollien et al. 2009). Splicing of Hac1/XBP1 mRNA and RIDD are mediated by distinct enzymatic activities (Li et al. 2018; Tam et al. 2014). Interestingly and in contrast to most other fungi, the ER stress response pathway in *Schizosaccharomyces pombe* and *Candida glabrata* is mediated independent of *HAC1* mRNA splicing suggesting that RIDD is not exclusively found in higher eukaryotes but an evolutionary ancient trait (Kimmig et al. 2012; Miyazaki et al. 2013). In *S. pombe*, the induction of ER stress results in cleavage and subsequent degradation of ER-associated mRNAs. Cleavage of *bip1* mRNA results in the loss of its polyA tail but also in increased stability and translation (Kimmig et al. 2012). *S. pombe* lacks a potential Hac1 homolog, and Hac1 is not involved in ER stress resistance in *C. glabrata*. Remarkably, expression of *C. glabrata* Hac1 in *S. cerevisiae* Δ*HAC1* strains restores ER stress resistance (Miyazaki et al. 2013). The observation that in most fungi, Δ*ire1* and Δ*hac1* strains display distinct phenotypes suggests additional roles for Ire1, independent of unconventional splicing of *HAC1* mRNA (Heimel 2015).

4. Hac1-Like Transcription Factors in Fungi

Hac1 represents the central transcriptional regulator that orchestrates the UPR by adapting gene expression to the demands imposed on the ER. Hac1 is a bZIP transcription factor that binds as a homodimer to **UPREs (unfolded protein response elements)**, in the promoters of regulated genes (Fig. 1), inducing or elevating their transcription. Cellular responses to ER stress are manifested by an increase of the ER folding capacity, restructuring of ER-associated transport processes, enlargement of the ER, and the targeted degradation of terminally misfolded proteins by the **ERAD pathway** (ER-associated degradation) (Amara et al. 1989; Chen et al. 1988; Ng et al. 2000; Travers et al. 2000). Genome-wide expression studies in different fungi identified a core set of UPR-regulated genes encoding **ER chaperones** and **foldases**, proteins involved in the synthesis of fatty acids and phospholipids, protein glycosylation, protein translocation, and protein degradation (Guillemette et al. 2007; Travers et al. 2000; Feng et al. 2011; Pinter et al. 2019). While several genes encoding the ER chaperones Bip1, Lhs1 (*L*umenal *H*sp *S*eventy 1), and the foldases and associated proteins Pdi1 (*P*rotein *D*isulfide *I*somerase 1), Mpd1 (*M*ulticopy suppressor of *PDI*1 *D*eletion 1), Ero1 (*ER O*xidation 1), and Cne1 (*Cal*n*e*xin 1) are universal UPR markers, a substantial amount of UPR regulated genes appears to be unique for certain fungal species.

In contrast to Ire1, Hac1-like proteins are only conserved in their bZIP domains (Fig. 3a) and display wide variations with respect to protein size. While *HAC1* in *S. cerevisiae* encodes a protein of 238 amino acids (aa), the molecular mass of Hac1-like proteins in filamentous fungi is increased and in *U. maydis* more than doubled (574 aa). The increased size of Hac1-like proteins provides platforms for interaction with other proteins and thus the incorporation and connection of other signaling pathway to the UPR. Experimentally validated binding to

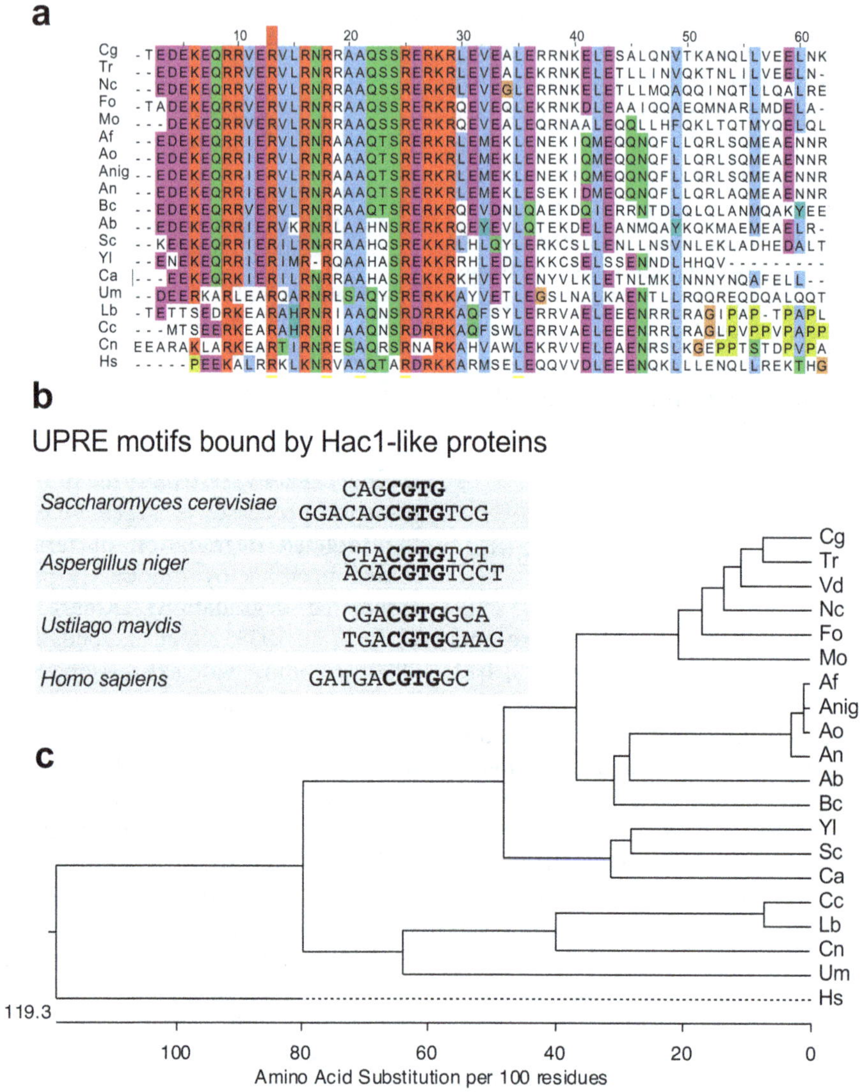

Fig. 3 Conservation of Hac1 and UPREs in fungi. (**a**) Alignment of bZIP domains of Hac1-like proteins in various fungi. *Cg Colletotrichum graminicola, Tr Trichoderma reesei, Nc Neurospora crassa, Fo Fusarium oxysporum, Mo Magnaporthe oryzae, Af Aspergillus fumigatus, Ao Aspergillus oryzae, Anig Aspergillus niger, An Aspergillus nidulans, Bc Botrytis cinerea, Ab Alternaria brassicicola, Sc Saccharomyces cerevisiae, Yl Yarrowia lipolytica, Ca Candida albicans, Um Ustilago maydis, Lb Laccaria bicolor, Cc Coprinopsis cinerea, Cn Cryptococcus neoformans, Hs Homo sapiens*. (**b**) Experimentally verified UPRE motifs bound by Hac1-like proteins in indicated fungi. (**c**) Phylogenetic relationship between Hac1-like proteins. The phylogenetic tree is based on multiple alignments of the bZIP domains in Hac1-like proteins from indicated species using ClustalW. *Cg Colletotrichum graminicola, Vd Verticillium dahliae, Tr Trichoderma reesei, Nc Neurospora crassa, Fo Fusarium oxysporum, Mo Magnaporthe oryzae, Af Aspergillus fumigatus, Ao Aspergillus oryzae, Anig Aspergillus niger, An Aspergillus nidulans, Bc Botrytis cinerea, Ab Alternaria brassicicola, Sc Saccharomyces cerevisiae, Yl Yarrowia lipolytica, Ca Candida albicans, Um Ustilago maydis, Lb Laccaria bicolor, Cc Coprinopsis cinerea, Cn Cryptococcus neoformans, Hs Homo sapiens*

the promoter regions of UPR-regulated genes identified similar UPREs in *S. cerevisiae* (Fordyce et al. 2012; Kohno et al. 1993; Mori et al. 1992), *Aspergillus niger* (Mulder et al. 2006), and *U. maydis* (Pinter et al. 2019) (Fig. 3b), the latter of which is highly similar to the

XBP1 binding site in humans (Yamamoto et al. 2004). Importantly, the Hac1 homolog Cib1 (Clp1 *i*nteracting *b*ZIP 1) in *U. maydis* can functionally replace Hac1 in *S. cerevisiae* (Heimel et al. 2013), and the C-terminal extension of Cib1 mediates the interaction with the developmental regulator Clp1 (*Cl*am*p*less 1), by which the developmental program and a modulated UPR interact to control fungal pathogenesis (Heimel et al. 2013, 2010a, b; Pinter et al. 2019).

In higher eukaryotes, XBP1 forms homodimers but also heterodimeric complexes with *ATF6* (Byrd and Brewer 2012). Gcn4 (*G*eneral *C*ontrol *N*onderepressible 4), the major regulator of the **amino acid starvation** responses (**Cross-Pathway Control**), is required for expression of the majority of Hac1-regulated UPR targets in *S. cerevisiae*, and both pathways are connected by bidirectional functional interactions (Herzog et al. 2013; Patil et al. 2004). Moreover, both proteins are regulated in their stability by Srb10 (*S*uppressor of *R*NA polymerase *B* 10)-mediated phosphorylation and Cdc4 (*C*ell *D*ivision *C*ycle 4)-mediated ubiquitylation (Pal et al. 2007; Chi et al. 2001). Similarly, the interaction between *U. maydis* Cib1 and Clp1 proteins reduces Cib1 phosphorylation but promotes increased stability of both proteins (Pinter et al. 2019). In contrast to the formation of the stable Cib1/Clp1 complex, a direct protein interaction between Gcn4 and Hac1 has not been observed.

III. The UPR in Fungal Pathogens

Human, animal, and plant hosts react to infectious threats by upregulating their defense mechanisms, including an increase of the body temperature, the focused generation of reactive oxygen species (ROS), as well as the induction of programmed cell death to prevent colonization by fungal pathogens. The ability of adapting intracellular pathways to the constant changes **fungal pathogens** are exposed to during interactions with challenging host environments is therefore a key determinant of fungal virulence. The UPR has been studied in different human and plant pathogenic fungi and is of central importance to infect and colonize the host organism and cause disease. A high capacity for protein secretion is pivotal for shaping the host environment in support of fungal infection including mobilization of nutrient sources and suppression of the plant immune system by secreted **effector proteins**. The underlying mechanism and functions of the UPR in fungal virulence appear to be common with respect to ER stress resistance, but also involve specific adaptations and functions that are connected to lifestyle and host organisms. Here, an overview on the distinct requirements for the UPR in disease development of human and plant pathogenic fungi will be provided.

A. The UPR in Human Pathogenic Fungi

Fungal pathogens infecting humans and animals rely on the ability to adapt their growth to the host environment like the body temperature, reduced oxygen supply, or altered nutrient availability and on mechanisms providing effective shielding from the host immune system. Functionality of the UPR is associated with various virulence-related traits including thermotolerance, stress resistance, **antifungal drug resistance**, and secretory activity in major human fungal pathogens such as *Aspergillus fumigatus*, *Candida albicans*, *Candida glabrata*, and the *Cryptococcus* species complex (Krishnan and Askew 2014, Richie et al. 2011b, a, Cheon et al. 2014, Glazier and Panepinto 2014, Miyazaki and Kohno 2014, Raj et al. 2015). In addition, the UPR is induced and important during morphological responses that are connected to disease development (Wimalasena et al. 2008; Jung et al. 2016; Monteiro et al. 2009).

1. *Aspergillus fumigatus*

Aspergillus fumigatus is an ascomycetous saprotrophic filamentous fungus commonly found in the soil and in compost heaps, growing on decaying organic matter and thus promoting flux of the carbon and nitrogen cycle (Fang and Latge 2018). Conidiospores, regularly inhaled by humans, are protected by surface hydrophobins from recognition by the human immune system (Aimanianda et al. 2009) and expelled or efficiently eliminated in immunocompetent individuals by neutrophil-induced programmed cell death (Shlezinger et al. 2017). By contrast, immunosuppressed individuals are at risk of developing invasive aspergillosis, which initially starts as a lung infection that can eventually disseminate through the blood stream and cause severe and potentially lethal cerebral infections. Investigations focusing on the role of the central UPR regulatory proteins IreA (Ire1) and HacA (Hac1) revealed important contributions of the UPR to fungal virulence. In infection models using immunosuppressed (triamcinolone acetonide) mice, deletion of *hacA* results in reduced virulence, and *ireA* deletion mutants were found to be completely avirulent (Feng et al. 2011; Richie et al. 2009). These observations were connected to reduced ER and cell wall stress resistance and decreased thermotolerance of both mutant strains. Differences between *ireA* and *hacA* mutants with respect to the ER stress response were investigated by genome-wide expression analysis, revealing 391 and 1157 differentially expressed genes in *hacA* and *ireA* deletion mutants (in comparison to the WT), respectively, of which 243 genes are shared between both mutant strains (Feng et al. 2011; Richie et al. 2011b). In higher eukaryotes, ER stress results in the global reduction of translation initiation by phosphorylation of eIF2α by *PERK* (PKR-like endoplasmic reticulum kinase) (Ron and Walter 2007). A similar mechanism contributing to phosphorylation of eIF2α under amino acid imbalance is mediated by Gcn2/CpcC and important for virulence in *A. fumigatus* (Krappmann et al. 2004). ER stress has not been reported to reduce translation initiation frequency in fungi. Conversely, analysis of the *A. fumigatus* translatome by **polysome profiling** under ER stress-inducing conditions and growth at increased temperatures revealed a predominantly increased ribosome association of mRNAs. Responses to the ER stress-inducing drugs DTT and tunicamycin are rapid and distinct with limited overlap between both treatments (DTT, total 717 mRNAs; TM total, 876 mRNAs; overlap, 233 mRNAs). The response to a temperature shift from 25 ° C to 37 ° C is associated with the increased translation of 372 mRNAs, overlapping with DTT and TM treatment in case of 44 mRNAs (Krishnan et al. 2014).

Degradation of terminally misfolded proteins is realized by the ER-associated degradation (ERAD) pathway, promoting the export of substrate proteins into the cytoplasm for ubiquitylation and subsequent degradation by the 26S proteasome. Deletion of *derA*, encoding a predicted Derlin ortholog, an ER membrane protein involved in retro-translocation during ERAD, or *hrdA* (*H*MG-coA *R*eductase *D*egradation), encoding the ER membrane-localized E3 ubiquitin ligase, did not affect ER stress resistance or virulence (Krishnan et al. 2013; Richie et al. 2011b). The parallel deletion of both genes leads to reduced resistance against cell wall and ER stress but does not affect virulence (Grahl et al. 2012; Krishnan et al. 2013). *derA* deletion mutants showed increased UPR activity, and the double deletion mutant of *derA* and *hacA* is severely affected in hyphal growth, antifungal drug resistance, and protease secretion and displayed strongly reduced virulence in triamcinolone acetonide immunosuppressed mice (Krishnan et al. 2013, Richie et al. 2011b). Similarly, UPR activity is increased upon conditional repression of the *Saccharomyces cerevisiae stt3* ortholog, encoding the catalytic subunit of the OST (*O*ligo*s*accharyl *T*ransferase) complex, mediating N-glycosylation in the ER (Li et al. 2011). Activity of the UPR during host colonization is apparently connected to growth at mammalian body temperature, nutrient deprivation and hypoxia (Krishnan and Askew 2014). The adaptation to hypoxic conditions is mediated by the **SREBP (sterol regulatory element-binding protein)** transcription factor SreA (Chung et al. 2014), which is activated by signal peptide peptidase-mediated cleavage of its ER transmembrane domain and required for full virulence of *A. fumigatus* (Bat-Ochir et al. 2016). The clinical relevance of the UPR is further corroborated via its role in **antifungal drug resistance.** *hacA* mutants showed increased sensitivity towards caspofungin, itraconazole, fluconazole, and amphotericin B (Richie et al. 2009), and resistance towards voriconazole was further reduced by the parallel deletion of *derA* (Richie et al. 2011a). Interestingly, *hrdA* deletion mutants show reduced voriconazole sensitivity (Krishnan et al. 2013). Deletion of *ormA*, encoding a protein involved in sphingolipid synthesis, results in increased susceptibility to **azoles**, whereas overexpression reduced azole sensitivity. Expression of *ormA* is induced by TM in a *hacA*-dependent manner, further corroborating the requirement of an intact ER stress response for antifungal drug resistance (Zhai et al. 2019).

2. *Cryptococcus neoformans*

Cryptococcus neoformans is a basidiomycetous yeast and an opportunistic pathogen of immunosuppressed patients. Basidiospores serve as predominant infectious propagules. After initially colonizing lung tissue *C. neoformans* can subsequently spread and cause cryptococcal meningitis, often as a secondary infection of HIV-positive individuals. In *C. neoformans*, the UPR has been studied with respect to virulence-associated traits, such as stress resistance and thermotolerance, but also concerning antifungal drug resistance and sexual mating.

While the ortholog of Ire1 in *C. neoformans* was identified by straightforward BLAST search, the identification of the first Hac1 ortholog in basidiomycetes required experimental validation (Cheon et al. 2011). Monitoring of potential unconventional splicing events of five candidate genes encoding bZIP transcription factors with significant homology of their bZIP domains to Hac1 identified Hxl1 (*H*ac1and *X*BP1-*l*ike gene 1) as the key UPR regulator in *C. neoformans*. Splicing of the unprocessed *HXL1* mRNA results in the Ire1-dependent removal of a 56 nt intron and expression of Hxl1. Expression of the intronless *HXL1* cDNA fully restores ER stress resistance of *IRE1* deletion mutants, confirming Hxl1 as the functional Hac1 ortholog. Interestingly, expression of Hxl1 in *S. cerevisiae* Δ*HAC1* mutants does not restore ER stress resistance, indicating divergent transcriptional machineries involved in Hac1- and Hxl1-dependent ER stress responses in either organism as causal for the functional incompatibility. Besides their requirement for ER stress resistance, *IRE1* and *HXL1* deletion mutants are strongly suppressed in thermotolerance and fail to grow at mammalian body temperature, resulting in avirulence in mice infection experiments (Cheon et al. 2011). Resolution of ER stress during growth at host temperature involves the major mRNA deadenylase Ccr4 (*C*arbon *C*atabolite *R*epression 4), which is a central component of CCR4-NOT complex mediating mRNA decay (Havel et al. 2011). In addition, the mRNA binding protein Puf4 (*Pu*milio homology domain *F*amily 4) destabilizes *HXL1* mRNA and therefore contributes to UPR dynamics and ER stress resistance. Deletion of *PUF4* alleviates growth at host temperature but does not affect virulence (Glazier et al. 2015). Overexpression of the central ER chaperone Kar2/Bip1 partly rescues ER stress resistance and thermotolerance of *HXL1* and *IRE1* deletion mutants (Jung et al. 2013). Ire1 and Hxl1 play conserved roles in ER stress resistance in *Cryptococcus* species. Interestingly, **antifungal drug resistance** towards **azole** derivatives and fludioxonil is in general decreased but differentially affected in Δ*IRE1* and Δ*HXL1* mutants, with divergent effects in the *C. neoformans* (H99), *C. deneoformans* (JEC21), and *C. deutrogattii* (R265) (Jung et al. 2018) backgrounds. In addition, *IRE1* but not *HXL1* deletion mutants are unable to undergo sexual mating, as reflected by reduced cell fusion and conjugation tube formation. This defect is connected to increased expression of the pheromone precursor gene MFα1 and an aberrant localization of the Ste6 pheromone transporter and in part suppressed by Kar2 overexpression in the Δ*IRE1* mutant background (Jung et al. 2016).

3. *Candida albicans*

The ascomycetous yeast *Candida albicans* is a commensal of the human mucosa and opportunistic pathogen in immunosuppressed individuals. The ability to colonize the host is connected to a yeast-to-hypha dimorphic transition. Although virulence has not been directly assayed, *HAC1* deletion mutants fail to induce hyphal growth, suggesting an important function of the UPR in *C. albicans* virulence (Wimalasena et al. 2008). Homozygous *HAC1* mutants are not impaired in vegetative growth but show reduced resistance towards drugs inducing ER or cell wall stress. Analysis of Hac1-dependent gene expression in response to DTT or TM treatment revealed only few genes regulated in common by Hac1 in *C. albicans* and *S. cerevisiae* (Travers et al. 2000). Differentially expressed and Hac1-dependent genes include several encoding for adhesins, such as *ALS4*, *ALS5*, *ALS12*, and *ALS10* (*A*gglutinin-*L*ike *S*equence) (Hoyer 2001), as well as *PMT1*, *PMT2*, and *PMT4-6* (*P*rotein O-*M*annosyl*T*ransferase), encoding O-mannosyltransferases important for adhesin function (Timpel et al. 1998; Timpel et al. 2000). Expression of UPR target genes upon ER stress is supported by the calcineurin effector Rta2 (*R*esistance *T*o *A*minocholesterol 2). Rta2 functions as plasma membrane-localized flippase involved in sphingolipid long-chain base release (Jia et al. 2009) and is transcriptionally induced upon TM treatment (Thomas et al. 2015). Rta2 does not influence abundance of the spliced *HAC1*^{i} mRNA and likely supports UPR target gene expression in a Hac1-independent manner. Consistently, the high susceptibility of *HAC1* and *RTA2* deletion mutants to TM is further increased upon parallel deletion of both genes (Thomas et al. 2015). The signaling mucin Msb2 (*M*ulticopy *S*uppression of a *B*udding defect 2) is a plasma membrane-localized surface receptor involved in sensing of and adaptation to hyperosmotic and cell wall stress (Roman et al. 2009) and required for **antifungal drug resistance**, ER stress resistance, and growth at elevated temperatures (Saraswat et al. 2016). It is moreover essential for activation of the CEK1 (*C*. albicans *E*RK-like *K*inase 1) MAPK (*M*itogen-*A*ctivated *P*rotein *K*inase) pathway (Roman et al. 2009) and for transcriptional induction of *IRE1*, *HAC1*, and *KAR2* (*BIP1*) during growth at elevated temperatures (Saraswat et al. 2016). Interestingly, in *S. cerevisiae*, Ire1 supports cleavage of Msb2 by regulating the transcriptional induction of the Msb2 processing protease Yps1 (*Ya*P*S*in 1) (Adhikari et al. 2015).

4. *Candida glabrata*

Similar to *C. albicans*, *Candida glabrata* is a commensal of human mucosal tissue that is prevalent in immunosuppressed patients and, based on the high resistance against antifungal **azoles**, represents a major threat in clinical settings (Fidel Jr. et al. 1999). In contrast to most other fungi, the UPR in *C. glabrata* is mediated in an Ire1-dependent but Hac1-independent manner (Miyazaki et al. 2013). Under ER stress-inducing conditions, no **unconventional splicing** of *HAC1* mRNA is observed, and although *IRE1* is required for ER stress resistance, the transcriptional response to treatment with ER stress-inducing drugs is independent of Ire1, but largely depends on calcineurin signaling and to a smaller extent on the Slt2 (*S*uppressor of the *Lyt*ic phenotype 2) MAPK signaling pathway (Cell Wall Integrity, CWI). Interestingly, expression of *CgIRE1* in an *S. cerevisiae* Δ*IRE1* mutant did not restore ER stress resistance, whereas expression of *CgHAC1* fully recovered ER stress resistance of *S. cerevisiae* Δ*IRE1* and Δ*HAC1* mutants (Miyazaki et al. 2013). The connection between ER stress resistance and functionality of the calcineurin- and Slt2-dependent MAPK signaling pathway is well established in *S. cerevisiae* (Bonilla and Cunningham 2003; Bonilla et al. 2002; Chen et al. 2005; Dudgeon et al. 2008). In *C. glabrata* Ire1, calcineurin and Slt2 are coordinately required for the ER stress response and display additive effects on ER stress susceptibility when individual mutations are combined. While calcineurin and Slt2 are important for the transcriptional response to ER stress (Miyazaki et al. 2013), Ire1 primarily regulates mRNA abundance via degradation of ER-associated mRNAs by **RIDD** (Hollien et al. 2009) to reduce ER protein synthesis and decrease the demands for protein folding in the ER (Miyazaki et al. 2013). *IRE1* is critical for virulence of *C. glabrata*, but in contrast to calcineurin not involved in resistance towards azole drugs (Miyazaki et al. 2013). Whereas individual mutants do not show increased sensitivity towards osmotic stress-inducing or cell wall-damaging drugs, calcineurin/*IRE1* double mutants display decreased resistance (Miyazaki et al. 2013; Tanaka et al. 2018).

B. The UPR in Plant Pathogenic Fungi

Plant pathogenic fungi use different strategies to enable the effective colonization of their host plants. While necrotrophic fungi induce plant cell death via secretion of proteinaceous effectors, plant cell wall-degrading enzymes or toxins, biotrophic fungi secrete effectors to establish compatible host-pathogen interactions. Effector molecules subvert the **plant immune system** and redirect metabolic fluxes for interfering with plant cell death-inducing pathways and keeping their host alive. Hemibiotrophic fungi typically establish an initial biotrophic interaction that with increasing time and disease progression is transformed to a necrotrophic interaction. UPR signaling in plant pathogens is connected to the ability of efficient protein secretion, regulation of **effector** genes, generation of **reactive oxygen species** (ROS), and resistance to plant defense compounds. Importantly, studies in the biotrophic corn smut fungus *Ustilago maydis* revealed an unexpected connection to regulatory pathways controlling pathogenic development and highlight the adaptation of conserved cellular pathways for lifestyle-specific functions in fungal pathogens.

1. *Alternaria brassicicola*

Alternaria brassicicola is a filamentous ascomycete and necrotrophic pathogen causing black spot disease in a wide range of plant species belonging to the Brassicaceae (Cho 2015; Guillemette et al. 2014). UPR signaling in *A. brassicicola* is controlled by the Hac1-like transcription factor AbHacA, which is crucial for ER stress resistance and important for growth and conidiation on complex carbon sources (potato dextrose medium). This phenotype is further accompanied by defects in cell wall construction, formation of aberrant enlarged and swollen hyphae, and hypersensitivity towards cell wall disturbing compounds. *AbhacA* is crucial for virulence in infection assays on *Arabidopsis thaliana* and *Brassica oleracea*, and deletion mutants are highly susceptible towards the indolic phytoalexin **camalexin**, a major plant defense compound of *A. thaliana* (Glawischnig 2007) and potent activator of the UPR in *A. brassicicola*. By contrast, *AbhacA*-mediated UPR activity is not required for conidia germination and efficient appressoria formation. Hence, the increased susceptibility towards plant defense compounds, the reduced vegetative growth, aberrant cell wall, and the reduced ability for protein secretion might be causal for the virulence defects in *A. brassicicola*.

2. *Botrytis cinerea*

The filamentous ascomycete *Botrytis cinerea* is the causal agent of grey mold disease and one of the most widespread plant pathogens, infecting more than 200 host species. The fungal-derived generation of ROS by the **Nox** (*N*ADPH *Ox*idase) complex is important for cellular differentiation, plant penetration, and sclerotia formation in *B. cinerea* (Roca et al. 2012; Segmuller et al. 2008; Siegmund et al. 2013). The protein disulfide isomerase BcPdi1, which mediates oxidative protein folding in the ER, was identified as an interactor of the NoxA complex. *BcPdi1* deletion mutants are reduced in virulence on *Phaseolus vulgaris* and display largely overlapping phenotypes with mutants of the NoxA signaling pathway, including reduced resistance to osmotic and redox stress and abolished formation of conidial anastomosis tubes (CAT) (Marschall and Tudzynski 2017). Phosphorylation of Ire1, as a key determinant of UPR activity, has been found specifically induced under host mimicking growth conditions (deproteinized tomato cell wall medium, TCW), suggesting that the UPR is activated during the infection process of *B. cinerea* (Escobar-Nino et al. 2019). The ER chaperone Bag1 (*B*cl-2 *a*ssociated athano*g*ene 1) is an important regulator of UPR signaling in

B. cinerea. Bag1 negatively regulates the central UPR components Ire1, Hac1, and Bip1, suggesting a protective and important function in promoting ER homeostasis. Absence of Bag1 results in increased sensitivity towards cell wall or ER stress-inducing compounds, and functionality of Bag1 is furthermore important for normal vegetative growth, formation of conidia, sclerotia, and full virulence in tomato infection assays (Zhang et al. 2019).

3. *Magnaporthe oryzae* (*Pyricularia oryzae*)

The filamentous ascomycete *Magnaporthe oryzae* is the causal agent of rice blast. *M. oryzae* exhibits a hemibiotrophic life cycle, starting with an initial biotrophic phase, in which plant penetration and initial colonization occurs that is followed by the necrotrophic stage featuring massive proliferation and host colonization culminating in the formation of conidia for further dissemination (Dean et al. 2012; Yan and Talbot 2016; Zhang et al. 2016a). The function of the UPR in virulence of *M. oryzae* has been investigated with respect to the individual contributions of the *S. cerevisiae* homologs of Hac1, Lhs1, Err1 (*E*R *r*etention *r*eceptor 1), and the MAPK Mps1. Deletion mutants of individual genes encoding these proteins show strongly reduced virulence, illustrating the central role of UPR in virulence of *M. oryzae*. Besides Kar2/Bip1, Lhs1 represents the major ER chaperone, being crucial for ER protein folding and ER protein homeostasis. Absence of Lhs1 results in increased UPR activity and reduced vegetative growth and conidiation. Defects in protein translocation across the ER membrane support a function of Lhs1 for efficient protein secretion, including effector proteins such as AVR-Pita. During pathogenic development Δ*lhs1* mutants are defective in plant penetration and biotrophic invasion of susceptible host plants. Interestingly, all defects of the Δ*lhs1* mutant are rescued by overexpression of Sil1, which functions as a nucleotide exchange factor for Kar2/Bip1 in *S. cerevisiae* (Steel et al. 2004). In genome-wide studies focusing on the role of bZIP transcription factors in *M. oryzae*, the homolog of Hac1 was identified (Kong et al. 2015; Tang et al. 2015). *MoHac1* mutants display strongly reduced virulence in leaf infection assays, which is connected to reduced vegetative growth, ER stress resistance, and the almost complete absence of conidiation. Expression of *Mohac1* is reported to be significantly reduced during invasive growth (72 hours post inoculation) in rice or barley leaves (Mathioni et al. 2011), suggesting that basal UPR activity is crucial for disease development. ER-resident proteins depend on their constant reimport into the ER after accidental transition into the Golgi, for proper subcellular localization and accumulation in the ER (Hardwick et al. 1990). C-terminal ER retention motifs (HDEL/KDEL and derivatives thereof) are bound by membrane-localized receptors preventing the transit of HDEL/KDEL-containing proteins through the secretory pathway. Mutants of the *M. oryzae* HDEL receptor Err1 show severe defects in vegetative growth and formation of aberrant conidia, which fail to adhere to the plant surface but produce functional appressoria and cause disease (Goh et al. 2017). The cell wall integrity (CWI) MAPK Mps1 is required for full UPR activity and thus represents a modulatory factor of the UPR in *M. oryzae*. CWI and UPR pathways show cross-regulation since phosphorylation of Mps1 is increased in response to cell wall stress and by endogenous and exogenous ER stress (Yin et al. 2016).

4. *Ustilago maydis*

Ustilago maydis is a facultatively biotrophic basidiomycete and the causal agent of corn smut disease, inducing tumor (gall) formation of its host plant maize (*Zea mays*) (Brefort et al. 2009). The analysis of UPR signaling in *U. maydis* revealed distinct adaptations of the ER stress response to the biotrophic lifestyle and fungal/plant interaction. *U. maydis* displays a dimorphic life cycle in which the pathogenic stage is induced after mating of saprotrophically growing and noninfectious haploid sporidia, to form the infectious biotrophic filament. The heterodikaryotic filament elongates by tip growth, as the cell cycle is arrested in the G2 phase until plant penetration has occurred and proliferation of the dikaryon is initiated (Flor-Parra et al. 2007; Heimel et al. 2010a; Perez-Martin and Castillo-Lluva 2008; Scherer et al. 2006). Concomitant with the massive fungal proliferation, plant tumors are formed in which fungal hyphae differentiate into diploid, melanized teliospores. Pathogenic development is controlled by the bE/bW heterodimer, encoded by the *b* **mating-type locus**. While activity of the b-heterodimer is sufficient for the activation of the pathogenic program including filament formation and appressoria-mediated plant penetration, proliferation of the dikaryon in the host plant is dependent on the functional modulation of the b-heterodimer by the Clp1 protein. Clp1 physically interacts with bW and with the master regulator downstream of the b-heterodimer, Rbf1, ultimately reducing activity of the mating-type signaling pathway to overcome the G2 cell cycle arrest and to induce mitotic growth of the dikaryon in planta (Heimel et al. 2010a; Scherer et al. 2006). The timing of cell cycle release is mediated by Clp1 accumulation, which is stabilized via the interaction with the Hac1 homolog Cib1. Cib1 is specifically expressed after successful plant penetration (Heimel et al. 2013; Heimel et al. 2010a), and the UPR is continuously active during the fungal/plant interaction. UPR activity is presumably connected to the strongly increased expression of large numbers of **effector genes** at this stage of the life cycle (Lanver et al. 2018) and required for efficient **secretion** and **processing** of effector proteins including the well-characterized Pit2, Tin2, Cmu1, and Pep1 effectors (Hampel et al. 2016; Lo Presti et al. 2016; Pinter et al. 2019). The stage-specific depletion of Cib1 during biotrophic development revealed that a functional UPR is not only important for the establishment of a compatible biotrophic interaction after plant penetration but is continuously required for suppressing the **plant defense** and maintenance of fungal **biotrophy** (Schmitz et al. 2019a). Interestingly, expression of *pit2* and *tin1-1* effector genes is controlled directly by Cib1 binding to their promoter regions, and abolishment of this regulation by targeted deletion of the Cib1 binding site in the *pit1/2* promoter leads to significantly reduced virulence of *U. maydis* (Hampel et al. 2016). The interaction between Clp1 and Cib1 affects phosphorylation and stability of Cib1 and confers strongly increased ER stress resistance by modulated UPR gene expression. Genome-wide analysis of Clp1-dependent effects on UPR genes revealed a set of 65 UPR core genes, which are induced by Cib1 and differentially modulated in their expression by Clp1. ChIP-seq (Chromatin Immunoprecipitation followed by massive parallel sequencing) analysis identified a consensus **UPRE** motif bound by Cib1 (TGACGTGG) (Hampel et al. 2016; Pinter et al. 2019), resembling the mammalian XBP1 binding motif (GATGACGTGG) (Yamamoto et al. 2004) (see also Fig. 3). Systematic deletion of UPR core genes identified the J-domain containing ER co-chaperone Dnj1 (*DnaJ*-like protein 1) as an important factor for fungal virulence, ER stress resistance, and effector secretion (Lo Presti et al. 2016). **Signal peptide peptidases** (SPP) are aspartic proteases catalyzing the intramembrane cleavage of type II oriented transmembrane (TM) domains in the ER membrane (Friedmann et al. 2004; Weihofen et al. 2002), including remnants of signal peptides (after initial cleavage by the signal peptidase), ER-TM containing transcription factors, and high-affinity nutrient transporters (Avci et al. 2014; Bat-Ochir et al. 2016). The signal peptide peptidase Spp1 is a UPR core gene in *U. maydis* and critical for fungal virulence. Spp1 is specifically required for plant defense suppression, as evidenced by massive accumulation of ROS in plant tissue infected by Δ*spp1* mutants and strongly induced expression of salicylic acid marker genes. The absence of Spp1 neither affects resistance to ER, cell wall, or oxidative stress nor pathogenic development prior plant infection, and the virulence-specific function of Spp1 is not connected to previously reported physiological roles of the conserved SPP protein family. Catalytic activity of Spp1 is essential for plant defense suppression but not required for secretion or processing of effectors (Pinter et al. 2019), pointing towards a central role of Spp1-dependently produced cleavage products that are directly or indirectly involved in manipulating the plant immune system. Although the UPR is required for proliferation in planta,

UPR activity inhibits pathogenic growth before plant infection by reducing phosphorylation of the pheromone MAPK Kpp2 (Heimel et al. 2013; Schmitz et al. 2019b). This effect is mediated by the dual-specificity phosphatase Rok1 (*R*egulator *o*f *K*pp2 1). Activity of Rok1 is inversely correlated with virulence of *U. maydis* as the absence of Rok1 activity results in hypervirulence and failure to sustain fungal biotrophy, whereas increased activity results in reduced virulence (Di Stasio et al. 2009). Since activity of Rok1 is regulated by the UPR, ER stress levels appear to be connected to the control of fungal biotrophy including negative feedback control of mating-type regulated effector gene expression (for overview see Fig. 4).

IV. Connections and Interplay Between the UPR and Other Signaling Pathways

The response to ER stress is functionally connected to a plethora of different pathways that altogether foster resistance to environmental changes and thereby contribute to fitness and resilience of fungi. The interaction between the UPR and **autophagy**, the **cell wall integrity (CWI) pathway**, **ER-associated degradation pathway (ERAD)**, **hypoxia response,** and **MAPK signaling pathways** is realized on different levels, including transcriptional control, posttranslational modifications, or protein-protein interactions. The constantly expanding data on these interactions suggest the existence of widely conserved crosstalk but also specific connections shared only between fungi with similar lifestyles or ecological niches. In the following we summarize the current knowledge on the interconnections of the different pathways and UPR.

A. UPR and ERAD

The ER-associated degradation (ERAD) pathway is an integral part of the ER quality control machinery, constantly removing terminally misfolded and potentially toxic proteins from the ER for proteasomal degradation in the cytoplasm. Several interrelated ERAD pathways exist in eukaryotes. The ERAD-L pathway specifically contributes to the degradation of substrates with luminal lesions, the ERAD-C pathway for degradation of substrates with cytosolic lesions, and the ERAD-M pathway mediates degradation of substrates with lesions in transmembrane domains (Thibault and Ng 2012). ERAD complexes are targeted to the ER membrane and consist of core E3 ubiquitin ligases (see also Fig. 1) (Doa10 in ERAD-C and Hrd1 in ERAD-L and M) and associated proteins. Doa10 (*D*egradation *O*f *A*lpha 10) forms two distinct complexes with Cue1 (*C*oupling of *U*biquitin conjugation to *E*R degradation 1) and Ubc7 (*Ub*iquitin-*C*onjugating enzyme 7) or with Cdc48 (*C*ell *D*ivision *C*ycle 48), Npl4 (*N*uclear *P*rotein *L*ocalization 4), Ufd1 (*U*biquitin *F*usion *D*egradation protein 1), and Ubx2 (*Ub*iquitin regulatory *X* 2). Hrd1 associates with Hrd3, Der1, Yos9 (*Y*east *OS-9* homolog), and Usa1 (*U*1-*S*np1 *A*ssociating 1) (Carvalho et al. 2006). Several ER and cytosolic chaperones are associated with ERAD function including Kar2/Bip1, Scj1 (*S. Cerevisiae DnaJ* 1), Jem1 (Dna*J*-like protein of the *E*R *M*embrane 1), and Pdi1 in the ER lumen (Gillece et al. 1999; Grubb et al. 2012; Nishikawa et al. 2001; Silberstein et al. 1998; Thibault et al. 2011) and Ydj1 (*Y*east *DnaJ* 1), Hlj1 (*H*omo*l*ogous to E. coli dna*J* protein 1), and Hsp70 in the cytosol (Huyer et al. 2004, Park et al. 2007, Vembar et al. 2009). Recognition of terminally misfolded protein for degradation via ERAD is mediated by a glycan timer cascade. If folding and glycosylation of substrate proteins is repetitively erroneous, mannosidase activity of the Htm1-Pdi1 complex generates a substrate recognition site for Yos9, which in turn delivers the substrate for Hrd1-dependent ubiquitylation and retro-translocation into the cytosol (Bhamidipati et al. 2005; Buschhorn et al. 2004; Gauss et al. 2011; Izawa et al. 2012). ERAD and UPR pathway components display compensatory interactions, and abrogation of both pathways leads to drastically increased stress sensitivity (Thibault et al. 2011). Several ERAD components are transcriptionally regu-

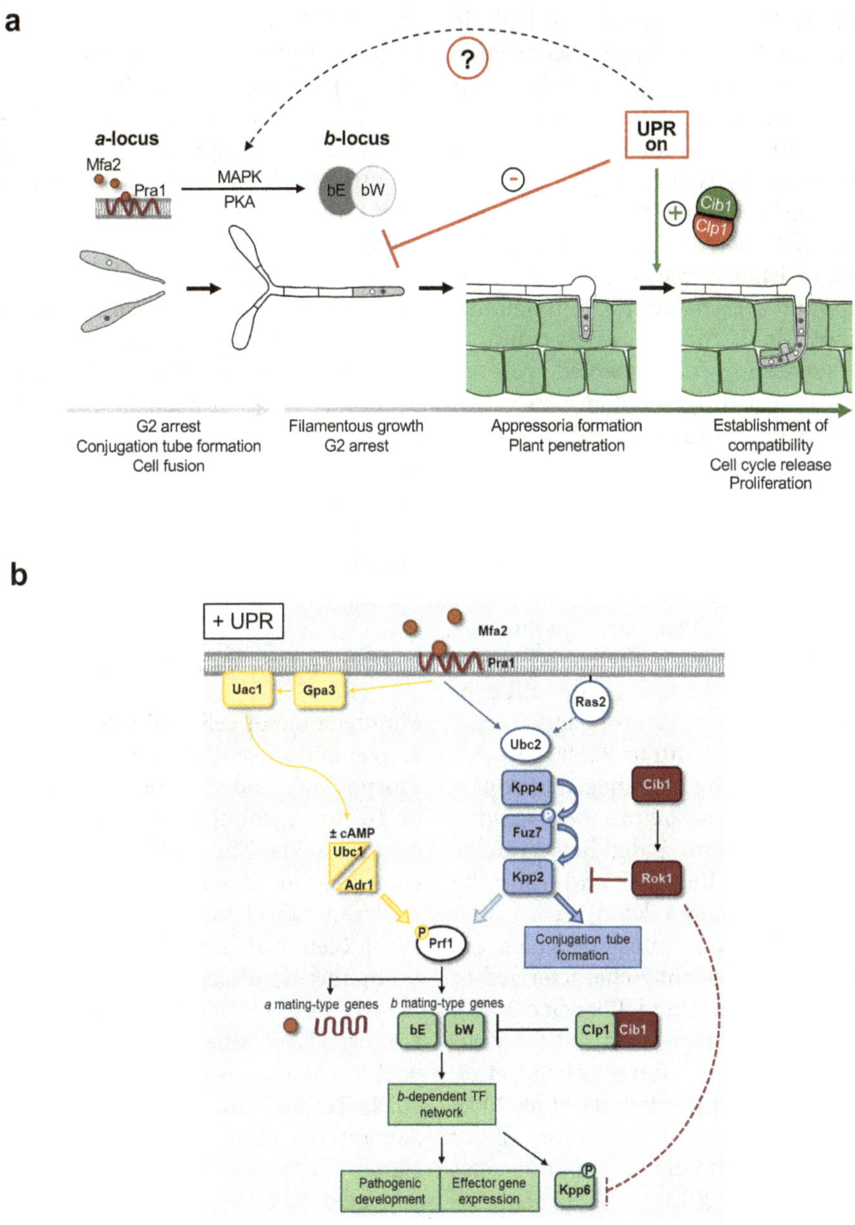

Fig. 4 UPR-mediated control of pathogenic development and mating-type signaling in *U. maydis*. Interaction between mating-type signaling and the UPR is realized on different levels. (**a**) UPR activity is specifically induced after penetration of the plant surface and facilitates fungal proliferation in planta via Clp1/Cib1 complex formation. UPR activity prior to plant penetration inhibits filamentous growth by Rok1-dependent dephosphorylation of the MAPK Kpp2. (**b**) Crosstalk between UPR and mating-type pathways via the Clp1/Cib1 interaction and by UPR-controlled Rok1 activity regulates the virulence potential and biotrophic development by balancing of mating-type signaling and UPR activity. This figure is adapted from published data (Schmitz et al. 2019b) (CC-BY 4.0)

lated by the UPR, whereas others are constitutively expressed (Feng et al. 2011; Pinter et al. 2019; Travers et al. 2000). Deletion of *derA* (*DER1*) in *A. niger* increased the yield of secreted glucoamylase-glucuronidase (Carvalho et al. 2011b). In *A. fumigatus* deletion of either *hrdA* (*HRD1*) or *derA* resulted in decreased virulence only upon parallel deletion of both genes or in combination with the *hacA* deletion (Krishnan et al. 2013; Richie et al. 2011b). By contrast, elimination of central ERAD components in single, double, or triple combinations did affect neither ER resistance nor virulence of *U. maydis* (Pinter et al. 2019). Interestingly, resistance against the novel antifungal compound sr7575 requires a functional ERAD pathway in both *S. cerevisiae* and *A. fumigatus*, suggesting a conserved role of ERAD in antifungal drug resistance in yeasts and filamentous fungi (Raj et al. 2015).

B. UPR and Autophagy

The response to ER stress results in restructuring of the secretory pathway and the enlargement of the ER to provide sufficient resources for accommodation of the upregulated ER protein folding machinery. Autophagy ("self-eating") is an essential intracellular pathway required for the recycling of damaged or superfluous intracellular macromolecules ranging from whole organelles to individual proteins (Pollack et al. 2009; Reggiori and Klionsky 2013). 42 *atg* (*Au*tophagy-related) genes were identified in *S. cerevisiae*, most of which are likely conserved but not fully characterized in filamentous fungi (Voigt and Pöggeler 2013). Functionality of autophagy is crucial for virulence in many pathogenic fungi (Khalid et al. 2019; Nadal and Gold 2010; Palmer et al. 2008; Tam et al. 2016; Yin et al. 2019) and for cellular differentiation of filamentous ascomycetes (Voigt and Pöggeler 2013). The process of **macroautophagy** mediates recycling of bulk cellular components, whereas **selective autophagy** accounts for the degradation and recycling of specific cellular components, including organelles. Autophagy-mediated degradation of the ER, referred to as **ER-phagy**, promotes cellular homeostasis by balancing ER synthesis and degradation. Both macroautophagy and selective autophagy contribute to ER-phagy and require the core autophagic machinery, including Atg8 proteins (Bernales et al. 2006; Lipatova and Segev 2015; Mochida et al. 2015). Selective ER-phagy is also reported to, at least in part, function independent of key *atg* genes (Schuck et al. 2014). In *S. cerevisiae*, expression of *ATG8* is induced by ER stress and the function of Atg8 and other Atg proteins is required for ER stress resistance (Bernales et al. 2006). By contrast, in *C. albicans* and *A. niger*, the absence of Atg8 does not impact ER resistance (Burggraaf and Ram 2016). Functional homologs of the *S. cerevisiae* ER-phagy receptors Atg39 and Atg40, recruiting Atg8 for selective ER-phagy (Mochida et al. 2015), have not been identified in filamentous fungi. Hence, connections between autophagy and the UPR appear to be conserved in fungi, whereas the extent of crosstalk and interactions between both pathways likely differs between individual fungal species.

C. UPR and the Cell Wall Integrity Pathway

Maintenance of cell wall integrity (CWI) is an active and essential process guiding fungal morphology and development, but also resistance to antifungal drugs, metabolites, or organic acids. The CWI is regulated by signal transduction cascades involving the most upstream mechanosensory proteins, termed Wsc1 (cell *W*all integrity and *S*tress response *C*omponent), Wsc2, Wsc3, and Mtl1 (*M*id-*T*wo *L*ike 1) in *S. cerevisiae*, which are functionally conserved in filamentous fungi in detecting cell wall stress (Beauvais et al. 2001; Dichtl et al. 2012; Lesage and Bussey 2006; Levin 2011; Samantaray et al. 2013). Transmission of the signal occurs via the Rho1 GTPase, which is activated by GTP exchange factors (GEFs) and in turn binds to and activates Pkc1 (*P*rotein *K*inase *C*). Pkc1 activity promotes MAPK-mediated phosphotransfer and activation of the terminal MAPK, termed Slt2 in *S. cerevisiae* and MpkA in *aspergilli*. MpkA activates the Rlm1/RlmA (*R*esistance to *L*ethality of *M*KK1P386 overexpression) transcription factor to alter expression of genes involved in cell wall synthesis and remodeling (Futagami et al. 2011; Jung and Levin 1999). Crosstalk between the UPR and CWI pathways has been observed in yeasts (Bonilla and Cunningham 2003; Chen et al. 2005; Scrimale et al. 2009) and filamentous fungi (Colabardini et al. 2010; Malavazi et al. 2014; Yin et al. 2016). Deletion mutants of the genes encoding homologs of Hac1 and Ire1 show increased susceptibility towards cell wall stress-inducing agents in *A. brassicicola*, *M. oryzae*, and *A. fumigatus* and *C. neoformans* (Cheon et al. 2011; Joubert et al. 2011; Kong et al. 2015; Tang et al. 2015), suggesting widely conserved interactions between UPR and CWI pathways. Conversely, key players involved in CWI signaling, including Slt2, are consistently required for ER stress resistance in *C. albicans* (Chen et al. 2005), *C. glabrata* (Miyazaki et al. 2013), and *S. cerevisiae* (Chen et al. 2005). Since UPR signaling affects vegetative growth and proper hyphal development and controls expression of genes important for cell wall synthesis (Joubert et al. 2011; Travers et al. 2000), the bidirectional activation of either pathway might establish a compensatory mechanism to generate distinct responses to various stress-inducing conditions.

D. UPR and MAPK Signaling

MAPK signaling pathways are key players in a wide variety of cellular adaptation processes and mediate the response to altered osmolality and light, cell wall integrity, nutrient deprivation, oxidative stress, and the presence of mating partners or compatible cells for fusion by anastomosis (Bahn and Jung 2013; Fleissner and Herzog 2016; Garrido-Bazan et al. 2018; Gonzalez-Rubio et al. 2019; Yu et al. 2016). Besides the previously elaborated crosstalk between UPR and CWI pathways (see Sect. IV. C.), additional regulatory interactions between the UPR and MAPK-controlled signaling pathways have been described. In *C. neoformans*, increased UPR activity is observed under osmotic stress, and the high-osmolarity glycerol (HOG) pathway is important for ER stress resistance upon exposure of increased TM concentrations (Cheon et al. 2011). In addition, deletion of IRE1 results in defects in sexual mating and a compensatory response of the pheromone MAPK cascade, leading to strongly increased expression of the pheromone precursor gene MFα1 (Jung et al. 2016). Since the mating defect is connected to erroneous localization of the pheromone transporter Ste6, the connection between Ire1 and the pheromone MAPK pathway is likely indirect (Jung et al. 2016). In *U. maydis*, expression of the intronless *cib1*^{s} mRNA leads to constitutive activation of the UPR and reduced expression of the *b* mating-type locus genes *bE* and *bW* and the major regulator downstream of the b-heterodimer Rbf1. This inhibitory effect of the UPR is reflected by reduced formation of infectious filaments and, as a consequence, fungal virulence (Heimel et al. 2013). Systematic analysis of crosstalk between the UPR and the pheromone MAPK pathway revealed strongly reduced phosphorylation of the MAPK Kpp2 in strains with constitutively active UPR, resulting in the UPR-dependent inhibition of the morphological and transcriptional response to pheromone (Schmitz et al. 2019b). The dual-specificity phosphatase Rok1 regulates activity of the mating-type signaling pathway by dephosphorylating Kpp2 and the partially redundant MAPK Kpp6. Deletion of *rok1* leads to hypervirulence of *U. maydis*, whereas overexpression reduces virulence (Di Stasio et al. 2009). The UPR promotes increased Rok1 activity, and deletion of *rok1* fully restores Kpp2 phosphorylation, formation of infectious filaments, and virulence, suggesting that the feedback regulation between the UPR and MAPK affects the virulence potential of *U. maydis* and ensures fungal biotrophy (Schmitz et al. 2019b) (Fig. 4).

E. UPR and Hypoxia

Hypoxic conditions suppress oxidation of membrane lipids and induce the activation of a compensatory response regulated by the **sterol regulatory element-binding protein** (SREBP) (Shimano and Sato 2017). Membrane aberrancies induced by inositol depletion activate the ER stress sensor Ire1 by a mechanism that is distinct from canonical activation via unfolded proteins (Promlek et al. 2011). The ability to grow under hypoxic conditions is critical for virulence in human pathogenic fungi such as *A. fumigatus*, *C. neoformans*, and *Paracoccidioides* spp. (Bien and Espenshade 2010; Chun et al. 2007; Chung et al. 2014; Grahl et al. 2012; Lima Pde et al. 2015). In *N. crassa*, the SREBP pathway is highly induced under lignocellulolytic conditions and negatively regulates expression of oxygen-consuming enzymes and protein secretion (Reilly et al. 2015; Qin et al. 2017). SREBP mediates suppression of the central ER stress response genes *bip-1* and *pdi-1* in a *hac-1* and *ire-1* independent manner. Importantly, the inability for efficient cellulase secretion in *hac-1* deletion strains is compensated by the concurrent deletion of the gene encoding the SREBP homolog SAH-2 (*S*hort *A*erial *H*yphae 2). In mammalian cells, the interplay between UPR and the SREBP-mediated hypoxia response is of high medical relevance. The Hac1 homolog XBP1 is required for tumor growth under hypoxic conditions and promotes tumor progression in triple-negative breast cancer cell lines by forming a transcriptional complex with HIF1 (*H*ypoxia-*I*nduced *F*actor 1) that regulates expression of HIF1 target genes (Chen et al. 2014).

V. The UPR in Biotechnology of Filamentous Fungi

The application of the UPR in white biotechnology is of long-standing interest as UPR activity provides strongly increased capacities for protein production and secretion (Heimel 2015). Comprehensive studies on the UPR in biotechnologically relevant filamentous fungi such as *A. niger*, *Trichoderma reesei*, and, more recently, *Neurospora crassa* substantially contributed to our current understanding of the UPR. Various approaches for harnessing the positive effects of the UPR have been described, aiming to increase yields of heterologous and endogenous secreted proteins. Genomic approaches detailing on the cellular and transcriptional responses to ER stress within the process of biotechnological protein production revealed distinct requirements for optimizing the production of different proteins of interest. So far, generally applicable strategies that fully exploit the potential of the UPR for high-level protein production and secretion remain to be established. In the following, UPR-related approaches for engineering the secretory capacity in important filamentous model fungi will be described.

A. *Aspergillus niger*

Aspergillus niger is used for more than 100 years in white biotechnology for the production of metabolites, proteins, and enzymes including citric acid, gluconic acid, glucoamylase, and heterologous proteins (Cairns et al. 2018). Several genes encoding central components of the ER folding machinery, such as the ER chaperone BipA, the foldases PdiA and CypB (*Cyclo*philin B), and the lectin chaperone calnexin ClxA, were identified as being induced under various stress conditions, including secretion stress (Derkx and Madrid 2001; Jeenes et al. 1997; Ngiam et al. 2000; van Gemeren et al. 1997). Transcriptional activation of these genes is mediated by the Hac1 homolog HacA (Mulder et al. 2004), recognizing the consensus sequence CAN(G/A)NTGT/GCCT (Mulder et al. 2006). Genome-wide expression studies identified a core set of 40 genes presumably required for high-level expression of *A. niger* glucoamylase (Kwon et al. 2012). In contrast to most yeasts and dimorphic fungi, deletion of *hacA* or expression of the intronless *hacA* mRNA (leading to constitutive UPR activity) reduces vegetative growth and induces formation of aberrant hyphae, complicating the functional characterization and application of HacA in biotechnological protein production (Heimel 2015; Mulder and Nikolaev 2009; Mulder et al. 2006; Mulder et al. 2004). Consequently, attempts for increasing the production of glucoamylase, manganese peroxidase, thaumatin, preprochymosin, or tissue plasminogen activator focused on the overexpression of genes encoding ER **chaperones** and **foldases**, including *bipA*, *pdiA*, and *clxA*, or the deletion of **ERAD** components *hrdC* and *derA* (Conesa et al. 2002; Lombrana et al. 2004; Ngiam et al.

2000), (Carvalho et al. 2011; Punt et al. 1998). Only in few cases, the yields of produced proteins were significantly increased. Overexpression of calnexin (*clxA*) resulted in a fivefold increased yield of manganese peroxidase and deletion of *hrdA* in a sixfold increased yield of glucoamylase-glucuronidase (Carvalho et al. 2011; Conesa et al. 2002). Interestingly, several genes encoding secretory proteins are **repressed under secretion stress (RESS)** (Al-Sheikh et al. 2004). A *cis*-acting element, located in the promoter of the secreted glucoamylase encoding gene *glaA*, is critical for RESS of *glaA*. A potential negative effect of the UPR on *glaA* expression might be indirect and connected to the repression of the AmyR transcription factor, a known regulator of *glaA* and other secretory proteins (Carvalho et al. 2012). An additional compensatory response to ER stress is deduced from the observation of reduced ribosome occupancy of mRNAs encoding secreted proteins under DTT-induced ER stress conditions (Guillemette et al. 2007), contrasting the results of a similar approach in *A. fumigatus* revealing an overall increase of ribosome association of mRNAs (Krishnan and Askew 2014).

B. *Trichoderma reesei*

The filamentous ascomycete *Trichoderma reesei* has an enormous intrinsic capacity for cellulose degradation and was identified as the causal organism degrading the cotton tents of GI stationed in Southeast Asia during the Second World War (Merivuori et al. 1985; Montenecourt et al. 1981). Similar to *A. niger*, the protein disulfide isomerase encoding gene *pdi1* and the ER chaperone encoding gene *bip1* were identified as ER stress induced (Saloheimo et al. 1999; Pakula et al. 2003). Genome-wide comparison of ER stress-mediated gene expression in *T. reesei* and *S. cerevisiae* revealed a set of commonly regulated UPR genes, but also identified the specific upregulation of *cpc1*, encoding the homolog of yeast Gcn4, the key regulator of the response to amino acid imbalance (cross-pathway control) and of the histone gene H4 (Arvas et al. 2006). High-level secretion of **cellulases** is connected to increased UPR activity, and overexpression of the UPR regulator Hac1 or the ER chaperone Bip1 supports increased cellulase production (Gao et al. 2018). Inhibition of protein transport through the secretory pathway by brefeldin A or DTT treatment results in Hac1-mediated induction of general UPR components but strongly decreased expression of genes encoding xylanases and cellulases by **RESS** (Pakula et al. 2003). Attempts to harness strains overexpressing the ER stress sensor Ire1 for increased production of *Phlebia radiata* laccase were unsuccessful, as expression of the UPR genes *pdi1* and *bip1* was only modestly increased. By contrast, deletion mutants of the Ire1 repressing phosphatase Ptc2 displayed strongly increased induction of the UPR. The resulting effects on the production of secreted proteins were not determined (Valkonen et al. 2004). The lectin chaperone calnexin, as an integral part of the ER quality control machinery, is required for efficient secretion of CBH1 (*C*ello*bio*hydrolase I). Targeted mutation of N-glycosylated aa residues in the catalytic domain reduces thermostability of CBH1 and activates UPR signaling (Qi et al. 2014).

C. *Neurospora crassa*

Neurospora crassa has a long-standing tradition as model organism for the investigation of fundamental genetic and cellular processes ("one gene-one enzyme" hypothesis, quelling and RNAi, mechanism of the circadian clock, and principles of epigenetic gene control) (Aramayo and Selker 2013; Roche et al. 2014). More recently, *N. crassa* has become of considerable interest with respect to fungal biotechnology, such as the production of **biofuels**. The production of biofuels requires the efficient degradation of plant biomass by carbohydrate active enzymes (**CAZymes**). System-wide analysis of the response to various complex polysaccharides found in plant cell walls identified the *hac-1* gene as being specifically induced during cultivation on medium with cellulose as sole carbon source, consistent with the induction of key UPR target genes, such as *bip-1*, *pdi-1*, and *calnexin* (Benz et al. 2014). Deletion of the *hac-1* gene results in the inability to efficiently degrade cellulose and suppression of growth on medium containing Avicel (microcrystalline cellulose) as carbon source (Montenegro-Montero et al. 2015). In contrast to *A. niger*, the absence of *hac-1* did affect neither vegetative growth nor resistance to cell wall stress-inducing drugs (Montenegro-Montero et al. 2015). Using a set of more 527 mutant strains, carrying individual deletions of genes implicated in the ER stress response, identified 249 genes important for ER stress resistance, of which 100 were so far uncharacterized, including several potential regulators important for lignocellulase secretion. Deletion of *hac-1* or *ire-1* does not affect expression of lignocellulase genes in response to sensing of cellulose, but severely affects the secretion of enzymatically active proteins (Fan et al. 2015). Hence, activation of the UPR provides the intracellular infrastructure important for efficient secretion of lignocellulolytic enzymes and utilization of cellulose as carbon source. Further profiling of single-gene deletion strains identified the sterol **regulatory element-binding protein** (SREBP) pathway as an important regulator of protein secretion under cellulolytic conditions. The SREBP pathway regulates ergosterol synthesis in response to hypoxic conditions (Bien and Espenshade 2010; Robichon and Dugail 2007). Deletion of the SREBP pathway components *tul-1* and *dsc-1*, the predicted homologs of *S. pombe dsc1* and *dsc2* (Defective in SREBP Cleavage) that mediate cleavage and activation of Sre1 in response to hypoxia, results in strongly increased protein secretion in response to cellulose but not sucrose or xylan (Reilly et al. 2015). Deletion of further SREBP pathway components consistently derepressed genes involved in the ER stress response and increased secretion of cellulolytic enzymes. Inactivation of the SREBP pathway suppressed the reduced **cellulase** production of *hac-1* mutant strains, revealing a regulatory interplay between both pathways in adapting the secretory pathway during cellulolytic growth (Qin et al. 2017).

VI. Conclusion and Outlook

The UPR is a central pathway in eukaryotes that counteracts ER stress but also compensates and interacts with a plethora of other stress-associated signaling pathways providing increased cellular robustness (Thibault et al. 2011). While the main theme of UPR function is widely conserved, we are at the beginning of understanding how the divergent lifestyles of fungi ranging from saprotrophically growing filamentous fungi over to biotrophic plant pathogens require specific adaptations of the UPR. The emerging picture of how signal integration and crosstalk between different cellular pathways such as the UPR, ERAD, autophagy, but also developmental control and mating-type signaling is achieved provides important means to uncover the detailed contribution of the UPR in fungal pathogenesis and to develop innovative strategies for unfolding the full potential of the UPR in biotechnological backgrounds.

Acknowledgements We acknowledge Niko Pinter for support in figure preparation. We acknowledge funding of our research by the Deutsche Forschungsgemeinschaft and the IRTG 2172 PRoTECT.

References

Adhikari H, Vadaie N, Chow J, Caccamise LM, Chavel CA, Li B, Bowitch A, Stefan CJ, Cullen PJ (2015) Role of the unfolded protein response in regulating the mucin-dependent filamentous-growth mitogen-activated protein kinase pathway. Mol Cell Biol 35:1414–1432

Aimanianda V, Bayry J, Bozza S, Kniemeyer O, Perruccio K, Elluru SR, Clavaud C, Paris S, Brakhage AA, Kaveri SV, Romani L, Latge JP (2009) Surface hydrophobin prevents immune recognition of airborne fungal spores. Nature 460:1117–1121

Al-Sheikh H, Watson AJ, Lacey GA, Punt PJ, MacKenzie DA, Jeenes DJ, Pakula T, Penttila M, Alcocer MJ, Archer DB (2004) Endoplasmic reticulum stress leads to the selective transcriptional downregulation of the glucoamylase gene in *Aspergillus niger*. Mol Microbiol 53:1731–1742

Amara JF, Lederkremer G, Lodish HF (1989) Intracellular degradation of unassembled asialoglycoprotein receptor subunits: a pre-Golgi, nonlysosomal endoproteolytic cleavage. J Cell Biol 109:3315–3324

Aramayo R, Selker EU (2013) *Neurospora crassa*, a model system for epigenetics research. Cold Spring Harb Perspect Biol 5:a017921

Arvas M, Pakula T, Lanthaler K, Saloheimo M, Valkonen M, Suortti T, Robson G, Penttila M (2006) Common features and interesting differences in transcriptional responses to secretion stress in the fungi *Trichoderma reesei* and *Saccharomyces cerevisiae*. BMC Genomics 7:32

Avci D, Fuchs S, Schrul B, Fukumori A, Breker M, Frumkin I, Chen CY, Biniossek ML, Kremmer E, Schilling O, Steiner H, Schuldiner M, Lemberg MK (2014) The yeast ER-intramembrane protease Ypf1 refines nutrient sensing by regulating transporter abundance. Mol Cell 56:630–640

Bahn YS, Jung KW (2013) Stress signaling pathways for the pathogenicity of *Cryptococcus*. Eukaryot Cell 12:1564–1577

Bat-Ochir C, Kwak JY, Koh SK, Jeon MH, Chung D, Lee YW, Chae SK (2016) The signal peptide peptidase SppA is involved in sterol regulatory element-binding protein cleavage and hypoxia adaptation in *Aspergillus nidulans*. Mol Microbiol 100:635–655

Beauvais A, Bruneau JM, Mol PC, Buitrago MJ, Legrand R, Latge JP (2001) Glucan synthase complex of *Aspergillus fumigatus*. J Bacteriol 183:2273–2279

Benz JP, Chau BH, Zheng D, Bauer S, Glass NL, Somerville CR (2014) A comparative systems analysis of polysaccharide-elicited responses in *Neurospora crassa* reveals carbon source-specific cellular adaptations. Mol Microbiol 91:275–299

Bernales S, McDonald KL, Walter P (2006) Autophagy counterbalances endoplasmic reticulum expansion during the unfolded protein response. PLoS Biol 4:e423

Bhamidipati A, Denic V, Quan EM, Weissman JS (2005) Exploration of the topological requirements of ERAD identifies Yos9p as a lectin sensor of misfolded glycoproteins in the ER lumen. Mol Cell 19:741–751

Bien CM, Espenshade PJ (2010) Sterol regulatory element binding proteins in fungi: hypoxic transcription factors linked to pathogenesis. Eukaryot Cell 9:352–359

Bonilla M, Cunningham KW (2003) Mitogen-activated protein kinase stimulation of Ca(2+) signaling is required for survival of endoplasmic reticulum stress in yeast. Mol Biol Cell 14:4296–4305

Bonilla M, Nastase KK, Cunningham KW (2002) Essential role of calcineurin in response to endoplasmic reticulum stress. EMBO J 21:2343–2353

Brefort T, Doehlemann G, Mendoza-Mendoza A, Reissmann S, Djamei A, Kahmann R (2009) *Ustilago maydis* as a pathogen. Annu Rev Phytopathol 47:423–445

Burggraaf AM, Ram AF (2016) Autophagy is dispensable to overcome ER stress in the filamentous fungus *Aspergillus niger*. Microbiology 5:647–658

Buschhorn BA, Kostova Z, Medicherla B, Wolf DH (2004) A genome-wide screen identifies Yos9p as essential for ER-associated degradation of glycoproteins. FEBS Lett 577:422–426

Byrd AE, Brewer JW (2012) Intricately regulated: a cellular toolbox for fine-tuning XBP1 expression and activity. Cell 1:738–753

Cairns TC, Nai C, Meyer V (2018) How a fungus shapes biotechnology: 100 years of *Aspergillus niger* research. Fungal Biol Biotechnol 5:13

Carvalho ND, Arentshorst M, Kooistra R, Stam H, Sagt CM, van den Hondel CA, Ram AF (2011) Effects of a defective ERAD pathway on growth and heterologous protein production in *Aspergillus niger*. Appl Microbiol Biotechnol 89:357–373

Carvalho P, Goder V, Rapoport TA (2006) Distinct ubiquitin-ligase complexes define convergent pathways for the degradation of ER proteins. Cell 126:361–373

Carvalho ND, Jorgensen TR, Arentshorst M, Nitsche BM, van den Hondel CA, Archer DB, Ram AF (2012) Genome-wide expression analysis upon constitutive activation of the HacA bZIP transcription factor in *Aspergillus niger* reveals a coordinated cellular response to counteract ER stress. BMC Genomics 13:350

Chaillot J, Tebbji F, Remmal A, Boone C, Brown GW, Bellaoui M, Sellam A (2015) The monoterpene carvacrol generates endoplasmic reticulum stress in the pathogenic fungus *Candida albicans*. Antimicrob Agents Chemother 59:4584–4592

Chawla A, Chakrabarti S, Ghosh G, Niwa M (2011) Attenuation of yeast UPR is essential for survival and is mediated by IRE1 kinase. J Cell Biol 193:41–50

Chen C, Bonifacino JS, Yuan LC, Klausner RD (1988) Selective degradation of T cell antigen receptor chains retained in a pre-Golgi compartment. J Cell Biol 107:2149–2161

Chen Y, Feldman DE, Deng C, Brown JA, De Giacomo AF, Gaw AF, Shi G, Le QT, Brown JM, Koong AC (2005) Identification of mitogen-activated protein kinase signaling pathways that confer resistance to endoplasmic reticulum stress in *Saccharomyces cerevisiae*. Mol Cancer Res 3:669–677

Chen X, Iliopoulos D, Zhang Q, Tang Q, Greenblatt MB, Hatziapostolou M, Lim E, Tam WL, Ni M, Chen Y, Mai J, Shen H, Hu DZ, Adoro S, Hu B, Song M, Tan C, Landis MD, Ferrari M, Shin SJ, Brown M, Chang JC, Liu XS, Glimcher LH (2014) XBP1 promotes triple-negative breast cancer by controlling the HIF1alpha pathway. Nature 508:103–107

Cheon SA, Jung KW, Bahn YS, Kang HA (2014) The unfolded protein response (UPR) pathway in *Cryptococcus*. Virulence 5:341–350

Cheon SA, Jung KW, Chen YL, Heitman J, Bahn YS, Kang HA (2011) Unique evolution of the UPR pathway with a novel bZIP transcription factor,

Hxl1, for controlling pathogenicity of *Cryptococcus neoformans*. PLoS Pathog 7:e1002177

Chi Y, Huddleston MJ, Zhang X, Young RA, Annan RS, Carr SA, Deshaies RJ (2001) Negative regulation of Gcn4 and Msn2 transcription factors by Srb10 cyclin-dependent kinase. Genes Dev 15:1078–1092

Cho Y (2015) How the necrotrophic fungus *Alternaria brassicicola* kills plant cells remains an enigma. Eukaryot Cell 14:335–344

Chun CD, Liu OW, Madhani HD (2007) A link between virulence and homeostatic responses to hypoxia during infection by the human fungal pathogen *Cryptococcus neoformans*. PLoS Pathog 3:e22

Chung D, Barker BM, Carey CC, Merriman B, Werner ER, Lechner BE, Dhingra S, Cheng C, Xu W, Blosser SJ, Morohashi K, Mazurie A, Mitchell TK, Haas H, Mitchell AP, Cramer RA (2014) ChIP-seq and in vivo transcriptome analyses of the *Aspergillus fumigatus* SREBP SrbA reveals a new regulator of the fungal hypoxia response and virulence. PLoS Pathog 10:e1004487

Cohen N, Breker M, Bakunts A, Pesek K, Chas A, Argemi J, Orsi A, Gal L, Chuartzman S, Wigelman Y, Jonas F, Walter P, Ernst R, Aragon T, van Anken E, Schuldiner M (2017) Iron affects Ire1 clustering propensity and the amplitude of endoplasmic reticulum stress signaling. J Cell Sci 130:3222–3233

Colabardini AC, De Castro PA, De Gouvea PF, Savoldi M, Malavazi I, Goldman MH, Goldman GH (2010) Involvement of the *Aspergillus nidulans* protein kinase C with farnesol tolerance is related to the unfolded protein response. Mol Microbiol 78:1259–1279

Conesa A, Jeenes D, Archer DB, van den Hondel CA, Punt PJ (2002) Calnexin overexpression increases manganese peroxidase production in *Aspergillus niger*. Appl Environ Microbiol 68:846–851

Dean R, Van Kan JA, Pretorius ZA, Hammond-Kosack KE, Di Pietro A, Spanu PD, Rudd JJ, Dickman M, Kahmann R, Ellis J, Foster GD (2012) The top 10 fungal pathogens in molecular plant pathology. Mol Plant Pathol 13:414–430

Derkx PM, Madrid SM (2001) The foldase CYPB is a component of the secretory pathway of *Aspergillus niger* and contains the endoplasmic reticulum retention signal HEEL. Mol Genet Genomics MGG 266:537–545

Di Stasio M, Brefort T, Mendoza-Mendoza A, Munch K, Kahmann R (2009) The dual specificity phosphatase Rok1 negatively regulates mating and pathogenicity in *Ustilago maydis*. Mol Microbiol 73:73–88

Dichtl K, Helmschrott C, Dirr F, Wagener J (2012) Deciphering cell wall integrity signalling in Aspergillus fumigatus: identification and functional characterization of cell wall stress sensors and relevant rho GTPases. Mol Microbiol 83:506–519

Dudgeon DD, Zhang N, Ositelu OO, Kim H, Cunningham KW (2008) Nonapoptotic death of *Saccharomyces cerevisiae* cells that is stimulated by Hsp90 and inhibited by calcineurin and Cmk2 in response to endoplasmic reticulum stresses. Eukaryot Cell 7:2037–2051

Escobar-Nino A, Lineiro E, Amil F, Carrasco R, Chiva C, Fuentes C, Blanco-Ulate B, Cantoral Fernandez JM, Sabido E, Fernandez-Acero FJ (2019) Proteomic study of the membrane components of signalling cascades of *Botrytis cinerea* controlled by phosphorylation. Sci Rep 9:9860

Fan F, Ma G, Li J, Liu Q, Benz JP, Tian C, Ma Y (2015) Genome-wide analysis of the endoplasmic reticulum stress response during lignocellulase production in *Neurospora crassa*. Biotechnol Biofuels 8:66

Fang W, Latge JP (2018) Microbe profile: *Aspergillus fumigatus*: a saprotrophic and opportunistic fungal pathogen. Microbiology 164:1009–1011

Feng X, Krishnan K, Richie DL, Aimanianda V, Hartl L, Grahl N, Powers-Fletcher MV, Zhang M, Fuller KK, Nierman WC, Lu LJ, Latge JP, Woollett L, Newman SL, Cramer RA Jr, Rhodes JC, Askew DS (2011) HacA-independent functions of the ER stress sensor IreA synergize with the canonical UPR to influence virulence traits in *Aspergillus fumigatus*. PLoS Pathog 7:e1002330

Fidel PL Jr, Vazquez JA, Sobel JD (1999) *Candida glabrata*: review of epidemiology, pathogenesis, and clinical disease with comparison to *C. albicans*. Clin Microbiol Rev 12:80–96

Fleissner A, Herzog S (2016) Signal exchange and integration during self-fusion in filamentous fungi. Semin Cell Dev Biol 57:76–83

Flor-Parra I, Castillo-Lluva S, Perez-Martin J (2007) Polar growth in the infectious hyphae of the phytopathogen *Ustilago maydis* depends on a virulence-specific cyclin. Plant Cell 19:3280–3296

Fordyce PM, Pincus D, Kimmig P, Nelson CS, El-Samad H, Walter P, Derisi JL (2012) Basic leucine zipper transcription factor Hac1 binds DNA in two distinct modes as revealed by microfluidic analyses. Proc Natl Acad Sci U S A 109:E3084–E3093

Friedmann E, Lemberg MK, Weihofen A, Dev KK, Dengler U, Rovelli G, Martoglio B (2004) Consensus analysis of signal peptide peptidase and homologous human aspartic proteases reveals opposite topology of catalytic domains compared with presenilins. J Biol Chem 279:50790–50798

Futagami T, Nakao S, Kido Y, Oka T, Kajiwara Y, Takashita H, Omori T, Furukawa K, Goto M (2011) Putative stress sensors WscA and WscB are involved in hypo-osmotic and acidic pH stress tolerance in *Aspergillus nidulans*. Eukaryot Cell 10:1504–1515

Gao F, Hao Z, Sun X, Qin L, Zhao T, Liu W, Luo H, Yao B, Su X (2018) A versatile system for fast screening and isolation of *Trichoderma reesei* cellulase hyperproducers based on DsRed and fluorescence-assisted cell sorting. Biotechnol Biofuels 11:261

Gardner BM, Pincus D, Gotthardt K, Gallagher CM, Walter P (2013) Endoplasmic reticulum stress sensing in the unfolded protein response. Cold Spring Harb Perspect Biol 5:a013169

Gardner BM, Walter P (2011) Unfolded proteins are Ire1-activating ligands that directly induce the unfolded protein response. Science 333:1891–1894

Garrido-Bazan V, Jaimes-Arroyo R, Sanchez O, Lara-Rojas F, Aguirre J (2018) SakA and MpkC stress MAPKs show opposite and common functions during stress responses and development in *Aspergillus nidulans*. Front Microbiol 9:2518

Gauss R, Kanehara K, Carvalho P, Ng DT, Aebi M (2011) A complex of Pdi1p and the mannosidase Htm1p initiates clearance of unfolded glycoproteins from the endoplasmic reticulum. Mol Cell 42:782–793

Gillece P, Luz JM, Lennarz WJ, de La Cruz FJ, Romisch K (1999) Export of a cysteine-free misfolded secretory protein from the endoplasmic reticulum for degradation requires interaction with protein disulfide isomerase. J Cell Biol 147:1443–1456

Glawischnig E (2007) Camalexin. Phytochemistry 68:401–406

Glazier VE, Kaur JN, Brown NT, Rivera AA, Panepinto JC (2015) Puf4 regulates both splicing and decay of HXL1 mRNA encoding the unfolded protein response transcription factor in *Cryptococcus neoformans*. Eukaryot Cell 14:385–395

Glazier VE, Panepinto JC (2014) The ER stress response and host temperature adaptation in the human fungal pathogen *Cryptococcus neoformans*. Virulence 5:351–356

Goh J, Jeon J, Lee YH (2017) ER retention receptor, MoERR1 is required for fungal development and pathogenicity in the rice blast fungus, *Magnaporthe oryzae*. Sci Rep 7:1259

Gonzalez-Rubio G, Fernandez-Acero T, Martin H, Molina M (2019) Mitogen-activated protein kinase phosphatases (MKPs) in fungal signaling: conservation, function, and regulation. Int J Mol Sci 20 (7):1709

Grahl N, Shepardson KM, Chung D, Cramer RA (2012) Hypoxia and fungal pathogenesis: to air or not to air? Eukaryot Cell 11:560–570

Grubb S, Guo L, Fisher EA, Brodsky JL (2012) Protein disulfide isomerases contribute differentially to the endoplasmic reticulum-associated degradation of apolipoprotein B and other substrates. Mol Biol Cell 23:520–532

Guillemette T, Calmes B, Simoneau P (2014) Impact of the UPR on the virulence of the plant fungal pathogen *A. brassicicola*. Virulence 5:357–364

Guillemette T, van Peij N, Goosen T, Lanthaler K, Robson GD, van den Hondel CA, Stam H, Archer DB (2007) Genomic analysis of the secretion stress response in the enzyme-producing cell factory *Aspergillus niger*. BMC Genomics 8:158

Hampel M, Jakobi M, Schmitz L, Meyer U, Finkernagel F, Doehlemann G, Heimel K (2016) Unfolded protein response (UPR) regulator Cib1 controls expression of genes encoding secreted virulence factors in *Ustilago maydis*. PLoS One 11:e0153861

Hardwick KG, Lewis MJ, Semenza J, Dean N, Pelham HR (1990) ERD1, a yeast gene required for the retention of luminal endoplasmic reticulum proteins, affects glycoprotein processing in the Golgi apparatus. EMBO J 9:623–630

Havel VE, Wool NK, Ayad D, Downey KM, Wilson CF, Larsen P, Djordjevic JT, Panepinto JC (2011) Ccr4 promotes resolution of the endoplasmic reticulum stress response during host temperature adaptation in *Cryptococcus neoformans*. Eukaryot Cell 10:895–901

Heimel K (2015) Unfolded protein response in filamentous fungi-implications in biotechnology. Appl Microbiol Biotechnol 99:121–132

Heimel K, Freitag J, Hampel M, Ast J, Bölker M, Kämper J (2013) Crosstalk between the unfolded protein response and pathways that regulate pathogenic development in *Ustilago maydis*. Plant Cell 25:4262–4277

Heimel K, Scherer M, Schuler D, Kämper J (2010a) The *Ustilago maydis* Clp1 protein orchestrates pheromone and b-dependent signaling pathways to coordinate the cell cycle and pathogenic development. Plant Cell 22:2908–2922

Heimel K, Scherer M, Vranes M, Wahl R, Pothiratana C, Schuler D, Vincon V, Finkernagel F, Flor-Parra I, Kämper J (2010b) The transcription factor Rbf1 is the master regulator for b-mating type controlled pathogenic development in *Ustilago maydis*. PLoS Pathog 6:e1001035

Herzog B, Popova B, Jakobshagen A, Shahpasandzadeh H, Braus GH (2013) Mutual crosstalk between the regulators Hac1 of the unfolded protein response and Gcn4 of the general amino acid control of *Saccharomyces cerevisiae*. Eukaryot Cell

Hinnebusch AG (1994) The eIF-2 alpha kinases: regulators of protein synthesis in starvation and stress. Semin Cell Biol 5:417–426

Hollien J, Lin JH, Li H, Stevens N, Walter P, Weissman JS (2009) Regulated Ire1-dependent decay of messenger RNAs in mammalian cells. J Cell Biol 186:323–331

Hooks KB, Griffiths-Jones S (2011) Conserved RNA structures in the non-canonical Hac1/Xbp1 intron. RNA Biol 8:552–556

Hoyer LL (2001) The ALS gene family of *Candida albicans*. Trends Microbiol 9:176–180

Huyer G, Piluek WF, Fansler Z, Kreft SG, Hochstrasser M, Brodsky JL, Michaelis S (2004) Distinct machinery is required in *Saccharomyces cerevisiae* for the endoplasmic reticulum-associated degradation of a multispanning membrane protein and a soluble luminal protein. J Biol Chem 279:38369–38378

Iracane E, Donovan PD, Ola M, Butler G, Holland LM (2018) Identification of an Exceptionally Long Intron in the HAC1 Gene of *Candida parapsilosis*. *mSphere* 3

Izawa T, Nagai H, Endo T, Nishikawa S (2012) Yos9p and Hrd1p mediate ER retention of misfolded proteins for ER-associated degradation. Mol Biol Cell 23:1283–1293

Jeenes DJ, Pfaller R, Archer DB (1997) Isolation and characterisation of a novel stress-inducible PDI-family gene from *Aspergillus niger*. Gene 193:151–156

Jia XM, Wang Y, Jia Y, Gao PH, Xu YG, Wang L, Cao YY, Cao YB, Zhang LX, Jiang YY (2009) RTA2 is involved in calcineurin-mediated azole resistance and sphingoid long-chain base release in *Candida albicans*. Cell Mol Life Sci 66:122–134

Joubert A, Simoneau P, Campion C, Bataille-Simoneau N, Iacomi-Vasilescu B, Poupard P, Francois JM, Georgeault S, Sellier E, Guillemette T (2011) Impact of the unfolded protein response on the pathogenicity of the necrotrophic fungus *Alternaria brassicicola*. Mol Microbiol 79:1305–1324

Jung KW, Kang HA, Bahn YS (2013) Essential roles of the Kar2/BiP molecular chaperone downstream of the UPR pathway in *Cryptococcus neoformans*. PLoS One 8:e58956

Jung KW, Lee KT, Averette AF, Hoy MJ, Everitt J, Heitman J, Bahn YS (2018) Evolutionarily conserved and divergent roles of unfolded protein response (UPR) in the pathogenic *Cryptococcus* species complex. Sci Rep 8:8132

Jung US, Levin DE (1999) Genome-wide analysis of gene expression regulated by the yeast cell wall integrity signalling pathway. Mol Microbiol 34:1049–1057

Jung KW, So YS, Bahn YS (2016) Unique roles of the unfolded protein response pathway in fungal development and differentiation. Sci Rep 6:33413

Khalid AR, Zhang S, Luo X, Mehmood K, Rahim J, Shaheen H, Dong P, Qiu D, Ren M (2019) Role of autophagy-related gene atg22 in developmental process and virulence of *Fusarium oxysporum*. Genes (Basel) 10(5):E365

Kimata Y, Ishiwata-Kimata Y, Ito T, Hirata A, Suzuki T, Oikawa D, Takeuchi M, Kohno K (2007) Two regulatory steps of ER-stress sensor Ire1 involving its cluster formation and interaction with unfolded proteins. J Cell Biol 179:75–86

Kimmig P, Diaz M, Zheng J, Williams CC, Lang A, Aragon T, Li H, Walter P (2012) The unfolded protein response in fission yeast modulates stability of select mRNAs to maintain protein homeostasis. elife 1:e00048

Kohno K, Normington K, Sambrook J, Gething MJ, Mori K (1993) The promoter region of the yeast KAR2 (BiP) gene contains a regulatory domain that responds to the presence of unfolded proteins in the endoplasmic reticulum. Mol Cell Biol 13:877–890

Kong S, Park SY, Lee YH (2015) Systematic characterization of the bZIP transcription factor gene family in the rice blast fungus, *Magnaporthe oryzae*. Environ Microbiol 17:1425–1443

Korennykh AV, Egea PF, Korostelev AA, Finer-Moore J, Zhang C, Shokat KM, Stroud RM, Walter P (2009) The unfolded protein response signals through high-order assembly of Ire1. Nature 457:687–693

Korennykh AV, Korostelev AA, Egea PF, Finer-Moore J, Stroud RM, Zhang C, Shokat KM, Walter P (2011) Structural and functional basis for RNA cleavage by Ire1. BMC Biol 9:47

Kozutsumi Y, Segal M, Normington K, Gething MJ, Sambrook J (1988) The presence of malfolded proteins in the endoplasmic reticulum signals the induction of glucose-regulated proteins. Nature 332:462–464

Krappmann S, Bignell EM, Reichard U, Rogers T, Haynes K, Braus GH (2004) The *Aspergillus fumigatus* transcriptional activator CpcA contributes significantly to the virulence of this fungal pathogen. Mol Microbiol 52:785–799

Krishnan K, Askew DS (2014) The fungal UPR: a regulatory hub for virulence traits in the mold pathogen *Aspergillus fumigatus*. Virulence 5:334–340

Krishnan K, Feng X, Powers-Fletcher MV, Bick G, Richie DL, Woollett LA, Askew DS (2013) Effects of a defective endoplasmic reticulum-associated degradation pathway on the stress response, virulence, and antifungal drug susceptibility of the mold pathogen *Aspergillus fumigatus*. Eukaryot Cell 12:512–519

Krishnan K, Ren Z, Losada L, Nierman WC, Lu LJ, Askew DS (2014) Polysome profiling reveals broad translatome remodeling during endoplasmic reticulum (ER) stress in the pathogenic fungus *Aspergillus fumigatus*. BMC Genomics 15:159

Kwon MJ, Jorgensen TR, Nitsche BM, Arentshorst M, Park J, Ram AF, Meyer V (2012) The transcriptomic fingerprint of glucoamylase over-expression in *Aspergillus niger*. BMC Genomics 13:701

Lanver D, Muller AN, Happel P, Schweizer G, Haas FB, Franitza M, Pellegrin C, Reissmann S, Altmuller J, Rensing SA, Kahmann R (2018) The biotrophic development of *Ustilago maydis* studied by RNA-Seq analysis. Plant Cell 30:300–323

Lesage G, Bussey H (2006) Cell wall assembly in *Saccharomyces cerevisiae*. Microbiol Mol Biol Rev 70:317–343

Levin DE (2011) Regulation of cell wall biogenesis in *Saccharomyces cerevisiae*: the cell wall integrity signaling pathway. Genetics 189:1145–1175

Li H, Korennykh AV, Behrman SL, Walter P (2010) Mammalian endoplasmic reticulum stress sensor IRE1 signals by dynamic clustering. Proc Natl Acad Sci U S A 107:16113–16118

Li W, Okreglak V, Peschek J, Kimmig P, Zubradt M, Weissman JS, Walter P (2018) Engineering ER-stress dependent non-conventional mRNA splicing. elife 7

Li K, Ouyang H, Lu Y, Liang J, Wilson IBH, Jin C (2011) Repression of N-glycosylation triggers the unfolded protein response (UPR) and overexpression of cell wall protein and chitin in *Aspergillus fumigatus*. Microbiology 157:1968–1979

Lima Pde S, Chung D, Bailao AM, Cramer RA, Soares CM (2015) Characterization of the *Paracoccidioides* hypoxia response reveals new insights into pathogenesis mechanisms of this important human pathogenic fungus. PLoS Negl Trop Dis 9: e0004282

Lipatova Z, Segev N (2015) A role for macro-ER-Phagy in ER quality control. PLoS Genet 11:e1005390

Lo Presti L, Lopez Diaz C, Turra D, Di Pietro A, Hampel M, Heimel K, Kahmann R (2016) A conserved co-chaperone is required for virulence in fungal plant pathogens. New Phytol 209:1135–1148

Lombrana M, Moralejo FJ, Pinto R, Martin JF (2004) Modulation of *Aspergillus awamori* thaumatin secretion by modification of bipA gene expression. Appl Environ Microbiol 70:5145–5152

Malavazi I, Goldman GH, Brown NA (2014) The importance of connections between the cell wall integrity pathway and the unfolded protein response in filamentous fungi. Brief Funct Genomics 13:456–470

Marschall R, Tudzynski P (2017) The protein disulfide Isomerase of *Botrytis cinerea*: an ER protein involved in protein folding and redox homeostasis influences NADPH oxidase signaling processes. Front Microbiol 8:960

Mathioni SM, Belo A, Rizzo CJ, Dean RA, Donofrio NM (2011) Transcriptome profiling of the rice blast fungus during invasive plant infection and in vitro stresses. BMC Genomics 12:49

Merivuori H, Siegler KM, Sands JA, Montenecourt BS (1985) Regulation of cellulase biosynthesis and secretion in fungi. Biochem Soc Trans 13:411–414

Miyazaki T, Kohno S (2014) ER stress response mechanisms in the pathogenic yeast *Candida glabrata* and their roles in virulence. Virulence 5:365–370

Miyazaki T, Nakayama H, Nagayoshi Y, Kakeya H, Kohno S (2013) Dissection of Ire1 functions reveals stress response mechanisms uniquely evolved in *Candida glabrata*. PLoS Pathog 9: e1003160

Mochida K, Oikawa Y, Kimura Y, Kirisako H, Hirano H, Ohsumi Y, Nakatogawa H (2015) Receptor-mediated selective autophagy degrades the endoplasmic reticulum and the nucleus. Nature 522:359–362

Monteiro JP, Clemons KV, Mirels LF, Coller JA Jr, Wu TD, Shankar J, Lopes CR, Stevens DA (2009) Genomic DNA microarray comparison of gene expression patterns in *Paracoccidioides brasiliensis* mycelia and yeasts in vitro. Microbiology 155:2795–2808

Montenecourt BS, Nhlapo SD, Trimino-Vazquez H, Cuskey S, Schamhart DH, Eveleigh DE (1981) Regulatory controls in relation to over-production of fungal cellulases. Basic Life Sci 18:33–53

Montenegro-Montero A, Goity A, Larrondo LF (2015) The bZIP transcription factor HAC-1 is involved in the unfolded protein response and is necessary for growth on cellulose in *Neurospora crassa*. PLoS One 10:e0131415

Mori K, Ma W, Gething MJ, Sambrook J (1993) A transmembrane protein with a cdc2+/CDC28-related kinase activity is required for signaling from the ER to the nucleus. Cell 74:743–756

Mori T, Ogasawara C, Inada T, Englert M, Beier H, Takezawa M, Endo T, Yoshihisa T (2010) Dual functions of yeast tRNA ligase in the unfolded protein response: unconventional cytoplasmic splicing of HAC1 pre-mRNA is not sufficient to release translational attenuation. Mol Biol Cell 21:3722–3734

Mori K, Sant A, Kohno K, Normington K, Gething MJ, Sambrook JF (1992) A 22 bp cis-acting element is necessary and sufficient for the induction of the yeast KAR2 (BiP) gene by unfolded proteins. EMBO J 11:2583–2593

Mulder HJ, Nikolaev I (2009) HacA-dependent transcriptional switch releases hacA mRNA from a translational block upon endoplasmic reticulum stress. Eukaryot Cell 8:665–675

Mulder HJ, Nikolaev I, Madrid SM (2006) HACA, the transcriptional activator of the unfolded protein response (UPR) in *Aspergillus niger*, binds to partly palindromic UPR elements of the consensus sequence 5'-CAN(G/a)NTGT/GCCT-3′. Fungal Genet Biol 43:560–572

Mulder HJ, Saloheimo M, Penttila M, Madrid SM (2004) The transcription factor HACA mediates the unfolded protein response in *Aspergillus niger*, and up-regulates its own transcription. Mol Genet Genomics MGG 271:130–140

Nadal M, Gold SE (2010) The autophagy genes ATG8 and ATG1 affect morphogenesis and pathogenicity in *Ustilago maydis*. Mol Plant Pathol 11:463–478

Ng DT, Spear ED, Walter P (2000) The unfolded protein response regulates multiple aspects of secretory and membrane protein biogenesis and endoplasmic reticulum quality control. J Cell Biol 150:77–88

Ngiam C, Jeenes DJ, Punt PJ, Van Den Hondel CA, Archer DB (2000) Characterization of a foldase, protein disulfide isomerase a, in the protein secretory pathway of *Aspergillus niger*. Appl Environ Microbiol 66:775–782

Nishikawa SI, Fewell SW, Kato Y, Brodsky JL, Endo T (2001) Molecular chaperones in the yeast endoplasmic reticulum maintain the solubility of proteins for retrotranslocation and degradation. J Cell Biol 153:1061–1070

Oikawa D, Kimata Y, Kohno K (2007) Self-association and BiP dissociation are not sufficient for activation of the ER stress sensor Ire1. J Cell Sci 120:1681–1688

Pakula TM, Laxell M, Huuskonen A, Uusitalo J, Saloheimo M, Penttila M (2003) The effects of drugs inhibiting protein secretion in the filamentous fungus *Trichoderma reesei*. Evidence for down-regulation of genes that encode secreted proteins in the stressed cells. J Biol Chem 278:45011–45020

Pal B, Chan NC, Helfenbaum L, Tan K, Tansey WP, Gething MJ (2007) SCFCdc4-mediated degradation of the Hac1p transcription factor regulates the unfolded protein response in *Saccharomyces cerevisiae*. Mol Biol Cell 18:426–440

Palmer GE, Askew DS, Williamson PR (2008) The diverse roles of autophagy in medically important fungi. Autophagy 4:982–988

Park SH, Bolender N, Eisele F, Kostova Z, Takeuchi J, Coffino P, Wolf DH (2007) The cytoplasmic Hsp70 chaperone machinery subjects misfolded and endoplasmic reticulum import-incompetent proteins to degradation via the ubiquitin-proteasome system. Mol Biol Cell 18:153–165

Patil CK, Li H, Walter P (2004) Gcn4p and novel upstream activating sequences regulate targets of the unfolded protein response. PLoS Biol 2:E246

Perez-Martin J, Castillo-Lluva S (2008) Connections between polar growth and cell cycle arrest during the induction of the virulence program in the phytopathogenic fungus *Ustilago maydis*. Plant Signal Behav 3:480–481

Pinter N, Hach CA, Hampel M, Rekhter D, Zienkiewicz K, Feussner I, Poehlein A, Daniel R, Finkernagel F, Heimel K (2019) Signal peptide peptidase activity connects the unfolded protein response to plant defense suppression by *Ustilago maydis*. PLoS Pathog 15:e1007734

Pollack JK, Harris SD, Marten MR (2009) Autophagy in filamentous fungi. Fungal Genet Biol 46:1–8

Promlek T, Ishiwata-Kimata Y, Shido M, Sakuramoto M, Kohno K, Kimata Y (2011) Membrane aberrancy and unfolded proteins activate the endoplasmic reticulum stress sensor Ire1 in different ways. Mol Biol Cell 22:3520–3532

Punt PJ, van Gemeren IA, Drint-Kuijvenhoven J, Hessing JG, van Muijlwijk-Harteveld GM, Beijersbergen A, Verrips CT, van den Hondel CA (1998) Analysis of the role of the gene bipA, encoding the major endoplasmic reticulum chaperone protein in the secretion of homologous and heterologous proteins in black *Aspergilli*. Appl Microbiol Biotechnol 50:447–454

Qi F, Zhang W, Zhang F, Chen G, Liu W (2014) Deciphering the effect of the different N-glycosylation sites on the secretion, activity, and stability of cellobiohydrolase I from *Trichoderma reesei*. Appl Environ Microbiol 80:3962–3971

Qin L, Wu VW, Glass NL (2017) Deciphering the regulatory network between the SREBP pathway and protein secretion in *Neurospora crassa*. MBio 8(2): e00233–e00217

Raj S, Krishnan K, Askew DS, Helynck O, Suzanne P, Lesnard A, Rault S, Zeidler U, d'Enfert C, Latge JP, Munier-Lehmann H, Saveanu C (2015) The toxicity of a novel antifungal compound is modulated by endoplasmic reticulum-associated protein degradation components. Antimicrob Agents Chemother 60:1438–1449

Reggiori F, Klionsky DJ (2013) Autophagic processes in yeast: mechanism, machinery and regulation. Genetics 194:341–361

Reilly MC, Qin L, Craig JP, Starr TL, Glass NL (2015) Deletion of homologs of the SREBP pathway results in hyper-production of cellulases in *Neurospora crassa* and *Trichoderma reesei*. Biotechnol Biofuels 8:121

Richie DL, Feng X, Hartl L, Aimanianda V, Krishnan K, Powers-Fletcher MV, Watson DS, Galande AK, White SM, Willett T, Latge JP, Rhodes JC, Askew DS (2011b) The virulence of the opportunistic fungal pathogen *Aspergillus fumigatus* requires cooperation between the endoplasmic reticulum-associated degradation pathway (ERAD) and the unfolded protein response (UPR). Virulence 2:12–21

Richie DL, Feng X, Krishnan K, Askew DS (2011a) Secretion stress and antifungal resistance: an Achilles' heel of *Aspergillus fumigatus*? Med Mycol 49(Suppl 1):S101–S106

Richie DL, Hartl L, Aimanianda V, Winters MS, Fuller KK, Miley MD, White S, McCarthy JW, Latge JP, Feldmesser M, Rhodes JC, Askew DS (2009) A role for the unfolded protein response (UPR) in virulence and antifungal susceptibility in *Aspergillus fumigatus*. PLoS Pathog 5:e1000258

Robichon C, Dugail I (2007) De novo cholesterol synthesis at the crossroads of adaptive response to extracellular stress through SREBP. Biochimie 89:260–264

Roca MG, Weichert M, Siegmund U, Tudzynski P, Fleissner A (2012) Germling fusion via conidial anastomosis tubes in the grey mould *Botrytis cinerea* requires NADPH oxidase activity. Fungal Biol 116:379–387

Roche CM, Loros JJ, McCluskey K, Glass NL (2014) *Neurospora crassa*: looking back and looking forward at a model microbe. Am J Bot 101:2022–2035

Roman E, Cottier F, Ernst JF, Pla J (2009) Msb2 signaling mucin controls activation of Cek1 mitogen-activated protein kinase in *Candida albicans*. Eukaryot Cell 8:1235–1249

Ron D, Walter P (2007) Signal integration in the endoplasmic reticulum unfolded protein response. Nat Rev Mol Cell Biol 8:519–529

Rose MD, Misra LM, Vogel JP (1989) KAR2, a karyogamy gene, is the yeast homolog of the mammalian BiP/GRP78 gene. Cell 57:1211–1221

Rubio C, Pincus D, Korennykh A, Schuck S, El-Samad H, Walter P (2011) Homeostatic adaptation to endoplasmic reticulum stress depends on Ire1 kinase activity. J Cell Biol 193:171–184

Saloheimo M, Lund M, Penttila ME (1999) The protein disulphide isomerase gene of the fungus *Trichoderma reesei* is induced by endoplasmic reticulum stress and regulated by the carbon source. Mol Gen Genet 262:35–45

Saloheimo M, Valkonen M, Penttila M (2003) Activation mechanisms of the HAC1-mediated unfolded protein response in filamentous fungi. Mol Microbiol 47:1149–1161

Samantaray S, Neubauer M, Helmschrott C, Wagener J (2013) Role of the guanine nucleotide exchange factor Rom2 in cell wall integrity maintenance of *Aspergillus fumigatus*. Eukaryot Cell 12:288–298

Saraswat D, Kumar R, Pande T, Edgerton M, Cullen PJ (2016) Signalling mucin Msb2 regulates adaptation to thermal stress in *Candida albicans*. Mol Microbiol 100:425–441

Scherer M, Heimel K, Starke V, Kämper J (2006) The Clp1 protein is required for clamp formation and pathogenic development of *Ustilago maydis*. Plant Cell 18:2388–2401

Schmitz L, Kronstad JW, Heimel K (2019a) Conditional gene expression reveals stage-specific functions of the unfolded protein response in the *Ustilago maydis*-maize pathosystem. Mol Plant Pathol 21 (2):258–271

Schmitz L, Schwier MA, Heimel K (2019b) The unfolded protein response regulates pathogenic development of *Ustilago maydis* by Rok1-dependent inhibition of mating-type signaling. *mBio* 10e

Schuck S, Gallagher CM, Walter P (2014) ER-phagy mediates selective degradation of endoplasmic reticulum independently of the core autophagy machinery. J Cell Sci 127:4078–4088

Scrimale T, Didone L, de Mesy Bentley KL, Krysan DJ (2009) The unfolded protein response is induced by the cell wall integrity mitogen-activated protein kinase signaling cascade and is required for cell wall integrity in *Saccharomyces cerevisiae*. Mol Biol Cell 20:164–175

Segmuller N, Kokkelink L, Giesbert S, Odinius D, van Kan J, Tudzynski P (2008) NADPH oxidases are involved in differentiation and pathogenicity in *Botrytis cinerea*. Mol Plant-Microbe Interact 21:808–819

Shimano H, Sato R (2017) SREBP-regulated lipid metabolism: convergent physiology - divergent pathophysiology. Nat Rev Endocrinol 13:710–730

Shlezinger N, Irmer H, Dhingra S, Beattie SR, Cramer RA, Braus GH, Sharon A, Hohl TM (2017) Sterilizing immunity in the lung relies on targeting fungal apoptosis-like programmed cell death. Science 357:1037–1041

Sidrauski C, Cox JS, Walter P (1996) tRNA ligase is required for regulated mRNA splicing in the unfolded protein response. Cell 87:405–413

Sidrauski C, Walter P (1997) The transmembrane kinase Ire1p is a site-specific endonuclease that initiates mRNA splicing in the unfolded protein response. Cell 90:1031–1039

Siegmund U, Heller J, van Kan JA, Tudzynski P (2013) The NADPH oxidase complexes in *Botrytis cinerea*: evidence for a close association with the ER and the tetraspanin Pls1. PLoS One 8:e55879

Silberstein S, Schlenstedt G, Silver PA, Gilmore R (1998) A role for the DnaJ homologue Scj1p in protein folding in the yeast endoplasmic reticulum. J Cell Biol 143:921–933

Steel GJ, Fullerton DM, Tyson JR, Stirling CJ (2004) Coordinated activation of Hsp70 chaperones. Science 303:98–101

Tam AB, Koong AC, Niwa M (2014) Ire1 has distinct catalytic mechanisms for XBP1/HAC1 splicing and RIDD. Cell Rep 9:850–858

Tam JM, Mansour MK, Acharya M, Sokolovska A, Timmons AK, Lacy-Hulbert A, Vyas JM (2016) The role of autophagy-related proteins in *Candida albicans* infections. Pathogens 5(2):E34

Tanaka Y, Sasaki M, Ito F, Aoyama T, Sato-Okamoto M, Takahashi-Nakaguchi A, Chibana H, Shibata N (2018) Cooperation between ER stress and calcineurin signaling contributes to the maintenance of cell wall integrity in *Candida glabrata*. Fungal Biol 122:19–33

Tang W, Ru Y, Hong L, Zhu Q, Zuo R, Guo X, Wang J, Zhang H, Zheng X, Wang P, Zhang Z (2015) System-wide characterization of bZIP transcription factor proteins involved in infection-related morphogenesis of *Magnaporthe oryzae*. Environ Microbiol 17:1377–1396

Thibault G, Ismail N, Ng DT (2011) The unfolded protein response supports cellular robustness as a broad-spectrum compensatory pathway. Proc Natl Acad Sci U S A 108:20597–20602

Thibault G, Ng DT (2012) The endoplasmic reticulum-associated degradation pathways of budding yeast. Cold Spring Harb Perspect Biol 4(12):a013193

Thomas E, Sircaik S, Roman E, Brunel JM, Johri AK, Pla J, Panwar SL (2015) The activity of RTA2, a downstream effector of the calcineurin pathway, is required during tunicamycin-induced ER stress response in *Candida albicans*. FEMS Yeast Res 15 (8):fov095

Timpel C, Strahl-Bolsinger S, Ziegelbauer K, Ernst JF (1998) Multiple functions of Pmt1p-mediated protein O-mannosylation in the fungal pathogen *Candida albicans*. J Biol Chem 273:20837–20846

Timpel C, Zink S, Strahl-Bolsinger S, Schroppel K, Ernst J (2000) Morphogenesis, adhesive properties, and antifungal resistance depend on the

Pmt6 protein mannosyltransferase in the fungal pathogen *Candida albicans*. J Bacteriol 182:3063–3071

Travers KJ, Patil CK, Wodicka L, Lockhart DJ, Weissman JS, Walter P (2000) Functional and genomic analyses reveal an essential coordination between the unfolded protein response and ER-associated degradation. Cell 101:249–258

Valkonen M, Penttila M, Saloheimo M (2004) The ire1 and ptc2 genes involved in the unfolded protein response pathway in the filamentous fungus *Trichoderma reesei*. Mol Genet Genomics MGG 272:443–451

van Gemeren IA, Punt PJ, Drint-Kuyvenhoven A, Broekhuijsen MP, van't Hoog A, Beijersbergen A, Verrips CT, van den Hondel CA (1997) The ER chaperone encoding bipA gene of black Aspergilli is induced by heat shock and unfolded proteins. Gene 198:43–52

Vembar SS, Jin Y, Brodsky JL, Hendershot LM (2009) The mammalian Hsp40 ERdj3 requires its Hsp70 interaction and substrate-binding properties to complement various yeast Hsp40-dependent functions. J Biol Chem 284:32462–32471

Voigt O, Pöggeler S (2013) Self-eating to grow and kill: autophagy in filamentous ascomycetes. Appl Microbiol Biotechnol 97:9277–9290

Weihofen A, Binns K, Lemberg MK, Ashman K, Martoglio B (2002) Identification of signal peptide peptidase, a presenilin-type aspartic protease. Science 296:2215–2218

Welihinda AA, Tirasophon W, Green SR, Kaufman RJ (1998) Protein serine/threonine phosphatase Ptc2p negatively regulates the unfolded-protein response by dephosphorylating Ire1p kinase. Mol Cell Biol 18:1967–1977

Wimalasena TT, Enjalbert B, Guillemette T, Plumridge A, Budge S, Yin Z, Brown AJ, Archer DB (2008) Impact of the unfolded protein response upon genome-wide expression patterns, and the role of Hac1 in the polarized growth, of *Candida albicans*. Fungal Genet Biol 45:1235–1247

Wiseman RL, Zhang Y, Lee KP, Harding HP, Haynes CM, Price J, Sicheri F, Ron D (2010) Flavonol activation defines an unanticipated ligand-binding site in the kinase-RNase domain of IRE1. Mol Cell 38:291–304

Yamamoto K, Yoshida H, Kokame K, Kaufman RJ, Mori K (2004) Differential contributions of ATF6 and XBP1 to the activation of endoplasmic reticulum stress-responsive cis-acting elements ERSE, UPRE and ERSE-II. J Biochem 136:343–350

Yan X, Talbot NJ (2016) Investigating the cell biology of plant infection by the rice blast fungus *Magnaporthe oryzae*. Curr Opin Microbiol 34:147–153

Yin Z, Feng W, Chen C, Xu J, Li Y, Yang L, Wang J, Liu X, Wang W, Gao C, Zhang H, Zheng X, Wang P, Zhang Z (2019) Shedding light on autophagy coordinating with cell wall integrity signaling to govern pathogenicity of *Magnaporthe oryzae*. Autophagy:1–17

Yin Z, Tang W, Wang J, Liu X, Yang L, Gao C, Zhang J, Zhang H, Zheng X, Wang P, Zhang Z (2016) Phosphodiesterase MoPdeH targets MoMck1 of the conserved mitogen-activated protein (MAP) kinase signalling pathway to regulate cell wall integrity in rice blast fungus *Magnaporthe oryzae*. Mol Plant Pathol 17:654–668

Yu Z, Armant O, Fischer R (2016) Fungi use the SakA (HogA) pathway for phytochrome-dependent light signalling. Nat Microbiol 1:16019

Zhai P, Song J, Gao L, Lu L (2019) A sphingolipid synthesis-related protein OrmA in *Aspergillus fumigatus* is responsible for azole susceptibility and virulence. Cell Microbiol 21:e13092

Zhang H, Li Y, Dickman MB, Wang Z (2019) Cytoprotective co-chaperone BcBAG1 is a component for fungal development, virulence, and unfolded protein response (UPR) of *Botrytis cinerea*. Front Microbiol 10:685

Zhang L, Zhang C, Wang A (2016b) Divergence and conservation of the major UPR branch IRE1-bZIP signaling pathway across eukaryotes. Sci Rep 6:27362

Zhang H, Zheng X, Zhang Z (2016a) The *Magnaporthe grisea* species complex and plant pathogenesis. Mol Plant Pathol 17:796–804

Zhou J, Liu CY, Back SH, Clark RL, Peisach D, Xu Z, Kaufman RJ (2006) The crystal structure of human IRE1 luminal domain reveals a conserved dimerization interface required for activation of the unfolded protein response. Proc Natl Acad Sci U S A 103:14343–14348

4 From Genetics to Molecular Oscillations: The Circadian Clock in *Neurospora crassa*

MEAGHAN S. JANKOWSKI[1], ZACHARY A. CHASE[1], JENNIFER M. HURLEY[1,2]

CONTENTS

I. Introduction: *Neurospora crassa* is an Important Model Organism for Research in Circadian Rhythms

Many organisms have evolved an anticipatory mechanism called the **circadian** clock to predict the consistent and dramatic daily changes in the levels of light, temperature, and UV radiation imposed by the Earth's rotation. Importantly, there is a strong correlation between the maintenance of a **molecular clock** that generates circadian (around 24 h) oscillations in a variety of behavioral and molecular processes and the biological fitness of an organism. For example, when different cyanobacteria strains are grown together, the strain whose circadian period is most closely aligned to an artificially generated light-dark cycle will outcompete the other strains to become the predominant organism in the environment (Ouyang et al. 1998). In addition, squirrels and chipmunks with functioning clocks have a higher survival rate against predators as compared to those with a dysfunctional circadian clock (DeCoursey et al. 1997, 2000; DeCoursey 2014). Proper circadian timing also increases the virulence of some fungi during plant infection (Hevia et al. 2015). Indeed, we know that innumerable biological functions are timed by the clock to occur at their optimal phase of the day (e.g., Hurley et al. 2014, 2018; Panda et al. 2002), and the benefits of these **circadian rhythms** have led them to be conserved in almost all branches of life (Dunlap 1999).

Authors "Meaghan S. Jankowski" and "Zachary A. Chase" contributed equally.

[1] Department of Biological Sciences, Rensselaer Polytechnic Institute, Troy, NY, USA; e-mail: hurlej2@rpi.edu
[2] Center for Biotechnology and Interdisciplinary Sciences, Rensselaer Polytechnic Institute, Troy, NY, USA

Genetics and Biotechnology, 3rd Edition
The Mycota II
J.P. Benz, K. Schipper (Eds.)

In identifying the behaviors that are regulated by the circadian clock, researchers have set four main criteria for an **oscillation** to be defined as truly circadian (Johnson et al. 2004). First, the **period** of the rhythm must be at or near 24 h, thereby coinciding with the external 24 h day/night cycle on Earth. Second, the rhythm must be entrainable, meaning that it is responsive to external cues, such as light, which can reset its phase relative to the current environment. For example, when traveling to a different time zone, a person will gradually begin to wake up at the local dawn time as a result of their internal clock adjusting to local lighting cues, rather than continue to wake up during the dawn phase of their previous time zone. Third, the rhythm must be able to persist without external cues. This characteristic demonstrates that the rhythm is an endogenous, anticipatory process and not simply a response to changes in the environment. Finally, the rhythm must be able to maintain a consistent period under different environmental conditions, a property called **compensation**. Since biochemical reactions speed up at higher temperatures, if the circadian clock were not buffered, then it would yield a shorter period and would not be a good pacemaker. While there are numerous biological oscillations in organisms, only those rhythms that meet these four criteria are considered circadian (Johnson et al. 2004).

As circadian clocks appear to have evolved independently in only a few cases, there is a high degree of similarity between the mechanisms that time the clocks of fungi and other higher eukaryotes, e.g., mammals and the fruit fly *Drosophila*, making *N. crassa* an excellent model system in which to study these clock mechanisms (Crosthwaite et al. 1997; Dunlap 1999; Bell-Pedersen et al. 2005; Dunlap and Loros 2017). Much of the foundational research into the molecular mechanism underlying circadian clocks in higher eukaryotes occurred in the filamentous fungus *N. crassa* (Dunlap 2008; Dunlap and Loros 2017; Loros 2019). A mutant in *N. crassa*, which displayed a "banding" developmental growth pattern on **race tubes** was

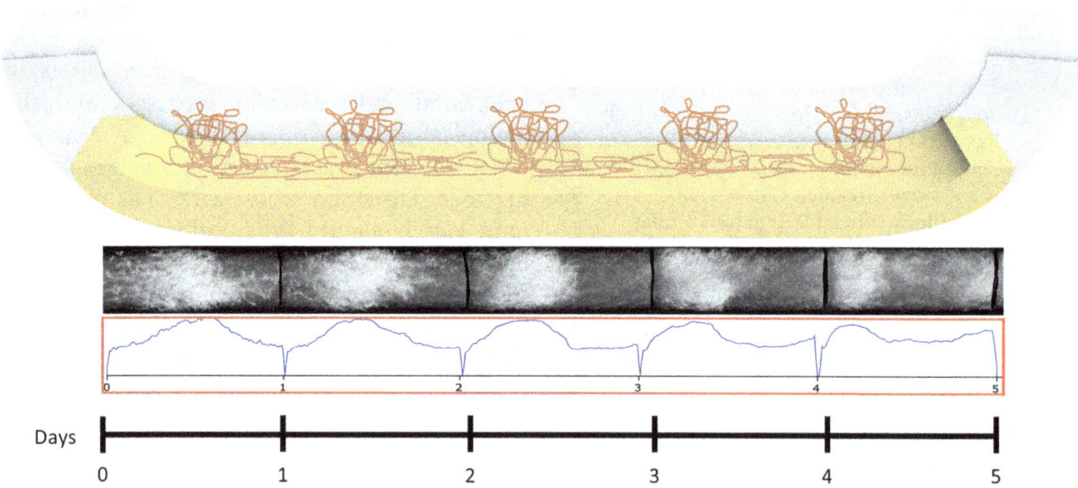

Fig. 1 Race tubes and the *bd* mutation enable period and phase analysis in *Neurospora crassa*. A race tube is a tool used to measure the overt phenotypic rhythm of banding strains of *N. crassa*. A hollow glass tube is bent upward at both ends and filled with molten agar growth media, as shown at top. Once the media has cooled and solidified, an *N. crassa* strain with a band (*bd*) background is inoculated at one end and then grows laterally down the tube. As the *bd* strain grows, it will periodically (in time with the circadian cycle) form aerial hyphae and conidia, shown here as the orange clumps. The race tube is marked once every 24 h at the growth front. The periodicity of the aerial hyphal formations is directly tied to the period of the clock. Race tubes can be scanned and densitometric analysis of these race tube images can be used to measure the time between successive bands to predict the period of the strain, as shown at bottom. Image courtesy of Joshua Thomas

among the earliest observable evidence that organisms have an internal circadian clock (Fig. 1) (Brandt 1953; Pittendrigh et al. 1959). This **band (*bd*)** strain is characterized by zones of hyphal growth alternating with daily "bands" or clumps of conidia during constant dark conditions. We now know that this banding phenotype is related to an increase in the expression of some genes (e.g., the conidial regulation protein *fluffy*) in response to an increase in reactive oxygen species (ROS), due to a mutation in the gene *ras-1* (Fig. 1) (Belden et al. 2007a).

A rapidly growing and non-pathogenic fungus that is easy to genetically manipulate, *N. crassa* has been a strong genetic model system since the work of George W. Beadle and Edward Tatum on metabolic mutants (Beadle and Tatum 1941). *N. crassa* can reproduce both sexually and asexually, depending on environmental cues such as light and nutrient availability, allowing for its efficient use in genetic research (Tan et al. 2004; Filippovich et al. 2015). With its strength as a genetic system, as well as the *bd* mutant phenotype allowing for easy detection of rhythms, *N. crassa* has served as an important model organism in the clocks field for over half a century (Dunlap et al. 2007; Dunlap and Loros 2017). In more recent years, technological advancements have yielded a fully sequenced genome and the development of further molecular tools will ensure that *N. crassa* remains an important model organism in the clocks field for years to come (Galagan et al. 2003; McCluskey et al. 2010).

The conserved circadian timekeeping apparatus in higher eukaryotes, uncovered in large part due to research in *N. crassa*, appears to be a transcriptional-translational negative feedback loop (TTFL) that forms a molecular oscillator or **core clock** (Dunlap and Loros 2018; Hurley et al. 2016a). At subjective dawn, the transcriptional apparatus of the TTFL, the positive arm protein complex (in *N. crassa* **White Collar-1 (WC-1)** and **White Collar-2 (WC-2)**), binds to the promoter region of one of the genes involved in the repressive part of the TTFL, the negative arm protein complex (in *N. crassa* **Frequency (FRQ)** and **Frequency-interacting RNA helicase (FRH)**), and activates its transcription (Fig. 2). The positive arm also activates a host of other gene promoters that are not involved in the regulation of the TTFL, generating oscillations in 10–40% or more of transcripts to effect the coordination of many cellular processes (Hurley et al. 2014, 2018; Sancar et al. 2015; Mure et al. 2018). Once translated, the negative arm enters the nucleus and interacts with the positive arm proteins to repress their transcriptional activity. As the day progresses, the negative arm proteins are phosphorylated by many different **kinases** (e.g., **Casein Kinase-1a (CK-1a)** and Casein Kinase-2 (CK-2)), which leads to their inactivation and **degradation**, allowing the positive arm proteins to resume transcriptional activity, starting the cycle anew. The clock is also able to incorporate environmental inputs (such as light, temperature, and nutrients) into the core oscillator to optimally time the circadian cycle, with the earliest examples of this integration in higher eukaryotes first demonstrated in *N. crassa* (Garceau et al. 1997; Crosthwaite et al. 1997; Liu et al. 1998; Sancar et al. 2012; Loros 2019).

In this review, we summarize the key investigations that use the genetic model organism *N. crassa* to create an understanding of the molecular underpinnings of circadian regulation. As the research in the clocks' field from *N. crassa* is substantial, and our space short, we focus our review into five broad contributions to clock research. The first three sections will describe what is known about *N. crassa* circadian timekeeping at the three basic levels of a molecular clock, the input, the core oscillator, and the output, and how these discoveries have informed an understanding of clocks in higher eukaryotes. The fourth section covers how circadian oscillations are measured in *N. crassa*, and the final section will address *N. crassa* as a model organism for clocks in other fungi. Where important research has occurred beyond the scope of these contributions, or at a depth our mandate did not allow for, we point the reader to previously published reviews focusing on that topic specifically.

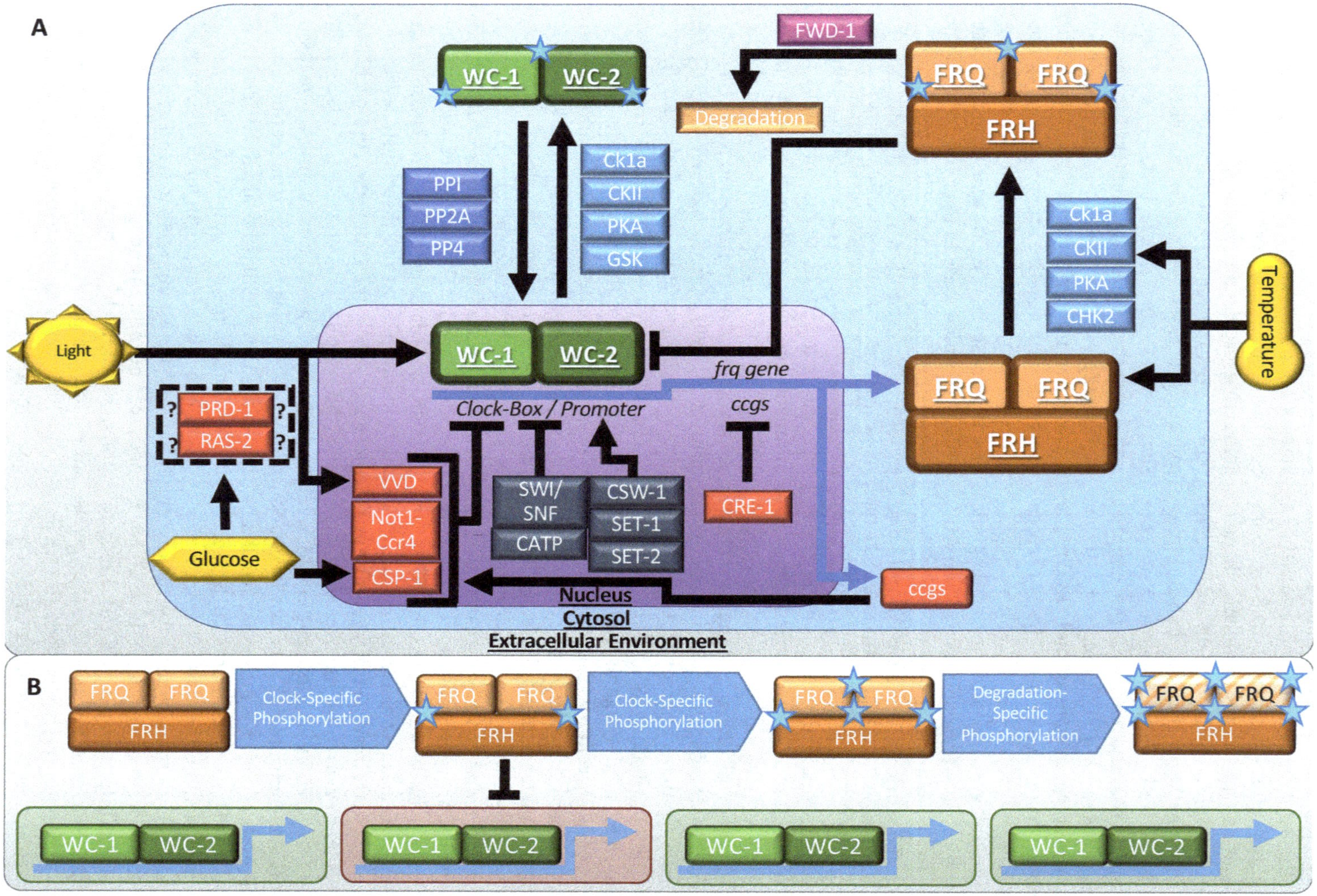

Fig. 2 The molecular components of the *Neurospora crassa* circadian clock. (**a**) The circadian cycle begins in the nucleus with the positive arm, made up of White Collar 1 (WC-1, light green) and White Collar 2 (WC-2, dark green), which form the White Collar Complex (WCC). The positive arm transcriptionally activates

II. Input: Various Inputs to the Core Clock Allow Phase Entrainment to the External Environment

One of the tenets of circadian rhythms is that their period is buffered against different constant conditions but is still able to incorporate signals from the environment, allowing the clock phase to align, or "entrain," to changes in the external environment. The most obvious of these signals is the light from the sun, a cue to which circadian rhythms were adapted (Hut and Beersma 2011). However, in recent years, the circadian field has discovered that the clock is responsive to many other environmental cues, such as temperature and glucose (Garceau et al. 1997; Liu et al. 1998; Sancar et al. 2012). For this review, we have grouped these signals, and the pathways that respond to them, into three basic categories: light, temperature, and nutrients.

A. *Neurospora crassa*, the Circadian Clock, and Photoreception

While fungi are not photosynthetic organisms, the ability to sense different colors and intensities of light still contributes to their overall fitness (Idnurm et al. 2010). *N. crassa* has conserved proteins that are essential for the detection of red light, green light, and blue light, though not all of these proteins are functional **photoreceptors** (for more complete reviews, see Fuller et al. 2015; Yu and Fischer 2019). The primary external timing cue for the clock in *N. crassa* is blue light, which is sensed through the positive arm protein WC-1 (Fig. 2a). WC-1 is a transcription factor that utilizes **Per-Arnt-Sim (PAS) domains** for transcriptional activation. The PAS domain located toward the protein's N-terminal end is of a special class called a **"light oxygen voltage" (LOV) domain** (Ballario and Macino 1997; Ballario et al. 1998). The LOV domain is able to form a transient cysteinyl adduct with a blue light-absorbing chromophore flavin adenine dinucleotide (FAD) when exposed to blue light (Ballario et al. 1998; Linden et al. 1997). Along with its partner protein, WC-2, WC-1 hetero-dimerizes to form the white collar complex (WCC) (Cheng et al. 2002). When the WCC is exposed to blue light, a conformational change occurs which creates a transcriptionally active WCC (Froehlich et al. 2002; He et al. 2002; Zoltowski et al. 2007; Malzahn et al. 2010). The WCC transcriptionally activates genes that have a **light-responsive element (LRE)** in their promoters, with hundreds of genes affected, either directly or indirectly, by WCC binding (Chen et al. 2009; Collett et al. 2002; Froehlich et al. 2003; Hurley et al. 2014; Smith et al. 2010; Wu et al. 2014). As the WCC contains both a light-sensitive element and forms the positive arm of the circadian clock (see Sect. III.A), research in *N. crassa* was foun-

Fig. 2 (continued) numerous genes, including *frequency* (*frq*). In order to do this, the promoter regions of these genes must be accessible and numerous chromatin remodeling factors (dark blue) work with the WCC to effect remodeling at these promoters. The transcriptional activation of *frq* leads to the transcription and translation of FRQ (light orange), which then dimerizes and associates with Frequency-interacting RNA helicase (FRH, dark orange), forming the FFC. FRQ is then phosphorylated (cyan stars) by numerous kinases (light blue) until the FFC can interact with the WCC, repressing WCC activation. The WCC is phosphorylated by several kinases, decreasing transcription at the *frq* locus and other loci regulated by the WCC, and then potentially removed from the nucleus. FRQ undergoes further phosphorylation until it can no longer interact with the WCC and it is recognized by FWD-1 (pink) for degradation. Without the FFC, the WCC is dephosphorylated by phosphatases (blue) and returns to the nucleus to restart the cycle. The genes under the influence of the WCC that do not play a role in the core clock are called clock-controlled genes (*ccgs*, red) and are the drivers of circadian behavior. Some of these *ccgs* can act back on the core circadian loop. In addition, inputs from the environment (yellow) are able to entrain the clock, such as light, temperature, and nutrients (e.g., glucose). (**b**) Negative arm turnover via phosphorylation triggered degradation was previously believed to relieve positive arm (WCC, green) repression. However, new evidence suggests that the negative arm (FFC, orange) undergoes two types of phosphorylation, a clock-specific phosphorylation, which alters the ability of the FFC to suppress WCC activity, and degradation-specific phosphorylation, which targets the negative arm proteins for degradation. This degradation occurs only after the FFC has lost its ability to repress the WCC, and this highly phosphorylated FFC does not impact the clock

dational in demonstrating that the positive arm serves the crucial function of integrating light cues into the clock.

While the WCC is the point protein for clock entrainment to light cues, the activity of another protein, **Vivid (VVD)**, is also crucial for proper clock phase timing based on light in the environment (Fig. 2a). A small protein, only containing an N-terminal cap and a LOV domain, VVD plays an important role in light integration into the clock by altering the activity of the WCC (Zoltowski et al. 2007; Hunt et al. 2010; Chen et al. 2010). The WCC induces *vvd* expression in response to light, after which VVD interacts with WC-1's LOV domain and blocks the transcriptional capabilities of WC-1 (Zoltowski et al. 2007; Hunt et al. 2010; Malzahn et al. 2010; Chen et al. 2010; Dasgupta et al. 2015). The action of VVD on the WCC is an important example of **photoadaptation**, ensuring that light-induced genes are only transiently upregulated (Hunt et al. 2010; Chen et al. 2010). VVD's photoadaptive activities are not only important for keeping the clock from resetting inappropriately, such as during nights with bright moonlight, but also in supporting the optimal timing of light responses by the WCC (Heintzen et al. 2001; Malzahn et al. 2010).

B. The Effect of Temperature on the Negative Arm of the Clock

Though it is logical to think of light as the primary environmental input for the clock, surprisingly many other factors, such as **temperature**, can strongly impact the circadian mechanism. For example, temperature changes can reset the phase of the clock, such as an increase from a low to a high temperature. While light input is chiefly transduced via the positive arm (see Sects. II.A and III.A), evidence from *N. crassa* shows that temperature input is primarily conveyed by changes in the molecular dynamics of the negative arm of the clock (Liu et al. 1997; Pregueiro et al. 2005, 2006; Diernfellner et al. 2005), particularly the dynamics of the levels of the core negative arm protein frequency (FRQ) (see Sect. III.B.1 and Fig. 2a). The current understanding of the response to temperature shifts is that while the level of *frq* mRNA remains stable at different temperatures, FRQ protein levels oscillate around a higher midline at higher temperatures, with the peak and trough of FRQ levels falling above the peak level of FRQ in a strain grown at a low enough temperature (Liu et al. 1997). Due to the difference in FRQ levels at different temperatures, the comparatively low levels of FRQ at a lower temperature will create an artificial "trough" from which FRQ levels rise when the organism is shifted to a higher temperature, thereby resetting the phase of the clock (Liu et al. 1998). A decrease in temperature can have the same resetting effect, as FRQ levels decrease from an artificial "peak" time in higher temperatures to start oscillating around a lower midline of protein levels (Liu et al. 1998). Extreme temperature changes (±10 °C) have a strong effect on setting the phase of the clock, so much so that it supersedes the effects of light, re-entraining the molecular clock even in a conflicting light-dark regime (Liu et al. 1998). This research paved the way for similar discoveries in higher eukaryotes. Temperature has subsequently been shown to be an important phase resetting cue in mammal peripheral tissues, such that daily temperature changes within physiological ranges (36–37.5 °C) can entrain mouse peripheral tissues (Buhr et al. 2010). Moreover, within *Drosophila*, the mutation of all detected phosphorylation sites to alanine on the positive arm clock protein dCLOCK led to compromised temperature resetting, but unaffected light resetting, suggesting that temperature is more important than light as a timing cue for *Drosophila* tissues as well (Lee et al. 2014).

Though temperature changes can shift the phase of the clock, one of the key tenets of a circadian rhythm is that its period is compensated and regulated at different growing temperatures. However, the underlying mechanism for temperature regulation in the clock is poorly understood. What is known is that temperature impacts the choice of FRQ start codons during translation (Colot et al. 2005; Diernfellner et al. 2007; Liu et al. 1997). A temperature-sensitive **alternate splicing** mechanism occurs in *frq* mRNA, leading to different ratios of the two isoforms of FRQ: long FRQ (L-FRQ; 1–989 aa) and short FRQ (S-FRQ; 100–989 aa). At higher temperatures, L-FRQ levels increase relative to S-FRQ (Liu et al. 1997; Colot et al. 2005; Diernfellner et al. 2007). L-FRQ-only strains have a shorter clock period than S-FRQ-only strains (Liu et al. 1997; Diernfellner et al. 2007). Therefore, the temperature-based ratio of L- to S-FRQ levels is believed to contribute to the robustness of the clock and the fine-tuning of the clock period at different ambient temperatures (Liu et al. 1997; Diernfellner et al. 2005, 2007). Evidence also suggests that post-translational mechanisms are involved in temperature compensation, e.g., the phosphorylation of FRQ. The mutation of **phosphorylation sites** that exist only on L-FRQ yield a strain with a longer period, similar to the period of a S-FRQ-only strain (Diernfellner et al. 2007; Baker et al. 2009). In addition, mutations in kinases that phosphorylate FRQ (i.e., CK2) have been shown to impact temperature compensation (Mehra et al. 2009b). While different FRQ isoform ratios and changes in FRQ phosphorylation contribute to temperature regulation, further work is necessary to uncover the underlying mechanisms that provide temperature compensation to the clock.

C. The Impact of Nutrient Sensing on the Circadian Clock

Though a great deal of research has gone into the understanding of how light and temperature align the molecular clock, recent data from *N. crassa* has shown that levels of **nutrients**, such as glucose and lipids, can also affect the clock and must be compensated for, termed metabolic or nutritional compensation (Hurley et al. 2016b; Sancar et al. 2012; Starkey et al. 2013). The best studied type of nutritional compensation is glucose compensation, which includes a known circadian-auxiliary loop that utilizes **Conidial Separation Protein 1 (CSP-1)** (Fig. 2a). When glucose levels are high, the core clock proteins and many of their downstream targets, including CSP-1, are translated at an increased rate (Sancar et al. 2012). To compensate for the increased clock protein levels, the resulting higher levels of CSP-1 interact with additional clock-regulated proteins, Regulation of Conidiation 1 (RCO-1) and Regulation of Conidiation and Morphology-1 (RCM-1), to repress *wc-1* expression and complete a negative feedback loop that provides glucose compensation (Sancar et al. 2012; Dovzhenok et al. 2015; Olivares-Yañez et al. 2016). Other proteins are also recognized for their impact on glucose compensation, including Period-1 (PRD-1/DBP-2) and Ras-like-2 (RAS-2). PRD-1 is an ATP-dependent DEAD-box RNA helicase, and the clock in *prd-1* mutants is not compensated against glucose changes, leading to longer periods when glucose is not limited (Starkey et al. 2013; Emerson et al. 2015; Adhvaryu et al. 2016). Like *prd-1* mutants, the clock in a Δ*ras2* strain displays longer periods and is not glucose-compensated (Gyöngyösi et al. 2017). Further work is needed to fully understand how glucose compensation is imparted, either by multiple feedback loops including CSP-1, RCO-1, RCM-1, PRD-1, and RAS-2, or through an underlying mechanism yet to be discovered. Nutrient compensation has long been shown to be an important element of the mammalian circadian clock and the research into the mechanisms that regulate nutrient compensation in *N. crassa* has enabled an appreciation of how these might work in higher eukaryotes (Peek et al. 2012).

Beyond glucose compensation, there is evidence that the clock can be affected by lipid levels. Longevity assurance gene (LAG-1) is thought to be part of a ceramide synthase complex and likely plays a role in metabolic clock regulation through stress responses involving sphingolipid metabolism (Case et al. 2014). In a *lag-1* mutant, the clock period lengthens to 41 h, suggesting that changes in sphingolipid levels can affect clock period. Further evidence that the clock is sensitive and responsive to lipid levels includes the relationship between high levels of DAG (sn-1,2-diacylglycerol) and longer clock periods (Ramsdale and Lakin-Thomas 2000; Case et al. 2014). With the extent of reported metabolic effects on the clock through glucose and lipid levels, it is likely that other cell components that play a role in clock nutritional compensation are still to be determined.

III. Core Oscillator: The *N. crassa* Core Circadian Timekeeping Mechanism

The circadian inputs described above feed into the architecturally-conserved circadian **transcriptional-translational negative feedback loop (TTFL)** (Hurley et al. 2016a; Dunlap and Loros 2017). In addition to the conservation of the underlying architecture of the TTFL across higher eukaryotes, there is sequence homology between WC-1 and WC-2, the positive arm **transcription factors** in *N. crassa*, and transcription factors in other fungal and mammalian systems (Dunlap and Loros 2017). Furthermore, conformational heterogeneity and post-translational regulation of the clock proteins by phosphorylation, especially of negative arm proteins such as FRQ, are also conserved (Lee et al. 2009; Chiu et al. 2011; Hurley et al. 2016a; Pelham et al. 2018). Finally, phosphorylation-dependent ubiquitination and degradation of the negative components supports the functioning of most eukaryotic clocks (Grima et al. 2002; Ko et al. 2002; He et al. 2003; Eide et al. 2005). All of these conserved features of the clock were, at least in part, discovered in *N. crassa*. In this section, we review the abovementioned key elements of the *N. crassa* core clock mechanism, which have contributed greatly to the understanding of the TTFL in other fungi and higher eukaryotes.

A. The Transcriptionally Active Positive Arm of the Clock

In *N. crassa*, the circadian cycle is initiated by the positive arm of the clock, a heterodimeric GATA-like transcription factor complex termed the White Collar Complex (WCC) comprising two proteins, WC-1 and WC-2 (Belden et al. 2007b; Hurley et al. 2014). WCC transcriptional activity peaks around subjective dawn, commencing the circadian cycle (Belden et al. 2007b; Hurley et al. 2014). WC-1 is a 1167 amino acid (aa) long protein that contains a circadian transactivation domain, three Per-Arnt-Sim (PAS) domains (including the light-responsive LOV domain and a "defective in binding" (DBD) region involved in DNA binding and mediating interactions with the negative arm), and a single zinc finger domain involved in DNA binding (Ballario et al. 1996; Lee et al. 2000; Cheng et al. 2002; Wang et al. 2014, 2016). As mentioned previously in Sect. II.A, WC-1 is the primary blue-light photoreceptor in *N. crassa* and is responsible for the integration of light signaling into the circadian clock.

WC-2 is a 530 aa-long protein that contains a single zinc finger domain, a single PAS domain and, unlike WC-1, is unable to sense light (Linden and Macino 1997; Collett et al. 2002). WC-1 and WC-2 exhibit strong oscillations at the transcriptional level, but neither show strong oscillations at the protein level (Hurley et al. 2018). Interestingly, the negative arm protein, FRQ, is needed for proper levels of WC-1 to accumulate (Schafmeier et al. 2006). WC-1 and WC-2 interact with each other through their PAS domains to form the White-Collar Complex (WCC), which promotes the expression of genes in the negative arm as well as other genes not involved in maintaining the core clock, termed **clock-controlled genes (*ccgs*)** (Fig. 2a) (Linden and Macino 1997; Crosthwaite et al. 1997; Cheng et al. 2002; Hurley et al. 2014; Wu et al. 2014).

The WCC exists in two conformations, a light-activated conformation and a dark conformation, which serve to control the DNA binding of the WCC. The light-activated complex is composed of two WC-1s and one WC-2, binding to a subset of genes that are considered to be light-responsive (Froehlich et al. 2002; Cheng et al. 2003). In contrast, the dark conformation comprises a single WC-1 and WC-2 and is essential for the expression of *frq* and other *ccgs* under "free-running" or constant dark conditions (Froehlich et al. 2002; Cheng et al. 2003). The zinc fingers of both WC-1 and WC-2 are essential for binding DNA in the dark, aided by the DBD motif on WC-1 (Wang et al.

2016). The ability of the WCC to bind DNA is also dependent on its phosphorylation state, as has been shown for *Drosophila* and mouse positive arm CLOCK protein, with daily rhythms in phosphorylation occurring (He et al. 2005; Baker et al. 2009; Yoshitane et al. 2009; Lee et al. 2014; Wang et al. 2019). There are 80 phosphorylation sites on WC-1, with phosphorylations near its zinc-finger domain playing an important role in closing the negative feedback loop (He et al. 2005; Baker et al. 2009; Sancar et al. 2009; Wang et al. 2019). WC-2 phosphorylation peaks around late subjective day, with 15 phosphorylation sites detected in 2 different clusters, with 1 cluster near its zinc finger domain, as in WC-1 (Schafmeier et al. 2005; Wang et al. 2019). While earlier work suggested that the mutation of 5 phosphorylation sites in WC-1 leads to arrhythmicity, it is now evident that phosphorylations on both WC-1 and WC-2 are needed to close the clock's negative feedback loop and support circadian rhythmicity (He et al. 2005; Wang et al. 2019). Plainly, both the light-dependent change in WCC structure and the overall phosphorylation state of the WCC play an important role in delineating the WCC's function in the light response pathway from its role in the positive arm of the circadian clock (Wang et al. 2016, 2019).

While the WCC is the primary transcriptional activator of the negative arm gene *frq*, there are other proteins that influence *frq* transcription through modifications of the chromatin structure at the *frq* locus. The *frq* promoter has two unique elements, the **proximal LRE (pLRE)** and the **Clock Box (c-box)** (Froehlich et al. 2003). The heterotrimeric, light-activated WCC binds to the pLRE in response to light, but the heterodimeric, dark-active WCC binds to the c-box to regulate the rhythmic expression of *frq* that is necessary for proper clock function in continual darkness (Froehlich et al. 2003; Gooch et al. 2014). This binding is facilitated by **chromatin remodelers**, including the positive regulators Clock ATPase (CATP), SWItch/SucroseNonFermentable (SWI-SNF), Chromodomain helicase DNA-binding (CHD-1), and Defective in methylation (DIM-2), and the negative regulators Clockswitch (CSW-1) and Su(var)3–9-enhancer-of-zeste-trithorax-1 and 2 (SET-1 and SET-2) (Fig. 2a). CATP positively regulates *frq* expression via its promotion of the removal of histones at the *frq* locus, allowing for WCC binding (Cha et al. 2013). Once the WCC binds to the c-box, it recruits the SWI-SNF complex, which removes a nucleosome that partially covers the c-box and likely loops the DNA, bringing the WCC closer to the transcription start site (Zhang et al. 2006; Wang et al. 2014). CHD-1 also contributes to necessary changes in *frq* chromatin structure to aid transcription, and DIM-2 catalyzes the removal of transient DNA methylation that helps set clock phase timing (Belden et al. 2011). The protein CSW-1 is involved in the final stages of closing the clock's feedback loop, compacting the chromatin structure at the c-box to down-regulate *frq* expression (Belden et al. 2007b). SET-1 and SET-2 further this process by methylating histones H3K4 and H3K36, respectively, making the chromatin less accessible (Raduwan et al. 2013; Sun et al. 2016). Only when these supporting proteins properly maintain the chromatin architecture of the *frq* locus can the WCC complete its positive activation of *frq* transcription to maintain rhythmicity. Circadian rhythms in histone modifications have also been detected in mammal systems, and chromatin-modifying proteins are important for the repression of positive arm clock proteins (Koike et al. 2012; Takahashi 2017).

B. The Negative Arm of the Clock

1. Oscillations in FRQ Protein and Phosphorylation Levels Determine Clock Period

Frequency's role in determining clock period was originally discovered through the isolation of "banding" (*bd*) mutants with different period lengths (Aronson et al. 1994; Feldman and Hoyle 1973; Gardner and Feldman 1980; Loros et al. 1986). In constant conditions (25 °C and constant darkness), the *bd* strain displays a conidial band once every 21.6 (± 0.5) h (Fig. 1). However, mutants in the *bd* strain were discovered to have periods ranging from

16.5–19.0 h (short-period mutants) to 24.0–29.0 h (long-period mutants) or to be completely arrhythmic (Feldman and Hoyle 1973; Gardner and Feldman 1980; Aronson et al. 1994). Many of the mutations in these short- and long-period strains mapped to a single locus, which was dubbed *frequency* (*frq*) (Feldman and Hoyle 1973; Gardner and Feldman 1980, 1981; Loros et al. 1986).

Full-length FRQ is a largely disordered 989 aa long protein containing several smaller potentially well-ordered domains that are involved in the formation of the negative arm protein complex. Due to the low proportion of in silico predicted structure, FRQ was believed to be an **intrinsically disordered protein (IDP)**, a protein that lacks a fixed or ordered three-dimensional structure instead sampling a hetergenous ensemble of conformations. This prediction of disorder was demonstrated biochemically as FRQ remained soluble after heat treatment and was more quickly digested by proteases (Hurley et al. 2013). While FRQ does not have sequence homologs amongst all higher eukaryotes, intrinsic disorder does appear to be conserved among negative arm proteins, suggesting that disorder is a key feature of negative arm proteins (Hurley et al. 2013).

There are two isoforms of FRQ that are the products of alternate translation start sites, Long-FRQ (989 aa) and Short-FRQ (889 aa, missing the first 100 amino acids), which are thought to contribute to temperature compensation of the clock (Garceau et al. 1997), as previously discussed in Sect. II.B. Both FRQ isoforms encompass several key protein-protein interaction domains, including the coiled-coil interaction domain that allows for the dimerization of FRQ (necessary for rhythmicity; Cheng et al. 2001a), the nuclear localization region (Luo et al. 1998), and the FRQ-CK1 interacting domains FCD1 and FCD2 (Querfurth et al. 2011). In addition, both isoforms contain two predicted proline, glutamic acid, serine, threonine (PEST) domains that may be important for protein turnover, PEST-1 and PEST-2 (Merrow and Dunlap 1994). Thus far, the only in vivo evidence that these domains act in protein turnover is that the PEST-1 region is phosphorylated before FRQ is degraded (Liu et al. 2000; Görl et al. 2001; Baker et al. 2009). Finally, the isoforms include an FRQ-FRH interacting domain called the FFD (Guo et al. 2010). These interaction domains contribute to the formation of the larger negative arm complex, the Frequency-FRH Complex (FFC) (Baker et al. 2009; Guo et al. 2010; Querfurth et al. 2011; Hurley et al. 2013).

FRQ transcript and protein levels oscillate on a daily basis as a result of the activity of the WCC. At the beginning of the circadian cycle, WCC activates *frq* expression, and approximately 4 h later there is a peak in translated FRQ levels (Garceau et al. 1997; Merrow et al. 1997). Shortly after translation, FRQ dimerizes with itself and binds to FRQ-interacting RNA Helicase (FRH) to form the FRQ/FRH complex (FFC); all FRQ is found in complex with FRH in a two to one ratio (Cheng et al. 2001a, 2005; Baker et al. 2009). This interaction helps to stabilize FRQ and is discussed further in Sect. III.B.2 (Cheng et al. 2005; Hurley et al. 2013). Many kinases act to phosphorylate FRQ (see Sect. III.D for more details) and this phosphorylation allows the FFC to enter the nucleus early on in the cycle to repress WCC activation of *frq* transcription, closing the negative feedback loop (Fig. 2) (Cheng et al. 2005; Hong et al. 2008; Baker et al. 2009; Cha et al. 2011). Phosphorylation eventually leads to FRQ degradation via the ubiquitin-proteasome pathway (Fig. 2), and a new wave of FRQ protein is translated as the circadian cycle continues (He and Liu 2005a; Larrondo et al. 2015).

In addition to direct WCC regulation, further transcriptional regulation supports robust oscillations in *frq* mRNA levels. The WCC activates transcription of *frq*'s long non-coding **antisense transcript,** ***qrf*** (*frq* spelled backwards) in response to light by utilizing a promoter in the 3′ UTR of *frq*, and this aids in light resetting of the clock (Kramer et al. 2003). In constant darkness, *qrf* expression is also regulated by another unknown mechanism (Xue et al. 2014). Data suggests that *qrf* silences *frq* expression to tune the timing and period of the clock, yielding more robust oscillations of *frq* by counteracting any "leaky" *frq* expression that would lead to dampening of the central oscillator. However, two competing models suggest two distinct mechanisms as to how this occurs. These include either the stalling/collision of the RNA polymerases transcribing *frq* and *qrf* (Xue et al. 2014) or transcriptional permissiveness at the *frq* locus (Li et al. 2015). As the mammalian clock also contains an antisense RNA to the negative arm proteins, the answer to this debate has implications for clock regulation in higher eukaryotes (Li et al. 2015; Xue et al. 2014).

In addition to the transcriptional regulation of *frq*, there is also regulation that occurs at the level of translation. One mechanism arises from the presence of non-optimal codons within *frq*'s open reading frame. Among the codons used to code for the same amino acid, some codons occur less often than others genome-wide and these are termed **non-optimal codons.** Non-optimal codons correlate to low copy numbers of a given tRNA and therefore are thought to result in slower translation elongation rates (Zhou et al. 2013). The sequence of *frq* is biased toward more non-optimal codons and evidence suggests that this codon bias is important for the stability of the resulting FRQ protein. When more frequently occurring codons were substituted into the N-terminus of the *frq* open reading frame, the clock no longer functioned (Zhou et al. 2013). As more optimal codons are thought to lead to faster rates of translation elongation, the loss of rhythmicity was attributed to the improper folding of FRQ, similar to results obtained from the codon optimization of the negative arm protein PER in *Drosophila* (Zhou et al. 2013; Fu et al. 2016). Alternatively, and to include the more recent data showing that FRQ is an IDP, if FRQ is translated too quickly, it may not have a chance to bind with its stabilizing "Nanny" FRH (see also Sect. III.B.2), which could lead to rapid degradation (Hurley et al. 2013). Regardless of the mechanism, the conservation of non-optimal codons within *frq* appears to be essential for circadian regulation in *N. crassa* and other organisms (Xu et al. 2013; Fu et al. 2016).

A second mechanism of FRQ translational regulation lies in the many **upstream Open Reading Frames (uORFs)** of *frq*. The *frq* transcript contains an approximate 1.5 kb untranslated region at its 5′ end (5′ UTR) that includes six uORFs (Garceau et al. 1997; Liu et al. 1997). When most of the 5′ UTR was deleted, FRQ levels increased, suggesting that ribosome scanning through uORFs could decrease levels of FRQ under some conditions (Liu et al. 1997). However, unpublished experiments suggest that the deletion of those six uORFs had no effect on the clock's overall rhythmicity and did not shorten the 4 h lag between peak *frq* mRNA and peak FRQ protein levels (Garceau et al. 1997). Interestingly, uORFs are also conserved in multiple clock proteins as a regulatory mechanism in mammals (Janich et al. 2015).

2. FRQ-Interacting RNA Helicase (FRH) Supports FRQ as a "Nanny" Protein

While all FRQ is associated with FRH, only ~40% of FRH within the cell is bound to FRQ, with the other ~60% presumably carrying out other functions within the cell (Cheng et al. 2005). These functions are likely related to RNA surveillance, as FRH is a DEAD-box-type RNA Helicase protein and is homologous to Mtr4p in *Saccharomyces cerevisiae,* which assists in RNA metabolism via its interaction with the exosome (Lykke-Andersen et al. 2011). FRH interacts with exosome proteins in *N. crassa* (e.g., RRP44), has been found to bind *frq* mRNA and influence its stability, and is also a functional RNA helicase (Guo et al. 2009; Morales et al. 2018).

FRH has many homologs among fungal and animal species, including some involved in the clock complex (Cheng et al. 2005; Padmanabhan et al. 2012), and has a well-characterized and conserved overall structure (Conrad et al. 2016; Morales et al. 2018). Four domains form a ring-like structure (two canonical RecA-like domains, a small winged helix domain, and a helical DSHCT domain) and a fifth domain forms an arch domain or "arm" that spans across one side of the ring and ends in a KOW or "fist" module (Conrad et al. 2016; Morales et al. 2018). A known binding region for FRQ is found in an additional domain (aa 100–150) within the largely unstructured N-terminus, and the KOW/fist domain includes a binding site for WCC interaction that can be further modulated by an interaction with VVD (Hunt et al. 2010; Shi et al. 2010; Hurley et al. 2013; Conrad et al. 2016; Morales et al. 2018).

Given the clear homology in structure with other RNA helicases and the demonstrated ATPase RNA Helicase function, it was surprising when a strain with a point mutation (R806H) in the KOW domain of FRH had healthy growth but an arrhythmic clock (Shi et al. 2010). This suggested that FRH's ATPase function did not explain its role in the core clock. Further support for FRH playing dual roles within the cell came from an additional mutant strain where FRH lost its ATPase function yet still supported a functional clock (Hurley et al. 2013). Moreover, the downregulation of FRH caused an ~80% reduction in FRQ protein levels (Cheng et al. 2005). In total, this suggested that FRH plays a structural function to support or act as a **"Nanny" protein** for the intrinsically disordered protein FRQ and protect it from premature turnover (Hurley et al. 2013). Without the stabilizing influence of FRH, FRQ phosphorylation profiles are impacted and FRQ becomes more prevalent in the nucleus rather than the cytoplasm, highlighting the importance of FRH's Nanny role for FRQ to function in the clock (Cheng et al. 2005; Guo et al. 2010; Cha et al. 2011; Hurley et al. 2013).

C. The Negative Arm Represses the Positive Arm to Close the TTFL

While a great deal is known about the activation that drives the transcriptional portion of the TTFL, the mechanistic underpinnings of WCC repression, the "closing of the loop," is not as well understood. What is known is that the FFC represses WCC transcriptional activity at the *frq* locus via a direct interaction which involves at least three key regions of FRQ (Aronson et al. 1994; Cheng et al. 2005; He and Liu 2005a; He et al. 2006; Guo et al. 2010). Within the clock complex, WC-2 can interact independently with WC-1 and FRQ and is thought to be required for the interaction between WC-1 and FRQ (Denault et al. 2001). In addition, recent evidence shows that in the dark, the "DBD" motif in WC-1 is also necessary for the interaction with the FFC (Wang et al. 2016). Finally, FRH has also been shown to interact independently with the WCC (Hunt et al. 2010), highlighting the complexity of the interactions within and between the core clock complexes.

However, direct interactions do not explain the functional mechanism of repression, and until recently there were two competing theories of how repression occurs. The first potential mechanism involves the deactivation of the WCC via phosphorylation by FRQ-associated kinases, which is thought to lead to the repression of WCC transcriptional activation of *frq* (Cheng et al. 2001b; Froehlich et al. 2003; Schafmeier et al. 2005; He et al. 2006; Wang et al. 2019). It is known that hyper-phosphorylated WCC has a weaker affinity for DNA and that the WCC is not phosphorylated unless it interacts with FRQ (Schafmeier et al. 2005; He et al. 2006; Wang et al. 2019). Recently, the specific clusters of phosphorylation sites that are important for decreased DNA binding have been identified, with phosphorylations near the zinc-finger binding domains of both WC-1 and WC-2 being important for repression of WCC activity (Wang et al. 2019). Additionally, there is some support for a second, "clearance-based" model of repression. Research has shown that oscillations in WCC phosphorylation and de-phosphorylation may also be important for the localization of the WCC complex (Schafmeier et al. 2008; Wang et al. 2016). There is evidence that FFC binding to the WCC in the nucleus leads to the export of the WCC to the cytoplasm (Hong et al. 2008). This would suggest that repression could occur via degradation or nuclear exclusion of the WCC due to the physical interaction between the WCC and the FFC (Hong et al. 2008). In the end, neither mechanism is mutually exclusive, and a two-step process has been suggested where the WCC is removed from the DNA (by phosphorylation) and then exported to the cytoplasm and thereby sequestered (Fig. 2) (Cha et al. 2008). Whether closure of the clock's negative feedback loop is restricted to WCC phosphorylation, or also includes nuclear export, the study of these repression mechanisms in *N. crassa* has, and will continue, to inform the understanding of similar mechanisms in higher eukaryotic clock systems (Yoshitane et al. 2009; Mahesh et al. 2014).

D. Post-translational Modifications in Clock Regulation

1. Kinases Play an Essential Role in Circadian Timing

Post-translational modifications, such as phosphorylation, are central to the functioning of the positive and negative arm of the core clock in *N. crassa* and other clock systems (for more complete reviews of the role of post-translational modifications in higher eukaryotic systems, see Mehra et al. 2009a; Weber et al. 2011; Hirano et al. 2016). The addition of phosphate groups by numerous kinases has been shown to affect circadian timing by altering the stability of the protein and its ability to form protein-protein interactions. In *N. crassa*, these modifications are carried out by the kinases CAMK1, CHK2 (PRD-4), CK1a, CK1b, CK2, GSK, and PKA, which act on the core proteins of the negative and positive arms of the clock (Fig. 2a).

As mentioned previously, FRQ is progressively phosphorylated in a circadian manner with over 85 different residues phosphorylated, making FRQ one of the most highly phosphorylated proteins ever documented (Baker et al. 2009; Tang et al. 2009). CK1a has been suggested to carry out much of this phosphorylation, as CK1a is found in complex with FRQ and FRH throughout a large proportion of the circadian day (Baker et al. 2009). Moreover, when FRQ is not able to interact with CK1a, FRQ is in a hypo-phosphorylated state and shows increased stability (He et al. 2006; Querfurth et al. 2011). In addition, other kinases such as CK2, PKA, CHK2 (PRD-4), CK1b, and CAMK-1 interact more transiently with FRQ and likely play a smaller role in its phosphorylation and circadian regulation (Yang et al. 2001, 2002; Pregueiro et al. 2006; He et al. 2006; Huang et al. 2007).

Throughout the daily cycle, different regions of FRQ are phosphorylated in a clustered manner by interacting kinases, and the phosphorylation of these different FRQ segments can variously stabilize or de-stabilize this protein (Baker et al. 2009). The kinase PKA increases FRQ stability, while CK1a, CK2, and CHK2 decrease FRQ stability over a circadian cycle (Pregueiro et al. 2006; Huang et al. 2007). Phosphorylation also regulates FRQ's ability to interact with other proteins. The extensive phosphorylation of FRQ affects its half-life through increased affinity with ubiquitin ligases, such as FWD-1 (He et al. 2003; also see Sect. III.D). In addition, FRQ's phosphorylation by CK2 decreases its ability to interact with the WCC, necessary for the start of another round of transcription in the circadian loop (Cheng et al. 2001b; Yang et al. 2002). Besides targeting FRQ for eventual degradation via the ubiquitin-proteasome pathway (Fig. 2a), these phosphorylations may affect FRQ's IDP conformation to regulate the time-specific binding of its different interaction partners to complete the negative feedback loop that determines clock period (He et al. 2006; Baker et al. 2009; Querfurth et al. 2011; Hurley et al. 2013; Pelham et al. 2018).

There is also evidence that phosphorylation regulates the positive arm of the clock, influencing the activity, interactions, and degradation of the WCC. The WCC is able to bind to the *frq* c-box promoter when hypo-phosphorylated, but its transcriptional activity is reduced when the WCC is phosphorylated first by PKA, which acts as a priming kinase, followed by CK1a and CK2 in a FRQ-dependent manner (He et al. 2006; Huang et al. 2007; Wang et al. 2019). Beyond the canonical circadian kinases, Glycogen synthase kinase (GSK) phosphorylates both WC-1 and WC-2, presumably promoting WCC degradation via phosphorylation, and the resultant decrease in WCC levels lengthens the circadian period (Tataroğlu et al. 2012). Like WC-1, WC-2 shows a circadian oscillation in phosphorylation (Schafmeier et al. 2005), with increasing WC-2 phosphorylation leading to a decrease in overall WCC activity.

2. Phosphatases Support the Proper Timing of Circadian Period and Phase

Phosphatases are a group of widely conserved post-transcriptional modifiers that remove phosphate groups and have demonstrated importance for proper clock function. Three protein phosphatases, PP1, PP2A, and PP4, all play a role in supporting the core clock (Cha et al. 2008; Yang et al. 2004). PP1 and PP4 support FRQ stability and therefore proper phase timing and period length, and PP2A is important for maintaining the proper *frq* mRNA level (Yang et al. 2004; Cha et al. 2008). However, the most detailed study of FRQ phosphorylation showed only progressive phosphorylation. Without in vivo evidence for de-phosphorylation, the effects of phosphatases on FRQ may thus be indirect (Baker et al. 2009). Indeed, some of these phosphatases also act on the WCC and thereby may affect *frq* levels and FRQ protein via the core feedback loop. For example, PP2A and PP4 de-phosphorylate the WCC, while it is in the cytosol. This is important as this de-phosphorylation potentially impacts the ability of the WCC to re-enter the nucleus and start a new round of transcription, suggesting that the major impact of phosphatases on FRQ is via the positive arm (Fig. 2a) (Cha et al. 2008; Schafmeier et al. 2005, 2008). Phosphatases also play an important role in the circadian clocks of *Drosophila* and mammals, modulating the phosphorylation state of clock proteins such that an approximately 24 h period is maintained (Reischl and Kramer 2011).

E. Ubiquitination and Degradation Contribute to a Robust Circadian Clock

While not all phosphorylation leads to degradation, the overall progressive phosphorylation of FRQ is associated with its regulated turnover by the **ubiquitin-proteasome pathway** (Garceau et al. 1997; Liu et al. 2000; Baker et al. 2009; Guo et al. 2010). Following the phosphorylation of FRQ, particularly in the PEST-1 region, FRQ is ubiquitinated by the F-box/WD40 repeat-containing protein FWD-1, which promotes FRQ's degradation via the proteasome (Fig. 2a) (He et al. 2003). FWD-1 is part of an SKP/Cullin/F-box (SCF)-type E3 ubiquitin ligase complex that can bind phosphorylated FRQ motifs, with more FRQ phosphorylation presenting more potential FWD-1 binding sites (He et al. 2003; He and Liu 2005a). When *fwd-1* is knocked out, the resulting strain is overtly arrhythmic; hyper-phosphorylated FRQ accumulates to high levels and conidial banding is lost (He et al. 2003; He and Liu 2005a). However, recent work using a **luciferase reporter** system has shown that FRQ degradation is not strictly required for the completion of the circadian feedback loop and the maintenance of period length (Fig. 2b) (Larrondo et al. 2015). By measuring in vivo luminescence from a strain expressing luciferase under the control of a segment from the promoter of *frq* that contains the c-box, circadian rhythms in *frq* transcriptional activation were found even when *fwd-1* was knocked out (Larrondo et al. 2015). This suggests that negative arm protein degradation is not required for the core circadian oscillator and is not actually a determinant of clock period, and has since been confirmed in a mammalian system (Larrondo et al. 2015; Ode et al. 2017).

Even though FRQ degradation may not be required for clock functionality, premature FRQ degradation still leads to arrhythmic strains (Guo et al. 2010; Hurley et al. 2013). As an intrinsically disordered protein (IDP), without its stabilizing FRH "Nanny," FRQ is rapidly degraded within 3 h through a passive "degradation by default" pathway that does not involve ubiquitination (Tsvetkov et al. 2009; Guo et al. 2010; Hurley et al. 2013). Only when protected by FRH can FRQ have a chance to mature and carry out its circadian functions before undergoing ubiquitination and targeted degradation. This degradation by default pathway also has implications for mammal clock systems, since the negative arm PER proteins are also predicted to be IDPs (Hurley et al. 2013).

IV. Output: Clock Output Regulates Cellular Physiology at the Transcriptional and Post-transcriptional Levels

The core circadian oscillator affects many processes within the cell via transcriptional and post-transcriptional regulation, termed the clock's "output" (Vitalini et al. 2006; Hurley et al. 2014, 2018; Sancar et al. 2015). Regulation occurs at multiple levels, including the regulation of transcriptional cascades by the positive arm of the clock and the more recent evidence of post-transcriptional regulation of protein levels (Smith et al. 2010; Hurley et al. 2014, 2018). For the purposes of this review, we will discuss output in two sections: transcriptional output and post-transcriptional output.

A. Transcriptional Regulation of Cellular Output via the Circadian Clock

1. Circadian Transcriptional Activation of Clock-Controlled Genes (*ccgs*)

The positive arm of the TTFL in animal and fungal clocks acts as a "pioneer-like" transcription factor to remodel chromatin and activate the transcription of the negative arm as well as a host of other genes, including other transcription factors (Hurley et al. 2014; Menet et al. 2014; Smith et al. 2010). Chromatin-immunoprecipitation (ChIP) sequencing identified that the WCC directly targets ~300 genes near circadian dawn (Smith et al. 2010; Hurley et al. 2014). Included in these WCC targets are additional transcription factors that regulate further groups of genes, creating a circadian transcription factor cascade. Among the detected WCC targets, a total of 28 transcription factors were found, including *ccg-9*, *sah-1*, *nit-2*, *hsf-2*, *adv-1*, *bek-1*, *sub-1*, *sah-2*, *sre*, *mip-1*, and *vad-2* (Smith et al. 2010). The binding motif for the WCC is "GATCGA," a sequence that is also found in the c-box and LRE of the *frq* promoter (Froehlich et al. 2002; Pavesi et al. 2004; He and Liu 2005b; Carlson et al. 2007; Smith et al. 2010).

Deep sequencing has shown that the transcriptional regulation stemming from the positive arm leads to up to 40% of genes within *N. crassa* undergoing rhythmic changes in transcript levels under constant conditions; these genes are termed clock-controlled genes or *ccgs* (Fig. 2a) (Hurley et al. 2014; Loros et al. 1989; Sancar et al. 2015). When the promoter sequences of all detected *ccgs* were analyzed, four motifs were found to be significantly enriched at certain times of day: STACASTA, GCRCTAAC, GRCGGGA, and GVCAGCCA, with each motif associated with essential cellular processes (Hurley et al. 2014). It is important to note that the set of *ccgs* varies to some extent depending on what growth media is used, showing that the clock can adjust its output to current metabolic conditions (Hurley et al. 2014). The extent and diversity of *ccgs* highlights the broad effect of circadian regulation in the cell and how the clock can fine-tune cellular output based on its metabolic environment. The diversity of output is also seen in higher eukaryotes, where as much as 80% of genes fell under circadian regulation when a wide variety of tissues were analyzed (Mure et al. 2018).

2. Circadian Transcriptional Repression of *ccgs* Is Vital to Clock Output

While the positive arm is generally considered a transcriptional activator, some of its direct or indirect targets are known transcriptional repressors. We have already reviewed some of these transcriptional repressors in the context of input signals to the clock (see Sect. II for discussions of VVD, CSP-1, RCO-1, and RCM-1). In addition to being a clock nutrient sensor and WC-1 transcriptional repressor, CSP-1 is also a direct target of the WCC and is expressed in a rhythmic fashion (Lambreghts et al. 2009; Smith et al. 2010). Forming a complex with RCO-1 and RCM-1, it modulates the expression of ~800, mostly evening-specific, genes in *N. crassa* that are predominantly involved in the metabolism of lipids and glucose (Sancar et al. 2011; see Sect. II). It is predicted that CSP-1 regulates evening-specific gene expression by morning-specific repression of CSP-1 target genes (Sancar et al. 2011).

Carbon catabolite regulation-1 (CRE-1) is also a WCC-regulated repressor that interacts with RCO-1 and RCM-1 to inhibit the transcription of genes related to the use of alternative carbon sources when glucose is present, otherwise known as **carbon catabolite repression** (Bailey and Arst 1975; Hurley et al. 2014; Cupertino et al. 2015; Adnan et al. 2017). Among other things, CRE-1 is an inhibitor of glycogen synthesis and represses glycogenic genes, such as *gsn*, *gbn*, and *gnn* (Cupertino et al. 2015). Interestingly, the WCC and other WCC-target transcription factors, such as Viability Of Spores-1 (VOS-1) and CSP-1, also target these glycogenic genes, highlighting the complexity of clock regulation of *ccgs* (Smith et al. 2010; Baek et al. 2019). An additional example, the NOT1-CCR4 (negative on TATA 1-carbon catabolite repression 4) complex regulates WCC stability and is important for circadian phase determination (Fig. 2a) (Huang et al. 2013). NOT1 acts directly on the WCC and its mRNA displays circadian oscillations in a similar phase to *frq* (Huang et al. 2013). When NOT1 or CCR4 levels are reduced, the clock's phase is delayed by 3–4 h (Huang et al. 2013). These examples show that not all clock output involves transcriptional activation but that transcriptional repressors are also important for the clock, including some clock-regulated repressors that feedback onto the core clock itself.

B. Post-transcriptional Regulation via the Circadian Clock

While transcriptional regulation has long been considered the primary output of the circadian clock, recent research suggests that clock output is also regulated post-transcriptionally. The initial evidence for this **post-transcriptional regulation** is the mismatch between promoter activation and steady state mRNA levels for a given gene (Fig. 3a). There are many genes whose promoters are rhythmically activated but whose steady-state mRNA levels remain constant and vice versa (Hurley et al. 2014). This post-transcriptional regulation occurs at the protein level as well, as approximately one quarter of reliably detected proteins oscillate with a circadian period but 40% of these arise from transcripts that are not oscillating (Fig. 3a) (Hurley et al. 2018). Moreover, post-transcriptional regulation appears to be circadianly directed. For example, differential peak phase timing of enzymes that are rhythmic only at the protein level within central metabolic pathways yields a coordination of peak timing within pathways such as glycolysis and the TCA cycle that peak anti-phase to the alternate pentose-phosphate pathway (Hurley et al. 2018). Many similar studies in higher eukaryotes demonstrate that circadian post-transcriptional regulation is conserved (Beckwith and Yanovsky 2014; Koike et al. 2012; Kojima et al. 2011).

There are several circadianly regulated mechanisms that impart circadian post-transcriptional output, including mRNA degradation, translation rates, and protein degradation. The FFC interacts with components of the exosome, which facilitate a circadian influence over mRNA degradation (Guo et al. 2009). The clock also regulates **MAPK pathways** that impact the phosphorylation state of translation factors, such as translation elongation factor eEF-2 (Lamb et al. 2011, 2012; Bennett et al. 2013; Caster et al. 2016). eEF-2 has lower phosphorylation levels and therefore higher activity near subjective dusk, a time when there seems to be more energy available for translation, as well as more translation initiation factors and ribosomes (Fig. 3b) (Caster et al. 2016; Hurley et al. 2018; Kafri et al. 2016). Finally, circadian regulation of E3 ubiquitin ligases could yield rhythms in protein degradation (Lück et al. 2014; Hurley et al. 2018). As clock regulation affects numerous additional cell regulatory systems, there may be many other avenues through which the clock can direct post-transcriptional regulation (Hurley et al. 2014, 2018; Sancar et al. 2015).

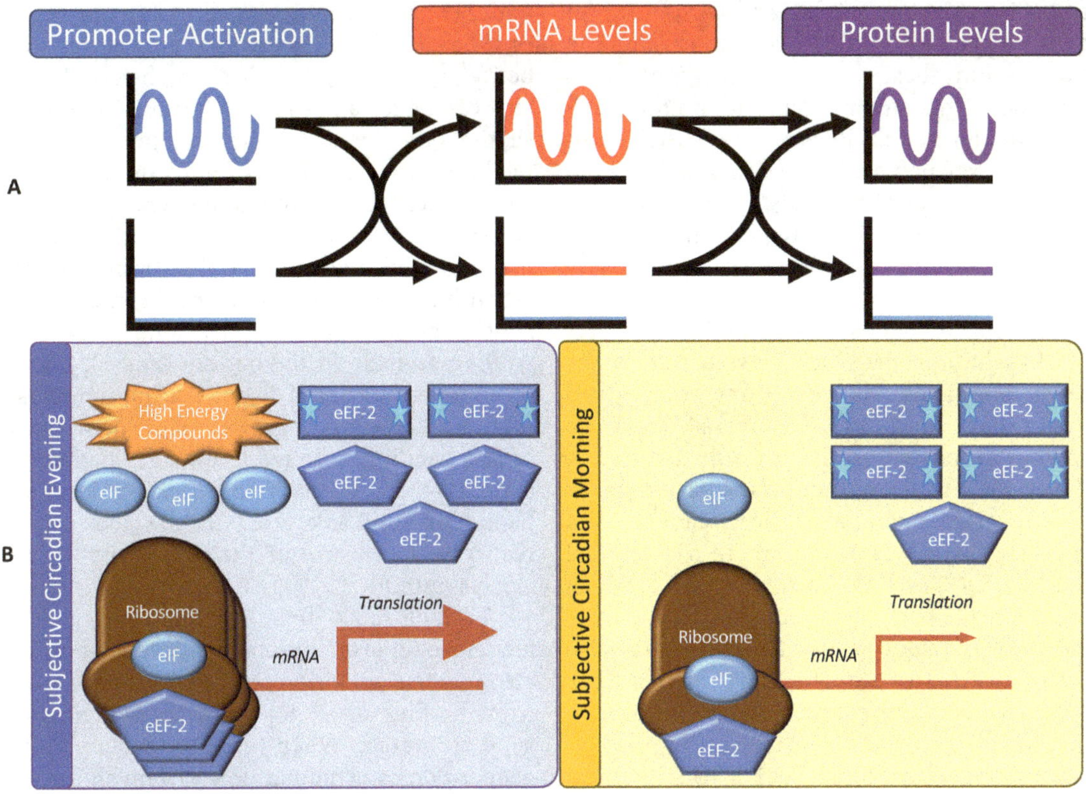

Fig. 3 Circadian output is highly impacted by post-transcriptional regulation. (**a**) Circadianly regulated/ rhythmic promoter activity or steady-state mRNA levels do not necessarily lead to oscillations at the level of the mRNA or protein. Similarly, oscillations at the level of mRNA or protein can happen despite constitutive promoter activity or steady-state mRNA levels. This suggests there is a significant amount of post-transcriptional regulation on circadian output. (**b**) There are some proposed mechanisms that may underlie circadian post-transcriptional regulation. First, excess amounts of high energy compounds, active ribosomes, and eukaryotic initiation factors (eIF) are available in the subjective circadian evening in *Neurospora crassa*, when rhythmic proteins reach peak levels. Second, the *N. crassa* translation factor eukaryotic elongation factor 2 (eEF-2) is less phosphorylated (cyan stars) in the circadian evening, which increases the translation elongation rate, likely yielding more proteins reaching peak levels at this time

C. Circadian Output Interfaces with Many Cellular Systems

The output generated from the circadian oscillator regulates many vital processes within the cell, the foremost of which is cellular **metabolism**. There are connections between the clock and metabolism that result in changes in clock output depending on the type of growth media, as discussed in Sect. IV.A.1, as well as some ion cycles that may in fact be circadianly-regulated (Hurley et al. 2014, 2016b; Feeney et al. 2016; Dunlap and Loros 2016). Overall, gene ontologies related to metabolic functions are highly enriched for *ccgs* at both the transcript and protein levels (Hurley et al. 2014, 2018; Sancar et al. 2015). These include, at the transcriptional level, the division between anabolic and catabolic pathways between subjective night and subjective day, while at the protein level, there is an overall peak in enzymes in many energy pathways near circadian dusk (Hurley et al. 2014, 2018; Sancar et al. 2015). Other studies have looked at oscillations within the cell that may occur outside the circadian feedback loop involving the WCC and the FFC, though many of these oscillations are not in fact compensated and so cannot be considered truly circadian (see Dunlap and Loros 2017 for a recent review). While there is much more to learn about these cellular oscillations, it is interesting that studies continue to find previously unknown links between these oscillations and the circadian clock, such as a widely conserved daily oscillation in Mg^{2+}, which can now be explained by clock regulation of genes involved in both active ion transport (via pumps) and passive ion channels (Feeney et al. 2016; Dunlap and Loros 2016).

In addition, the clock plays a role in the regulation of the cell cycle. Hong et al. (2014) found that there was an overall circadian rhythm in **mitosis**, with most cells dividing in the evening, while morning is the peak time for cells to be in interphase. The circadian clock rhythmically regulates the kinase STK-29 and the G1 and G2 cyclins, CLN-1 and CLB-1, which in turn regulate entry into mitosis (Hong et al. 2014). The clock and cell cycle are also interconnected by Chk2 (PRD-4 in *N. crassa*), a key kinase in both processes, and it is through this kinase that the cell cycle can impact the clock (Pregueiro et al. 2006). Chk2 binds to FRQ and phosphorylates it in a DNA damage-dependent manner, triggering destabilization and leading to FRQ degradation, shifting the phase of the clock and enabling circadian responsiveness to DNA damage (Gamsby et al. 2009; Pregueiro et al. 2006). This research in *N. crassa* demonstrates that the cross-talk between the circadian clock and cellular systems contributes to the optimal timing of cellular growth and development, important relationships that are conserved and are now being elucidated in higher eukaryotes (e.g., Gaucher et al. 2018; Reinke and Asher 2019).

V. Methods for Detecting Circadian Rhythms

As noted earlier (see Sect. I), for something to be defined as a circadian rhythm, the period must be free running and at or near 24 h. Therefore, to identify circadian rhythms in *Neurospora crassa*, it is standard procedure to collect a time series of regularly recorded observations under constant conditions, after a strong resetting cue is given. From this time series, the period, amplitude, and phase can be calculated, as it can be for any oscillatory wave. To look at rhythms in vivo in *N. crassa*, two common methods are utilized: the "band" or *bd* mutant strain background and a luciferase reporter system. To track the period of the clock using the *bd* mutant background, the strain of interest with a *bd* background is grown in constant conditions on a race-tube, a long glass tube with agar media in the bottom (Fig. 1). Bands of conidia and aerial hyphae form at regular intervals as the strain grows and the periodicity of this banding is linked to the circadian clock (Brandt 1953; Pittendrigh et al. 1959). This banding is used as a proxy for the clock but in reality is a measure of the periodicity of the clock's output.

To look more directly at the core clock, a luciferase reporter system was developed using an *N. crassa* codon-optimized luciferase gene from fireflies fused to a promoter region or gene of interest. When fused to the core clock gene *frq* or its promoter, this system allows for the tracking of changes in steady-state protein levels or transcriptional activity, respectively, over the course of multiple days (Gooch et al. 2008; Larrondo et al. 2012; Morgan et al. 2003). With the advent of RNA microarrays, RNA-sequencing and mass spectrometry, ex vivo methods of tracking rhythms in mRNA or protein levels have allowed further investigation of circadian regulation within the cell. Samples are extracted from tissue grown under constant conditions, which has been shown to preserve the oscillations in clock genes and proteins in *N. crassa* (Loros et al. 1989). From the extracted samples, a variety of assays and computational approaches can be performed to measure rhythms in the core clock and identify clock output, including Western blots, qPCR, RNA-seq, and proteomics (e.g., Hurley et al. 2014, 2018). Novel techniques and technologies are developed in *N. crassa* on a regular basis to allow circadian biologists to learn more about the core clock and its regulation of cellular output (e.g., De los Santos et al. 2017; Pelham et al. 2018).

VI. Clock Conservation in Fungi

Neurospora crassa is not only a model for clocks in higher eukaryotes but is a model for clocks in other fungi. An analysis of 64 fungal genomes based on the *N. crassa* circadian genes *frq*, *wc-1*, *wc-2*, *frh*, and *fwd-1* (Salichos and Rokas 2010) showed that FRH and FWD-1 were the earliest conserved clock proteins, not surprising based on their other essential functions in the organism (Jonkers and Rep 2009; Putnam and Jankowsky 2013). WC-1 and WC-2 are further along the evolutionary tree and FRQ is the most recently evolved protein, with sequential homologs appearing in only three classes within the Ascomycetes: Sordariomycetes, Leotiomycetes, and Dothideomycetes (Salichos and Rokas 2010). However, as FRQ is a known IDP, and it has been documented that there is little conservation of sequence among IDPs, it is likely that the functional homologs of FRQ may not have sequence homology and that other negative arm proteins exist in the wider fungal families (Hurley et al. 2013). Though many species contain homologs of clock proteins, and some even display oscillatory phenotypes (e.g., *Podospora anserina*), so far only a few fungi that display measurable phenotypic or molecular rhythms have been shown to be truly circadian, such as the rhythms in Ascomycetes *Sordaria fimicola*, *Pyronema confluens*, *Botrytis cinerea*, *Aspergillus flavus*, and *Cercospora kikuchii* and the Basidiomycete *Neonothopanus gardneri* (Brandt 1953; Austin 1968; Lysek and Esser 1970; Greene et al. 2003; Bluhm et al. 2010; Hevia et al. 2015, 2016; Traeger and Nowrousian 2015).

B. cinerea has the best molecular and phenotypic evidence for a fungal circadian clock outside of *N. crassa* (Montenegro-Montero et al. 2015). It has orthologues of *frq*, *wc*-1, and *wc*-2 (*bcfrq1*, *bcwcl1*, and *bcwcl2*, respectively), and the clock has been shown to maximize virulence through the control of its sexual and asexual cycle (Schumacher and Tudzynski 2012; Canessa et al. 2013; Hevia et al. 2015). In contrast, while the *Aspergillus* family does not contain a recognized *frq* homolog, they do have *wc-1* and *wc-2* homologs and evidence of circadian regulation. The level of the *gpdA* gene may oscillate under free running conditions in *A. nidulans*, and *A. flavus* has rhythms in sclerotia formation (Greene et al. 2003). *Cercospora kikuchii*, of the Dothideomycetes subclass, displays phenotypic rhythms through concentric hyphal rings which persist on Petri dishes under free-running conditions for several days that are both temperature-compensated and dependent on a white collar-like gene (Bluhm et al. 2010). Through these examples it can be seen that the clock of *N. crassa* can be used to gain further insights into the clocks of other fungi.

VII. Summary

Neurospora crassa is a key model organism for research into the molecular mechanism of the circadian clock. All of the above discussed elements of clock function were, at least in part, discovered through the use of molecular and genetic techniques in *N. crassa*. The more that is understood about the clock in *N. crassa*, including the similarity of its basic architecture and circadian output to the clocks in higher eukaryotes, the more we understand what this organism has to contribute as a model system to studies of clocks in other eukaryotes. It is therefore likely that *N. crassa* will continue to contribute as a model for circadian and fungal research for years to come.

Acknowledgments This work was supported by a NIH-National Institute of General Medical Sciences T32 Training Grant GM067545 (M.S.J. and Z.A.C), a NIH-National Institute of Biomedical Imaging and Bioengineering Grant EB022546 (J.M.H.), a NIH-National Institute of General Medical Sciences Grant GM128687 (J.M.H.), and Rensselaer Polytechnic Startup funds (J.M.H.).

References

Adhvaryu K, Firoozi G, Motavaze K, Lakin-Thomas P (2016) PRD-1, a component of the circadian system of *Neurospora crassa*, is a member of the DEAD-box RNA helicase family. J Biol Rhythm 31:258–271. https://doi.org/10.1177/0748730416639717

Adnan M, Zheng W, Islam W et al (2017) Carbon catabolite repression in filamentous Fungi. Int J Mol Sci 19:48. https://doi.org/10.3390/ijms19010048

Aronson BD, Johnson KA, Dunlap JC (1994) Circadian clock locus frequency: protein encoded by a single open reading frame defines period length and temperature compensation. Proc Natl Acad Sci U S A 91:7683–7687

Austin B (1968) An endogenous rhythm of spore discharge in *Sordaria fimicola*. Ann Bot 32:262–278

Baek M, Virgilio S, Lamb TM et al (2019) Circadian clock regulation of the glycogen synthase (*gsn*) gene by WCC is critical for rhythmic glycogen metabolism in *Neurospora crassa*. Proc Natl Acad Sci 116:10435–10440. https://doi.org/10.1073/pnas.1815360116

Bailey C, Arst HNJ (1975) Carbon catabolite repression in *Aspergillus nidulans*. Eur J Biochem 51:573–577. https://doi.org/10.1111/j.1432-1033.1975.tb03958.x

Baker CL, Kettenbach AN, Loros JJ et al (2009) Quantitative proteomics reveals a dynamic Interactome and phase-specific phosphorylation in the *Neurospora* circadian clock. Mol Cell 34:354–363. https://doi.org/10.1016/j.molcel.2009.04.023

Ballario P, Macino G (1997) White collar proteins: PASsing the light signal in *Neurospora crassa*. Trends Microbiol 5:458–462. https://doi.org/10.1016/S0966-842X(97)01144-X

Ballario P, Vittorioso P, Magrelli A et al (1996) White collar-1, a central regulator of blue light responses in *Neurospora*, is a zinc finger protein. EMBO J 15:1650–1657. https://doi.org/10.1002/j.1460-2075.1996.tb00510.x

Ballario P, Talora C, Galli D et al (1998) Roles in dimerization and blue light photoresponse of the PAS and LOV domains of *Neurospora crassa* white collar proteins. Mol Microbiol 29:719–729. https://doi.org/10.1046/j.1365-2958.1998.00955.x

Beadle GW, Tatum EL (1941) Genetic control of biochemical reactions in *Neurospora*. Proc Natl Acad Sci 27:499

Beckwith EJ, Yanovsky MJ (2014) Circadian regulation of gene expression: at the crossroads of transcriptional and post-transcriptional regulatory networks. Curr Opin Genet Dev 27:35–42. https://doi.org/10.1016/j.gde.2014.03.007

Belden WJ, Larrondo LF, Froehlich AC et al (2007a) The band mutation in *Neurospora crassa* is a dominant allele of *ras-1* implicating RAS signaling in circadian output. Genes Dev 21:1494–1505. https://doi.org/10.1101/gad.1551707

Belden WJ, Loros JJ, Dunlap JC (2007b) Execution of the circadian negative feedback loop in *Neurospora* requires the ATP-dependent chromatin-remodeling enzyme CLOCKSWITCH. Mol Cell 25:587–600. https://doi.org/10.1016/j.molcel.2007.01.010

Belden WJ, Lewis ZA, Selker EU et al (2011) CHD1 remodels chromatin and influences transient DNA methylation at the clock gene frequency. PLoS Genet 7:e1002166. https://doi.org/10.1371/journal.pgen.1002166

Bell-Pedersen D, Cassone VM, Earnest DJ et al (2005) Circadian rhythms from multiple oscillators: lessons from diverse organisms. Nat Rev Genet 6:544–556. https://doi.org/10.1038/nrg1633

Bennett LD, Beremand P, Thomas TL, Bell-Pedersen D (2013) Circadian activation of the mitogen-activated protein kinase MAK-1 facilitates rhythms in clock-controlled genes in *Neurospora crassa*. Eukaryot Cell 12:59–69. https://doi.org/10.1128/EC.00207-12

Bluhm BH, Burnham AM, Dunkle LD (2010) A circadian rhythm regulating hyphal melanization in *Cercospora kikuchii*. Mycologia 102:1221–1228. https://doi.org/10.3852/09-041

Brandt WH (1953) Zonation in a prolineless strain of *Neurospora*. Mycologia 45:194–208

Buhr ED, Yoo S-H, Takahashi JS (2010) Temperature as a universal resetting cue for mammalian circadian oscillators. Science 330:379–385. https://doi.org/10.1126/science.1195262

Canessa P, Schumacher J, Hevia MA et al (2013) Assessing the effects of light on differentiation and virulence of the plant pathogen *Botrytis cinerea*: characterization of the white collar complex. PLoS One 8:e84223. https://doi.org/10.1371/journal.pone.0084223

Carlson JM, Chakravarty A, DeZiel CE, Gross RH (2007) SCOPE: a web server for practical de novo motif discovery. Nucleic Acids Res 35:W259–W264

Case ME, Griffith J, Dong W et al (2014) The aging biological clock in *Neurospora crassa*. Ecol Evol 4:3494–3507. https://doi.org/10.1002/ece3.1202

Caster SZ, Castillo K, Sachs MS, Bell-Pedersen D (2016) Circadian clock regulation of mRNA translation through eukaryotic elongation factor eEF-2. Proc Natl Acad Sci 113:9605–9610. https://doi.org/10.1073/pnas.1525268113

Cha J, Chang SS, Huang G et al (2008) Control of WHITE COLLAR localization by phosphorylation is a critical step in the circadian negative feedback process. EMBO J 27:3246–3255. https://doi.org/10.1038/emboj.2008.245

Cha J, Yuan H, Liu Y (2011) Regulation of the activity and cellular localization of the circadian clock protein FRQ. J Biol Chem 286:11469–11478. https://doi.org/10.1074/jbc.M111.219782

Cha J, Zhou M, Liu Y (2013) CATP is a critical component of the *Neurospora* circadian clock by regulating the nucleosome occupancy rhythm at the frequency locus. EMBO Rep 14:923–930. https://doi.org/10.1038/embor.2013.131

Chen CH, Ringelberg CS, Gross RH et al (2009) Genome-wide analysis of light-inducible responses reveals hierarchical light signalling in

Neurospora. EMBO J 28:1029–1042. https://doi.org/10.1038/emboj.2009.54

Chen C-H, DeMay BS, Gladfelter AS et al (2010) Physical interaction between VIVID and white collar complex regulates photoadaptation in *Neurospora*. Proc Natl Acad Sci 107:16715–16720

Cheng P, Yang Y, Heintzen C, Liu Y (2001a) Coiled-coil domain-mediated FRQ-FRQ interaction is essential for its circadian clock function in *Neurospora*. EMBO J 20:101–108. https://doi.org/10.1093/emboj/20.1.101

Cheng P, Yang Y, Liu Y (2001b) Interlocked feedback loops contribute to the robustness of the *Neurospora* circadian clock. Proc Natl Acad Sci U S A 98:7408–7413. https://doi.org/10.1073/pnas.121170298

Cheng P, Yang Y, Gardner KH, Liu Y (2002) PAS domain-mediated WC-1/WC-2 interaction is essential for maintaining the steady-state level of WC-1 and the function of both proteins in circadian clock and light responses of *Neurospora*. Mol Cell Biol 22:517–524. https://doi.org/10.1128/MCB.22.2.517-524.2002

Cheng P, Yang Y, Wang L et al (2003) WHITE COLLAR-1, a multifunctional *Neurospora* protein involved in the circadian feedback loops, light sensing, and transcription repression of *wc-2*. J Biol Chem 278:3801–3808. https://doi.org/10.1074/jbc.M209592200

Cheng P, He Q, He Q et al (2005) Regulation of the *Neurospora* circadian clock by an RNA helicase. Genes Dev 19:234–241. https://doi.org/10.1101/gad.1266805.234

Chiu JC, Ko HW, Edery I (2011) NEMO/NLK phosphorylates PERIOD to initiate a time-delay phosphorylation circuit that sets circadian clock speed. Cell 145:357–370. https://doi.org/10.1016/j.cell.2011.04.002

Collett MA, Garceau N, Dunlap JC, Loros JJ (2002) Light and clock expression of the *Neurospora* clock gene frequency is differentially driven by but dependent on WHITE COLLAR-2. Genetics 160:149–158

Colot HV, Loros JJ, Dunlap JC (2005) Temperature-modulated alternative splicing and promoter use in the circadian clock gene frequency. Mol Biol Cell 16:5563–5571. https://doi.org/10.1091/mbc.E05-08-0756

Conrad KS, Hurley JM, Widom J et al (2016) Structure of the frequency - interacting RNA helicase : a protein interaction hub for the circadian clock. EMBO J 35:4–7. https://doi.org/10.15252/embj.201694327

Crosthwaite SK, Dunlap JC, Loros JJ (1997) *Neurospora wc-1* and *wc-2*: transcription, Photoresponses, and the origins of circadian rhythmicity. Science 276 (80):763–769

Cupertino FB, Virgilio S, Freitas FZ et al (2015) Regulation of glycogen metabolism by the CRE-1, RCO-1 and RCM-1 proteins in *Neurospora crassa*. The role of CRE-1 as the central transcriptional regulator. Fungal Genet Biol 77:82–94. https://doi.org/10.1016/j.fgb.2015.03.011

Dasgupta A, Chen C-H, Lee C et al (2015) Biological significance of photoreceptor photocycle length: VIVID photocycle governs the dynamic VIVID-white collar complex pool mediating photo-adaptation and response to changes in light intensity. PLoS Genet 11:e1005215

De los Santos H, Collins EJ, Hurley JM, Bennett KP (2017) Circadian rhythms in *Neurospora* exhibit biologically relevant driven and damped harmonic oscillations. In: Proceedings of the 8th ACM international conference on bioinformatics, computational biology, and health informatics. ACM, New York, NY, pp 455–463

DeCoursey PJ (2014) Survival value of suprachiasmatic nuclei (SCN) in four wild sciurid rodents. Behav Neurosci 128:240–249

DeCoursey PJ, Krulas JR, Mele G, Holley DC (1997) Circadian performance of suprachiasmatic nuclei (SCN)-lesioned antelope ground squirrels in a desert enclosure. Physiol Behav 62:1099–1108. https://doi.org/10.1016/S0031-9384(97)00263-1

DeCoursey PJ, Walker JK, Smith SA (2000) A circadian pacemaker in free-living chipmunks: essential for survival? J Comp Physiol A 186:169–180. https://doi.org/10.1007/s003590050017

Denault DL, Loros JJ, Dunlap JC (2001) WC-2 mediates WC-1–FRQ interaction within the PAS protein-linked circadian feedback loop of *Neurospora*. EMBO J 20:109

Diernfellner ACR, Schafmeier T, Merrow MW, Brunner M (2005) Molecular mechanism of temperature sensing by the circadian clock of *Neurospora crassa*. Genes Dev 19:1968–1973. https://doi.org/10.1101/gad.345905

Diernfellner A, Colot HV, Dintsis O et al (2007) Long and short isoforms of *Neurospora* clock protein FRQ support temperature-compensated circadian rhythms. FEBS Lett 581:5759–5764. https://doi.org/10.1016/j.febslet.2007.11.043

Dovzhenok AA, Baek M, Lim S, Hong CI (2015) Mathematical modeling and validation of glucose compensation of the *Neurospora* circadian clock. Biophys J 108:1830–1839. https://doi.org/10.1016/j.bpj.2015.01.043

Dunlap JC (1999) Molecular bases for circadian clocks. Cell 96:271–290. https://doi.org/10.1016/S0092-8674(00)80566-8

Dunlap JC (2008) Salad days in the rhythms trade. Genetics 178:1–13

Dunlap JC, Loros JJ (2016) Yes, circadian rhythms actually do affect almost everything. Cell Res 26:759–760. https://doi.org/10.1038/cr.2016.65

Dunlap JC, Loros JJ (2017) Making time: conservation of biological clocks from fungi to animals. Microbiol Spectr 5(3). https://doi.org/10.1128/microbiolspec.FUNK-0039-2016

Dunlap JC, Loros JJ (2018) Just-so stories and origin myths: phosphorylation and structural disorder in circadian clock proteins. Mol Cell 69:165–168. https://doi.org/10.1016/j.molcel.2017.11.028

Dunlap JC, Borkovich KA, Henn MR et al (2007) Enabling a community to dissect an organism: overview of the *Neurospora* functional genomics project. Adv Genet 57:49–96. https://doi.org/10.1016/S0065-2660(06)57002-6

Eide EJ, Woolf MF, Kang H et al (2005) Control of mammalian circadian rhythm by CKI-regulated proteasome-mediated PER2 degradation. Mol Cell Biol 25:2795–2807. https://doi.org/10.1128/MCB.25.7.2795-2807.2005

Emerson JM, Bartholomai BM, Ringelberg CS et al (2015) Period-1 encodes an ATP-dependent RNA helicase that influences nutritional compensation of the *Neurospora* circadian clock. Proc Natl Acad Sci U S A 112:15707–15712. https://doi.org/10.1073/pnas.1521918112

Feeney KA, Hansen LL, Putker M et al (2016) Daily magnesium fluxes regulate cellular timekeeping and energy balance. Nature 532:375

Feldman JF, Hoyle MN (1973) Isolation of circadian clock mutants of *Neurospora crassa*. Genetics 75:605–613

Filippovich SY, Bachurina GP, Shcherbakov DL (2015) Quantitative assessment of the combined effect of the nitrogen status, light and dehydration of mycelium on conidiation in *Neurospora crassa*. Appl Biochem Microbiol 51:342–349. https://doi.org/10.1134/S0003683815030060

Froehlich AC, Liu Y, Loros JJ, Dunlap JC (2002) White Collar-1, a circadian blue light photoreceptor, binding to the frequency promoter. Science 297 (80):815–819

Froehlich AC, Loros JJ, Dunlap JC (2003) Rhythmic binding of a WHITE COLLAR-containing complex to the frequency promoter is inhibited by FREQUENCY. Proc Natl Acad Sci U S A 100:5914–5919. https://doi.org/10.1073/pnas.1030057100

Fu J, Murphy KA, Zhou M et al (2016) Codon usage affects the structure and function of the *Drosophila* circadian clock protein PERIOD. Genes Dev 30:1761–1775. https://doi.org/10.1101/gad.281030.116

Fuller KK, Loros JJ, Dunlap JC (2015) Fungal photobiology: visible light as a signal for stress, space and time. Curr Genet 61:275. https://doi.org/10.1007/s00294-014-0451-0

Galagan JE, Calvo SE, Borkovich KA et al (2003) The genome sequence of the filamentous fungus *Neurospora crassa*. Nature 422:859

Gamsby JJ, Loros JJ, Dunlap JC (2009) A phylogenetically conserved DNA damage response resets the circadian clock. J Biol Rhythm 24:193–202. https://doi.org/10.1177/0748730409334748

Garceau NY, Liu Y, Loros JJ, Dunlap JC (1997) Alternative initiation of translation and time-specific phosphorylation yield multiple forms of the essential clock protein FREQUENCY. Cell 89:469. https://doi.org/10.1016/S0092-8674(00)80227-5

Gardner GF, Feldman JF (1980) The frq locus in *Neurospora crassa*: a key element in circadian clock organization. Genetics 96:877–886

Gardner GF, Feldman JF (1981) Temperature compensation of circadian period length in clock mutants of *Neurospora crassa*. Plant Physiol 68:1244–1248

Gaucher J, Montellier E, Sassone-Corsi P (2018) Molecular cogs: interplay between circadian clock and cell cycle. Trends Cell Biol 28:368–379. https://doi.org/10.1016/j.tcb.2018.01.006

Gooch V, Mehra A, Larrondo L et al (2008) Fully codon-optimized luciferase uncovers novel temperature characteristics of the *Neurospora* clock. Eukaryot Cell 7(1):28–37

Gooch VD, Johnson AE, Bourne BJ et al (2014) A kinetic study of the effects of light on circadian rhythmicity of the frq promoter of *Neurospora crassa*. J Biol Rhythm 29:38–48. https://doi.org/10.1177/0748730413517981

Görl M, Merrow M, Huttner B et al (2001) A PEST-like element in FREQUENCY determines the length of the circadian period in *Neurospora crassa*. EMBO J 20:7074–7084. https://doi.org/10.1093/emboj/20.24.7074

Greene AV, Keller N, Haas H, Bell-Pedersen D (2003) A circadian oscillator in *Aspergillus* spp. regulates daily development and gene expression. Eukaryot Cell 2:231–237. https://doi.org/10.1128/EC.2.2.231-237.2003

Grima B, Lamouroux A, Chélot E et al (2002) The F-box protein Slimb controls the levels of clock proteins period and timeless. Nature 420:178

Guo J, Cheng P, Yuan H, Liu Y (2009) The exosome regulates circadian gene expression in a posttranscriptional negative feedback loop. Cell 138:1236–1246. https://doi.org/10.1016/j.cell.2009.06.043

Guo J, Cheng P, Liu Y (2010) Functional significance of FRH in regulating the phosphorylation and stability of *Neurospora* circadian clock protein FRQ. J Biol Chem 285:11508–11515. https://doi.org/10.1074/jbc.M109.071688

Gyöngyösi N, Szőke A, Ella K, Káldi K (2017) The small G protein RAS2 is involved in the metabolic compensation of the circadian clock in the circadian model *Neurospora crassa*. J Biol Chem 292:14929–14939. https://doi.org/10.1074/jbc.M117.804922

He Q, Liu Y (2005a) Degradation of the *Neurospora* circadian clock protein FREQUENCY through the ubiquitin–proteasome pathway. Biochem Soc Trans 33:953

He Q, Liu Y (2005b) Molecular mechanism of light responses in *Neurospora*: from light-induced transcription to photoadaptation. Genes Dev 19:2888–2899. https://doi.org/10.1101/gad.1369605

He Q, Cheng P, Yang Y et al (2002) White Collar-1, a DNA binding transcription factor and a light sensor. Science 297(80):840–843

He Q, Cheng P, Yang Y et al (2003) FWD1-mediated degradation of FREQUENCY in *Neurospora* establishes a conserved mechanism for circadian clock regulation. EMBO J 22:4421–4430. https://doi.org/10.1093/emboj/cdg425

He Q, Shu H, Cheng P et al (2005) Light-independent phosphorylation of WHITE COLLAR-1 regulates its function in the *Neurospora* circadian negative feedback loop. J Biol Chem 280:17526–17532. https://doi.org/10.1074/jbc.M414010200

He Q, Cha J, He Q et al (2006) CKI and CKII mediate the FREQUENCY-dependent phosphorylation of the WHITE COLLAR complex to close the *Neurospora* circadian negative feedback loop. Genes Dev 20:2552–2565. https://doi.org/10.1101/gad.1463506

Heintzen C, Loros JJ, Dunlap JC (2001) The PAS protein VIVID defines a clock-associated feedback loop that represses light input, modulates gating, and regulates clock resetting. Cell 104:453–464. https://doi.org/10.1016/S0092-8674(01)00232-X

Hevia MA, Canessa P, Müller-Esparza H, Larrondo LF (2015) A circadian oscillator in the fungus Botrytis cinerea regulates virulence when infecting *Arabidopsis thaliana*. Proc Natl Acad Sci 112:8744–8749

Hevia MA, Canessa P, Larrondo LF (2016) Circadian clocks and the regulation of virulence in fungi: getting up to speed. Semin Cell Dev Biol 57:147–155. https://doi.org/10.1016/j.semcdb.2016.03.021

Hirano A, Fu Y-H, Ptáček LJ (2016) The intricate dance of post-translational modifications in the rhythm of life. Nat Struct Mol Biol 23:1053

Hong CI, Ruoff P, Loros JJ, Dunlap JC (2008) Closing the circadian negative feedback loop: FRQ-dependent clearance of WC-1 from the nucleus. Genes Dev 22:3196–3204. https://doi.org/10.1101/gad.1706908

Hong CI, Zamborszky J, Baek M et al (2014) Circadian rhythms synchronize mitosis in *Neurospora crassa*. Proc Natl Acad Sci 111:1397–1402. https://doi.org/10.1073/pnas.1319399111

Huang G, Chen S, Li S et al (2007) Protein kinase a and casein kinases mediate sequential phosphorylation events in the circadian negative feedback loop. Genes Dev 21:3283–3295. https://doi.org/10.1101/gad.1610207

Huang G, He Q, Guo J et al (2013) The Ccr4-not protein complex regulates the phase of the *Neurospora* circadian clock by controlling WHITE COLLAR protein stability and activity. J Biol Chem 288:31002–31009. https://doi.org/10.1074/jbc.M113.494120

Hunt SM, Thompson S, Elvin M, Heintzen C (2010) VIVID interacts with the WHITE COLLAR complex and FREQUENCY-interacting RNA helicase to alter light and clock responses in *Neurospora*. Proc Natl Acad Sci 107:16709–16714. https://doi.org/10.1073/pnas.1009474107

Hurley JM, Larrondo LF, Loros JJ, Dunlap JC (2013) Conserved RNA helicase FRH acts nonenzymatically to support the intrinsically disordered *Neurospora* clock protein FRQ. Mol Cell 52:832–843. https://doi.org/10.1016/j.molcel.2013.11.005

Hurley JM, Dasgupta A, Emerson JM et al (2014) Analysis of clock-regulated genes in *Neurospora* reveals widespread posttranscriptional control of metabolic potential. Proc Natl Acad Sci 111:16995–17002. https://doi.org/10.1073/pnas.1418963111

Hurley JM, Loros JJ, Dunlap JC (2016a) Circadian oscillators: around the transcription–translation feedback loop and on to output. Trends Biochem Sci 41:834–846. https://doi.org/10.1016/j.tibs.2016.07.009

Hurley JM, Loros JJ, Dunlap JC (2016b) The circadian system as an organizer of metabolism. Fungal Genet Biol 90:39–43. https://doi.org/10.1016/j.fgb.2015.10.002

Hurley JM, Jankowski MS, De los Santos H et al (2018) Circadian proteomic analysis uncovers mechanisms of post-transcriptional regulation in metabolic pathways. Cell Syst 7:613. https://doi.org/10.1016/j.cels.2018.10.014

Hut RA, Beersma DGM (2011) Evolution of timekeeping mechanisms: early emergence and adaptation to photoperiod. Philos Trans R Soc Lond Ser B Biol Sci 366:2141–2154. https://doi.org/10.1098/rstb.2010.0409

Idnurm A, Verma S, Corrochano LM (2010) A glimpse into the basis of vision in the kingdom Mycota. Fungal Genet Biol 47:881–892. https://doi.org/10.1016/j.fgb.2010.04.009

Janich P, Arpat AB, Castelo-Szekely V et al (2015) Ribosome profiling reveals the rhythmic liver translatome and circadian clock regulation by upstream open reading frames. Genome Res 25:1848–1859. https://doi.org/10.1101/gr.195404.115

Johnson CH, Elliott J, Russell F et al (2004) Fundamental properties of circadian rhythms. In: Dunlap JC, Loros JJ, DeCoursey PJ (eds) Chronobiology: biological timekeeping. Sinauer Associates, Sunderland, MA, pp 67–106

Jonkers W, Rep M (2009) Lessons from fungal F-box proteins. Eukaryot Cell 8:677–695. https://doi.org/10.1128/EC.00386-08

Kafri M, Metzl-Raz E, Jona G, Barkai N (2016) The cost of protein production. Cell Rep 14:22–31. https://doi.org/10.1016/j.celrep.2015.12.015

Ko HW, Jiang J, Edery I (2002) Role for Slimb in the degradation of *Drosophila* period protein phosphorylated by Doubletime. Nature 420:673

Koike N, Yoo S-H, Huang H-C et al (2012) Transcriptional architecture and chromatin landscape of the core circadian clock in mammals. Science 338:349–354. https://doi.org/10.1126/science.1226339

Kojima S, Shingle DL, Green CB (2011) Post-transcriptional control of circadian rhythms. J Cell Sci 124:311–320. https://doi.org/10.1242/jcs.065771

Kramer C, Loros JJ, Dunlap JC, Crosthwaite SK (2003) Role for antisense RNA in regulating circadian clock function in *Neurospora crassa*. Nature 421:948

Lamb TM, Goldsmith CS, Bennett L et al (2011) Direct transcriptional control of a p38 MAPK pathway by the circadian clock in *Neurospora crassa*. PLoS One 6:e27149–e27149. https://doi.org/10.1371/journal.pone.0027149

Lamb TM, Finch KE, Bell-Pedersen D (2012) The *Neurospora crassa* OS MAPK pathway-activated transcription factor ASL-1 contributes to circadian rhythms in pathway responsive clock-controlled genes. Fungal Genet Biol 49:180–188. https://doi.org/10.1016/j.fgb.2011.12.006

Lambreghts R, Shi M, Belden WJ et al (2009) A high-density single nucleotide polymorphism map for *Neurospora crassa*. Genetics 181:767–781. https://doi.org/10.1534/genetics.108.089292

Larrondo LF, Loros JJ, Dunlap JC (2012) High-resolution spatiotemporal analysis of gene expression in real time: in vivo analysis of circadian rhythms in *Neurospora crassa* using a FREQUENCY-luciferase translational reporter. Fungal Genet Biol 49:681–683. https://doi.org/10.1016/j.fgb.2012.06.001

Larrondo LF, Olivares-Yañez C, Baker CL et al (2015) Decoupling circadian clock protein turnover from circadian period determination. Science 347:1257277. https://doi.org/10.1126/science.1257277

Lee K, Loros JJ, Dunlap JC (2000) Interconnected feedback loops in the *Neurospora* circadian system. Science 289(80):107–110. https://doi.org/10.1126/science.289.5476.107

Lee H, Chen R, Lee Y et al (2009) Essential roles of CKIδ and CKIε in the mammalian circadian clock. Proc Natl Acad Sci 106:21359. https://doi.org/10.1073/pnas.0906651106

Lee E, Jeong EH, Jeong H-J et al (2014) Phosphorylation of a central clock transcription factor is required for thermal but not photic entrainment. PLoS Genet 10:e1004545–e1004545. https://doi.org/10.1371/journal.pgen.1004545

Li N, Joska TM, Ruesch CE, Coster SJ, Belden WJ (2015) The natural antisense transcript first promotes, then represses, gene expression via facultative heterochromatin. Proc Natl Acad Sci USA 112(14):4357–4362. https://doi.org/10.1073/pnas.1406130112

Linden H, Macino G (1997) White collar 2, a partner in blue-light signal transduction, controlling expression of light–regulated genes in *Neurospora crassa*. EMBO J 16:98–109. https://doi.org/10.1093/emboj/16.1.98

Linden H, Ballario P, Macino G (1997) Blue light regulation in *Neurospora crassa*. Fungal Genet Biol 22:141–150. https://doi.org/10.1006/fgbi.1997.1013

Liu Y, Garceau NY, Loros JJ, Dunlap JC (1997) Thermally regulated translational control of FRQ mediates aspects of temperature responses in the *Neurospora* circadian clock. Cell 89:477–486. https://doi.org/10.1016/S0092-8674(00)80228-7

Liu Y, Merrow M, Loros JJ, Dunlap JC (1998) How temperature changes reset a circadian oscillator. Science 281(80):825

Liu Y, Loros J, Dunlap JC (2000) Phosphorylation of the *Neurospora* clock protein FREQUENCY determines its degradation rate and strongly influences the period length of the circadian clock. Proc Natl Acad Sci U S A 97:234–239

Loros JJ (2019) Principles of the animal molecular clock learned from *Neurospora*. Eur J Neurosci 51:19. https://doi.org/10.1111/ejn.14354

Loros JJ, Richman A, Feldman J (1986) A recessive circadian clock mutation at the *frq* locus of *Neurospora crassa*. Genetics 114(4):1095–1110

Loros JJ, Denome SA, Dunlap JC (1989) Molecular cloning of genes under control of the circadian clock in *Neurospora*. Science 243(80):385–388. https://doi.org/10.1126/science.2563175

Lück S, Thurley K, Thaben PF, Westermark PO (2014) Rhythmic degradation explains and unifies circadian transcriptome and proteome data. Cell Rep 9:741–751. https://doi.org/10.1016/j.celrep.2014.09.021

Luo C, Loros JJ, Dunlap JC (1998) Nuclear localization is required for function of the essential clock protein FRQ. Embo 17:1228–1235

Lykke-Andersen S, Tomecki R, Jensen TH, Dziembowski A (2011) The eukaryotic RNA exosome: same scaffold but variable catalytic subunits. RNA Biol 8:61–66

Lysek G, Esser K (1970) Rhythmic mycelial growth in *Podospora anserina*. Arch Mikrobiol 73:224–230. https://doi.org/10.1007/BF00410623

Mahesh G, Jeong E, Ng FS et al (2014) Phosphorylation of the transcription activator CLOCK regulates progression through a ~ 24-h feedback loop to influence the circadian period in *Drosophila*. J Biol Chem 289:19681–19693. https://doi.org/10.1074/jbc.M114.568493

Malzahn E, Ciprianidis S, Káldi K et al (2010) Photoadaptation in *Neurospora* by competitive interaction of activating and inhibitory LOV domains. Cell 142:762–772. https://doi.org/10.1016/j.cell.2010.08.010

McCluskey K, Wiest A, Plamann M (2010) The fungal genetics stock center: a repository for 50 years of fungal genetics research. J Biosci 35:119

Mehra A, Baker C, Loros J, Dunlap J (2009a) Post-translational modification in circadian rhythms. Trends Biochem Sci 34:483–490. https://doi.org/10.1016/j.tibs.2009.06.006.Post-Translational

Mehra A, Shi M, Baker CL et al (2009b) A role for casein kinase 2 in the mechanism underlying circadian temperature compensation. Cell 137:749–760. https://doi.org/10.1016/j.cell.2009.03.019

Menet JS, Pescatore S, Rosbash M (2014) CLOCK: BMAL1 is a pioneer-like transcription factor. Genes Dev 28:8–13. https://doi.org/10.1101/gad.228536.113

Merrow MW, Dunlap JC (1994) Intergeneric complementation of a circadian rhythmicity defect: phylogenetic conservation of structure and function of the clock gene frequency. EMBO J 13:2257–2266. https://doi.org/10.1002/j.1460-2075.1994.tb06507.x

Merrow MW, Garceau NY, Dunlap JC (1997) Dissection of a circadian oscillation into discrete domains. Proc Natl Acad Sci 94:3877–3882. https://doi.org/10.1073/pnas.94.8.3877

Montenegro-Montero A, Canessa P, Larrondo LF (2015) Chapter four - Around the fungal clock: recent advances in the molecular study of circadian clocks in *Neurospora* and other Fungi. In: Friedmann T, Dunlap JC, Goodwin SF (eds) Advances in genetics. Academic Press, Cambridge, MA, pp 107–184

Morales Y, Olsen KJ, Bulcher JM, Johnson SJ (2018) Structure of frequency-interacting RNA helicase from *Neurospora crassa* reveals high flexibility in a domain critical for circadian rhythm and RNA surveillance. PLoS One 13:e0196642

Morgan LW, Greene AV, Bell-Pedersen D (2003) Circadian and light-induced expression of luciferase in *Neurospora crassa*. Fungal Genet Biol 38:327–332. https://doi.org/10.1016/S1087-1845(02)00562-5

Mure LS, Le HD, Benegiamo G et al (2018) Diurnal transcriptome atlas of a primate across major neural and peripheral tissues. Science 359(80): eaao0318. https://doi.org/10.1126/science.aao0318

Ode KL, Ukai H, Susaki EA et al (2017) Knockout-rescue embryonic stem cell-derived mouse reveals circadian-period control by quality and quantity of CRY1. Mol Cell 65:176–190. https://doi.org/10.1016/j.molcel.2016.11.022

Olivares-Yañez C, Emerson J, Kettenbach A et al (2016) Modulation of circadian gene expression and metabolic compensation by the RCO-1 corepressor of *Neurospora crassa*. Genetics 204:163–176. https://doi.org/10.1534/genetics.116.191064

Ouyang Y, Andersson CR, Kondo T et al (1998) Resonating circadian clocks enhance fitness in cyanobacteria. Proc Natl Acad Sci 95:8660–8664. https://doi.org/10.1073/pnas.95.15.8660

Padmanabhan K, Robles MS, Westerling T, Weitz CJ (2012) Feedback regulation of transcriptional termination by the mammalian circadian clock PERIOD complex. Science 337(80):599–602. https://doi.org/10.1126/science.1221592

Panda S, Hogenesch JB, Kay SA (2002) Circadian rhythms from flies to human. Nature 417:329

Pavesi G, Mereghetti P, Mauri G, Pesole G (2004) Weeder web: discovery of transcription factor binding sites in a set of sequences from co-regulated genes. Nucleic Acids Res 32:W199–W203. https://doi.org/10.1093/nar/gkh465

Peek CB, Ramsey KM, Marcheva B, Bass J (2012) Nutrient sensing and the circadian clock. Trends Endocrinol Metab 23:312–318. https://doi.org/10.1016/j.tem.2012.02.003

Pelham JF, Mosier AE, Hurley JM (2018) Characterizing time-of-day conformational changes in the intrinsically disordered proteins of the circadian clock, 1st edn. Elsevier, Cambridge

Pittendrigh CS, Bruce VG, Rosenweig NS, Rubin ML (1959) A biological clock in *Neurospora*. Nature 184:169–170. https://doi.org/10.1038/184169a0

Pregueiro AM, Price-Lloyd N, Bell-Pedersen D et al (2005) Assignment of an essential role for the *Neurospora frequency* gene in circadian entrainment to temperature cycles. Proc Natl Acad Sci U S A 102:2210

Pregueiro AM, Liu Q, Baker CL et al (2006) The *Neurospora* checkpoint kinase 2: a regulatory link between the circadian and cell cycles. Science 313 (80):644–649

Putnam AA, Jankowsky E (2013) DEAD-box helicases as integrators of RNA, nucleotide and protein binding. Biochim Biophys Acta 1829:884–893. https://doi.org/10.1016/j.bbagrm.2013.02.002

Querfurth C, Diernfellner ACR, Gin E et al (2011) Circadian conformational change of the *Neurospora* clock protein FREQUENCY triggered by clustered hyperphosphorylation of a basic domain. Mol Cell 43:713–722. https://doi.org/10.1016/j.molcel.2011.06.033

Raduwan H, Isola AL, Belden WJ (2013) Methylation of histone H3 on lysine 4 by the lysine methyltransferase SET1 protein is needed for normal clock gene expression. J Biol Chem 288:8380–8390. https://doi.org/10.1074/jbc.M112.359935

Ramsdale M, Lakin-Thomas PL (2000) sn-1,2-diacylglycerol levels in the fungus *Neurospora crassa* display circadian rhythmicity. J Biol Chem 275 (36):27541–27550

Reinke H, Asher G (2019) Crosstalk between metabolism and circadian clocks. Nat Rev Mol Cell Biol 20:227–241. https://doi.org/10.1038/s41580-018-0096-9

Reischl S, Kramer A (2011) Kinases and phosphatases in the mammalian circadian clock. FEBS Lett 585:1393–1399. https://doi.org/10.1016/j.febslet.2011.02.038

Salichos L, Rokas A (2010) The diversity and evolution of circadian clock proteins in fungi. Mycologia 102:269–278

Sancar G, Sancar C, Brunner M, Schafmeier T (2009) Activity of the circadian transcription factor white collar complex is modulated by phosphorylation of SP-motifs. FEBS Lett 583:1833–1840. https://doi.org/10.1016/j.febslet.2009.04.042

Sancar G, Sancar C, Brügger B et al (2011) A global circadian repressor controls antiphasic expression of metabolic genes in neurospora. Mol Cell 44:687–697. https://doi.org/10.1016/j.molcel.2011.10.019

Sancar G, Sancar C, Brunner M (2012) Metabolic compensation of the *Neurospora* clock by a glucose-dependent feedback of the circadian repressor

CSP1 on the core oscillator. Genes Dev 26:2435–2442. https://doi.org/10.1101/gad.199547.112

Sancar C, Sancar G, Ha N et al (2015) Dawn- and dusk-phased circadian transcription rhythms coordinate anabolic and catabolic functions in *Neurospora*. BMC Biol 13:17–19. https://doi.org/10.1186/s12915-015-0126-4

Schafmeier T, Haase A, Káldi K et al (2005) Transcriptional feedback of *Neurospora* circadian clock gene by phosphorylation-dependent inactivation of its transcription factor. Cell 122:235–246. https://doi.org/10.1016/j.cell.2005.05.032

Schafmeier T, Káldi K, Diernfellner A et al (2006) Phosphorylation-dependent maturation of *Neurospora* circadian clock protein from a nuclear repressor toward a cytoplasmic activator. Genes Dev 20:297–306. https://doi.org/10.1101/gad.360906

Schafmeier T, Diernfellner A, Schäfer A et al (2008) Circadian activity and abundance rhythms of the *Neurospora* clock transcription factor WCC associated with rapid nucleo-cytoplasmic shuttling. Genes Dev 22:3397–3402. https://doi.org/10.1101/gad.507408

Schumacher J, Tudzynski P (2012) In: Pérez-Martín J, Di Pietro A (eds) Morphogenesis and infection in *Botrytis cinerea* BT - morphogenesis and pathogenicity in Fungi. Springer, Berlin, pp 225–241

Shi M, Collett M, Loros JJ, Dunlap JC (2010) FRQ-interacting RNA helicase mediates negative and positive feedback in the *Neurospora* circadian clock. Genetics 184:351–361. https://doi.org/10.1534/genetics.109.111393

Smith KM, Sancar G, Dekhang R et al (2010) Transcription factors in light and circadian clock signaling networks revealed by genomewide mapping of direct targets for *Neurospora* white collar complex. Eukaryot Cell 9:1549–1556. https://doi.org/10.1128/EC.00154-10

Starkey N, McDonald J, Lee K (2013) PERIOD-1 is a component of the *Neurospora* metabolic oscillator. Lett Gen Microbiol 1:13–15

Sun G, Zhou Z, Liu X et al (2016) Suppression of WHITE COLLAR-independent frequency transcription by histone H3 lysine 36 methyltransferase SET-2 is necessary for clock function in *Neurospora*. J Biol Chem 291:11055–11063. https://doi.org/10.1074/jbc.M115.711333

Takahashi JS (2017) Transcriptional architecture of the mammalian circadian clock. Nat Rev Genet 18:164–179. https://doi.org/10.1038/nrg.2016.150

Tan Y, Merrow M, Roenneberg T (2004) Photoperiodism in *Neurospora crassa*. J Biol Rhythm 19:135–143. https://doi.org/10.1177/0748730404263015

Tang C-T, Li S, Long C et al (2009) Setting the pace of the *Neurospora* circadian clock by multiple independent FRQ phosphorylation events. Proc Natl Acad Sci U S A 106:10722–10727. https://doi.org/10.1073/pnas.0904898106

Tataroğlu Ö, Lauinger L, Sancar G et al (2012) Glycogen synthase kinase is a regulator of the circadian clock of *Neurospora crassa*. J Biol Chem 287:36936–36943. https://doi.org/10.1074/jbc.M112.396622

Traeger S, Nowrousian M (2015) Analysis of circadian rhythms in the basal filamentous ascomycete *Pyronema confluens*. G3 Genes|Genomes|Genetics 5:2061–2071. https://doi.org/10.1534/g3.115.020461

Tsvetkov P, Reuven N, Shaul Y (2009) The nanny model for IDPs. Nat Chem Biol 5:778

Vitalini MW, de Paula RM, Park WD, Bell-Pedersen D (2006) The rhythms of life: circadian output pathways in *Neurospora*. J Biol Rhythm 21:432–444. https://doi.org/10.1177/0748730406294396

Wang B, Kettenbach AN, Gerber SA et al (2014) *Neurospora* WC-1 recruits SWI/SNF to remodel frequency and initiate a circadian cycle. PLoS Genet 10:e1004599

Wang B, Zhou XY, Loros JJ, Dunlap JC (2016) Alternative use of DNA binding domains by the *Neurospora* white collar complex dictates circadian regulation and light responses. Mol Cell Biol 36:781–793. https://doi.org/10.1128/MCB.00841-15

Wang B, Kettenbach AN, Zhou X et al (2019) The Phospho-code determining circadian feedback loop closure and output in *Neurospora*. Mol Cell 74:771–784.e3. https://doi.org/10.1016/j.molcel.2019.03.003

Weber F, Zorn D, Rademacher C, Hung H-C (2011) Post-translational timing mechanisms of the *Drosophila* circadian clock. FEBS Lett 585:1443–1449. https://doi.org/10.1016/j.febslet.2011.04.008

Wu C, Yang F, Smith KM et al (2014) Genome-wide characterization of light-regulated genes in *Neurospora crassa*. G3 Genes|Genomes|Genetics 4:1731–1745. https://doi.org/10.1534/g3.114.012617

Xu Y, Ma P, Shah P et al (2013) Non-optimal codon usage is a mechanism to achieve circadian clock conditionality. Nature 495:116–120. https://doi.org/10.1038/nature11942

Xue Z, Ye Q, Anson SR et al (2014) Transcriptional interference by antisense RNA is required for circadian clock function. Nature 514:650–653. https://doi.org/10.1038/nature13671

Yang Y, Cheng P, Zhi G, Liu Y (2001) Identification of a calcium/calmodulin-dependent protein kinase that phosphorylates the *Neurospora* circadian clock protein FREQUENCY. J Biol Chem 276:41064–41072. https://doi.org/10.1074/jbc.M106905200

Yang Y, Cheng P, Liu Y (2002) Regulation of the *Neurospora* circadian clock by casein kinase II. Genes Dev 16:994–1006. https://doi.org/10.1101/gad.965102

Yang Y, He Q, Cheng P (2004) Distinct roles for PP1 and PP2A in the *Neurospora* circadian clock. Genes Dev 18:255–260. https://doi.org/10.1101/gad.1152604

Yoshitane H, Takao T, Satomi Y et al (2009) Roles of CLOCK phosphorylation in suppression of E-box-dependent transcription. Mol Cell Biol 29:3675–3686. https://doi.org/10.1128/MCB.01864-08

Yu Z, Fischer R (2019) Light sensing and responses in fungi. Nat Rev Microbiol 17:25–36. https://doi.org/10.1038/s41579-018-0109-x

Zhang Y, Smith CL, Saha A et al (2006) DNA translocation and loop formation mechanism of chromatin remodeling by SWI/SNF and RSC. Mol Cell 24:559–568. https://doi.org/10.1016/j.molcel.2006.10.025

Zhou M, Guo J, Cha J et al (2013) Non-optimal codon usage affects expression, structure and function of clock protein FRQ. Nature 494:111–115. https://doi.org/10.1038/nature11833

Zoltowski BD, Schwerdtfeger C, Widom J et al (2007) Conformational switching in the fungal light sensor Vivid. Science 316:1054–1057. https://doi.org/10.1126/science.1137128

5 Small RNAs in Fungi

Francisco E. Nicolás[1], Laura Murcia[1], Eusebio Navarro[1], José T. Cánovas-Márquez[1], Victoriano Garre[1]

CONTENTS

I. Introduction

Fungi are one of the oldest groups of organisms on the Earth, where they play pivotal roles in maintaining life by participating in nutrient recycling and symbiosis with plants. Despite their benefits to the planet, they also put in risk the survival of other species, including humans, when they are genuine or opportunistic pathogens. Besides, we use several fungi for our own benefit in different aspects from food to the treatment of diseases. Thereby, understanding the biology of fungi at different levels ranging from ecology to molecular details is essential, although this kingdom received little attention in the past apart from some classical models such as yeasts, *Neurospora crassa*, and *Aspergillus nidulans*. Fortunately, the study of fungi is living now an exciting period with spectacular advances in our knowledge about the molecular mechanisms underlying the fascinating responses of these organisms to their faced environmental challenges. Several examples are showing how fungi can adapt rapidly to stressful situations that can put their lives at risk, such as the presence of antifungal drugs and high temperature (Calo et al. 2014; Chang et al. 2019; Kronholm et al. 2016; Noble et al. 2016; Slepecky and Starmer 2009). In this context, non-coding small RNAs (sRNAs) play essential roles in genome integrity preservation, gene expression, phenotypic plasticity, and the ability to interact with other organisms via the conserved eukaryotic RNA interference (RNAi) pathway, known also as the RNA-mediated gene silencing mechanism, which mainly represses the expression of target genes at the transcriptional or post-transcriptional level (Torres-Martínez and Ruiz-Vázquez 2017).

The phenomenon of RNAi was initially discovered in plants when the introduction of the chalcone synthase gene in petunia suppressed the expression of both the transgene and the endogenous gene, a phenomenon called co-suppression (Napoli et al. 1990). **Several RNAi pathways have been described in fungi** (Chang et al. 2012; Torres-Martínez and Ruiz-Vázquez 2017; Villalobos-Escobedo et al. 2016) and other species, but the **basic conserved machinery** (Fig. 1) comprises an RNase III protein, called Dicer, which produces the sRNA molecules from a double-stranded RNA (dsRNA) precursor. These sRNAs, which receive differ-

[1] Departamento de Genética y Microbiología, Facultad de Biología, Universidad de Murcia, Murcia, Spain; e-mail: fnicolas@um.es; lauramur@um.es; sebi@um.es; josetomas.canovas@um.es; vgarre@um.es

Genetics and Biotechnology, 3rd Edition
The Mycota II
J.P. Benz, K. Schipper (Eds.)

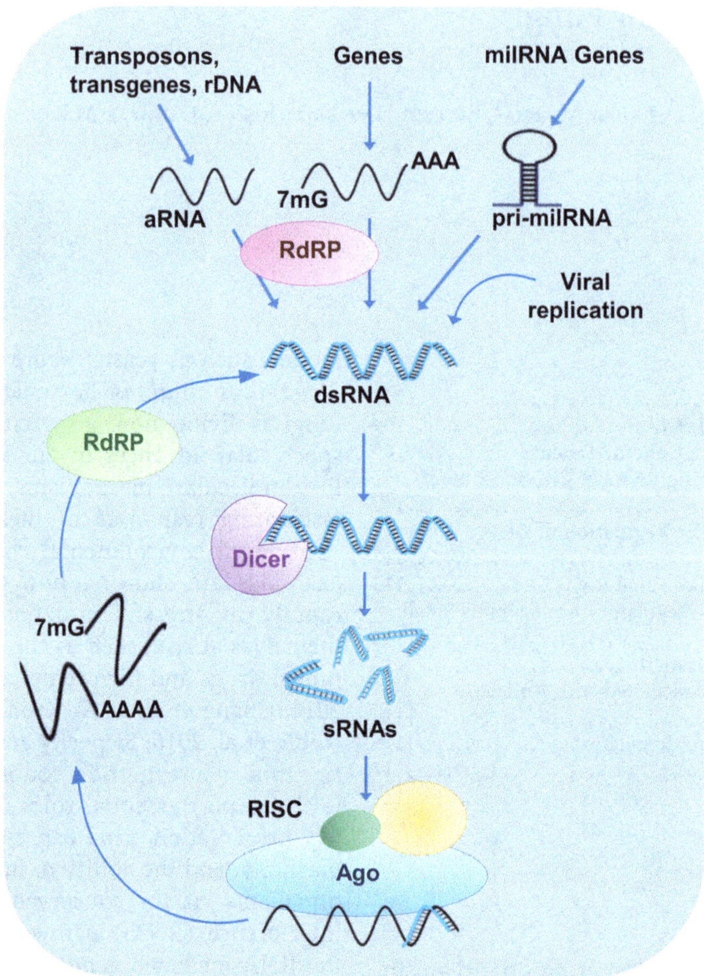

Fig. 1 A simplified model for RNAi pathways in fungi. Core components of the RNAi pathways are depicted (Table 2). Fungi show a wide diversity of specific RNAi-related pathways that carry out particular functions, which have been previously outlined in great detail (Calo et al. 2017; Chang et al. 2012; Torres-Martínez and Ruiz-Vázquez 2017; Villalobos-Escobedo et al. 2016)

ent names depending on their function and features (Table 1), are bound to an Argonaute (Ago) protein that uses one strand of these sRNAs to guide selective destruction, translational repression, or transcriptional suppression of target RNAs. Fungi, as plants and nematodes, present an additional RNA-dependent RNA polymerase (RdRP) protein that generates dsRNA from single-strand RNA (ssRNA) acting as inducers or amplifiers of sRNA signals (Torres-Martínez and Ruiz-Vázquez 2017). Genes coding for the key RNAi components are present in the genomes of most fungi, except for some relevant examples, such as *Cryptococcus deuterogattii* (Feretzaki et al. 2016), some species of the phylum Microsporidia and subphylum Ustilaginomycotina, and those of the subphylum Saccharomycotina and the class Wallemiomycetes (Choi et al. 2014; Nakayashiki et al. 2006; Laurie et al. 2008). Besides the core RNAi components, several accessories proteins with secondary functions have been described (Table 2). The **high number of fungi that may have a functional**

Table 1 sRNAs in fungi

Acronym	Name	Function	Features	References
siRNAs	Small-interfering RNAs	Genome integrity protection against integrative transgenes, non-integrative transgenes, transposons, and viruses	21–25 nt long with uracil at the 5′ end	Catalanotto et al. (2002) and Nicolás et al. (2003)
masiRNAs	MSUD-associated small-interfering RNAs	Genome integrity protection by silencing of unpaired DNA during meiosis	~25 nt long with uracil at the 5′ end	Hammond et al. (2013b)
qiRNAs	QDE-2-interacting small RNAs	Genome integrity protection in response to DNA damage	21–22 nt long with a strong preference for uracil at the 5′ end	Lee et al. (2009)
rasiRNAs	Repeat-associated small interfering RNA	Epigenetic silencing of transposons	21 nt long	Borgognone et al. (2018)
nat-siRNAs	Natural antisense transcript-derived siRNAs	Regulation of gene expression by targeting specific mRNAs	22–23 nt long	Drinnenberg et al. (2009)
disiRNAs	Dicer-independent esRNAs	Regulation of gene expression by targeting specific mRNAs	22 nt long with uracil at the 5′ end	Lee et al. (2010)
Ex-siRNAs	Exon- derived siRNAs	Regulation of gene expression by targeting specific mRNAs	Classes I and II, 23–24 nt long with a preference for uracil at the 5′ end Classes III and IV, different sizes with a strong preference for uracil in the penultimate position	Nicolas et al. (2010)
milRNA	miRNA-like small RNAs	Regulation of gene expression by targeting specific mRNAs	25 and 19 nt long with a strong preference for uracil at the 5'position	Lee et al. (2010)
rdRNAs	Rdrp-dependent degraded RNAs	Regulation of gene expression by a non-canonical dicer-independent mechanism?	Different sizes with uracil at the penultimate position	Trieu et al. (2015)
vsRNAs	Virus-derived siRNAs	Control of virus	20–22 nt with a strong preference for A at the 3′ position	Zhang et al. (2008)
priRNAs	Primal RNAs	Initiation of heterochromatin formation	22–23 nt long with uracil at the 5′ end	Halic and Moazed (2010)

Table 2 RNAi proteins in fungi

Core silencing proteins		
Protein family	Protein name	Function
Argonaute	AgoAGLQDE-2 (*N. crassa*)SMS2 (specific for MSUD in *N. crassa*)	They use one strand of sRNAs as a guide to bind target RNAs. They have endonuclease activity
RNase III protein	DicerDCRDCL	Riboendonuclease that produce sRNA from a dsRNA
RNA-dependent RNA polymerase	RdRPQDE-1 (*N. crassa*)SAD-1 (specific for MSUD in *N. crassa*)	They generate dsRNA from ssRNA. A DNA-dependent RNA polymerase activity has described for some of them
Accessory silencing proteins		
Protein family	**Protein name**	**Function**
RecQ DNA helicase	QDE-3 (*N. crassa*)	They promote aRNA production
RNA/DNA helicase	SAD-3 (specific for MSUD in *N. crassa*)Hrr1 (specific for heterochromatin formation in *S. pombe*)RnhA (*M. circinelloides*)	They act in unwinding nucleic-acid strands
SNF2 helicase-related proteins	SAD-6 (specific for MSUD in *N. crassa*)	Unpaired DNA detection during MSUD
Atypical RNase III protein	R3B2 (*M. circinelloides*)	Ribonuclease involved in the non-canonical RNAi pathway
Protein with RNase III domain	MRPL3 (*N. crassa*)	Dicer-independent biogenesis of some milRNAs
Exonuclease	QIP	They remove the passenger strand from the double-stranded siRNAs bound by ago proteins
Exoribonuclease	Dhp (*S. pombe*)	Dicer-independent biogenesis of sRNAs
Exonuclease	ERI-1 (*N. crassa*)	Biogenesis of disiRNAs
SAD-2	SAD-2 (specific for MSUD in *N. crassa*)	Constitutes a scaffold protein in MSUD
SAD-4	SAD-4 (specific for MSUD in *N. crassa*)	Production of masiRNAs
SAD-5	SAD-5 (specific for MSUD in *N. crassa*)	Production of masiRNAs

RNAi mechanism suggests that it plays an essential role in their biology, although in some particular cases its absence might provide some advantage (Drinnenberg et al. 2011; Nicolás et al. 2013; Wang et al. 2010). Despite the **original function of RNAi was to protect the genome** from deleterious movements of transposable elements (TEs) and viruses, the basic machinery has evolved in all groups of eukaryotes to generate specific pathways with **endogenous regulatory functions**. These pathways produce particular types of sRNAs that are involved in the regulation of gene expression (Table 1) (Cerutti and Casas-Mollano 2006). The most representative example of these regulatory sRNAs is the animal and plant microRNA (miRNA) (Ghildiyal and Zamore 2009). In fungi, various endogenous regulatory RNAi pathways have also been described (Calo et al. 2017; Lee et al. 2010; Nicolas et al. 2010; Yu et al. 2018). These pathways interact with each other at several levels, competing for and sharing substrates, components, and cross-regulating each other (Fig. 1) (Calo et al. 2017).

Several excellent reviews have examined the RNAi-related pathways, types of sRNAs and their associated functions in fungi (Chang et al. 2012; Torres-Martínez and Ruiz-Vázquez 2017; Villalobos-Escobedo et al. 2016). In this Chapter of The Mycota, we have updated the most recent discoveries and contributions to the field with particular emphasis on functional aspects. Readers interested in mechanistic details specific for each RNAi-related pathway are advised to go through the indicated reviews or the references in this Chapter.

II. Protective Small RNAs

Soon after the discovery of co-suppression in plants (Napoli et al. 1990), a similar RNAi phenomenon in response to transgenes, named **quelling**, was described in *Neurospora crassa*. In this mechanism, the expression of genes harbored in the genome was suppressed when copies of the same gene were introduced (Romano and Macino 1992). Similar responses to transgenes were observed in other fungi (Nicolás et al. 2003; Wang et al. 2010). The RNAi mechanism was well-characterized in *N. crassa* (Chang et al. 2012), which belongs to the Ascomycota phylum, and subsequently this characterization was extended to other significant fungal phyla such as Mucoromycota and Basidiomycota, represented by *Mucor circinelloides* (Torres-Martínez and Ruiz-Vázquez 2017) and *Cryptococcus neoformans* (Feretzaki et al. 2016; Janbon et al. 2010; Wang et al. 2012), respectively. These analyses and others described below revealed that RNAi pathways have a critical role in the maintenance of the genome integrity in response to TEs and other invading nucleic acids. Moreover, the RNAi mechanism as a guard of the genome soon acquired a secondary role in the function and evolution of centromeres, probably due to the high abundance of TEs in these specialized regions of the chromosomes (Friedman and Freitag 2017; Yadav et al. 2018).

A. Small RNAs in Genome Defense

RNAi-meditated protection mechanisms that participate in the maintenance of genome integrity have been identified in different stages of fungal life cycles, reflecting their importance.

1. Defense Responses During Vegetative Growth

The RNAi mechanism that acts in the defense against exogenous nucleic acids during vegetative growth received different names such as quelling in *N. crassa* (Romano and Macino 1992), mitotic-induced silencing (MIS) in *C. neoformans* (Wang et al. 2012) or gene silencing in *M. circinelloides* (Nicolás et al. 2003). However, they constitute basically the same mechanism with particular details in each fungus that affect the accessory RNAi proteins (Table 2) and sRNAs features (Table 1) (Chang et al. 2012; Torres-Martínez and Ruiz-Vázquez 2017). In these pathways, the long dsRNA is cleaved into double-stranded small-interfering RNAs (siRNAs) of 21–25 nucleotides, which bind to an Ago protein in the RNA-induced silencing complex (RISC) (Fig. 1). In most studied fungal models so far, the silencing is associated with a dramatic decrease of mature mRNA levels from target genes, whereas those of primary transcripts are not affected, indicating that it is mainly a **post-transcriptional gene silencing (PTGS)** phenomenon (Nicolás et al. 2003; Pickford and Cogoni 2003).

Full understanding of the function of the silencing mechanism in genome defense requires the **identification of the triggering signal.** Quelling and MIS are initiated when multiple transgene copies are integrated in tandem repeats in the genome (Cogoni et al. 1996; Romano and Macino 1992; Wang et al. 2010). This may also occur in *M. circinelloides* although transgenes are maintained in self-replicative plasmids, since in Mucorales these plasmids are prone to form concatemers and rearrange (Meussen et al. 2012; Michielse et al. 2004; Papp et al. 2013), which might favor the formation of tandem repeats. How the siRNA production is initiated in fungi is best characterized for quelling in *N. crassa*. Tandem repeats alone are not sufficient to induce siRNA production since it also requires double-stranded breaks and homologous recombination (HR) (Yang et al. 2015). Repetitive sequences are regions of genome instability due to hyperrecombination events caused by replication stress (Bzymek and Lovett 2001; Vader et al. 2011). Thus, the proposed model suggests that recombination intermediates produced during HR can be recognized by QDE-3, a putative RecQ DNA helicase involved in RNAi. This helicase might resolve the recombination intermediates into single-stranded DNA (ssDNA), after recruiting the ssDNA-binding complex (RPA) and the dual RdRP enzyme

QDE-1 with RNA/DNA-dependent RNA polymerase activities, which produces an aberrant single-stranded RNA (aRNA) and subsequently the dsRNA that activates the RNAi pathway (Fig. 1) (Liu et al. 2010).

In addition to transgenes, fungal genomes naturally contain two main types of tandem repetitive sequences, TE arrays, and ribosomal genes (rDNA), which are known to be a major source of genome instability (Butler 1992; Bzymek and Lovett 2001; Vader et al. 2011). In fact, several studies have revealed large amounts of sRNAs derived from both types of loci in different fungi (Dumesic et al. 2013; Janbon et al. 2010; Nicolas et al. 2010), including qiRNA (QDE-2-interacting small RNAs) and rasiRNAs (repeat-associated small-interfering RNA) originated from rDNA in *N. crassa* (Lee et al. 2009) and transposons in the basidiomycete *Pleurotus ostreatus* (Borgognone et al. 2018), respectively. Even though qiRNA and the transgenes-induced siRNAs have different origins, they require the same RNAi components and are the result of DNA damage (Chang et al. 2012; Zhang et al. 2013). In addition, the RNAi pathway suppresses transposon proliferation in several fungi (Borgognone et al. 2018; Janbon et al. 2010; Nolan et al. 2005; Wang et al. 2010), and it is required to maintain transgene tandem repeats (Yang et al. 2015) and retrotransposon arrays found in centromeres (Yadav et al. 2018).

An alternative mechanism to control TEs during vegetative growth has been described in *C. neoformans*. This mechanism identifies transposon transcripts due to the presence of suboptimal introns, which provokes that they stall on spliceosomes. A spliceosome-coupled and nuclear RNAi (SCANR) complex is able to recognize these transposons transcripts stuck on spliceosomes promoting siRNA production that leads to control of transposons (Dumesic et al. 2013).

Together, these observations suggest that the **principal function of the siRNA produced from repeat regions is to maintain genome stability**, which is supported by the identification of DNA damage-induced sRNAs and the involvement of Dicer enzymes in the maintenance of genome stability in plants and animals (Bonath et al. 2018; Francia et al. 2013; Lu et al. 2018; Michalik et al. 2012; Wei et al. 2012).

2. Defense Responses in the Sexual Cycle

In addition to the RNAi mechanisms that operate in the vegetative growth to protect the genome integrity, other RNAi pathways carry out this function during sexual reproduction. One of such RNAi mechanisms, called **sex-induced silencing (SIS)**, was described in *C. neoformans*. This is a PTGS mechanism that is triggered by tandem integration of a transgene array both in opposite-sex mating and unisexual reproduction. It shares the basic RNAi machinery required for MIS, including Ago, Dicer, and RdRP proteins (Wang et al. 2010, 2013). Beyond silencing exogenous transgenes, SIS plays a critical role in transposon control because RNAi mutants show an increase in transposition/mutation rate and elevated levels of siRNAs derived from repetitive TEs (Wang et al. 2010, 2013). Interestingly, the higher robustness of SIS compared to MIS might be related to the fact that transposons in *C. neoformans* are more active during mating than during vegetative growth (Wang et al. 2010, 2013).

An additional RNAi mechanism involved in defense of the genome during the sexual cycle has been described in species of the Ascomycota phylum, such as *N. crassa*, *Neurospora tetrasperma*, and *Gibberella zeae* (anamorph *Fusarium graminearum*) (Ramakrishnan et al. 2011; Shiu et al. 2001; Son et al. 2011), suggesting that it has evolved recently (Hammond 2017). This RNAi mechanism protects genome integrity through silencing of all those DNA sequences that remain unpaired during the meiotic prophase I. Consequently, it was named **Meiotic Silencing by Unpaired DNA (MSUD)** (Shiu et al. 2001). This mechanism seems to be mechanistically distinct from the SIS of *C. neoformans* because SIS is not triggered by unpaired DNA structures (Wang et al. 2010). The molecular mechanism of MSUD has been studied in detail in *N. crassa*, where it can be divided into a detection stage of unpaired DNA between homologous chromosomes and a silencing stage of unpaired DNA and any related sequences found at other locations in the genome, regardless of their pairing state. Nine proteins are involved in MSUD, three of which correspond to the core components of the canonical RNAi pathway and the rest perform specific functions of this mechanism (Hammond 2017). In the latter group are included SAD-5 and SAD-6, which are the only proteins known to participate in the identification of unpaired sequences (Samarajeewa et al. 2014). Despite the poor characterization of the mechanism of unpaired DNA detection, this process is thought to trigger the production of an aRNA, which is recognized in the perinuclear region by the proteins involved in the silencing stage. Thus, a MSUD-specific RdRP SAD-1 and the helicase SAD-3 (Hammond et al. 2011, 2013a; Shiu and Metzenberg 2002; Shiu et al. 2001) generate a dsRNA molecule that is processed by Dicer DCL-1 into MSUD-associated small-interfering RNAs (masiRNAs) (Alexander et al. 2008). These are then bound by AGO SMS-2, and the passenger is removed by the exonuclease QIP (Lee et al. 2003; Xiao et al. 2010). The identification of masiRNAs derived from an unpaired transposon in a sexual cross suggests that MSUD protects the genome from transposons (Wang et al. 2015), although another hypothesis points to the protection of the genome from meiotic drive elements, called spore killers, which rearrange large genomic segments as part of the driving mechanism (Hammond 2017).

Spore killers, described in Ascomycota phylum, are single genes or complexes of genes that favor their own propagation through meiosis and/or gametogenesis by killing the meiotic products not containing them. The best characterized spore killer is *sk-2* of *N. crassa* (Turner and Perkins 1979). The *sk-2* drive mechanism requires that at least two distantly located genes, a resistance gene called *rsk* and a killer gene called *rfk*, inherit together during meiosis (Campbell and Turner 1987). Tight linkage between these two genes appears to be derived from chromosome rearrangements within the *sk-2* element (Harvey et al. 2014). The *sk-2*, and also *sk-3*, suppresses MSUD (Raju et al. 2007), suggesting that this blockage allows the evolution of chromosome rearrangements, which would be MSUD targets (Hammond 2017).

3. Antiviral Small-Interfering RNAs

The use of the RNAi pathway as an innate immune system against viruses was one of the first roles associated with this mechanism. Triggering of siRNA production by dsRNA or ssRNA viruses has been found in plants, flies, worms, mammals, and fungi (Harvey et al. 2011; Jeang 2012; Segers et al. 2007; Wilkins et al. 2005; Zambon et al. 2006). Among fungi, this response has been intensively studied in the ascomycete *Cryphonectria parasitica*, a filamentous fungus that is the causal agent of chestnut blight (Segers et al. 2007), and also described in *Colletotrichum higginsianum* (Campo et al. 2016), *Aspergillus nidulans* (Hammond et al. 2008), *Fusarium graminearum* (Yu et al. 2018), and *Sclerotinia sclerotiorum* (Mochama et al. 2018). *C. parasitica* produces an RNAi-mediated antiviral response by the production of virus-derived siRNAs (vsRNAs) (Table 1) that target and destroy viral sequences. Although this fungus has two *dicer*-like genes, four *ago*-like genes, and four *rdrp*-like genes, only genes *dcl2* and *agl2* are involved in the RNAi-mediated antiviral response. Thus, single deletion mutants in any of these two genes were defective in the production of vsRNAs and highly susceptible to mycovirus infections, resulting in a severely debilitated growth (Segers et al. 2007; Sun et al. 2009). Interestingly, the *C. parasitica* **RNAi mechanism also promotes recombination of viral genomic RNA**, a central component of virus evolution that contributes to the emergence of new viruses (Sun et al. 2009; Zhang and Nuss 2008).

The importance of the RNAi-mediated antiviral defense is reinforced by the fact that some **viruses present mechanisms of RNAi suppression**. Thus, the hypovirus *Cryphonectria* hypovirus 1 (CHV1-EP173), a mycovirus that infects *C. parasitica*, expresses the protein p29, a papain-like protease, that inhibits the RNAi pathway by repressing the transcriptional induction of *agl2* in response to virus infections (Sun et al. 2009). Likewise, the existence of a viral suppressor has also been demonstrated in *A. nidulans* (Hammond et al. 2008).

B. Small-Interfering RNAs in Chromosome Function

In previous sections, we have described several mechanisms that use siRNAs to protect the genome from TEs. These protective mechanisms seem to be adopted by eukaryotic cells during evolution to play other relevant functions in the cells. This might be the case for the **formation of heterochromatin** in different regions of the chromosomes, particularly in regional centromeres, where siRNAs play an essential role (Pidoux and Allshire 2005; Volpe and Martienssen 2011). Regional centromeres, contrary to point centromeres, are long ranging from few kilobases to megabases and made of repetitive DNA that consists of either arrays of satellite DNA or TEs or both (Roy and Sanyal 2011). TEs are proposed to play a more significant role in the evolution of this type of centromeres, and their domestication may have given rise to the *dh*/*dg* and α-satellite repeats present in the centromeres of fission yeast (*Schizosaccharomyces pombe*) and humans, respectively (Gao et al. 2015; Wong and Choo 2004).

In fission yeast, RNAi plays a critical role in heterochromatin formation, which it is required for normal centromere function leading to chromosome segregation (Pidoux and Allshire 2005). Dicer-independent sRNAs, named primal RNAs or priRNAs (Table 1), are the main effectors in the initiation of heterochromatin formation (Halic and Moazed 2010). The priRNAs are generated from ssRNAs to later form a complex with Ago1 that targets long non-coding centromeric transcripts. This complex recruits the RdRP to synthesize dsRNA, which is subsequently processed by Dicer into secondary siRNAs. These secondary siRNAs are loaded onto the RNA-induced transcriptional silencing complex (RITS), which includes Ago1. The binding of siRNA-containing RITS to the nascent transcripts in the centromeric regions recruits the Clr4 methyltransferase complex, which deposits the H3K9me mark and HP1 family proteins leading to heterochromatin formation (Martienssen and Moazed 2015; Ugolini and Halic 2018).

In addition to the establishment of heterochromatin formation in the centromeres, RNAi also seems to play an essential function in **centromere evolution** in species belonging to the phylum Basidiomycota. Here, RNAi and cytosine methylation are proposed to maintain repetitive transposon-rich centromeres, since RNAi-deficient species or mutants from RNAi-proficient species have shorter centromeres (Yadav et al. 2018). This agrees with the fact that fungi with RNAi systems have more total DNA corresponding to TEs than RNAi-deficient species, although fewer elements are functional, reflecting stringent control over transposition (Muszewska et al. 2017).

III. Small RNAs in the Regulation of Gene Expression

Soon after the discovery of RNAi, the focus on its defensive role gave way to a whole new research field based on the endogenous regulatory functions of sRNAs. It was initiated when miRNAs, previously identified and studied in *Caenorhabditis elegans* (Lee et al. 1993), were found to be conserved in most plants and animals. These discoveries opened an extensive period in which the regulatory roles of RNAi and **endogenous sRNAs (esRNAs)** were the hot topic in molecular biology for several years. Hundreds of miRNAs were found creating a regulatory network that might involve more than 60% of total genes in humans (Friedman et al. 2009) and regulate most of the complex biological processes in living cells such as development, differentiation, maintenance, cell death, and diseases associated with the misregulation of these molecules (Esteller 2011; López-Camarillo and Marchat 2013; Stefani and Slack 2008). During this time, miRNAs and other regulatory esRNAs were considered to be absent in fungi, until molecules similar to miRNAs and regulatory esRNAs were discovered in *Neurospora crassa* and *Mucor circinelloides*, respectively (Lee et al. 2010; Nicolas et al. 2010). However, the regulatory role of esRNAs has been scarcely studied in fungi because few

RNAi-deficient mutants show altered phenotypes in development or physiology. The most remarkable exceptions to this general rule are *M. circinelloides* and *Trichoderma atroviride*, since mutants in core components of the RNAi pathway show clear phenotypes associated with deregulation of large numbers of genes (Carreras-Villaseñor et al. 2013; Cervantes et al. 2013; de Haro et al. 2009; Nicolás et al. 2007, 2015; Trieu et al. 2015).

A. MicroRNA-Like Small RNAs

miRNAs are the most important small regulatory RNAs found in animals and plants as they negatively regulate the expression of hundreds of genes by either inducing degradation or repressing translation of the target mRNAs (Carthew and Sontheimer 2009; Ghildiyal and Zamore 2009). In fungi, an in-depth sequencing study of sRNAs interacting with Ago proteins in *N. crassa* revealed a complex system containing four different types of sRNAs that shared several similarities with miRNA, thereby named as miRNA-like small RNAs (milRNA) (Lee et al. 2010). These four classes of milRNAs were defined according to the RNAi components that were required for their biogenesis.

The synthesis of milRNAs type 1 (milR-1) depends on the activity of Dicer, QDE-2, QIP, and MRPL3 (Table 2), whereas the synthesis of milRNAs type 2 (milR-2) only relies on the catalytic activity of QDE-2, being totally independent of Dicer enzymes. \It is a conserved protein harboring a predicted RNase III domain similar to the well-known domains found in Dicer enzymes. The role of MRPL3 in other organisms is unknown, although it might be related to the fate of mRNAs during the synthesis of proteins, considering its association to the large subunit of the mitochondrial ribosomes (Smits et al. 2007). Biogenesis of milRNAs belonging to the type 3(milR-3) is quite similar to a pathway found in plants that only depends on the activity of Dicer enzymes (Carthew and Sontheimer 2009). The fourth type of milRNAs (milR-4) is the most unusual class as it requires MRPL3 and is partially dependent on Dicer enzymes (Lee et al. 2010).

Regardless of specific differences in the biogenesis of each type of milRNAs, all of them share three main features with the synthesis of miRNAs. First, they are produced from RNA precursors of a single strand that forms a typical hairpin structure. Second, the dsRNA region of the typical hairpin structure is processed to generate two complementary sRNAs, although only one of them is predominantly preserved while the other is degraded. Finally, a strong bias for uracil bases at the 5′ termini of the milRNA is commonly found in these sRNAs and shared with canonical miRNAs (Lee et al. 2010). Despite these similarities, the processing and the final features of milRNAs in *N. crassa* also present several fundamental differences compared with miRNAs in animals and plants. Thus, milRNA genes of *N. crassa* are not highly conserved in other species of fungi, indicating an early evolution of milRNAs independent of canonical miRNAs, which usually are highly conserved and share sequence, function and even tissue-specific expression in several species (Bartel 2004; Carthew and Sontheimer 2009). Additional divergences between milRNAs and miRNAs can be found at the beginning of their biogenesis. Thus, cleavage of the milRNA precursors (pri-milRNAs or pre-milRNAs) might be processed by the putative ribonuclease MRPL3, a conserved protein harboring a predicted RNase III domain similar to the well-known domains found in Dicer enzymes that has not been involved in miRNA biogenesis in plant and animals (Bartel 2004; Carthew and Sontheimer 2009). Also, pri-milRNAs are transcribed by RNA polymerase III, representing an additional significant difference compared to the well-established role of RNA polymerase II in the transcription of pri-miRNAs in plants and animals (Yang et al. 2013).

The function of **milRNAs has been linked to the regulation of the expression of target mRNAs** similarly to miRNAs (Lee et al. 2010). The first result supporting the functionality of milRNAs was the overexpression of target mRNAs in Dicer mutants of *N. crassa*, suggesting that target mRNAs are repressed by milRNAs (Lee et al. 2010). Correspondingly, mutants devoid in the Ago protein QDE2 (Table 2) also displayed increased levels of predicted target mRNAs and QDE2 co-precipitates with these target mRNAs. More direct evidence showed that mutants lacking specific milRNA genes exhibited elevated levels of the mRNAs harboring predicted target sequences, indicating that these target sequences are regulated by milRNAs (Lee et al. 2010).

The discovery of milRNAs in *N. crassa*, a fungus of the Ascomycota phylum, proved that regulatory sRNAs are also produced in the lower branches of the Eukarya domain. Subsequently, other studies revealed the presence of similar regulatory sRNAs in fungi belonging to both the Ascomycota (*Fusarium oxysporum* (Chen et al. 2014), *Penicillium chrysogenum* (Dahlmann and Kück 2015), and *Trichoderma reesei* (Kang et al. 2013)) and Basidiomycota (*Coprinopsis cinerea* (Lau et al. 2013) and *Antrodia cinnamomea* (Lin et al. 2015)) phyla, suggesting that milRNAs have functional roles in these fungal groups. However, the thoroughly accomplished studies in *N. crassa* and other fungi could not identify phenotypes or specific functions associated with mutants enable to produce all milRNAs, such as deletion mutants lacking *dicer* or *ago* genes, or particular milRNA. Despite the absence of phenotypes, these studies uncovered several cases of expression profiles associated with specific developmental stages and groups of milRNAs targeting predicted mRNAs involved in the same functional roles (Jiang et al. 2017; Lau et al. 2013, 2018; Lin et al. 2015). Other studies have also identified **milRNAs that could control the production of compounds with industrial applications**, such as hydrolases and antibiotics (Dahlmann and Kück 2015; Kang et al. 2013).

The diversity of milRNAs across fungi, including sequence, number, expression pattern, and especially the lack of particular phenotypes associated with mutations in these regulators, represents a major obstacle that is preventing the unveiling of the different roles of milRNAs in fungal physiology. Nevertheless, all the studies describing fungal milRNAs have contributed to enlighten a heterogeneous world of endogenous regulatory sRNAs in Ascomycota and Basidiomycota phyla. The only common feature among all these fungi is the **lack of a unique or universal pathway for the production of milRNAs,** showing a scenario where the diversification of regulatory silencing mechanisms is the chosen strategy rather than the conserved pathways found in animals and plants (Bartel 2004; Carthew and Sontheimer 2009).

B. Exon-Derived Regulatory Endogenous Small RNAs

M. circinelloides accumulates several types of regulatory esRNAs generated by both Dicer-dependent and Dicer-independent RNAi pathways (Cervantes et al. 2013; Nicolas et al. 2010; Trieu et al. 2015). The Dicer-dependent esRNAs, named ex-siRNAs, are derived from exons of protein-coding genes and regulate the expression of the producing genes by degradation of the corresponding mRNAs (Nicolas et al. 2010). In addition to Dicer, the rest of components of the RNAi pathway involved in siRNA-mediated genome defense are also implicated in the biogenesis of ex-siRNAs, although they are combined to originate four different classes of ex-siRNAs (Nicolas et al. 2010). The four ex-siRNA classes not only differ in the biogenesis pathway but also in their binding to the main Ago proteins (Ago1) and structural characteristics, such as length, strand bias and 5′ nucleotide (Cervantes et al. 2013; Nicolas et al. 2010). The ex-siRNA-mediated regulation extends beyond the genes that produce them since hundreds of likely second targets genes are differentially expressed in deletion mutants for genes coding for RNAi proteins involved in ex-siRNA biogenesis (Nicolás et al. 2015). The massive alteration in gene expression in these mutants should affect genes that participate in the developmental and physiological processes affected in these mutants, which include vegetative growth, response to nutrient starvation and oxidative stress, asexual sporulation, mating, and development (Cervantes et al. 2013; de Haro et al. 2009; Nicolás et al. 2007, 2015; Trieu et al. 2015).

Analysis of esRNA accumulating in Dicer mutants in *M. circinelloides* revealed the existence of a new type of esRNAs, named rdRNAs (*rdrp*-dependent degraded RNAs). This led to the discovery of a **non-canonical RNAi mechanism** in which participate the three RdRP enzymes of *M. circinelloides* and the Sad-3-like helicase *rnhA* (Calo et al. 2017; Trieu et al. 2015). The Dicer function is carried out by an atypical RNase III-like enzyme, named R3B2 (Table 2), which has been found only in Mucorales. The rdRNAs have characteristics of degradation products (Table 1), suggesting that **this mechanism is likely a degradation pathway** that controls the levels of specific mRNAs (Trieu et al. 2015). This pathway regulates the expression of hundreds of highly expressed genes involved in metabolism, regular cellular processes, and signaling; consequently, the deletion mutant lacking the key gene *r3b2* shows resistance to oxidative stress and defects in asexual sporulation, response to nutrient, and mating (Trieu et al. 2017). These kinds of Dicer-independent mechanisms are not restricted to *M. circinelloides,* but have been described in other fungi. In *Schizosaccharomyces pombe,* Dhp exoribonuclease drives a novel RNAi and exosome-independent pathway of epigenetic silencing and also plays a role in PTGS (Tucker et al. 2016). Similarly, R3B2 of *M. circinelloides* plays a relevant role in the canonical silencing mechanism in parallel to its function in the Dicer-independent degradation pathway (Trieu et al. 2015). Besides, Dicer-independent small non-coding RNAs (disiRNAs), described in *N. crassa,* are also produced by a Dicer-independent RNAi pathway from loci that generate overlapping sense and antisense transcripts as a result of convergent transcription (Lee et al. 2010). Although these sRNAs show structural characteristics corresponding to siRNAs, none of the components of the RNAi machinery is involved in their biogenesis (Lee et al. 2010).

The presence of esRNAs derived from protein-coding genes has also been detected in several fungi belonging to Ascomycota phylum, but only RNAi mutants of some species show a phenotype that suggests the involvement of esRNAs in regulation. Different components of the RNAi machinery are involved in light-dependent asexual reproduction and light-independent hyphal growth in *Trichoderma atroviride* (Carreras-Villaseñor et al. 2013), in vegetative growth in *Magnaporthe oryzae* (Kadotani et al. 2004), and formation of sexual spores after sexual interaction in *Fusarium graminearum* (Son et al. 2017). Despite these remarkable examples, the number of phenotypes observed in *M. circinelloides* and their strengths suggest that the **RNAi pathway plays a more relevant role in early-divergent fungi** than in more evolved fungi (Ascomycota and Basidiomycota), where milRNAs and other regulatory sRNAs may fine-tune gene expression (Villalobos-Escobedo et al. 2016).

C. Phenotypic Plasticity

In the previous sections, we described general RNAi-mediated regulatory mechanisms that affect the expression of hundreds of genes, but a specific regulatory RNAi mechanism has been revealed in *M. circinelloides* that target the mRNA of a particular gene. This mechanism was initially discovered in the screening for spontaneous **resistants to the antifungal drug** FK506 (tacrolimus). In addition to mutants in the gene *fkbA*, which encodes the FK506 target protein FKBP12, epimutant strains were isolated that transiently silenced the expression of *fkbA* (Calo et al. 2014). Interestingly, *M. circinelloides* isolated from humans or other animal hosts are able to produce a high number of FK506-resistant epimutants (Calo et al. 2014, 2017), suggesting that this mechanism may enable this opportunistic pathogen to readily adapt to different environments. Silencing of *fkbA* was accompanied by elevated levels of sRNAs and required the core components of the RNAi machinery (Calo et al. 2014, 2017). The analysis of RNAi proteins involved in this regulatory mechanism revealed the existence of an epimutational RNAi pathway in *M. circinelloides*, which is very similar to the ex-siRNA-producing pathways (Calo et al. 2017), but the set of RNAi proteins and their relevancy is different, suggesting that a distinct sRNA class is involved in this pathway (Calo et al. 2014, 2017). The generation of sRNAs depends on the production of a dsRNA generated by the action of one of the RdRPs (RdRP2) present in *M. circinelloides*, and not from antisense transcription in *fkbA* locus (Calo et al. 2014). An unanswered question is why the accumulation of sRNAs from the *fkbA* locus is induced in response to the drug. One possible explanation is based on the high level of expression of *fkbA* that might lead to the formation of aRNA, triggering the silencing as it has been observed in plants (Gazzani et al. 2008).

The isolation of *M. circinelloides* epimutants exhibiting resistance to 5-fluoroorotic acid (5-FOA) has revealed that **RNAi-dependent epimutation plays a broad role** enabling rapid and reversible responses of this fungus (Chang et al. 2019). The distribution of RNAi-related mechanisms involved in the phenotypic plasticity in other fungi is unknown and requires further studies, but it is very promising that one of such mechanism could operate in response to multiple environments in *N. crassa* (Kronholm et al. 2016).

D. Small RNAs Associated with Antisense Transcripts

Noncoding RNAs that overlap with protein-coding regions as a consequence of convergent transcription, named natural antisense transcripts (NATs), have been identified in a high number in animals and plants (Faghihi and Wahlestedt 2009) and in a lesser extent in phylogenetically diverse fungi (Donaldson and Saville 2012). These NATs use different mechanisms to control gene expression, including transcriptional interference, chromatin remodeling, and dsRNA formation. The dsRNA can be a substrate of the RNAi machinery to produce nat-siRNAs, which has been observed only in *S. pombe* within the fungal kingdom (Donaldson and Saville 2012). In several fungi, the presence of NATs is not associated with the formation of nat-siRNAs, although NATs regulates gene expression and cellular processes (Nevers et al. 2018; Shao et al. 2017), including pathogenesis (Donaldson and Saville 2013).

The biogenesis of disiRNAs in *N. crassa* is related to nat-siRNAs because they derive from genes in which convergent transcription causes stalling of RNA polymerase II (Dang et al. 2016). The slow progression of polymerase II allows binding of the exonuclease ERI-1 (Table 2) to nascent mRNA provoking the generation of disiRNAs, although the precise role ERI-1 in disiRNA biogenesis is unknown (Dang et al. 2016). Like siRNAs, disiRNAs map to both strands of the genome and are bound to Argonaute protein QDE-2 (Lee et al. 2010). Interestingly, disiRNA loci are associated with DNA methylation and K3K9me3 (Dang et al. 2013), which are dependent on ERI-1 binding on mRNA and antisense transcription, whereas disiRNAs only contribute to these epigenetic modifications (Dang et al. 2016). The role of disiRNA in QDE-2-mediated DNA methylation is not clear, although a mechanism similar to siRNA-mediated heterochromatin formation in *S. pombe* by recognizing the nascent RNA and recruiting histone-modifying enzymes has been suggested (Dang et al. 2016).

E. Cross-Kingdom Regulation

Recent pieces of evidence have shown that sRNAs can move between interacting organisms to silence the gene expression *in trans* in the non-related species, a phenomenon called cross-kingdom or trans-kingdom RNAi (Kuan et al. 2016; Weiberg and Jin 2015; Weiberg et al. 2015). In fungi, it was observed for the first time in the aggressive plant pathogen *Botrytis cinerea*, which can transfer siRNAs during the infection to the plant host that hijack components of the plant RNAi pathway, **suppressing the expression of host immunity genes** (Weiberg et al. 2013). These sRNAs are derived from long-terminal repeat (LTR) retrotransposons and require the two Dicer-like enzymes of *B. cinerea* for their production. Consequently, deletion of both *dicer* genes leads to reduced virulence in plants (Weiberg et al. 2013). *B. cinerea* sRNAs are translocated by an unknown mechanism to the plant cell where they bind to one of the plant Ago proteins (AGO1 in *Arabidopsis thaliana*) to silence genes involved in plant immunity (Weiberg et al. 2013). Interestingly, the *A. thaliana* mutant in AGO1 is also resistant to another plant fungal pathogen, *Verticillium dahlia* (Ellendorff et al. 2009), and *V. dahlia* sRNAs with predicted plant targets have been associated with plant AGO1, suggesting that this fungus also employs sRNAs and cross-kingdom RNAi for successful infection (Wang et al. 2016). Moreover, **sRNAs have been found in extracellular vesicles** secreted by *Malassezia sympodialis*, *Saccharomyces cerevisiae*, *Candida albicans*, *Paracoccidioides brasiliensis*, and *Cryptococcus neoformans* (Peres da Silva et al. 2015; Rayner et al. 2017). These proofs suggest that an unknown cell-to-cell transport mechanism must exist to translocate sRNAs to the plant that could be used by other pathogens to suppress host immune systems (Weiberg et al. 2015).

In addition to sRNA translocation from fungal pathogens into plants, **plants can also transfer sRNAs into fungi**. This was initially revealed when dsRNA targeting virulence genes of fungal pathogens produced by engineered plants were able to trigger silencing in the fungi and confer resistance to the infection (Koch et al. 2013; Nowara et al. 2010; Panwar et al. 2013; Tinoco et al. 2010). The use of this strategy, name host-induced gene silencing (HIGS), to control invading fungi has the potential to become an important disease-protection method (Cai et al. 2018a; Nunes and Dean 2012). Interestingly, the delivery of own plant sRNAs was later confirmed by the identification of specific plant sRNAs and miRNAs in fungal cells infecting plants (Cai et al. 2018b; Zhang et al. 2016). The delivery mechanism of sRNAs by the plant and taking up by fungi is basically unknown, but exosome-like vesicles have proven to play a critical role in the secretion of sRNAs by *A. thaliana* infected with *B. cinerea*. The fungus takes up these vesicles resulting in the silencing of fungal virulence-related genes, including genes involved in vesicle-trafficking (Cai et al. 2018b). Together these results suggest that cross-kingdom RNAi might play an essential role in the arms race between pathogens and host with bidirectional translocation of sRNAs. Moreover, the ability of the fungi to incorporate extracellular RNAs has been devised as a strategy **to control pathogens by direct application of dsRNAs or sRNAs** onto host plants or post-harvest products to silence target fungal genes and confers efficient disease control (Cai et al. 2018a).

IV. Conclusions

The study of RNAi in the fungal kingdom has revealed that it is a conserved mechanism present in most fungi with a primary function of genome protection against foreign invaders and TEs during vegetative and sexual cycles. More impressively, recent studies have additionally unveiled a whole regulatory layer composed of different classes of regulatory esRNAs that can control several target genes and biological processes. In some fungi, illustrated by *Mucor circinelloides*, the relevancy of the RNAi-related regulation is huge controlling physiological and developmental processes. In contrast, milRNAs seems to fine-tune gene expression, and more in-depth studies are required to validate their functional roles in particular biological processes experimentally. The regulation of these processes by RNAi-related mechanisms is transient and reversible, conferring substantial phenotypic plasticity for rapid adaptation, as it has been demonstrated for antifungal drug response. In addition to the regulation of endogenous processes, plant fungal pathogens produce sRNAs to regulate the expression of plant genes involved in plant immunity. On the whole, all the different studies describing esRNAs in fungi have contributed to enlighten a heterogeneous world of new regulatory mechanisms in Ascomycota, Basidiomycota, and Mucoromycota. The only common feature among all these fungi is the lack of a unique or universal pathway for the production of esRNAs, showing a scenario where the diversification of regulatory silencing mechanisms is the chosen strategy rather than the conserved pathways found in animals and plants.

References

Alexander WG, Raju NB, Xiao H, Hammond TM, Perdue TD, Metzenberg RL, Pukkila PJ, Shiu PKT (2008) DCL-1 colocalizes with other components of the MSUD machinery and is required for silencing. Fungal Genet Biol 45:719–727

Bartel DP (2004) MicroRNAs: genomics, biogenesis, mechanism, and function. Cell 116:281–297

Bonath F, Domingo-Prim J, Tarbier M, Friedländer MR, Visa N (2018) Next-generation sequencing reveals two populations of damage-induced small RNAs at endogenous DNA double-strand breaks. Nucleic Acids Res 46:11869–11882

Borgognone A, Castanera R, Morselli M, López-Varas L, Rubbi L, Pisabarro AG, Pellegrini M, Ramírez L (2018) Transposon-associated epigenetic silencing during *Pleurotus ostreatus* life cycle. DNA Res 25:451–464

Butler DK (1992) Ribosomal DNA is a site of chromosome breakage in aneuploid strains of *Neurospora*. Genetics 131:581–592

Bzymek M, Lovett ST (2001) Instability of repetitive DNA sequences: the role of replication in multiple mechanisms. Proc Natl Acad Sci 98:8319–8325

Cai Q, He B, Kogel K-H, Jin H (2018a) Cross-kingdom RNA trafficking and environmental RNAi — nature's blueprint for modern crop protection strategies. Curr Opin Microbiol 46:58–64

Cai Q, Qiao L, Wang M, He B, Lin FM, Palmquist J, Da Huang S, Jin H (2018b) Plants send small RNAs in extracellular vesicles to fungal pathogen to silence virulence genes. Science 360:1126–1129

Calo S, Shertz-Wall C, Lee SC, Bastidas RJ, Nicolás FE, Granek JA, Mieczkowski P, Torres-Martínez S, Ruiz-Vázquez RM, Cardenas ME et al (2014) Antifungal drug resistance evoked via RNAi-dependent epimutations. Nature 513:555–558

Calo S, Nicolás FE, Lee SC, Vila A, Cervantes M, Torres-Martinez S, Ruiz-Vazquez RM, Cardenas ME, Heitman J (2017) A non-canonical RNA degradation pathway suppresses RNAi-dependent epimutations in the human fungal pathogen *Mucor circinelloides*. PLoS Genet 13:e1006686

Campbell JL, Turner BC (1987) Recombination block in the spore killer region of *Neurospora*. Genome 29:129–135

Campo S, Gilbert KB, Carrington JC (2016) Small RNA-based antiviral defense in the phytopathogenic fungus *Colletotrichum higginsianum*. PLoS Pathog 12:e1005640

Carreras-Villaseñor N, Esquivel-Naranjo EU, Villalobos-Escobedo JM, Abreu-Goodger C, Herrera-Estrella A (2013) The RNAi machinery regulates growth and development in the filamentous fungus *Trichoderma atroviride*. Mol Microbiol 89:96–112

Carthew RW, Sontheimer EJ (2009) Origins and mechanisms of miRNAs and siRNAs. Cell 136:642–655

Catalanotto C, Azzalin G, Macino G, Cogoni C (2002) Involvement of small RNAs and role of the *qde* genes in the gene silencing pathway in *Neurospora*. Genes Dev 16:790–795

Cerutti H, Casas-Mollano JA (2006) On the origin and functions of RNA-mediated silencing: from protists to man. Curr Genet 50:81–99

Cervantes M, Vila A, Nicolás FE, Moxon S, de Haro JP, Dalmay T, Torres-Martínez S, Ruiz-Vázquez RM (2013) A single argonaute gene participates in exogenous and endogenous RNAi and controls cellular functions in the basal fungus *Mucor circinelloides*. PLoS One 8:e69283

Chang S-S, Zhang Z, Liu Y (2012) RNA interference pathways in fungi: mechanisms and functions. Annu Rev Microbiol 66:305–323

Chang Z, Billmyre RB, Lee SC, Heitman J (2019) Broad antifungal resistance mediated by RNAi-dependent epimutation in the basal human fungal pathogen *Mucor circinelloides*. PLoS Genet 15:e1007957

Chen R, Jiang N, Jiang Q, Sun X, Wang Y, Zhang H, Hu Z (2014) Exploring microRNA-like small RNAs in the filamentous fungus *Fusarium oxysporum*. PLoS One 9:e104956

Choi J, Kim KT, Jeon J, Wu J, Song H, Asiegbu FO, Lee YH (2014) FunRNA: a fungi-centered genomics platform for genes encoding key components of RNAi. BMC Genomics 15:S14

Cogoni C, Irelan JT, Schumacher M, Schmidhauser TJ, Selkerl EU, Macino G, Umana B, Cellulare B (1996) Transgene silencing of the *al-1* gene in vegetative cells of *Neurospora* is mediated by a cytoplasmic effector and does not depend on DNA-DNA interactions or DNA methylation. EMBO J 1:3153–3163

Dahlmann TA, Kück U (2015) Dicer-dependent biogenesis of small RNAs and evidence for microRNA-like RNAs in the penicillin producing fungus *Penicillium chrysogenum*. PLoS One 10:e0125989

Dang Y, Li L, Guo W, Xue Z, Liu Y (2013) Convergent transcription induces dynamic DNA methylation at disiRNA loci. PLoS Genet 9:e1003761

Dang Y, Cheng J, Sun X, Zhou Z, Liu Y (2016) Antisense transcription licenses nascent transcripts to mediate transcriptional gene silencing. Genes Dev 30:2417–2432

de Haro JP, Calo S, Cervantes M, Nicolás FE, Torres-Martínez S, Ruiz-Vázquez RM (2009) A single *dicer* gene is required for efficient gene silencing associated with two classes of small antisense RNAs in *Mucor circinelloides*. Eukaryot Cell 8:1486–1497

Donaldson ME, Saville BJ (2012) Natural antisense transcripts in fungi. Mol Microbiol 85:405–417

Donaldson ME, Saville BJ (2013) *Ustilago maydis* natural antisense transcript expression alters mRNA stability and pathogenesis. Mol Microbiol 89:29–51

Drinnenberg IA, Weinberg DE, Xie KT, Mower JP, Wolfe KH, Fink GR, Bartel DP (2009) RNAi in budding yeast. Science 326:544–550

Drinnenberg IA, Fink GR, Bartel DP (2011) Compatibility with killer explains the rise of RNAi-deficient fungi. Science 333:1592

Dumesic PA, Natarajan P, Chen C, Drinnenberg IA, Schiller BJ, Thompson J, Moresco JJ, Yates JR, Bartel DP, Madhani HD (2013) Stalled spliceosomes are a signal for RNAi-mediated genome defense. Cell 152:957–968

Ellendorff U, Fradin EF, De Jonge R, Thomma BPHJ (2009) RNA silencing is required for *Arabidopsis* defence against *Verticillium* wilt disease. J Exp Bot 60:591–602

Esteller M (2011) Non-coding RNAs in human disease. Nat Rev Genet 12:861–874

Faghihi MA, Wahlestedt C (2009) Regulatory roles of natural antisense transcripts. Nat Rev Mol Cell Biol 10:637–643

Feretzaki M, Billmyre RB, Clancey SA, Wang X, Heitman J (2016) Gene network polymorphism illuminates loss and retention of novel RNAi silencing components in the *Cryptococcus* pathogenic species complex. PLoS Genet 12:1005868

Francia S, Michelini F, Saxena A, Tang D, De Hoon M, Mione M, Carninci P (2013) Site-specific DICER and DROSHA RNA products control the DNA damage response. Nature 488:231–235

Friedman S, Freitag M (2017) Centrochromatin of Fungi. In: Black BE (ed) Centromeres and kinetochores. Springer International Publishing AG, Cham, pp 85–109

Friedman RC, Farh KKH, Burge CB, Bartel DP (2009) Most mammalian mRNAs are conserved targets of microRNAs. Genome Res 19:92–105

Gao D, Jiang N, Wing RA, Jiang J, Jackson SA (2015) Transposons play an important role in the evolution and diversification of centromeres among closely related species. Front Plant Sci 6:216

Gazzani S, Lawrenson T, Woodward C, Headon D, Sablowski R (2008) A link between mRNA turnover and RNA interference in *Arabidopsis*. Science 1046:1046–1049

Ghildiyal M, Zamore PD (2009) Small silencing RNAs: an expanding universe. Nat Rev Genet 10:94–108

Halic M, Moazed D (2010) Dicer-independent primal RNAs trigger RNAi and heterochromatin formation. Cell 140:504–516

Hammond TM (2017) Sixteen years of meiotic silencing by unpaired DNA. Adv Genet 97:1–42

Hammond TM, Andrewski MD, Roossinck MJ, Keller NP (2008) *Aspergillus* mycoviruses are targets and suppressors of RNA silencing. Eukaryot Cell 7:350–357

Hammond TM, Xiao H, Boone EC, Perdue TD, Pukkila PJ, Shiu PKT (2011) SAD-3, a putative helicase required for meiotic silencing by unpaired DNA, interacts with other components of the silencing machinery. G3 (Bethesda) 1:369–376

Hammond TM, Xiao H, Boone EC, Decker LM, Lee SA, Perdue TD, Pukkila PJ, Shiu PKT (2013a) Novel proteins required for meiotic silencing by unpaired DNA and siRNA generation in *Neurospora crassa*. Genetics 194:91–100

Hammond TM, Spollen WG, Decker LM, Blake SM, Springer GK, Shiu PKT (2013b) Identification of small RNAs associated with meiotic silencing by unpaired DNA. Genetics 194:279–284

Harvey JJW, Lewsey MG, Patel K, Westwood J, Heimstädt S, Carr JP, Baulcombe DC (2011) An antiviral defense role of AGO2 in plants. PLoS One 6:e14639

Harvey AM, Rehard DG, Groskreutz KM, Kuntz DR, Sharp K, Shiu PK, Hammond TM (2014) A critical component of meiotic drive in *Neurospora* is located near a chromosome rearrangement. Genetics 197:1165–1174

Janbon G, Maeng S, Yang DH, Ko YJ, Jung KW, Moyrand F, Floyd A, Heitman J, Bahn YS (2010) Characterizing the role of RNA silencing components in *Cryptococcus neoformans*. Fungal Genet Biol 47:1070–1080

Jeang KT (2012) RNAi in the regulation of mammalian viral infections. BMC Biol 10:58

Jiang X, Qiao F, Long Y, Cong H, Sun H (2017) MicroRNA-like RNAs in plant pathogenic fungus *Fusarium oxysporum* f. sp. *niveum* are involved in toxin gene expression fine tuning. 3 Biotech 7:354

Kadotani N, Nakayashiki H, Tosa Y, Mayama S (2004) One of the two dicer-like proteins in the filamentous fungi *Magnaporthe oryzae* genome is responsible for hairpin RNA-triggered RNA silencing and related small interfering RNA accumulation. J Biol Chem 279:44467–44474

Kang K, Zhong J, Jiang L, Liu G, Gou CY, Wu Q, Wang Y, Luo J, Gou D (2013) Identification of microRNA-like RNAs in the filamentous fungus *Trichoderma reesei* by Solexa sequencing. PLoS One 8:e76288

Koch A, Kumar N, Weber L, Keller H, Imani J, Kogel K-H (2013) Host-induced gene silencing of cytochrome P450 lanosterol C14 -demethylase-encoding genes confers strong resistance to *Fusarium* species. Proc Natl Acad Sci 110:19324–19329

Kronholm I, Johannesson H, Ketola T (2016) Epigenetic control of phenotypic plasticity in the filamentous fungus *Neurospora crassa*. G3 (Bethesda) 6:4009–4022

Kuan T, Zhai Y, Ma W (2016) Small RNAs regulate plant responses to filamentous pathogens. Semin Cell Dev Biol 56:190–200

Lau SKP, Chow WN, Wong AYP, Yeung JMY, Bao J, Zhang N, Lok S, Woo PCY, Yuen KY (2013) Identification of microRNA-like RNAs in mycelial and yeast phases of the thermal dimorphic fungus *Penicillium marneffei*. PLoS Negl Trop Dis 7:e2398

Lau AYT, Cheng X, Cheng CK, Nong W, Cheung MK, Chan RHF, Hui JHL, Kwan HS (2018) Discovery of microRNA-like RNAs during early fruiting body development in the model mushroom *Coprinopsis cinerea*. PLoS One 13:e0198234

Laurie JD, Linning R, Bakkeren G (2008) Hallmarks of RNA silencing are found in the smut fungus *Ustilago hordei* but not in its close relative *Ustilago maydis*. Curr Genet 53:49–58

Lee RC, Feinbaum RL, Ambros V (1993) The *C. elegans* heterochronic gene *lin-4* encodes small RNAs with antisense complementarity to *lin-14*. Cell 75:843–854

Lee DW, Pratt RJ, McLaughlin M, Aramayo R (2003) An Argonaute-like protein is required for meiotic silencing. Genetics 164:821–828

Lee HC, Chang S-S, Choudhary S, Aalto AP, Maiti M, Bamford DH, Liu Y (2009) qiRNA, a new type of small interfering RNA induced by DNA damage. Nature 459:274–277

Lee HC, Li L, Gu W, Xue Z, Crosthwaite SK, Pertsemlidis A, Lewis ZA, Freitag M, Selker EU, Mello CC, Liu Y (2010) Diverse pathways generate microRNA-like RNAs and dicer-independent small interfering RNAs in fungi. Mol Cell 38:803–814

Lin YL, Ma LT, Lee YR, Lin SS, Wang SY, Chang TT, Shaw JF, Li WH, Chu FH (2015) MicroRNA-like small RNAs prediction in the development of *Antrodia cinnamomea*. PLoS One 10:e0123245

Liu Y, Lee HC, Aalto AP, Yang Q, Chang SS, Huang G, Fisher D, Cha J, Poranen MM, Bamford DH (2010) The DNA/RNA-dependent RNA polymerase QDE-1 generates aberrant RNA and dsRNA for RNAi in a process requiring replication protein a and a DNA helicase. PLoS Biol 8:e1000496

López-Camarillo C, Marchat LA (2013) MicroRNAs in cancer. Annu Rev Pathol 9:287–314

Lu WT, Hawley BR, Skalka GL, Baldock RA, Smith EM, Bader AS, Malewicz M, Watts FZ, Wilczynska A, Bushell M (2018) Drosha drives the formation of DNA:RNA hybrids around DNA break sites to facilitate DNA repair. Nat Commun 9:532

Martienssen R, Moazed D (2015) RNAi and heterochromatin. Cold Spring Harb Perspect Biol 7:a019323

Meussen BJ, De Graaff LH, Sanders JPM, Weusthuis RA (2012) Metabolic engineering of *Rhizopus oryzae* for the production of platform chemicals. Appl Microbiol Biotechnol 94:875–886

Michalik KM, Böttcher R, Förstemann K (2012) A small RNA response at DNA ends in *Drosophila*. Nucleic Acids Res 40:9596–9603

Michielse CB, Salim K, Ragas P, Ram AFJ, Kudla B, Jarry B, Punt PJ, Van Den Hondel CAMJJ (2004) Development of a system for integrative and stable transformation of the zygomycete *Rhizopus oryzae* by agrobacterium-mediated DNA transfer. Mol Gen Genomics 271:499–510

Mochama P, Jadhav P, Neupane A, Marzano SYL (2018) Mycoviruses as triggers and targets of RNA silencing in white mold fungus *Sclerotinia sclerotiorum*. Viruses 10:214

Muszewska A, Steczkiewicz K, Stepniewska-Dziubinska M, Ginalski K (2017) Cut-and-paste transposons in fungi with diverse lifestyles. Genome Biol Evol 9:3463–3477

Nakayashiki H, Kadotani N, Mayama S (2006) Evolution and diversification of RNA silencing proteins in fungi. J Mol Evol 63:127–135

Napoli C, Lemieux C, Jorgensen R (1990) Introduction of a chimeric chalcone synthase gene into Petunia results in reversible co-suppression of homologous genes in trans. Plant Cell 2:279–289

Nevers A, Doyen A, Malabat C, Néron B, Kergrohen T, Jacquier A, Badis G (2018) Antisense transcriptional interference mediates condition-specific gene repression in budding yeast. Nucleic Acids Res 46:6009–6025

Nicolás FE, Torres-Martínez S, Ruiz-Vázquez RM (2003) Two classes of small antisense RNAs in fungal RNA silencing triggered by non-integrative transgenes. EMBO J 22:3983–3991

Nicolás FE, de Haro JP, Torres-Martínez S, Ruiz-Vázquez RM (2007) Mutants defective in a *Mucor circinelloides dicer*-like gene are not compromised in siRNA silencing but display developmental defects. Fungal Genet Biol 44:504–516

Nicolas FE, Moxon S, de Haro JP, Calo S, Grigoriev IV, Torres-Martínez S, Moulton V, Ruiz-Vázquez RM, Dalmay T (2010) Endogenous short RNAs generated by dicer 2 and RNA-dependent RNA polymerase 1 regulate mRNAs in the basal fungus *Mucor circinelloides*. Nucleic Acids Res 38:5535–5541

Nicolás FE, Torres-Martínez S, Ruiz-Vázquez RM (2013) Loss and retention of RNA interference in fungi and parasites. PLoS Pathog 9:e1003089

Nicolás FE, Vila A, Moxon S, Cascales MD, Torres-Martínez S, Ruiz-Vázquez RM, Garre V (2015) The RNAi machinery controls distinct responses to environmental signals in the basal fungus *Mucor circinelloides*. BMC Genomics 16:237

Noble LM, Holland LM, McLauchlan AJ, Andrianopoulos A (2016) A plastic vegetative growth threshold governs reproductive capacity in *Aspergillus nidulans*. Genetics 204:1161–1175

Nolan T, Braccini L, Azzalin G, De Toni A, Macino G, Cogoni C (2005) The post-transcriptional gene silencing machinery functions independently of DNA methylation to repress a LINE1-like retrotransposon in *Neurospora crassa*. Nucleic Acids Res 33:1564–1573

Nowara D, Gay A, Lacomme C, Shaw J, Ridout C, Douchkov D, Hensel G, Kumlehn J, Schweizer P (2010) HIGS: host-induced gene silencing in the obligate biotrophic fungal pathogen *Blumeria graminis*. Plant Cell 22:3130–3141

Nunes CC, Dean RA (2012) Host-induced gene silencing: a tool for understanding fungal host interaction and for developing novel disease control strategies. Mol Plant Pathol 13:519–529

Panwar V, McCallum B, Bakkeren G (2013) Host-induced gene silencing of wheat leaf rust fungus *Puccinia triticina* pathogenicity genes mediated by the barley stripe mosaic virus. Plant Mol Biol 81:595–608

Papp T, Csernetics Á, Nagy G, Bencsik O, Iturriaga EA, Eslava AP, Vágvölgyi C (2013) Canthaxanthin production with modified *Mucor circinelloides* strains. Appl Microbiol Biotechnol 97:4937–4950

Peres da Silva R, Puccia R, Rodrigues ML, Oliveira DL, Joffe LS, César GV, Nimrichter L, Goldenberg S, Alves LR (2015) Extracellular vesicle-mediated export of fungal RNA. Sci Rep 5:7763

Pickford AS, Cogoni C (2003) RNA-mediated gene silencing. Cell Mol Life Sci 60:871–882

Pidoux AL, Allshire RC (2005) The role of heterochromatin in centromere function. Philos Trans R Soc Lond Ser B Biol Sci 360:569–579

Raju NB, Metzenberg RL, Shiu PK (2007) *Neurospora* spore killers *Sk-2* and *Sk-3* suppress meiotic silencing by unpaired DNA. Genetics 176:43–52

Ramakrishnan M, Naga Sowjanya T, Raj KB, Kasbekar DP (2011) Meiotic silencing by unpaired DNA is expressed more strongly in the early than the late perithecia of crosses involving most wild-isolated *Neurospora crassa* strains and in self-crosses of *N. tetrasperma*. Fungal Genet Biol 48:1146–1152

Rayner S, Bruhn S, Vallhov H, Andersson A, Billmyre RB, Scheynius A (2017) Identification of small RNAs in extracellular vesicles from the commensal yeast *Malassezia sympodialis*. Sci Rep 7:39742

Romano N, Macino G (1992) Quelling: transient inactivation of gene expression in *Neurospora crassa* by transformation with homologous sequences. Mol Microbiol 6:3343–3353

Roy B, Sanyal K (2011) Diversity in requirement of genetic and epigenetic factors for centromere function in fungi. Eukaryot Cell 10:1384–1395

Samarajeewa DA, Sauls PA, Sharp KJ, Smith ZJ, Xiao H, Groskreutz KM, Malone TL, Boone EC, Edwards KA, Shiu PKT et al (2014) Efficient detection of unpaired DNA requires a member of the rad54-like family of homologous recombination proteins. Genetics 198:895–904

Segers GC, Zhang X, Deng F, Sun Q, Nuss DL (2007) Evidence that RNA silencing functions as an antiviral defense mechanism in fungi. Proc Natl Acad Sci 104:12902–12906

Shao J, Chen H, Yang D, Jiang M, Zhang H, Wu B, Li J, Yuan L, Liu C (2017) Genome-wide identification and characterization of natural antisense transcripts by strand-specific RNA sequencing in *Ganoderma lucidum*. Sci Rep 7:5711

Shiu PKT, Metzenberg RL (2002) Meiotic silencing by unpaired DNA: properties, regulation and suppression. Genetics 161:1483–1495

Shiu PKT, Raju NB, Zickler D, Metzenberg RL (2001) Meiotic silencing by unpaired DNA. Cell 107:905–916

Slepecky RA, Starmer WT (2009) Phenotypic plasticity in fungi: a review with observations on *Aureobasidium pullulans*. Mycologia 101:823–832

Smits P, Smeitink JAM, van den Heuvel LP, Huynen MA, Ettema TJG (2007) Reconstructing the evolution of the mitochondrial ribosomal proteome. Nucleic Acids Res 35:4686–4703

Son H, Min K, Lee J, Raju NB, Lee YW (2011) Meiotic silencing in the homothallic fungus *Gibberella zeae*. Fungal Biol 115:1290–1302

Son H, Park AR, Lim JY, Shin C, Lee YW (2017) Genome-wide exonic small interference RNA-mediated gene silencing regulates sexual reproduction in the homothallic fungus *Fusarium graminearum*. PLoS Genet 13:e1006595

Stefani G, Slack FJ (2008) Small non-coding RNAs in animal development. Nat Rev Mol Cell Biol 9:219–230

Sun Q, Choi GH, Nuss DL (2009) A single Argonaute gene is required for induction of RNA silencing antiviral defense and promotes viral RNA recombination. Proc Natl Acad Sci 106:17927–17932

Tinoco MLP, Dias BBA, Dall'Astta RC, Pamphile JA, Aragão FJL (2010) In vivo trans-specific gene silencing in fungal cells by in planta expression of a double-stranded RNA. BMC Biol 8:27

Torres-Martínez S, Ruiz-Vázquez RM (2017) The RNAi universe in fungi: a varied landscape of small rnas and biological functions. Annu Rev Microbiol 71:371–391

Trieu TA, Calo S, Nicolás FE, Vila A, Moxon S, Dalmay T, Torres-Martínez S, Garre V, Ruiz-Vázquez RM (2015) A non-canonical rna silencing pathway promotes mRNA degradation in basal fungi. PLoS Genet 11:1005168

Trieu TA, Navarro-Mendoza MI, Pérez-Arques C, Sanchis M, Capilla J, Navarro-Rodriguez P, Lopez-Fernandez L, Torres-Martínez S, Garre V, Ruiz-Vázquez RM et al (2017) RNAi-based functional genomics identifies new virulence determinants in mucormycosis. PLoS Pathog 13:e1006150

Tucker JF, Ohle C, Schermann G, Bendrin K, Zhang W, Fischer T, Zhang K (2016) A novel epigenetic silencing pathway involving the highly conserved 5′-3′ exoribonuclease Dhp1/Rat1/Xrn2 in *Schizosaccharomyces pombe*. PLoS Genet 12: e1005873

Turner BC, Perkins DD (1979) Spore killer, a chromosomal factor in neurospora that kills meiotic products not containing it. Genetics 93:587–606

Ugolini I, Halic M (2018) Fidelity in RNA-based recognition of transposable elements. Philos Trans R Soc B Biol Sci 373:20180168

Vader G, Blitzblau HG, Tame MA, Falk JE, Curtin L, Hochwagen A (2011) Protection of repetitive DNA borders from self-induced meiotic instability. Nature 477:115–121

Villalobos-Escobedo JM, Herrera-Estrella A, Carreras-Villaseñor N (2016) The interaction of fungi with the environment orchestrated by RNAi. Mycologia 108:556–571

Volpe T, Martienssen RA (2011) RNA interference and heterochromatin assembly. Cold Spring Harb Perspect Biol 3:a003731

Wang X, Hsueh Y-P, Li W, Floyd A, Skalsky R, Heitman J (2010) Sex-induced silencing defends the genome of *Cryptococcus neoformans* via RNAi. Genes Dev 24:2566–2582

Wang X, Wang P, Sun S, Darwiche S, Idnurm A, Heitman J (2012) Transgene induced co-suppression during vegetative growth in *Cryptococcus neoformans*. PLoS Genet 8:e1002885

Wang X, Darwiche S, Heitman J (2013) Sex-induced silencing operates during opposite-sex and unisexual reproduction in *Cryptococcus neoformans*. Genetics 193:1163–1174

Wang Y, Smith KM, Taylor JW, Freitag M, Stajich JE (2015) Endogenous small RNA mediates meiotic silencing of a novel DNA transposon. G3 (Bethesda) 5:1949–1960

Wang M, Weiberg A, Lin FM, Thomma BPHJ, Da Huang H, Jin H (2016) Bidirectional cross-kingdom RNAi and fungal uptake of external RNAs confer plant protection. Nat Plants 2:1–10

Wei W, Ba Z, Gao M, Wu Y, Ma Y, Amiard S, White CI, Danielsen JMR, Yang YG, Qi Y (2012) A role for small RNAs in DNA double-strand break repair. Cell 149:101–112

Weiberg A, Jin H (2015) Small RNAs-the secret agents in the plant-pathogen interactions. Curr Opin Plant Biol 26:87–94

Weiberg A, Wang M, Lin FM, Zhao H, Zhang Z, Kaloshian I, Da Huang H, Jin H (2013) Fungal small RNAs suppress plant immunity by hijacking host RNA interference pathways. Science 342:118–123

Weiberg A, Bellinger M, Jin H (2015) Conversations between kingdoms: small RNAs. Curr Opin Biotechnol 32:207–215

Wilkins C, Dishongh R, Moore SC, Whitt MA, Chow M, Machaca K (2005) RNA interference is an antiviral defence mechanism in *Caenorhabditis elegans*. Nature 436:1044–1047

Wong LH, Choo KHA (2004) Evolutionary dynamics of transposable elements at the centromere. Trends Genet 20:611–616

Xiao H, Alexander WG, Hammond TM, Boone EC, Perdue TD, Pukkila PJ, Shiu PKT (2010) QIP, a protein that converts duplex siRNA into single strands, is required for meiotic silencing by unpaired DNA. Genetics 186:119–126

Yadav V, Sun S, Billmyre RB, Thimmappa BC, Shea T, Lintner R, Bakkeren G, Cuomo CA, Heitman J, Sanyal K (2018) RNAi is a critical determinant of centromere evolution in closely related fungi. Proc Natl Acad Sci 115:201713725

Yang Q, Li L, Xue Z, Ye Q, Zhang L, Li S, Liu Y (2013) Transcription of the major *Neurospora crassa* microRNA-like small RNAs relies on RNA polymerase III. PLoS Genet 9:e1003227

Yang Q, Ye QA, Liu Y (2015) Mechanism of siRNA production from repetitive DNA. Genes Dev 29:526–537

Yu J, Lee K-M, Cho WK, Park JY, Kim K-H (2018) Differential contribution of RNAi components in response to distinct *Fusarium graminearum* virus infections. J Virol 92:e01756–e01717

Zambon RA, Vakharia VN, Wu LP (2006) RNAi is an antiviral immune response against a dsRNA virus in *Drosophila melanogaster*. Cell Microbiol 8:880–889

Zhang X, Nuss DL (2008) A host dicer is required for defective viral RNA production and recombinant virus vector RNA instability for a positive sense RNA virus. Proc Natl Acad Sci 105:16749–16754

Zhang X, Segers GC, Sun Q, Deng F, Nuss DL (2008) Characterization of hypovirus-derived small RNAs generated in the chestnut blight fungus by an inducible DCL-2-dependent pathway. J Virol 82:2613–2619

Zhang Z, Chang SS, Zhang Z, Xue Z, Zhang H, Li S, Liu Y (2013) Homologous recombination as a mechanism to recognize repetitive DNA sequences in an RNAi pathway. Genes Dev 27:145–150

Zhang T, Zhao YL, Zhao JH, Wang S, Jin Y, Chen ZQ, Fang YY, Hua CL, Ding SW, Guo HS (2016) Cotton plants export microRNAs to inhibit virulence gene expression in a fungal pathogen. Nat Plants 2:16153

6 NLR Function in Fungi as Revealed by the Study of Self/Non-self Recognition Systems

Asen Daskalov[1,2], Witold Dyrka[3], Sven J. Saupe[4]

CONTENTS

I. Introduction

All organisms rely on innate immunity to cope appropriately with challenges imposed by a variety of biotic interactions. In particular both plants and animals employ different types of germ-line-encoded *p*attern *r*ecognition *r*eceptors (PRR) to mount a scaled response to various forms of microbial infection. In addition, some of these receptors also appear to play a role in establishing beneficial relationships making it increasingly clear that biotic interactions align along a continuum ranging from pathogenicity to symbiosis (Lipinski and Rosenstiel 2013; Sellge and Kufer 2015). Among these PRRs, the cytoplasmic NLR receptors represent a large and highly diversified family (NLR for *N*od-*l*ike *r*eceptors or *N*BS-*LRR*, nucleotide binding site and leucine-rich repeats) (Duxbury et al. 2016; Jones et al. 2016; Meunier and Broz 2017; Zhang et al. 2017). Plant and animal NLRs share a similar domain architecture and function. This similarity in domain organization is currently proposed to result from convergent evolution in plants and animals (Urbach and Ausubel 2017). These intracellular receptors have a typical tripartite domain architecture with a central nucleotide-binding and oligomerization domain (NOD) (Leipe et al. 2004) (most often of the NB-ARC type in plants and NACHT type in animals) flanked C-terminally by a LRR domain and N-terminally by an effector or signaling domain, thought to be responsible for mediating the downstream host responses including programmed cell death. In plant systems, one distinguishes PAMP-triggered immunity (PTI) where the immune response relies on the detection of pathogen-associated molecular patterns (PAMPs), from effector-triggered immunity (ETI), where the cell response is dependent on detection of modifications of the host state by pathogen-secreted effectors (Jones and Dangl 2006; Jones et al. 2016). The "guard model" was proposed as a frame for this form of immune response. The receptors ("guards") monitor the integrity of host proteins ("guardees") which are the target of pathogen effectors that either degrade or modify these targets as part of their host invasion scheme. As further

[1] The Plant and Microbial Biology Department, University of California, California, USA
[2] Institut Européen de Chimie et Biologie (IECB), Pessac, France
[3] Politechnika Wrocławska, Wydział Podstawowych Problemów Techniki, Katedra Inżynierii Biomedycznej, Wrocław, Poland
[4] Non-self Recognition in Fungi, UMR5095 Institut de Biochimie et Génétique Cellulaires, CNRS – Université de Bordeaux, Bordeaux Cedex, France; e-mail: sven.saupe@ibgc.cnrs.fr

Genetics and Biotechnology, 3rd Edition
The Mycota II
J.P. Benz, K. Schipper (Eds.)

refinement of this defense strategy, the plant immune system includes decoys representing mimics of effector targets that serve solely to detect pathogen intrusion. Although in many cases the mechanistic details of NLR function are still unknown (in particular in the case of plant NLRs), the general working paradigm for these receptors corresponds to a ligand-induced oligomerization model. The receptors are thought to recognize PAMPs or DAMPs (damage-associated molecular patterns) via the C-terminal LRR domain. Ligand binding then leads to receptor oligomerization mediated by the central nucleotide-binding domain. The oligomerization step into discrete-sized or open-ended oligomers then leads to activation of the N-terminal effector or signaling domain. This general scheme probably shows many variations and exceptions, for instance, elicitor binding was shown to involve the NOD domain rather than the C-terminal LRR domain in certain mouse NLRs (Tenthorey et al. 2014). The effector or signaling domains occurring at the N-terminus of NLRs differ in plant and animal lineages. Two main types are found in plants, the CC (for coiled coil) and TIR domains, while in animal lineages one finds predominantly death-fold domains (such as CARD, PYD, and DD), which engage into higher-order homotypic interactions (Jones et al. 2016; Vajjhala et al. 2017).

While this family of proteins is intensively studied in plant and animal lineages, less is known about fungal NLR homologs. A significant fraction of the work devoted to the characterization of NLR in fungi was developed in connection with the mechanistic dissection of a self/non-self recognition process known as heterokaryon incompatibility (Chevanne et al. 2009, 2010; Choi et al. 2012; Daskalov et al. 2015b; Heller et al. 2018; Paoletti et al. 2007; Saupe et al. 1995a; Zhang et al. 2014). Incompatibility occurs both in ascomycetes and basidiomycetes fungi when strains of unlike genotype undergo anastomosis (somatic cell fusion). The phenomenon was first described in the plant pathogen *Diaporthe perniciosa* in 1923 by Dorothy M. Cayley (a British mycologist coworker of W. Bateson in the "ladies lab" at the John Innes Center) and termed "mutual aversion" (Cayley 1923). Incompatibility is genetically controlled by so-called heterokaryon incompatibility loci (Aanen et al. 2010; Daskalov et al. 2017; Paoletti 2016; Saupe 2000). Incompatibility systems can either be allelic (incompatible alleles of the same locus) or non-allelic (incompatibility is caused by gene interactions at unlinked loci) (Pinan-Lucarre et al. 2007). Karl Esser, the historical editor of this compendium, was a pioneer in the biochemical and genetic dissection of incompatibility in *Podospora* (Esser 1965). Over the years it has become clear that in several species, incompatibility genes encode proteins displaying an NLR-like domain architecture. Although direct functional evidence for an immune function of NLRs in fungi is currently still lacking, a role for this class of proteins in non-self recognition processes and the control of programmed cell death is established in different species (Choi et al. 2012; Daskalov et al. 2015b; Heller et al. 2018; Saupe et al. 1995a). Further indirect evidence supports the hypothesis that fungal NLR-type proteins could play a similar role as in animals and plants (Paoletti and Saupe 2009; Uehling et al. 2017), raising the question of the evolutionary scenario explaining the possible common use for host defense of this type of receptor in these three lineages.

The aim of this brief chapter is to review the role of NLRs in fungal incompatibility and to propose a general description of NLR repertoires in fungal genomes. Similarities and differences of this protein family in fungal as compared to plant and animal lineages will be discussed. We will also review a specific mechanism of NLR-mediated signal transduction common in fungi and involving prion propagation of an amyloid fold. We discuss the importance of taking into account the proposed immune role of fungal NLRs to interpret the evolutionary trajectory of this protein architecture in plants, animals, and fungi (and bacteria).

II. NLRs in Fungal Incompatibility

Heterokaryon incompatibility manifests itself as a programmed cell death reaction occurring rapidly after cell fusion between genetically unlike strains differing at one or more *het* loci (Saupe 2000). Although incompatibility appears to be ubiquitous in filamentous fungi, the species *Neurospora crassa*, *Cryphonectria parasitica*, and *Podospora anserina* stand out so far as the only species in which heterokaryon incompatibility genes have been characterized at the molecular level. In each of these species, at least one of the known *het* systems involves an NLR protein. *het-d*, *het-e*, *het-r*, and *nwd2* in *Podospora* (Chevanne et al. 2009; Daskalov et al. 2015b; Espagne et al. 2002; Saupe et al. 1995a), *plp-1* in *Neurospora* (Heller et al. 2018), and *vic2* and *vic4* in *Cryphonectria* (Choi et al. 2012; Zhang et al. 2014) all correspond to genes encoding proteins with an NLR architecture (Fig. 1a).

In *P. anserina*, *het-e* and *het-c* form a non-allelic incompatibility system. The *het-e* gene encodes an NLR with a NACHT domain (in fact HET-e is one of the proteins used to initially define the NACHT domain; the H in the acronym stands for HET-e) (Koonin and Aravind 2000; Saupe et al. 1995a). The N-terminal domain of HET-e corresponds to a **HET domain,** a protein domain present in many

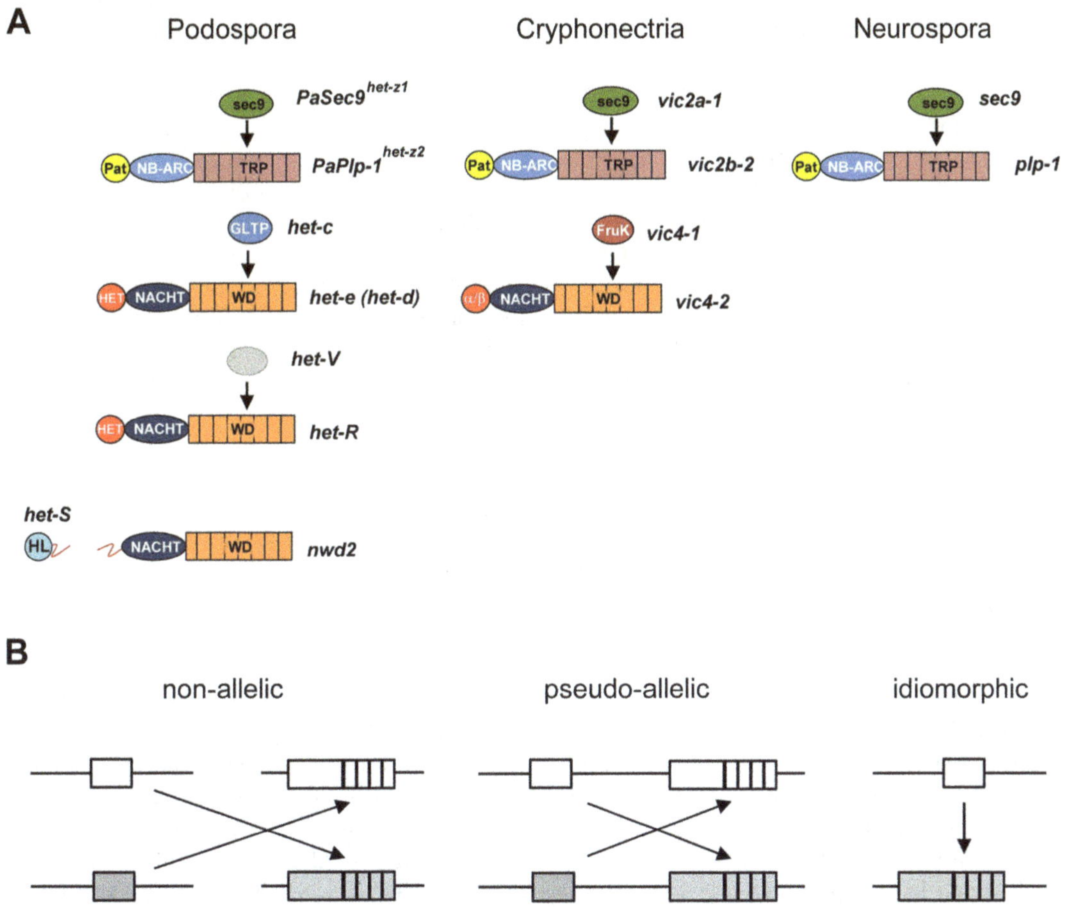

Fig. 1 Fungal heterokaryon incompatibility systems involving NLRs. (**a**) For each of the species, the incompatibility systems involving a NLR protein are represented with the corresponding gene designation and the protein domain annotation, Pat for patatin-like phospholipase, α/β for α/β-hydrolase, FruK for fructosamine kinase, GLTP for glycolipid transfer protein, HL for HeLo. (**b**) Categories of genetic interaction occurring in fungal incompatibility systems involving NLRs. The interacting genes are represented as boxes, the "guardee" genes as simple boxes, and the NLR "guard" gene as multiple boxes, the dashed part representing the repeat domain involved in recognition

proteins involved in heterokaryon incompatibility (Smith et al. 2000; Zhao et al. 2015) and found to show a remote homology to plant and mammalian TIR domains (Dyrka et al. 2014). The C-terminal domain of HET-e is a WD-repeat domain. *het-e* and *het-c* both exist as numerous alleles, and incompatibility is triggered when a specific allele of *het-e* interacts with a specific allele of *het-c* (Bastiaans et al. 2014; Saupe et al. 1995b). *het-c* encodes a glycolipid transfer protein that is conserved in plants and mammals and involved in transfer of glycolipids between cellular compartments. Recognition of proteins encoded by *het-c* alleles is mediated by the C-terminal **WD-repeat domain.** *het-e* presents many paralogs in *Podospora*, two of which are also *het*-genes. *het-d* defines a second incompatibility system in interaction with the *het-c* locus (Espagne et al. 2002), and *Het-r* defines an incompatibility system with a second gene termed *het-v* which has not been molecularly characterized at present (Chevanne et al. 2009). Two types of *het-r* alleles have been described: *het-R* alleles active in incompatibility and *het-r* alleles which are inactive. *het-R* and *het-r* alleles differ by the number of WD-repeats in their C-terminal domain. Inactivity of *het-r* alleles results from loss of repeats in the WD-repeat domain (Chevanne et al. 2009, 2010). NLRs are also involved, although indirectly, in allelic incompatibility systems in *Podospora*. The allelic *het-s/het-S* prion-dependent incompatibility system derives from an NLR-based signaling pathway controlled by a protein termed NWD2, which is encoded by a gene belonging to the same gene family as *het-d*, *het-e*, and *het-r*. NWD2 displays a NACHT domain and WD-repeats, but unlike the above described examples of incompatibility systems, this NLR is not required for the incompatibility reaction per se as will be developed in a later section (Daskalov et al. 2015b).

N. crassa is a major model system for the study of incompatibility, and incompatibility loci have been genetically identified in that species, and several of the corresponding genes have been cloned and functionally characterized (Daskalov et al. 2017; Garnjobst and Wilson 1956). In addition, in that species, **a programmed cell death reaction** has been found to occur after fusion of germ tubes from germinating conidia of unlike genotype (Heller et al. 2018). This **programmed germling death** is conceptually analogous to heterokaryon incompatibility although it was not identified using the classical assays employed to characterize *het* genes, namely, confrontation tests (barrage tests) or forced heterokaryons. Germling death is regulated by a two linked highly polymorphic loci with four haplotypes in *N. crassa* population samples (Heller et al. 2018). More specifically, death is mediated by non-allelic interactions between two linked genes, *plp-1* and *sec-9*. *Plp-1* encodes an NLR with a NB-ARC NOD domain, C-terminal TPR-repeats, and an N-terminal patatin-like phospholipase domain. *sec-9* codes for a **SNARE protein** involved in the membrane fusion process during exocytosis. Germling death is triggered by the physical interaction of PLP-1 and SEC-9 and involves PLP-1 oligomerization. Death is dependent on the predicted phospholipase activity of PLP-1 since a catalytic mutant of PLP-1 fails to induce cell death but retains the ability to recognize SEC-9. Remarkably, the orthologs of these genes in *Cryphonectria parasitica* correspond to the *vic2a* and *vic2b* incompatibility genes and to the *het-z* incompatibility locus of *P. anserina*, which also encodes *plp-1* and *sec-9* orthologs (Choi et al. 2012; Heller et al. 2018). This gene pair thus functions in non-self recognition and programmed cell death control in three distinct species. Furthermore, functional studies in *P. anserina* suggest that cell death induced by *het-z* during the incompatibility reaction shares many mechanistic similarities with germling-regulated death controlled by *plp-1* in *Neurospora*. Notably, in both cases, recognition specificity is mediated by the TPR domain, relies on a functional patatin-like domain, and is suppressed by a mutation of the P-loop motif of the NOD domain (Heller et al. 2018). The occurrence of an allorecognition system involving the same pair of genetic partners in these three species could either be explained by a scenario of long-term conservation of the allorecognition function or by convergent evolution and re-occurring recruitment of these genes as allorecognition genes. Phylogenetic studies support a scenario of convergent evolution, implying that **this locus has been repeatedly co-opted as a non-self recognition locus** (Heller et al. 2018).

In *C. parasitica*, the *vic4* locus displays two incompatible alleles, *vic4-1* and *vic4-2* (Fig. 1a). *vic4-2* encodes a NLR with an NACHT NOD domain and WD-repeats and a N-terminal signaling/effector domain with an α-/β-hydrolase fold, while *vic4-1* encodes a predicted small molecule kinase of the fructosamine kinase family (Pfam PF03881) (Zhang et al. 2014). Further functional studies will be required to determine whether this system shares mechanistic similarities with the other incompatibility systems involving NLRs, that is, whether the WD-repeats are involved in the interaction with the putative kinase and whether the predicted hydrolase represents a cell death effector domain.

Although not all incompatibility systems involve NLRs (Paoletti 2016), these studies (in three different and distantly related fungal species) reveal a frequent occurrence of members of this protein family among the products of *het* genes and related genes. All **these systems involve a protein partner that appears to be surveyed by the NLR.** Still, the genetic architecture of these systems is variable (Fig. 1b). In the *het-e/het-c* system, unlinked genes encode the NLR and the putative "guardee," leading to a classical non-allelic incompatibility system. In the case of *het-z* (*PaPlp1/PaSec9*), interactions between the NLR and the "guardee" are non-allelic, but since the gene pair is encoded at the same locus as a haplotype, classical genetics identified this system as allelic. Finally at *C. parasitica vic4* locus, interactions appear allelic, but the NLR and the "guardee" are encoded by idiomorphs (sequences encoded at the same locus but totally unrelated, a situation already described for mating-type incompatibility in *N. crassa*), (Glass et al. 1988).

III. In Silico Survey of Fungal NLR Repertoires

The functional studies reported above show that in several fungal species, proteins with a tripartite domain architecture (effector or signaling domain/NOD/superstructure-forming repeats) resembling plant and animal NLRs operate in allorecognition and the control of programmed cell death. These observations support the hypothesis that **NLR function is conserved from fungi to animals and plants,** prompting the description of the general occurrence of NLR-like proteins in fungal genomes. A survey of NLR occurrence, domain architecture, and variability was reported in 2014 (Dyrka et al. 2014). Here, we have repeated this survey with a similar methodology but with a larger dataset of complete fungal genomes now available (identified by a NCBI BioProject ID) (882 strains of 561 species versus 198 strains in 2014) (Fig. 2). **Fungal genomes harbor a variable number of NLR-encoding genes.** Considering only species with at least one hit (487 strains out of 882), the median number of NLRs per genome is 41 in all fungi (with 42 in Ascomycota and 49 in Basidiomycota). The mean number is 57 (52 in Ascomycota and 83 in Basidiomycota). As described for NLRs in plants, the number of NLRs varies greatly between species, reaching 602 in *Fibularhizoctonia* (the so-called cuckoo fungus producing sclerotia mimicking termite eggs) (Matsuura et al. 2009). Table 1 provides a list of the species with the highest NLR repertoire content, and Fig. 2b shows the abundance of NLR hits by phylogenetic class. The top-ranked groups are Basidiomycota, but NLR count can also be very high in Ascomycota in particular in species displaying symbiotic interactions with plants such as endophytic species like *Phialocephala subalpina* or forming ectomycorrhizal associations like *Cenococcum geophilum*. The NLR repertoires in fungi are generally a bit smaller than in plants and possibly larger than in most metazoan lineages (although large variations in repertoire size between and within lineages are also common) (Jones et al. 2016). NLRs were found in a large majority of the Pezizomycotina and Agaricomycetes genomes but were absent from Ascomycota and Basidiomycota yeast genomes (Saccharomycetes, Schizosaccharomycetes, and Tremellomycetes, which were dominated in the dataset by Cryptococcaceae) suggesting that presence of NLRs is associated with multicellularity (Fig. 2b).

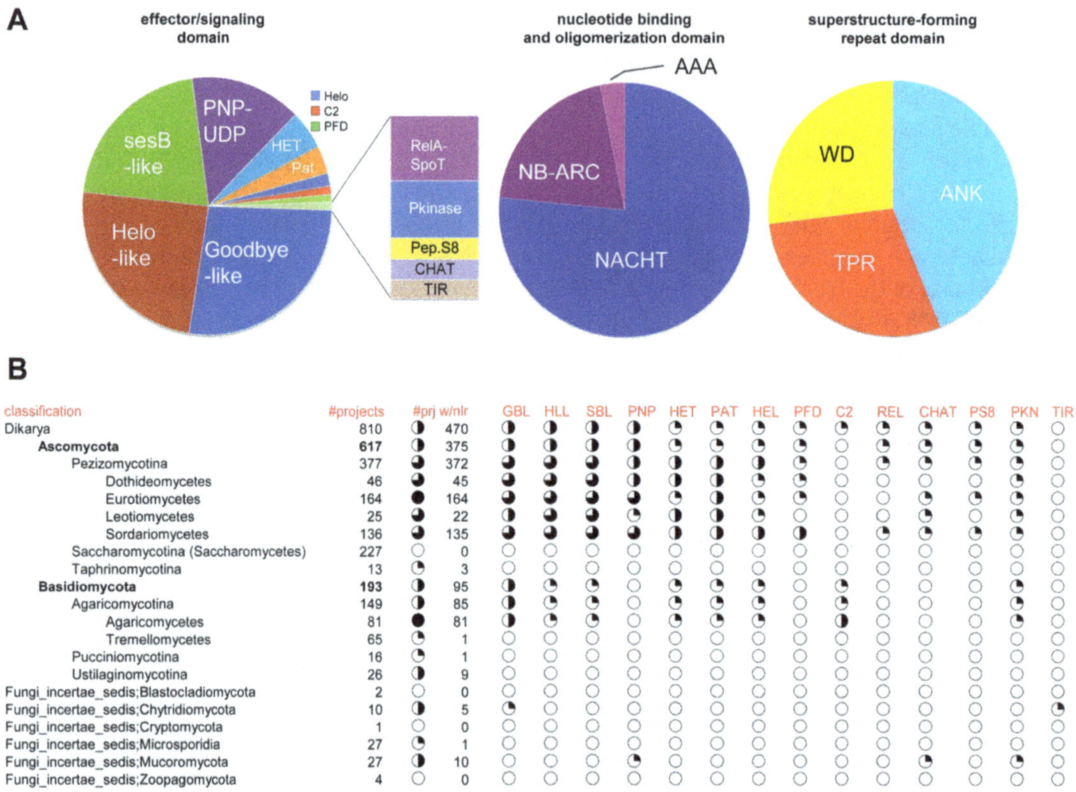

Fig. 2 Domain annotation of NLRs from fungal genomes. (**a**) Each of the pie charts shows the proportions of the indicated domain annotations in a total of 36,141 NLR proteins identified in fungal genomes. Proportions are given as a fraction of the domain that has received a given annotation over the total number of annotated domains in the N-terminal (left chart), NBD (middle chart), and C-terminal domain (right chart). (**b**) For each phylogenetic group, the total number of analyzed genomes (#projects) and the number of genomes with at least one NLR (#prj w/nlr) are given, as well as a semi-quantitative representation of the NLR abundance and specific N-terminal domain annotation in each phylogenetic group. GBL, Goodbye-like (overlaps with Pfam Goodbye (PF17109) and NACHT_N (PF17100)); HLL, HeLo-like (overlaps with Pfam STAND_N clan (CL0587) including HeLo_like_N (PF17111), SesA (PF17107)); SBL, sesb-like (overlaps with Pfam AB_hydrolase clan (CL0028)) (Dyrka et al. 2014); PNP, PNP_UDP_1 (PF01048); HET (PF06985); PAT, Pfam Patatin clan (CL0323); HEL, HeLo (PF14479); PFD, prion-forming domain like (overlaps with Pfam HET-S (PF17108), NACHT_sigma (PF17106), and ses_B (PF17046)) (Dyrka et al. 2014); C2 (PF00168); REL, RelA_SpoT (PF04607); CHAT (PF12770); PS8, Peptidase_S8 (PF00082); PKN, Pfam Pkinase clan (CL0016); and TIR, TIR_2 (PF01582, PF13676). The open circle denotes a fraction of genomes with NLR/domain annotation = 0, the first quadrant a fraction >0 and < 0.25, half-filled symbol a fraction ≥0.25 and < 0.75, third quadrant a fraction ≥0.75 and < 1, and the filled symbol a fraction = 1

The large majority of NOD domains of fungal NLRs are either of the NACHT or NB-ARC type, with NACHT being more frequent than NB-ARC (in a ~ 3:1 ratio) (Fig. 2a, middle chart) (Dyrka et al. 2014). Superstructure-forming repeats (SSFRs) are not LRRs as found in most plant and animal NLRs but of the ANK, TPR, and WD type in about equal prevalence (Dyrka et al. 2014) (Fig. 2a, right chart). ANK-repeats are more abundant in Ascomycota and WD-repeats in Basidiomycota. There is a preferential association of NB-ARC with TPR-repeats and of NACHT with ANK or WD-repeats. A variety of effector

Table 1 Fungal species with the highest NLR gene content

Species and strain	Number of NLRs	Phylum	Lifestyle
Fibularhizoctonia sp. *CBS 109695*	602	Basidiomycota	Termite-associated
Serendipita vermifera MAFF 305830	401	Basidiomycota	Orchid mycorrhizal
Amanita muscaria Koide BX008	390	Basidiomycota	Ectomycorrhizal
Gymnopus luxurians FD-317 M1	336	Basidiomycota	Saprotroph
Galerina marginata CBS 339.88	334	Basidiomycota	Saprotroph
Sphaerobolus stellatus SS14	333	Basidiomycota	Saprotroph
Laccaria amethystina LaAM-08-1	320	Basidiomycota	Ectomycorrhizal
Serendipita vermifera 'subsp. *bescii' NFPB0129*	306	Basidiomycota	Mycorrhizal
Phialocephala subalpina UAMH11012	302	Ascomycota	Endophytic
Serendipita indica DSM 11827	272	Basidiomycota	Mycorrhizal, endophytic
Cenococcum geophilum 1.58	262	Ascomycota	Ectomycorrhizal
Rhizoctonia solani AG2-2IIIB	242	Basidiomycota	Plant pathogenic
Piloderma croceum F 1598	242	Basidiomycota	Ectomycorrhizal
Meliniomyces bicolor E	234	Ascomycota	Ericoid-ectomycorrhizal
Cadophora sp. *DSE1049*	205	Ascomycota	Endophytic
Leucoagaricus sp. *SymC.cos*	203	Basidiomycota	Ectomycorrhizal
Meliniomyces variabilis F	186	Ascomycota	Ericoid, endophytic
Oidiodendron maius Zn	178	Ascomycota	Ericoid
Pezoloma ericae strain:UAMH 7357	176	Ascomycota	Ericoid, mycorrhizal
Trichoderma virens Gv29-8	169	Ascomycota	Mycoparasitic

domains can occupy the N-terminal position, the majority of which have no functional annotation (in particular in Basidiomycota) (Dyrka et al. 2014). Figure 2a (left chart) presents the 14 most frequent effector/signaling domains found in the fungal NLR set. The annotated domains fall into several broad categories. Several domains display enzymatic activities as lipases (sesB-like, patatin), proteases (peptidase S8, CHAT), or purine nucleoside phosphorylases (PNP-UDP). A second category corresponds to HeLo-related membrane-targeting cell death-inducing domains (HeLo-like and HeLo). Different types of amyloid signaling motifs (PFD-like as in prion-forming domain) were identified that function by amyloid templating and activation of downstream effectors as will be developed below. Finally, the HET domain, which shows a remote homology with the TIR domain prevalent in plant NLRs and animal immune receptors, is likely to function (as described for TIR domains) as a signaling domain engaging in homotypic interaction with downstream effectors also bearing a HET domain (Dyrka et al. 2014; Nimma et al. 2017). In comparison with the 2014 survey, two newcomers to the list of common N-terminal domains are the TIR and CHAT domains. TIR domains were found in a group of NLRs from the flagellated Chytridiomycota, *Rhizoclosmatium globosum*. The CHAT domain is a protease domain evolutionarily related to caspases (Aravind and Koonin 2002).

Considering these different annotation classes, one might propose that these N-terminal domains could either be termed signaling domains when their putative function is as adaptors that transmit the receptor activation signal to further downstream effector proteins (typically HET/TIR domains or prion amyloid signaling motifs), or alternatively as effector domains, where it is presumed or shown that the domain functions as a terminal execution module (as is, for instance, the case for the HeLo cell death-inducing domain which directly targets membrane integrity). Of course, due to the scarcity of functional studies, this distinction cannot be clearly made at present.

The functional annotation of the fungal NLR repertoires revealed a greater diversity of the N-terminal domain but also point to some **similarities with plant and animal NLRs** and also NLR-related proteins from bacteria. The HET and HeLo-like domains have been shown to exhibit a remote homology with the TIR domain (as already mentioned) and the RPW8 subclass of CC-type domains found in plant NLRs (Daskalov et al. 2016; Dyrka et al. 2014). Some fungal NLRs thus show domain architectures analogous to their plant counterparts (except that WD/ANK/TPR-repeats are found in place of the LRR). Similarly, nucleoside phosphorylase domains frequent in fungal NLRs have also been described as N-terminal domain of NLRs in the coral *Acropora digitifera* (Hamada et al. 2013), and Pkinase and a hydrolase domains (DUF676, which overlaps with our sesB-like annotation) were found as N-terminal domain of NLRs (with LRR-repeats) in the mosses *Physcomitrella patens* and *Sphagnum fallax*, respectively (Gao et al. 2018). Additionally, a number of NACHT or NB-ARC domain STAND proteins from multicellular bacteria (Cyanobacteria and Actinobacteria) display domain architectures similar to fungal NLRs (Dunin-Horkawicz et al. 2014; Koonin and Aravind 2002; Leipe et al. 2004; Urbach and Ausubel 2017) raising those question about the evolutionary history of the protein family as will be detailed in a following section.

IV. Evolution and Variability of NLR Repertoires in Fungi

Work on the *nwd* gene family of *P. anserina* which includes the *het-d*, *het-e*, and *het-r* incompatibility loci and *nwd2* has shown that the **WD-repeat units undergo concerted evolution within and between genes of the family** (Chevanne et al. 2010; Daskalov et al. 2015b; Paoletti et al. 2007). Genes of the *nwd* family show repeat number polymorphism and undergo frequent rearrangements, repeat loss, gain, and shuffling. This concerted evolution process is both a cause and consequence of the very high internal conservation of the repeats (homology between repeat unit ranges from 85 to 95%). At the same time, specific positions of the WD-repeat units are hypervariable and under positive diversifying selection (Paoletti et al. 2007). These positions are predicted by homology modeling to correspond to residues forming the interaction surface of the β-propeller structure formed cooperatively by the WD-repeats. It was suggested that this mechanism of **concerted evolution allows for hypervariability of the WD-repeat recognition domain** in the NWD proteins in relation with their non-self recognition function. NLRs from *P. anserina* with ANK- and TPR-repeats with high internal conservation were also analyzed for repeat number polymorphism and as observed for the *nwd* family, and high internal repeat conservation was correlated with frequent repeat number polymorphism in natural isolates (Dyrka et al. 2014; Marold et al. 2015). Again, as in the case of the WD-repeats, specific positions of the ANK- and TPR-repeats were hypervariable, under diversifying selection and mapped to residues forming the interaction surface of the ANK or TPR superstructures in homology modeling approaches. Bioinformatic analysis of the repeat domains in a collection of fungal NLRs revealed that high internal repeat conservation is also found in many other species suggesting that the proposed mechanism of recognition domain diversification might also be operating in a wide range of Basidiomycota and Ascomycota species (Dyrka et al. 2014). WD- and TPR-repeats with high internal conservation are also abundant in NACHT and NB-ARC proteins from Actinobacteria and Cyanobacteria (Table 2). In fact, WD-repeats with **high internal conservation are almost exclusively found in fungi and bacteria** (Cyanobacteria and Actinobacteria in particular) (Hu et al. 2017). It is likely that, akin to the situation described in *P. anserina*, concerted evolution in WD- and TPR-repeats provides adaptive hypervariability to other NLR-type proteins not only in fungi but also in Actinobacteria and Cyanobacteria. High internal conservation of WD-repeats is also observed in an NLR-type gene family from the red algae *Chondrus crispus* where 27 out of 50 members of the gene family display highly conserved repeats.

Of note here is the description of a highly expanded gene family encoding NACHT-ANK proteins in *Tuber melanosporum*. This gene family characterized by the presence of a variable number of introns (from 17 to up to 91 per gene) gives rise to codon-sized microexons (Iotti et al. 2012). Through alternative splicing, intron skipping, and retention, a variety of transcripts are generated that differ in sequence at the end of the

Table 2 Fraction of NLR architecture proteins with high internal conservation repeats in different phylogenetic classes

Group	NB-ARC TPR[a]	HiC	%	NACHT WD	HiC	%	NACHT ANK	HiC	%	NB-ARC LRR	HiC	%	NACHT LRR	HiC	%	NB-ARC WD	HiC	%
Actinobacteria	1136	27	2.4	46	7	**15.2**	0	0	n.a.	0	0	n.a.	0	0	n.a.	35	4	**11.4**
Cyanobacteria	198	88	**44.4**	118	9	7.6	0	0	n.a.	0	0	n.a.	0	0	n.a.	229	7	3
Ascomycota	169	47	**27.8**	429	82	**19.1**	740	43	5.8	0	0	n.a.	0	0	n.a.	1	0	0
Basidiomycota	32	6	**18.7**	270	39	**14.4**	15	0	0	0	0	n.a.	0	0	n.a.	4	1	**25**
Metazoa	12	0	0	370	0	0	17	1	5.8	0	0	n.a.	2654	12	0.4	219	0	0
Viridiplantae	0	0	n.a	0	0	n.a	0	0	n.a.	1230	12	0.9	53	1	1.8	13	0	0

[a]Domain architecture searches were performed with SMART (Letunic and Bork 2018) and repeat annotation with Xstream (Newman and Cooper 2007) with the following setting: i and $j > 0.8$; minimum period 15 aa; maximum period 45 aa; minimum copy number 5. (n.a., non-applicable)

ANK coding region at the 3′ end of the coding sequence. Together these observations suggest that there is an evolutionary necessity associated with the generation of variability in these fungal NLR-encoding genes. In the brown algae *Ectocarpus siliculosus*, NB-ARC TPR-encoding genes were also found to display an unusual gene organization with exons encoding one single TPR-repeat and exon shuffling taking place (Zambounis et al. 2012). Apparently, different evolutionary mechanisms ensure high functional variations in superstructure-forming repeat domains in NLR-type proteins.

Quite in line with what is known from the evolution of NLR repertoires in plant models, genes encoding proteins of this family in fungi apparently undergo **death-and-birth evolution** with common, lineage-, species-, or even strain-specific expansions (Dyrka et al. 2014; Fedorova et al. 2008; Iotti et al. 2012; Kubicek et al. 2011; Martin et al. 2008; Van der Nest et al. 2014; Zuccaro et al. 2011). For example, this phenomenon is apparent when the NLR repertoires are compared in species for which genomics data for several strains are available. A significant fraction of the NLRs are restricted to one strain, and NLR genes are more variable than the rest of the genome (Dyrka et al. 2014). Importantly, this gene family appears plastic in the sense that phylogenetic analyses suggest that domain architecture re-inventions frequently occur. Different effector/signaling domains are re-shuffled with distinct NOD and repeat domains. NLR domain modularity is illustrated by the fact that of the 60 possible domain architectures that can theoretically be achieved by free combinatorial association of the 10 most common N-terminal domain types with 2 classes of NBDs and 3 classes of repeat domains, 39 are actually found in our current dataset.

Considering the specific mechanisms of repeat domain diversification, the death-and-birth evolution with lineage-specific expansion, and the common domain architecture re-invention events, the NLR family in fungi stands out as highly dynamic and variable. A variety of comparative genomic approaches have pointed to the fact that **NLR-encoding genes show expansions in specific lineages.** Significant NLR gene expansion was reported in mycoparasitic, lichen-forming, mycorrhizal, and endophytic species (Kubicek et al. 2011; Martino et al. 2018; Peter et al. 2016; Wang et al. 2014). This observation is mirrored in Table 1 where many species with very high numbers of NLR genes show complex symbiotic/parasitic lifestyles. It is tempting to propose, as was suggested in the case of the coral *A. digitifera* (Hamada et al. 2013), that at least a part of the extensive NLR sets in these fungi has a role in the establishment and maintenance of a symbiotic lifestyle.

V. Prion-Forming Domains in NLR Signaling

A subset of N-terminal signaling domains of the fungal NLRs forms β-sheet-rich amyloid structures, whose inherent **prion properties play a crucial role in the signaling process** (Daskalov et al. 2012, 2015a, b, 2016; Loquet and Saupe 2017). Prion amyloids are strictly cooperative protein aggregates (fibers) exposing a supersecondary cross-β conformation, in which the β-sheets are perpendicular to the axis of the amyloid fiber. The high cooperativity and repetitiveness of the amyloid fold bring forth a remarkable feature of "self-propagation" (the prion aspect), where the amyloid serves as a molecular template for the unstructured primary sequence of a protein able to adopt a similar amyloid fold (Riek and Eisenberg 2016). The amyloid signal transduction is based on the controlled emergence and transmission (as templating) of structural information from the activated NLR receptor to a downstream executioner protein. In the current model, the oligomerization of an activated NLR protein brings the N-terminal unstructured amyloid motifs of the monomeric NLR molecules in close proximity (Daskalov et al. 2015b). The spatial clustering of these motifs induces **the cooperative folding of an amyloid structure**, which then templates the amyloid folding of an unstructured amyloid domain situated on the executioner protein. The NLR and its downstream effector protein represent a functional unit and are frequently encoded by adjacent genes in the genomes of various filamentous ascomycetes (Daskalov et al. 2012, 2015a).

The first NLR protein identified to carry an accessory amyloid domain was NWD2 (NACHT-WD) from *P. anserina* (Daskalov et al. 2015b). The *nwd2* gene is located adjacent to the *het-S* gene, which encodes the downstream effector of the NLR (Fig. 3a). HET-S is a 289 amino acids pore-forming protein consisting of a cytotoxic N-terminal HeLo domain and C-terminal prion forming domain (PFD) (Seuring et al. 2012). The PFD, or HET-S (218–289), transitions from an unstructured state in the inactive monomer to a left-handed β-helix (or β-solenoid) in the activated cytotoxic state. The adoption of the amyloid fold by the PFD triggers a conformational change in the HeLo domain resulting in the release of an N-terminal transmembrane α-helix, which induces plasma membrane damage and causes cell death. The HET-S PFD is used as a model to study the fundamental properties of amyloids and a solid-state NMR structure of the β-solenoid fold has been obtained more than a decade ago (Wasmer et al. 2008). The PFD consists of two 21 amino acid pseudo-repeats (termed R1 and R2) connected by a flexible 15 amino acid long loop (Fig. 3b, c). The repeats are alternately stacked along the axis of the β-solenoid, each contributing with four β-strands per ring (Fig. 3c). The resulting β-sheets delimit a highly hydrophobic triangular β-solenoid core. The core is tightly packed with predominantly hydrophobic residues while polar and charged residues are found on the exterior, solvent-facing side of the amyloid (Fig. 3d). Three salt bridges and two asparagine ladders are part of the β-solenoid and impact its stability and prion properties. While the PFD of HET-S contains two pseudo-repeats, NWD2 contains only one repeat, termed R0 (Fig. 3b). This R0 motif (NWD2(3-24)) is essential for the NLR signaling process, shows sequence homology to the R1 and R2 pseudo-repeats and adopts a related amyloid fold (Daskalov et al. 2015b) (Fig. 3b, d). It is proposed, based on the accumulated experimental evidence, that the oligomerization of NWD2 brings the R0 motifs of different NWD2 molecules in proximity so that these motifs cooperatively adopt a HET-S-like amyloid fold and subsequently template the PFD of HET-S to trigger the cytotoxic activity of the HeLo domain (Fig. 3a).

NLR-mediated amyloid signaling appears to be widespread in the fungal kingdom, and several other putative signal-transducing amyloid domains have been identified (Daskalov et al. 2012, 2015a, 2016). Noteworthy, some of these signal-transducing amyloids are evolutionarily related to the PFD of HET-S/s and are named HRAMs (HET-s-related amyloid motifs). The HRAMs have diversified into

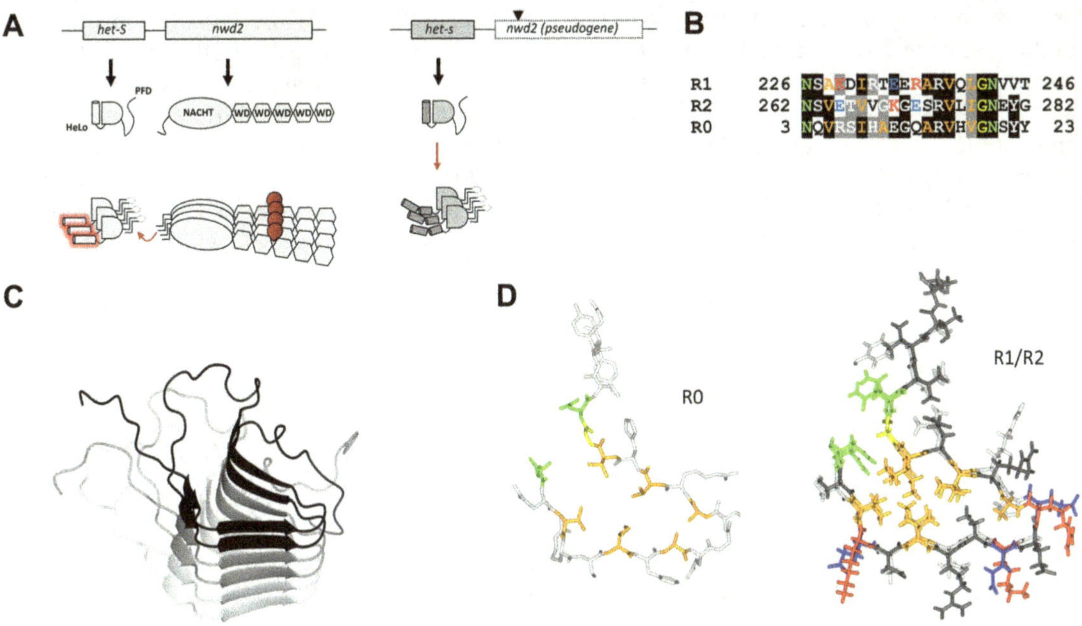

Fig. 3 Amyloid signaling cascade involving the NWD2 NLR and HET-S in *Podospora anserina*. (**a**) A schematic representation of the *nwd2/het-S* gene architecture in *het-S* (left) and *het-s* strains (right) in *Podospora anserina* as well as the domain organization of the corresponding proteins. The mode of activation of the HeLo domain of HET-S by amyloid templating of the C-terminal PFD domain by the N-terminal region of NWD2 is represented. In the current model, NWD2 undergoes ligand-induced oligomerization upon binding of a specific ligand (red shape) in the WD-repeat domain. Oligomerization is proposed to induce cooperative amyloid folding of the N-terminal region. The amyloid fold then serves as a template for conversion of the PFD region of HET-S. Amyloid folding of the PFD region in turn induces transconformation of the HeLo domain which exposes a N-terminal hydrophobic a-helix responsible for membrane-targeting and pore-forming activity. In *het-s* strains, *nwd2* is a pseudogene, and the gene is inactivated through integration of a transposable element (depicted by a black arrowhead). The HeLo domain of HET-s in altered in the N-terminal region and lacks pore-forming activity (in contrast to HET-S), HET-s can thus switch to the prion state without inducing cell death. (**b**) Alignment of the R1 and R2 repeat region of HET-s and the R0 region of NWD2. The color code highlights specific residues depicted in panel D. (**c**) Structure of HET-s(218-289) in the prion conformation. Each of the three monomers in the structural model is presented in a different shading (pdb:2kj3). (**d**) The predicted amyloid structure of the NWD2 RO region modeled after HET-s(218-289) is given together with the structure of the stacked R1 and R2 repeats of HET-s(218-289) (R1 layer in dark gray, R2 layer in light gray). The color code is as follows: the asparagine forming the two asparagine ladder (226/262 and 243/279) are given in green, the hydrophobic residues in the core in orange, the conserved glycine in the kink in yellow, and charged residues forming the three salt bridges in HET-s(218-289) in red and blue, respectively, for positively and negatively charged residues. Sequences in panel B use the same coloring code

discrete phylogenetic subtypes which include the R0 motifs of the corresponding NLR-like proteins, suggesting that a signal-transducing specificity exists for each subtype. Five different HRAM subtypes have been described, which share a pattern of alternating hydrophobic/polar (charged) residues (Daskalov et al. 2015a). The common primary sequence pattern and conservation of functionally important residues for the β-solenoid fold of HET-S suggest that the HRAMs adopt an overall similar amyloid fold. The discovery of the HRAMs establishes a superfamily of signal-transducing amyloids, which has prompted comparisons with the death domains superfamily (CARD, PYD, DD, and DED), involved in the control of apoptosis and inflammatory cell death in metazoans (Park et al. 2007). The death-fold domains form structurally similar α-helical bundles yet have distinct signaling specificities dependent also on homotypic interactions. In line with this conceptual analogy, it was demonstrated that the PFD of HET-S and the R0 NWD2 amyloid motif can replace the PYD domains of NLRP3 and ASC, underlying that prion-like polymerization is equally a feature of the death domains superfamily, although not based on controlled amyloid folding or amyloid propagation (Cai et al. 2014).

Besides the HRAMs, at least **two other signal-transducing amyloids have been identified** and named PP (pseudo-palindromic) and σ (sigma). The σ putative signaling amyloid was identified on the N-terminus of an NLR protein encoded by the gene *het-eN* in the genome of *Nectria haematococca* (Daskalov et al. 2012; Graziani et al. 2004). Adjacent to *het-eN* two other genes are found, *sesA* and *sesB*, encoding proteins carrying at their C-termini the σ putative amyloid domain. The length of the σ-motif is about 25 amino acid residues. The amino acid composition of the motif is biased toward predominantly G, N/Q, and aromatic residues. Unlike the HRAMs and their associated NLRs, the σ-motif found on the HET-eN NLR and its downstream effectors (SESA and SESB) are of equivalent length. Nonetheless, a common point with the HRAMs is the identification of sub-motifs composing the various σ-motifs. These sub-motifs, called A and B, are usually composed of eight or nine amino acids and arranged in three different patterns (AAA, ABA, ABB). There is currently no direct evidence for the amyloidogenic prion properties of the σ-motif; however the gene cluster (*het-eN/sesA/sesB*) has been previously associated in *N. haematococca* with an infectious phenomenon termed "secteur" caused by the σ infectious element (Graziani et al. 2004). The hypothesis that this σ infectious cytoplasmic element corresponds to a prion form of SESA and SESB is a plausible one.

More direct experimental evidence supports the amyloid prion properties of the PP signaling domain (Daskalov et al. 2016). In one reported case, the PP domains of an NLR and its effector protein have been shown to form prion amyloids. The NLR protein is PNT1 (PP-NACHT-TPR), and the effector protein is HELLP (HeLo-like PP), both encoded by a gene cluster in the genome of *Chaetomium globosum*, a species in the same taxonomic order as *P. anserina*. Like the *nwd2/het-S* cluster, the *pnt1/hellp* and *het-eN/sesA/sesB* gene clusters are present in the genomes of a variety of other ascomycete species. The distinct prion specificity of the PP amyloids, revealed in the lack of detected heterospecific interactions (cross-seeding) with the PFD of HET-S, offers experimental evidence that supports the existence of different signaling specificity for each amyloid domain. The pseudo-palindromic consensus sequence NxGϕQϕGxN, centered on the Q residue, is the most highly conserved central part of the 17/18 amino acids PP-motif. Remarkably, the **PP-motif shows strong sequence similarity to RHIM** (RIP homotypic interaction motif)—an amyloid controlling a pro-inflammatory programmed cell death reaction in metazoans termed necroptosis, which plays a role in innate immunity (Li et al. 2012). A solid-state NMR structure of a RHIM hetero-amyloid has been recently reported, where the RHIM motifs from two different proteins (RIPK1 and RIPK3) alternate to form two fronting β-sheets with a unique serpentine fold and a common hydrophobic core (Mompean et al. 2018). The amyloid core is structured around the association of the two central GϕQϕG motifs in which the similarity to the PP-motif lies, making it likely that fungal PP amyloids adopt a similar fold, which in turn would strengthen a model of long-term evolutionary conservation of programmed cell death-related amyloid signaling from fungi to animals.

VI. Evolutionary Origin of NLRs: Parallel Versus Convergent Evolution

When it was shown that the NOD1 protein in mammals detects bacterial lipopolysaccharides and thus functions as an immune receptor, the striking conclusion that plants and animals employ analogous receptors in immune functions soon followed (Inohara et al. 2001). The common NBS-LRR architecture of plant and animal immune receptors could either be explained **by parallel evolution or result from convergent evolution.** Several authors have attempted to resolve this issue by phylogenetic analyses in particular by tracing the phylogenies of NBS-SSFR proteins (proteins with NACHT or NB-ARC nucleotide-binding domains followed by superstructure-forming repeats) in different phyla. A recent study supports the convergent evolution model for emergence of the NBS-LRR architecture (Urbach and Ausubel 2017). The conclusion is based on the observation that proteins ancestral to plant R-proteins and animal NLRs did not have an NBS-LRR architecture. The NBS-LRR architecture is in fact rare and occurs almost exclusively in plants and metazoans. It is proposed that the last common ancestor of R-proteins and animal NLRs had a NBS-TPR structure and that the path to the NBS-LRR architecture probably involved intermediates of a NBS-WD architecture (for R-proteins) and non-repeat-associated NBS in the case of animal NLRs. The authors conclude that "it follows logically that the NBS-LRR architecture of plant R-proteins and metazoans NLRs evolved in independent events" (Urbach and Ausubel 2017). However, when the authors ask "what is so special about the NBS-LRR combination in the context of immune receptors" and discuss "what makes LRR-repeats any more suited to ligand-binding roles in immune receptors relative to other repeat domains," we feel that their perspective does not take into account the mounting and yet still partial evidence that NBS-SSFR proteins might have an immune-related function in fungi (and possibly also in multicellular bacteria). The central point here is that the existence of fungal proteins with NBS-TPR and NBS-WD architectures which induce programmed cell death in response to non-self makes it reasonable to assume that proposed more ancestral architectures (like NBS-TPR or NBS-WD) *already* displayed immune-related functions in a common ancestor. Urbach and Ausubel claim that NBS-TPR and NBS-WD proteins in eukaryotes are poorly characterized and that it is not known if they have an immune function or play a role in programmed cell death (Urbach and Ausubel 2017). We feel that the work on fungal NLRs challenges this view: fungal NACHT and NB-ARC STAND proteins controlling programmed cell death and non-self recognition are found in fungi. Similarly in the case of prokaryotic STAND proteins, Koonin has noted that the abundance of genes with NBS-TPR and NBS-WD architecture specifically in bacterial lineages with complex multicellular structures (Actinobacteria, Cyanobacteria, and some Proteobacteria) might be explained by the fact that some of these proteins have roles in host defense and programmed cell death (Koonin and Aravind 2002). Likewise, NBS-TPR proteins were proposed to act as immune receptors in brown algae (Zambounis et al. 2012). Urbach et al. show that acquisition of the NBS-LRR architecture in plants and animals is a relatively late event and occurred independently in plants and animal lineages (Urbach and Ausubel 2017). However if proteins with the other, possibly more ancestral E-NBS-SSFR architectures (effector/nucleotide-binding site/superstructure-forming repeats) function as immune receptors in fungi, multicellular bacteria, basal metazoans, basal plants, stramenopiles, and red algae (*Chondrus crispus*), then we are more likely dealing with parallel evolution of immune receptors with E-NBS-SSFR architectures (even though the late event of acquisition of LRR occurred as a convergent event). The existence of NBS-TPR and NBS-WD with immune-related functions represents—we feel—**a strong argument in favor of a global parallel evolutionary history for all E-NBS-SSFR architectures,** characterized here and there by convergent domain architecture re-inventions, lineage-specific losses or expansions, and possibly horizontal gene transfer events. The perspective adopted on NLR evolution comes down in some respect to a matter of semantics and depends on how NLRs are defined. We suggested, based on the functional studies in fungi and the genomic information on NLR-related proteins in different branches, that the NLR designation can be safely extended to

NACHT and NB-ARC domain proteins containing other types of SSFRs (Dyrka et al. 2014), while other authors insist that the NLR term should be reserved for proteins with the NBS-LRR architecture (Yuen et al. 2014). If NBS-SSFR proteins (with non-LRR-repeats) indeed act as non-self receptors in a variety of organisms, notably in fungi, then it appears advisable to frame the receptors under a common term. This nomenclature would then account for a possible long-term, parallel evolution of this protein family throughout the tree of life. This proposed parallel evolution apparently involved combinatorial recruitment and reassortment of effector, NBS and SSFR domains. As previously mentioned the resemblance of animal, fungal, plant, and bacterial NLR-related genes extends to the N-terminal effector/signaling domains. For instance, kinase domains and the DUF676 domain are found as N-terminal domains of NB-SSFR protein s in both fungi and bryophytes, and PNP_UDP_1 in fungi and corals, TIR domains are found in plants and bacteria and the related HET domain in fungi, and the HeLo-like domain is related to RPW8 domain. In this context, it has been proposed that STAND proteins (the class to which NLRs belong) have been acquired in eukaryotes through horizontal transfer from prokaryotic origin (Koonin and Aravind 2000, 2002). Therefore, in addition to repeated domain architecture re-invention events (as the ones identified in fungi), evolution of NLRs might involve horizontal transfer events.

The study by Urbach and Ausubel (2017) reveals the existence of a limited number of NBS-LRR proteins in non-plant and non-metazoan eukaryote lineages (6 proteins) and bacteria (12 proteins). Of those, one is of particular interest and found in the Planctomycetes *Gemmata obscuriglobus*. This predicted protein (GenBank: WP_010049624.1) displays a NACHT NBD domain and 13 C-terminal LRR-repeats with high internal repeat conservation and an N-terminal domain showing remote homology to death-fold domains (as detected by HHPred (Zimmermann et al. 2018)). The bacterial protein thus displays the same domain architecture as typical metazoan NLRs. Thus, although the NBS-LRR architecture is exceedingly rare outside of plant and animal lineages (Urbach and Ausubel 2017), it does exist, and the possibility that this architecture has also a long evolutionary history cannot be totally excluded.

VII. The Exaptation Model: How to Craft an Allorecognition System

If NLRs have a general function as immune receptors in fungi, then one needs to explain why genes of this family have repeatedly been identified in the context of fungal incompatibility. One plausible explanation for this trend is that fungal NLRs and other molecular determinants of **fungal incompatibility have been evolutionary co-opted to control heterokaryon formation from other molecular pathways, notably those controlling organismal defense**. This evolutionary co-optation, also termed exaptation (Gould and Vrba 1982), would result in some NLR genes being recruited to play a role in two distinct biological processes (allo- and hetero-recognition, understood as intra- and interspecific recognition) and/or to switch between these processes on the basis of the evolutionary pressures exercised on the locus. One tempting speculative model for the emergence of NLR-based incompatibility would be, in the context of the “guard” model, to co-express a “guard”—say *het-e* in *P. anserina*—with an independently diverged “guardee” allele of *het-c* from a different *Podospora* strain. Such “foreign” forms of *het-c* alleles would occasionally mimic an active state of the “guardee” for an independently evolved *het-e* and lead to the activation of the NLR, resulting in the emergence of an allorecognition system (Bastiaans et al. 2014; Paoletti and Saupe 2009). Selective forces driving gene diversification would accelerate the divergence between an independently evolving “guard” and a “guardee.” In support of this hypothesis, it has been reported that molecular marks of positive selection are found on *het-c* and the *nwd* genes in *Podospora*, although it is currently not clear if gene diversification is solely a consequence of the adaptive value of the allorecognition process itself (Bastiaans et al. 2014). The model of independently evolving “guard-guardee” pairs could be applied to the three NLR-encoding genes—*plp-1*, *het-z*, and *vic2*—triggering heterokaryon incompatibility with different *sec-9* homologs in *Neurospora*, *Podospora*, and *Cryphonectria*, respectively (Choi et al. 2012; Heller et al. 2018).

A parallel model linking NLRs and allorecognition would consist in bypassing the recognition step for the activation of the NLR and directly relying on the recruitment of its downstream executioner protein for the purposes of conspecific discrimination. One such incompatibility system is the *nwd2/het-S* gene cluster. In this example, mutations are proposed to have occurred in the *het-S* gene, generating a new allele termed *het-s* (small s) (Daskalov and Saupe 2015). A critical mutation at position 33 in the HeLo domain, replacing a histidine residue with a proline residue (H33P), renders the pore-forming protein unable to induce cell death. As a consequence, the *het-s* allele encodes a protein (the HET-s protein) that is nearly identical to HET-S but only with the prion-forming domain (PFD) remaining functional. This HET-s variant can exist as a soluble monomer, in a state termed [Het-s*] (star state), or as infectious aggregates (a prion state), termed [Het-s]. The appearance of the [Het-s] prion occurs sporadically at a low rate and corresponds to a conformational transition of the PFD of HET-s from unstructured—in the monomeric form of the protein—to the specific cross-β amyloid fold in the prion form. As the HeLo domain of the protein is non-functional, the prion state spreads in the colony, converting all HET-s monomers to adopt the amyloid prion fold, so that strains of *het-s* genotype are either prion-infected or prion-free. Only prion-infected [Het-s] strains trigger incompatibility with strains of the *het-S* genotype. The cell death reaction occurs when the amyloid death fold carried by the [Het-s] prion templates the PFD of the HET-S allelic variant, with a functional cytotoxic HeLo domain. Thus, the [Het-s] prion plays an analogous role to an activated NWD2, serving as a cell death trigger for HET-S. The *het-S* gene has been co-opted in the *het-S/het-s* incompatibility system, which likely operates only in *Podospora*, while the *nwd2/het-S* gene cluster is conserved in various other ascomycetes.

One way to frame **fungal incompatibility** is to consider it with an **autoimmune reaction**, analogous to what has been described as hybrid necrosis in plants (Bomblies et al. 2007; Bomblies and Weigel 2007). Hybrid incompatibilities in plants are post-mating genetic incompatibilities, characterized by dwarfism and tissue necrosis of the F1 offspring between genetically distinct parents. The phenomenon has been consistently linked to innate immunity genes in plants and notably R (resistance) genes, encoding the repertoire of plant NLRs (Chae et al. 2014). One striking example of plant autoimmunity is a lesion-mimicking mutant of *Arabidopsis thaliana*—*acd11* (*accelerated cell death 11*). Autoimmunity is caused by a mutation in the *ACD11* gene, which is a homolog of *het-c*, the allorecognition determinant from *P. anserina* (Brodersen et al. 2002). Remarkably, the autoimmune necrotic phenotype is dependent on an NLR receptor encoded by the gene *LAZ5* (Palma et al. 2010). While this situation is reminiscent of the *het-e/het-c* and *het-d/het-c* incompatibility systems in *Podospora*, where inappropriate NLR activation may be caused by disruption/perturbation of the "guardian-guardee" pair, other autoimmune reactions in plants seem to implicate different mechanisms. This is true, for example, for the hybrid necrosis triggered by co-expression of DM1 (DANGEROUS MIX 1) and DM2d, two distinct NLR proteins, an incompatibility mechanism not yet encountered in fungi (Tran et al. 2017).

Hybrid necrosis in plants and heterokaryon incompatibility in fungi may thus represent very similar phenomena, emerging from anomalous activation of genes evolving primarily to mediate heterospecific non-self recognition and detection of modified self. However, the adaptive value of the two phenomena is likely different. Notably, the fungal specific lifestyle, where colony establishment relies on anastomosis, occasionally between genetically distinct individuals, would integrate some of the sporadically emerging allorecognition systems to prevent conspecific parasitism and the spread of deleterious elements like mycoviruses. In that regard, one could argue that some of the co-opted NLRs are (re)utilized as immune receptors.

VIII. Conclusions

In this chapter, we recapitulate the state of knowledge on the large family of fungal NOD-like receptors, identified and investigated as allorecognition determinants in several filamentous ascomycete species. Subsequent in silico analyses revealed that NLR proteins are widespread in the fungal kingdom, display remarkable architectural variability, and carry accessory signaling/executioner domains, which appear in some cases homologous to counterparts in metazoan and plant NLRs. The accumulation of observational and experimental data has led to the hypothesis that the fungal NLRs are an integral part of a complex and yet to be fully described fungal immune system that may play roles both in host defense and symbiotic interactions. Future approaches will need to demonstrate the involvement of a fungal NLR protein in direct mediation of heterospecific non-self recognition to establish a broader functional equivalence between the fungal NLR type and the NLR receptors that have been shown to control innate immunity in other major eukaryotic lineages. This achievement would impact not only NLR biology but also our general understanding of the molecular evolution of non-self discrimination in eukaryotes and possibly beyond, in some prokaryotic species. The ascertainment that these proteins belong to the same family, have a shared evolutionary origin, and mediate heterospecific non-self discrimination in three of the main branches in Eukaryota, would validate these use of a common conceptual framework for the investigation of commonalities and particularities in the molecular mechanics and evolutionary strategies of diversification of the NOD-like receptors in plants, metazoans, and fungi (and perhaps also in multicellular prokaryotes). The likely prokaryotic origin of programmed cell death and innate immunity machinery has been repeatedly stressed (Aravind and Koonin 2002; Dunin-Horkawicz et al. 2014; Koonin and Aravind 2002). Some commonality appears in the genome content of NLR-type gene architectures in fungi and bacteria. In both cases, the architecture is prevalent in filamentous/multicellular species and rare or absent in unicellular species. NB-ARC TPR and NACHT WD architectures are abundant, and SSFRs with high internal repeat conservation occur frequently in both branches. It will be of great interest in the future to gain some insight into the evolutionary and functional basis of this resemblance in the NLR-type gene content in prokaryotic and eukaryotic multicellular microbes.

Many aspects of fungal biology are currently contemplated in the context of inter-organismal interactions (endophytic and mycorrhizal lifestyles, lichen formation, pathogenicity, mycoparasitism, fungal-bacterial interactions, control of secondary metabolism). If indeed the large and diversified repertoires of NLRs in fungi have a general role in sensing and responding to a variety of biotic cues, then it can safely be predicted that this family of proteins will turn out to be of some importance for the future of our understanding of the myriad of fungal lifestyle modalities.

Acknowledgments The authors thank Corinne Clavé for critical reading of the manuscript. This work was supported in part by the National Science Centre of Poland (grant no. 2015/17/D/ST6/04054) and an ANR grant (SFAS, ANR-17-CE11-0035-01). AD was supported by a Laboratory Directed Research and Development Program of Lawrence Berkeley National Laboratory under the US Department of Energy Contract No. DE-AC02-05CH11231.

References

Aanen DK, Debets AJM, Glass NL, Saupe SJ (2010) Biology and genetics of vegetative incompatibility in fungi. In: Borkovich KA, Ebbole DJ (eds) Cellular and molecular biology of filamentous fungi. Washington, DC, ASM Press, pp 274–288

Aravind L, Koonin EV (2002) Classification of the caspase-hemoglobinase fold: detection of new families and implications for the origin of the eukaryotic separins. Proteins 46:355–367

Bastiaans E, Debets AJ, Aanen DK, van Diepeningen AD, Saupe SJ, Paoletti M (2014) Natural variation of heterokaryon incompatibility gene het-c in *Podospora anserina* reveals diversifying selection. Mol Biol Evol 31:962–974

Bomblies K, Weigel D (2007) Hybrid necrosis: autoimmunity as a potential gene-flow barrier in plant species. Nat Rev Genet 8:382–393

Bomblies K, Lempe J, Epple P, Warthmann N, Lanz C, Dangl JL, Weigel D (2007) Autoimmune response as a mechanism for a Dobzhansky-Muller-type incompatibility syndrome in plants. PLoS Biol 5: e236

Brodersen P, Petersen M, Pike HM, Olszak B, Skov S, Odum N, Jorgensen LB, Brown RE, Mundy J (2002) Knockout of Arabidopsis accelerated-cell-death11 encoding a sphingosine transfer protein causes activation of programmed cell death and defense. Genes Dev 16:490–502

Cai X, Chen J, Xu H, Liu S, Jiang QX, Halfmann R, Chen ZJ (2014) Prion-like polymerization underlies signal transduction in antiviral immune defense and inflammasome activation. Cell 156:1207–1222

Cayley DM (1923) The phenonenon of mutual aversion between mono-spore mycelia of the same fungus (*Diaporthe perniciosa* march.) with a discussion on sex heterothallism in fungi. Jour Gen 13:353–370

Chae E, Bomblies K, Kim ST, Karelina D, Zaidem M, Ossowski S, Martin-Pizarro C, Laitinen RA, Rowan BA, Tenenboim H et al (2014) Species-wide genetic incompatibility analysis identifies immune genes as hot spots of deleterious epistasis. Cell 159:1341–1351

Chevanne D, Bastiaans E, Debets A, Saupe SJ, Clave C, Paoletti M (2009) Identification of the het-r vegetative incompatibility gene of *Podospora anserina* as a member of the fast evolving HNWD gene family. Curr Genet 55:93–102

Chevanne D, Saupe SJ, Clave C, Paoletti M (2010) WD-repeat instability and diversification of the *Podospora anserina* hnwd non-self recognition gene family. BMC Evol Biol 10:134

Choi GH, Dawe AL, Churbanov A, Smith ML, Milgroom MG, Nuss DL (2012) Molecular characterization of vegetative incompatibility genes that restrict hypovirus transmission in the chestnut blight fungus *Cryphonectria parasitica*. Genetics 190:113–127

Daskalov A, Saupe SJ (2015) As a toxin dies a prion comes to life: a tentative natural history of the [Het-s] prion. Prion 9:184–189

Daskalov A, Paoletti M, Ness F, Saupe SJ (2012) Genomic clustering and homology between HET-S and the NWD2 STAND protein in various fungal genomes. PLoS One 7:e34854

Daskalov A, Dyrka W, Saupe SJ (2015a) Theme and variations: evolutionary diversification of the HET-s functional amyloid motif. Sci Rep 5:12494

Daskalov A, Habenstein B, Martinez D, Debets AJ, Sabate R, Loquet A, Saupe SJ (2015b) Signal transduction by a fungal NOD-like receptor based on propagation of a prion amyloid fold. PLoS Biol 13: e1002059

Daskalov A, Habenstein B, Sabate R, Berbon M, Martinez D, Chaignepain S, Coulary-Salin B, Hofmann K, Loquet A, Saupe SJ (2016) Identification of a novel cell death-inducing domain reveals that fungal amyloid-controlled programmed cell death is related to necroptosis. Proc Natl Acad Sci U S A 113:2720–2725

Daskalov A, Heller J, Herzog S, Fleissner A, Glass NL (2017) Molecular mechanisms regulating cell fusion and heterokaryon formation in filamentous fungi. Microbiol Spectr 5(2). https://doi.org/10.1128/microbiolspec.FUNK-0015-2016

Dunin-Horkawicz S, Kopec KO, Lupas AN (2014) Prokaryotic ancestry of eukaryotic protein networks mediating innate immunity and apoptosis. J Mol Biol 426:1568–1582

Duxbury Z, Ma Y, Furzer OJ, Huh SU, Cevik V, Jones JD, Sarris PF (2016) Pathogen perception by NLRs in plants and animals: parallel worlds. BioEssays 38:769–781

Dyrka W, Lamacchia M, Durrens P, Kobe B, Daskalov A, Paoletti M, Sherman DJ, Saupe SJ (2014) Diversity and variability of NOD-like receptors in fungi. Genome Biol Evol 6:3137–3158

Espagne E, Balhadere P, Penin ML, Barreau C, Turcq B (2002) HET-E and HET-D belong to a new subfamily of WD40 proteins involved in vegetative incompatibility specificity in the fungus *Podospora anserina*. Genetics 161:71–81

Esser K (1965) Heterogenic incompatibility. In: Esser K, Raper JR (eds) Incompatibility in fungi. Springer, New York, pp 6–12

Fedorova ND, Khaldi N, Joardar VS, Maiti R, Amedeo P, Anderson MJ, Crabtree J, Silva JC, Badger JH, Albarraq A et al (2008) Genomic islands in the pathogenic filamentous fungus *Aspergillus fumigatus*. PLoS Genet 4:e1000046

Gao Y, Wang W, Zhang T, Gong Z, Zhao H, Han GZ (2018) Out of water: the origin and early diversification of plant R-genes. Plant Physiol 177:82–89

Garnjobst L, Wilson JF (1956) Heterocaryosis and protoplasmic incompatibility in *Neurospora crassa*. Proc Natl Acad Sci U S A 42:613–618

Glass NL, Vollmer SJ, Staben C, Grotelueschen J, Metzenberg RL, Yanofsky C (1988) DNAs of the two mating-type alleles of *Neurospora crassa* are highly dissimilar. Science 241:570–573

Gould JS, Vrba ES (1982) Exaptation-a missing term in the science of form. Paleobiology 8:4–15

Graziani S, Silar P, Daboussi MJ (2004) Bistability and hysteresis of the 'Secteur' differentiation are controlled by a two-gene locus in *Nectria haematococca*. BMC Biol 2:18

Hamada M, Shoguchi E, Shinzato C, Kawashima T, Miller DJ, Satoh N (2013) The complex NOD-like receptor repertoire of the coral *Acropora digitifera* includes novel domain combinations. Mol Biol Evol 30:167–176

Heller J, Clave C, Gladieux P, Saupe SJ, Glass NL (2018) NLR surveillance of essential SEC-9 SNARE proteins induces programmed cell death upon allorecognition in filamentous fungi. Proc Natl Acad Sci U S A 115:E2292

Hu XJ, Li T, Wang Y, Xiong Y, Wu XH, Zhang DL, Ye ZQ, Wu YD (2017) Prokaryotic and highly-repetitive WD40 proteins: a systematic study. Sci Rep 7:10585

Inohara N, Ogura Y, Chen FF, Muto A, Nunez G (2001) Human Nod1 confers responsiveness to bacterial lipopolysaccharides. J Biol Chem 276:2551–2554

Iotti M, Rubini A, Tisserant E, Kholer A, Paolocci F, Zambonelli A (2012) Self/nonself recognition in *Tuber melanosporum* is not mediated by a heterokaryon incompatibility system. Fungal Biol 116:261–275

Jones JD, Dangl JL (2006) The plant immune system. Nature 444:323–329

Jones JD, Vance RE, Dangl JL (2016) Intracellular innate immune surveillance devices in plants and animals. Science 354:aaf6395

Koonin EV, Aravind L (2000) The NACHT family – a new group of predicted NTPases implicated in apoptosis and MHC transcription activation. Trends Biochem Sci 25:223–224

Koonin EV, Aravind L (2002) Origin and evolution of eukaryotic apoptosis: the bacterial connection. Cell Death Differ 9:394–404

Kubicek CP, Herrera-Estrella A, Seidl-Seiboth V, Martinez DA, Druzhinina IS, Thon M, Zeilinger S, Casas-Flores S, Horwitz BA, Mukherjee PK et al (2011) Comparative genome sequence analysis underscores mycoparasitism as the ancestral life style of Trichoderma. Genome Biol 12:R40

Leipe DD, Koonin EV, Aravind L (2004) STAND, a class of P-loop NTPases including animal and plant regulators of programmed cell death: multiple, complex domain architectures, unusual phyletic patterns, and evolution by horizontal gene transfer. J Mol Biol 343:1–28

Letunic I, Bork P (2018) 20 years of the SMART protein domain annotation resource. Nucleic Acids Res 46:D493–D496

Li J, McQuade T, Siemer AB, Napetschnig J, Moriwaki K, Hsiao YS, Damko E, Moquin D, Walz T, McDermott A et al (2012) The RIP1/RIP3 necrosome forms a functional amyloid signaling complex required for programmed necrosis. Cell 150:339–350

Lipinski S, Rosenstiel P (2013) Debug your bugs – how NLRs shape intestinal host-microbe interactions. Front Immunol 4:479

Loquet A, Saupe SJ (2017) Diversity of amyloid motifs in NLR signaling in Fungi. Biomol Ther 7(2):E38

Marold JD, Kavran JM, Bowman GD, Barrick D (2015) A naturally occurring repeat protein with high internal sequence identity defines a new class of TPR-like proteins. Structure 23:2055–2065

Martin F, Aerts A, Ahren D, Brun A, Danchin EG, Duchaussoy F, Gibon J, Kohler A, Lindquist E, Pereda V et al (2008) The genome of Laccaria bicolor provides insights into mycorrhizal symbiosis. Nature 452:88–92

Martino E, Morin E, Grelet GA, Kuo A, Kohler A, Daghino S, Barry KW, Cichocki N, Clum A, Dockter RB et al (2018) Comparative genomics and transcriptomics depict ericoid mycorrhizal fungi as versatile saprotrophs and plant mutualists. New Phytol 217:1213–1229

Matsuura K, Yashiro T, Shimizu K, Tatsumi S, Tamura T (2009) Cuckoo fungus mimics termite eggs by producing the cellulose-digesting enzyme beta-glucosidase. Curr Biol 19:30–36

Meunier E, Broz P (2017) Evolutionary convergence and divergence in NLR function and structure. Trends Immunol 38:744–757

Mompean M, Li W, Li J, Laage S, Siemer AB, Bozkurt G, Wu H, McDermott AE (2018) The structure of the necrosome RIPK1-RIPK3 core, a human hetero-amyloid signaling complex. Cell 173(1244–1253):e1210

Newman AM, Cooper JB (2007) XSTREAM: a practical algorithm for identification and architecture modeling of tandem repeats in protein sequences. BMC Bioinform 8:382

Nimma S, Ve T, Williams SJ, Kobe B (2017) Towards the structure of the TIR-domain signalosome. Curr Opin Struct Biol 43:122–130

Palma K, Thorgrimsen S, Malinovsky FG, Fiil BK, Nielsen HB, Brodersen P, Hofius D, Petersen M, Mundy J (2010) Autoimmunity in Arabidopsis acd11 is mediated by epigenetic regulation of an immune receptor. PLoS Pathog 6(10):e1001137

Paoletti M (2016) Vegetative incompatibility in fungi: from recognition to cell death, whatever does the trick. Fungal Biol Rev 30:152–162

Paoletti M, Saupe SJ (2009) Fungal incompatibility: evolutionary origin in pathogen defense? BioEssays 31:1201–1210

Paoletti M, Saupe SJ, Clave C (2007) Genesis of a fungal non-self recognition repertoire. PLoS One 2:e283

Park HH, Lo YC, Lin SC, Wang L, Yang JK, Wu H (2007) The death domain superfamily in intracellular signaling of apoptosis and inflammation. Annu Rev Immunol 25:561–586

Peter M, Kohler A, Ohm RA, Kuo A, Krutzmann J, Morin E, Arend M, Barry KW, Binder M, Choi C et al (2016) Ectomycorrhizal ecology is imprinted in the genome of the dominant symbiotic fungus *Cenococcum geophilum*. Nat Commun 7:12662

Pinan-Lucarre B, Paoletti M, Clave C (2007) Cell death by incompatibility in the fungus Podospora. Semin Cancer Biol 17:101–111

Riek R, Eisenberg DS (2016) The activities of amyloids from a structural perspective. Nature 539:227–235

Saupe SJ (2000) Molecular genetics of heterokaryon incompatibility in filamentous ascomycetes. Microbiol Mol Biol Rev 64:489–502

Saupe S, Turcq B, Begueret J (1995a) A gene responsible for vegetative incompatibility in the fungus *Podospora anserina* encodes a protein with a GTP-binding motif and G beta homologous domain. Gene 162:135–139

Saupe S, Turcq B, Begueret J (1995b) Sequence diversity and unusual variability at the het-c locus involved in vegetative incompatibility in the fungus *Podospora anserina*. Curr Genet 27:466–471

Sellge G, Kufer TA (2015) PRR-signaling pathways: learning from microbial tactics. Semin Immunol 27:75–84

Seuring C, Greenwald J, Wasmer C, Wepf R, Saupe SJ, Meier BH, Riek R (2012) The mechanism of toxicity in HET-S/HET-s prion incompatibility. PLoS Biol 10:e1001451

Smith ML, Micali OC, Hubbard SP, Mir-Rashed N, Jacobson DJ, Glass NL (2000) Vegetative incompatibility in the het-6 region of *Neurospora crassa* is mediated by two linked genes. Genetics 155:1095–1104

Tenthorey JL, Kofoed EM, Daugherty MD, Malik HS, Vance RE (2014) Molecular basis for specific recognition of bacterial ligands by NAIP/NLRC4 inflammasomes. Mol Cell 54:17–29

Tran DTN, Chung EH, Habring-Muller A, Demar M, Schwab R, Dangl JL, Weigel D, Chae E (2017) Activation of a plant NLR complex through Heteromeric association with an autoimmune risk variant of another NLR. Curr Biol 27:1148–1160

Uehling J, Deveau A, Paoletti M (2017) Do fungi have an innate immune response? An NLR-based comparison to plant and animal immune systems. PLoS Pathog 13:e1006578

Urbach JM, Ausubel FM (2017) The NBS-LRR architectures of plant R-proteins and metazoan NLRs evolved in independent events. Proc Natl Acad Sci U S A 114:1063–1068

Vajjhala PR, Ve T, Bentham A, Stacey KJ, Kobe B (2017) The molecular mechanisms of signaling by cooperative assembly formation in innate immunity pathways. Mol Immunol 86:23–37

Van der Nest MA, Olson A, Lind M, Velez H, Dalman K, Durling MB, Karlsson M, Stenlid J (2014) Distribution and evolution of het gene homologs in the basidiomycota. Fungal Genet Biol 64:45–57

Wang YY, Liu B, Zhang XY, Zhou QM, Zhang T, Li H, Yu YF, Zhang XL, Hao XY, Wang M et al (2014) Genome characteristics reveal the impact of lichenization on lichen-forming fungus *Endocarpon pusillum* Hedwig (Verrucariales, Ascomycota). BMC Genomics 15:34

Wasmer C, Lange A, Van Melckebeke H, Siemer AB, Riek R, Meier BH (2008) Amyloid fibrils of the HET-s(218-289) prion form a beta solenoid with a triangular hydrophobic core. Science 319:1523–1526

Yuen B, Bayes JM, Degnan SM (2014) The characterization of sponge NLRs provides insight into the origin and evolution of this innate immune gene family in animals. Mol Biol Evol 31:106–120

Zambounis A, Elias M, Sterck L, Maumus F, Gachon CM (2012) Highly dynamic exon shuffling in candidate pathogen receptors ... what if brown algae were capable of adaptive immunity? Mol Biol Evol 29:1263–1276

Zhang DX, Spiering MJ, Dawe AL, Nuss DL (2014) Vegetative incompatibility loci with dedicated roles in allorecognition restrict Mycovirus transmission in chestnut blight fungus. Genetics 197:701–714

Zhang X, Dodds PN, Bernoux M (2017) What do we know about NOD-like receptors in plant immunity? Annu Rev Phytopathol 55:205–229

Zhao J, Gladieux P, Hutchison E, Bueche J, Hall C, Perraudeau F, Glass NL (2015) Identification of allorecognition loci in *Neurospora crassa* by genomics and evolutionary approaches. Mol Biol Evol 32:2417–2432

Zimmermann L, Stephens A, Nam SZ, Rau D, Kubler J, Lozajic M, Gabler F, Soding J, Lupas AN, Alva V (2018) A completely reimplemented MPI bioinformatics toolkit with a new HHpred server at its core. J Mol Biol 430:2237–2243

Zuccaro A, Lahrmann U, Guldener U, Langen G, Pfiffi S, Biedenkopf D, Wong P, Samans B, Grimm C, Basiewicz M et al (2011) Endophytic life strategies decoded by genome and transcriptome analyses of the mutualistic root symbiont *Piriformospora indica*. PLoS Pathog 7:e1002290

7 Genetics and Genomics Decipher Partner Biology in Arbuscular Mycorrhizas

LUISA LANFRANCO[1], GENNARO CAROTENUTO[1], ANDREA GENRE[1], PAOLA BONFANTE[1]

CONTENTS

I. Introduction

Arbuscular mycorrhizas (AMs) are often defined as the most widespread plant symbiosis: 72% of vascular plants (Brundrett and Tedersoo 2018) interact at root level with a group of early-diverging fungi, **Glomeromycotina,** originating a symbiosis which is not detectable at naked eye, but has deep consequences on a global scale, from nutrient cycles and soil structure to plant health, photosynthesis, and productivity. In addition, several nonvascular plants, including many bryophytes, also host Glomeromycotina in the cells of their haploid thalli. Being able to colonize both sporophytes and gametophytes, AM fungi are therefore central for land plant biology, making AM symbiosis a major scientific topic in diverse fields, from mycology to botany, microbiology, ecology, agronomy, and bioinformatics, also involving modeling and economics studies. For this reason it is not surprising to obtain millions of hits when entering the keywords *arbuscular mycorrhizas* in any search engine on the World Wide Web (December 2018). However, adding the term *Genetics* strongly reduces the number of hits, and the scientific papers that have both *Genetics and AMs* among their keywords are only a few. This scenario mirrors the history of mycorrhiza studies: while the so-called endotrophic mycorrhizas have been discovered and then studied since the end of the nineteenth century (Bonfante 2018), application of molecular techniques to AMs required the advent of PCR (Mullis 1990), almost exactly 100 years later. The first papers reporting the application of this powerful tool to mycorrhizal research were focused on the development of molecular probes based on RNA ribosomal genes and aimed at the identification of fungal symbionts in ectomycorrhizas (White et al. 1990; Gardes and Bruns 1993). Only later, molecular tools as well as -omics approaches were successfully applied to AM fungi, originating in the two main trajectories (Ferlian et al. 2018) that still characterize mycorrhizal studies: on the one hand, molecular approaches have provided a wealth of information on the ecology, distribution, and diversity of AM fungi; on the other hand, they have represented the starting point to decipher the molecular mechanisms underlying plant-fungal interactions.

[1] Dipartimento di Scienze della Vita e Biologia dei Sistemi, Università di Torino, Torino, Italy; e-mail: luisa.lanfranco@unito.it; paola.bonfante@unito.it

Genetics and Biotechnology, 3rd Edition
The Mycota II
J.P. Benz, K. Schipper (Eds.)

Differently from model fungi (*Neurospora*, *Aspergillus*) and plants (*Arabidopsis*), the **genetics of arbuscular mycorrhizas**, intended as the sum of interacting plants and fungi, **is** therefore **a very recent domain of science**. Parniske (2004) was one of the first to use the term genetics in his highly quoted review "Molecular genetics of the arbuscular mycorrhizal symbiosis."

In this context, the aim of this chapter is to provide a review of the multiple interactions that are included in the term "arbuscular mycorrhizas" and present an updated view of our knowledge on the molecular genetics of AMs, covering the genomes of AM fungi, the cellular and molecular responses of the host plant, as well as the fungal and plant natural variation that contributes to the outcome of this fascinating interaction.

II. A New Look at the Interacting Partners: From Two to Many

Arbuscular mycorrhizas (AMs) are traditionally described as the symbiosis resulting from the interaction between the roots of land plants and soil fungi. Even if AM host plants can survive if deprived of their fungal symbionts, this condition is virtually unknown in natural ecosystems, where AM fungi are associated as helper microorganisms in most of the environments so far investigated (Davison et al. 2015, 2018). On the other hand, **AM fungi are still considered to be unculturable in the absence of their host: being auxotrophic for lipids** (Jiang et al. 2017; Luginbuehl et al. 2017; Keymer et al. 2017), they strictly depend on their green hosts for growth and reproduction, which gives them the status of obligate biotrophs. From an evolutionary point of view, the ecological success of AM fungi demonstrates that the advantages of such a strict association with plants have overcome the risks arising from the loss of saprotrophic capabilities (Bonfante and Genre 2010). Recent findings have however demonstrated that the interaction is more complex than a *one-to-one* relationship, since **roots may simultaneously host many AM fungi and also other microbes.** In addition to the Glomeromycotina, some members of the related subphylum of Mucoromycotina (Spatafora et al. 2016) have been demonstrated to colonize early-diverging plants, i.e. liverworts and hornworts, and lycophytes (Bidartondo et al. 2011; Desirò et al. 2013; Rimington et al. 2015). In addition, a typology of AM fungi, usually identified as *Glomus tenuis* (the fine endophyte), has been recently suggested to be related with the Mucoromycotina (Orchard et al. 2017), broadening therefore the taxonomic spectrum of aseptate fungi which colonize land plants.

The colonization process of AM fungi belonging to the Glomeromycotina has been described in detail by using light and electron microscopy (Bonfante 2018), while more recent information is based on the use of in vivo confocal microscopy (McLean et al. 2017; Lanfranco et al. 2018; Pimprikar and Gutjahr 2018). By contrast, the colonization processes by Mucoromycotina are still to be defined: these fungi may establish different interactions with plants; some of them also establish ectomycorrhizas (Fassi et al. 1969) and have been detected mostly by using molecular tools, while morphology suggests that they form characteristic intracellular swellings (Bidartondo et al. 2011). In the light of the hypothesis that *G. tenuis* is actually belonging to Mucoromycotina, a few observations dating back to the 1980s described how the cellular features of these fungi and their interaction with the plant host are impressively similar to those of Glomeromycotina (Gianinazzi-Pearson et al. 1981). Since many reports revealed that **a single plant may be simultaneously colonized by both Mucoromycotina and Glomeromycotina** (Desirò et al. 2013), it will be crucial to develop cellular tools to clearly distinguish between the two fungal subphyla during their growth *in planta*. Lastly, **many AM fungi host obligate endobacteria** which live inside their cytoplasm and have an impact on fungal biology (Bonfante and Desirò 2017; Salvioli et al. 2016). **The bacteria living in Glomeromycotina appear to be common to many Mucoromycota,** since related microbes have been detected in *Rhizopus*, *Mortierella*, as well as *Endogone* species, suggesting that their presence predates the

Mucoromycota divergence (Bonfante and Desirò 2017).

On the basis of the current data, Glomeromycotina can therefore be defined as a stable component of the plant microbiota, since they are found in most of the environments so far investigated (Davison et al. 2015, 2018), but on the other hand, they also host their own microbiota, given by the intracellular endobacteria as well as by the bacteria which are commonly associated to the surface of their extraradical hyphae (Turrini et al. 2018).

Interestingly, the molecular investigations, including the exploitation of transcriptomics data, have also allowed the description of **viral sequences hosted within AM fungi** (Turina et al. 2018). Mycoviruses can therefore be considered an additional component of the AM microbiome with the potential to influence the biology of AM fungi and their host plant (Ikeda et al. 2012).

III. Lessons from the Genome Sequencing of AM Fungi

Our knowledge of the AM symbiosis mainly mirrors a plant-centric view. This is due to (1) the obligate biotrophic status of Glomeromycotina, which cannot be cultivated in the absence of their host plants; (2) their multinuclear condition, i.e., hundreds of nuclei coexist within one continuous cytoplasm; and (3) the absence of observable sexual reproduction and a uninucleated life stage (Chen et al. 2018). All these aspects hamper the use of the classical genetic tools which have, by contrast, allowed to study model fungi like *Neurospora* or *Aspergillus*, or their host plants which offer genetically tractable systems. In addition, protocols to obtain a stable genetic transformation of AM fungi are not yet available.

In the first decade of the new century, the development of -omics approaches and the first sequencing of an AM fungal genome have offered novel groundbreaking insights in their biology. However, achieving **the first sequenced genome of a Glomeromycotina, *Rhizophagus irregularis***, was not an easy task and required many years (Martin et al. 2004; Tisserant et al. 2013). The strain DAOM-197198 was selected for several reasons: it had been hypothesized to possess a very small genome; it easily grows in association with root organ cultures, producing a large amount of non-contaminated fungal material; and – as a last key feature – it does not host endobacteria, thus representing a potentially more amenable scenario. The sequence of its 153-Mb haploid genome showed a repertoire of about 30,000 genes and revealed a low level of polymorphism offering for the first time a reply to the crucial question: do the nuclei of AM fungi possess multiple, highly diverged genomes? The data strongly suggested the inconsistency of such a hypothesis, which was also elegantly refuted by the whole sequence of isolated single nuclei (Lin et al. 2014). Mating (MAT)-related genes were found to be expanded, suggesting the **existence of cryptic sex-related processes** and opening the possibility that a non-observable mating does not mean absence of sex. Genomic analyses of several *R. irregularis* isolates clearly proved that some strains are **homokaryotic** (containing genetically identical nuclei with one putative MAT locus) while other strains are **dikaryotic** (harboring two distinct nuclear genotypes each with a different MAT locus; Ropars et al. 2016; Corradi and Brachmann 2017). Moreover, Chen et al. (2018), following the single-nucleus sequencing approach (Fig. 1), demonstrated that nuclei with distinct genotypes in their MAT alleles can undergo recombination, originating genetic diversity. Despite evidence of recombination, however, **clonality still appears to be the prevalent mode of reproduction** (Chen et al. 2018).

A. The Biotrophism of AM Fungi

The expectations of the researchers involved in genome sequencing of *Rhizophagus irregularis* were first focused on another crucial question: why are AM fungi unculturable? At a first glance, their obligate biotrophy was not explained by genome erosion or any related loss of metabolic complexity in central metabolism. Only later it was clear that **AM fungi do**

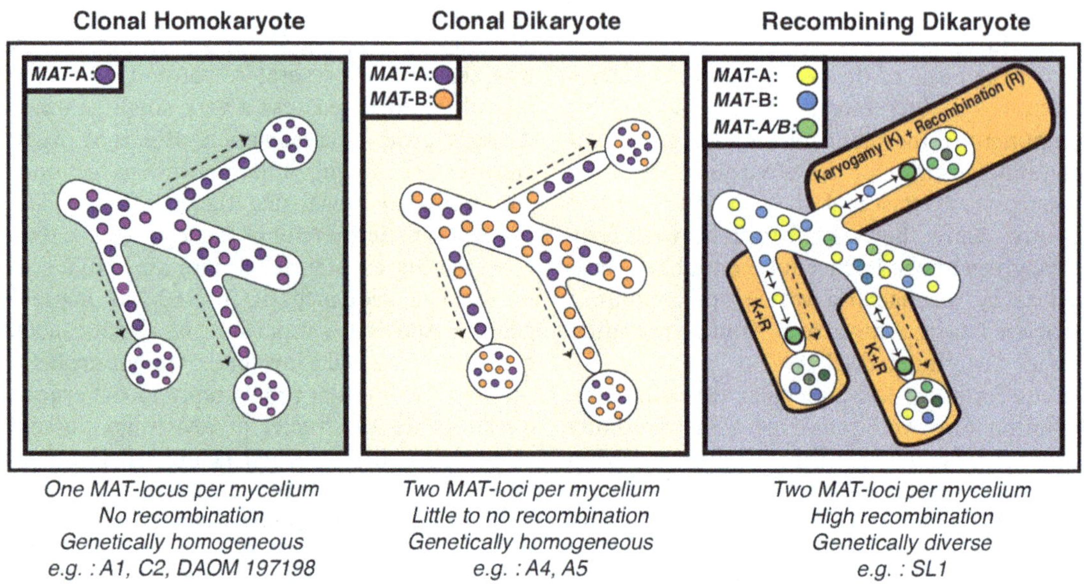

Fig. 1 Schematic representation of the three genome organizations found to date in the model AM fungus *Rhizophagus irregularis*. Left: Most isolates analyzed using genome analysis and PCR targeted to the MAT locus have been found to carry nuclei with the same MAT locus. In these isolates, genetic variability is lower than dikaryotic relatives, and recombination is undetectable. Middle: The *R. irregularis* isolates A4 and A5 carry nuclei with two distinct MAT loci. Evidence for recombination is very rare, and two divergent genotypes appear to coexist in the cytoplasm. Right: In some cases, strains can harbor nuclei with two distinct MAT loci that undergo frequent karyogamy. The frequency of karyogamy increases nuclear diversity within the mycelium. From Chen et al. (2018), licensed under CC BY 2.0 (https://creativecommons.org/licenses/by/2.0/)

not possess fatty acid synthetases (FAS; Wewer et al. 2014), resulting to be auxotroph for this essential group of compounds which are totally received from the host plant (Luginbuehl et al. 2017; Jiang et al. 2017; Keymer et al. 2017). Interestingly, Tisserant et al. (2013) firstly pointed to the **almost total lack of genes encoding plant cell wall-degrading enzymes** as well as of genes involved in toxin and thiamine synthesis. The limited number of cell wall-degrading enzymes is currently considered one of the signatures of AM fungal genomes, since this feature has been found in all other genomes sequenced later: *R. clarus* (Kobayashi et al. 2018), *Rhizophagus cerebriforme*, and *Rhizophagus diaphanum* (Morin et al. 2018) all belonging to the same *Rhizophagus* genus, as well as members of the Diversisporales, *Diversispora epigea* (Sun et al. 2018), *Gigaspora rosea* (Morin et al. 2018), and *Gigaspora margarita* (Venice et al. 2019). Taken in the whole, these results highlight a still unsolved question: how do AM fungi penetrate within their host tissue, being for a large part of their life cycle intracellular endophytes? How do they cross the plant cell wall? On one hand, the limited number of such hydrolytic enzymes well fits with the life strategy of AM fungi: they do not activate plant defenses, since they have to keep their host alive and prone to accommodate the fungus, but, on the other hand, they enter the root cells probably thanks to a finely tuned regulation of the plant metabolism. Among the battery of mycorrhiza-induced secreted proteins which are mainly expressed in symbiotic tissues (Tisserant et al. 2013; Sędzielewska Toro and Brachmann 2016), some of them can act as effectors which might manipulate plant regulatory pathways (see next paragraph). It can be hypothesized that some effectors may induce the weakening and swelling of plant cell walls, which may compensate for the absence of hydrolytic enzymes and is indeed observed by morphological analyses (Balestrini and Bonfante 2014).

The repertoire of *R. irregularis* genes has therefore provided an excellent basis for understanding not only the fungal biology but also the genetics mechanisms underlying the AM symbiosis. The genomes of AM fungi which have been recently sequenced thanks to the impressive advances in DNA sequencing technologies have provided the opportunity to develop **comparative genomics analyses** (Morin et al. 2018; Venice et al. 2019). The genome size of AM fungi can be very diverse, ranging from about 150 Mb of *Rhizophagus* species (Chen et al. 2018) to 600 Mb of *G. rosea* and 784 Mb of *G. margarita*. The genome expansion seems to be strictly correlated with the presence of **transposable elements** (TE) which, in the case of *G. margarita*, represent more than 80% of the whole genome. Interestingly, also ectomycorrhizal fungi belonging to the truffle groups and characterized by the production of hypogeous fruitbodies have genomes very large and rich in TE (Murat et al. 2018).

Differently from Glomeromycotina, the ecological role of their sister group, Mucoromycotina, is much more enigmatic. In the order Endogonales, the family Endogonaceae contains some ectomycorrhizal fungi producing small hypogeous truffle-like fruitbodies (Fassi et al. 1969; Yamamoto et al. 2017), while the other family, Densosporaceae, groups fungi that live associated to early-diverging plants (Desirò et al. 2013; Rimington et al. 2015) as well as the fine root endophytes (Orchard et al. 2017). Remarkably, the genome sequencing of four Endogonaceae fungi (Chang et al. 2018) has detected the symbiotic signatures already identified in the other mycorrhizal fungi such as large genome size, high repetitive DNA content, and low diversity of plant cell wall-degrading enzymes but without elevated small secreted proteins/secretome ratios. Notwithstanding the absence of fungi belonging to the Archeosporaceae, all the genomes of these mycorrhizal Mucoromycota fungi share similar features. AM fungi, however, have some further specific traits which well explain their unculturability (lipid auxotrophism), their TE abundance (TE burst), their high compatibility with host plants (strong reduction of cell wall-degrading CAZymes), and expansion of some metabolic pathways (chitin synthesis and degradation) allowing therefore a fine-tuning of the molecular dialogue with their host (Venice et al. 2019).

The genomes of *D. epigea* (Sun et al. 2018), *G. margarita* (Venice et al. 2019), and Endogonales (Chen et al. 2018) also allowed to gain new information on **endobacteria** living in their cytoplasm. The genome sequence led to the detection of Mollicutes-related endobacteria (MREs) in *D. epigea* and in three of the four sequenced Endogonaceae. Their genomes can therefore be read as "metagenomes." By contrast, *G. margarita* genome confirmed the presence of *Candidatus* Glomeribacter gigasporarum, which was already sequenced (Ghignone et al. 2012). The presence of such endobacteria, which have also been discovered in the phylogenetically related Mortierellomycotina (Uehling et al. 2017), strongly suggests that endobacteria may be an evolutionary marker of Mucoromycota. The intimate contact between bacteria and fungi may have favored horizontal gene transfer (Torres-Cortés et al. 2015; Naito et al. 2015; Sun et al. 2018), potentially leading to an impact on the fungal biology (Salvioli et al. 2016).

B. From Structure to Function

The genome sequencing of AM fungi has so far provided relevant information concerning their genome structure and evolution, even if data from some more distantly related members, such as *Archeospora*, would be essential to better define their ancient relationships. By contrast, functional genomics study of AM fungi is still at its infancy. Many genes, and mostly those expressed during the symbiotic phase, are orphan genes (i.e., do not show similarities with genes listed in databases), and the lack of genetic transformation procedures further hampers their characterization.

Following the studies on pathogenic interactions, attention has been given to the **secretome, the pool of proteins characterized by the**

presence of a signal peptide that guides proteins toward the endomembrane system for secretion (Kloppholz et al. 2011; Tisserant et al. 2013; Lin et al. 2014; Sędzielewska Toro and Brachmann 2016; Kamel et al. 2017; Zeng et al. 2018) since the secretome includes proteins called effectors which are of crucial relevance in host-microbe interactions. **Effectors** are microbial molecules that, once delivered to the host cells, can manipulate cellular mechanisms often leading to an attenuation of innate immune response or a promotion of nutrient exchange and thus favoring host colonization (Lo Presti et al. 2015).

AM fungi possess a rather rich secretome with hundreds of candidate **secreted proteins** which, in the different publications, may vary in number according to the criteria used to define them. Comparative analyses of genomes and transcriptomic data from AM fungi showed that many secreted proteins are conserved in phylogenetically related AM species; however, in analogy to other fungal groups with different lifestyles/nutritional strategies (Schirawski et al. 2010; Heard et al. 2015; Pellegrin et al. 2015), there is a prevalence of lineage-specific proteins, suggesting specific biological roles. Indeed, AM effectors have also been hypothesized to be important factors to control symbiotic efficiency and/or host preferences (Zeng et al. 2018), two aspects of AM fungi biology whose molecular mechanisms are still largely unknown.

Interestingly, two studies have clearly demonstrated that, while some secreted proteins showed similar gene expression levels in different host plants, suggesting that they fulfill conserved roles, a subset of them were differentially expressed depending on the host species (Kamel et al. 2017; Zeng et al. 2018). Host-specifically expressed secreted proteins, candidate effectors, also have been observed for the endophyte *Piriformospora indica* (Lahrmann et al. 2013). Evidence that these secreted proteins can play a significant role in host specificity also comes from plant pathogens where their evolution seems to be under host-directed selection (Zhong et al. 2016).

On the other hand, a small set of secreted proteins, also shared by distantly related AM fungi (*Rhizophagus irregularis* and *Gigaspora rosea*), showed similar expression patterns in different host plants (Kamel et al. 2017). These genes, described as the AM symbiotic core secretome, encode proteins with unknown function or proteases. The proteolytic activity may play a role in the production of oligopeptides and amino acids with nutritional roles, the inactivation of plant defense proteins (Jashni et al. 2015), or the generation or turnover of fungal/plant signaling proteins.

Induced expression *in planta* is a commonly applied additional predictive criterion to identify effectors among secreted proteins. Gene expression profiles from laser microdissected cells even allowed to identify a set of genes most specifically expressed at the arbuscule stage (Zeng et al. 2018). Although the majority of them are orphan genes, some secreted proteins could be associated to lipid signaling (which is of particular interest considering the finding of fatty acid auxotrophy of AM fungi) or show homology, again, to endopeptidases (Zeng et al. 2018). But, so far, **only three AM effectors have been characterized in detail.** The first, called SP7, was shown to target the host cell nucleus where it counteracts the function of the pathogenesis-related transcription factor MtERF19 (Kloppholz et al. 2011). The putative secreted protein SIS1 from *R. irregularis* was found among those genes upregulated in strigolactones-treated germinating spores (Tsuzuki et al. 2016) and strongly expressed in the intraradical mycelium, including arbuscules (Zeng et al. 2018), in line with its predicted role in intraradical colonization (Tsuzuki et al. 2016). Recently, a crinkler (CRN) effector (RiCRN1) that belongs to a subfamily of secreted CRN proteins from *R. irregularis* was also characterized (Voß et al. 2018). As CRNs were originally described in plant pathogenic oomycetes (Schornack et al. 2009), this finding extends the similarity between AM fungi and plant pathogens also outside the fungal kingdom. Although not yet defined, the mechanism of action of RiCRN1 does not involve cell death processes as often described for CRNs from oomycetes.

In all these three abovementioned cases, **host-induced gene silencing** (HIGS) has been used to specifically silence the fungal genes during the symbiotic phase allowing the description of an impaired colonization pattern. These examples also highlight how, in the absence of protocols for stable genetic transformation for AM fungi, genetic manipulation tools developed for the host plants can be successfully applied to study the function of AM fungal genes at least in the *in planta* phase.

The availability of genome and transcriptomic data also allowed the large-scale analysis of Cu, Fe, and Zn transporters: beside an expansion of some gene families, it has been observed that some genes were upregulated in the intraradical phase suggesting that metals are important for plant colonization (Tamayo et al. 2014), in analogy to what has been observed for the endophytic fungus *Epichloë festucae* (Johnson et al. 2013).

Analogous investigations led to an inventory of conserved proteins of the **RNA interference machinery (RNAi)** in *Rhizoglomus irregulare* with the discovery of putative ancient events of horizontal gene transfer involving two class I ribonuclease III protein-coding genes possibly from cyanobacterial genomes (Lee et al. 2018). From an evolutionary perspective, this finding may reflect an ancient symbiosis history of AM fungi with cyanobacteria. Remarkably, this type of interaction can still be observed today between *Geosiphon pyriforme*, an AM fungus assigned to the basal order Archaeosporales and the cyanobacterium *Nostoc punctiforme* (Gehrig et al. 1996). The presence of a RNA silencing machinery suggests that *R. irregulare* has the potential to produce small RNAs and, hypothetically, to use them also as effectors, in a process so called cross-kingdom RNAi, as it has been shown in the pathogenic interaction of *Botrytis cinerea* and *Arabidopsis* (Wang et al. 2017). Some evidences in this direction have been recently obtained (Silvestri et al. 2019). Further investigations are needed to verify whether also AM fungi use RNA effectors to regulate plant processes; on the other hand, the inverse phenomenon, that is, the delivery of small RNA with gene silencing purposes from the plant to the fungus, is also likely to occur as again it has been demonstrated in the *Botrytis cinerea-Arabidopsis* interaction (Cai et al. 2018). The success of the HIGS approach as a tool to silence fungal genes in the AM symbiosis (Helber et al. 2011; Tsuzuki et al. 2016) is a strong clue toward the occurrence of such a process.

IV. Molecular Tools Reveal Plant Responses to AM Fungi

A. Cellular and Molecular Changes in the Host Plant

In analogy to most root-microbe interactions, AM establishment depends on finely tuned **recognition processes** (Bonfante and Genre 2015) through signal release and perception between both partners before their physical contact (Gianinazzi-Pearson 1996).

Root exudates (Fig. 2) contain several bioactive, low molecular weight compounds. Among them, strigolactones are a class of terpenic compounds deriving from the carotenoid metabolism. They are biosynthesized in the root and actively transported to the shoot or released in the rhizosphere (Gomez-Roldan et al. 2008). **Strigolactones**, whose release increases under phosphate starvation, rapidly undergo spontaneous hydrolysis in water solutions, which limits their diffusion in the rhizosphere and makes them reliable signals of root vicinity for several root-interacting organisms. Indeed, strigolactones were originally identified in plant root exudates as germination stimulants for parasitic plants of the family Orobanchaceae (Matusova et al. 2005). Later, strigolactones have been reported to be indispensable for the establishment of AM (Akiyama et al. 2005, 2010; Besserer et al. 2006, 2008) and more recently also to be involved in symbiotic nitrogen fixation (Soto et al. 2009; Foo and Davies 2011).

Strigolactone perception triggers a cascade of molecular and cellular events in AM fungi, such as nuclear multiplication, mitochondrial growth, and a fast increase of cytotoxic NADH and ATP content, indicating respiration as one of the primary metabolic targets (Akiyama and Hayashi 2006). Such cellular responses associ-

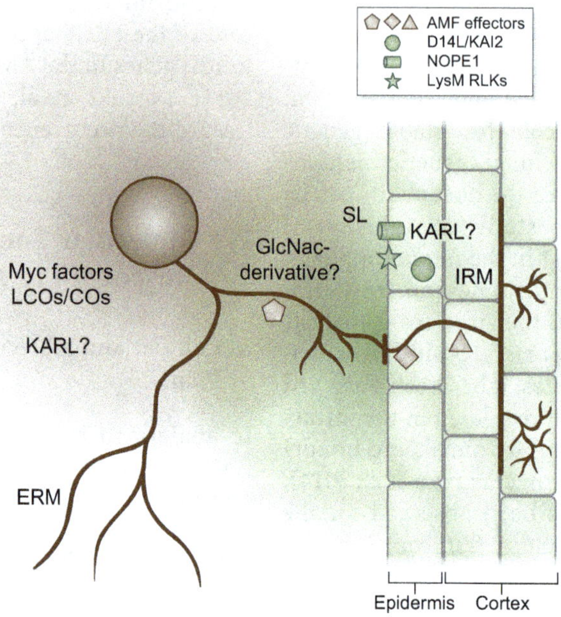

Fig. 2 Molecules involved in the communication between AM fungi and host plants. Plant roots release strigolactones (SL) which stimulate fungal metabolism and hyphal branching to promote colonization (Akiyama et al. 2005; Besserer et al. 2006, 2008). A rice mutant deficient for the D14L gene is characterized by an absence of hyphopodia (Gutjahr et al. 2015). The D14L/KAI2 protein localizes to the nucleus and cytoplasm. It is yet unclear whether the karrikin-like (KARL) ligand of D14L/KAI2 relevant for AM symbiosis is of plant or fungal origin. The recent finding that a plasma membrane-resident plant N-acetylglucosamine (GlcNAc) transporter (NOPE1) is required for early signaling in AM suggests the existence of GlcNAc-based diffusible plant molecules, which may trigger pre-symbiotic fungal reprogramming (Nadal et al. 2017). Also AM fungi use GlcNAc-based molecules, which include lipo-chito-oligosaccharides (LCOs; Maillet et al. 2011) and short chitin tetra- and pentamers (COs; Genre et al. 2013); these are perceived by plant LysM-RLKs (Zipfel and Oldroyd 2017) and activate plant symbiotic responses. AM fungal effector candidates, thought to interfere with host cellular processes to favor colonization at early and/or late stages of the AM symbiosis, have been predicted from fungal genomes and transcriptomes (Sędzielewska Toro and Brachmann 2016; Kamel et al. 2017). SLs stimulate the production of chitin oligomers (Genre et al. 2013) and secreted proteins (Tsuzuki et al. 2016; Kamel et al. 2017) by AM fungi. Note that the tissue-specific expression of D14L/KAI2 and NOPE1 is currently unknown. IRM, intraradical mycelium; ERM, extraradical mycelium. From Lanfranco et al. (2018) with permission

ate with hyphal proliferation and repeated branching close to the root, facilitating contact with the host surface (Besserer et al. 2006, 2008).

The role of strigolactones as fungus-directed signals in AM interactions has indirectly been confirmed by the observation that rice mutants defective in the SLs receptor D14 are not perturbed in AM colonization (Yoshida et al. 2012; Gutjahr et al. 2015). The study of strigolactone perception by the host plant has anyway revealed intriguing crosstalk mechanisms with other signaling pathways. For strigolactone perception, D14 forms a receptor complex with the F-box protein MAX2/D3/RMS4 (Hamiaux et al. 2012). In turn, MAX2 was shown to also be involved with KAI2/D14LIKE in the receptor complex for karrikins, the butenolide molecules found in smoke extracts that promote seed germination of many plant species (Flematti et al. 2004; Nelson et al. 2010; Waters et al. 2012). Interestingly, rice *d3* and pea *rms4* mutants displayed important defects in AM colonization and arbuscule formation, respectively (Yoshida et al. 2012; Foo et al. 2013; Gutjahr et al. 2015); furthermore, a *d14l/kai2* rice mutant does not stimulate the formation of hyphopodia (Gutjahr et al.

2015) and does not respond transcriptionally to AM germinating spore exudates. Together these results suggest that the karrikin receptor complex plays a role in symbiotic signaling even if the involvement of karrikin-like molecules of fungal or plant origin remains to be investigated (Gutjahr et al. 2015; Waters et al. 2017).

The discovery of a rice and maize N-acetylglucosamine transporter (NOPE1) required for AM signaling and colonization points to the role of additional diffusible plant molecules in the activation of pre-symbiotic fungal reprogramming (Nadal et al. 2017). Elucidating the exact molecular function of NOPE1 and its substrate will shed light on long predicted new molecular actors in AM signaling (Bonfante and Requena 2011).

Even if fungal receptors for strigolactones remain unknown (Waters et al. 2017), recent data suggest the activation of a calcium-mediated pathway (Moscatiello et al. 2014) and – intriguingly – the release of pre-symbiotic fungal signals (Genre et al. 2013).

In fact, AM fungi release water-soluble molecules collectively known as **Myc factors** (mycorrhizal factors) (Bonfante and Genre 2015). Myc factor perception triggers plant symbiotic responses (Bonfante and Requena 2011) through a Ca^{2+}-mediated **signal transduction pathway**. Responses include transcriptional regulation, starch accumulation in roots, and lateral root formation (Kosuta et al. 2003, 2008; Oláh et al. 2005; Kuhn et al. 2010; Chabaud et al. 2011; Mukherjee and Ané 2011; Maillet et al. 2011; Genre et al. 2013), overall preparing the plant to a successful symbiotic association.

The first evidence of diffusible Myc factors was found in the exudates of germinated spores (GSE), which triggered a transient increase in cytosolic calcium concentration of soybean cultured cells (Navazio et al. 2007). The GSE was later shown to contain different chitin-related oligomers that are responsible for such plant responses: these include **lipo-chito-oligosaccharides** (Myc-LCOs; Maillet et al. 2011) and **tetra- and penta-chito-oligosaccharides** (Myc-COs; Genre et al. 2013).

Myc-LCOs were purified from sterile exudates of mycorrhizal carrot roots and identified as putative Myc factors by their induction of the early symbiotic gene *MtENOD11* in *Medicago* plants. Furthermore, such LCOs stimulate root hair branching (a typical Nod factor-related response) in *Vicia sativa* (Maillet et al. 2011). In fact, LCOs have a striking structural similarity to rhizobial Nod factors (Dénarié et al. 1996).

Myc-COs have been isolated from distantly related AM fungi, *Rhizophagus irregularis* and *Gigaspora rosea* (Genre et al. 2013). They are the most effective elicitors of **nuclear Ca^{2+} spiking** patterns that resemble the irregular spiking triggered by GSE. Myc-COs are active in both legumes and non-legumes at very low concentration, down to 10^{-8} M (Genre et al. 2013; Sun et al. 2015), and can be considered as universal AM-specific elicitors.

1. The Common Symbiotic Signaling Pathway

The study of Myc factor signaling mechanisms in legumes such as *Medicago truncatula* and *Lotus japonicus* has mostly come as a follow-up of analogous research on **symbiotic nitrogen fixation** (SNF; Dénarié and Cullimore 1993; Maillet et al. 2011). Since the latter evolved almost 400 million years later than the oldest known AM-related interactions, it is currently acknowledged that legumes and rhizobia have adapted the pre-existing AM signaling pathway to control the new interaction. Indeed, several legume mutants that cannot transduce Nod factor signals are equally impaired in the early development of SNF and AM (Sagan et al. 1995; Catoira et al. 2000; Oldroyd and Downie 2006; Kosuta et al. 2008; Parniske 2008). Once characterized, the corresponding genes have been associated in the so-called common symbiotic signaling pathway (CSSP), a signal transduction pathway that mediates AM and – in legumes – SNF establishment (Oldroyd 2013; Gobbato 2015). Phylogenetic studies have demonstrated that key CSSP genes are present throughout eudicots, monocots, basal land plants, and charophytes (Banba et al. 2008; Gutjahr et al. 2008; Chen et al. 2009; Wang et al. 2010; Delaux et al. 2013, 2015).

The CSSP mediates **fungal and bacterial signal transduction** from plasma membrane-bound receptors into the nucleus, where gene expression is modulated. The LRR receptor-like kinase *Lj*SYMRK is believed to interact with so far unidentified Myc factor receptor(s) (Gobbato 2015). In its cytoplasmic domain, the *M. truncatula* homolog of *Lj*SYMRK, *Mt*DMI2, interacts with a 3-hydroxy-3-methylglutaryl-CoA reductase (HMGR1) involved in mevalonate synthesis (Kevei et al. 2007). Mevalonate is therefore believed to be produced in the vicinity of the cytoplasmic face of the plasma membrane (Venkateshwaran et al. 2015) and acts as an intermediate messenger. The following group of known CSSP proteins are localized on the nuclear envelope. They include three nucleoporins (NUP85, NUP133, and NENA; Kanamori et al. 2006; Saito et al. 2007; Groth et al. 2010), the ATP-powered Ca^{2+} pump (*Mt*MCA8; Capoen et al. 2011), a cyclic nucleotide-gated channel (*Mt*CNGC15s; Charpentier et al. 2016), and the cationic channels *Lj*CASTOR (*Mt*DMI1) and *Lj*POLLUX (Ané et al. 2004; Charpentier et al. 2008). Nucleoporins were proposed to be involved in the targeting of CSSP channels and pumps to the inner nuclear membrane (Kanamori et al. 2006; Saito et al. 2007; Groth et al. 2010), whereas the latter are directly involved in generating nuclear Ca^{2+} spiking: a series of repeated oscillations in Ca^{2+} concentration (Ané et al. 2004; Imaizumi-Anraku et al. 2005; Capoen et al. 2011; Venkateshwaran et al. 2012). In more detail, CNGC15 channels are predicted to release Ca^{2+} from the nuclear envelope lumen, a release compensated by the opposite flow of K^+ ions through CASTOR (Parniske 2008; Venkateshwaran et al. 2012). MCA8 activity contributes to the restoration of Ca^{2+} concentration at the end of each peak.

The last group of CSSP proteins resides in the nucleoplasm: Ca^{2+} spiking is supposed to activate a Ca^{2+}- and calmodulin-dependent protein kinase (*Lj*CCaMK; Miller et al. 2013), which in turn phosphorylates its interacting partner, CYCLOPS (Yano et al. 2008). Activated CYCLOPS regulates gene expression either directly or through the action of other transcription factors like NSP1, NSP2, and RAM1 (Oldroyd 2013). Importantly, the identified CSSP actors only constitute subsets of the whole transduction pathway, leaving gaps that still hamper the definition of a complete picture of the signaling process (Genre and Russo 2016).

Beside canonical CSSP members, recent evidence showed a role for additional players: the Nod factor receptor *Mt*LYK3 (*Lj*NFR1) – but not *Mt*NFP (*Lj*NFR5) – is required for AM-specific activation of the CSSP and subsequent colonization (Miyata et al. 2014; Zhang et al. 2015). Furthermore, Myc-LCO-induced responses were shown to be NFP-dependent, indicating the involvement of this receptor in both Nod- and Myc-LCO perception (Op den Camp et al. 2011). Although *nfp* mutants exhibit normal Ca^{2+} spiking and AM colonization (Maillet et al. 2011; Genre et al. 2013; Zhang et al. 2015), such mutants do not display nuclear Ca^{2+} spiking in response to Myc-LCOs (Sun et al. 2015). Taken together, these findings point to possible **partial overlaps and crosstalks between Nod and Myc signaling in legumes.**

By contrast, studies in non-legumes are opening new perspectives in the characterization of AM-specific CSSP activation. A **dual role** has been demonstrated for the rice receptor-like kinase CERK1. This gene is required for both chitin-triggered immunity and AM colonization (Miya et al. 2007; Shimizu et al. 2010). In defense-related chitin perception, OsCERK1 acts with its LysM co-receptor *Os*CEBiP (Kaku et al. 2006; Shimizu et al. 2010), but *Os*CEBiP is not required for AM symbiosis (Miyata et al. 2014), suggesting the involvement of additional players in Myc factor perception. In short, legumes and non-legumes appear to differ in their perception of Myc-LCO and Myc-CO signals, outlining a more complex scenario, where different plant species respond to different components in the mix of signals produced by AM fungi (Sun et al. 2015).

Since CSSP activation is required for the transcription of genes involved in both AM and SNF, it is not fully clear how this pathway can **discriminate between the two signals** and induce different developmental programs. Evidence points to a differential regulation of CCaMK by Ca^{2+} and calmodulin. In fact, the CaM-binding domain is redundant for AM but essential for SNF establishment (Shimoda et al. 2012). Moreover, different calcium signatures have been proposed to act in the two symbioses, with specific spiking patterns and differential responding cell types (Russo et al. 2013).

2. Symbiosis Establishment, Functioning, and Senescence

IPD3/CYCLOPS targets a number of genes through a cis-element called AM-CYC box (Favre et al. 2014; Pimprikar et al. 2016). Interestingly, CYCLOPS also interacts with the gibberellin signaling repressor DELLA (Pimprikar et al. 2016), and *Medicago truncatula della* mutants are impaired in AM and SNF colonization (Floss et al. 2013; Jin et al. 2016). DELLA proteins are repressors of gibberellic acid (GA) signaling (Alvey and Harberd 2005), and GA was shown to accumulate in *Lotus japonicus* roots during mycorrhization (Takeda et al. 2015). Furthermore, DELLA proteins also control key developmental processes such as the cell cycle (Gallego-Bartolome et al. 2012; Daviere and Achard 2013). In *M. truncatula*, *DELLA1* and *DELLA2* act redundantly to promote arbuscule development (Floss et al. 2013; Foo et al. 2013; Yu et al. 2014). In short, the CCaMK-CYCLOPS-DELLA pathway may act in the adjustment of AM establishment based on plant developmental and environmental stimuli (Carbonnel and Gutjahr 2014; Breuillin-Sessoms et al. 2015; Nagata et al. 2015; Konvalinkova and Jansa 2016).

Among the CCaMK/CYCLOPS-regulated transcription factors, the GRAS-domain proteins NSP1 (Nodulation Signalling Pathway 1) and NSP2 play an essential role in Nod factor signaling (Catoira et al. 2000; Kaló et al. 2005; Smit et al. 2005). Evidence suggests direct roles of NSP1 and NSP2 also in Myc factor signaling: with NSP2 being involved in NS-LCO-induced lateral root growth (Maillet et al. 2011) and NSP1 being required for the induction of three mycorrhizal genes in response to NS-LCO (Delaux et al. 2013). NSP1 and NSP2 interaction is required for the induction of nodulation-specific promoters (Hirsch et al. 2009; Jin et al. 2016) but not crucial for AM symbiosis, suggesting that different GRAS transcription factor complexes regulate distinct groups of genes (Pimprikar and Gutjahr 2018).

RAM1 is an AM-specific transcription factor regulated by the CYCLOPS/DELLA complex (Gobbato et al. 2012). RAM1 expression restores arbuscule formation in *cyclops* mutants (Pimprikar et al. 2016), and both RAM1 expression and arbuscule formation are rescued in *cyclops* mutants by either overexpression of DELLA1 (Floss et al. 2013; Park et al. 2015) or GA treatment (Pimprikar et al. 2016). RAM1 forms a complex with NSP2 that was proposed to regulate the expression of AM-specific genes, similarly to the NSP1/NSP2 complex in SNF (Gobbato et al. 2012).

Additional GRAS proteins were shown to have a role in fungal colonization and arbuscule development, such as RAD1 (Required for Arbuscule Development 1), an interactor of RAM1 and NSP2 (Xue et al. 2015), or DIP1 (Della Interacting Protein 1), a rice GRAS family interactor of the DELLA protein SLR1 (Yu et al. 2014). Furthermore, two CAAT-box transcription factors, *MtCbf3* and *MtCbf4*, were involved in the pre-symbiotic stage (Hogekamp et al. 2011; Czaja et al. 2012; Hogekamp and Kuster 2013).

Lastly, recent studies revealed that the CSSP also activates several other effectors, including miRNA and small interfering RNA (siRNA) with a critical role in transcriptional regulation (Lelandais-Briére et al. 2016; Bazin et al. 2017), mRNA splicing, RNA-directed DNA methylation, and epigenetic functions (Ariel et al. 2015; Chekanova 2015).

Following this chemical courtship, the **pre-symbiotic phase of AM development** culminates in physical contact between symbionts, with a hyphal tip touching the root surface (Bonfante and Genre 2010). This crucial step in AM symbiosis consists in the formation of a large, swollen, and often branched **hyphopodium**, attached to the root epidermal surface. Epidermal cell wall was proposed to stimulate hyphopodium differentiation (Nagahashi and Douds 1997) through specific physicochemical signals. This idea arose from observations in rice – where hyphopodia form on large lateral roots but never on fine lateral roots (Gutjahr et al. 2009) – and seminal studies of Giovannetti et al. (1993) and Nagahashi and Douds (1997) pointing at the presence of **wall-associated hyphopodium-stimulating signals**. More recently, one such signal has been identified in monomeric cutin. This deduction came from the observation of mutants in *ram2* (Gobbato et al. 2012), a glycerol-3-phosphate acyltransferase that is highly induced by RAM1 during AM symbiosis (Harrison 2012) and is involved in the biosynthesis of cutin precursors (Wang et al. 2012; Vijayakumar et al. 2015). In fact, root-bound cuticle monomers (Wang et al. 2012) that have also been reported to stimulate AM hyphal branching (Nagahashi and Douds 2011) are less abundant in *ram2* mutants.

Hyphopodium development is followed by hyphal penetration in the sub-hyphopodial epidermal cell (Genre et al. 2005; Bonfante and Genre 2010). **Intracellular fungal accommodation** is the central feature of AM symbiosis, and plant cells have to change their architecture and molecular composition in a process referred to as host cell reprogramming (Dörmann et al. 2014). Epidermal cells reorganize to accommodate the fungal symbiont with precise **nuclear movements, cytoplasm aggregation, and cytoskeleton remodeling**. This cellular reorganization allows the assembling of a subcellular column-shaped structure, the so-called **prepenetration apparatus** (PPA), that the plant cell forms in anticipation of fungal infection (Genre et al. 2005). PPA assembly may require 4–6 h and starts with the movement of the epidermal cell nucleus toward the hyphopodium. The nucleus then moves away from the contact site and traverses the plant cell vacuole inside a broad cytoplasmic bridge. The resulting columnar **cytoplasmic aggregation** includes numerous Golgi stacks, extensive trans-Golgi network, endoplasmic reticulum, cytoskeleton, and secretory vesicles (Genre et al. 2005, 2008, 2012). Only at this stage a hyphopodium-derived hypha starts penetrating the epidermal cell.

Endoplasmic reticulum and Golgi membranes that surround the penetrating hypha are ideally positioned for the synthesis of the **perifungal membrane**, which is believed to be the main function of the PPA. In fact, intense exocytic activity and the accumulation of SNARE and exocyst proteins have been observed around the penetrating hyphal tip (Genre et al. 2008, 2012; Ivanov et al. 2012), alongside the upregulation of the corresponding genes (Ivanov et al. 2012; Zhang et al. 2015).

PPA formation was shown to be CSSP-dependent (Genre et al. 2005; Gutjahr and Parniske 2013), and several GRAS and CAAT-box transcription factors are active during this stage (Hogekamp et al. 2011; Hogekamp and Kuster 2013), regulating a large number of genes (Diédhiou and Diouf 2018). *Mt*ENOD11 is an atypical cell wall-associated protein presumed to limit cross-linking between other wall components (Journet et al. 2001). As such, *Mt*ENOD11 expression during AM colonization may contribute to **cell wall plasticity**, especially considering the lack of cell wall-degrading enzymes in glomeromycotan genomes (Tisserant et al. 2012, 2013). Additional cell wall remodeling enzymes are expressed in roots during AM colonization: a xyloglucan endotransglycosidase (van Buuren et al. 1999) and cellulose synthase-like and expansin-like proteins (Balestrini and Bonfante 2005; Siciliano et al. 2007). Vapyrin, a VAMP-associated protein, is also expressed during early AM and SNF establishment (Pumplin et al. 2010; Murray et al. 2011).

Penetrating hyphae cross the epidermal cell lumen strictly following the route traced by the PPA and reach the root cortex. Overall, prepenetration responses in outer cortical cells resemble those observed in epidermal cells (Genre et al. 2008). By contrast, as AM hyphae reach the inner cortex, a substantial change is observed in both fungal growth pattern and host cell responses: the fungus switches from **radial to longitudinal growth,** and inner cortical cells develop broad PPA-like structures in preparation of **arbuscule accommodation** (Genre et al. 2008).

Arbuscule accommodation in cortical cells involves the biogenesis of an extensive apoplastic compartment, the **symbiotic interface** (Bonfante 2001; Balestrini and Bonfante 2014), which consists of the periarbuscular space, containing plant cell wall material and directly outlining the fungal cell wall (Balestrini and Bonfante 2014), and **periarbuscular membrane** (PAM; Harrison 2012), continuous with the host cell plasmalemma.

The process of arbuscule accommodation in cortical cells is the most striking feature of AM development and requires a broad reorganization of the host cells in strict coordination with fungal development: hyphal penetration associates with **nuclear movement** at the center of the cell (Bonfante 2001), engulfed by a broad PPA (Genre et al. 2008). This anticipates the formation of the arbuscule trunk and the PAM trunk domain (Pumplin and Harrison 2009), characterized by a set of proteins that is analogous to that of the plasma membrane. Later on, smaller PPA-like aggregates organize in the areas where the **arbuscule branches** and their associated PAM branch domain develop (Genre et al. 2008), harboring a **specific set of proteins** devoted to nutrient exchange (Pumplin and Harrison 2009), whose genes are only expressed during this phase of arbuscule development (McLean et al. 2017).

The extensive, repeated branching of arbuscule hyphae requires a very intense membrane synthesis (Pumplin and Harrison 2009) and a **polarized exocytic process** that dwarfs the analogous mechanisms described in outer cell layers and positions AM-specific membrane proteins in the PAM. An important transcriptional response drives such cellular changes, and the roles of individual genes are gradually being revealed (McLean et al. 2017; Hogekamp et al. 2011; Gaude et al. 2012; Hogekamp and Kuster 2013).

In response to CSSP activation, several transcription factors are expressed during either early or later stages of arbuscule formation (Bucher et al. 2014; Luginbuehl and Oldroyd 2017; Diédhiou and Diouf 2018; Pimprikar and Gutjahr 2018), in turn regulating the expression of genes involved in nutrient transfer, primary and specialized metabolism, membrane and cell wall modifications, secretion, and signal transduction (Hohnjec et al. 2005; Gaude et al. 2012; Hogekamp and Kuster 2013; Handa et al. 2015). The BLUE COPPER-BINDING PROTEIN 1 localizes to the peripheral plasma membrane and PAM trunk domain (Pumplin and Harrison 2009; Pumplin et al. 2012; Ivanov and Harrison 2014). By contrast, the GRAS-domain transcription factor, RAM1, is regulated by DELLA proteins and required for arbuscule branch development (Gobbato et al. 2013; Park et al. 2015; Rich et al. 2015; Pimprikar et al. 2016). RAM1 regulates the expression of exocytic markers such as the EXO70I subunit of the exocyst complex (Zhang et al. 2015b). Indeed, several proteins involved in membrane dynamics are expressed during the PAM branch domain development, such as the symbiosis-specific t-SNARE SYP132A (Huisman et al. 2016; Pan et al. 2016), VAPYRIN, involved in membrane fusion processes (Feddermann et al. 2010; Pumplin et al. 2010; Murray et al. 2011), and two symbiosis-specific v-SNARES of the VAMP721 group (Ivanov et al. 2012).

Other GRAS-domain proteins, such as RAD1 (Xue et al. 2015) and DIP1 (Yu et al. 2014), play a role in arbuscule development by interacting with RAM1 and DELLAs, suggesting the existence of a large transcription factor complex (Floss et al. 2016). Furthermore, another GRAS protein, MIG1 (MYCORRHIZA-INDUCED GRAS 1), was proposed to control radial expansion of cortical cells during arbuscule formation and interact with DELLA1 to regulate AM root development (Heck et al. 2016; Luginbuehl and Oldroyd 2017). AM-upregulated transcription factors also include AP2 (APETALA2)-EREBP, an ethylene-responsive element binding protein with a role in arbuscule development (Devers et al. 2013), and *Mt*ERF1, specifically expressed in arbusculated cells and required for arbuscule maturation (Devers et al. 2013).

AM fungi provide the host plant with a **more efficient access to soil mineral nutrients**, in particular **phosphate** and **ammonium**. After absorption, phosphorus (P) and nitrogen (N) are translocated along the extraradical and intraradical mycelium in the form of polyphosphate and arginine. From arbuscules, they are released into the periarbuscular space as phosphate (Javot et al. 2007) and ammonium (Tanaka and Yano 2005), respectively.

AM fungal **exporters** for nutrients bound to the periarbuscular space have not yet been identified. By contrast, symbiosis-specific phosphate **importers** - PT4 in *Medicago truncatula* and PT11 in rice (Javot et al. 2007; Pumplin and Harrison 2009; Kobae et al. 2010) - have been localized to the PAM. *M. truncatula pt4* deletion induces a premature arbuscule collapse and symbiosis abortion (Javot et al. 2007), suggesting that host cells monitor phosphate delivery from arbuscules and induce arbuscule degeneration if it is not sufficient. Similarly, nitrogen has also proven to act as a signal supporting arbuscule survival (Javot et al. 2011). Ammonium transporters are transcriptionally induced in mycorrhized roots and localize in the periarbuscular membrane (Kobae et al. 2010; Koegel et al. 2013). In fact, an ammonium transporter 2 family protein AMT2-3 has been identified in *M. truncatula*, whose mutation induces premature arbuscule degeneration (Yang et al. 2012; Breuillin-Sessoms et al. 2015). In the same context, PAM-associated H^+-ATPases of the MtHA1 family (Krajinski et al. 2014; Hogekamp et al. 2011; Gaude et al. 2012) are believed to generate the proton gradient required for the import of different nutrients from the periarbuscular space.

A few studies have analyzed the molecular bases of **carbon transfer from the host plant to the AM fungus**. The sugar transporter *Mt*ST1 was found to be expressed in *Medicago truncatula* root tissues colonized by AM fungi (Harrison 1996). Radiolabelling studies suggest that glucose can be absorbed by intraradical but not extraradical hyphae (Bago et al. 2000; Douds et al. 2000). Glucose transfer requires the expression of invertases and sucrose synthases in mycorrhizal roots, suggesting sucrose as a possible source of the hexoses delivered to the fungus. In parallel, a high-affinity monosaccharide transporter (MST2) has been identified in arbuscules and intraradical hyphae of *Rhizophagus irregularis* (Helber et al. 2011). Once the assimilated **sugars** reach the fungal cytoplasm, they are converted into glycogen and trehalose and exported to the extraradical mycelium (Pfeffer et al. 1999).

Even if the main form of carbon storage in AM fungi is represented by triacylglycerols (TAGs), extraradical hyphae and fungal spores are not capable of de novo fatty acid synthesis (Pfeffer et al. 1999; Gobbato et al. 2013). Indeed, studies revealed that **host plants also provide fatty acids to AM fungi**, likely in the form of palmitic acid (Wewer et al. 2014). This scenario is supported by the upregulation of several genes involved in lipid biosynthesis and secretion in arbusculated cells. They include the acyl-ACP (acyl carrier protein) thioesterase FatM and the glycerol-3-phosphate acyltransferase RAM2, both required for the establishment of a functional AM symbiosis (Gobbato et al. 2013; Luginbuehl et al. 2017). FatM has been suggested to produce palmitic acid from palmitoyl-ACP in the chloroplast (Bravo et al. 2017), whereas RAM2 has been shown to preferentially use palmitoyl-coenzyme A as a substrate to produce 2-monopalmitin (Luginbuehl et al. 2017). Interestingly, RAM1-dependent expression has been reported for the ABCG lipid exporters STR and STR2, which localize to the branch domain of the PAM (Gutjahr et al. 2012; Bravo et al. 2017; Luginbuehl et al. 2017).

The functional **lifetime** of arbuscules has been estimated a few days (Kobae and Hata 2010) after which senescence processes are initiated. Coupled with their non-synchronous formation, this relatively short time of activity leads to the **coexistence of symbiotic structures at different stages** within the same area of the root. Arbuscule collapse and degeneration involves the rapid shrinkage of arbuscule branches and PAM, including the dismantling of its associated proteins (Kobae and Hata 2010). Abundant vesicles and endoplasmic reticulum cisternae persist around the collapsing PAM, alongside peroxisomes, possibly assisting lipid breakdown or protecting the host cell from oxidative damage (Pumplin and Harrison 2009). As senescence proceeds, progressively larger branches and the arbuscule trunk become septate and collapse, eventually leading to the disappearance of the arbuscule from its host cell (Luginbuehl and Oldroyd 2017). The MYB1 transcription factor is required for the expression of several *M. truncatula* genes associated with arbuscule collapse. MYB1 was shown to interact with DELLAs and NSP1 to form a transcription factor complex that is believed to trigger arbuscule degeneration by inducing hydrolase genes (Floss et al. 2017). In fact, arbuscule senescence is a regulated process where the host cell remains active during and after arbuscule collapse and maintains the ability to be colonized again by a new arbuscule.

B. Mycorrhizal Omics: From Local to Systemic Responses

The AM symbiosis develops in roots where extensive cellular reorganizations and specific metabolic changes occur, which are mirrored by local changes in the transcript profiles as it has been demonstrated by **transcriptomic** analyses carried out on several plant species. The root metabolome is also reprogrammed upon mycorrhization. Laparre et al. (2014) detected 71 compounds exclusively present or more abundant in *M. truncatula* roots colonized by *R. irregularis*, including propionyl- and butyryl-carnitines. Remarkably, the accumulation of carnitine, which is known to be involved in lipid metabolism, could reflect changes in the fungal metabolism and can be linked to the described fatty acid auxotrophy of AM fungi.

An untargeted **metabolomic** analysis was also recently performed on tomato mycorrhizal roots with the aim to identify key metabolites involved in the mycorrhiza-induced protection against osmotic stresses (Rivero et al. 2018). AM-colonized roots accumulated some amino acids, lignans, oxylipins, and carotenoids, among which some compounds are known to be involved in plant stress adaptation (Havaux 2014). Interestingly, the protective effect of the mycorrhizal symbiosis was higher than that observed upon exogenous application of purified compounds highlighting that the AM symbiosis could be considered a more versatile strategy for plant protection.

Even if AM colonization is physically confined to root tissues, epigeous portions of the plants also experience physiological and metabolic changes. Indeed, transcriptomic analyses revealed significant gene modulation in shoots (Liu et al. 2007; Fiorilli et al. 2009; Cervantes-Gámez et al. 2016) and even fruits (Zouari et al. 2014) of mycorrhizal plants, indicating the occurrence of a long-distance systemic response. Reorganization of the metabolic profiles was also observed in leaves of mycorrhizal plants. A comparative multi-species metabolomic approach carried out on plants inoculated with the same AM fungus revealed that, although a core metabolome could be identified, leaves metabolic responses to arbuscular mycorrhiza showed strong species specificity (Schweiger et al. 2014).

It has been envisaged that these **AM-induced changes at systemic level may have an impact on the outcome of biotic and abiotic interactions.** Martinez-Medina et al. (2016) have described how mycorrhizal fungi as beneficial microbes induce a priming status (i.e., the induction of a physiological state in which a plant is conditioned for the activation of defenses against environmental challenges). Interestingly, such priming status is raised also by native microbiota which are associated to tomato, including also AM fungi (Chialva et al. 2018). As a consequence, mycorrhizal plants acquire the so-called **mycorrhiza-induced resistance** (MIR; Jung et al. 2012; Cameron et al. 2013), thanks to which they have been shown to alleviate the damage caused by pathogen attacks. The dissection of the tripartite interaction among wheat, the AM fungus *Funneliformis mosseae*, and the bacterial pathogen *Xanthomonas translucens* by using a combined transcriptomic-proteomic-metabolomic approach revealed that AM symbiosis does exert a positive effect on wheat growth and productivity but also does provide protection against *X. translucens* (Fiorilli et al. 2018). Indeed, induction of genes involved in a general defense line (e.g., coding for pathogenesis-related proteins, or leading to ROS formation) was the result of the AM fungus presence at local and systemic level, while specific defense genes (encoding, e.g., a cytochrome P450 enzyme, involved in iron binding and with oxidoreductase activity) were detected exclusively after the pathogen attack (Fiorilli et al. 2018).

It is worth to mention that studies on leaf metabolome have also been instrumental for the identification of blumenol-derived compounds which were detected in leaves of several dicot and monocotyledonous plants so that they have been proposed as foliar markers of the AM association (Wang et al. 2018). **Blumenols** therefore represent an extremely powerful tool for high-throughput screening for a functional AM symbiosis and can be used instead of laborious measurements of AM-induced transcripts or microscopic analyses.

Small RNA (sRNA) molecules are also attractive candidates for long-distance signaling and have been targeted by -omics analyses in arbuscular mycorrhizas. So far investigations focused on roots revealed a large-scale reprograming of microRNAs (miRNAs) upon AM colonization (Devers et al. 2011; Wu et al. 2016; Pandey et al. 2018). In a recent work, the analysis of putative targets of selected miRNAs revealed an involvement in P starvation, phytohormone signaling, and defense (Pandey et al. 2018).

In conclusion, all these studies convincingly demonstrate that AM fungi have a local and systemic influence on their host plant, since they lead to a deep reorganization of the plant biology acting on multiple transcriptomic, regulatory, and metabolomic pathways.

V. The AM Symbiosis in the Light of Natural Variation

Despite the low morphological variation and their large host range, **AM fungal species and isolates show different efficiency in promoting plant performance**; on the other hand, the plant genotype has an important role in determining the extent of plant responsiveness to the AM symbiosis (Smith et al. 2004). AM fungi can present high functional diversity: even isolates belonging to the same species can exert on a specific host plant different growth effects, which can vary in amplitude and direction (promotion or inhibition) (Hart and Reader 2002; Munkvold et al. 2004; Feddermann et al. 2008; Antunes et al. 2011; Hong et al. 2012). An extensive comparative study, considering 56 AM isolates belonging to 6 different families and 17 genera inoculated on 3 different host plants, revealed that the plant growth response could not be predicted from AM species identity or morphological traits, such as extra- and intraradical fungal volumes (Koch et al. 2017). It can be hypothesized that the functional variability of the fungal symbionts may rely on other factors such as qualitative and/or quantitative differences in the production of signaling molecules, such as chitin oligomers, and/or effectors, which are crucial for triggering the symbiotic program and dampening the plant immune response, and in the expression and functioning of transporters and/or metabolic enzymes which control nutrient exchanges and may guarantee, in the end, a highly compatible and efficient mutualistic symbiosis. Since the mycorrhizal growth effect is also dependent on the host plant, it is likely that the host plant contributes, to some extent, to the regulation of these fungal genetic determinants. The differential expression of some secreted proteins, candidate effectors, on different host plants (Kamel et al. 2017; Zeng et al. 2018) already provides a support to this hypothesis (Fig. 3).

Mateus et al. (2019) analyzed transcriptomes from five genetically different cultivars of another important crop worldwide, cassava (*Manihot esculenta*), each inoculated with two different *R. irregularis* isolates to explore how plant and fungal gene expression profiles are affected by the intra-specific variability of the partner organism. Interestingly, expression of most plant genes responded in a different direction or magnitude depending on the plant genotype. The abundance of several fungal transcripts was also strongly influenced by the genotype of the plant.

Comparative genomics and functional genomics studies on several AM strains and species will be instrumental to determine the genetic, and possibly epigenetic, polymorphisms controlling the impact of specific AM inocula on host plant performance. Hopefully, evaluation of plant performance should not be limited to the growth effect but also to other traits provided by AM fungi, such as enhanced tolerance to abiotic and abiotic stresses to cover all the potential benefits of the symbiosis.

Exploring the genomic variations of AM fungi is already giving important insights on

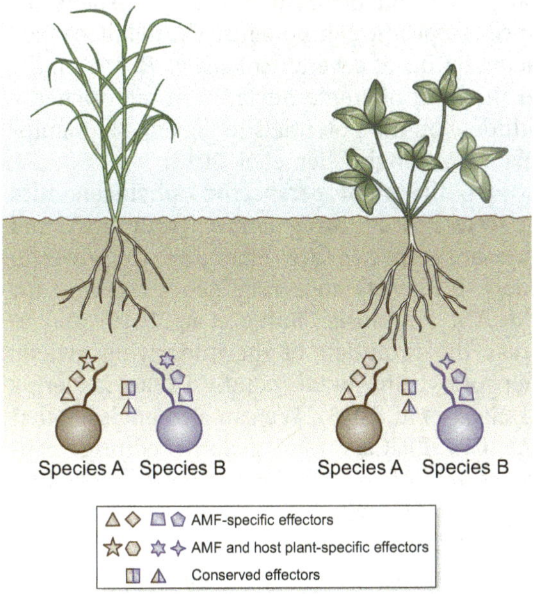

Fig. 3 Scheme of the variety of symbiotic effectors produced by AM fungi during the interaction with host plants. For a single fungal species, some effectors are expressed in association with all plant species, whereas others are expressed in a host plant-specific manner. Some effectors are conserved among AM fungi and may play core symbiotic functions. From Lanfranco et al. (2018) with permission

the nature of their genome organization: the discovery of homokaryotic and dikaryotic strains of *R. irregularis* (Ropars et al. 2016; Corradi and Brachmann 2017) and the finding that distinct nuclear genotypes can undergo recombination events (possibly through a meiotic process) in dikaryotic mycelia (Chen et al. 2018b) highlight the **potential of AM fungi for sexual reproduction and** offer **perspectives for genetic strain improvement.** Together with functional analyses, this knowledge will be fundamental to allow a selection of AM fungi with specific impact on plant performance.

On the plant side, susceptibility to AM fungi, evaluated by the measurement of colonized root length, depends on several environmental factors, among which soil nutrient availability, in particular that of P, is a crucial parameter, with high fertilization having generally a negative effect (Sawers et al. 2010; Chu et al. 2013). In addition, not only susceptibility to AM fungi but also the mycorrhizal growth response depends on the plant genotype. It is worth to note that current literature data, however, could not highlight a clear correlation between the amount of colonization and plant performance (Koch et al. 2017; Sawers et al. 2017; Lekberg and Koide 2005).

There is increasing interest in exploring the variations in AM susceptibility and responsiveness in cultivated accessions and, through genome-wide association analyses, in identifying the genetic determinants associated to those traits. One of the first large-scale studies considering **genetic variation in AM fungi susceptibility was carried out on 94 wheat (*Triticum aestivum*) genotypes** inoculated by a mixed inoculum of 3 AM species (Lehnert et al. 2017). Interestingly, six quantitative trait loci (QTLs) associated with colonization level could be identified: they contained genes related to cell wall metabolism and defense, suggesting that they may be involved in controlling root colonization.

A recent work analyzed a large collection of wild, domesticated, and cultivated lines of *Triticum turgidum* ssp. *durum* colonized by two AM fungi (*Funneliformis mosseae* and *Rhizoglomus irregulare*). Seven QTLs were linked to mycorrhizal susceptibility, and candidate proteins with roles in host-parasite interactions, degradation of cellular proteins, homeostasis regulation, plant growth, and disease/defense were identified (De Vita et al. 2018).

Concerning the **responsiveness to AM fungi,** Sawers et al. (2017) analyzed the growth response of 30 maize lines upon colonization by *F. mosseae*; variations in shoot dry weight and shoot Pi content were observed, and, interestingly, these correlated with the amount of extraradical mycelium, suggesting a plant-fungus reciprocal effect on growth performances. The molecular bases for this effect are completely unknown and may rely on differential regulation of genes involved in nutrient transport in both partners. In addition, the concentration of 19 elements was also determined in roots and leaves of the same maize lines (Ramirez-Flores et al. 2017): a number of other elements, besides P, responded significantly to inoculation, and the impact of AM symbiosis on the concentration of these ions was genotype specific, indicating again the relevance of plant genetics.

In the attempt to map the genetic bases of AM symbiotic variations, also wild relatives and old varieties are often analyzed since they represent important genetic resources for breeding (Singh et al. 2012; Lehmann et al. 2012). It has been hypothesized that the selection of modern varieties, which was likely carried out under highly fertilized conditions, may have decreased the susceptibility/responsiveness to the AM symbiosis (Zhu et al. 2001; Lehmann et al. 2012). Depending on the crop species diverse mycorrhizal response patterns were observed (Kapulnik & Kushnir 1991; Koltai & Kapulnik 2010; Steinkellner et al. 2012; Xing et al. 2012; Turrini et al. 2016). A recent work, through a comparative investigation on 27 crop species and their wild progenitors, showed that **the growth benefits exerted by the AM symbiosis were dependent on P availability**; while wild progenitors positively responded to the AM symbiosis irrespective of P availability, in domesticated plants the growth effect observed at low P became negligible when P availability increased (Martín-Robles et al. 2018). In addition, domesticated plants reduced AM fungal colonization more strongly than did wild progenitors in response to increased P availability.

On the whole these studies indicate **a strong fungal genotype X plant genotype interaction in the mycorrhizal symbiosis.** This variation may have profound impact in natural populations and has to be considered in agricultural practices where AM fungi are exploited to improve plant health and productivity.

VI. Conclusions

Genetics and genomics have recently provided crucial novel information on the biology of arbuscular mycorrhizas. The genome sequencing of a number of AM fungal species is allowing to identify common features such as the fatty acid auxotrophy but also dispensable species-specific components. The detailed characterization of several isolates of *R. irregularis* at the level of single nuclei has even opened a window on the potentials to genetically manipulate AM fungi (Chen et al. 2018).

On the plant perspective, phylogenomics analyses based on genomes from host and non-host species are emerging as powerful tools to identify conserved genes required for the AM symbiosis (Bravo et al. 2016) and to trace the evolution of the underlying genetic network from basal plants to angiosperms (Delaux et al. 2015). We can also envisage that the CRISPR/Cas-based genome editing technique will offer an efficient strategy for producing plant genotypes with mutations in genes of interest. These genes could be selected among those responsible of the molecular dialogue between partners (also considering the pre-symbiotic steps) and among those which regulate AM functionality. In the frame of a more friendly agriculture, these plant genes could be the targets for the development of new crop varieties more susceptible and responsive to the beneficial AM fungi.

Acknowledgments Research is supported by the Italian Ministry for University and Research (MIUR - UNITO Ricerca Locale 2016) and by TOMRES from the European Union's Horizon 2020 research and innovation program under grant agreement no. 727929 to L.L. and by the Italian Ministry for University and Research (MIUR - UNITO Ricerca Locale 2016) and Fondazione Cassa di Risparmio di Cuneo (AMforQuality - Bando Ricerca Scientifica 2015) to A.G.

References

Akiyama K, Hayashi H (2006) Strigolactones: chemical signals for fungal symbionts and parasitic weeds in plant roots. Ann Bot 97(6):925–931

Akiyama K, Matsuzaki K, Hayashi H (2005) Plant sesquiterpenes induce hyphal branching in arbuscular mycorrhizal fungi. Nature 435(7043):824–827

Akiyama K, Ogasawara S, Ito S, Hayashi H (2010) Structural requirements of strigolactones for hyphal branching in AM fungi. Plant Cell Physiol 51(7):1104–1117

Alvey L, Harberd NP (2005) DELLA proteins: integrators of multiple plant growth regulatory inputs? Physiol Plant 123:153–160

Ané JM, Kiss GB, Riely BK, Penmetsa RV, Oldroyd GE, Ayax C, Lévy J, Debellé F, Baek JM, Kalo P, Rosenberg C, Roe BA, Long SR, Dénarié J, Cook DR (2004) *Medicago truncatula* DMI1 required for bacterial and fungal symbioses in legumes. Science 303:1364–1367

Antunes PM, Koch AM, Morton JB, Rillig MC, Klironomos JN (2011) Evidence for functional divergence in arbuscular mycorrhizal fungi from contrasting climatic origins. New Phytol 189:507–514

Ariel F, Romero-Barrios N, Jégu T, Benhamed M, Crespi M (2015) Battles and hijacks: noncoding transcription in plants. Trends Plant Sci 20:362–371

Bago B, Pfeffer PE, Shachar-Hill Y (2000) Carbon metabolism and transport in arbuscular mycorrhizas. Plant Physiol 124:949–957

Balestrini R, Bonfante P (2005) The interface compartment in arbuscular mycorrhizae: a special type of plant cell wall? Plant Biosystems 139:8–15

Balestrini R, Bonfante P (2014) Cell wall remodeling in mycorrhizal symbiosis: a way towards biotrophism. Front Plant Sci 5:237

Banba M, Gutjahr C, Miyao A, Hirochika H, Paszkowski U, Kouchi H, Imaizumi-Anraku H (2008) Divergence of evolutionary ways among common sym genes: CASTOR and CCaMK show functional conservation between two symbiosis systems and constitute the root of a common signalling pathway. Plant Cell Physiol 49(11):1659–1671

Bazin J, Baerenfaller K, Gosai SJ, Gregory BD, Crespi M, Bailey-Serres J (2017) Global analysis of ribosome-associated noncoding RNAs unveils new modes of translational regulation. Proc Natl Acad Sci U S A 114(46):E10018–E10027

Besserer A, Puech-Pagès V, Kiefer P, Gomez-Roldan V, Jauneau A, Roy S, Portais JC, Roux C, Bécard G, Séjalon-Delmas N (2006) Strigolactones stimulate arbuscular mycorrhizal fungi by activating mitochondria. PLoS Biol 4(7):1239–1247

Besserer A, Bécard G, Jauneau A, Roux C, Séjalon-Delmas N (2008) GR24, a synthetic analog of strigolactones, stimulates the mitosis and growth of the arbuscular mycorrhizal fungus *Gigaspora rosea* by boosting its energy metabolism. Plant Physiol 148(1):402–413

Bidartondo MI, Read DJ, Trappe JM, Merckx V, Ligrone R, Duckett JG (2011) The dawn of symbiosis between plants and fungi. Biol Lett 7:574–577

Bonfante P (2001) At the interface between mycorrhizal fungi and plants: the structural organization of cell wall, plasma membrane and cytoskeleton. In: Hock B (ed) The mycota, IX: fungal associations. Springer, Berlin, pp 45–61

Bonfante P (2018) The future has roots in the past: the ideas and scientists that shaped mycorrhizal research. New Phytol 220:982–995

Bonfante P, Desirò A (2017) Who lives in a fungus? The diversity, origins and functions of fungal endobacteria living in Mucoromycota. ISME J 11:1727–1735

Bonfante P, Genre A (2010) Mechanisms underlying beneficial plant-fungus interactions in mycorrhizal symbiosis. Nature Comm 1:48

Bonfante P, Genre A (2015) Arbuscular mycorrhizal dialogues: do you speak 'plantish' or 'fungish'? Trends Plant Sci 20:150–154

Bonfante P, Requena N (2011) Dating in the dark: how roots respond to fungal signals to establish arbuscular mycorrhizal symbiosis. Curr Opin Plant Biol 14:451–457

Bravo A, York T, Pumplin N, Mueller LA, Harrison MJ (2016) Genes conserved for arbuscular mycorrhizal symbiosis identified through phylogenomics. Nat Plants 2:15208

Bravo A, Brands M, Wewer V, Dormann P, Harrison MJ (2017) Arbuscular mycorrhiza-specific enzymes FatM and RAM2 finetune lipid biosynthesis to promote development of arbuscular mycorrhiza. New Phytol 214:1631–1645

Breuillin-Sessoms F, Floss DS, Gomez SK, Pumplin N, Ding Y, Levesque-Tremblay V, Noar RD, Daniels DA, Bravo A, Eaglesham JB, Benedito VA, Udvardi MK, Harrison MJ (2015) Suppression of arbuscule degeneration in *Medicago truncatula* phosphate transporter4 mutants is dependent on the ammonium transporter 2 family protein AMT2;3. Plant Cell 27:1352–1366

Brundrett MC, Tedersoo L (2018) Evolutionary history of mycorrhizal symbioses and global host plant diversity. New Phytol 220:1108–1115. https://doi.org/10.1111/nph.14976

Bucher M, Hause B, Krajinski F, Kuster H (2014) Through the doors of perception to function in arbuscular mycorrhizal symbioses. New Phytol 204:833–840

Cai Q, Qiao LL, Wang M, He B-Y, Lin FM, Palmquist J, Jin H-L (2018) Plants send small RNAs in extracellular vesicles to fungal pathogen to silence virulence genes. Science 360:1126–1129

Cameron DD, Neal AL, Van Wees SCM, Ton J (2013) Mycorrhiza-induced resistance: more than the sum of its parts? Trends Plant Sci 18:539–545

Capoen W, Sun J, Wysham D, Otegui MS, Venkateshwaran M, Hirsch S, Miwa H, Downie JA, Morris RJ, Ané JM, Oldroyd GE (2011) Nuclear membranes control symbiotic calcium signaling of legumes. Proc Natl Acad Sci U S A 108:14348–14353

Carbonnel S, Gutjahr C (2014) Control of arbuscular mycorrhiza development by nutrient signals. Front Plant Sci 5:462

Catoira R, Galera C, de Billy F, Penmetsa RV, Journet EP, Maillet F, Rosenberg C, Cook D, Gough C, Dénarié J (2000) Four genes of *Medicago truncatula* controlling components of a nod factor transduction pathway. Plant Cell 12:1647–1665

Cervantes-Gámez RG, Bueno-Ibarra MA, Cruz-Mendívil A, Calderón-Vázquez CL, Ramírez-Douriet CM,

Maldonado-Mendoza IE, Villalobos-López MA, Valdez-Ortíz A, López-Meyer M (2016) Arbuscular mycorrhizal symbiosis-induced expression changes in *Solanum lycopersicum* leaves revealed by RNA-seq analysis. Plant Mol Biol Rep 34:89–102

Chabaud M, Genre A, Sieberer BJ, Faccio A, Fournier J, Novero M, Barker DG, Bonfante P (2011) Arbuscular mycorrhizal hyphopodia and germinated spore exudates trigger Ca^{2+} spiking in the legume and nonlegume root epidermis. New Phytol 189:347–355

Chang Y, Desirò A, Na H, Sandor L, Lipzen A, Clum A, Barry K, Grigoriev IV, Martin FM, Stajich JE, Smith ME, Bonito G, Spatafora JW (2018) Phylogenomics of Endogonaceae and evolution of mycorrhizae within Mucoromycota. New Phytol 222(1):511–525. https://doi.org/10.1111/nph.15613

Charpentier M, Bredemeier R, Wanner G, Takeda N, Schleiff E, Parniske M (2008) *Lotus japonicus* CASTOR and POLLUX are ion channels essential for perinuclear calcium spiking in legume root endosymbiosis. Plant Cell 20:3467–3479

Charpentier M, Sun J, Vaz Martins T, Radhakrishnan GV, Findlay K, Soumpourou E, Thouin J, Véry AA, Sanders D, Morris RJ, Oldroyd GED (2016) Nuclear-localized cyclic nucleotide-gated channels mediate symbiotic calcium oscillations. Science 352:1102–1105

Chekanova JA (2015) Long non-coding RNAs and their functions in plants. Curr Opin Plant Biol 27:207–216

Chen C, Zou J, Zhang S, Zaitlin D, Zhu L (2009) Strigolactones are a new-defined class of plant hormones which inhibit shoot branching and mediate the interaction of plant-AM fungi and plant-parasitic weeds. Sci Chin C Life Sci/Chin Acad Sci 52 (8):693–700

Chen EC, Mathieu S, Hoffrichter A, Sedzielewska-Toro K, Peart M, Pelin A, Ndikumana S, Ropars J, Dreissig S, Fuchs J, Brachmann A, Corradi N (2018) Single nucleus sequencing reveals evidence of inter-nucleus recombination in arbuscular mycorrhizal fungi. elife 7:pii:e39813. https://doi.org/10.7554/eLife.39813

Chen ECH, Morin E, Beaudet D, Noel J, Yildirir G, Ndikumana S, Charron P, St-Onge C, Giorgi J, Krüger M, Marton T, Ropars J, Grigoriev IV, Hainaut M, Henrissat B, Roux C, Martin F, Corradi N (2018b) High intraspecific genome diversity in the model arbuscular mycorrhizal symbiont *Rhizophagus irregularis*. New Phytol 220:1161–1171

Chialva M, Salvioli A, Daghino S, Ghignone S, Bagnaresi P, Chiapello M, Novero M, Spadaro D, Perotto S, Bonfante P (2018) Native soils with their microbiotas elicit a state of alert in tomato plants. New Phytol 220:1296–1308

Chu Q, Wang XX, Yang Y, Chen FJ, Zhang FS, Feng G (2013) Mycorrhizal responsiveness of maize (*Zea mays* L.) genotypes as related to releasing date and available P content in soil. Mycorrhiza 23:497–505

Corradi C, Brachmann A (2017) Fungal mating in the most widespread plant symbionts? Trends Plant Sci 22:175–183

Czaja LF, Hogekamp C, Lamm P, Maillet F, Andres Martinez E, Samain E, Dénarié J, Kuster H, Hohnjec N (2012) Transcriptional responses towards diffusible signals from symbiotic microbes reveal MtNFP-and MtDMI3-dependent reprogramming of host gene expression by arbuscular mycorrhizal fungal lipochitooligosaccharides. Plant Physiol 159:1671–1685

Daviere JM, Achard P (2013) Gibberellin signaling in plants. Development 140:1147–1151

Davison J, Moora M, Öpik M, Adholeya A, Ainsaar L, Bâ A, Burla S, Diedhiou AG, Hiiesalu I, Jairus T, Johnson NC, Kane A, Koorem K, Kochar M, Ndiaye C, Pärtel M, Reier U, Saks U, Singh R, Vasar M, Zobel M (2015) Global assessment of arbuscular mycorrhizal fungus diversity reveals very low endemism. Science 349:970–973

Davison J, Moora M, Öpik M, Ainsaar L, Ducousso M, Hiiesalu I, Jairus T, Johnson N, Jourand P, Kalamees R, Koorem K, Meyer JY, Püssa K, Reier U, Pärtel M, Semchenko M, Traveset A, Vasar M, Zobel M (2018) Microbial island biogeography: isolation shapes the life history characteristics but not diversity of root symbiotic fungal communities. ISME J12:2211–2224

Delaux PM, Séjalon-Delmas N, Bécard G, Ané M (2013) Evolution of the plant-microbe symbiotic "toolkit.". Trends Plant Sci 18(6):298–304

Delaux PM, Radhakrishnan GV, Jayaraman D, Cheema J, Malbreil M, Volkening JD, Sekimoto H, Nishiyama T, Melkonian M, Pokorny L, Rothfels CJ, Sederoff HW, Stevenson DW, Surek B, Zhang Y, Sussman MR, Dunand C, Morris RJ, Roux C, Wong GK-S, Oldroyd GED, Ané J-M (2015) Algal ancestor of land plants was preadapted for symbiosis. Proc Natl Acad Sci U S A 112:13390–13395

Dénarié J, Cullimore J (1993) Lipo-oligosaccharide nodulation factors: a minireview new class of signaling molecules mediating recognition and morphogenesis. Cell 74:951–954

Dénarié J, Debellé F, Promé JC (1996) *Rhizobium* lipochitooligosaccharide nodulation factors: signaling molecules mediating recognition and morphogenesis. Annu Rev Biochem 65:503–535

Desirò A, Duckett JG, Pressel S, Villarreal JC, Bidartondo MI (2013) Fungal symbioses in hornworts: a chequered history. Proc Biol Sci 280 (1759):20130207

De Vita P, Avio L, Sbrana C, Laidò G, Marone D, Mastrangelo AM, Cattivelli L, Giovannetti M (2018) Genetic markers associated to arbuscular mycorrhizal colonization in durum wheat. Sci Rep 8:10612

Devers EA, Branscheid A, May P, Krajinski F (2011) Stars and symbiosis: microRNA- and microRNA*-mediated transcript cleavage involved in arbuscular mycorrhizal symbiosis. Plant Physiol 156:1990–2010

Devers E, Teply J, Reinert A, Gaude N, Krajinski F (2013) An endogenous artificial microRNA system for unraveling the function of root endosymbioses related genes in *Medicago truncatula*. BMC Plant Biol 13:82

Diédhiou I, Diouf D (2018) Transcription factors network in root endosymbiosis establishment and development. World J Microbiol Biotechnol 34 (3):37

Dörmann P, Kim H, Ott T, Schulze-Lefert P, Trujillo M, Wewer V, Hückelhoven R (2014) Cell-autonomous defense, reorganization and trafficking of membranes in plant-microbe interactions. New Phytol 204(4):815–822

Douds DD, Pfeffer PE, Shachar-Hill Y (2000) Carbon partitioning, cost and metabolism of arbuscular mycorrhizae in arbuscular mycorrhizas: physiology and function. In: Kapulnick Y, Douds DD Jr (eds) Arbuscular mycorrhizas: molecular biology and physiology. Kluwer Academic, Dordrecht

Fassi B, Fontana A, Trappe JM (1969) Ectomycorrhizae formed by *Endogone lactiflua* with species of *Pinus* and *Pseudotsuga*. Mycologia 61:412–414

Favre P, Bapaume L, Bossolini E, Delorenzi M, Falquet L, Reinhardt D (2014) A novel bioinformatics pipeline to discover genes related to arbuscular mycorrhizal symbiosis based on their evolutionary conservation pattern among higher plants. BMC Plant Biol 14:333

Feddermann N, Boller T, Salzer P, Elfstrand S, Wiemken A, Elfstrand M (2008) *Medicago truncatula* shows distinct patterns of mycorrhiza-related gene expression after inoculation with three different arbuscular mycorrhizal fungi. Planta 227:671–680

Feddermann N, Muni RRD, Zeier T, Stuurman J, Ercolin F, Schorderet M, Reinhardt D (2010) The PAM1 gene of petunia, required for intracellular accommodation and morphogenesis of arbuscular mycorrhizal fungi, encodes a homologue of VAPYRIN. Plant J 64:470–481

Ferlian O, Eisenhauer N, Aguirrebengoa M, Camara M, Ramirez-Rojas I, Santos F, Thakur MP (2018) Invasive earthworms erode soil biodiversity: a meta-analysis. J Anim Ecol 87:162–172

Fiorilli V, Catoni M, Miozzi L, Novero M, Accotto GP, Lanfranco L (2009) Global and cell-type gene expression profiles in tomato plants colonized by an arbuscular mycorrhizal fungus. New Phytol 184:975–987

Fiorilli V, Vannini C, Ortolani F, Garcia-Seco D, Chiapello M, Novero M, Domingo G, Terzi V, Morcia C, Bagnaresi P, Moulin L, Bracale M, Bonfante P (2018) Omics approaches revealed how arbuscular mycorrhizal symbiosis enhances yield and resistance to leaf pathogen in wheat. Sci Rep 8:9625

Flematti GR, Ghisalberti EL, Dixon KW, Trengove RD (2004) A compound from smoke that promotes seed germination. Science 305:977

Floss DS, Levy JG, Lévesque-Tremblay V, Pumplin N, Harrison MJ (2013) DELLA proteins regulate arbuscule formation in arbuscular mycorrhizal symbiosis. Proc Natl Acad Sci U S A 110:E5025–E5034

Floss DS, Lévesque-Tremblay V, Park H-J, Harrison MJ (2016) DELLA proteins regulate expression of a subset of AM symbiosis- induced genes in *Medicago truncatula*. Plant Signal Behav 11:e1162369

Floss DS, Gomez SK, Park HJ, McLean AM, Muller LM, Bhattarai KK, Lévesque-Tremblay V, Maldonado-Mendoza IE, Harrison MJ (2017) A transcriptional program for arbuscule degeneration during AM symbiosis is regulated by MYB1. Curr Biol 27:1206–1212

Foo E, Davies NW (2011) Strigolactones promote nodulation in pea. Planta 234(5):1073–1081

Foo E, Ross JJ, Jones WT, Reid JB (2013) Plant hormones in arbuscular mycorrhizal symbioses: an emerging role for gibberellins. Ann Bot 111:769–779

Gallego-Bartolome J, Minguet EG, Grau-Enguix F, Abbas M, Locascio A, Thomas SG, Alabadi D, Blazquez MA (2012) Molecular mechanism for the interaction between gibberellin and brassinosteroid signaling pathways in *Arabidopsis*. Proc Natl Acad Sci U S A 109:13446–13451

Gardes M, Bruns TD (1993) ITS primers with enhanced specificity for basidiomycetes application to the identification of mycorrhizae and rusts. Mol Ecol 2:113–118

Gaude N, Bortfeld S, Duensing N, Lohse M, Krajinski F (2012) Arbuscule-containing and non-colonized cortical cells of mycorrhizal roots undergo extensive and specific reprogramming during arbuscular mycorrhizal development. Plant J 69:510–528

Gehrig H, Schüßler A, Kluge M (1996) *Geosiphon pyriformis*, a fungus forming endocytobiosis with Nostoc (cyanobacteria), is an ancestral member of the Glomerales: evidence by SSU rRNA analysis. J Mol Evol 43:71–81

Genre A, Russo G (2016) Does a common pathway transduce symbiotic signals in plant-microbe interactions? Front Plant Sci 7:96

Genre A, Chabaud M, Timmers T, Bonfante P, Barker DG (2005) Arbuscular mycorrhizal fungi elicit a novel intracellular apparatus in *Medicago truncatula* root epidermal cells before infection. Plant Cell 17:3489–3499

Genre A, Chabaud M, Faccio A, Barker DG, Bonfante P (2008) Prepenetration apparatus assembly precedes and predicts the colonization patterns of arbuscular mycorrhizal fungi within the root cortex of both *Medicago truncatula* and *Daucus carota*. Plant Cell 20:1407–1420

Genre A, Ivanov S, Fendrych M, Faccio A, Zársky V, Bisseling T, Bonfante P (2012) Multiple exocytotic markers accumulate at the sites of perifungal membrane biogenesis in arbuscular mycorrhizas. Plant Cell Physiol 53(1):244–255

Genre A, Chabaud M, Balzergue C, Puech-Pagès V, Novero M, Rey T, Fournier J, Rochange S, Bécard G, Bonfante P, Barker DG (2013) Short-chain chitin oligomers from arbuscular mycorrhizal fungi trigger nuclear Ca^{2+} spiking in *Medicago truncatula* roots and their production is enhanced by strigolactone. New Phytol 198:190–202

Ghignone S, Salvioli A, Anca I, Lumini E, Ortu G, Petiti L, Cruveiller S, Bianciotto V, Piffanelli P, Lanfranco L, Bonfante P (2012) The genome of the obligate endobacterium of an AM fungus reveals an interphylum network of nutritional interactions. ISME J 6:136–145

Gianinazzi-Pearson V (1996) Plant cell responses to arbuscular mycorrhizal fungi: getting to the roots of the symbiosis. Plant Cell 8(10):1871–1883

Gianinazzi-Pearson V, Morandi D, Dexheimer J, Gianinazzi S (1981) Ultrastructural and ultracytochemical features of a *Glomus tenuis* mycorrhiza. New Phytol 88:633–639

Giovannetti M, Ayio L, Sbrana C, Citernesi AS (1993) Factors affecting appressorium development in the vesicular-arbuscular mycorrhizal fungus *Glomus mosseae* (Nicol. and Gerd.) Gerd. And Trappe. New Phytol 123:115–122

Gobbato E (2015) Recent developments in arbuscular mycorrhizal signaling. Curr Opin Plant Biol 26:1–7

Gobbato E, Marsh JF, Vernié T, Wang E, Maillet F, Kim J, Miller JB, Sun J, Bano SA, Ratet P, Mysore KS, Dénarié J, Schultze M, Oldroyd GED (2012) A GRAS-type transcription factor with a specific function in mycorrhizal signaling. Curr Biol 22:2236–2241

Gobbato E, Wang E, Higgins G, Bano SA, Henry C, Schultze M, Oldroyd GED (2013) RAM1 and RAM2 function and expression during arbuscular mycorrhizal symbiosis and *Aphanomyces euteiches* colonization. Plant Signal Behav 8(10): pii:e26049. https://doi.org/10.4161/psb.26049

Gomez-Roldan V, Fermas S, Brewer PB, Puech-Pagès V, Dun EA, Pillot JP, Letisse F, Matusova R, Danoun S, Portais JC, Bouwmeester H, Bécard G, Beveridge CA, Rameau C, Rochange SF (2008) Strigolactone inhibition of shoot branching. Nature 455 (7210):189–194

Groth M, Takeda N, Perry J, Uchida H, Dräxl S, Brachmann A, Sato S, Tabata S, Kawaguchi M, Wang TL, Parniske M (2010) NENA, a Lotus japonicus homolog of Sec13, is required for rhizodermal infection by arbuscular mycorrhiza fungi and rhizobia but dispensable for cortical endosymbiotic development. Plant Cell 22(7):2509–2526

Gutjahr C, Parniske M (2013) Cell and developmental biology of arbuscular mycorrhiza symbiosis. Anna Rev Cell Dev Biol 29:593–617

Gutjahr C, Banba M, Croset V, An K, Miyao A, An G, Hirochika H, Imaizumi-Anraku H, Paszkowski U (2008) Arbuscular mycorrhiza-specific signaling in rice transcends the common symbiosis signaling pathway. Plant Cell 20:2989–3005

Gutjahr C, Novero M, Guether M, Montanari O, Udvardi M, Bonfante P (2009) Presymbiotic factors released by the arbuscular mycorrhizal fungus *Gigaspora margarita* induce starch accumulation in *Lotus japonicus* roots. New Phytol 183(1):53–61

Gutjahr C, Radovanovic D, Geoffroy J, Zhang Q, Siegler H, Chiapello M, Casieri L, An K, An G, Guiderdoni E, Kumar CS, Sundaresan V, Harrison MJ, Paszkowski U (2012) The half- size ABC transporters STR1 and STR2 are indispensable for mycorrhizal arbuscule formation in rice. Plant J 69:906–920

Gutjahr C, Gobbato E, Choi J, Riemann M, Johnston MG, Summers W, Carbonnel S, Mansfield C, Yang SY, Nadal M, Acosta I, Takano M, Jiao WB, Schneeberger K, Kelly KA, Paszkowski U (2015) Rice perception of symbiotic arbuscular mycorrhizal fungi requires the karrikin receptor complex. Science 350:1521–1524

Hamiaux C, Drummond RS, Janssen BJ, Ledger SE, Cooney JM, Newcomb RD, Snowden KC (2012) DAD2 is an a/b hydrolase likely to be involved in the perception of the plant branching hormone strigolactone. Curr Biol 22:2032–2036

Handa Y, Nishide H, Takeda N, Suzuki Y, Kawaguchi M, Saito K (2015) RNA-seq transcriptional profiling of an arbuscular mycorrhiza provides insights into regulated and coordinated gene expression in *Lotus japonicus* and *Rhizophagus irregularis*. Plant Cell Physiol 56:1490–1511

Harrison MJ (1996) A sugar transporter from *Medicago truncatula*: altered expression pattern in roots during vesicular-arbuscular (VA) mycorrhizal associations. Plant J 9:491–503

Harrison MJ (2012) Cellular programs for arbuscular mycorrhizal symbiosis. Curr Opin Plant Biol 15 (6):691–698

Hart MM, Reader RJ (2002) Host plant benefit from association with arbuscular mycorrhizal fungi: variation due to differences in size of mycelium. Biol Fertil Soils 36:357–366

Havaux M (2014) Carotenoid oxidation products as stress signals in plants. Plant J 79:597–606

Heard S, Brown NA, Hammond-Kosack K (2015) An interspecies comparative analysis of the predicted secretomes of the necrotrophic plant pathogens *Sclerotinia sclerotiorum* and *Botrytis cinerea*. PLoS One 10:e0130534. https://doi.org/10.1371/journal.pone.0130534

Heck C, Kuhn H, Heidt S, Walter S, Rieger N, Requena N (2016) Symbiotic fungi control plant root cortex development through the novel GRAS transcription factor MIG1. Curr Biol 26:2770–2778

Helber N, Wippel K, Sauer N, Schaarschmidt S, Hause B, Requena N (2011) A versatile monosaccharide transporter that operates in the arbuscular mycorrhizal fungus *Glomus* sp is crucial for the symbiotic relationship with plants. Plant Cell 23:3812–3823

Hirsch S, Kim J, Munoz A, Heckmann AB, Downie JA, Oldroyd GED (2009) GRAS proteins form a DNA binding complex to induce gene expression during nodulation signaling in *Medicago truncatula*. Plant Cell 21:545–557

Hogekamp C, Kuster H (2013) A roadmap of cell-type specific gene expression during sequential stages of the arbuscular mycorrhiza symbiosis. BMC Genomics 14:306

Hogekamp C, Arndt D, Pereira PA, Becker JD, Hohnjec N, Küster H (2011) Laser microdissection unravels cell-type- specific transcription in arbuscular mycorrhizal roots, including CAAT-box transcription factor gene expression correlating with fungal contact and spread. Plant Physiol 157(4):2023–2043

Hohnjec N, Vieweg M, Puhler A, Becker A, Kuster H (2005) Overlaps in the transcriptional profiles of *Medicago truncatula* roots inoculated with two different *Glomus* fungi provide insights in the genetic program activated during arbuscular mycorrhiza. Plant Physiol 137:1283–1301

Hong JJ, Park YS, Bravo A, Bhattarai KK, Daniels DA, Harrison MJ (2012) Diversity of morphology and function in arbuscular mycorrhizal symbioses in *Brachypodium distachyon*. Planta 236:851–865

Huisman R, Hontelez J, Mysore KS, Wen JQ, Bisseling T, Limpens E (2016) A symbiosis- dedicated SYNTAXIN OF PLANTS 13II isoform controls the formation of a stable host-microbe interface in symbiosis. New Phytol 211:1338–1351

Ikeda Y, Shimura H, Kitahara R, Masuta C, Ezawa (2012) A novel virus-like double-stranded RNA in an obligate biotroph arbuscular mycorrhizal fungus: a hidden player in mycorrhizal symbiosis. Mol Plant-Microbe Interact 25:1005–1012

Imaizumi-Anraku H, Takeda N, Charpentier M, Perry J, Miwa H, Umehara Y, Kouchi H, Murakami Y, Mulder L, Vickers K, Pike J, Downie JA, Wang T, Sato S, Asamizu E, Tabata S, Yoshikawa M, Murooka Y, Wu GJ, Kawaguchi M, Kawasaki S, Parniske M, Hayashi M (2005) Plastid proteins crucial for symbiotic fungal and bacterial entry into plant roots. Nature 433:527–531

Ivanov S, Harrison MJ (2014) A set of fluorescent protein-based markers expressed from constitutive and arbuscular mycorrhiza-inducible promoters to label organelles, membranes and cytoskeletal elements in *Medicago truncatula*. Plant J 80:1151–1163

Ivanov S, Fedorova EE, Limpens E, De Mita S, Genre A, Bonfante P, Bisseling T (2012) Rhizobium-legume symbiosis shares an exocytotic pathway required for arbuscule formation. Proc Natl Acad Sci U S A 109(21):8316–8321

Jashni MK, Mehrabi R, Collemare J, Mesarich CH, de Wit PJ (2015) The battle in the apoplast: further insights into the roles of proteases and their inhibitors in plant-pathogen interactions. Front Plant Sci 6:584

Javot H, Penmetsa RV, Terzaghi N, Cook DR, Harrison MJ (2007) A *Medicago truncatula* phosphate transporter indispensable for the arbuscular mycorrhizal symbiosis. Proc Natl Acad Sci U S A 104(5):1720–1725

Javot H, Penmetsa RV, Breuillin F, Bhattarai KK, Noar RD, Gomez SK, Zhang Q, Cook DR, Harrison MJ (2011) *Medicago truncatula* mtpt4 mutants reveal a role for nitrogen in the regulation of arbuscule degeneration in arbuscular mycorrhizal symbiosis. Plant J 68:954–965

Jiang Y, Wang W, Xie O, Liu N, Liu L, Wang D, Zhang X, Yang C, Chen X, Tang D, Wang E (2017) Plants transfer lipids to sustain colonization by mutualistic mycorrhizal and parasitic fungi. Science 356:1172–1175

Jin Y, Liu H, Luo DX, Yu N, Dong WT, Wang C, Zhang XW, Dai HL, Yang J, Wang ET (2016) DELLA proteins are common components of symbiotic rhizobial and mycorrhizal signalling pathways. Nat Commun 7:12433

Johnson LJ, Koulman A, Christensen M, Lane GA, Fraser K, Forester N, Johnson RD, Bryan GT, Rasmussen S (2013) An extracellular siderophore is required to maintain the mutualistic interaction of *Epichloë festucae* with *Lolium perenne*. PLoS Pathog 9(5):e1003332

Journet EP, El-Gachtouli N, Vernoud V, de Billy F, Pichon M, Dedieu A, Arnould C, Morandi D, Barker DG, Gianinazzi-Pearson V (2001) *Medicago truncatula* ENOD11: a novel RPRP-encoding early nodulin gene expressed during mycorrhization in arbuscule-containing cells. Mol Plant-Microbe Interact 14(6):737–748

Jung SC, Martinez-Medina A, Lopez-Raez JA, Pozo MJ (2012) Mycorrhiza-induced resistance and priming of plant defenses. J Chem Ecol 38:651–664

Kaku H, Nishizawa Y, Ishii-Minami N, Akimoto-Tomiyama C, Dohmae N, Takio K, Minami E, Shibuya N (2006) Plant cells recognize chitin fragments for defense signalling through a plasma membrane receptor. Proc Natl Acad Sci U S A 103(29):11086–11091

Kaló P, Gleason C, Edwards A, Marsh J, Mitra RM, Hirsch S, Jakab J, Sims S, Long SR, Rogers J, Kiss GB, Downie JA, Oldroyd GED (2005) Nodulation signalling in legumes requires NSP2, a member of the GRAS family of transcriptional regulators. Science 308(5729):1786–1789

Kamel L, Tang NW, Malbreil M, San Clemente H, Le Marquer M, Roux C, Frei Dit Frey N (2017) The comparison of expressed candidate secreted proteins from two arbuscular mycorrhizal fungi unravels common and specific molecular tools to invade different host plants. Front Plant Sci 8:124

Kanamori N, Madsen LH, Radutoiu S, Frantescu M, Quistgaard EM, Miwa H, Downie JA, James EK, Felle HH, Haaning LL, Jensen TH, Sato S, Nakamura Y, Tabata S, Sandal N, Stougaard J (2006) A nucleoporin is required for induction of Ca^{2+} spiking in legume nodule development and essential for rhizobial and fungal symbiosis. Proc Natl Acad Sci U S A 103:359–336

Kapulnik Y, Kushnir U (1991) Growth dependency of wild, primitive and modern cultivated wheat lines on vesicular-arbuscular mycorrhiza fungi. Euphytica 56:27–36

Kevei Z, Lougnon G, Mergaert P, Horváth GV, Kereszt A, Jayaraman D, Zaman N, Marcel F, Regulski K, Kiss GB, Kondorosi A, Endre G, Kondorosi E, Ané JM (2007) 3-hydroxy-3-methylglutaryl coenzyme a reductase 1 interacts with NORK and is crucial for nodulation in *Medicago truncatula*. Plant Cell 19 (12):3974–3989

Keymer A, Pimprikar P, Wewer V, Huber C, Brands M, Bucerius SL, Delaux PM, Klingl V, von Röpenack-Lahaye E, Wang TL, Eisenreich W, Dörmann P, Parniske M, Gutjahr C (2017) Lipid transfer from plants to arbuscular mycorrhiza fungi. elife 6: e29107

Kloppholz S, Kuhn H, Requena N (2011) A secreted fungal effector of *Glomus intraradices* promotes symbiotic biotrophy. Curr Biol 21:1204–1209

Kobae Y, Hata S (2010) Dynamics of periarbuscular membranes visualized with a fluorescent phosphate transporter in arbuscular mycorrhizal roots of rice. Plant Cell Physiol 51:341–353

Kobae Y, Tamura Y, Takai S, Banba M, Hata S (2010) Localized expression of arbuscular mycorrhiza-inducible ammonium transporters in soybean. Plant Cell Physiol 51:1411–1415

Kobayashi Y, Maeda T, Yamaguchi K, Kameoka H, Tanaka S, Ezawa T, Shigenobu S, Kawaguchi M (2018) The genome of *Rhizophagus clarus* HR1 reveals a common genetic basis for auxotrophy among arbuscular mycorrhizal fungi. BMC Genomics 19:465

Koch AM, Antunes PM, Maherali H, Hart MM, Klironomos JN (2017) Evolutionary asymmetry in the arbuscular mycorrhizal symbiosis; conservatism in the fungal morphology does not predict host plant growth. New Phytol 214:1330–1337

Koegel S, Ait Lahmidi N, Arnould C, Chatagnier O, Walder F, Ineichen K, Boller T, Wipf D, Wiemken A, Courty PE (2013) The family of ammonium transporters (AMT) in *Sorghum bicolor*: two AMT members are induced locally, but not systemically in roots colonized by arbuscular mycorrhizal fungi. New Phytol 198:853–865

Koltai H, Kapulnik Y (2010) Arbuscular mycorrhizas: physiology and function. Springer, Heidelberg, pp 209–236

Konvalinkova T, Jansa J (2016) Lights off for arbuscular mycorrhiza: on its symbiotic functioning under light deprivation. Front Plant Sci 7:782

Kosuta S, Chabaud M, Gough C, De J, Barker DG, Bécard G (2003) A diffusible factor from arbuscular mycorrhizal fungi induces symbiosis-specific MtENOD11 expression in roots of *Medicago truncatula*. Plant Physiol 131:952–962

Kosuta S, Hazledine S, Sun J, Miwa H, Morris RJ, Downie JA, Oldroyd GE (2008) Differential and chaotic calcium signatures in the symbiosis signaling pathway of legumes. Proc Natl Acad Sci U S A 105:9823–9828

Krajinski F, Courty PE, Sieh D, Franken P, Zhang H, Bucher M, Gerlach N, Kryvoruchko I, Zoeller D, Udvardi M, Hause B (2014) The H^{+}-ATPase HA1 of *Medicago truncatula* is essential for phosphate transport and plant growth during arbuscular mycorrhizal symbiosis. Plant Cell 26:1808–1817

Kuhn H, Küster H, Requena N (2010) Membrane steroid-binding protein 1 induced by a diffusible fungal signal is critical for mycorrhization in *Medicago truncatula*. New Phytol 185:716–733

Lahrmann U, Ding Y, Banhara A, Rath M, Hajirezaei MR, Döhlemann S, von Wirén N, Parniske M, Zuccaro A (2013) Host-related metabolic cues affect colonization strategies of a root endophyte. Proc Natl Acad Sci U S A 110:13965–13970

Lanfranco L, Fiorilli V, Gutjahr C (2018) Partner communication and role of nutrients in the arbuscular mycorrhizal symbiosis. New Phytol 220:1031–1046

Laparre J, Malbreil M, Letisse F, Portais JC, Roux C, Becard G, Puech-Pagès V (2014) Combining metabolomics and gene expression analysis reveals that propionyland butyryl-carnitines are involved in late stages of arbuscular mycorrhizal symbiosis. Mol Plant 7:554–566

Lee SJ, Kong M, Harrison P, Hijri M (2018) Conserved proteins of the RNA interference system in the arbuscular mycorrhizal fungus *Rhizoglomus irregulare* provide new insight into the evolutionary history of glomeromycota. Genome Biol Evol 10:328–343

Lehmann A, Barto EK, Powell JR, Rillig MC (2012) Mycorrhizal responsiveness trends in annual crop plants and their wild relatives - a meta-analysis on studies from 1981 to 2010. Plant Soil 355:231–250

Lehnert H, Serfling A, Enders M, Friedt W, Ordon F (2017) Genetics of mycorrhizal symbiosis in winter wheat (*Triticum aestivum*). New Phytol 215:779–791

Lekberg Y, Koide RT (2005) Is plant performance limited by abundance of arbuscular mycorrhizal fungi? A meta analysis of studies published between 1988 and 2003. New Phytol 168:189–204

Lelandais-Brière C, Moreau J, Hartmann C, Crespi M (2016) Non-coding RNAs, emerging regulators in root endosymbioses. Mol Plant-Microbe Interact 29:170–180

Lin K, Limpens E, Zhang Z, Ivanov S, Saunders DGO, Mu D, Pang E, Cao H, Cha H, Lin T, Zhou Q, Shang Y, Li Y, Sharma T, van Velzen R, de Ruijter N, Aanen DK, Win J, Kamoun S, Bisseling T, Geurts R, Huang S (2014) Single nucleus genome sequencing reveals high similarity among nuclei of an endomycorrhizal fungus. PLoS Genet 10: e1004078

Liu J, Maldonado-Mendoza I, Lopez-Meyer M, Cheung F, Town CD, Harrison MJ (2007) Arbuscular mycorrhizal symbiosis is accompanied by local and systemic alterations in gene expression and an increase in disease resistance in the shoots. Plant J 50:529–544

Lo Presti L, Lanver D, Schweizer G, Tanaka S, Liang L, Tollot M, Zuccaro A, Reissmann S, Kahmann R (2015) Fungal effectors and plant susceptibility. Annu Rev Plant Biol 66:513–545

Luginbuehl LH, Oldroyd GE (2017) Understanding the arbuscule at the heart of endomycorrhizal symbioses in plants. Curr Biol 27:R952–R963

Luginbuehl LH, Menard GN, Kurup S, Van Erp H, Radhakrishnan GV, Breakspear A, Oldroyd GED, Eastmond PJ (2017) Fatty acids in arbuscular mycorrhizal fungi are synthesized by the host plant. Science 356:1175–1178

Maillet F, Poinsot V, André O, Puech-Pagès V, Haouy A, Gueurnier M, Cromer L, Giraudet D, Formey D, Niebel A, Andres Martinez E, Driguez H, Bécard G, Dénarié J (2011) Fungal lipochitooligosaccharide symbiotic signals in arbuscular mycorrhiza. Nature 469:58–64

Martin FM, Tuskan GA, DiFazio SP, Lammers P, Newcombe G, Podila GK (2004) Symbiotic sequencing for the Populus mesocosm. New Phytol 161:330–335

Martín-Robles N, Lehmann A, Seco E, Aroca R, Rillig MC, Milla R (2018) Impacts of domestication on the arbuscular mycorrhizal symbiosis of 27 crop species. New Phytol 218(1):322–334

Martinez-Medina A, Flors V, Heil M, Mauch-Mani B, Pieterse C, Pozo M, Ton J, van Dam N, Conrath U (2016) Recognizing plant defense priming. Trends Plant Sci 21:818–822

Mateus ID, Masclaux FG, Aletti C, Rojas EC, Savary R, Dupuis C, Sanders IR (2019) Dual RNA-seq reveals large-scale non-conserved genotype×genotype-specific genetic reprograming and molecular crosstalk in the mycorrhizal symbiosis. ISME J 13(5):1226–1238. https://doi.org/10.1038/s41396-018-0342-3

Matusova R, Rani K, Verstappen W, Franssen M, Beale M, Bouwmeester H (2005) The strigolactone germination stimulants of the plant-parasitic *Striga* and *Orobanche* spp. are derived from the carotenoid pathway. Plant Physiol 139:920–934

McLean AM, Bravo A, Harrison MJ (2017) Plant signaling and metabolic pathways enabling arbuscular symbiosis. Plant Cell 29:2319–2335

Miller JB, Pratap A, Miyahara A, Zhou L, Bornemann S, Morris RJ, Oldroyd GED (2013) Calcium/calmodulin-dependent protein kinase is negatively and positively regulated by calcium, providing a mechanism for decoding calcium responses during symbiosis signaling. Plant Cell 25:5053–5066

Miya A, Albert P, Shinya T, Desaki Y, Ichimura K, Shirasu K, Narusaka Y, Kawakami N, Kaku H, Shibuya N (2007) CERK1, a LysM receptor kinase, is essential for chitin elicitor signaling in *Arabidopsis*. Proc Natl Acad Sci U S A 104:19613–19618

Miyata K, Kozaki T, Kouzai Y, Ozawa K, Ishii K, Asamizu E, Okabe Y, Umehara Y, Miyamoto A, Kobae Y, Akiyama K, Kaku H, Nishizawa Y, Shibuya N, Nakagawa T (2014) The bifunctional plant receptor, OsCERK1, regulates both chitin-triggered immunity and arbuscular mycorrhizal symbiosis in rice. Plant Cell Physiol 55:1864–1872

Morin E, Miyauchi S, San Clemente H, Chen ECH, Pelin A, de la Providencia I, Ndikumana S, Beaudet D, Hainaut M, Drula E, Kuo A, Tang N, Roy S, Viala J, Henrissat B, Grigoriev IV, Corradi N, Roux C, Martin FM (2018) Comparative genomics of *Rhizophagus irregularis, R. cerebriforme, R. diaphanus* and *Gigaspora rosea* highlights specific genetic features in Glomeromycotina. New Phytol 222 (3):1584–1598. https://doi.org/10.1111/nph.15687

Moscatiello R, Sello S, Novero M, Negro A, Bonfante P, Navazio L (2014) The intracellular delivery of TAT-aequorin reveals calcium-mediated sensing of environmental and symbiotic signals by the arbuscular mycorrhizal fungus *Gigaspora margarita*. New Phytol 203(3):1012–1020

Mukherjee A, Ané J-M (2011) Germinating spore exudates from arbuscular mycorrhizal fungi: molecular and developmental responses in plants and their regulation by ethylene. Mol Plant-Microbe Interact 24:260–270

Mullis KB (1990) The unusual origin of the polymerase chain reaction. Sci Am 262(4):56–61. 64–65

Munkvold L, Kjøller R, Vestberg M, Rosendahl S, Jakobsen I (2004) High functional diversity within species of arbuscular mycorrhizal fungi. New Phytol 164:357–364

Murat C, Payen T, Noel B, Kuo A, Morin E, Chen J, Kohler A, Krizsán K, Balestrini R, Da Silva C, Montanini B (2018) Pezizomycetes genomes reveal the molecular basis of ectomycorrhizal truffle lifestyle. Nat Ecol Evol 2:1956–1965

Murray JD, Muni RR, Torres-Jerez I, Tang Y, Allen S, Andriankaja M, Li G, Laxmi A, Cheng X, Wen J, Vaughan D, Schultze M, Sun J, Charpentier M, Oldroyd G, Tadege M, Ratet P, Mysore KS, Chen R, Udvardi MK (2011) Vapyrin, a gene essential for intracellular progression of arbuscular mycorrhizal symbiosis, is also essential for infection by rhizobia in the nodule symbiosis of *Medicago truncatula*. Plant J 65(2):244–252

Nadal M, Sawers R, Naseem S, Bassin B, Kulicke C, Sharman A, An G, An K, Ahern KR, Romag A, Brutnell TP, Gutjahr C, Geldner N, Roux C, Martinoia E, Konopka JB, Paszkowski U (2017) An N-acetylglucosamine transporter required for arbuscular mycorrhizal symbioses in rice and maize. Nat Plants 26:17073

Nagahashi G, Douds DD (2011) The effects of hydroxy fatty acids on the hyphal branching of germinated spores of AM fungi. Fungal Biol 115(4–5):351–358

Nagahashi G, Douds DD Jr (1997) Appressorium formation by AM fungi on isolated cell walls of carrot roots. New Phytol 136:299–304

Nagata M, Yamamoto N, Shigeyama T, Terasawa Y, Anai T, Sakai T, Inada S, Arima S, Hashiguchi M, Akashi R, Nakayama H, Ueno D, Hirsch AM, Suzuki A (2015) Red/far red light controls arbuscular mycorrhizal colonization via jasmonic acid and strigolactone signaling. Plant Cell Physiol 56:2100–2109

Naito M, Morton JB, Pawlowska TE (2015) Minimal genomes of mycoplasma-related endobacteria are plastic and contain host-derived genes for sustained life within Glomeromycota. Proc Natl Acad Sci U S A 112:7791–7796

Navazio L, Moscatiello R, Genre A, Novero M, Baldan B, Bonfante P, Mariani P (2007) A diffusible signal from arbuscular mycorrhizal fungi elicits a transient cytosolic calcium elevation in host plant cells. Plant Physiol 144:673–681

Nelson DC, Flematti GR, Riseborough JA, Ghisalberti EL, Dixon KW, Smith SM (2010) Karrikins enhance light responses during germination and seedling development in *Arabidopsis thaliana*. Proc Natl Acad Sci U S A 107:7095–7100

Oláh B, Brière C, Bécard G, Dénarié J, Gough C (2005) Nod factors and a diffusible factor from arbuscular mycorrhizal fungi stimulate lateral root formation in *Medicago truncatula* via the DMI1/DMI2 signalling pathway. Plant J 44:195–207

Oldroyd GED (2013) Speak, friend, and enter: signalling systems that promote beneficial symbiotic associations in plants. Nat Rev Microbiol 11(4):252–263

Oldroyd GED, Downie JA (2006) Nuclear calcium changes at the core of symbiosis signalling. Curr Opin Plant Biol 9:351–357

Op Den Camp R, Streng A, De Mita S, Cao Q, Polone E, Liu W, Ammiraju JS, Kudrna D, Wing R, Untergasser A, Bisseling T, Geurts R (2011) LysM-type mycorrhizal receptor recruited for rhizobium symbiosis in non legume *Parasponia*. Science 331:909–912

Orchard S, Hilton S, Bending GD, Dickie IA, Standish RJ, Gleeson RD, Jeffery P, Powell JR, Walker C, Bass D, Monk J, Simonin A, Ryan MH (2017) Fine endophytes (*Glomus tenue*) are related to Mucoromycotina not Glomeromycota. New Phytol 213:481–486

Pan HR, Oztas O, Zhang XW, Wu XY, Stonoha C, Wang E, Wang B, Wang D (2016) A symbiotic SNARE protein generated by alternative termination of transcription. Nat Plants 2:15197

Pandey P, Wang M, Baldwin IT, Pandey SP, Groten K (2018) Complex regulation of microRNAs in roots of competitively-grown isogenic *Nicotiana attenuata* plants with different capacities to interact with arbuscular mycorrhizal fungi. BMC Genomics 19 (1):937

Park H-J, Floss DS, Levesque-Tremblay V, Bravo A, Harrison MJ (2015) Hyphal branching during arbuscule development requires RAM1. Plant Physiol 169:2774–2788

Parniske M (2004) Molecular genetics of the arbuscular mycorrhizal symbiosis. Curr Opin Plant Biol 7:414–421

Parniske M (2008) Arbuscular mycorrhiza: the mother of plant root endosymbioses. Nat Rev Microbiol 6 (10):763–775

Pellegrin C, Morin E, Martin FM, Veneault-Fourrey C (2015) Comparative analysis of secretomes from ectomycorrhizal fungi with an emphasis on small-secreted proteins. Front Microbiol 6:1278

Pfeffer PE, Douds DD, Bécard G, Shachar-Hill Y (1999) Carbon uptake and the metabolism and transport of lipids in an arbuscular mycorrhiza. Plant Physiol 120:587–598

Pimprikar P, Gutjahr C (2018) Transcriptional regulation of arbuscular mycorrhiza development. Plant Cell Physiol 59:673–690

Pimprikar P, Carbonnel S, Paries M, Katzer K, Klingl V, Bohmer MJ, Karl L, Floss DS, Harrison MJ, Parniske M, Gutjahr C (2016) A CCaMK-CYCLOPS-DELLA complex activates transcription of RAM1 to regulate arbuscule branching. Curr Biol 26:987–998

Pumplin N, Harrison MJ (2009) Live-cell imaging reveals periarbuscular membrane domains and organelle location in *Medicago truncatula* roots during arbuscular mycorrhizal symbiosis. Plant Physiol 151:809–819

Pumplin N, Mondo SJ, Topp S, Starker CG, Gantt JS, Harrison MJ (2010) *Medicago truncatula* Vapyrin is a novel protein required for arbuscular mycorrhizal symbiosis. Plant J 61:482–494

Pumplin N, Zhang X, Noar RD, Harrison MJ (2012) Polar localization of a symbiosis-specific phosphate transporter is mediated by a transient reorientation of secretion. Proc Natl Acad Sci USA 109: E665–E672

Ramirez-Flores MR, Rellan-Alvarez R, Wozniak B, Gebreselassie MN, Jakobsen I, Olalde-Portugal V, Baxter I, Paszkowski U, RJH S (2017) Coordinated changes in the accumulation of metal ions in maize (*Zea mays* ssp. mays L.) in response to inoculation with the arbuscular mycorrhizal fungus *Funneliformis mosseae*. Plant Cell Physiol 58:1689–1699

Rich MK, Schorderet M, Bapaume L, Falquet L, Morel P, Vandenbussche M, Reinhardt D (2015) The Petunia GRAS transcription factor ATA/RAM1 regulates symbiotic gene expression and fungal morphogenesis in arbuscular mycorrhiza. Plant Physiol 168:788–797

Rimington WR, Pressel S, Duckett JG, Bidartondo MI (2015) Fungal associations of basal vascular plants: reopening a closed book? New Phytol 205:1394–1398

Rivero J, Álvarez D, Flors V, Azcón-Aguilar C, Pozo MJ (2018) Root metabolic plasticity underlies functional diversity in mycorrhiza-enhanced stress tolerance in tomato. New Phytol 220 (4):1322–1336

Ropars J, Lo Y-C, Dumas E, Snirc A, Begerow D, Rollnik T, Lacoste S, Dupont J, Giraud T, López-Villavicencio M (2016) Fertility depression among cheese-making *Penicillium roqueforti* strains suggests degeneration during domestication. Evolution 70(9):2099–2109

Russo G, Spinella S, Sciacca E, Bonfante P, Genre A (2013) Automate analysis of calcium spiking profiles with CaSA software: two case studies from root-microbe symbioses. BMC Plant Biol 13:224

Sagan M, Morandi D, Tarenghi E, Duc G (1995) Selection of nodulation and mycorrhizal mutants in the model plant *Medicago truncatula* (Gaertn.) after c-ray mutagenesis. Plant Sci 111:63–71

Saito K, Yoshikawa M, Yano K, Miwa H, Uchida H, Asamizu E, Sato S, Tabata S, Imaizumi-Anraku H, Umehara Y, Kouchi H, Murooka Y, Szczyglowski K, Downie JA, Parniske M, Hayashi M, Kawaguchi M (2007) NUCLEOPORIN85 is required for calcium spiking, fungal and bacterial symbioses, and seed production in *Lotus japonicus*. Plant Cell 19:610–624

Salvioli A, Ghignone S, Novero M, Navazio L, Venice F, Bagnaresi P, Bonfante P (2016) Symbiosis with an endobacterium increases the fitness of a mycorrhizal fungus, raising its bioenergetic potential. ISME J 10:130–144

Sawers RJH, Gebreselassie MN, Janos DP, Paszkowski U (2010) Characterizing variation in mycorrhiza effect among diverse plant varieties. Theor Appl Genet 120:1029–1039

Sawers RJH, Svane SF, Quan C, Grønlund M, Wozniak B, Gebreselassie MN, González Muñoz E, Chávez Montes RA, Baxter I, Goudet J, Jakobsen I, Paszkowski U (2017) Phosphorus acquisition efficiency in arbuscular mycorrhizal maize is correlated with the abundance of root external hyphae and the accumulation of transcripts encoding PHT1 phosphate transporters. New Phytol 214:632–643

Schirawski J, Mannhaupt G, Munch K, Brefort T, Schipper K, Doehlemann G, Di Stasio M, Rossel N, Mendoza-Mendoza A, Pester D, Muller O, Winterberg B, Meyer E, Ghareeb H, Wollenberg T, Munsterkotter M, Wong P, Walter M, Stukenbrock E, Guldener U, Kahmann R (2010) Pathogenicity determinants in smut fungi revealed by genome comparison. Science 330:1546–1548

Schornack S, Huitema E, Cano LM, Bozkurt TO, Oliva R, Van Damme M et al (2009) Ten things to know about oomycete effectors. Mol Plant Pathol 10:795–803

Schweiger R, Baier MC, Persicke M, Muller C (2014) High specificity in plant metabolic responses to arbuscular mycorrhiza. Nat Commun 5:3886

Sędzielewska Toro K, Brachmann A (2016) The effector candidate repertoire of the arbuscular mycorrhizal fungus *Rhizophagus clarus*. BMC Genomics 17:101. https://doi.org/10.1186/s12864-016-2422-y

Shimizu T, Nakano T, Takamizawa D, Desaki Y, Ishii-Minami N, Nishizawa Y, Minami E, Okada K, Yamane H, Kaku H, Shibuya N (2010) Two LysM receptor molecules, CEBiP and OsCERK1, cooperatively regulate chitin elicitor signaling in rice. Plant J 64:204–214

Shimoda Y, Han L, Yamazaki T, Suzuki R, Hayashi M, Imaizumi-Anraku H (2012) Rhizobial and fungal symbioses show different requirements for calmodulin binding to calcium calmodulin-dependent protein kinase in *Lotus japonicus*. Plant Cell 24(1):304–321

Siciliano V, Genre A, Balestrini R, Cappellazzo G, Dewit PJGM, Bonfante P (2007) Transcriptome analysis of arbuscular mycorrhizal roots during development of the prepenetration apparatus. Plant Physiol 144:1455–1466

Silvestri A, Fiorilli V, Miozzi L, Accotto GP, Turina M, Lanfranco L (2019) *In silico* analysis of fungal small RNA accumulation reveals putative plant mRNA targets in the symbiosis between an arbuscular mycorrhizal fungus and its host plant. BMC Genomics 20:169

Singh AK, Hamel C, DePauw RM, Knox RE (2012) Genetic variability in arbuscular mycorrhizal fungi compatibility supports the selection of durum wheat genotypes for enhancing soil ecological services and cropping systems. Can J Microbiol 58(2012):293–302

Smit P, Raedts J, Portyanko V, Debellé F, Gough C, Bisseling T, Geurts R (2005) NSP1 of the GRAS protein family is essential for rhizobial nod factor-induced transcription. Science 308(5729):1789–1791

Smith SE, Smith FA, Jakobsen I (2004) Functional diversity in arbuscular mycorrhizal (AM) symbioses: the contribution of the mycorrhizal P uptake pathway is not correlated with mycorrhizal responses in growth or total P uptake. New Phytol 162:511–524

Soto MJ, Domínguez-Ferreras A, Pérez-Mendoza D, Sanjuán J, Olivares J (2009) Mutualism versus pathogenesis: the give-and- take in plant-bacteria interactions. Cell Microbiol 11(3):381–388

Spatafora JW, Chang Y, Benny GL, Lazarus K, Smith ME, Berbee ML, Bonito G, Corradi N, Grigoriev I, Gryganskyi A, James TY, O'Donnell K, Roberson RW, Taylor TN, Uehling J, Vilgalys R, White MM, Stajich JE (2016) A phylum-level phylogenetic classification of zygomycete fungi based on genome-scale data. Mycologia 108:1028–1046

Steinkellner S, Hage-Ahmed K, Garcia-Garrido JM, Illana A, Ocampo JA, Vierheilig H (2012) A comparison of wild-type, old and modern tomato cultivars in the interaction with the arbuscular mycorrhizal fungus *Glomus mosseae* and the tomato pathogen *Fusarium oxysporum* f. sp. *lycopersici*. Mycorrhiza 22:189–194

Sun J, Miller JB, Granqvist E, Wiley-Kalil A, Gobbato E, Maillet F, Cottaz S, Samain E, Venkateshwaran M, Fort S, Morris RJ, Ané JM, Dénarié J, Oldroyd GED (2015) Activation of symbiosis signaling by arbuscular mycorrhizal fungi in legumes and rice. Plant Cell 27:823–838

Sun X, Chen W, Ivanov S, MacLean AM, Wight H, Ramaraj T, Mudge J, Harrison MJ, Fei Z (2018) Genome and evolution of the arbuscular mycorrhizal fungus *Diversispora epigaea* (formerly *Glomus versiforme*) and its bacterial endosymbionts. New Phytol 221(3):1556–1573. https://doi.org/10.1111/NPH.15472

Takeda N, Handa Y, Tsuzuki S, Kojima M, Sakakibara H, Kawaguchi M (2015) Gibberellins interfere with symbiosis signaling and gene expression and alter colonization by arbuscular mycorrhizal fungi in *Lotus japonicus*. Plant Physiol 167:545–557

Tamayo E, Gómez-Gallego T, Azcón-Aguilar C, Ferrol N (2014) Genome-wide analysis of copper, iron and zinc transporters in the arbuscular mycorrhizal fungus *Rhizophagus irregularis*. Front Plant Sci 5:547

Tanaka Y, Yano K (2005) Nitrogen delivery to maize via mycorrhizal hyphae depends on the form of N supplied. Plant Cell Environ 28:1247–1254

Tisserant E, Kohler A, Dozolme-Seddas P, Balestrini R, Benabdellah K, Colard A, Croll D, Da Silva C, Gomez SK, Koul R, Ferrol N, Fiorilli V, Formey D, Franken P, Helber N, Hijri M, Lanfranco L, Lindquist E, Liu Y, Malbreil M, Morin E, Poulain J, Shapiro H, van Tuinen D, Waschke A, Azcón-Aguilar C, Bécard G, Bonfante P, Harrison MJ, Küster H, Lammers P, Paszkowski U, Requena N, Rensing SA, Roux C, Sanders IR, Shachar-Hill Y, Tuskan G, Young JP, Gianinazzi-Pearson V, Martin F (2012) The transcriptome of the arbuscular mycorrhizal fungus *Glomus intraradices* (DAOM 197198) reveals functional tradeoffs in an obligate symbiont. New Phytol 193(3):755–769

Tisserant E, Malbreil M, Kuo A, Kohler A, Symeonidi A, Balestrini R, Charron P, Duensing N, Frei dit Frey N, Gianinazzi-Pearson V, Gilbert LB, Handa Y, Herr JR, Hijri M, Koul R, Kawaguchi M, Krajinski F, Lammers PJ, Masclaux FG, Murat C, Morin E, Ndikumana S, Pagni M, Petitpierre D, Requena N, Rosikiewicz P, Riley R, Saito K, San Clemente H, Shapiro H, van Tuinen D, Bécard G, Bonfante P, Paszkowski U, Shachar-Hill YY, Tuskan GA, Young JP, Sanders IR, Henrissat B, Rensing SA, Grigoriev IV, Corradi N, Roux C, Martin F (2013) Genome of an arbuscular mycorrhizal fungus provides insight into the oldest plant symbiosis. Proc Natl Acad Sci U S A 110(50):20117–20122

Torres-Cortés G, Ghignone S, Bonfante P, Schüßler A (2015) Mosaic genome of endobacteria in arbuscular mycorrhizal fungi: transkingdom gene transfer in an ancient mycoplasma-fungus association. Proc Natl Acad Sci U S A 112:7785–7790

Tsuzuki S, Handa Y, Takeda N, Kawaguchi M (2016) Strigolactone-induced putative secreted protein 1 is required for the establishment of symbiosis by the arbuscular mycorrhizal fungus *Rhizophagus irregularis*. Mol Plant-Microbe Interact 29:277–286

Turina M, Ghignone S, Astolfi N, Silvestri A, Bonfante P, Lanfranco L (2018) The virome of the arbuscular mycorrhizal fungus *Gigaspora margarita* reveals the first report of DNA fragments corresponding to replicating nonretroviral RNA viruses in fungi. Environ Microbiol 20(6):2012–2025. https://doi.org/10.1111/1462-2920.14060

Turrini A, Sbrana C, Avio L, Mugendi Njeru E, Bocci G, Bàrberi P, Giovannetti M (2016) Changes in the composition of native root arbuscular mycorrhizal fungal communities during a short-term cover crop-maize succession. Biol Fertil Soils 52:643–653

Turrini A, Avio L, Giovannetti M, Agnolucci M (2018) Functional complementarity of arbuscular mycorrhizal fungi and associated microbiota: the challenge of translational research. Front Plant Sci 9:1407

Uehling J, Gryganskyi A, Hameed K, Tschaplinski T, Misztal PK, Wu S, Desirò A, Vande Pol N, Du Z, Zienkiewicz A, Zienkiewicz K, Morin E, Tisserant E, Splivallo R, Hainaut M, Henrissat B, Ohm R, Kuo A, Yan J, Lipzen A, Nolan M, LaButti K, Barry K, Goldstein AH, Labbé J, Schadt C, Tuskan G, Grigoriev I, Martin F, Vilgalys R, Bonito G (2017) Comparative genomics of *Mortierella elongata* and its bacterial endosymbiont *Mycoavidus cysteinexigens*. Environ Microbiol 19(8):2964–2983

Van Buuren ML, Maldonado-Mendoza IE, Trieu AT, Blaylock LA, Harrison MJ (1999) Novel genes induced during an arbuscular mycorrhizal (AM) symbiosis formed between *Medicago truncatula* and *Glomus versiforme*. Mol Plant-Microbe Interact 12(3):171–181

Venkateshwaran M, Cosme A, Han L, Banba M, Satyshur KA, Schleiff E, Parniske M, Imaizumi-Anraku H, Ané JM (2012) The recent evolution of a symbiotic ion channel in the legume family altered ion conductance and improved functionality in calcium signaling. Plant Cell 24:2528–2545

Venkateshwaran M, Jayaraman D, Chabaud M, Genre A, Balloon AJ, Maeda J, Forshey K, den Os D, Kwiecien NW, Coon JJ, Barker DG, Ané J-M (2015) A role for the mevalonate pathway in early plant symbiotic signaling. Proc Natl Acad Sci U S A 112:9781–9786

Venice F, Ghignone S, Salvioli A, Amselem J, Novero M, Xianan X, Sedzielewska Toro K, Morin E, Lipzen A, Grigoriev IV, Henrissat B, Martin F, Bonfante P (2019) At the nexus of three kingdoms: the genome of the mycorrhizal fungus *Gigaspora margarita* provides insights into plant, endobacterial and fungal interactions. Environ Microbiol 22:122–141

Vijayakumar V, Liebisch G, Buer B, Xue L, Gerlach N, Blau S, Schmitz J, Bucher M (2015) Integrated multi-omics analysis supports role of lysophosphatidylcholine and related glycerophospholipids in the *Lotus japonicus*-*Glomus intraradices* mycorrhizal symbiosis. Plant Cell Environ 39 (2):393–415

Voß S, Betz R, Heidt S, Corradi N, Requena N (2018) RiCRN1, a crinkler fffector from the arbuscular mycorrhizal fungus *Rhizophagus irregularis*, functions in arbuscule development. Front Microbiol 9:2068. https://doi.org/10.3389/fmicb.2018.02068. eCollection 2018

Wang B, Yeun LH, Xue JY, Liu Y, Ané JM, Qiu YL (2010) Presence of three mycorrhizal genes in the common ancestor of land plants suggests a key role of mycorrhizas in the colonization of land by plants. New Phytol 186:514–525

Wang E, Schornack S, Marsh JF, Gobbato E, Schwessinger B, Eastmond P, Schultze M, Kamoun S, Oldroyd GED (2012) A common signalling process that promotes mycorrhizal and oomycete colonization of plants. Curr Biol 22(23):2242–2246

Wang M, Weiberg A, Dellota E, Yamane D, Jin H (2017) Botrytis smallRNABcsiR37 suppresses plant defense genes by cross-kingdom RNAi. RNA Biol 14:421–428

Wang M, Schäfer M, Li D, Halitschke R, Dong C, McGale E, Paetz C, Song Y, Li S, Dong J, Heiling S, Groten K, Franken P, Bitterlich M, Harrison MJ, Paszkowski U, Baldwin IT (2018) Blumenols as shoot markers of root symbiosis with arbuscular mycorrhizal fungi. elife 7:e37093. https://doi.org/10.7554/eLife.37093

Waters MT, Brewer PB, Bussell JD, Smith SM, Beveridge CA (2012) The *Arabidopsis* ortholog of rice DWARF27 acts upstream of MAX1 in the control of plant development by strigolactones. Plant Physiol 159(3):1073–1085

Waters MT, Gutjahr C, Bennett T, Nelson DC (2017) Strigolactone signaling and evolution. Annu Rev Plant Biol 68:291–322

Wewer V, Brands M, Dörmann P (2014) Fatty acid synthesis and lipid metabolism in the obligate biotrophic fungus *Rhizophagus irregularis* during mycorrhization of *Lotus japonicus*. Plant J 79:398–412

White TJ, Bruns T, Lee S, Taylor J (1990) Amplification and direct sequencing of fungal ribosomal RNA genes for phylogenetics. In: Innis MA, Gelfand DH, Sninsky JJ, White TJ (eds) PCR protocols: a guide to methods and applications. Academic Press, New York, pp 315–322

Wu P, Wu Y, Liu C-C, Liu L-W, Ma F-F, Wu X-Y, Wu M, Hang Y-Y, Chen J-Q, Shao Z-Q, Wang B (2016) Identification of arbuscular mycorrhiza (AM)-responsive microRNAs in tomato. Front Plant Sci 7:429

Xing X, Koch AM, Jones AMP, Ragone D, Murch S, Hart MM (2012) Mutualism breakdown in breadfruit domestication. Proc Biol Sci 279:1122–1130

Xue L, Cui H, Buer B, Vijayakumar V, Delaux P-M, Junkermann S, Bucher M (2015) Network of GRAS transcription factors involved in the control of arbuscule development in *Lotus japonicus*. Plant Physiol 167:854–871

Yamamoto K, Endo N, Degawa Y, Fukuda M, Yamada A (2017) First detection of Endogone ectomycorrhizas in natural oak forests. Mycorrhiza 27:295–301

Yang S-Y, Grønlund M, Jakobsen I, Grotemeyer MS, Rentsch D, Miyao A, Hirochika H, Santhosh Kumar C, Sundaresan V, Salamin N, Catausan S, Mattes N, Heuer S, Paszkowski U (2012) Nonredundant regulation of rice arbuscular mycorrhizal symbiosis by two members of the phosphate transporter1 gene family. Plant Cell 24:4236–4251

Yano K, Yoshida S, Müller J, Singh S, Banba M, Vickers K, Markmann K, White C, Schuller B, Sato S, Asamizu E, Tabata S, Murooka Y, Perry J, Wang TL, Kawaguchi M, Imaizumi-Anraku H, Hayashi M, Parniske M (2008) CYCLOPS, a mediator of symbiotic intracellular accommodation. Proc Natl Acad Sci U S A 105(51):20540–20545

Yoshida S, Kameoka H, Tempo M, Akiyama K, Umehara M, Yamaguchi S, Hayashi H, Kyozuka J, Shirasu K (2012) The D3 F-box protein is a key component in host strigolactone responses essential for arbuscular mycorrhizal symbiosis. New Phytol 196:1208–1216

Yu N, Luo D, Zhang X, Liu J, Wang W, Jin Y, Dong W, Liu J, Liu H, Yang W, Zeng L, Li Q, He Z, Oldroyd GED, Wang E (2014) A DELLA protein complex controls the arbuscular mycorrhizal symbiosis in plants. Cell Res 24:130–133

Zeng T, Holmer R, Hontelez J, Te Lintel HB, Marufu L, de Zeeuw T, Wu F, Schijlen E, Bisseling T, Limpens E (2018) Host- and stage-dependent secretome of the arbuscular mycorrhizal fungus *Rhizophagus irregularis*. Plant J 94:411–425

Zhang X, Dong W, Sun J, Feng F, Deng Y, He Z, Oldroyd GE, Wang E (2015) The receptor kinase CERK1

has dual functions in symbiosis and immunity signalling. Plant J 81:258–267
Zhang X, Pumplin N, Ivanov S, Harrison MJ (2015b) EXO70I is essential for development of a sub-domain of the periarbuscular membrane during arbuscular mycorrhizal symbiosis. Curr Biol 25:2189–2195
Zhong Z, Norvienyeku J, Chen M, Bao J, Lin L, Chen L, Lin Y, Wu X, Cai Z, Zhang Q, Lin X, Hong Y, Huang J, Xu L, Zhang H, Chen L, Tang W, Zheng H, Chen X, Wang Y, Lian B, Zhang L, Tang H, Lu G, Ebbole DJ, Wang B, Wang Z (2016) Directional selection from host plants is a major force driving host specificity in *Magnaporthe* species. Sci Rep 6:25591
Zhu YG, Smith SE, Barritt AR, Smith FA (2001) Phosphorus (P) efficiencies and mycorrhizal responsiveness of old and modern wheat cultivars. Plant Soil 237:249–255
Zipfel C, Oldroyd GE (2017) Plant signalling in symbiosis and immunity. Nature 15:328–336
Zouari I, Salvioli A, Chialva M, Novero M, Miozzi L, Tenore G, Bagnaresi P, Bonfante P (2014) From root to fruit: RNA-Seq analysis shows that arbuscular mycorrhizal symbiosis may affect tomato fruit metabolism. BMC Genomics 15:221

8 Coordination of Fungal Secondary Metabolism and Development

JENNIFER GERKE[1], ANNA M. KÖHLER[1], CINDY MEISTER[1], KARL G. THIEME[1], HUGO AMOEDO[1], GERHARD H. BRAUS[1]

CONTENTS

I. Introduction

Fungi are ubiquitous lifeforms that can grow in a wide range of diverse environments. Based on current phylogenetic analyses, the fungal kingdom is divided into 8 phyla with 12 subphyla and 46 classes (Spatafora et al. 2017). The two monophyletic groups of Ascomycota and Basidiomycota form together the subkingdom of *Dikarya*. Many fungi have a saprophytic lifestyle. They live on decaying organic matter as principal decomposers of our ecosystem. Fungi play important roles in our food industry to make cheese, ferment soybeans, or brew beer and in the pharma industry to produce drugs (Gerke and Braus 2014). But their appearance can also be harmful to us. Fungal spores are dispersed through the air and can be inhaled into our lungs where most of them are inactivated as long as the immune system is not compromised (Shlezinger et al. 2017). Fungi are responsible for many diseases in humans and animals, ranging from allergies to life-threatening intoxications and mycoses. The infection of plants and contamination of harvest products by fungi lead to high economic losses and are a threat for food supply and safety (Meyer et al. 2016). In fungi, differentiation processes, including the formation of infection structures, are closely linked to the production of specific chemicals. These interconnected processes and their regulations are the main focus of this chapter.

[1] Institut für Mikrobiologie und Genetik, Abteilung Molekulare Mikrobiologie und Genetik, Göttinger Zentrum für Molekulare Biowissenschaften (GZMB), Georg-August-Universität Göttingen, Göttingen, Germany; e-mail: gbraus@gwdg.de

Genetics and Biotechnology, 3rd Edition
The Mycota II
J.P. Benz, K. Schipper (Eds.)

A. Fungal Differentiation

Filamentous fungi of the *Dikarya* germinate from single spores and initially grow in long, tubular vegetative hyphae by tip extension (Riquelme et al. 2018). The Spitzenkörper, which is located at the hyphal tip, acts as a dynamic center for the organization and supply of the vesicles required for material transport. Hyphae fuse through anastomosis tubes at the tips and form branched two dimensional networks, called mycelia. When a certain state of developmental competence is reached, the fungus can respond to external signals to proliferate through either asexual or sexual differentiation (Fig. 1a). In asexual reproduction, the fungus produces spore-forming structures called conidiophores, which carry the uninuclear, mitotic-derived asexual spores called conidia (Adams et al. 1998). These airborne spores are distributed by the wind. In sexual reproduction, ascomycetes produce sexual spores called ascospores in a sac-shaped ascus.[1] In most species, each ascus carries eight spores which are formed by meiosis and subsequent mitotic division. The asci usually form within the protecting fruiting bodies, which can be closed and spherical (cleistothecia), closed and flask-like (perithecia), or open and cup-shaped (apothecia) and which serve as overwintering structures in the soil (Pöggeler et al. 2018). Whether asexual or sexual structures are formed depends strongly on the environmental conditions, such as nutrients, light, temperature, or oxygen availability. Fungi change their lifestyle also during the infection process. Whereas some fungi can gain entry into the host without forming specialized structures, there are many plant pathogenic fungi that produce adhesion and penetration structures, such as appressoria and hyphopodia. These penetration organs form tiny infection hooks and penetrate the host using turgor pressure and/or by secreting large amounts of plant cell wall-degrading enzymes (Lo Presti et al. 2015).

[1] Ascus = skin bag.

B. Fungal Secondary Metabolism

Whether a fungus initiates a differential program depends not only on environmental conditions but also on endogenous factors such as the formation of primary or secondary metabolites including pheromones. Metabolic programs are tightly connected with morphogenic differentiation through sophisticated signal sensing and transduction mechanisms as well as transcriptional networks. Whereas primary metabolites (also called central metabolites) are essential for the growth of an organism, secondary metabolites (also called specialized metabolites or natural products) are dispensable but offer advantages in the natural habitat of their producer. Usually, primary metabolites are the precursors for the biosynthesis of secondary metabolites, and the biosynthesis occurs during developmental or aging processes (Bayram et al. 2016). Several secondary metabolites directly related to development are known and show the close connection between these two processes in fungi (Fig. 1a). These secondary metabolites are involved, for example, in the initiation and regulation of development and in protection and survival. They are also required for communication, competition, and defense against other microorganisms as well as for virulence in plant and animal infections (Macheleidt et al. 2016; Künzler 2018).

Although the biological benefit of most secondary metabolites has not yet been understood, their biological activities have been used for a very long time. Already our ancestors used the healing extracts of fungi and plants as medicine and nowadays many fungal compounds serve as lead structures for the synthesis of new drugs (Newman and Cragg 2012). Examples of secondary metabolites with pharmaceutical relevance from ascomycetes are the antibiotic penicillin, the anticancer drug taxol, the cholesterol-lowering drug lovastatin, or the immunosuppressant ciclosporin. However, the biological activities of secondary metabolites can also be harmful for us. Fungi synthesize strong carcinogens and mycotoxins such as aflatoxins produced by *Aspergillus* and trichothecenes produced by *Fusarium* (Gerke and Braus 2014).

A

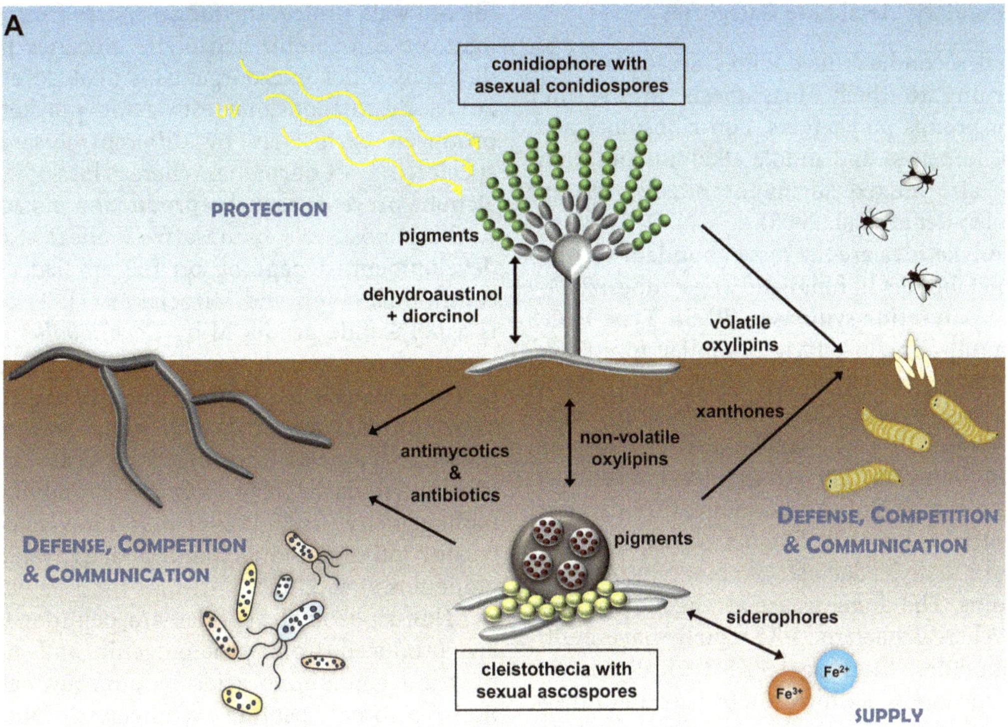

B

Main Classes	Precursors	Enzymes	Examples
Polyketides	malonyl-CoA, acetyl-CoA, propionyl-CoA, methylmalonyl-CoA	polyketide synthases (PKS)	aflatoxins, sterigmatocystin, lovastatin, orsellinic acid, F-9775A, F-9775B, xanthones, amphotericin B
Nonribosomal peptides	proteinogenic and non-proteinogenic amino acids	non-ribosomal peptide synthetases (NRPS)	penicillin, ciclosporin, gliotoxin
Terpenes	isoprene units	terpene synthases (terpene cyclases)	aristolochenes, carotenoids, gibberellins, trichothecenes, taxol
Indole alkaloids	tryptophan, indole-3-glycerol-phosphate, tryptamine	dimethylallyl tryptophan synthases (DMATS)	paxilline, communesin F, lysergic acid, psilocybin
Subclasses			
Peptide-polyketide hybrids	malonyl-CoA, acetyl-CoA, propionyl-CoA, methylmalonyl-CoA, proteinogenic and non-proteinogenic amino acids	PKS-NRPS hybrids	emericellamides, aspyridone, pseurotin A
Meroterpenoids	isoprene units, malonyl-CoA, acetyl-CoA, propionyl-CoA, methylmalonyl-CoA	terpene synthases (terpene cyclases) & PKS	austinol, dehydroaustinol, fumagillin
	isoprene units, proteinogenic and non-proteinogenic amino acids	terpene synthases (terpene cyclases) & NRPS	terpendole E
Isocyanides	proteinogenic and non-proteinogenic amino acids	isocyanide synthases, isocyanide synthase-NRPS hybrids	xanthocillin, xanthoascin
Oxylipins	polyunsaturated fatty acids	oxygenases	hydroxylated oleic (18:1) acid, hydroxylated linoleic (18:2) acid

Fig. 1 Functions of fungal secondary metabolites during life of an ascomycetous filamentous fungus living in the soil. (**a**) Examples of secondary metabolites contributing to growth, differentiation, communication, competition, protection, and survival of *Aspergillus nidulans*. (**b**) Classification of fungal secondary metabolites into the most common classes

1. Secondary Metabolite Categories

Fungal secondary metabolites are categorized according to their biosynthesis in the four major groups polyketides, non-ribosomal peptides, terpenes, and indole alkaloids but comprise also mixed forms or smaller groups (Fig. 1b; Keller et al. 2005).

Polyketides are the most abundant secondary metabolites in fungi and are synthesized by type I *p*oly*k*etide *s*ynthases (PKS). Type I PKS are multi-domain enzymes similar to eukaryotic fatty acid synthases, consisting of the essential *k*etoacyl CoA *s*ynthase (KS), *a*cyl*t*ransferase (AT), *a*cyl *c*arrier *p*rotein (ACP) domains, and a termination domain (Keller et al. 2005). Additionally, they can harbor variable domains like dehydratase, ketoreductase, enoylreductase, or methyltransferase domains. The domains are arranged in a module. Whereas bacterial PKS usually have multiple modules, the fungal PKS typically carries only one module, which can be used iteratively. The precursors for the biosynthesis of polyketides are usually acetyl *co*enzyme A (acetyl-CoA) and malonyl-CoA, but also propionyl-CoA or methylmalonyl-CoA are used. During the starting stage of polyketide synthesis, the precursors are loaded onto the starter domains KS and ACP, catalyzed by the AT domain. In the elongation stage, these loaded precursors are condensed by decarboxylative condensation similar to the fatty acid synthesis, resulting in a polyketide chain. By loading of another starter unit, the next cycle starts, and the polyketide chain is elongated by the new starter unit similarly. The control of how many cycles are performed is not yet understood. Finally, the polyketide chain is released from the enzyme by the termination domain (Keller et al. 2005).

Among the class of polyketides are several products that demonstrate the tight link between developmental processes and secondary metabolism by fulfilling survival or longevity tasks, inducing sexual development as hormones or conferring virulence. Typical examples are melanins, zearalenone, and T-toxin. Melanins are pigments that can be incorporated into the fungal cell wall or secreted to the environment. They strengthen the cell wall, protect the fungal spores from UV light, or can inhibit hydrolytic enzymes produced by other microorganisms (Toledo et al. 2017). Zearalenone, an estrogenic polyketide produced exclusively by different Fusarium species, is a sex hormone. Whereas loss of zearalenone prevents sexual reproduction, its addition can positively or negatively affect sexual development, depending on the applied concentrations (Wolf and Mirocha 1977). T-toxin is a polyketide produced by *Cochliobolus heterostrophus*, a necrotrophic fungal plant pathogen that causes southern corn leaf blight in maize. It is connected with high virulence[2] in maize cultivars carrying the "Texas male sterile cytoplasm", but it is not required for pathogenicity[3] itself. It contributes to virulence by disrupting the mitochondrial activity (Stergiopoulos et al. 2013).

Non-ribosomal peptides are peptides that are produced from proteinogenic and non-proteinogenic amino acids by enzymes called *n*on-*r*ibosomal *p*eptide *s*ynthetases (NRPS) without the use of ribosomes. NRPS are multi-domain and multi-modular enzymes. Each module contains an adenylation domain for the recognition of the amino acid, a peptidyl carrier domain for the activation and covalent binding of the amino acid to the *p*hospho*p*antetheine *t*ransferase cofactor (PPTase) and a condensation domain for peptide formation. The final peptide is usually released by a thioesterase domain at the C-terminus of the enzyme. The diversity of non-ribosomal peptides derives from cyclization or branching of peptides. Among them are, for instance, the antibiotic penicillin, siderophores, and gliotoxin. Siderophores are iron-chelating compounds produced for the uptake and storage of iron ions. Usually they contain hydroxamate groups that form strong iron(III) binding bidentates. Most filamentous fungi and some yeasts, except for the model organisms *S. cerevisiae* or *C. albicans*, produce siderophores. A fine-tuned system between iron uptake and storage is necessary to prevent the fungus from iron starva-

[2] Virulence = severity of a disease caused by a pathogen.
[3] Pathogenicity = ability of a pathogen to cause a disease.

tion or toxicity (Haas 2014). Gliotoxin is another prominent example of non-ribosomal peptides. It is a mycotoxin produced by human pathogens such as *Aspergillus fumigatus*, *Trichoderma* spp., and *Penicillium* spp. It inhibits the proteasome and prevents NF-κB activation (see Sect. V.A) as presumed virulence factor of *A. fumigatus* by targeting primarily the activity of neutrophils or other phagocytes of the host immune system (Scharf et al. 2016).

Terpenes are built up from isoprene units by terpene synthases (terpene cyclases). The subgroup of oxygenated forms of terpenes is called terpenoids. Most of known terpenes derive from plants, for example, the major constituents of essential oils. In terpene synthesizing fungi, they are produced by the mevalonate pathway. Prominent examples are the carotenoid pigments, the gibberellin phytohormones or the trichothecene mycotoxins (Schmidt-Dannert 2014).

Indole alkaloids are nitrogen-containing secondary metabolites incorporating one or more indole or indoline moieties (Xu et al. 2014). Mostly, the indole precursors are tryptophan or related indole-3-glycerol-phosphate. Tryptophan is prenylated with dimethylallyl pyrophosphate by *dim*ethyl*a*llyl *t*ryptophan *s*ynthases (DMATS). Ergotamine from *Claviceps purpurea* represents a prominent example for an indole alkaloid. It works as vasoconstrictor with structural similarity to several neurotransmitters and is medicinally used against acute migraine attacks (Schardl et al. 2006).

Besides the four main groups, there are additional classes of fungal secondary metabolites (Fig. 1b). Metabolites consisting of combinations of peptides and polyketides (**peptide-polyketide hybrids**), such as emericellamides or aspyridone, are formed by PKS-NRPS hybrids that contain all typical domains of the single PKS and NRPS (Du et al. 2001). **Meroterpenoids** are products with a partial terpenoid structure plus a polyketide or a peptide moiety (Matsuda and Abe 2016). An example for a meroterpenoid with a polyketide moiety is dehydroaustinol produced by *A. nidulans*. The addition of dehydroaustinol together with the polyketide diorcinol restores spore formation of *A. nidulans* mutants defective in sporulation, suggesting that these two compounds or derivatives serve as developmental signal for asexual sporulation (Rodríguez-Urra et al. 2012; Fig. 1a). **Oxylipins** (collectively termed psi-factor (*p*recocious *s*exual *i*nducer)) are oxygenated polyunsaturated fatty acids produced by oxygenases (Brodhun and Feussner 2011). In *Aspergillus nidulans*, three *p*si factor *p*roducing *o*xygenases PpoA, PpoB, and PpoC have been characterized producing mainly hydroxylated oleic (18:1) and linoleic (18:2) acid. A careful regulated system of oxylipins is necessary for a proper balance between asexual and sexual development (Tsitsigiannis et al. 2004). Besides these non-volatile oxylipins, also volatile oxylipins exist. They share an eight-carbon scaffold and are suggested to be involved in fungus-invertebrate interactions (Holighaus and Rohlfs 2018). **Isocyanides**, which function as chalkophores, contain nitrile groups and are synthesized by isocyanide synthases that can also occur as hybrids with NRPS. The first characterized isocyanide was xanthocillin, which is produced, e.g., by *Penicillium notatum* and by *A. fumigatus* (Lim et al. 2018).

2. Secondary Metabolite Clusters Are Often Silenced and Have to Be Activated for Analysis in the Laboratory

The large diversity of secondary metabolites is due to modifications that are performed after the initial step of building the skeleton. These include oxidations, cyclizations, methylations, dehydrations, or reductions. Fungal genes that are necessary for the biosynthesis of the final secondary metabolites as well as genes encoding proteins necessary for regulation of the synthesis or transport of the final metabolite product are usually, and similarly to genes for specialized or accessory primary metabolism, clustered together on one chromosomal locus (Rokas et al. 2018). The **gene clusters** enable an economical production of secondary metabolites due to transcriptional co-regulation. Most secondary metabolite biosynthetic gene clusters are silenced and are only expressed under very specific conditions. Especially under laboratory conditions, where the fungus does not naturally meet environmental triggers such as stressors or nutrient limitations, the analysis of secondary metabolism is difficult. The rapidly growing number of genome studies has shown that there are many more secondary metabolite clusters in the genome of fungi than metabolites have been identified so far (Andersen et al. 2013; Inglis et al. 2013; van der Lee and Medema 2016). In order to eliminate this discrepancy and exploit the full potential of fungal metabolite production, strategies have been developed to activate silenced gene clusters and thus enable their analysis in the laboratory (Gerke and Braus 2014; Ren et al. 2017).

The easiest way to induce secondary metabolite production is to change the growth parameters as described by the **OSMAC** (*o*ne *s*train *ma*ny *c*ompounds) approach (Bode et al. 2002). Parameters such as temperature, media components, pH, air supply, and light can be varied to force the fungus to adapt its secondary metabolite repertoire. Since secondary metabolites are often used in nature for communication, competition, and defense, their production can also be stimulated by co-cultivation with other microorganisms (Netzker et al. 2015). Additionally, genes involved in the metabolite biosynthesis can be heterologously introduced and expressed in suitable hosts. **Heterologous expression** systems for fungal biosynthetic genes have been designed, e.g., for *A. niger* (Gressler et al. 2015), *A. nidulans* (Chiang et al. 2013), or *A. oryzae* (Sakai et al. 2012). Silenced biosynthetic gene clusters can also be awakened by manipulating the transcriptional machinery, the epigenetic status, or the degradation machinery of the cell.

In this chapter we focus on the molecular regulatory links of the coordinated and specific formation of secondary metabolites in different phases of fungal developmental programs, with a particular focus on the mold *Aspergillus nidulans*. This includes the control and interaction of different genetic networks, which can be organized chronologically and hierarchically and can contain several feedback functions. Genetic transcriptional control is linked to post-translational histone modifications as epigenetic control and specific signal transduction pathways. Several additional post-translational control mechanisms, such as the attachment and removal of ubiquitin, link fungal differentiation to the corresponding secondary metabolism by altering protein function and cellular localization and by controlling protein stability through the ubiquitin 26S proteasome as well as autophagy degradation pathways.

II. Transcriptional Networks Linked to Signal Transduction Pathways Control Development and Secondary Metabolism

Fungal growth and differentiation and the concomitant secondary metabolism occur in response to internal and external signals that are sensed through receptors and transported by highly controlled signal transduction pathways. This leads to a choreography of changes in transcription, translation, post-translational histone modifications and protein stability followed by proteomic changes. Several examples of well-studied signal transduction pathways and responding transcriptional regulatory circuits are summarized in the following section.

A. Transcriptional Networks Interact in Fungal Morphogenic Transitions

Transcriptional reprogramming plays a crucial role in the morphological transition from vegetative fungal growth to developmental programs. Besides chromatin modifications (see Sect. III), changes in transcriptomes are mediated by approximately 80 families of currently classified fungal DNA binding transcription factors including a few examples of dual factors with two distinct DNA binding domains. Ascomycetous **transcription factors** of the largest group carry a Zn_2Cys_6 zinc cluster domain. This domain is generally found in all fungi but also in a few additional non-fungal organisms. In contrast, four types of transcription factors carrying an APSES-type DNA binding domain (named after fungal developmental transcription factors *A*sm1, *P*hd1, *S*ok2, *E*fg1, and *S*tuA), mating-type MAT α1, copper fist

DNA-binding, or velvet domains are fungal specific. Notably, the mating-type protein is exclusively present in ascomycetes (Ahmed et al. 2013; Shelest 2017). The velvet domain is structurally similar to the immunoglobulin-like Rel homology domain of the mammalian transcription factor NF-κB p50 (Ahmed et al. 2013). The four velvet domain proteins of *Aspergilli* steer large hierarchical networks and interconnect developmental programs with secondary metabolism triggered by chemical, oxidative, light, or temperature sensing and subsequent signal transduction pathways (Bayram et al. 2008b; Bayram and Braus 2012; Lind et al. 2016).

1. Chemical Sensing and Oxidative Stress as Developmental Signal: How Fungi Smell Their Environment

Chemical sensing is accomplished by *G p*rotein-*c*oupled *r*eceptors (GPCRs) with a common seven transmembrane domain architecture that initiate signal transduction through binding to the heterotrimeric Gα, Gβ, and Gγ protein complex. In *Aspergilli*, three different Gα, one Gβ, and one Gγ protein exist (de Vries et al. 2017; Brown et al. 2018). GPRCs are the largest fungal group of surface receptors important for sensing of pheromones, nutrients, and host cells in pathogenic interactions. For instance, the binding of a pheromone to the receptor initiates GDP-GTP exchange on the Gα protein that then dissociates from the βγ dimer and typically activates a three-step *m*itogen-*a*ctivated *p*rotein (MAP) kinase cascade, which finally changes transcription. The result of this well conserved cascade in fungi is a cell cycle arrest and the cellular fusion of both mating partners (Bahn et al. 2007). Further, similar sensing systems through G protein-coupled receptors with different transcriptional outcomes exist for different macro- and micronutrients including glucose as main carbon energy source. Glucose is sensed through the sugar receptor Gpr1 (*G p*rotein-coupled *r*eceptor 1) that stimulates Gpa2 resulting in increased *c*yclic *a*denosine *m*ono*p*hosphate (cAMP) levels through activation of adenylyl cyclase and *p*rotein *k*inase A (PKA). Active PKA inhibits the Atg1 (*a*u*t*ophagy-specific *g*ene 1) initiation kinase of autophagy degradation pathways (see Sect. V.B). Specialized sugar transporters such as yeast Snf3 (*s*ucrose *n*on*f*ermenting 3) and Rgt2 (*r*estores *g*lucose *t*ransport 2) act as sensors inducing signal transduction and changed transcription (Bahn et al. 2007).

The regulation of the mycotoxin sterigmatocystin in *A. nidulans* works through a Gα-PKA pathway. The Gα protein FadA (*f*luffy *a*utolytic *d*ominant A) activates PkaA (*p*rotein *k*inase *A* A), which in turn is a repressor of the specific transcription factor of the sterigmatocystin biosynthesis cluster, AflR (*afl*atoxin *r*egulator). The activation of the Gα protein FadA is inhibited by the *fluffy low brlA* protein FlbA (Shimizu et al. 2003; Fig. 2).

R*eactive* **o***xygen* **s***pecies* (ROS) function as signalling molecules in light perception and redox biology and thereby induce fungal development (Gessler et al. 2007). Therefore, an impact on ROS homeostasis is often linked to distorted development in filamentous fungi (Nahlik et al. 2010; Kolog Gulko et al. 2018). Fungi respond to oxidative stress induced by ROS through activation of a MAP kinase pathway (Yu and Fischer 2019). The MAP kinase Hog1 (*h*igh *o*smolarity *g*lycerol response 1) of *S. cerevisiae* or its homologue SakA (*s*tress *a*ctivated *k*inase A) in *A. nidulans* are activated by ROS and in turn induce transcription factors needed for the induction of the oxidative stress response, such as the *A. nidulans* AtfA (*a*ctivating *t*ranscription *f*actor A). These transcription factors regulate different antioxidants, such as superoxide dismutases, catalases, peroxidases, glutathione peroxidases, peroxiredoxins, and antioxidative secondary metabolites. The stress-activated kinase SakA is not only induced by ROS but also by light, mediated through the red light photoreceptor FphA (*f*ungal *ph*ytochrome A), which represses fruiting body formation in *A. nidulans* (Yu and Fischer 2019). SakA additionally mediates repression of the *A. nidulans* NADPH *ox*idase gene *noxA*, which is essential for different steps during sexual fruiting body formation (Lara-Ortíz et al. 2003).

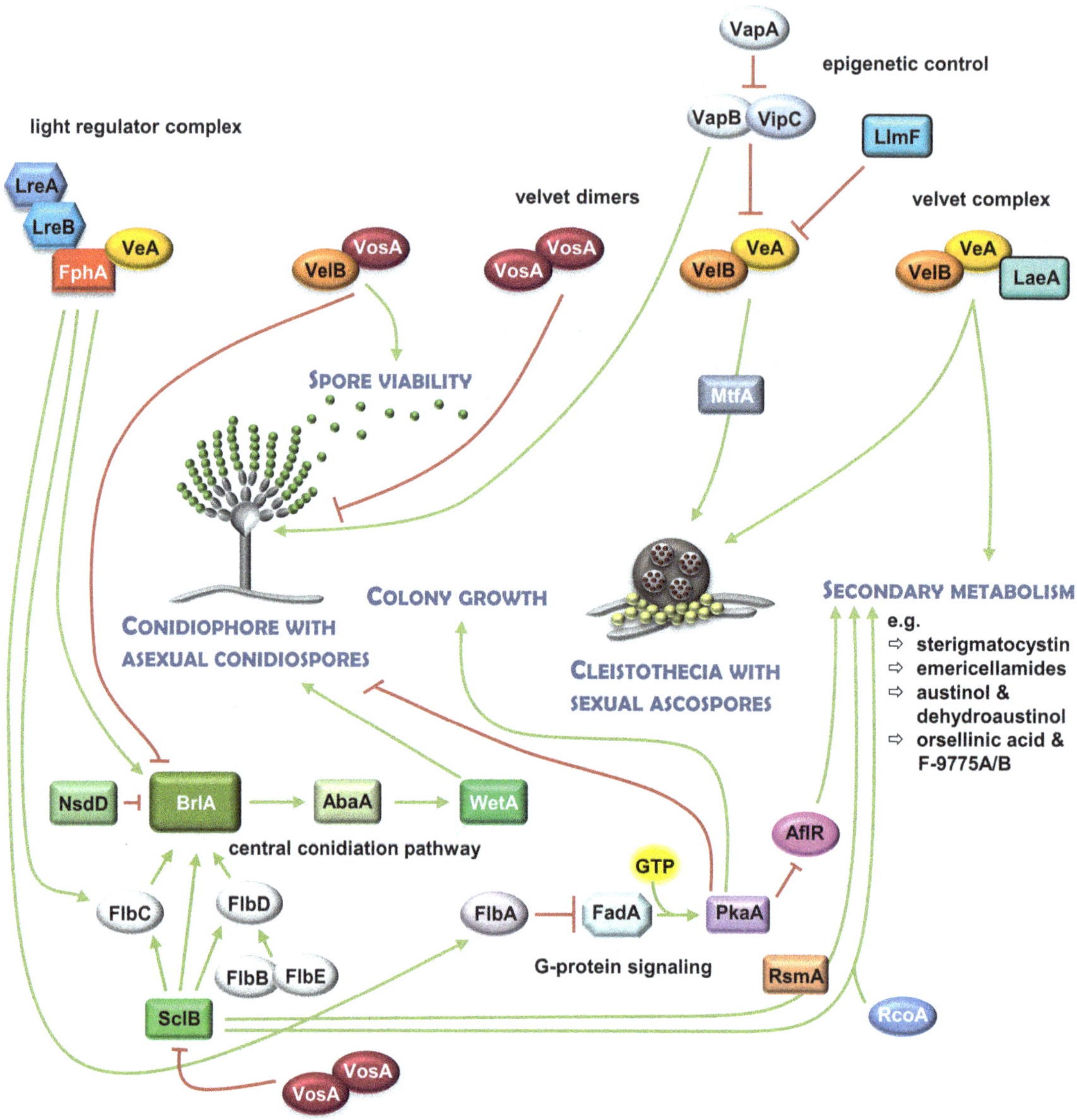

Fig. 2 Interconnection of genetic networks and subnetworks for light and developmental control of *A. nidulans*. Simplified model depicting examples of major molecular control lines to support sexual or asexual development and the appropriate secondary metabolism. The velvet domain protein dimer VeA-VelB supports sexual development in dark and VeA is connected to blue (LreA, B) and red (FphA) light receptors. VeA is controlled by epigenetic signal transduction (VapA, VipC-VapB) and bridges DNA binding with epigenetic control of secondary metabolism (LaeA: *l*oss of *a*flR expression A). Velvet proteins (VelB-VosA, VosA-VosA) as well as NsdD and different signal transduction pathways (e.g., the cAMP PkaA pathway) inhibit the expression of *brlA* as part of the central conidiation pathway (with AbaA and WetA), whereas different Flb factors, SclB, and the epigenetic VapB-VipC support asexual development

2. Light Response: The Fungal Vision

Almost all fungi are able to perceive changes in light or temperature conditions. Light is perceived as a signal and also as a stressor and induces a wide range of responses by inducing different genetic networks resulting in changes in development (photomorphogenesis), secondary metabolism, circadian clock, phototropism, or DNA repair (Bayram et al. 2016; Cetz-

Chel et al. 2016; Fischer et al. 2016). The developmental light responses vary considerably between different fungi. Blue or red light can promote asexual spore formation in *A. nidulans* (Fig. 1a) but inhibit sporulation in *Botrytis cinerea* (Tan 1974). Similarly, *A. nidulans* sexual fruiting bodies are preferentially formed in darkness, but *Trichoderma reesei* forms its corresponding sexual structures in light (Pöggeler et al. 2018).

Typical fungal photosensory systems include light sensors specific for different wavelengths from 350 to 650 nm distributed into four types of light receptor proteins. The white collar proteins perceive blue light, phytochromes sense red light, opsins are specific for green light, and cryptochromes react to blue and *u*ltra*v*iolet (UV) light. The light perception occurs through binding of the photoreceptor to a chromophore, which are flavin for blue and UV light, linear tetrapyrrole for red light and retinal for green light.

White collar (WC) proteins are blue-light regulated transcription factors with a GATA-type zinc finger DNA binding domain and three PAS (*P*er, *A*rnt, *S*im) domains directly controlling gene expression. One of the PAS domains is a LOV (*l*ight *o*xygen *v*oltage) domain, which is able to bind the chromophore flavin. In *Neurospora crassa*, two WC proteins are present, WC-1 and WC-2 (Wu et al. 2014). They form a heterodimeric complex through the PAS domains, which binds to light response elements in the dark and activates light-induced transcription after phosphorylation and dissociation of WC-1. The complex regulates approximately 20% of the *N. crassa* genes encoding transcription factors as large hierarchical network responding to light and acting downstream of the WC complex in light and circadian clock control (Wu et al. 2014). Besides the typical WC proteins, blue light photoreceptors with only one LOV domain (*vivid*, VVD) exist that act in fine-tuning the light response (Malzahn et al. 2010).

N. crassa biosynthesis of carotenoid secondary metabolites and asexual spore formation are dependent on light and controlled by the circadian clock (Rodriguez-Romero et al. 2010). WC-1 comprises a flavin mononucleotide binding LOV domain for environmental sensing of light, oxygen, and voltage. The WC-1 and WC-2 zinc fingers dimerize to form the white collar complex transcription factor for DNA binding. WC-1 and WC-2 complexes are conserved in numerous fungi with sometimes several orthologs. WC-1 homologs of pathogenic species are connected to virulence, probably through secondary metabolite production (Idnurm et al. 2010; Fischer et al. 2016). The white collar complex of *Aspergilli*, LreA/LreB (*l*ight *re*sponse), activates the expression of the *brlA* (*b*ristle A) gene encoding the central transcription factor of the conidiation pathway in response to light (Ruger-Herreros et al. 2011, Fig. 2). The *brlA* promoter is repressed by binding factors such as NsdD (*n*ever in *s*exual *d*evelopment D) or the velvet domain protein VosA (*v*iability *o*f *s*pores A) (Ni and Yu 2007). *brlA* must be activated through a complex cascade of early genes for proteins which produce small signal molecules (*fluG*: *fluffy* G) or act as transcription factors (*flbB-E*: *fluffy* *l*ow *b*rlA B-E; *sclB*: *s*clerotia-*l*ike B). Activation of *brlA* by light then activates the downstream pathway with the genes *abaA* (*aba*cus) and *wetA* (*wet*-white; Fig. 2; Thieme et al. 2018).

Phytochromes are multi-domain photoreceptors that bind linear tetrapyrrole as chromophore at their N-terminal photosensory domain and have red/far-red light absorbance properties. The tetrapyrrole undergoes a conformational change upon red light induction and thereby changes its spectroscopic properties by shifting the absorption maximum toward far-red light. The photosensory domain consists of a PAS, a GAF (vertebrate *c*GMP-specific phosphodiesterases, cyanobacterial *a*denylate cyclases, transcription activator *F*hlA) and a PHY (*phy*tochrome-specific PAS-related) domain. The further domains are a histidine kinase domain and a response regulator domain. Phytochromes seem to be absent from *Mucoromycotina* fungi but might be present in some *Chytridiomycetes* and have been extensively studied in *Aspergilli* (Fischer et al. 2016). The *A. nidulans* phytochrome FphA interacts with the white collar orthologs LreA and LreB, corresponding to WC-1 and WC-2, and the velvet transcription factor VeA (see

Sect. II.B.2) that coordinates development and secondary metabolism (Purschwitz et al. 2008; Bayram et al. 2008b; Fig. 2). LreA binds to its target sequences in the dark in dependency of FphA while being released after illumination with blue or red light (Blumenstein et al. 2005; Hedtke et al. 2015). LreA modulates gene expression together with FphA through modification of histone H3 by interacting with the acetyltransferase GcnE (*g*eneral *c*ontrol *n*on-derepressible E) and the *h*istone *dea*cetylase HdaA (Grimaldi et al. 2006; Hedtke et al. 2015). FphA of *A. nidulans* is a phytochrome with light-driven histidine kinase activity and presumably transmits the white collar and phytochrome light signal directly to the velvet protein VeA by phosphorylation (Rauscher et al. 2016). This results in an enhancement of asexual development in light and a delay in dark, where sexual development is favored. After 30 min of illumination as minimum time required to initiate conidiation, 19% of the transcriptome of competent *A. nidulans* mycelia reacts to light (Bayram et al. 2010, 2016; Hedtke et al. 2015; Macheleidt et al. 2016). Austinol and dehydroaustinol (see Sect. I.B.1) are secondary metabolites connected to conidiospore production that are only produced in light but not in dark (Rodríguez-Urra et al. 2012).

Cryptochromes are blue and UV light receptors with similarities to blue light-dependent DNA repairing photolyases, carrying a photolyase domain (Idnurm et al. 2010; Fischer et al. 2016). They bind non-covalently to *f*lavin *a*denine *d*inucleotide (FAD) as well as other chromophores such as pterin or deazaflavin. The *cry*ptochromes CryA of *A. nidulans* and Cry1 of *T. reesei* are so far the only known dual-function cryptochrome/photolyase proteins. *A. nidulans* CryA inhibits sexual development in UV light and has a DNA repair function (Bayram et al. 2008a), whereas *T. reesei* Cry1 is needed for light-induced transcription besides its DNA repair function in conidia (García-Esquivel et al. 2016). In *N. crassa*, Cry1 acts as transcriptional repressor of the white collar blue light complex without photolyase activity (Nsa et al. 2015).

Opsins, the green light receptors, are membrane-bound proteins associated with a retinal chromophore. They are related to bacterial and archaeal rhodopsins that, in their activated form, channel ions across the membrane (Yu and Fischer 2019). In filamentous fungi, opsins are poorly characterized. In *F. fujikuroi*, the opsin CarO (*car*otenoid O) was described as green light-driven proton pump that is involved in spore germination (García-Martínez et al. 2015). In contrast, the *N. crassa* NOP-1 (*N*eurospora *op*sin-1) lacks proton pump activity but is involved in the regulation of the switch between asexual and sexual development in response to light and ROS levels (Wang et al. 2018).

B. Gene Expression for Secondary Metabolite Production Is Interconnected with Morphological Differentiation

Environmental stimuli such as light, oxygen, pH, and nutrients affect fungal morphological programs as well as the tightly interconnected specific secondary metabolite production. Accordingly, defects in light control affect secondary metabolism. One example is *Fusarium fujikuroi*, which is impaired in secondary metabolite production when the white collar blue light sensor WC-1 is defective, and which uses its *cry*ptochrome CryD to repress the production of the antibiotic bikaverin during growth in light (Estrada and Avalos 2008; Castrillo et al. 2013).

1. Expression of Silenced Secondary Metabolite Clusters by Specific and Global Regulators

Filamentous fungi are a vast reservoir of yet undescribed secondary metabolites, carrying dozens of usually clustered but only specifically expressed biosynthetic gene clusters. Many of these clusters are controlled by cluster-specific, poorly conserved transcription factors (Alberti et al. 2017; Keller 2018). In addition, there are several master regulators of secondary metabolism such as the originally in *A. nidulans* described methyltransferase LaeA or the *m*ulti-*c*luster *r*egulator A (McrA). LaeA is encoded by a conserved regulatory gene with homologues in filamentous fungi such as *Fusar-*

ium, *Penicillium*, *Trichoderma*, and *Aspergillus* (Bok and Keller 2016). Absence of *laeA* or overexpression of *mcrA* results in silencing, whereas *laeA* overexpression or *mcrA* deletion leads to increased production of several secondary metabolites (Bok and Keller 2016; Oakley et al. 2016). The fact that only some global regulators are conserved, whereas the targets of individual regulators differ considerably, correlates with the finding that secondary metabolite genes are more variable and significantly less conserved than genes of the primary metabolism (Lind et al. 2015).

2. The Fungal Velvet Complex Physically Connects Transcriptional and Heterochromatin Control to Coordinate Secondary Metabolism and Development

Velvet proteins with their characteristic velvet domain, such as VeA (*v*elvet A), VelB (*v*elvet-*l*ike B) VelC (*v*elvet-*l*ike C), and VosA in *A. nidulans*, form complex regulatory networks by direct DNA binding to thousands of target gene promoters in numerous fungi (Ahmed et al. 2013; Becker et al. 2016; Fig. 2). The master regulator of secondary metabolism LaeA is physically linked to the heterodimer VeA-VelB to coordinate secondary metabolite biosynthesis with developmental programs. It was originally shown in *A. nidulans* that this VelB-VeA-LaeA complex is required for the appropriate formation of fruiting bodies and concomitant production of the aflatoxin family metabolite sterigmatocystin (Bayram et al. 2008b). LaeA is a methyltransferase and acts as epigenetic control element by counteracting the silencing heterochromatic lysine 9 methylation marks at histone H3 in secondary metabolite clusters (Strauss and Reyes-Dominguez 2011; see Sect. III.A). VeA, which provides the interphase of the trimeric complex, therefore physically links transcription to post-translational epigenetic control of histone modifications (Sarikaya-Bayram et al. 2015). Velvet proteins and LaeA contribute to the virulence of several fungi, possibly through mycotoxin production (Wiemann et al. 2010; Kumar et al. 2016; López-Díaz et al. 2018). The trimeric velvet complex is conserved in the fungal kingdom and can physically interact through VeA with the phytochrome FphA as part of the light and presumably temperature control machinery (Lind et al. 2016; Yu and Fischer 2019). The light reception through VeA-FphA-LreA-LreB is presumably less stable and rather transient in comparison to the velvet complex VelB-VeA-LaeA (Bayram et al. 2010).

3. Velvet Domain Transcription Factors Expand Their Transcriptional Networks Through Formation of Sub-networks

Velvet domain proteins control several genetic networks either acting as homo- or heterodimers (Fig. 2). VelB does not only support as VelB-VeA heterodimer sexual development linked to its specific secondary metabolism but also represses asexual development in combination with the velvet protein VosA as VelB-VosA together with VosA-VosA (Park and Yu 2012). Additionally, VelB-VosA is a positive regulator for spore viability (Park and Yu 2012). How the formation of the different VosA complexes is regulated is still unknown. Velvet proteins bind to promoters of hundreds of genes, including genes for additional transcription factors, which in turn control their own genetic network (Ahmed et al. 2013).

VosA directly represses the central asexual regulator gene *brlA* and the gene for the Zn_2Cys_6 zinc cluster domain transcriptional activator SclB. The SclB network induces early developmental genes including *brlA* to promote asexual sporulation and to support spore viability. SclB links asexual spore formation to its specific secondary metabolism, including the putative signal molecule dehydroaustinol, and to the fungal oxidative stress response (Thieme et al. 2018). This illustrates a convoluted surveillance apparatus, including sub-networks with feedback mechanisms and overlapping as well as antagonistic functions, to perform fungal development and its accurate linkage to the appropriate secondary metabolism (Fig. 2). Developmental functions of velvet domain proteins are rewired in different fungi. Whereas *veA* is dispensable for conidiation in *A. nidu-*

lans, the *veA* homolog of *N. crassa ve-1* regulates asexual sporulation (Bayram et al. 2008c). Rewiring of transcriptional regulation is presumably, together with gene mutations and gene transfer, an important driving force to evolve fungal divergences and to adapt to different lifestyles (Nocedal et al. 2017).

III. Epigenetics and Fungal Secondary Metabolism and Development

Secondary metabolite gene clusters evolved either from gene relocations with sometimes prior gene duplication or, in rarer cases, by horizontal gene cluster transfer from bacteria to fungal ancestors (Rokas et al. 2018). Clustering offers the economic advantage that these genes can be transcriptionally co-regulated by changes in the chromatin structure through covalent chromatin modifications (Fig. 3). These chromatin regulations can be inherited to the next generation of spores and are therefore epigenetic (epi-: "over, outside of"), which describes heritable phenotypic changes that are not caused by a change in the DNA sequence. In chromatin, the DNA is wrapped twice around eight histones (two H2A, two H2B, two H3, and two H4) to form the nucleosome. With the help of non-histone proteins, the nucleosomes are tightly packed up to the chromatin fiber and then condensed further yielding the chromosome structure. Two different types of chromatin exist which can be distinguished by their condensation state. Euchromatin is the less condensed active form, which allows gene transcription. Heterochromatin is the highly

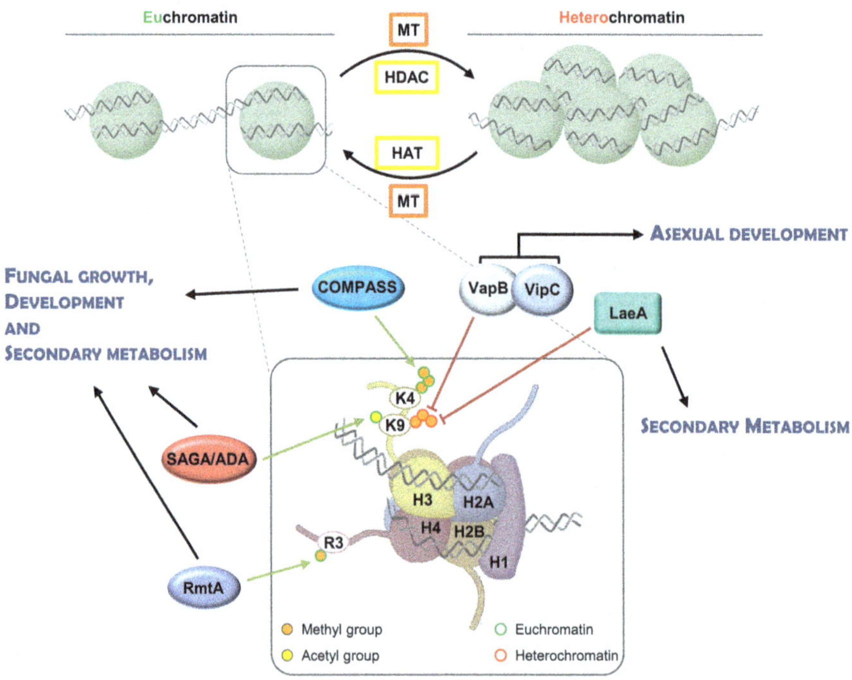

Fig. 3 Epigenetic control of chromatin dynamics for the coordination of fungal development and secondary metabolism. Secondary metabolite gene clusters are mostly silenced during vegetative growth by negative histone tags (red) resulting in heterochromatin and have to be activated by positive tags (green) to allow transition to euchromatin. *M*ethyl*t*ransferases (MT), *h*istone *a*cetyl*t*ransferases (HAT), and *h*istone *de*acetylases *c*omplexes (HDAC) add (writer) or remove (eraser) these tags. Histone H3K4Me3 represents a positive (euchromatin) and H3K9Me3 a negative (heterochromatin) example for a hallmark chromatin tag. Fungal gene expression can be induced by methylation of H3K4 (COMPASS), demethylation of H3K9 (VapB, LaeA), methylation of H4R3 (RmtA), and acetylation of H3K9 (SAGA/ADA) resulting in increased transcription supporting fungal growth, development, and the corresponding secondary metabolism

condensed and transcriptionally inactive form that can either be constitutive and serve as structural element of the chromosomal centro- and telomeres or be facultative and switch to transcriptionally active euchromatin. Covalent modifications of DNA or histones determine whether hetero- or euchromatin is formed. DNA can be reversibly modified by methylation, whereas lysine residues of histones can be methylated, acetylated, ubiquitinated, or sumoylated, and serine residues can be phosphorylated (Kouzarides 2007). Writers and erasers, which attach and remove modifications, modify chromatin in different possible combinations. This results in a histone code in a certain genomic region, which is recognized by readers. Readers induce the restructuring of chromatin into the open euchromatin or the closed heterochromatin form and by this regulate fungal development and secondary metabolism (Pfannenstiel et al. 2018; Keller 2018). In the following paragraphs, we will discuss some writers and erasers in more detail with the focus on consequences of histone or DNA modification on secondary metabolism and development.

A. DNA and Histone Methylation and Demethylation

Methylation of chromatin can occur at DNA and histones. Cytosine modification of DNA is conserved between plants, mammals, and the fungus *N. crassa*, but has not been found in all fungi. Histone methylations and demethylations at lysine and arginine residues are present in plants, mammals, and fungi and important for fungal development and the associated secondary metabolism (Liu et al. 2010a; Greer and Shi 2012; Nie et al. 2018; Fig. 3). Arginines can be mono- or dimethylated (symmetric or asymmetric) and are less well studied than lysine modifications, which include mono-, di-, or trimethylations. *P*rotein arginine (*R*) *m*ethyl*t*ransferases (PRMTs) are divided into four major classes according to the methylation pattern they provide (Bachand 2007; Stopa et al. 2015). Three of nine human PRMTs (PRMT1, PRMT3, PRMT5) are conserved in yeast or filamentous fungi (Bachand 2007). The corresponding *A. nidulans* proteins RmtA and RmtC possess H4R3[4] specificity, whereas RmtB can methylate the histones H4, H3, and H2A in vitro and all three methylate additional non-histone substrates that affect transcriptional regulation (Lee and Stallcup 2009; Bauer et al. 2010). RmtA and RmtC can affect mycelial growth, oxidative stress response, development, or secondary metabolism in different *Aspergilli*, whereas the role and the non-histone substrates of RmtB still remain unclear (Bauer et al. 2010; Satterlee et al. 2016).

Lysine methylation usually occurs at histones 3 and 4. The methyltransferase Set1 (*SET* domain-containing 1) is part of the eight subunit COMPASS (*COMP*lex *AS*sociated with Set1) complex involved as writer in mono-, di-, and trimethylation of histone H3K4 euchromatin marks in fungi and in humans. Whereas dysfunction in humans has been linked to several types of human cancer (Meeks and Shilatifard 2017), the degree of H3K4 methylation in fungi has important implications on development and secondary metabolism. Defects in COMPASS subunits, which significantly reduce H3K4Me3[5] euchromatin marks, change significantly the secondary metabolite production profile in *Aspergillus* and *Fusarium*, and the development of *Aspergillus* (Palmer et al. 2013a; Studt et al. 2017). COMPASS-dependent methylation of H3K4 is connected to the methylation of H3K79 by the non-Set domain-containing enzyme Dot1 (*d*isruptor *o*f *t*elomeric silencing 1) and dependent on histone H2B ubiquitination (Shilatifard 2012). Dot1 is to date the only non-SET domain-containing enzyme for histone H3K79 methylation and controls, e.g., production of aflatoxin, conidiation, sclerotia formation, and pathogenicity of *A. flavus* (Liang et al. 2017).

Another epigenetic mark commonly associated with "active" chromatin is the trimethylation of H3K36 performed by the methyltransferase Set2, which is able to mono-,

[4] H3R4 = histone H4 modified at arginine residue 3 (R3).

[5] H3K4Me3 = histone H3 trimethylated (Me3) at lysine residue 4 (K4).

di-, and trimethylate this residue (Wagner and Carpenter 2012). Methylation of H3K36 controls *N. crassa* development as well as vegetative growth, fungal virulence, and secondary metabolism of *F. verticillioides* (Adhvaryu et al. 2005; Gu et al. 2017). Ash1 (*a*symmetric *s*ynthesis of *HO* 1) represents a second histone methyltransferase for H3K36 in *N. crassa* or *F. fujikuroi*. *N. crassa* methyl tags deposited by Set2 mark actively transcribed genes, whereas the H3K36 methylation by Ash1 occurs in inactive genes. *F. fujikuroi* Ash1 is involved in developmental processes and secondary metabolism (Janevska et al. 2018; Bicocca et al. 2018).

Repressive marks associated with gene silencing and heterochromatic regions include the trimethylation of H3K9 or H3K27, which are enriched in silenced fungal secondary metabolite gene clusters. The H3K27 mark is not used in *A. nidulans*, but in *F. graminearum*, where the methyltransferase Kmt6 is responsible for the establishment of the H3K27Me3 and regulates production of many secondary metabolites as well as development (Connolly et al. 2013).

Lysine demethylation is performed by histone demethylases, such as KdmA (lysine (*K*)-*dem*ethylase A) and KdmB of *A. nidulans*. Both are erasers that remove methyl marks from histone H3. KdmA demethylates H3K36Me3 and has a dual role in transcriptional regulation as co-repressor of primary metabolism genes and activator of secondary metabolite genes. KdmB demethylates H3K4Me3 and promotes transcriptional downregulation as prerequisite for accurate induction of *A. nidulans* secondary metabolism (Gacek-Matthews et al. 2015, 2016).

One of the first described *Aspergillus* proteins with a methyltransferase domain is LaeA, a master regulator of secondary metabolism. LaeA, which is part of the velvet complex (see Sect. II.A.2), possesses automethylation activity (Patananan et al. 2013). Additionally, it counteracts H3K9Me3 marks in repressive heterochromatin, activating many secondary metabolite biosynthetic gene clusters (Reyes-Dominguez et al. 2010, Fig. 3). However, the exact molecular mechanism of LaeA function is yet unknown. There are nine additional *A. nidulans* *L*aeA-*l*ike *m*ethyltransferases (LlmA-LlmG, LlmI, LlmJ) that share sequence similarity with LaeA. LlmF interacts with VeA in the cytoplasm and reduces its nuclear import, resulting in altered development and secondary metabolism, whereas the interaction of VeA with LaeA takes place in the nucleus (Palmer et al. 2013b, Fig. 2). VeA interacts with at least two additional methyltransferases in both cellular compartments, the *V*eA *i*nteracting *p*rotein C (VipC) and the *V*ipC-*a*ssociated *p*rotein B (VapB). The nuclear heterodimeric methyltransferases VipC-VapB are, together with the membrane tethering zinc finger domain protein VapA (*V*ipC *a*ssociated *p*rotein A), part of a novel type of epigenetic signal transduction pathway (Sarikaya-Bayram et al. 2014, Fig. 2). VapA can exclude VipC-VapB from the nucleus by forming the membrane-bound trimeric VapA-VipC-VapB complex. Release of the VipC-VapB heterodimer from VapA is induced by a yet elusive molecular trigger and leads to its transport from the membrane to the nucleus. VipC-VapB interaction with VeA in the cytoplasm inhibits its nuclear accumulation, resulting in decreased sexual development and corresponding secondary metabolism. Without VeA interaction, VipC-VapB enters the nucleus and activates the light-promoted *brlA* master regulatory gene of conidiation and thereby the asexual differentiation program, for instance, by decreasing heterochromatin through VapB, which was shown to reduce H3K9Me3 marks (Sarikaya-Bayram et al. 2014; Figs. 2 and 3).

B. Histone Acetylation and Deacetylation

Histone hyperacetylation at amino groups of lysine residues is, in most cases, associated with euchromatin, whereas deacetylation results in heterochromatin formation (Fig. 3). Promoter spreading of histone H3 or H4 acetylation results in higher transcription and production of secondary metabolites of the aflatoxin family in *Aspergilli* (Roze et al. 2007; Reyes-Dominguez et al. 2010).

Histone acetylation is performed by type A or type B histone acetyltransferases (HATs) as writers. Cytoplasmic type B HATs acetylate newly synthesized histones prior to their assembly into nucleosomes, whereas nuclear type A HATs acetylate nucleosomal histones.

A well-studied group of type A HATs are the *G*cn5-related *N*-*a*cetyl*t*ransferases (GNAT). They catalyze the transfer of acetyl groups to lysine residues on the histones H2B, H3, or H4 from acetyl-CoA donors. Founding member of the family is the yeast Gcn5 (*g*eneral *c*ontrol *n*on-derepressed 5), a transcriptional cofactor associating as HAT with several complexes (e.g. SAGA: *S*pt-*A*da-*G*cn5-*A*cetyltransferase). The *A. nidulans* SAGA complex acetylates the histone sites H3K9 and H3K14 and regulates the biosynthesis of penicillin, sterigmatocystin, terrequinone, and orsellinic acid (Nützmann et al. 2011). Gcn5 is important for the control of the mycotoxin deoxynivalenol biosynthesis in *Fusarium graminearum* (Kong et al. 2018). Numerous homologs of *S. cerevisiae* Gcn5 in filamentous fungi are involved in growth, development, regulation of secondary metabolite production, and pathogenicity.

The MYST (*M*OZ, *Y*bf2, *S*as2, and *T*ip60) family represents another type A HAT group with a characteristic zinc finger in the highly conserved MYST domain, which acetylates H2A, H3, and H4 lysine residues. EsaA (*e*ssential *S*AS2-related *a*cetyltransferase A), a MYST HAT of *A. nidulans*, is an activator of secondary metabolism by acetylating H4K12 in the sterigmatocystin, penicillin, terrequinone, and orsellinic acid gene clusters (Soukup et al. 2012). MYST3 of *A. parasiticus* is required for aflatoxin production (Roze et al. 2011b).

Histone deacetylation in fungi is performed by erasers such as the classical histone deacetylases (HDACs) of class I (Rpd3/Hos2-type) or class II (HDA1-type), with a zinc ion in their catalytic site, or by non-conventional class III SIR2-type sirtuin HDACs, which require NAD^+ as cofactor (Brosch et al. 2008).

Yeast Rpd3 (*r*educed *p*otassium *d*ependency factor 3) as name-giving class I HDAC regulates transcription by RNA polymerases I and II through chromatin silencing and controls mitosis, meiosis, aging, or macroautophagy (see Sect. V.B). The corresponding *A. nidulans* RpdA deacetylates H3 and H4 and is essential for growth and conidiation (Tribus et al. 2010). Trichostatin A, an inhibitor of classical HDACs such as RpdA, is a promising anticancer drug (Bauer et al. 2016) and inhibits appressorium formation and decreases pathogenicity of the rice blast fungus *Magnaporthe oryzae* (Izawa et al. 2009). *A. nidulans* HosA (corresponding to yeast Hos2 (*H*da *O*ne *s*imilar 2)) is another class I HDAC, which deacetylates H4 and represses orsellinic acid production but has inducing effects on other secondary metabolites by overriding the regulatory effects of other HDACs (Pidroni et al. 2018). The corresponding *A. oryzae* protein (HdaD/Hos2) is involved in the regulation of growth, asexual development, stress response, and the biosynthesis of the industrially important chelator agent kojic acid (Kawauchi and Iwashita 2014).

The class II Hda1-type HDACs include HdaA of *A. nidulans*, which represses sterigmatocystin and penicillin gene clusters as well as additional clusters located close to telomeres. HdaA, which can be overridden by class I HDAC HosA, is therefore an antagonist of the positive secondary metabolite gene cluster regulator LaeA (see Sects. II.B.2 and III.A). Analysis of the corresponding proteins of *Alternaria alternata* or *Penicillium expansum* further support HDAC-mediated repression of secondary metabolism as conserved function in fungi. One exception is the *A. fumigatus* gliotoxin cluster, which is activated by HdaA. The protein is additionally required for growth and germination but not for virulence in this fungus (Shwab et al. 2007; Lee et al. 2009).

Class III SIR2-type sirtuins HDACs include SirA, which deacetylates H4K16 in the promotor regions of the sterigmatocystin and penicillin gene clusters and controls together with the sirtuin HstA (*h*omolog of *S*IR *T*wo A) the formation of these metabolites. Some secondary metabolite genes are repressed, whereas others are activated, which reflects the complex control of histone modifications (Shwab et al. 2007; Shimizu et al. 2012; Itoh et al. 2017). This is further illustrated by the *Aspergillus oryzae* sirtuin HstD, which controls the expression of the velvet complex member LaeA and thus the kojic acid production and conidiation (Kawauchi et al. 2013).

C. Histone Phosphorylation, Ubiquitination, and Sumoylation

In addition to methylation and acetylation, further post-translational histone modifications comprise phosphorylation, ubiquitination, or sumoylation. These modifications were rather analyzed in yeast than in filamentous fungi. Histone phosphorylations by kinases that transfer the phosphate group from ATP to the hydroxyl group of the modified amino acid take place at serines, threonines, or tyrosines and are reversed by phosphatases. Phosphorylation of H3S10 is crucial for yeast chromosome condensation and cell cycle progression during mitosis and meiosis (Nowak and Corces 2004).

Ubiquitination of histones by ubiquitin ligases is the covalent attachment of the small modifier ubiquitin to lysine residues and is reversed by deubiquitinating enzymes (DUBs, see Sect. IV.B). Methylation of H3K4 is mediated by the ubiquitination of H2BK123 in yeast (Sun and Allis 2002). These concerted histone modifications on distinct histone tails are referred to as *trans*-tail regulation (Zheng et al. 2010). Similar to ubiquitination, sumoylation is the covalent attachment of the *s*mall *u*biquitin-like *mo*difier (SUMO) to lysines of proteins through the activity of an E1-E2-E3 enzyme cascade. Sumoylation has been found for all four histones and is associated with transcriptional repression. SUMO, which is required for multicellular fungal development in *A. nidulans*, most likely blocks the histone modification sites to prevent acetylation or ubiquitination (Nathan et al. 2006; Harting et al. 2013). Co-purification experiments identified SUMO-associated proteins connected to the ubiquitin network involved in *A. nidulans* histone modification. They include one subunit of the COMPASS complex required for histone methylation as well as subunits of the SAGA complex for histone acetylation (Harting et al. 2013). Like other post-translational modifications, ubiquitin and ubiquitin-like modifications do not only modify histones but also other proteins and thereby can influence their activity, function, localization, or stability and affect fungal secondary metabolism and development (see Sects. III and IV).

D. Epigenetic Tools to Activate Silenced Gene Clusters

The crucial role of histone modifications such as methylation or acetylation in the regulation of secondary metabolism can be utilized in the laboratory for the identification of new secondary metabolites. Epigenetically silenced clusters can be reactivated by changes in chromatin modifications. Deletion of the genes encoding HDACs, which are generally associated with transcriptional repression, can lead to derepression of secondary metabolite gene clusters. Accordingly, penicillin and sterigmatocystin gene clusters of *A. nidulans* are activated when the HDAC gene *hdaA* had been deleted (Shwab et al. 2007). These types of experiments have resulted in chemical epigenetics as an emerging new research field (Okada and Seyedsayamdost 2017). The addition of HDAC or methyltransferase inhibitors to fungal cultures alters the epigenetic status of the cells to activate, identify, and study new secondary metabolite gene clusters and their products. The advantage of this method is that the fungus does not have to be genetically modified, and it can therefore be applied to any organism. Examples for such inhibiting chemicals are the HDAC inhibitors valproic acid, trichostatin A (see Sect. III.B) and its synthetic derivative *s*ub*e*roylanilide *h*ydroxamic *a*cid (SAHA), or the DNA methyltransferase inhibitor 5-*aza*cytidine (5-AZA). Successful chemical epigenetics approaches were applied in many different fungi, such as *A. alternata* and *P. expansum* (Shwab et al. 2007), *Cladosporium cladosporioides* and *Diatrype disciformis* (Williams et al. 2008), *A. niger* (Fisch et al. 2009; Henrikson et al. 2009), and *A. fumigatus* (Magotra et al. 2017).

IV. The Role of Ubiquitination and Deubiquitination in Fungal Development and Secondary Metabolism

The importance of post-translational modifications for coordinated fungal development and secondary metabolism is not restricted to histone modifications such as the epigenetic control in connection with transcriptional networks. Target protein modifications by phosphate, ubiquitin, or ubiquitin-like proteins are also essential for the accurate adjustment of the fungal proteome. They alter the physical and chemical properties of a protein in response to external or internal stimuli in order to regulate and control its activity, half-life, transport, and subsequent cellular localization. These processes, which include the highly conserved *u*biquitin-26S *p*roteasome *p*athway (UPP) for protein stability control, are prerequisites for accurate fungal differentiation and physiology.

A. The Ubiquitin Attachment Machinery Influences Fungal Development and Secondary Metabolism on Several Layers

Ubiquitin is the most prominent modifier of the family of *ub*iquitin-*l*ike proteins (UBL), which includes SUMO, the cullin modifier *ne*ural precursor cell expressed, *d*evelopmentally *d*own-regulated *8* (Nedd8) or autophagy-related modifiers (Atg8, Atg12; see Sect. V). Ubiquitin consists of 76 amino acids and is encoded as fusion to ribosomal proteins or as head-to-tail fusions of many ubiquitin moieties encompassing two ubiquitin genes in filamentous fungi such as *A. nidulans* or four in *S. cerevisiae* (Noventa-Jordão et al. 2000; Lee et al. 2017). These ubiquitin fusion proteins have to be cleaved by *deub*iquitinating enzymes (DUBs) to create a pool of ubiquitin monomers that can be used for the actual ubiquitination reaction (Grou et al. 2015). Ubiquitin is essential for fungal growth, although single deletion strains can grow but have strong defects in fungal growth, stress response, and development (Leach et al. 2011; Oh et al. 2012).

About 20% of all *A. nidulans* proteins are ubiquitinated during hyphal growth and are located in the nucleus, whereas in *S. cerevisiae* the biggest portion is contained in the transmembrane protein fraction (Peng et al. 2003; Chu et al. 2016). Ubiquitin itself contains seven conserved lysine residues (K6, K11, K27, K29, K33, K48, K63), which can be used for ubiquitin chain formation. The attachment of ubiquitin chains linked through a certain lysine residue can alter activity, localization, or stability of the substrate. The K48-linked polyubiquitin chains are the most abundant in yeast and usually mark the modified protein for degradation by the 26S proteasome (Spasser and Brik 2012; Zuin et al. 2014).

The attachment of ubiquitin as well as other ubiquitin-like proteins involves an enzyme cascade consisting of E1 activating, E2 conjugating, and E3 ligase enzymes (Fig. 4). An E1 enzyme activates ubiquitin in an ATP-dependent reaction by formation of a thioester bond. The corresponding E1 encoding *UBA1* (*ub*iquitin *a*ctivating 1) gene of *S. cerevisiae* is essential for spore formation and vegetative growth (McGrath et al. 1991). The activated ubiquitin molecule is transferred to E2, which can physically interact with E3 ubiquitin ligases. Several yeast genes encoding E2 enzymes as well as the polyubiquitin encoding locus are induced during heat stress or starvation conditions (McGrath et al. 1991; Hiraishi et al. 2006). **E3 ubiquitin ligases** are classified into the *c*ullin-*R*ING *l*igase (CRLs) and the HECT (*h*omologous to *E*6AP *c*arboxyl *t*erminus) ubiquitin ligase families.

Fungi such as *A. nidulans* express three *cul*lin proteins (CulA, CulC and CulD) and humans even eight (Marín 2009; von Zeska Kress et al. 2012). CulA, corresponding to human cullin-1, is part of the largest group of CRLs, the SCF (*S*kpA, *C*ullinA, *F*box) complexes. They are activated by the attachment of Nedd8 to CulA, a process called neddylation. Nedd8 and proteins of the neddylation cascade and the SCF complex are essential for *A. nidulans*, whereas SUMO-deficient mutants can grow but are impaired in multicellular development and secondary metabolite production (von Zeska Kress et al. 2012; Harting et al. 2013). In *S. cerevisiae*, it is the other way

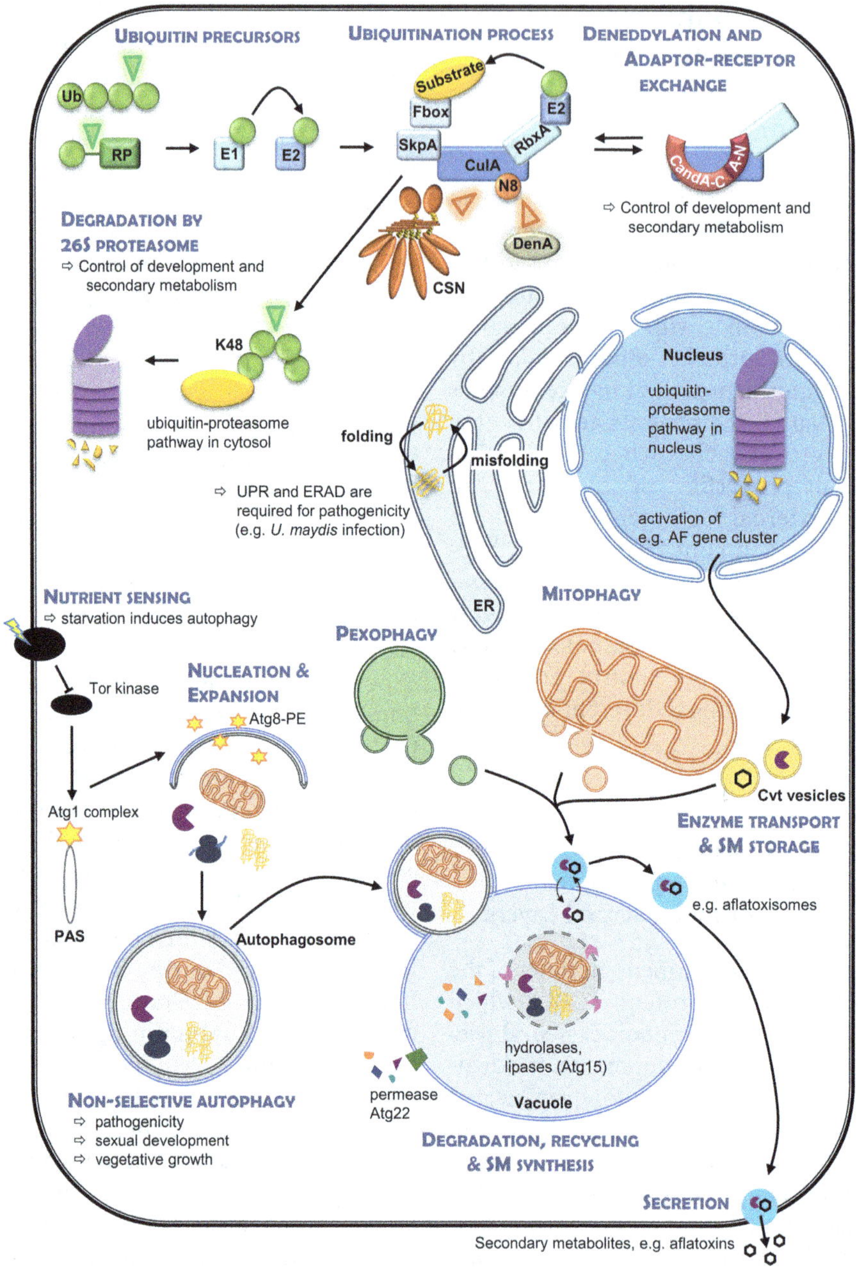

Fig. 4 Protein degradation pathways and their involvement in growth, development, pathogenicity, and secondary metabolite production. Free ubiquitin is generated from ubiquitin precursors and used to post-translationally label substrates by the concerted action of E1 activating, E2 conjugating, and E3 SCF (*S*kpA-*C*ulA-*F*box) ligase enzymes. E3 SCF is activated by neddylation and disassembled by deneddylation. Deneddylation is primarily catalyzed by the COP9 signalosome or in addition by DenA. Exchange of the adaptor/receptor complex (SkpA/Fbox) requires binding of the CandA receptor exchange factor. E3 SCF activity results primarily in K48 polyubiquitinated substrates, which are tagged for degradation by the 26S proteasome. Damaged proteasomes, organelles, or aggregated cellular material of misfolded proteins are enclosed by autophagosomes for vacuolar degradation. Fungal proteins that are damaged, misfolded, or only required for a certain timespan are targets for degradation by the 26S proteasome in the nucleus or cytosol. Misfolded ER proteins are recognized by the *u*nfolded *p*rotein *r*esponse (UPR) and the *ER*-*a*ssociated *d*egradation (ERAD) pathway and exported to the cytoplasm for proteasomal degradation. Signal transduction path-

round, SUMO is essential but not Nedd8 (Liakopoulos et al. 1998).

The active **SCF complex** requires at one side of the post-translational neddylated cullin scaffold a small RING protein RbxA, which binds to E2-ubiquitin, and at the other side the target substrate, which is connected to the cullin by the general SkpA (*s*uppressor of *k*inetochore *p*rotein mutant A) adaptor. SkpA binds the specific substrate receptors, the **Fbox proteins**, which differ considerably to interact with different substrates, but share a characteristic N-terminal 40–50 amino acids F-box motif as SkpA interaction domain. The genome of *A. nidulans* encompasses approximately 70 different Fbox proteins and the recruited target proteins are often primed by phosphorylation prior to ubiquitination (Draht et al. 2007). Fungal adaptation to changing environmental conditions requires the rapid exchange of Fbox proteins at the SCF complexes in order to label and degrade different substrates, what requires the removal of Nedd8 from CulA (Flick and Kaiser 2013). Several Fbox proteins are involved in fungal development. Fbox protein GrrA (*g*lucose *r*epression-*r*esistant A) is needed for the induction of meiosis and thus for the development of mature ascospores in *A. nidulans* (Krappmann et al. 2006). The corresponding orthologs of the fungal pathogens *Cryptococcus neoformans F-b*ox *p*rotein 1 (Fbp1) and *Gibberella zeae* FBP1 are also involved in sexual reproduction and virulence (Han et al. 2007; Liu et al. 2011). *A. nidulans* Fbx15 is required for asexual and sexual and Fbx23 for light-dependent development (von Zeska Kress et al. 2012). The homologous Fbx15 of the human pathogen *A. fumigatus* represses the formation of the secondary metabolite gliotoxin and provides a general stress response and virulence to the fungus (Jöhnk et al. 2016).

The disruption of the UPP leads to an imbalanced protein degradation of, e.g., secondary metabolite cluster-specific or global transcription factors. This can be used for the identification of novel interesting secondary metabolites (Gerke et al. 2012; Zheng et al. 2017). For instance, the deletion of *csnE* (*C*OP9 *sig*nalosome E), encoding the catalytically active subunit of the COP9 signalosome, interferes with the deneddylation of cullins, which mediate the ubiquitination of proteins (Beckmann et al. 2015; see Sect. IV.B). In *A. nidulans*, this deneddylation defect leads to the accumulation of 100 metabolites and resulted in the identification of the antimicrobial 2,4-*dih*ydroxy-3-*m*ethyl-6-(2-oxopropyl) *b*enz*a*ldehyde (DHMBA) (Nahlik et al. 2010; Gerke et al. 2012).

B. Controlled Removal of Ubiquitin Family Proteins from Substrates Is Important for Fungal Growth and Development

The attachment of ubiquitin and UBLs such as Nedd8 or SUMO to proteins is a reversible process. Two desumoylating enzymes UlpA and UlpB were characterized in *A. nidulans*, which are required for asexual conidiospore production and the formation of mature sexual fruiting bodies (Harting et al. 2013).

The removal of Nedd8 (deneddylation) from SCF complexes leads to a conformational change of the cullin protein (Duda et al. 2008)

Fig. 4 (continued) ways (e.g., the target of rapamycin Tor kinase) inhibit the formation of the fungal Atg1 autophagy initiation complex for *p*hagophore *a*ssembly *s*ite (PAS) in the presence of nutrients. Atg1 is activated by starvation or during development or interaction with other organisms and results in coupling of the ubiquitin-like Atg8 to the lipid *p*hosphatidyl*e*thanolamine (PE). Atg8-PE is recruited to PAS, which promotes nucleation and elongation of the phagophore. The phagophore engulfes proteins, protein aggregates or organelles like mitochondria or ribosomes resulting in the double membraned autophagosome. Autophagosomes fuse with the vacuole. The membrane of the resulting autophagic body is degraded by Atg15 lipase and the contents by vacuolar hydrolases. Permeases such as Atg22 can release degraded material back to the cytoplasm for recycling. The budding of vesicles from mitochondria is called mitophagy and the budding from peroxisomes is called pexophagy. Aflatoxin producing secondary metabolite enzymes and their substrates can be transported by *c*ytoplasm-to-*v*acuole *t*ransport (Cvt) vesicles and exported from the vacuole out of the fungal cell by aflatoxisomes

and is performed by deneddylases such as the **COP9 signalosome** (*co*nstitutive *p*hotomorphogenesis 9, CSN) (Beckmann et al. 2015). In most fungi, plants or in humans, it is a highly conserved complex consisting of eight CSN subunits including an intrinsic JAMM motif with Nedd8 isopeptidase activity in one subunit. *Neurospora crassa* and *S. pombe* possess seven- and six-subunit CSNs, respectively, and in *S. cerevisiae* there is only the conserved deneddylase subunit as part of an alternative CSN complex. Dysfunction of CSN results in lethality in plants or mammals but not in fungi, which makes them an attractive model organism to study CSN complex assembly and the function of the different subunits (Braus et al. 2010). CSN recognizes substrate free CRLs (Lingaraju et al. 2014) and physically interacts within the nucleus with a second deneddylase Den1/A. This interaction is found in *A. nidulans* as well as in human cells (Christmann et al. 2013; Schinke et al. 2016). Fungal CSN is composed of an inactive seven-subunit pre-complex, which incorporates the deneddylase subunit in a final step, resulting in a catalytically active complex (Beckmann et al. 2015). In *N. crassa*, CSN mutant strains result in stabilization of the circadian clock regulator FRQ, defects in light control (see Sect. II.A.2), and impaired asexual development (Wang et al. 2010). Loss of any of the eight CSN subunits of *A. nidulans* leads to impaired light control, a block in sexual development and a brownish colony with more than a hundred accumulated secondary metabolites including orsellinic acid derivatives (Busch et al. 2003, 2007; Nahlik et al. 2010).

COP9-mediated deneddylation of CulA deactivates E3 SCF ubiquitin ligases and results in disassembly. **Cand1** (*c*ullin *a*ssociated *N*edd8 *d*issociated 1) acts as Fbox receptor exchange factor by blocking the neddylation site of CulA. Multiple cycles of disassembly and assembly ensure the appropriate exchange of different Fbox proteins to degrade different target proteins in different environments or during development (Choo et al. 2011; Wu et al. 2013). Cand1/A is encoded as single protein in higher eukaryotes and fungi like *N. crassa*, *Verticillium dahliae*, or the opportunistic pathogen *C. albicans*, whereas the *A. nidulans* gene has been split into two separate genes, which encoded proteins are interacting (Helmstaedt et al. 2011; Sela et al. 2012). Deletion of one or both *cand* genes in *A. nidulans* leads to impaired conidiospore and fruiting body formation and to a similarly altered secondary metabolism with a brownish colony color as for the defective CSN (Helmstaedt et al. 2011; Nahlik et al. 2010). The deneddylation of cullins, the dissociation of the SkpA adaptor/Fbox receptor complex, the binding of the Cand protein, and the following reactivation of the SCF complex are highly dynamic processes that require tight regulation (Liu et al. 2018).

Deubiquitinases (DUBs) are required for the processing of ubiquitin precursors and for the removal of ubiquitin molecules from target proteins in order to generate the free ubiquitin pool for E1 ubiquitin-activating enzymes (Fig. 4). DUBs are classified into seven different families (Komander et al. 2009; Rehman et al. 2016). (1) The JAMM domain metalloprotease DUBs, such as the Rpn11 (*r*egulatory *p*article *n*on-ATPase 11) of the proteasomal lid,[6] contain an MPN+ domain and are similar to the deneddylase subunit of CSN (Meister et al. 2016). All other DUBs are cysteine proteases: (2) *u*biquitin *C*-terminal *h*ydrolases (UCH); (3) *M*achado-*J*oseph *d*omain (Josephin-domain) containing proteases (MJD), which are missing in some yeasts (Hutchins et al. 2013); (4) *o*varian *tu*mor proteases (OTU); (5) *u*biquitin-specific *p*roteases (USP); (6) the *m*otif *i*nteracting with Ub-containing *n*ovel *D*UB famil*y* (MINDY) (Rehman et al. 2016); and (7) the *z*inc finger with *UF*M1-*s*pecific *p*eptidase domain protein/C6orf113/ZUP1 (ZUFSP) family (Haahr et al. 2018).

The DUB distribution to the different families is quite conserved between fungi and humans with most members in the ubiquitin-specific proteases family with 17 USPs in *S. cerevisiae* and more than 50 in humans (Hutchins et al. 2013). Any single deletion of the genes for the 17 USP DUBs of *S. cerevisiae* is viable

[6] lid = removable or hinged cover of a container, here: the cover of the proteasome core.

with only partially moderate growth defects (Amerik et al. 2000). The ubiquitin-specific protease Doa4 (*d*egradation *o*f *A*lpha 4) is associated to the 26S proteasome and recycles ubiquitin chains from proteins that were targeted for degradation to the 26S proteasome or from membrane proteins that are targeted to the vacuole (Swaminathan et al. 1999). The *ub*iquitin-specific *p*rotease Ubp14 hydrolyzes rather free polyubiquitin chains, which are not degraded by the 26S proteasome as its unfolding would require too much energy (Amerik et al. 2000). The *C. neoformans* ubiquitin-specific protease Ubp5 is required for virulence and controls melanin and capsule formation (Fang et al. 2012).

V. Protein Degradation Pathways

Clearance of misfolded, damaged, or no longer needed proteins and organelles directs time controlled developmental transitions, multicellular development, pathogenicity, and secondary metabolism in filamentous fungi and leads to recycling of building bricks for the synthesis of new proteins and organelles. Protein degradation is mediated through different pathways (Fig. 4) including the nuclear and cytoplasmic *u*biquitin-26S *p*roteasome *p*athway (UPP) and autophagy pathways, which target defective cellular compartments and 26S proteasomes for degradation in the fungal vacuole. Misfolded proteins of the *e*ndoplasmic *r*eticulum (ER) activate the unfolded protein response and the *ER a*ssociated *d*egradation (ERAD) pathway, which finally results in degradation of these proteins in 26S proteasomes of the cytoplasm. The plant pathogenic fungus *Ustilago maydis* requires the unfolded protein response of the ER for pathogenicity (Heimel et al. 2013; Hampel et al. 2016).

A. Protein Degradation by the 26S Proteasome

The conserved 2.5 MDa **26S proteasome** is one of the major degradation machineries in eukaryotes and has approximately half the size of a ribosome. This multi-protease complex consists of the 20S barrel-like core particle and on one or both sides associated 19S regulatory particles consisting of lid and base subcomplexes (Tomko and Hochstrasser 2013). The lid receives ubiquitinated substrates and the base, containing six subunits with ATPase activity, unfolds substrates by ATP hydrolysis, and translocates them into the barrel of the core particle, while the JAMM domain metalloprotease DUB Rpn11 cleaves the ubiquitin chains (Lander et al. 2012; de la Peña et al. 2018). The core particle contains four rings. Each ring is built of seven subunits providing gate and catalytic activity with different trypsin-, chymotrypsin-, or caspase-like peptidases. These degrade substrates processively, resulting in recyclable amino acids (Budenholzer et al. 2017).

26S proteasomal-mediated degradation is initiated by the recognition of K48-polyubiquitinated substrates by Rpn1, Rpn10, or Rpn13, which possess ubiquitin-binding domains (Finley 2009). After capturing these substrates, ubiquitin tags are released by the intrinsic lid Rpn11 deubiquitinase, which is embedded in a similar protein subunit architecture as the intrinsic deneddylase of the COP9 signalosome (Meister et al. 2016; see Sect. IV. A). The eight CSN subunits correspond to eight Rpn (Rpn3, 5–9, 11, 12) subunits of the lid of the regulatory particle. The lid contains as ninth additional subunit the versatile small multifunctional and intrinsically disordered Sem1/Dss1 (*s*uppressor of *e*xocyst *m*utations 1; *d*eletion of *s*plit hand/*s*plit foot 1), and a ninth COP9 signalosome subunit (CSN acidic protein) is also present in metazoa and some plants and fungi (Barth et al. 2016).

Sem1 stabilizes the assembly of 26S proteasomes by recruiting the receptors Rpn13 and Rpn10 as well as the tethering factor Ecm29 (*e*xtra*c*ellular *m*utant 29) to support complex formation. The *A. nidulans sem1* deletion strain is delayed in conidiophore development, has less conidiospores, and produces immature fruiting bodies without ascospores. Sem1 of *A. nidulans* links fungal stress response to development and controls secondary metabolism similar to CSN or CandA (Kolog Gulko et al. 2018; see Sect. IV.A).

A link between the 26S proteasome of filamentous fungi and development and control of secondary metabolism is further corroborated by the application of proteasome inhibitors such as bortezomib. Proteasome inhibitors delay germination, appressoria formation, and host infection processes of the rice blast fungus *M. oryzae* (Wang et al. 2011; Oh et al. 2012). Inhibition of the proteasome resulted in the identification of several new compounds in *Pleosporales* or in the plant pathogen *Pestalotiopsis sydowiana* (VanderMolen et al. 2014; Xia et al. 2016). Although the targets of the 26S proteasome, which coordinate fungal development and secondary metabolism in these specific cases, are still unknown, it seems likely that they include regulatory proteins with a limited half-life that interconnect different cellular pathways.

B. Degradation by Autophagy

High protein turnover results in damaged fungal proteasomes, which are degraded in yeasts and presumably also in filamentous fungi by *self-eating* **autophagy** (Waite et al. 2016; Hoeller and Dikic 2016). This is, besides the proteasomal degradation, the second major eukaryotic degradation system. It is a highly organized membrane-trafficking pathway, specialized for long-lived proteins as well as quality control for large and heterogeneous cellular material including protein aggregates or organelles. Malfunction of this conserved process is, e.g., associated to neurodegenerative diseases in humans as well as impairment of accurate fungal secondary metabolism and differentiation. Autophagy provides nutrients during stress conditions, starvation, or transition phases of developmental programs (Voigt and Pöggeler 2013; Popova et al. 2018, Fig. 4). Fungi produce secondary metabolites, which directly affect autophagy such as the beneficial rasfonin from *Talaromyces*, which reduces different cancer cells by inducing autophagy (Xiao et al. 2014; Sun et al. 2016).

The targeted engulfment of cellular material by membrane invaginations of the vacuole is termed microautophagy. **Macroautophagy** corresponds to the sequestration of bulk cellular material such as damaged organelles or protein aggregates by de novo formed phagophore, e.g., during yeast starvation (Li et al. 2012; Reggiori et al. 2012). Cargo proteins are engulfed during this process by the double membrane phagophore, which closes to form the autophagosome. This compartment fuses with the fungal vacuole at its outer membrane, while the inner autophagic body is digested by the hydrolytic vacuolar milieu. Autophagy can be either non-selective or selective. **Selective autophagy** for the degradation of different cellular organelles, e.g., the pexophagy for defective peroxisomes in *Sordaria macrospora*, requires specific cargo receptors (Werner et al. 2019).

Instead of degradation, selective autophagy pathways can also be used for hydrolytic enzyme transportation to the vacuole as the yeast *c*ytoplasm-to-*v*acuole *t*ransport (Cvt) pathway (Lynch-Day and Klionsky 2010). Fungal secondary metabolism takes place in different cellular compartments such as the cytoplasm, the peroxisomes, vesicles, or vacuoles. The transport of enzymes and metabolic compounds or precursors is therefore often mediated through autophagic pathways. For instance, aflatoxin production requires a complex interplay of different autophagic processes. Vesicles bud from mitochondria and peroxisomes, which deliver precursors and enzymes to Cvt vesicles containing other enzymes of the biosynthetic pathway to form the aflatoxisomes. **Aflatoxisomes** mediate the biosynthesis by the transport of active enzymes as well as compartmentalization by storing different aflatoxin intermediates and the aflatoxin export to the cell exterior (Chanda et al. 2009; Roze et al. 2011a).

Autophagy-associated *atg* genes were first identified in the unicellular fungus *S. cerevisiae*. Until 2016, 42 yeast *atg* genes were identified including 18 core genes for autophagosome formation, which are mostly conserved in filamentous fungi (Wen and Klionsky 2016; Parzych et al. 2018).

In the initiation phase, starvation-induced autophagy requires an active Atg1 Ser/Thr kinase complex, which is inhibited under non-starvation conditions by the active Tor (*t*arget *o*f *r*apamycin) kinase complex, sensing nitrogen sources, and the cAMP-dependent protein kinase A complex, sensing glucose (Budovskaya et al. 2004; González and Hall 2017). In the following nucleation phase, Atg1 complex kinase activity results in the assembly of several other Atg proteins and the phosphatidylinositol kinase Vps34, which action recruits transmembrane protein Atg9-containing vesicles for the transport to the phagophore assembly site (He et al. 2008; Stjepanovic et al. 2017). Phagophore expansion as the third phase includes two ubiquitin-like conjugation systems (Atg8, Atg12), which cooperate to transfer Atg8 to the lipid *p*hosphatidyl*e*thanolamine (PE). Conjugation of Atg8 to Atg8-PE requires a similar mechanism of E1 activating, E2 conjugating, and E3 ligase cascade as previously described for other ubiquitin-like proteins (Liu et al. 2010b; see Sect. IV.A). After the fusion of the outer phagosomal membrane with the vacuolar membrane, vacuolar lipases such as Atg15 degrade the membrane of the released autophagic body. The permease Atg22 exports the degradation products into the cytoplasm for recycling (Sugimoto et al. 2011; Ramya and Rajasekharan 2016; Fig. 4).

Numerous fungal ***atg* genes** are essential for growth or are required for development, pathogenicity, or the formation of secondary metabolites. An *A. fumigatus atg1* deletion strain is impaired in vegetative growth in the presence of different nitrogen sources and produces less asexual spores, but is not affected in pathogenicity (Richie et al. 2007; Richie and Askew 2008). A *Penicillium chrysogenum atg1* mutant strain shows increased penicillin production due to an increased number of peroxisomes as a result of impaired pexophagy (Bartoszewska et al. 2011). Penicillin production takes place in the cytosol and in peroxisomes (Meijer et al. 2010). The relocation of AcvA, the first enzyme of the penicillin biosynthesis (δ-(L-α-*a*minoadipyl)-L-*c*ysteinyl-D-*v*aline synthetase), from the cytosol to the peroxisomes significantly increased the penicillin production of *A. nidulans* by more than three-fold (Herr and Fischer 2014).

The genes for the ubiquitin-like Atg12 from different yeasts are required for ascosporogenesis under nitrogen starvation, and the corresponding gene of *N. crassa* supports efficient fruiting body formation (Mukaiyama et al. 2010; Chinnici et al. 2014). In contrast, a *M. oryzae atg12*-deficient strain can undergo sexual development but is impaired in host infection. Whereas the *atg* genes participating in selective autophagy are dispensable, 22 non-selective *M. oryzae atg* genes are required for pathogenicity (Kershaw and Talbot 2009). This includes non-selective autophagy mutants, which are unable to transport the cellular content from the conidiospore into the appressoria, impairing leaf penetration.

Defects in the UBL encoding gene *atg8* as well as in several other *atg* genes result in impaired vegetative growth and a block in sexual development in *S. macrospora* (Nolting et al. 2009; Voigt and Pöggeler 2013). Both autophagy UBLs are involved in aging of hyphae, which change in their thickness from young thin to old thick hyphae. Deletion of *atg8* for the UBL or of *atg10* encoding the E2 conjugating enzyme for the Atg12 UBL shortened this aging process in *A. niger* (Nitsche et al. 2013). Aging of hyphae by autophagy might be due to degradation of organelles in older hyphae, which provides nutrients and building bricks for tip growth (Shoji et al. 2010). The degradation by **mitophagy** of damaged mitochondria increases presumably also fungal life span, because it prevents the accumulation of ROS or proapoptotic factors (Tyler and Johnson 2018).

Autophagy ensures survival during starvation, mediates nutrient supply for ascospore formation, and transports hydrolytic enzymes into the vacuoles in yeast. Autophagy protects cells against toxic metabolites or macromolecules and damaged organelles by their degradation and export. In ascomycetes, autophagy is required for development under starvation and nutrient-rich conditions. Furthermore, pathogenicity and secondary metabolism rely on autophagy.

VI. Conclusion

The burden of infectious diseases caused by bacteria and fungi, which are resistant to the usual antibiotics or antimycotics such as azoles, is growing (Cassini et al. 2019; Lestrade et al. 2019). Filamentous fungi form a vast reservoir of yet unknown beneficial or toxic secondary metabolites, which are only produced as responses to environmental biotic or abiotic stimuli. A more comprehensive picture of the complex and subtle control of transcriptional networks, which are nested within each other, are more and more emerging. They interconnect distinct fungal developmental programs with the production of specific secondary metabolites. This includes the genetic networks of the fungal velvet domain proteins and their sub-networks, which link transcriptional to epigenetic control of gene expression, but also protein degradation machineries like the 26S proteasome and the autophagy membrane trafficking pathway. They control maintenance or changes of fungal proteomes as well as cellular localization and transport of proteins as additional levels of coordinating secondary metabolite production in response to environmental parameters or during development, aging, or pathogenesis. Manipulation of gene expression including chemical epigenetics as well as genetic or chemical reprogramming of fungal protein degradation are promising approaches to find new biologically active secondary metabolites acting as antibiotics or as antifungal drugs.

Acknowledgments We thank Fruzsina Bakti and Christoph Sasse for critical reading of the manuscript. Funding of this research was provided by the Deutsche Forschungsgemeinschaft (DFG) and the SFB860.

References

Adams TH, Wieser JK, Yu JH (1998) Asexual sporulation in *Aspergillus nidulans*. Microbiol Mol Biol Rev 62:35–54

Adhvaryu KK, Morris SA, Strahl BD, Selker EU (2005) Methylation of histone H3 lysine 36 is required for normal development in *Neurospora crassa*. Eukaryot Cell 4:1455–1464

Ahmed YL, Gerke J, Park H-S, Bayram Ö, Neumann P, Ni M, Dickmanns A, Kim SC, Yu J-H, Braus GH, Ficner R (2013) The velvet family of fungal regulators contains a DNA-binding domain structurally similar to NF-κB. PLoS Biol 11:e1001750

Alberti F, Foster GD, Bailey AM (2017) Natural products from filamentous fungi and production by heterologous expression. Appl Microbiol Biotechnol 101:493–500

Amerik AY, Li S, Hochstrasser M (2000) Analysis of the deubiquitinating enzymes of the yeast *Saccharomyces cerevisiae*. Biol Chem 381:981–992

Andersen MR, Nielsen JB, Klitgaard A, Petersen LM, Zachariasen M, Hansen TJ, Blicher LH, Gotfredsen CH, Larsen TO, Nielsen KF, Mortensen UH (2013) Accurate prediction of secondary metabolite gene clusters in filamentous fungi. Proc Natl Acad Sci USA 110:E99–E107

Bachand F (2007) Protein arginine methyltransferases: from unicellular eukaryotes to humans. Eukaryot Cell 6:889–898

Bahn Y-S, Xue C, Idnurm A, Rutherford JC, Heitman J, Cardenas ME (2007) Sensing the environment: lessons from fungi. Nat Rev Microbiol 5:57–69

Barth E, Hübler R, Baniahmad A, Marz M (2016) The evolution of COP9 signalosome in unicellular and multicellular organisms. Genome Biol Evol 8:1279–1289

Bartoszewska M, Kiel JAKW, Bovenberg RAL, Veenhuis M, van der Klei IJ (2011) Autophagy deficiency promotes beta-lactam production in *Penicillium chrysogenum*. Appl Environ Microbiol 77:1413–1422

Bauer I, Graessle S, Loidl P, Hohenstein K, Brosch G (2010) Novel insights into the functional role of three protein arginine methyltransferases in *Aspergillus nidulans*. Fungal Genet Biol 47:551–561

Bauer I, Varadarajan D, Pidroni A, Gross S, Vergeiner S, Faber B, Hermann M, Tribus M, Brosch G, Graessle S, Turgeon EBG (2016) A class 1 histone deacetylase with potential as an antifungal target. MBio 7:e00831–e00816

Bayram Ö, Braus GH (2012) Coordination of secondary metabolism and development in fungi: the velvet family of regulatory proteins. FEMS Microbiol Rev 36:1–24

Bayram O, Biesemann C, Krappmann S, Galland P, Braus GH (2008a) More than a repair enzyme: *Aspergillus nidulans* photolyase-like CryA is a regulator of sexual development. Mol Biol Cell 19:3254–3262

Bayram Ö, Krappmann S, Ni M, Bok JW, Helmstaedt K, Valerius O, Braus-Stromeyer S, Kwon NJ, Keller NP, Yu J, Braus GH (2008b) VelB/VeA/LaeA complex coordinates light signal with fungal development and secondary metabolism. Science 320:1504–1506

Bayram Ö, Krappmann S, Seiler S, Vogt N, Braus GH (2008c) *Neurospora crassa ve-1* affects asexual conidiation. Fungal Genet Biol 45:127–138

Bayram Ö, Braus GH, Fischer R, Rodriguez-Romero J (2010) Spotlight on *Aspergillus nidulans* photosensory systems. Fungal Genet Biol 47:900–908

Bayram Ö, Feussner K, Dumkow M, Herrfurth C, Feussner I, Braus GH (2016) Changes of global gene expression and secondary metabolite accumulation during light-dependent *Aspergillus nidulans* development. Fungal Genet Biol 87:30–53

Becker K, Ziemons S, Lentz K, Freitag M, Kück U (2016) Genome-wide chromatin immunoprecipitation sequencing analysis of the *Penicillium chrysogenum* velvet protein PcVelA identifies methyltransferase PcLlmA as a novel downstream regulator of fungal development. mSphere 1: e00149–e00116

Beckmann EA, Köhler AM, Meister C, Christmann M, Draht OW, Rakebrandt N, Valerius O, Braus GH (2015) Integration of the catalytic subunit activates deneddylase activity in vivo as final step in fungal COP9 signalosome assembly. Mol Microbiol 97:110–124

Bicocca VT, Ormsby T, Adhvaryu KK, Honda S, Selker EU (2018) ASH1-catalyzed H3K36 methylation drives gene repression and marks H3K27me2/3-competent chromatin. Elife 7:e41497

Blumenstein A, Vienken K, Tasler R, Purschwitz J, Veith D, Frankenberg-Dinkel N, Fischer R (2005) The *Aspergillus nidulans* phytochrome FphA represses sexual development in red light. Curr Biol 15:1833–1838

Bode HB, Bethe B, Höfs R, Zeeck A (2002) Big effects from small changes: possible ways to explore nature's chemical diversity. ChemBioChem 3:619–627

Bok JW, Keller NP (2016) Insight into fungal secondary metabolism from ten years of LaeA research. In: Hoffmeister D (ed) Biochemistry and molecular biology: the Mycota (A comprehensive treatise on fungi as experimental systems for basic and applied research), vol III. Springer International Publishing, Cham, pp 21–29

Braus GH, Irniger S, Bayram Ö (2010) Fungal development and the COP9 signalosome. Curr Opin Microbiol 13:672–676

Brodhun F, Feussner I (2011) Oxylipins in fungi. FEBS J 278:1047–1063

Brosch G, Loidl P, Graessle S (2008) Histone modifications and chromatin dynamics: a focus on filamentous fungi. FEMS Microbiol Rev 32:409–439

Brown NA, Schrevens S, van Dijck P, Goldman GH (2018) Fungal G-protein-coupled receptors: mediators of pathogenesis and targets for disease control. Nat Microbiol 3:402–414

Budenholzer L, Cheng CL, Li Y, Hochstrasser M (2017) Proteasome structure and assembly. J Mol Biol 429:3500–3524

Budovskaya Y, Stephan J, Reggiori F, Klionsky DJ, Herman PK (2004) The Ras/PKA signaling pathway regulates an early step of the autophagy process in *Saccharomyces cerevisiae*. J Biol Chem 279:20663

Busch S, Eckert SE, Krappmann S, Braus GH (2003) The COP9 signalosome is an essential regulator of development in the filamentous fungus *Aspergillus nidulans*. Mol Microbiol 49:717–730

Busch S, Schwier EU, Nahlik K, Bayram Ö, Helmstaedt K, Draht OW, Krappmann S, Valerius O, Lipscomb WN, Braus GH (2007) An eight-subunit COP9 signalosome with an intact JAMM motif is required for fungal fruit body formation. Proc Natl Acad Sci USA 104:8089–8094

Cassini A, Diaz Högberg L, Plachouras D, Quattrocchi A, Hoxha A, Skov Simonsen G (2019) Attributable deaths and disability-adjusted life-years caused by infections with antibiotic-resistant bacteria in the EU and the European Economic Area in 2015: a population-level modelling analysis. Lancet Infect Dis 19:56–66

Castrillo M, García-Martínez J, Avalos J (2013) Light-dependent functions of the *Fusarium fujikuroi* CryD DASH cryptochrome in development and secondary metabolism. Appl Environ Microbiol 79:2777–2788

Cetz-Chel JE, Balcázar-López E, Esquivel-Naranjo EU, Herrera-Estrella A (2016) The *Trichoderma atroviride* putative transcription factor Blu7 controls light responsiveness and tolerance. BMC Genomics 17:327

Chanda A, Roze L, Kang S, Artymovich KA, Hicks GR, Raikhel N, Calvo AM, Linz JE (2009) A key role for vesicles in fungal secondary metabolism. Proc Natl Acad Sci USA 106:19533–19538

Chiang YM, Oakley CE, Ahuja M, Entwistle R, Schultz A, Chang S-L, Sung CT, Wang CCC, Oakley BR (2013) An efficient system for heterologous expression of secondary metabolite genes in *Aspergillus nidulans*. J Am Chem Soc 135:7720–7731

Chinnici JL, Fu C, Caccamise LM, Arnold JW, Free SJ (2014) *Neurospora crassa* female development requires the PACC and other signal transduction pathways, transcription factors, chromatin remodeling, cell-to-cell fusion, and autophagy. PLoS One 9:e110603

Choo YY, Boh BK, Lou JJW, Eng J, Leck YC, Anders B, Smith PG, Hagen T (2011) Characterization of the role of COP9 signalosome in regulating cullin E3 ubiquitin ligase activity. Mol Biol Cell 22:4706–4715

Christmann M, Schmaler T, Gordon C, Huang X, Bayram Ö, Schinke J, Stumpf S, Dubiel W, Braus GH (2013) Control of multicellular development by the physically interacting deneddylases DEN1/DenA and COP9 signalosome. PLoS Genet 9: e1003275

Chu X-L, Feng M-G, Ying S-H (2016) Qualitative ubiquitome unveils the potential significances of protein lysine ubiquitination in hyphal growth of *Aspergillus nidulans*. Curr Genet 62:191–201

Connolly LR, Smith KM, Freitag M (2013) The *Fusarium graminearum* histone H3 K27 methyltransferase KMT6 regulates development and expression of secondary metabolite gene clusters. PLoS Genet 9:e1003916 de la Peña AH, Goodall EA, Gates SN, Lander GC, Martin A (2018) Substrate-engaged 26S proteasome structures reveal mechanisms for ATP-hydrolysis-driven translocation. Science 362: eaav0725

de Vries RP, Riley R, Wiebenga A, Aguilar-Osorio G, Amillis S, Uchima CA, Anderluh G, Asadollahi M, Askin M, Barry K, Battaglia E, Bayram Ö, Benocci T, Braus-Stromeyer SA, Caldana C, Cánovas D, Cerqueira GC, Chen F, Chen W, Choi C, Clum A, dos Santos RAC, de Damásio ARL, Diallinas G, Emri T, Fekete E, Flipphi M, Freyberg S, Gallo A, Gournas C, Habgood R, Hainaut M, Harispe ML, Henrissat B, Hildén KS, Hope R, Hossain A, Karabika E, Karaffa L, Karányi Z, Kraševec N, Kuo A, Kusch H, LaButti K, Lagendijk EL, Lapidus A, Levasseur A, Lindquist E, Lipzen A, Logrieco AF, MacCabe A, Mäkelä MR, Malavazi I, Melin P, Meyer V, Mielnichuk N, Miskei M, Molnár ÁP, Mulé G, Ngan CY, Orejas M, Orosz E, Ouedraogo JP, Overkamp KM, Park H-S, Perrone G, Piumi F, Punt PJ, Ram AFJ, Ramón A, Rauscher S, Record E, Riaño-Pachón DM, Robert V, Röhrig J, Ruller R, Salamov A, Salih NS, Samson RA, Sándor E, Sanguinetti M, Schütze T, Sepčić K, Shelest E, Sherlock G, Sophianopoulou V, Squina FM, Sun H, Susca A, Todd RB, Tsang A, Unkles SE, van de Wiele N, van Rossen-Uffink D, de Oliveira JVC, Vesth TC, Visser J, Yu J-H, Zhou M, Andersen MR, Archer DB, Baker SE, Benoit I, Brakhage AA, Braus GH, Fischer R, Frisvad JC, Goldman GH, Houbraken J, Oakley B, Pócsi I, Scazzocchio C, Seiboth B, vanKuyk PA, Wortman J, Dyer PS, Grigoriev IV (2017) Comparative genomics reveals high biological diversity and specific adaptations in the industrially and medically important fungal genus *Aspergillus*. Genome Biol 18:28

Draht O, Busch S, Hofmann K, Braus-Stromeyer S, Helmstaedt K, Goldman G, Braus G (2007) Amino acid supply of *Aspergillus*. In: Osmani SA, Goldman GH (eds) The Aspergilli: genomics, medical aspects, biotechnology, and research methods. CRC Press/Taylor & Francis Group, Boca Raton, FL, pp 143–175

Du L, Sánchez C, Shen B (2001) Hybrid peptide–polyketide natural products: biosynthesis and prospects toward engineering novel molecules. Metab Eng 3:78–95

Duda DM, Borg LA, Scott DC, Hunt HW, Hammel M, Schulman BA (2008) Structural insights into NEDD8 activation of cullin-RING ligases: conformational control of conjugation. Cell 134:995–1006

Estrada AF, Avalos J (2008) The white collar protein WcoA of *Fusarium fujikuroi* is not essential for photocarotenogenesis, but is involved in the regulation of secondary metabolism and conidiation. Fungal Genet Biol 45:705–718

Fang W, Price MS, Toffaletti DL, Tenor J, Betancourt-Quiroz M, Price JL, Pan W, Liao W, Perfect JR (2012) Pleiotropic effects of deubiquitinating enzyme Ubp5 on growth and pathogenesis of *Cryptococcus neoformans*. PLoS One 7:38326

Finley D (2009) Recognition and processing of ubiquitin-protein conjugates by the proteasome. Annu Rev Biochem 78:477–513

Fisch KM, Gillaspy AF, Gipson M, Henrikson JC, Hoover AR, Jackson L, Najar FZ, Wägele H, Cichewicz RH (2009) Biosynthetic pathway transcription in *Aspergillus niger*. J Ind Microbiol Biotechnol 36:1199–1213

Fischer R, Aguirre J, Herrera-Estrella A, Corrochano LM (2016) The complexity of fungal vision. Microbiol Spectr 4:1–22

Flick K, Kaiser P (2013) Set them free: F-box protein exchange by Cand1. Cell Res 23:870–871

Gacek-Matthews A, Noble LM, Gruber C, Berger H, Sulyok M, Marcos AT, Strauss J, Andrianopoulos A (2015) KdmA, a histone H3 demethylase with bipartite function, differentially regulates primary and secondary metabolism in *Aspergillus nidulans*. Mol Microbiol 96:839–860

Gacek-Matthews A, Berger H, Sasaki T, Wittstein K, Gruber C, Lewis ZA, Strauss J (2016) KdmB, a jumonji histone H3 demethylase, regulates genome-wide H3K4 trimethylation and is required for normal induction of secondary metabolism in *Aspergillus nidulans*. PLoS Genet 12:e1006222

García-Esquivel M, Esquivel-Naranjo EU, Hernández-Oñate MA, Ibarra-Laclette E, Herrera-Estrella A (2016) The *Trichoderma atroviride* cryptochrome/photolyase genes regulate the expression of blr1-independent genes both in red and blue light. Fungal Biol 120:500–512

García-Martínez J, Brunk M, Avalos J, Terpitz U (2015) The CarO rhodopsin of the fungus *Fusarium fujikuroi* is a light-driven proton pump that retards spore germination. Sci Rep 5:7798

Gerke J, Braus GH (2014) Manipulation of fungal development as source of novel secondary metabolites for biotechnology. Appl Microbiol Biotechnol 98:8443–8455

Gerke J, Bayram Ö, Feussner K, Landesfeind M, Shelest E, Feussner I, Braus GH (2012) Breaking the silence: protein stabilization uncovers silenced biosynthetic gene clusters in the fungus *Aspergillus nidulans*. Appl Environ Microbiol 78:8234–8244

Gessler NN, Aver AA, Belozerskaya TA (2007) Reactive oxygen species in regulation of fungal development. Biochemist 72:1091–1109

González A, Hall MN (2017) Nutrient sensing and TOR signaling in yeast and mammals. EMBO J 36:397–408

Greer EL, Shi Y (2012) Histone methylation: a dynamic mark in health, disease and inheritance. Nat Rev Genet 13:343–357

Gressler M, Hortschansky P, Geib E, Brock M (2015) A new high-performance heterologous fungal expression system based on regulatory elements from the *Aspergillus terreus* terrein gene cluster. Front Microbiol 6:184

Grimaldi B, Coiro P, Filetici P, Berge E, Dobosy JR, Freitag M, Selker EU, Ballario P (2006) The *Neurospora crassa* white collar-1 dependent blue light response requires acetylation of histone H3 lysine 14 by NGF-1. Mol Biol Cell 17:4576–4583

Grou CP, Pinto MP, Mendes AV, Domingues P, Azevedo JE (2015) The *de novo* synthesis of ubiquitin: identification of deubiquitinases acting on ubiquitin precursors. Sci Rep 5:12836

Gu Q, Wang Z, Sun X, Ji T, Huang H, Yang Y, Zhang H, Tahir HAS, Wu L, Wu H, Gao X (2017) FvSet2 regulates fungal growth, pathogenicity, and secondary metabolism in *Fusarium verticillioides*. Fungal Genet Biol 107:24–30

Haahr P, Borgermann N, Guo X, Typas D, Achuthankutty D, Hoffmann S, Shearer R, Sixma TK, Mailand N (2018) ZUFSP deubiquitylates K63-linked polyubiquitin chains to promote genome stability. Mol Cell 70:165–174.e6

Haas H (2014) Fungal siderophore metabolism with a focus on *Aspergillus fumigatus*. Nat Prod Rep 31:1266–1276

Hampel M, Jakobi M, Schmitz L, Meyer U, Finkernagel F, Doehlemann G, Heimel K (2016) Unfolded protein response (UPR) regulator Cib1 controls expression of genes encoding secreted virulence factors in *Ustilago maydis*. PLoS One 11:e0153861

Han Y, Kim M, Lee S, Yun S, Lee Y (2007) A novel F-box protein involved in sexual development and pathogenesis in *Gibberella zeae*. Mol Microbiol 63:768–779

Harting R, Bayram Ö, Laubinger K, Valerius O, Braus GH (2013) Interplay of the fungal sumoylation network for control of multicellular development. Mol Microbiol 90:1125–1145

He C, Baba M, Cao Y, Klionsky DJ (2008) Self-interaction is critical for Atg9 transport and function at the phagophore assembly site during autophagy. Mol Biol Cell 19:5506–5516

Hedtke M, Rauscher S, Röhrig J, Rodríguez-Romero J, Yu Z, Fischer R (2015) Light-dependent gene activation in *Aspergillus nidulans* is strictly dependent on phytochrome and involves the interplay of phytochrome and white collar-regulated histone H3 acetylation. Mol Microbiol 97:733–745

Heimel K, Freitag J, Hampel M, Ast J, Bolker M, Kamper J (2013) Crosstalk between the unfolded protein response and pathways that regulate pathogenic development in *Ustilago maydis*. Plant Cell 25:4262–4277

Helmstaedt K, Schwier EU, Christmann M, Nahlik K, Tansey WP (2011) Recruitment of the inhibitor Cand1 to the cullin substrate adaptor site mediates interaction to the neddylation site. Mol Biol Cell 22:153–164

Henrikson JC, Hoover AR, Joyner PM, Cichewicz RH (2009) A chemical epigenetics approach for engineering the in situ biosynthesis of a cryptic natural product from *Aspergillus niger*. Org Biomol Chem 7:435–438

Herr A, Fischer R (2014) Improvement of *Aspergillus nidulans* penicillin production by targeting AcvA to peroxisomes. Metab Eng 25:131–139

Hiraishi H, Mochizuki M, Takagi H (2006) Enhancement of stress tolerance in *Saccharomyces cerevisiae* by overexpression of ubiquitin ligase Rsp5 and ubiquitin-conjugating enzymes. Biosci Biotechnol Biochem 70:2762–2765

Hoeller D, Dikic I (2016) How the proteasome is degraded. Proc Natl Acad Sci USA 113:13266–13268

Holighaus G, Rohlfs M (2018) Volatile and non-volatile fungal oxylipins in fungus-invertebrate interactions. Fungal Ecol 38:28–36

Hutchins AP, Liu S, Diez D, Miranda-Saavedra D (2013) The repertoires of ubiquitinating and deubiquitinating enzymes in eukaryotic genomes. Mol Biol Evol 30:1172–1187

Idnurm A, Verma S, Corrochano LM (2010) A glimpse into the basis of vision in the kingdom Mycota. Fungal Genet Biol 47:881–892

Inglis DO, Binkley J, Skrzypek MS, Arnaud MB, Cerqueira GC, Shah P, Wymore F, Wortman JR, Sherlock G (2013) Comprehensive annotation of secondary metabolite biosynthetic genes and gene clusters of *Aspergillus nidulans*, *A. fumigatus*, *A. niger* and *A. oryzae*. BMC Microbiol 13:91

Itoh E, Shigemoto R, Oinuma K-I, Shimizu M, Masuo S, Takaya N (2017) Sirtuin A regulates secondary metabolite production by *Aspergillus nidulans*. J Gen Appl Microbiol 63:228–235

Izawa M, Takekawa O, Arie T, Teraoka T, Yoshida M, Kimura M, Kamakura T (2009) Inhibition of histone deacetylase causes reduction of appressorium formation in the rice blast fungus *Magnaporthe oryzae*. J Gen Appl Microbiol 55:489–498

Janevska S, Baumann L, Sieber CMK, Münsterkötter M, Ulrich J, Kämper J, Güldener U, Tudzynski B (2018) Elucidation of the two H3K36me3 histone methyltransferases Set2 and Ash1 in *Fusarium fujikuroi* unravels their different chromosomal targets and a major impact of Ash1 on genome stability. Genetics 208:153–171

Jöhnk B, Bayram Ö, Abelmann A, Heinekamp T, Mattern DJ, Brakhage AA, Jacobsen ID, Valerius O, Braus GH (2016) SCF ubiquitin ligase F-box protein Fbx15 controls nuclear co-repressor localization, stress response and virulence of the human pathogen *Aspergillus fumigatus*. PLoS Pathog 12:1–10

Kawauchi M, Iwashita K (2014) Functional analysis of histone deacetylase and its role in stress response, drug resistance and solid-state cultivation in *Aspergillus oryzae*. J Biosci Bioeng 118:172–176

Kawauchi M, Nishiura M, Iwashita K (2013) Fungal-specific sirtuin HstD coordinates secondary metabolism and development through control of LaeA. Eukaryot Cell 12:1087–1096

Keller NP (2018) Fungal secondary metabolism: regulation, function and drug discovery. Nat Rev Microbiol 17:167–180

Keller NP, Turner G, Bennett JW (2005) Fungal secondary metabolism—from biochemistry to genomics. Nat Rev Microbiol 3:937–947

Kershaw MJ, Talbot NJ (2009) Genome-wide functional analysis reveals that infection-associated fungal autophagy is necessary for rice blast disease. Proc Natl Acad Sci USA 106:15967–15972

Kolog Gulko M, Heinrich G, Gross C, Popova B, Valerius O, Neumann P, Ficner R, Braus GH (2018) Sem1 links proteasome stability and specificity to multicellular development. PLoS Genet 14: e1007141

Komander D, Clague MJ, Urbé S (2009) Breaking the chains: structure and function of the deubiquitinases. Nat Rev Mol Cell Biol 10:550–563

Kong X, van Diepeningen AD, van der Lee TAJ, Waalwijk C, Xu J, Xu J, Zhang H, Chen W, Feng J (2018) The *Fusarium graminearum* histone acetyltransferases are important for morphogenesis, DON biosynthesis, and pathogenicity. Front Microbiol 9:654

Kouzarides T (2007) Chromatin modifications and their function. Cell 128:693–705

Krappmann S, Jung N, Medic B, Busch S, Prade RA, Braus GH (2006) The *Aspergillus nidulans* F-box protein GrrA links SCF activity to meiosis. Mol Microbiol 61:76–88

Kumar D, Barad S, Chen Y, Luo X, Tannous J, Dubey A, Glam Matana N, Tian S, Li B, Keller N, Prusky D (2016) LaeA regulation of secondary metabolism modulates virulence in *Penicillium expansum* and is mediated by sucrose. Mol Plant Pathol 18:1150–1163

Künzler M (2018) How fungi defend themselves against microbial competitors and animal predators. PLoS Pathog 14:e1007184

Lander GC, Estrin E, Matyskiela ME, Bashore C, Nogales E, Martin A (2012) Complete subunit architecture of the proteasome regulatory particle. Nature 482:186–191

Lara-Ortíz T, Riveros-Rosas H, Aguirre J (2003) Reactive oxygen species generated by microbial NADPH oxidase NoxA regulate sexual development in *Aspergillus nidulans*. Mol Microbiol 50:1241–1255

Leach MD, Stead DA, Argo E, Maccallum DM, Brown AJP (2011) Molecular and proteomic analyses highlight the importance of ubiquitination for the stress resistance, metabolic adaptation, morphogenetic regulation and virulence of *Candida albicans*. Mol Microbiol 79:1574–1593

Lee Y-H, Stallcup MR (2009) Minireview: protein arginine methylation of nonhistone proteins in transcriptional regulation. Mol Endocrinol 23:425–433

Lee I, Oh J-H, Keats Shwab E, Dagenais TRT, Andes D, Keller NP (2009) HdaA, a class 2 histone deacetylase of *Aspergillus fumigatus*, affects germination and secondary metabolite production. Fungal Genet Biol 46:782–790

Lee S, Tumolo JM, Ehlinger AC, Jernigan KK, Qualls-Histed SJ, Hsu P-C, Mcdonald H, Chazin WJ, Macgurn JA (2017) Ubiquitin turnover and endocytic trafficking in yeast are regulated by Ser57 phosphorylation of ubiquitin. Elife 6:e29176

Lestrade PP, Bentvelsen RG, Schauwvlieghe AFAD, Schalekamp S, van der Velden WJFM, Kuiper EJ, van Paassen J, van der Hoven B, van der Lee HA, Melchers WJG, de Haan AF, van der Hoeven HL, Rijnders BJA, van der Beek MT, Verweij PE (2019) Voriconazole resistance and mortality in invasive aspergillosis: a multicenter retrospective cohort study. Clin Infect Dis 68:1463–1471

Li W, Li J, Bao J (2012) Microautophagy: lesser-known self-eating. Cell Mol Life Sci 69:1125–1136

Liakopoulos D, Doenges G, Matuschewski K, Jentsch S (1998) A novel protein modification pathway related to the ubiquitin system. EMBO J 17:2208–2214

Liang L, Liu Y, Yang K, Lin G, Xu Z, Lan H, Wang X, Wang S (2017) The putative histone methyltransferase DOT1 regulates aflatoxin and pathogenicity attributes in *Aspergillus flavus*. Toxins 9:232

Lim FY, Won TH, Raffa N, Baccile JA, Wisecaver J, Rokas A, Schroeder FC, Keller NP (2018) Fungal isocyanide synthases and xanthocillin biosynthesis in *Aspergillus fumigatus*. MBio 9:e00785–e00718

Lind AL, Wisecaver JH, Smith TD, Feng X, Calvo AM, Rokas A (2015) Examining the evolution of the regulatory circuit controlling secondary metabolism and development in the fungal genus *Aspergillus*. PLoS Genet 11:e1005096

Lind AL, Smith TD, Saterlee T, Calvo AM, Rokas A (2016) Regulation of secondary metabolism by the velvet complex is temperature-responsive in *Aspergillus*. G3 6:4023–4033

Lingaraju GM, Bunker RD, Cavadini S, Hess D, Hassiepen U, Renatus M, Fischer ES, Thomä NH (2014) Crystal structure of the human COP9 signalosome. Nature 512:161–165

Liu C, Lu F, Cui X, Cao X (2010a) Histone methylation in higher plants. Annu Rev Plant Biol 61:395–420

Liu T-B, Liu X-H, Lu J-P, Zhang L, Min H, Lin F-C (2010b) The cysteine protease MoAtg4 interacts with MoAtg8 and is required for differentiation and pathogenesis in *Magnaporthe oryzae*. Autophagy 6:74–85

Liu T, Wang Y, Stukes S, Chen Q, Casadevall A, Xue C (2011) The F-Box protein Fbp1 regulates sexual reproduction and virulence in *Cryptococcus neoformans*. Eukaryot Cell 10:791–802

Liu X, Reitsma JM, Mamrosh JL, Zhang Y, Straube R, Deshaies RJ (2018) Cand1-mediated adaptive exchange mechanism enables variation in F-box protein expression. Mol Cell 69:773–786.e6

Lo Presti L, Lanver D, Schweizer G, Tanaka S, Liang L, Tollot M, Zuccaro A, Reissmann S, Kahmann R (2015) Fungal effectors and plant susceptibility. Annu Rev Plant Biol 66:513–545

López-Díaz C, Rahjoo V, Sulyok M, Ghionna V, Martín-Vicente A, Capilla J, Di Pietro A, López-Berges MS (2018) Fusaric acid contributes to virulence of *Fusarium oxysporum* on plant and mammalian hosts. Mol Plant Pathol 19:440–453

Lynch-Day MA, Klionsky DJ (2010) The Cvt pathway as a model for selective autophagy. FEBS Lett 584:1359–1366

Macheleidt J, Mattern DJ, Fischer J, Netzker T, Weber J, Schroeckh V, Valiante V, Brakhage AA (2016) Regulation and role of fungal secondary metabolites. Annu Rev Genet 50:371–392

Magotra A, Kumar M, Kushwaha M, Awasthi P, Raina C, Gupta AP, Shah BA, Gandhi SG, Chaubey A (2017) Epigenetic modifier induced enhancement of fumiquinazoline C production in *Aspergillus fumigatus* (GA-L7): an endophytic fungus from *Grewia asiatica* L. AMB Express 7:43

Malzahn E, Ciprianidis S, Káldi K, Schafmeier T, Brunner M (2010) Photoadaptation in *Neurospora* by competitive interaction of activating and inhibitory LOV domains. Cell 142:762–772

Marín I (2009) Diversification of the cullin family. BMC Evol Biol 11:1–11

Matsuda Y, Abe I (2016) Biosynthesis of fungal meroterpenoids. Nat Prod Rep 33:26–53

McGrath JP, Jentsch S, Varshavsky A (1991) *Uba1*: an essential yeast gene encoding ubiquitin-activating enzyme. EMBO J 10:227–236

Meeks JJ, Shilatifard A (2017) Multiple roles for the MLL/COMPASS family in the epigenetic regulation of gene expression and in cancer. Annu Rev Cancer Biol 1:425–446

Meijer WH, Gidijala L, Fekken S, Kiel JAKW, van den Berg MA, Lascaris R, Bovenberg RAL, van der Klei IJ (2010) Peroxisomes are required for efficient penicillin biosynthesis in *Penicillium chrysogenum*. Appl Environ Microbiol 76:5702–5709

Meister C, Gulko MK, Köhler AM, Braus GH (2016) The devil is in the details: comparison between COP9 signalosome (CSN) and the LID of the 26S proteasome. Curr Genet 62:129–136

Meyer V, Andersen MR, Brakhage AA, Braus GH, Caddick MX, Cairns TC, de Vries RP, Haarmann T, Hansen K, Hertz-Fowler C, Krappmann S, Mortensen UH, Peñalva MA, Ram AFJ, Head RM (2016) Current challenges of research on filamentous fungi in relation to human welfare and a sustainable bio-economy: a white paper. Fungal Biol Biotechnol 3:6

Mukaiyama H, Nakase M, Nakamura T, Kakinuma Y, Takegawa K (2010) Autophagy in the fission yeast *Schizosaccharomyces pombe*. FEBS Lett 584:1327–1334

Nahlik K, Dumkow M, Bayram Ö, Helmstaedt K, Busch S, Valerius O, Gerke J, Hoppert M, Schwier E, Opitz L, Westermann M, Grond S, Feussner K, Goebel C, Kaever A, Meinicke P, Feussner I, Braus GH (2010) The COP9 signalosome mediates transcriptional and metabolic response to hormones, oxidative stress protection and cell wall rearrangement during fungal development. Mol Microbiol 78:964–979

Nathan D, Ingvarsdottir K, Sterner DE, Bylebyl GR, Dokmanovic M, Dorsey JA, Whelan KA, Krsmanovic M, Lane WS, Meluh PB, Johnson ES, Berger SL (2006) Histone sumoylation is a negative regulator in *Saccharomyces cerevisiae* and shows dynamic interplay with positive-acting histone modifications. Genes Dev 20:966–976

Netzker T, Fischer J, Weber J, Mattern DJ, König CC, Valiante V, Schroeckh V, Brakhage AA (2015) Microbial communication leading to the activation of silent fungal secondary metabolite gene clusters. Front Microbiol 6:299

Newman DJ, Cragg GM (2012) Natural products as sources of new drugs over the 30 years from 1981 to 2010. J Nat Prod 75:311–335

Ni M, Yu JH (2007) A novel regulator couples sporogenesis and trehalose biogenesis in *Aspergillus nidulans*. PLoS One 2:e970

Nie X, Li B, Wang S (2018) Epigenetic and posttranslational modifications in regulating the biology of *Aspergillus* species. In: Gadd GM, Sariaslani S (eds) Advances in applied microbiology. Academic Press, London, pp 191–226

Nitsche BM, Burggraaf-van Welzen A-M, Lamers G, Meyer V, Ram AFJ (2013) Autophagy promotes survival in aging submerged cultures of the filamentous fungus *Aspergillus niger*. Appl Microbiol Biotechnol 97:8205–8218

Nocedal I, Mancera E, Johnson AD (2017) Gene regulatory network plasticity predates a switch in function of a conserved transcription regulator. Elife 6: e23250

Nolting N, Bernhards Y, Pöggeler S (2009) SmATG7 is required for viability in the homothallic ascomycete *Sordaria macrospora*. Fungal Genet Biol 46:531–542

Noventa-Jordão MA, Mendes do Nascimento A, Goldman MHS, Terenzi HF, Goldman GH (2000) Molecular characterization of ubiquitin genes from *Aspergillus nidulans*: mRNA expression on different stress and growth conditions. Biochim Biophys Acta Mol Cell Res 1490:237–244

Nowak SJ, Corces VG (2004) Phosphorylation of histone H3: a balancing act between chromosome condensation and transcriptional activation. Trends Genet 20:214–220

Nsa IY, Karunarathna N, Liu X, Huang H, Boetteger B, Bell-Pedersen D (2015) A novel cryptochrome-dependent oscillator in *Neurospora crassa*. Genetics 199:233–245

Nützmann H-W, Reyes-Dominguez Y, Scherlach K, Schroeckh V, Horn F, Gacek A, Schümann J, Hertweck C, Strauss J, Brakhage AA (2011) Bacteria-induced natural product formation in the fungus *Aspergillus nidulans* requires Saga/Ada-mediated histone acetylation. Proc Natl Acad Sci USA 108:14282–14287

Oakley CE, Ahuja M, Sun W, Entwistle R, Akashi T, Yaegashi J, Guo C, Cerqueira GC, Wortman JR, Clay C, Wang C, Chiang Y, Oakley BR (2016) Discovery of McrA, a master regulator of *Aspergillus* secondary metabolism. Mol Microbiol 103:347–365

Oh Y, Franck WL, Han SO, Shows A, Gokce E, Muddiman DC, Dean RA (2012) Polyubiquitin is required for growth, development and pathogenicity in the rice blast fungus *Magnaporthe oryzae*. PLoS One 7:e42868

Okada BK, Seyedsayamdost MR (2017) Antibiotic dialogues: induction of silent biosynthetic gene clusters by exogenous small molecules. FEMS Microbiol Rev 41:19–33

Palmer JM, Bok JW, Lee S, Dagenais TRT, Andes DR, Kontoyiannis DP, Keller NP (2013a) Loss of CclA, required for histone 3 lysine 4 methylation, decreases growth but increases secondary metabolite production in *Aspergillus fumigatus*. PeerJ 1: e4

Palmer JM, Theisen JM, Duran RM, Grayburn WS, Calvo AM, Keller NP (2013b) Secondary metabolism and development is mediated by LlmF control of VeA subcellular localization in *Aspergillus nidulans*. PLoS Genet 9:e1003193

Park HS, Yu JH (2012) Genetic control of asexual sporulation in filamentous fungi. Curr Opin Microbiol 15:669–677

Parzych KR, Ariosa A, Mari M, Klionsky DJ (2018) A newly characterized vacuolar serine carboxypeptidase, Atg42/Ybr139w, is required for normal vacuole function and the terminal steps of autophagy in the yeast *Saccharomyces cerevisiae*. Mol Biol Cell 29:1089–1099

Patananan AN, Palmer JM, Garvey GS, Keller NP, Clarke SG (2013) A novel automethylation reaction in the *Aspergillus nidulans* LaeA protein generates S-methylmethionine. J Biol Chem 288:14032–14045

Peng J, Schwartz D, Elias JE, Thoreen CC, Cheng D, Marsischky G, Roelofs J, Finley D, Gygi SP (2003) A proteomics approach to understanding protein ubiquitination. Nat Biotechnol 21:921–926

Pfannenstiel BT, Greco C, Sukowaty AT, Keller NP (2018) The epigenetic reader SntB regulates secondary metabolism, development and global histone modifications in *Aspergillus flavus*. Fungal Genet Biol 120:9–18

Pidroni A, Faber B, Brosch G, Bauer I, Graessle S (2018) A class 1 histone deacetylase as major regulator of secondary metabolite production in *Aspergillus nidulans*. Front Microbiol 9:2212

Pöggeler S, Nowrousian M, Teichert I, Beier A, Kück U (2018) Fruiting-body development in ascomycetes. In: Anke T, Schüffler A (eds) The Mycota XV. Physiology and genetics, 2nd edn. Springer International Publishing, Cham, pp 1–56

Popova B, Kleinknecht A, Arendarski P, Mischke J, Wang D, Braus GH (2018) Sumoylation protects against β-synuclein toxicity in yeast. Front Mol Neurosci 11:94

Purschwitz J, Müller S, Kastner C, Schöser M, Haas H, Espeso EA, Atoui A, Calvo AM, Fischer R (2008) Functional and physical interaction of blue- and red-light sensors in *Aspergillus nidulans*. Curr Biol 18:255–259

Ramya V, Rajasekharan R (2016) *ATG15* encodes a phospholipase and is transcriptionally regulated by YAP1 in *Saccharomyces cerevisiae*. FEBS Lett 590:3155–3167

Rauscher S, Pacher S, Hedtke M, Kniemeyer O, Fischer R (2016) A phosphorylation code of the *Aspergillus nidulans* global regulator VelvetA (VeA) determines specific functions. Mol Microbiol 99:909–924

Reggiori F, Komatsu M, Finley K, Simonsen A (2012) Autophagy: more than a nonselective pathway. Int J Cell Biol 2012:1–18

Rehman SAA, Kristariyanto YA, Choi S, Labib K, Hofmann K, Kulathu Y, Arif S, Rehman A, Kristariyanto YA, Choi S, Nkosi PJ, Weidlich S (2016) MINDY-1 is a member of an evolutionarily conserved and structurally distinct new family of deubiquitinating enzymes. Mol Cell 63:146–155

Ren H, Wang B, Zhao H (2017) Breaking the silence: new strategies for discovering novel natural products. Curr Opin Biotechnol 48:21–27

Reyes-Dominguez Y, Bok JW, Berger H, Shwab EK, Basheer A, Gallmetzer A, Scazzocchio C, Keller N, Strauss J (2010) Heterochromatic marks are associated with the repression of secondary metabolism clusters in *Aspergillus nidulans*. Mol Microbiol 76:1376–1386

Richie DL, Askew DS (2008) Autophagy: a role in metal ion homeostasis? Autophagy 4:115–117

Richie DL, Fuller KK, Fortwendel J, Miley MD, McCarthy JW, Feldmesser M, Rhodes JC, Askew DS (2007) Unexpected link between metal ion deficiency and autophagy in *Aspergillus fumigatus*. Eukaryot Cell 6:2437–2447

Riquelme M, Aguirre J, Bartnicki-García S, Braus GH, Feldbrügge M, Fleig U, Hansberg W, Herrera-Estrella A, Kämper J, Kück U, Mouriño-Pérez RR, Takeshita N, Fischer R (2018) Fungal morphogenesis, from the polarized growth of hyphae to complex reproduction and infection structures. Microbiol Mol Biol Rev 82:e00068–e00017

Rodriguez-Romero J, Hedtke M, Kastner C, Müller S, Fischer R (2010) Fungi, hidden in soil or up in the air: light makes a difference. Annu Rev Microbiol 64:585–610

Rodríguez-Urra AB, Jiménez C, Nieto MI, Rodríguez J, Hayashi H, Ugalde U (2012) Signaling the induction of sporulation involves the interaction of two secondary metabolites in *Aspergillus nidulans*. ACS Chem Biol 7:599–606

Rokas A, Wisecaver JH, Lind AL (2018) The birth, evolution and death of metabolic gene clusters in fungi. Nat Rev Microbiol 16:731–744

Roze LV, Arthur AE, Hong S-Y, Chanda A, Linz JE (2007) The initiation and pattern of spread of histone H4 acetylation parallel the order of transcriptional activation of genes in the aflatoxin cluster. Mol Microbiol 66:713–726

Roze LV, Chanda A, Linz JE (2011a) Compartmentalization and molecular traffic in secondary metabolism: a new understanding of established cellular processes. Fungal Genet Biol 48:35–48

Roze LV, Koptina AV, Laivenieks M, Beaudry RM, Jones DA, Kanarsky AV, Linz JE, Roze LV, Linz JE, Koptina AV, Kanarsky AV, Laivenieks M, Beaudry RM, Jones DA (2011b) Willow volatiles influence growth, development, and secondary metabolism in *Aspergillus parasiticus*. Appl Microbiol Biotechnol 92:359–370

Ruger-Herreros C, Rodríguez-Romero J, Fernández-Barranco R, Olmedo M, Fischer R, Corrochano LM, Canovas D (2011) Regulation of conidiation by light in *Aspergillus nidulans*. Genetics 188:809–822

Sakai K, Kinoshita H, Nihira T (2012) Heterologous expression system in *Aspergillus oryzae* for fungal biosynthetic gene clusters of secondary metabolites. Appl Microbiol Biotechnol 93:2011–2022

Sarikaya-Bayram Ö, Bayram Ö, Feussner K, Kim JH, Kim HS, Kaever A, Feussner I, Chae KS, Han DM, Han KH, Braus GH (2014) Membrane-bound methyltransferase complex VapA-VipC-VapB guides epigenetic control of fungal development. Dev Cell 29:406–420

Sarikaya-Bayram Ö, Palmer JM, Keller N, Braus GH, Bayram Ö (2015) One Juliet and four Romeos: VeA and its methyltransferases. Front Microbiol 6:1

Satterlee T, Cary JW, Calvo AM (2016) RmtA, a putative arginine methyltransferase, regulates secondary metabolism and development in *Aspergillus flavus*. PLoS One 11:e0155575

Schardl CL, Panaccione DG, Tudzynski P (2006) Ergot alkaloids-biology and molecular biology. In: Cordell GA (ed) The alkaloids.chemistry and biology, vol 63, 1st edn. Academic Press, Cambridge, pp 45–86

Scharf DH, Brakhage AA, Mukherjee PK (2016) Gliotoxin—bane or boon? Environ Microbiol 18:1096–1109

Schinke J, Gulko MK, Christmann M, Valerius O, Stumpf K, Stirz M, Braus GH (2016) The DenA/DEN1 interacting phosphatase DipA controls septa positioning and phosphorylation-dependent stability of cytoplasmatic DenA/DEN1 during fungal development. PLoS Genet 12:e1005949

Schmidt-Dannert C (2014) Biosynthesis of terpenoid natural products in fungi. In: Schrader JBJ (ed) Biotechnology of isoprenoids, Advances in biochemical engineering/biotechnology, vol 148. Springer, Cham, pp 19–61

Sela N, Atir-lande A, Kornitzer D (2012) Neddylation and CAND1 independently stimulate SCF ubiquitin ligase activity in *Candida albicans*. Eukaryot Cell 11:42–52

Shelest E (2017) Transcription factors in fungi: TFome dynamics, three major families, and dual-specificity TFs. Front Genet 8:53

Shilatifard A (2012) The COMPASS family of histone H3K4 methylases: mechanisms of regulation in development and disease pathogenesis. Annu Rev Biochem 81:65–95

Shimizu K, Hicks JK, Huang T, Keller NP (2003) Pka, Ras and RGS protein interactions regulate activity of AflR, a Zn(II)2Cys6 transcription factor in *Aspergillus nidulans*. Genetics 1104:1095–1104

Shimizu M, Masuo S, Fujita T, Doi Y, Kamimura Y, Takaya N (2012) Hydrolase controls cellular NAD, sirtuin, and secondary metabolites. Mol Cell Biol 32:3743–3755

Shlezinger N, Irmer H, Dhingra S, Beattie SR, Cramer RA, Braus GH, Sharon A, Hohl TM (2017) Sterilizing immunity in the lung relies on targeting fungal apoptosis-like programmed cell death. Science 357:1037–1041

Shoji J, Kikuma T, Arioka M, Kitamoto K (2010) Macroautophagy-mediated degradation of whole nuclei in the filamentous fungus *Aspergillus oryzae*. PLoS One 5:e15650

Shwab EK, Bok JW, Tribus M, Galehr J, Graessle S, Keller NP (2007) Histone deacetylase activity regulates chemical diversity in *Aspergillus*. Eukaryot Cell 6:1656–1664

Soukup AA, Chiang Y-M, Woo Bok J, Reyes-Dominguez Y, Oakley BR, Wang CCC, Strauss J, Keller NP (2012) Overexpression of the *Aspergillus nidulans* histone 4 acetyltransferase EsaA increases activation of secondary metabolite production. Mol Microbiol 86:314–330

Spasser L, Brik A (2012) Chemistry and biology of the ubiquitin signal. Angew Chem 51:6840–6862

Spatafora JW, Aime MC, Grigoriev IV, Martin F, Stajich JE, Blackwell M (2017) The fungal tree of life: from molecular systematics to genome-scale phylogenies. Microbiol Spectr 5:5

Stergiopoulos I, Collemare J, Mehrabi R, de Wit PJGM (2013) Phytotoxic secondary metabolites and peptides produced by plant pathogenic *Dothideomycete* fungi. FEMS Microbiol Rev 37:67–93

Stjepanovic G, Baskaran S, Lin MG, Hurley JH (2017) Vps34 kinase domain dynamics regulate the autophagic PI 3-kinase complex. Mol Cell 67:528–534

Stopa N, Krebs JE, Shechter D (2015) The PRMT5 arginine methyltransferase: many roles in development, cancer and beyond. Cell Mol Life Sci 72:2041–2059

Strauss J, Reyes-Dominguez Y (2011) Regulation of secondary metabolism by chromatin structure and epigenetic codes. Fungal Genet Biol 48:62–69

Studt L, Janevska S, Arndt B, Boedi S, Sulyok M, Humpf HU, Tudzynski B, Strauss J (2017) Lack of the COMPASS component Ccl1 reduces H3K4 trimethylation levels and affects transcription of secondary metabolite genes in two plant-pathogenic *Fusarium* species. Front Microbiol 7:2144

Sugimoto N, Iwaki T, Chardwiriyapreecha S, Shimazu M, Kawano M, Sekito T, Takegawa K, Kakinuma Y (2011) Atg22p, a vacuolar membrane protein involved in the amino acid compartmentalization of *Schizosaccharomyces pombe*. Biosci Biotechnol Biochem 75:385–387

Sun ZW, Allis CD (2002) Ubiquitination of histone H2B regulates H3 methylation and gene silencing in yeast. Nature 418:104–108

Sun H, Wang W, Che Y, Jiang X (2016) Fungal secondary metabolites rasfonin induces autophagy, apoptosis and necroptosis in renal cancer cell line. Mycology 7:81–87

Swaminathan S, Amerik AY, Hochstrasser M (1999) The Doa4 deubiquitinating enzyme is required for ubiquitin homeostasis in yeast. Mol Biol Cell 10:2583–2594

Tan KK (1974) Blue-light inhibition of sporulation in *Botrytis cinerea*. J Gen Microbiol 82:191–200

Thieme KG, Gerke J, Sasse C, Valerius O, Thieme S, Karimi R, Heinrich AK, Finkernagel F, Smith K, Bode HB, Freitag M, Ram AFJ, Braus GH (2018) Velvet domain protein VosA represses the zinc cluster transcription factor SclB regulatory network for *Aspergillus nidulans* asexual development, oxidative stress response and secondary metabolism. PLoS Genet 14:e1007511

Toledo AV, Franco MEE, Yanil Lopez SM, Troncozo MI, Saparrat MCN, Balatti PA (2017) Melanins in fungi: types, localization and putative biological roles. Physiol Mol Plant Pathol 99:2–6

Tomko RJ, Hochstrasser M (2013) Molecular architecture and assembly of the eukaryotic proteasome. Annu Rev Biochem 82:415–445

Tribus M, Bauer I, Galehr J, Rieser G, Trojer P, Brosch G, Loidl P, Haas H, Graessle S (2010) A novel motif in fungal class 1 histone deacetylases is essential for growth and development of *Aspergillus*. Mol Biol Cell 21:345–353

Tsitsigiannis DI, Zarnowski R, Keller NP (2004) The lipid body protein, PpoA, coordinates sexual and asexual sporulation in *Aspergillus nidulans*. J Biol Chem 279:11344–11353

Tyler JK, Johnson JE (2018) The role of autophagy in the regulation of yeast life span. Ann N Y Acad Sci 1418:31–43

van der Lee TAJ, Medema MH (2016) Computational strategies for genome-based natural product discovery and engineering in fungi. Fungal Genet Biol 89:29–36

VanderMolen KM, Darveaux BA, Chen W-L, Swanson SM, Pearce CJ, Oberlies NH (2014) Epigenetic manipulation of a filamentous fungus by the proteasome-inhibitor bortezomib induces the production of an additional secondary metabolite. RSC Adv 4:18329–18335

Voigt O, Pöggeler S (2013) Autophagy genes *Smatg8* and *Smatg4* are required for fruiting-body development, vegetative growth and ascospore germination in the filamentous ascomycete *Sordaria macrospora*. Autophagy 9:33–49

von Zeska Kress MR, Harting R, Bayram Ö, Christmann M, Irmer H, Valerius O, Schinke J, Goldman GH, Braus GH (2012) The COP9 signalosome counteracts the accumulation of cullin SCF ubiquitin E3 RING ligases during fungal development. Mol Microbiol 83:1162–1177

Wagner EJ, Carpenter PB (2012) Understanding the language of Lys36 methylation at histone H3. Nat Rev Mol Cell Biol 13:115–126

Waite KA, De-La Mota-Peynado A, Vontz G, Roelofs J (2016) Starvation induces proteasome autophagy with different pathways for core and regulatory particles. J Biol Chem 291:3239–3253

Wang J, Hu Q, Chen H, Zhou Z, Li W, Wang Y, Li S, He Q (2010) Role of individual subunits of the *Neurospora crassa* CSN complex in regulation of deneddylation and stability of cullin proteins. PLoS Genet 6:12–16

Wang Y, Kim S-G, Wu J, Yu S, Kang K-Y, Kim S-T (2011) Proteasome inhibitors affect appressorium formation and pathogenicity of the rice blast fungus, *Magnaporthe oryzae*. Plant Pathol J 27:225–231

Wang Z, Wang J, Li N, Li J, Trail F, Dunlap JC, Townsend JP (2018) Light sensing by opsins and fungal ecology: NOP-1 modulates entry into sexual reproduction in response to environmental cues. Mol Ecol 27:216–232

Wen X, Klionsky DJ (2016) An overview of macroautophagy in yeast. J Mol Biol 428:1681–1699

Werner A, Herzog B, Voigt O, Valerius O, Braus GH, Pöggeler S (2019) NBR1 is involved in selective pexophagy in filamentous ascomycetes and can be functionally replaced by a tagged version of its human homolog. Autophagy 15:78–97

Wiemann P, Brown DW, Kleigrewe K, Bok JW, Keller NP, Humpf HU, Tudzynski B (2010) FfVel1 and FfLae1, components of a *velvet*-like complex in *Fusarium fujikuroi*, affect differentiation, secondary metabolism and virulence. Mol Microbiol 77:972–994

Williams RB, Henrikson JC, Hoover AR, Lee AE, Cichewicz RH (2008) Epigenetic remodeling of the fungal secondary metabolome. Org Biomol Chem 6:1895–1897

Wolf JC, Mirocha CJ (1977) Control of sexual reproduction in *Gibberella zeae* (*Fusarium roseum* "Graminearum"). Appl Environ Microbiol 33:546–550

Wu S, Zhu W, Nhan T, Toth JI, Petroski MD, Wolf DA (2013) CAND1 controls *in vivo* dynamics of the cullin 1-RING ubiquitin ligase repertoire. Nat Commun 4:1642

Wu C, Yang F, Smith KM, Peterson M, Dekhang R, Zhang Y, Zucker J, Bredeweg EL, Mallappa C, Zhou X, Lyubetskaya A, Townsend JP, Galagan JE, Freitag M, Dunlap JC, Bell-Pedersen D, Sachs MS (2014) Genome-wide characterization of light-regulated genes in *Neurospora crassa*. G3 4:1731–1745

Xia X, Kim S, Liu C, Shim S, Xia X, Kim S, Liu C, Shim SH (2016) Secondary metabolites produced by an endophytic fungus *Pestalotiopsis sydowiana* and their 20S proteasome inhibitory activities. Molecules 21:944

Xiao Z, Li L, Li Y, Zhou W, Cheng J, Liu F, Zheng P, Zhang Y, Che Y (2014) Rasfonin, a novel 2-pyrone derivative, induces ras-mutated Panc-1 pancreatic tumor cell death in nude mice. Cell Death Dis 5: e1241

Xu W, Gavia DJ, Tang Y (2014) Biosynthesis of fungal indole alkaloids. Nat Prod Rep 31:1474–1487

Yu Z, Fischer R (2019) Light sensing and responses in fungi. Nat Rev Microbiol 17:25–36

Zheng S, Wyrick JJ, Reese JC (2010) Novel *trans*-tail regulation of H2B ubiquitylation and H3K4 methylation by the N terminus of histone H2A. Mol Cell Biol 30:3635–3645

Zheng Y, Ma K, Lyu H, Huang Y, Liu H, Liu L, Che Y, Liu X, Zou H, Yin W-B (2017) Genetic manipulation of the COP9 signalosome subunit PfCsnE leads to the discovery of pestaloficins in *Pestalotiopsis fici*. Org Lett 19:4700–4703

Zuin A, Isasa M, Crosas B (2014) Ubiquitin signaling: extreme conservation as a source of diversity. Cells 3:690–701

9 Fungal Genomics

Robin A. Ohm[1]

CONTENTS

I. Introduction

In the past two decades, genomics has developed into a formidable tool to study various aspects of fungi. In the year 1996, *Saccharomyces cerevisiae* was the first fungal genome to be sequenced (Goffeau et al. 1996), and since then the number of publicly available genome sequences has increased to 2128 in GenBank and 1398 in MycoCosm (at the time of writing in September 2019) (Grigoriev et al. 2014; Clark et al. 2016). Fungal genomics as a research field was kick-started by the sequencing efforts of institutes and consortia, including the Fungal Genome Initiative of the BROAD Institute of MIT and Harvard (Cuomo and Birren 2010) and the Fungal Program of the US DOE Joint Genome Institute (Grigoriev et al. 2011). In recent years, sequencing costs have decreased considerably, placing genome sequencing and analysis well within the reach of smaller labs.

After the first fungal genome of *S. cerevisiae* (Goffeau et al. 1996) was published, genome sequencing efforts initially focused on other previously established model systems. Examples include *Neurospora crassa* (Galagan et al. 2003), various species of *Aspergillus* (Galagan et al. 2005), the human pathogen *Cryptococcus neoformans* (Loftus et al. 2005), the plant pathogen *Fusarium graminearum* (Cuomo et al. 2007), and *Trichoderma reesei* (Martinez et al. 2008). These genome sequences are still an indispensable tool for studying these important model systems. Among many other things, they facilitate high-throughput experiments such as RNA-Seq to study genome-wide gene expression or ChIP-Seq to study various aspects of epigenetic regulation. In combination, these approaches aim to assign functions to regions of the genome and are called Functional Genomics.

Moreover, the increasing number of available genome sequences (including those of non-model organisms) allowed for a comparative genomics approach. By comparing genomes of related species, new insights can be gained into genome evolution, gene evolution, gene association with a particular lifestyle, as well as phylogeny (examples of this are described below).

In general, a genome sequencing project starts with sequencing the genomic DNA using next-generation sequencing technologies. This is followed by genome assembly, which aims to computationally reconstruct the

[1] Microbiology, Department of Biology, Utrecht University, Utrecht, the Netherlands; e-mail: r.a.ohm@uu.nl

Genetics and Biotechnology, 3rd Edition
The Mycota II
J.P. Benz, K. Schipper (Eds.)

genome from the (relatively short) sequence reads. Next, the genome assembly is annotated. This entails the identification of repetitive sequences, genes, and other functional elements in the genome. The predicted genes are subsequently annotated by assigning a putative function, usually based on homology to known genes and domains.

This chapter will describe advances in technologies underlying fungal genome sequencing, annotation, and analysis. Furthermore, the impact of fungal genome sequencing is illustrated using examples from several fields of biotechnology.

II. Advances in Genome Sequencing Technologies

In the past decade, sequencing technologies have improved dramatically, radically changing the landscape of fungal genome sequencing. **Sanger** sequencing was the first sequencing technique that was used for genome sequencing (Sanger et al. 1977b). It was used to sequence landmark genomes such as the first bacteriophage φX174 (Sanger et al. 1977a), the first bacterium *Haemophilus influenza* (Fleischmann et al. 1995), the first eukaryote (and first fungus) *Saccharomyces cerevisiae* (Goffeau et al. 1996), the first plant *Arabidopsis thaliana* (Arabidopsis Genome Initiative 2000), the first animal *Caenorhabditis elegans* (The *C. elegans* Sequencing Consortium 1998), as well as the human genome (International Human Genome Sequencing Consortium 2001). The sequencing of these genomes was generally a multi-year undertaking and was performed by large consortia of collaborating labs. Sequencing reads that were obtained with Sanger technology were relatively long (up to approximately 1500 bp) and were relatively straightforward to assemble using assembly software such as Jazz (Aparicio et al. 2002) or Arachne (Batzoglou et al. 2002).

More recently, since the mid-2000s, several new sequencing platforms were developed that are collectively known as "**next-generation sequencing**" (NGS). Initially, these techniques included the now mostly defunct technologies Roche 454 (Margulies et al. 2005), IonTorrent (Life Technologies) and SOLiD (Applied Biosystems). Currently the most prominent short read sequencing technology, however, is **Illumina** (Bennett 2004). Although the sequences generated by Illumina technology were initially too short for efficient genome sequencing (up to 25 bp), this has increased to currently 2 × 300 bp on an Illumina MiSeq machine. New assembly approaches and software were developed for these short reads, such as Velvet (Zerbino and Birney 2008), ABySS (Simpson et al. 2009), SOAPdenovo (Luo et al. 2012), and SPAdes (Bankevich et al. 2012).

Rather paradoxically, the assemblies generated from early NGS techniques were not nearly as good as the ones generated from Sanger reads, with respect to assembly fragmentation. Especially repetitive genomic regions (e.g., originating from transposable elements) were challenging to assembly using short reads. However, crucial advantages of NGS technologies are that they are considerably faster and cheaper than Sanger sequencing (Ghurye and Pop 2019). This meant that genome sequencing became affordable to core facilities and even individual researchers, as opposed to the large sequencing consortia that were required for Sanger-based genome sequencing. This is illustrated by the following back-of-the-envelope calculation: sequencing a typical fungal genome of 30 Mbp with 100-fold coverage (each bp sequenced on average 100 times) currently costs less than 250 euro per genome on an Illumina NextSeq500 machine (if 35 genomes are pooled onto one lane). This is a stark difference with the multi-million euro Sanger sequencing efforts of the past (Goffeau et al. 1996).

Since the early 2010s, new technologies have become commercially available that produce considerably longer reads than Illumina. **Pacific Biosciences** (PacBio) is based on single-molecule sequencing and can produce reads of on average 5 kbp and a maximum of 20 kbp (Eid et al. 2009). **Oxford Nanopore** further revolutionized sequencing by vastly reducing the size of the machine to a mere USB flash drive (Jain et al. 2016). This MinION machine produces reads of over 100 kbp. However, both

PacBio and Oxford Nanopore reads have a considerably higher error rate (up to 15% errors) than Illumina technology (Mardis 2017). Although this error rate will likely improve as new protocols become available, it is problematic for accurate genome sequencing and assembly. Approaches for assembly using these long reads are either high coverage sequencing (Chin et al. 2013) or a hybrid approach that uses Illumina reads to correct sequencing and assembly errors (Walker et al. 2014). As these long-read NGS technologies mature further, it is likely that obtaining genome assemblies with telomere-to-telomere chromosomes will become trivial and affordable within a few years.

III. Genome Annotation

Sequencing a genome is only the first step, and the even more important next step is to annotate the genome. This process generally includes the identification of regions of repetitive DNA, the prediction of genes, and a function prediction for these genes and domains. These individual steps can be strung together into a **pipeline**. Several pipelines exist for eukaryotic genome annotation, and two examples of frequently used pipelines for fungal genome annotation are MAKER (Cantarel et al. 2008) and the pipeline used by the US DOE Joint Genome Institute (Haridas et al. 2018). This section describes the steps of genome annotation in more detail.

A. Repeats

The term "repeat" may refer to various types of sequences: "low-complexity regions" (sometimes called "simple repeats") such as a homopolymeric run of nucleotides, as well as transposable (mobile) elements (transposons) (Kapitonov and Jurka 2008). These transposable elements can essentially copy themselves and thus spread throughout the genome. They can be subdivided into two classes, depending on their mode of proliferation (Wicker et al. 2007). Class I elements use an RNA-intermediate (reminiscent of a retrovirus) and move via a "copy-paste" mechanism. They include long interspersed nuclear elements (LINEs), short interspersed nuclear elements (SINEs), and long terminal repeats (LTRs). Class II elements move via a DNA intermediate and include helitrons and terminal inverted repeats (Kapitonov and Jurka 2001, 2008).

Since repetitive regions are markedly different from gene-coding regions, it is common practice to "mask" the repetitive regions before commencing gene prediction. Masking ensures that any spurious open reading frames that may be present in the repeats will not confound (the training of) the gene predictor. Repeats can be identified in a newly sequenced genome using either homology-based or de novo tools. Homology-based tools rely on a database of known repetitive elements such as Repbase (Jurka et al. 2005) and a search algorithm such as RepeatMasker (Smit et al. 2015). Novel or genome-specific repeats can be identified using de novo tools such as Repeatscout (Price et al. 2005), which looks for sequences that occur repeatedly throughout the genome. Since transposable elements tend to be relatively AT-rich, their proliferation can result in large AT-rich regions. Those regions can be distinguished from gene-coding GC-rich regions by tools such as OcculterCut (Testa et al. 2016).

The repetitive content of the genome varies widely between fungi. For example, the very compact 13.6 Mbp genome of the fern pathogen *Mixia osmundae* has a repetitive content of <1% (Toome et al. 2014), whereas the 177.6 Mbp genome of the mycorrhizal ascomycete *Cenococcum geophilum* consist for 81% of repetitive sequences (Peter et al. 2016). Repetitive sequences are usually predominantly found in centromeric and sub-telomeric regions but may be spread throughout the assembly. Generally, self-replicating repeats are considered deleterious since their spread may interrupt genes. Fungi have evolved a defense mechanism that recognizes repeats and inactivates these by causing point mutations (repeat-induced point mutations, or RIP) (Clutterbuck 2011; Castanera et al. 2016). Intriguingly, genome sequencing of several plant pathogens has revealed that pathogenesis-related genes frequently co-localize with repetitive sequences in these species. A potential evolutionary benefit of this colocalization is a higher rate of mutation due to RIP, which in turn may lead to a higher rate of evolution. This may allow these pathogens to adapt more quickly to the host plant's defenses (Rouxel et al. 2011; Ohm et al. 2012).

B. Gene Prediction

Genes are (arguably) the most important functional elements in a fungal genome. However, their accurate identification is non-trivial. The presence of introns in fungal genes precludes simple scanning for open reading frames (ORFs), which is a common initial approach in gene prediction in prokaryotes. The structure of protein-coding genes varies widely between eukaryotes (Yandell and Ence 2012) and even between fungi. Differences include GC content of the coding regions, splicing acceptor and donor sites, intron length, number of introns per gene, gene length, etc. For example, the ascomycete yeast *S. cerevisiae* has 6576 predicted genes with a median gene length of 1071 bp, of which 4.2% contain an intron (Goffeau et al. 1996). In contrast, the basidiomycete mushroom-forming fungus *Schizophyllum commune* has 16204 predicted genes with a median gene length of 1517 bp, of which 86.3% contain an intron (Ohm et al. 2010). Therefore, the gene-finding approach needs to be tailored to each organism individually.

Gene prediction algorithms can be divided into two categories: evidence-driven and ab initio approaches. Evidence-driven predictors take external evidence to identify the locations of protein-coding genes. This evidence usually takes the form of sequenced cDNA (Haas et al. 2003) or homology with known proteins of related species (Birney et al. 2004). Sequenced cDNA (in this context usually referred to as Expressed Sequence Tags or ESTs) are aligned to the assembly, and exons and intron splice sites are inferred. This approach has the advantage that it uses evidence specific to the organism but has the disadvantage that unexpressed genes are less likely to be identified correctly. Homology-based gene predictors rely on the alignment of known proteins from related organisms to identify exons. Advantages of this approach are that it is cheap (since no cDNA sequencing is required) but has the disadvantage that organism-specific genes are less likely to be identified correctly. An ab initio approach uses a mathematical model of the gene structure to predict genes. These algorithms require training, which means that they need to learn what a gene looks like (e.g., typical gene length, intron length, GC content of coding regions, etc.) from a subset of known genes. This poses a problem, since for most fungal genomes there is no prior knowledge available. Modern approaches use a hybrid strategy in which RNA-Seq data is used as evidence to train an ab initio gene predictor. The algorithm **BRAKER**, for example, only requires aligned RNA-Seq reads and a genome assembly and no other prior knowledge (Hoff et al. 2016). It uses these data to train the ab initio predictors GeneMark (Lomsadze et al. 2014) and Augustus (Stanke and Waack 2003) and subsequently generates a high-quality gene prediction.

Various gene prediction algorithms may predict different genes at the same locus. Although these sometimes represent alternative splicing variants (especially when the gene predictor uses expression data as evidence), it is more likely that only one variant is correct. Various methods have been published that aim to select the correct gene prediction at each locus; examples include MAKER (Cantarel et al. 2008), the US DOE Joint Genome Institute pipeline (Haridas et al. 2018), and FunGAP (Min et al. 2017).

The quality and completeness of the set of predicted genes can be assessed by determining the percentage of highly conserved eukaryotic genes that are found in the predicted gene set. Since these highly conserved genes (histones, DNA polymerase, etc.) are expected to be present among the genes of the newly sequenced fungus, their absence can be indicative of an incompleteness of the genome assembly or the gene prediction. CEGMA (Core Eukaryotic Genes Mapping Approach) was initially a popular tool to determine completeness (Parra et al. 2007). However, a key issue with CEGMA was that the conserved genes were identified from only six eukaryotic species. BUSCO takes a clade-specific approach that is based on more eukaryotic genomes, and fungi-specific conserved gene sets are available (Simão et al. 2015). More recently, FGMP (Fungal Genome Mapping Project) was developed that provides a computational framework and sequence resource specifically designed to assess the completeness of fungal genomes (Cissé and Stajich 2019). It is based on 246 fungal genomes and can be used to assess assembly and annotation completeness as well as suggest assembly improvements.

C. Functional Annotation of the Predicted Genes

Once a reliable set of genes has been predicted, the next step is to determine the putative role of the encoded proteins. This is referred to as functional annotation of the predicted proteins. It is important to note, however, that automated function predictions should be interpreted with care. Lab experiments may be required to definitively confirm the function of individual genes (e.g., an enzyme activity assay to confirm the predicted activity of a putative enzyme).

Functional annotation usually starts with homology searches in a database of known proteins, for example, using Blast (Altschul et al. 1990) to search for homologs in GenBank (Clark et al. 2016) or UniProt/Swiss-Prot (Bateman et al. 2017). Moreover, conserved protein domains can be identified using InterPro (Hunter et al. 2009), which comprises a collection of domain databases that includes PFAM (Finn et al. 2016). Cellular localization of the proteins can be predicted using SignalP (Petersen et al. 2011), TMHMM (Krogh et al. 2001), and WoLF PSORT (Horton et al. 2007). Proteases/peptidases can be identified by homology to known enzymes in the MEROPS database (Rawlings et al. 2014). More generally, Gene Ontology (GO) aims to provide a hierarchical functional annotation of the predicted proteins, based on their molecular function, cellular localization, and the biological process they are involved in Ashburner et al. (2000). Similarly, KEGG (Kyoto Encyclopedia of Genes and Genomes) provides a classification system into metabolic pathways, including predicted enzyme activities based on the Enzyme Commission (EC) system (Kanehisa and Goto 2000).

Several functional annotation approaches have been developed that aim to identify genes that are involved in the lifestyle of fungi. The **CAZy** (carbohydrate-active enzymes) database focuses on enzymes that assemble, modify, or break down polysaccharides (Lombard et al. 2014). CAZymes are especially important in the context of plant biomass breakdown, for example, in lignocellulose degradation and plant disease (further discussed below). Fungi are known to produce a wide variety of secondary metabolites and other natural products (further discussed below). The genes involved in this process are frequently clustered in the genome, and these biosynthetic gene clusters can be identified by tools like AntiSMASH (Blin et al. 2017) or SMURF (Khaldi et al. 2010).

D. Data Visualization, Analysis, and Manual Curation

Large amounts of data are generated by genome sequencing and annotation. These can be challenging to interpret unless they are visualized. Genome sequencing consortia and/or institutes generally make the data accessible to the public by means of a centrally hosted web database, which allows users to analyze the genome sequence, gene predictions, and functional annotations. Examples include the genus-specific websites Saccharomyces Genome Database (SGD) and the Aspergillus Genome Database (AspGD) (Cherry et al. 2012; Cerqueira et al. 2014). MycoCosm hosts all fungal genome portals of the US DOE Joint Genome Institute (Grigoriev et al. 2014). FungiDB hosts numerous published fungi (Basenko et al. 2018). Upon publication of a genome, the data is generally submitted to NCBI GenBank, which has therefore amassed a large collection of fungal genome data (Clark et al. 2016).

Typically, a genome portal contains tools to visualize and analyze the genome data. These tools include Blast to search for homology (Altschul et al. 1990), a search function for functional annotations and a genome browser. Since the generated data will likely contain errors in gene prediction, it is important that these predictions can be fixed manually, based on external evidence. Genome browsers can facilitate this in an intuitive way. Data of various origins can be displayed, evaluated, and (if needed) manually corrected. This process is referred to as manual curation. Data that can be visualized include gene predictions, expression data, regions of homology (e.g., blast hits), genome synteny, etc.

An early example of a web-based genome browser was the UCSC (University of California, Santa Cruz) Genome Browser, which was originally developed to visualize the human genome (Kent et al. 2002) and is also used in MycoCosm. Later, GBrowse was developed (Stein 2013), which was designed to integrate well with the Generic Model Organism Database suite (www.gmod.org). Its successor JBrowse (Buels et al. 2016) offers an intuitive and flexible genome browser that can be easily installed and used in small-scale genome sequencing initiatives. Web Apollo (later renamed to Apollo) (Lee et al. 2013) is a plugin for JBrowse that facilitates the manual curation (correction) of gene predictions as well as other genomic features, making it a valuable tool for genome visualization, analysis, and curation. All corrections are stored in a centralized database, allowing collaborators from all over the world to simultaneously work on the same genome.

IV. Genomics and Biotechnology

Fungi play important roles in a wide range of fields that are interesting from a biotechnological perspective. Genome sequencing and annotation has greatly facilitated the development of these fields by revealing the genes involved in these processes. Examples of biotechnologically relevant topics include secondary metabolites, carbohydrate-active enzymes, mushroom development, and plant interactions. Obviously, this is by no means an exhaustive list of biotechnological topics. This section will discuss the roles genome sequencing and analysis have played in these important fields of study.

A. Secondary Metabolites or Natural Products

Fungi can produce a wide range of secondary metabolites, which are relatively small molecules that are not directly encoded by genes. In the context of biotechnology, they are frequently referred to as natural products. These metabolites can play an important role in processes such as pathogenesis, defense, interactions, pigmentation, etc. Often, they play an ecological role and help the fungi to colonize a niche. From a biotechnology perspective, they are interesting for their antibacterial, antifungal, and antitumor activities. Some well-known examples of natural products are the antibiotic penicillin, which is produced by species of *Penicillium* (Bennett and Chung 2001), and the cholesterol-lowering drug lovastatin (Downs et al. 1998).

Secondary metabolites are not directly encoded by genes, but instead they are generally produced by a set of enzymes that synthesize the metabolite in a conveyor belt-like fashion. These enzymes include polyketide synthases (PKS), non-ribosomal peptide synthetases (NRPS), terpene cyclases (TC), dimethyl-allyl-tryptophan synthetases (DMATS), and a range of accessory enzymes including methyltransferases (Keller et al. 2005; Keller 2019). Intriguingly, the genes encoding these enzymes are frequently clustered in the genome, which makes them relatively easy to identify (Nützmann et al. 2018). These gene clusters are known as biosynthetic gene clusters. AntiSMASH is a commonly used tool to identify these clusters (Blin et al. 2017). It first identifies core genes (PKS, NRPS, TC, and DMATS) and then looks for putative accessory genes involved in the production of the secondary metabolite. Moreover, the identified putative clusters can be compared to known clusters in other organisms. This homology and the gene families in the cluster are used to predict the type of secondary metabolite that may be produced, although this is currently still rather inaccurate.

The wide diversity among members of the fungal kingdom is also reflected in the wide range of natural products they produce, making fungi an interesting source for novel drugs. Genome sequencing has resulted in a large catalog of biosynthetic gene clusters (Keller 2019). Unfortunately, most natural products are not produced under lab conditions, complicating their identification in high-throughput screens (Keller et al. 2005). Several companies (e.g., Hexagon Bio, USA) are currently using high-throughput genome sequencing to identify novel natural products, purely based on their gene content. Interesting candidate gene clusters are then heterologously expressed in production species using a synthetic biology approach, thus circumventing the problem of low production of the natural products in their natural host. *S. cerevisiae* and *Aspergillus nidulans* are examples of production species (Billingsley et al. 2016; Clevenger et al. 2017; Harvey et al. 2018). This approach illustrates the power of large-scale genome sequencing and analysis.

B. Carbohydrate-Active Enzymes

Fungi are heterotrophs: they feed on organic matter. A large source of organic matter is plant biomass, or, more specifically, polysaccharides in lignocellulose (including cellulose, hemicellulose, and pectin). Fungi have evolved a wide range of extracellular enzymes to break down these recalcitrant polysaccharides into smaller compounds (monosaccharides and oligosaccharides) that can be transported over the cell membrane. Collectively, these enzymes are known as carbohydrate-active enzymes (CAZymes) and are organized in a special database, the CAZy database. CAZy describes the families of structurally related catalytic and carbohydrate-binding modules (or functional domains) of enzymes that degrade, modify, or create glycosidic bonds (Lombard et al. 2014). More generally, CAZymes are enzymes involved in the breakdown, biosynthesis, and modification of carbohydrates.

Based on their domain structure CAZymes are classified into glycoside hydrolases (GH), glycosyl transferases (GT), polysaccharide lyases (PL), carbohydrate esterases (CE), and enzymes with auxiliary activities (AA). Each of these categories is subdivided into numerous families with predicted enzyme activities (Lombard et al. 2014). Although their identification is based on sequence homology (and therefore relatively straightforward), it is important to note that even within families there can be a range of predicted enzyme activities. It may therefore be necessary to confirm the enzyme activity of the predicted CAZyme with lab experiments.

From a biotechnology perspective, CAZymes are interesting due to their ability to break down (unfermentable) polysaccharides into oligosaccharides and monosaccharides that can be fermented into ethanol by *S. cerevisiae*. As such, CAZymes play an important role in converting plant biomass into biofuel. Moreover, fungal pathogens of plants use CAZymes as an important weapon in their arsenal to attack their host. In the case of pathogens of important agricultural crops, the CAZyme content of fungal genome can lead to important insights (discussed below).

Initial genome sequencing efforts focused on established model systems used to study CAZymes. Examples include *Aspergillus niger* (Pel et al. 2007) and *Neurospora crassa* (Galagan et al. 2003). This resulted in a wide range of well-characterized enzymes (Coutinho et al. 2009). Furthermore, several key regulators involved in the regulation of CAZyme gene expression were identified (Benocci et al. 2017). Later, large-scale sequencing efforts focused on fungi that break down plant polysaccharides. An important sequencing effort is the 1000 Fungal Genomes Project by the Joint Genome Institute (Grigoriev et al. 2014), resulting in a large number of genomes from across the fungal kingdom, including many plant biomass degrading fungi. More targeted sequencing efforts have focused on groups of fungi, such as the genus *Aspergillus* (Vesth et al. 2018) or the class *Agaricomycetes* (Floudas et al. 2012; Ohm et al. 2014), which includes potent degraders of lignocellulose.

The combined genome sequencing efforts have resulted in a large catalog of putative CAZymes, maintained in the CAZy database (Lombard et al. 2014). Similar to how it was described above for secondary metabolism, this catalog can be screened using a high-throughput synthetic biology approach. Putatively interesting CAZymes can be expressed in a production host, and the enzyme activity can be assayed. This approach precludes the need to grow the original host fungus.

C. Mushroom Development

Mushrooms are the sexual reproductive structures of fungi (predominantly) of the phylum *Basidiomycota* or, more specifically, the class *Agaricomycetes* (Kües and Liu 2000; Kües and Navarro-González 2015). Mushrooms are a nutritious and sustainable food source for a growing world population. They can be cultivated on low-quality agricultural waste streams (e.g., manure, saw dust or straw), which they convert into high quality food. As such, they contribute to a circular economy (Grimm and Wösten 2018) and are interesting from a biotechnology perspective. Examples of edible mushrooms include the white button mushroom (*Agaricus bisporus*), the oyster mushroom (*Pleurotus ostreatus*), and shiitake mushroom (*Lentinula edodes*).

Few mushroom-forming fungi are genetically accessible, but notable exceptions are *Schizophyllum commune* and *Coprinopsis cinerea*, both of which have been used as model systems for decades (Kües and Navarro-González 2015). This has resulted in the identification of structural proteins involved in mushroom development, such as hydrophobins (Wösten 2001), as well as multiple developmental regulators (Terashima et al. 2005; Ohm et al. 2011, 2013).

The number of available genomes of mushroom-forming fungi has dramatically increased in recent years, although it should be noted that most mushroom-forming fungi were sequenced due to their capacity to degrade lignocellulose (Ohm et al. 2014). Comparative genomics studies have given important new insights into the phylogeny of mushroom-forming fungi (Varga et al. 2019), showing that morphological diversification occurred especially in the Cretaceous and Paleocene. Moreover, numerous novel gene families have been identified that may be involved in mushroom development (Sipos et al. 2017; Krizsán et al. 2019; Almási et al. 2019), based on their conservation in mushroom-forming species as well as their gene expression profile during mushroom development. These genes are currently studied in more detail, which is facilitated by the recent development of CRISPR/Cas9 genome editing tools (Sugano et al. 2017; Vonk et al. 2019).

D. Plant Interactions

Many fungi interact with plants in one way or another. This can be beneficial for the host plant, for example, in the case of **mycorrhizal fungi** that form a symbiosis with plant roots. In contrast, **fungal pathogens** can be detrimental to plant health. Both these fungal lifestyles are important from a biotechnology perspective, since they can strongly impact the yield of agricultural crops.

Although plant pathogens are found across the fungal kingdom, many destructive pathogens belong to the phylum *Ascomycota*. Examples include various species of *Fusarium* and *Verticillium*, which were early targets of genome sequencing (Cuomo et al. 2007; Ma et al. 2010; Klosterman et al. 2011). Comparative genome analysis allowed the reconstruction of gene evolution of pathogenesis-related genes, which are generally called effector genes. More recently all *Verticillium* species were sequenced, and the subsequent analysis revealed frequent chromosomal rearrangements as well as gene family losses. Moreover, in these species only about 200–600 species-specific genes occurred, which are markedly different from the conserved genes and are likely candidates for host specificity (Shi-Kunne et al. 2018). The class *Dothideomycetes* harbors many pathogens, including the wheat pathogen *Zymoseptoria tritici* (formerly known as *Mycosphaerella graminicola*), tomato pathogen *Passalora fulva* (formerly known as *Clados-*

porium fulvum), pine pathogen *Dothistroma septosporum*, and maize pathogen *Bipolaris maydis* (formerly known as *Cochliobolus heterostrophus*) (Goodwin et al. 2011; de Wit et al. 2012; Condon et al. 2013). *P. fulva* and *D. septosporum* are closely related but have very different host plants (tomato and pine, respectively) and lifestyles (hemibiotroph and necrotroph, respectively). Genome sequencing revealed the evolution of a gene cluster involved in the production of dothistromin toxin by *D. septosporum*, as well as effector genes specific to *P. fulva*. Comparing the two genomes suggests that these pathogens had a common ancestral host but have since diverged into different hosts and lifestyles by a differentiation in gene content, pseudogenization, as well as gene regulation (de Wit et al. 2012). More generally, a comparative analysis of members of the class *Dothideomycetes* showed that genome evolution follows a pattern of frequent short intra-chromosomal inversions and few inter-chromosomal rearrangements (Hane et al. 2011; Ohm et al. 2012).

In contrast to plant pathogens, mycorrhizal fungi form a symbiosis that is beneficial to the plant host. Generally, during this symbiosis the fungus provides micronutrients to the plant, while the plant provides carbohydrates (sugars produced by photosynthesis) to the fungus. This mycorrhizal lifestyle evolved independently several times across the fungal kingdom, in species as diverse as the mushroom-forming *Basidiomycete Laccaria bicolor*, the *Dothideomycete Cenococcum geophilum*, and the Périgord black truffle *Tuber melanosporum* (Martin et al. 2008, 2010; Peter et al. 2016). Although there are many differences between these mycorrhizal fungi, a general pattern is that (compared to their non-mycorrhizal relatives) the number of plant cell wall degrading CAZymes decreased, while the number of lineage-specific genes increased (especially genes that were differentially expressed during symbiosis). Nevertheless, mycorrhizal fungi have retained a unique set of CAZymes, which suggests that they are still capable of degrading lignocellulose and therefore are not fully reliant on their plant host (Kohler et al. 2015; Martino et al. 2018).

The genus *Trichoderma* contains several mycoparasitic species that promote plant growth. To some extent this can be explained by the fact that they parasitise on deleterious plant pathogens. However, several strains also induce root branching and increase shoot biomass (Kubicek et al. 2011; Druzhinina et al. 2011; Contreras-Cornejo et al. 2016).

V. Conclusions

This chapter described recent improvements in sequencing technologies that are used to sequence fungal genomes. As these sequencing technologies mature further, it will soon be trivial and affordable to obtain a high-quality telomere-to-telomere assembly. Accurate gene prediction and data analysis is still a challenge, although algorithms and pipelines continue to improve. Currently fungal genome sequencing is already affordable to small labs and individual researchers. For those who are interested in starting with fungal genome sequencing, the following pipeline has proven to work very well in my lab (as an example): we routinely sequence fungal genomes using Illumina (occasionally supplemented with long read from Oxford Nanopore) and genome assembly is done with SPAdes (Bankevich et al. 2012). Gene prediction is preferably done with BRAKER in combination with RNA-Seq expression data (Hoff et al. 2016). Basic functional annotation is done with InterProScan (Hunter et al. 2009) and supplemented with other algorithms, depending on the scientific questions.

Important next steps include a functional genomics approach, which relies heavily on an accurate genome sequence. In functional genomics, high-throughput (sequencing-based) techniques are used in an effort to assign function to elements of the genome (usually genes). These techniques may include RNA-Seq (to study gene expression), ChIP-Seq (to study various aspects of epigenetics), as well as high-throughput gene inactivations. Gene inactivations and other genome editing approaches have been greatly facilitated by the development of CRISPR/Cas9 across the fungal kingdom (Shi et al. 2017). I expect that in the coming years, large improvements will be made in techniques related to functional genomics, further accelerating discoveries across the fungal kingdom.

Fungal genome sequencing, comparative genomics and functional genomics are Big Data sciences and require a specific skill set: most bioinformatics tools run in a Linux environment and programming skills (e.g. in Python and R) are essential for advanced analyses. Most universities now include these aspects in their curriculum, ensuring that the next generation of researchers will be skilled in both experimental lab work and computational biology.

References

Almási É, Sahu N, Krizsán K, Bálint B, Kovács GM, Kiss B, Cseklye J, Drula E, Henrissat B, Nagy I, Chovatia M, Adam C, LaButti K, Lipzen A, Riley R, Grigoriev IV, Nagy LG (2019) Comparative genomics reveals unique wood-decay strategies and fruiting body development in the Schizophyllaceae. New Phytol. https://doi.org/10.1111/nph.16032

Altschul SF, Gish W, Miller W, Myers EW, Lipman DJ (1990) Basic local alignment search tool. J Mol Biol 215:403–410. https://doi.org/10.1016/S0022-2836(05)80360-2

Aparicio S, Chapman J, Stupka E, Putnam N, Chia J-M, Dehal P, Christoffels A, Rash S, Hoon S, Smit A, Gelpke MDS, Roach J, Oh T, Ho IY, Wong M, Detter C, Verhoef F, Predki P, Tay A, Lucas S, Richardson P, Smith SF, Clark MS, Edwards YJK, Doggett N, Zharkikh A, Tavtigian SV, Pruss D, Barnstead M, Evans C, Baden H, Powell J, Glusman G, Rowen L, Hood L, Tan YH, Elgar G, Hawkins T, Venkatesh B, Rokhsar D, Brenner S (2002) Whole-genome shotgun assembly and analysis of the genome of *Fugu rubripes*. Science 297:1301–1310. https://doi.org/10.1126/science.1072104

Arabidopsis Genome Initiative (2000) Analysis of the genome sequence of the flowering plant *Arabidopsis thaliana*. Nature 408:796–815. https://doi.org/10.1038/35048692

Ashburner M, Ball CA, Blake JA, Botstein D, Butler H, Cherry JM, Davis AP, Dolinski K, Dwight SS, Eppig JT, Harris MA, Hill DP, Issel-Tarver L, Kasarskis A, Lewis S, Matese JC, Richardson JE, Ringwald M, Rubin GM, Sherlock G (2000) Gene ontology: tool for the unification of biology. Nat Genet 25:25–29. https://doi.org/10.1038/75556

Bankevich A, Nurk S, Antipov D, Gurevich AA, Dvorkin M, Kulikov AS, Lesin VM, Nikolenko SI, Pham S, Prjibelski AD, Pyshkin AV, Sirotkin AV, Vyahhi N, Tesler G, Alekseyev MA, Pevzner PA (2012) SPAdes: a new genome assembly algorithm and its applications to single-cell sequencing. J Comput Biol 19:455–477. https://doi.org/10.1089/cmb.2012.0021

Basenko EY, Pulman JA, Shanmugasundram A, Harb OS, Crouch K, Starns D, Warrenfeltz S, Aurrecoechea C, Stoeckert CJ, Kissinger JC, Roos DS, Hertz-Fowler C (2018) FungiDB: an integrated bioinformatic resource for fungi and oomycetes. J Fungi 4:39. https://doi.org/10.3390/jof4010039

Bateman A, Martin MJ, O'Donovan C, Magrane M, Alpi E, Antunes R, Bely B, Bingley M, Bonilla C, Britto R, Bursteinas B, Bye-A-Jee H, Cowley A, Da Silva A, De Giorgi M, Dogan T, Fazzini F, Castro LG, Figueira L, Garmiri P, Georghiou G, Gonzalez D, Hatton-Ellis E, Li W, Liu W, Lopez R, Luo J, Lussi Y, MacDougall A, Nightingale A, Palka B, Pichler K, Poggioli D, Pundir S, Pureza L, Qi G, Renaux A, Rosanoff S, Saidi R, Sawford T, Shypitsyna A, Speretta E, Turner E, Tyagi N, Volynkin V, Wardell T, Warner K, Watkins X, Zaru R, Zellner H, Xenarios I, Bougueleret L, Bridge A, Poux S, Redaschi N, Aimo L, Argoud-Puy G, Auchincloss A, Axelsen K, Bansal P, Baratin D, Blatter M-C, Boeckmann B, Bolleman J, Boutet E, Breuza L, Casal-Casas C, de Castro E, Coudert E, Cuche B, Doche M, Dornevil D, Duvaud S, Estreicher A, Famiglietti L, Feuermann M, Gasteiger E, Gehant S, Gerritsen V, Gos A, Gruaz-Gumowski N, Hinz U, Hulo C, Jungo F, Keller G, Lara V, Lemercier P, Lieberherr D, Lombardot T, Martin X, Masson P, Morgat A, Neto T, Nouspikel N, Paesano S, Pedruzzi I, Pilbout S, Pozzato M, Pruess M, Rivoire C, Roechert B, Schneider M, Sigrist C, Sonesson K, Staehli S, Stutz A, Sundaram S, Tognolli M, Verbregue L, Veuthey A-L, Wu CH, Arighi CN, Arminski L, Chen C, Chen Y, Garavelli JS, Huang H, Laiho K, McGarvey P, Natale DA, Ross K, Vinayaka CR, Wang Q, Wang Y, Yeh L-S, Zhang J (2017) UniProt: the universal protein knowledgebase. Nucleic Acids Res 45:D158–D169. https://doi.org/10.1093/nar/gkw1099

Batzoglou S, Jaffe DB, Stanley K, Butler J, Gnerre S, Mauceli E, Berger B, Mesirov JP, Lander ES (2002) ARACHNE: a whole-genome shotgun assembler. Genome Res 12:177–189. https://doi.org/10.1101/gr.208902

Bennett S (2004) Solexa Ltd. Pharmacogenomics 5:433–438. https://doi.org/10.1517/14622416.5.4.433

Bennett JW, Chung KT (2001) Alexander Fleming and the discovery of penicillin. Adv Appl Microbiol 49:163–184

Benocci T, Aguilar-Pontes MV, Zhou M, Seiboth B, de Vries RP (2017) Regulators of plant biomass degradation in ascomycetous fungi. Biotechnol Biofuels 10:152. https://doi.org/10.1186/s13068-017-0841-x

Billingsley JM, DeNicola AB, Tang Y (2016) Technology development for natural product biosynthesis in *Saccharomyces cerevisiae*. Curr Opin Biotechnol 42:74–83. https://doi.org/10.1016/j.copbio.2016.02.033

Birney E, Clamp M, Durbin R (2004) GeneWise and genomewise. Genome Res 14:988–995. https://doi.org/10.1101/gr.1865504

Blin K, Wolf T, Chevrette MG, Lu X, Schwalen CJ, Kautsar SA, Suarez Duran HG, de Los Santos ELC, Kim HU, Nave M, Dickschat JS, Mitchell DA, Shelest E, Breitling R, Takano E, Lee SY, Weber T, Medema MH (2017) AntiSMASH 4.0—improvements in chemistry prediction and gene cluster boundary identification. Nucleic Acids Res 45:W36–W41. https://doi.org/10.1093/nar/gkx319

Buels R, Yao E, Diesh CM, Hayes RD, Munoz-Torres M, Helt G, Goodstein DM, Elsik CG, Lewis SE, Stein L, Holmes IH (2016) JBrowse: a dynamic web platform for genome visualization and analysis. Genome Biol 17:66. https://doi.org/10.1186/s13059-016-0924-1

Cantarel BL, Korf I, Robb SMC, Parra G, Ross E, Moore B, Holt C, Sánchez Alvarado A, Yandell M (2008) MAKER: an easy-to-use annotation pipeline designed for emerging model organism genomes. Genome Res 18:188–196. https://doi.org/10.1101/gr.6743907

Castanera R, López-Varas L, Borgognone A, LaButti K, Lapidus A, Schmutz J, Grimwood J, Pérez G, Pisabarro AG, Grigoriev IV, Stajich JE, Ramírez L (2016) Transposable elements versus the fungal genome: impact on whole-genome architecture and transcriptional profiles. PLoS Genet 12: e1006108. https://doi.org/10.1371/journal.pgen.1006108

Cerqueira GC, Arnaud MB, Inglis DO, Skrzypek MS, Binkley G, Simison M, Miyasato SR, Binkley J, Orvis J, Shah P, Wymore F, Sherlock G, Wortman JR (2014) The *Aspergillus* Genome Database: multispecies curation and incorporation of RNA-Seq data to improve structural gene annotations. Nucleic Acids Res 42:D705–D710. https://doi.org/10.1093/nar/gkt1029

Cherry JM, Hong EL, Amundsen C, Balakrishnan R, Binkley G, Chan ET, Christie KR, Costanzo MC, Dwight SS, Engel SR, Fisk DG, Hirschman JE, Hitz BC, Karra K, Krieger CJ, Miyasato SR, Nash RS, Park J, Skrzypek MS, Simison M, Weng S, Wong ED (2012) *Saccharomyces* Genome Database: the genomics resource of budding yeast. Nucleic Acids Res 40:D700–D705. https://doi.org/10.1093/nar/gkr1029

Chin C-S, Alexander DH, Marks P, Klammer AA, Drake J, Heiner C, Clum A, Copeland A, Huddleston J, Eichler EE, Turner SW, Korlach J (2013) Nonhybrid, finished microbial genome assemblies from long-read SMRT sequencing data. Nat Methods 10:563–569. https://doi.org/10.1038/nmeth.2474

Cissé OH, Stajich JE (2019) FGMP: assessing fungal genome completeness. BMC Bioinformatics 20:184. https://doi.org/10.1186/s12859-019-2782-9

Clark K, Karsch-Mizrachi I, Lipman DJ, Ostell J, Sayers EW (2016) GenBank. Nucleic Acids Res 44:D67–D72. https://doi.org/10.1093/nar/gkv1276

Clevenger KD, Bok JW, Ye R, Miley GP, Verdan MH, Velk T, Chen C, Yang K, Robey MT, Gao P, Lamprecht M, Thomas PM, Islam MN, Palmer JM, Wu CC, Keller NP, Kelleher NL (2017) A scalable platform to identify fungal secondary metabolites and their gene clusters. Nat Chem Biol 13:895–901. https://doi.org/10.1038/nchembio.2408

Clutterbuck AJ (2011) Genomic evidence of repeat-induced point mutation (RIP) in filamentous ascomycetes. Fungal Genet Biol 48:306–326. https://doi.org/10.1016/J.FGB.2010.09.002

Condon BJ, Leng Y, Wu D, Bushley KE, Ohm RA, Otillar R, Martin J, Schackwitz W, Grimwood J, MohdZainudin N, Xue C, Wang R, Manning VA, Dhillon B, Tu ZJ, Steffenson BJ, Salamov A, Sun H, Lowry S, LaButti K, Han J, Copeland A, Lindquist E, Barry K, Schmutz J, Baker SE, Ciuffetti LM, Grigoriev IV, Zhong S, Turgeon BG (2013) Comparative genome structure, secondary metabolite, and effector coding capacity across *Cochliobolus* pathogens. PLoS Genet 9:e1003233. https://doi.org/10.1371/journal.pgen.1003233

Contreras-Cornejo HA, Macías-Rodríguez L, del-Val E, Larsen J (2016) Ecological functions of *Trichoderma* spp. and their secondary metabolites in the rhizosphere: interactions with plants. FEMS Microbiol Ecol 92:fiw036. https://doi.org/10.1093/femsec/fiw036

Coutinho PM, Andersen MR, Kolenova K, vanKuyk PA, Benoit I, Gruben BS, Trejo-Aguilar B, Visser H, van Solingen P, Pakula T, Seiboth B, Battaglia E, Aguilar-Osorio G, de Jong JF, Ohm RA, Aguilar M, Henrissat B, Nielsen J, Stålbrand H, de Vries RP (2009) Post-genomic insights into the plant polysaccharide degradation potential of *Aspergillus nidulans* and comparison to *Aspergillus niger* and *Aspergillus oryzae*. Fungal Genet Biol 46(Suppl 1): S161–S169

Cuomo CA, Birren BW (2010) The fungal genome initiative and lessons learned from genome sequencing. Methods Enzymol 470:833–855. https://doi.org/10.1016/S0076-6879(10)70034-3

Cuomo CA, Guldener U, Xu J-R, Trail F, Turgeon BG, Di Pietro A, Walton JD, Ma L-J, Baker SE, Rep M, Adam G, Antoniw J, Baldwin T, Calvo S, Chang Y-L, DeCaprio D, Gale LR, Gnerre S, Goswami RS, Hammond-Kosack K, Harris LJ, Hilburn K, Kennell JC, Kroken S, Magnuson JK, Mannhaupt G, Mauceli E, Mewes H-W, Mitterbauer R, Muehlbauer G, Munsterkotter M, Nelson D, O'Donnell K, Ouellet T, Qi W, Quesneville H, Roncero MIG,

Seong K-Y, Tetko IV, Urban M, Waalwijk C, Ward TJ, Yao J, Birren BW, Kistler HC (2007) The *Fusarium graminearum* genome reveals a link between localized polymorphism and pathogen specialization. Science 317:1400–1402. https://doi.org/10.1126/science.1143708

de Wit PJGM, van der Burgt A, Ökmen B, Stergiopoulos I, Abd-Elsalam KA, Aerts AL, Bahkali AH, Beenen HG, Chettri P, Cox MP, Datema E, de Vries RP, Dhillon B, Ganley AR, Griffiths SA, Guo Y, Hamelin RC, Henrissat B, Kabir MS, Jashni MK, Kema G, Klaubauf S, Lapidus A, Levasseur A, Lindquist E, Mehrabi R, Ohm RA, Owen TJ, Salamov A, Schwelm A, Schijlen E, Sun H, van den Burg HA, van Ham RCHJ, Zhang S, Goodwin SB, Grigoriev IV, Collemare J, Bradshaw RE (2012) The genomes of the fungal plant pathogens *Cladosporium fulvum* and *Dothistroma septosporum* reveal adaptation to different hosts and lifestyles but also signatures of common ancestry. PLoS Genet 8: e1003088. https://doi.org/10.1371/journal.pgen.1003088

Downs JR, Clearfield M, Weis S, Whitney E, Shapiro DR, Beere PA, Langendorfer A, Stein EA, Kruyer W, Gotto AM Jr, for the AFCAPS/TexCAPS Research Group for the AR (1998) Primary prevention of acute coronary events with lovastatin in men and women with average cholesterol levels. JAMA 279:1615. https://doi.org/10.1001/jama.279.20.1615

Druzhinina IS, Seidl-Seiboth V, Herrera-Estrella A, Horwitz BA, Kenerley CM, Monte E, Mukherjee PK, Zeilinger S, Grigoriev IV, Kubicek CP (2011) Trichoderma: the genomics of opportunistic success. Nat Rev Microbiol 9:749–759. https://doi.org/10.1038/nrmicro2637

Eid J, Fehr A, Gray J, Luong K, Lyle J, Otto G, Peluso P, Rank D, Baybayan P, Bettman B, Bibillo A, Bjornson K, Chaudhuri B, Christians F, Cicero R, Clark S, Dalal R, Dewinter A, Dixon J, Foquet M, Gaertner A, Hardenbol P, Heiner C, Hester K, Holden D, Kearns G, Kong X, Kuse R, Lacroix Y, Lin S, Lundquist P, Ma C, Marks P, Maxham M, Murphy D, Park I, Pham T, Phillips M, Roy J, Sebra R, Shen G, Sorenson J, Tomaney A, Travers K, Trulson M, Vieceli J, Wegener J, Wu D, Yang A, Zaccarin D, Zhao P, Zhong F, Korlach J, Turner S (2009) Real-time DNA sequencing from single polymerase molecules. Science 323:133–138. https://doi.org/10.1126/science.1162986

Finn RD, Coggill P, Eberhardt RY, Eddy SR, Mistry J, Mitchell AL, Potter SC, Punta M, Qureshi M, Sangrador-Vegas A, Salazar GA, Tate J, Bateman A (2016) The Pfam protein families database: towards a more sustainable future. Nucleic Acids Res 44:D279–D285. https://doi.org/10.1093/nar/gkv1344

Fleischmann R, Adams M, White O, Clayton R, Kirkness E, Kerlavage A, Bult C, Tomb J, Dougherty B, Merrick J, Mewes HW, Murakami Y, Philippsen P, Tettelin H, Oliver SG et al (1995) Whole-genome random sequencing and assembly of *Haemophilus influenzae* Rd. Science 269:496–512. https://doi.org/10.1126/science.7542800

Floudas D, Binder M, Riley R, Barry K, Blanchette RA, Henrissat B, Martínez AT, Otillar R, Spatafora JW, Yadav JS, Aerts A, Benoit I, Boyd A, Carlson A, Copeland A, Coutinho PM, de Vries RP, Ferreira P, Findley K, Foster B, Gaskell J, Glotzer D, Górecki P, Heitman J, Hesse C, Hori C, Igarashi K, Jurgens JA, Kallen N, Kersten P, Kohler A, Kües U, Kumar TKA, Kuo A, LaButti K, Larrondo LF, Lindquist E, Ling A, Lombard V, Lucas S, Lundell T, Martin R, McLaughlin DJ, Morgenstern I, Morin E, Murat C, Nagy LG, Nolan M, Ohm RA, Patyshakuliyeva A, Rokas A, Ruiz-Dueñas FJ, Sabat G, Salamov A, Samejima M, Schmutz J, Slot JC, St John F, Stenlid J, Sun H, Sun S, Syed K, Tsang A, Wiebenga A, Young D, Pisabarro A, Eastwood DC, Martin F, Cullen D, Grigoriev IV, Hibbett DS (2012) The Paleozoic origin of enzymatic lignin decomposition reconstructed from 31 fungal genomes. Science 336:1715–1719. https://doi.org/10.1126/science.1221748

Galagan JE, Calvo SE, Borkovich KA, Selker EU, Read ND, Jaffe D, FitzHugh W, Ma L-J, Smirnov S, Purcell S, Rehman B, Elkins T, Engels R, Wang S, Nielsen CB, Butler J, Endrizzi M, Qui D, Ianakiev P, Bell-Pedersen D, Nelson MA, Werner-Washburne M, Selitrennikoff CP, Kinsey JA, Braun EL, Zelter A, Schulte U, Kothe GO, Jedd G, Mewes W, Staben C, Marcotte E, Greenberg D, Roy A, Foley K, Naylor J, Stange-Thomann N, Barrett R, Gnerre S, Kamal M, Kamvysselis M, Mauceli E, Bielke C, Rudd S, Frishman D, Krystofova S, Rasmussen C, Metzenberg RL, Perkins DD, Kroken S, Cogoni C, Macino G, Catcheside D, Li W, Pratt RJ, Osmani SA, DeSouza CPC, Glass L, Orbach MJ, Berglund JA, Voelker R, Yarden O, Plamann M, Seiler S, Dunlap J, Radford A, Aramayo R, Natvig DO, Alex LA, Mannhaupt G, Ebbole DJ, Freitag M, Paulsen I, Sachs MS, Lander ES, Nusbaum C, Birren B (2003) The genome sequence of the filamentous fungus *Neurospora crassa*. Nature 422:859–868. https://doi.org/10.1038/nature01554

Galagan JE, Calvo SE, Cuomo C, Ma L-J, Wortman JR, Batzoglou S, Lee S-I, Baştürkmen M, Spevak CC, Clutterbuck J, Kapitonov V, Jurka J, Scazzocchio C, Farman M, Butler J, Purcell S, Harris S, Braus GH, Draht O, Busch S, D'Enfert C, Bouchier C, Goldman GH, Bell-Pedersen D, Griffiths-Jones S, Doonan JH, Yu J, Vienken K, Pain A, Freitag M, Selker EU, Archer DB, Peñalva MÁ, Oakley BR, Momany M, Tanaka T, Kumagai T, Asai K, Machida M, Nierman WC, Denning DW, Caddick M, Hynes M, Paoletti M, Fischer R, Miller B, Dyer P, Sachs MS, Osmani SA, Birren BW (2005) Sequencing of *Aspergillus nidulans* and comparative analysis with *A. fumigatus* and *A. oryzae*. Nature 438:1105–1115. https://doi.org/10.1038/nature04341

Ghurye J, Pop M (2019) Modern technologies and algorithms for scaffolding assembled genomes. PLoS Comput Biol 15:e1006994. https://doi.org/10.1371/journal.pcbi.1006994

Goffeau A, Barrell BG, Bussey H, Davis RW, Dujon B, Feldmann H, Galibert F, Hoheisel JD, Jacq C, Johnston M, Louis EJ, Mewes HW, Murakami Y, Philippsen P, Tettelin H, Oliver SG (1996) Life with 6000 genes. Science 274(546):563–567. https://doi.org/10.1126/SCIENCE.274.5287.546

Goodwin SB, Ben M'Barek S, Dhillon B, Wittenberg AHJ, Crane CF, Hane JK, Foster AJ, Van der Lee TAJ, Grimwood J, Aerts A, Antoniw J, Bailey A, Bluhm B, Bowler J, Bristow J, van der Burgt A, Canto-Canché B, Churchill ACL, Conde-Ferràez L, Cools HJ, Coutinho PM, Csukai M, Dehal P, De Wit P, Donzelli B, van de Geest HC, van Ham RCHJ, Hammond-Kosack KE, Henrissat B, Kilian A, Kobayashi AK, Koopmann E, Kourmpetis Y, Kuzniar A, Lindquist E, Lombard V, Maliepaard C, Martins N, Mehrabi R, Nap JPH, Ponomarenko A, Rudd JJ, Salamov A, Schmutz J, Schouten HJ, Shapiro H, Stergiopoulos I, Torriani SFF, Tu H, de Vries RP, Waalwijk C, Ware SB, Wiebenga A, Zwiers L-H, Oliver RP, Grigoriev IV, Kema GHJ (2011) Finished genome of the fungal wheat pathogen *Mycosphaerella graminicola* reveals dispensome structure, chromosome plasticity, and stealth pathogeness. PLoS Genet 7:e1002070. https://doi.org/10.1371/journal.pgen.1002070

Grigoriev IV, Cullen D, Goodwin SB, Hibbett D, Jeffries TW, Kubicek CP, Kuske C, Magnuson JK, Martin F, Spatafora JW, Tsang A, Baker SE (2011) Fueling the future with fungal genomics. Mycology 2:192–209. https://doi.org/10.1080/21501203.2011.584577

Grigoriev IV, Nikitin R, Haridas S, Kuo A, Ohm R, Otillar R, Riley R, Salamov A, Zhao X, Korzeniewski F, Smirnova T, Nordberg H, Dubchak I, Shabalov I (2014) MycoCosm portal: gearing up for 1000 fungal genomes. Nucleic Acids Res 42: D699–D704. https://doi.org/10.1093/nar/gkt1183

Grimm D, Wösten HAB (2018) Mushroom cultivation in the circular economy. Appl Microbiol Biotechnol 102:7795–7803. https://doi.org/10.1007/s00253-018-9226-8

Haas BJ, Delcher AL, Mount SM, Wortman JR, Smith RK Jr, Hannick LI, Maiti R, Ronning CM, Rusch DB, Town CD, Salzberg SL, White O (2003) Improving the *Arabidopsis* genome annotation using maximal transcript alignment assemblies. Nucleic Acids Res 31:5654–5666. https://doi.org/10.1093/nar/gkg770

Hane JK, Rouxel T, Howlett BJ, Kema GHJ, Goodwin SB, Oliver RP (2011) A novel mode of chromosomal evolution peculiar to filamentous Ascomycete fungi. Genome Biol 12:R45. https://doi.org/10.1186/gb-2011-12-5-r45

Haridas S, Salamov A, Grigoriev IV (2018) Fungal genome annotation. Methods Mol Biol 1775:171–184

Harvey CJB, Tang M, Schlecht U, Horecka J, Fischer CR, Lin H-C, Li J, Naughton B, Cherry J, Miranda M, Li YF, Chu AM, Hennessy JR, Vandova GA, Inglis D, Aiyar RS, Steinmetz LM, Davis RW, Medema MH, Sattely E, Khosla C, St Onge RP, Tang Y, Hillenmeyer ME (2018) HEx: a heterologous expression platform for the discovery of fungal natural products. Sci Adv 4:eaar5459. https://doi.org/10.1126/sciadv.aar5459

Hoff KJ, Lange S, Lomsadze A, Borodovsky M, Stanke M (2016) BRAKER1: unsupervised RNA-Seq-based genome annotation with GeneMark-ET and AUGUSTUS. Bioinformatics 32:767–769. https://doi.org/10.1093/bioinformatics/btv661

Horton P, Park K-J, Obayashi T, Fujita N, Harada H, Adams-Collier CJ, Nakai K (2007) WoLF PSORT: protein localization predictor. Nucleic Acids Res 35:W585–W587. https://doi.org/10.1093/nar/gkm259

Hunter S, Apweiler R, Attwood TK, Bairoch A, Bateman A, Binns D, Bork P, Das U, Daugherty L, Duquenne L, Finn RD, Gough J, Haft D, Hulo N, Kahn D, Kelly E, Laugraud A, Letunic I, Lonsdale D, Lopez R, Madera M, Maslen J, McAnulla C, McDowall J, Mistry J, Mitchell A, Mulder N, Natale D, Orengo C, Quinn AF, Selengut JD, Sigrist CJA, Thimma M, Thomas PD, Valentin F, Wilson D, Wu CH, Yeats C (2009) InterPro: the integrative protein signature database. Nucleic Acids Res 37:D211–D215. https://doi.org/10.1093/nar/gkn785

International Human Genome Sequencing Consortium (2001) Initial sequencing and analysis of the human genome. Nature 409:860–921. https://doi.org/10.1038/35057062

Jain M, Olsen HE, Paten B, Akeson M (2016) The Oxford Nanopore MinION: delivery of nanopore sequencing to the genomics community. Genome Biol 17:239. https://doi.org/10.1186/s13059-016-1103-0

Jurka J, Kapitonov VV, Pavlicek A, Klonowski P, Kohany O, Walichiewicz J (2005) Repbase update, a database of eukaryotic repetitive elements. Cytogenet Genome Res 110:462–467. https://doi.org/10.1159/000084979

Kanehisa M, Goto S (2000) KEGG: Kyoto encyclopedia of genes and genomes. Nucleic Acids Res 28:27–30. https://doi.org/10.1093/nar/28.1.27

Kapitonov VV, Jurka J (2001) Rolling-circle transposons in eukaryotes. Proc Natl Acad Sci USA 98:8714–8719. https://doi.org/10.1073/pnas.151269298

Kapitonov VV, Jurka J (2008) A universal classification of eukaryotic transposable elements implemented in Repbase. Nat Rev Genet 9:411–412. https://doi.org/10.1038/nrg2165-c1

Keller NP (2019) Fungal secondary metabolism: regulation, function and drug discovery. Nat Rev Microbiol 17:167–180. https://doi.org/10.1038/s41579-018-0121-1

Keller NP, Turner G, Bennett JW (2005) Fungal secondary metabolism—from biochemistry to genomics. Nat Rev Microbiol 3:937–947. https://doi.org/10.1038/nrmicro1286

Kent WJ, Sugnet CW, Furey TS, Roskin KM, Pringle TH, Zahler AM, Haussler D (2002) The human genome browser at UCSC. Genome Res 12:996–1006. https://doi.org/10.1101/gr.229102

Khaldi N, Seifuddin FT, Turner G, Haft D, Nierman WC, Wolfe KH, Fedorova ND (2010) SMURF: genomic mapping of fungal secondary metabolite clusters. Fungal Genet Biol 47:736–741. https://doi.org/10.1016/j.fgb.2010.06.003

Klosterman SJ, Subbarao KV, Kang S, Veronese P, Gold SE, Thomma BPHJ, Chen Z, Henrissat B, Lee Y-H, Park J, Garcia-Pedrajas MD, Barbara DJ, Anchieta A, de Jonge R, Santhanam P, Maruthachalam K, Atallah Z, Amyotte SG, Paz Z, Inderbitzin P, Hayes RJ, Heiman DI, Young S, Zeng Q, Engels R, Galagan J, Cuomo CA, Dobinson KF, Ma L-J (2011) Comparative genomics yields insights into niche adaptation of plant vascular wilt pathogens. PLoS Pathog 7:e1002137. https://doi.org/10.1371/journal.ppat.1002137

Kohler A, Kuo A, Nagy LG, Morin E, Barry KW, Buscot F, Canbäck B, Choi C, Cichocki N, Clum A, Colpaert J, Copeland A, Costa MD, Doré J, Floudas D, Gay G, Girlanda M, Henrissat B, Herrmann S, Hess J, Högberg N, Johansson T, Khouja H-R, LaButti K, Lahrmann U, Levasseur A, Lindquist EA, Lipzen A, Marmeisse R, Martino E, Murat C, Ngan CY, Nehls U, Plett JM, Pringle A, Ohm RA, Perotto S, Peter M, Riley R, Rineau F, Ruytinx J, Salamov A, Shah F, Sun H, Tarkka M, Tritt A, Veneault-Fourrey C, Zuccaro A, Tunlid A, Grigoriev IV, Hibbett DS, Martin F, Martin F (2015) Convergent losses of decay mechanisms and rapid turnover of symbiosis genes in mycorrhizal mutualists. Nat Genet 47:410–415. https://doi.org/10.1038/ng.3223

Krizsán K, Almási É, Merényi Z, Sahu N, Virágh M, Kószó T, Mondo S, Kiss B, Bálint B, Kües U, Barry K, Cseklye J, Hegedüs B, Henrissat B, Johnson J, Lipzen A, Ohm RA, Nagy I, Pangilinan J, Yan J, Xiong Y, Grigoriev IV, Hibbett DS, Nagy LG (2019) Transcriptomic atlas of mushroom development reveals conserved genes behind complex multicellularity in fungi. Proc Natl Acad Sci USA 116:7409–7418. https://doi.org/10.1073/PNAS.1817822116

Krogh A, Larsson B, von Heijne G, Sonnhammer EL (2001) Predicting transmembrane protein topology with a hidden Markov model: application to complete genomes. J Mol Biol 305:567–580. https://doi.org/10.1006/JMBI.2000.4315

Kubicek CP, Herrera-Estrella A, Seidl-Seiboth V, Martinez DA, Druzhinina IS, Thon M, Zeilinger S, Casas-Flores S, Horwitz BA, Mukherjee PK, Mukherjee M, Kredics L, Alcaraz LD, Aerts A, Antal Z, Atanasova L, Cervantes-Badillo MG, Challacombe J, Chertkov O, McCluskey K, Coulpier F, Deshpande N, von Döhren H, Ebbole DJ, Esquivel-Naranjo EU, Fekete E, Flipphi M, Glaser F, Gómez-Rodríguez EY, Gruber S, Han C, Henrissat B, Hermosa R, Hernández-Oñate M, Karaffa L, Kosti I, Le Crom S, Lindquist E, Lucas S, Lübeck M, Lübeck PS, Margeot A, Metz B, Misra M, Nevalainen H, Omann M, Packer N, Perrone G, Uresti-Rivera EE, Salamov A, Schmoll M, Seiboth B, Shapiro H, Sukno S, Tamayo-Ramos JA, Tisch D, Wiest A, Wilkinson HH, Zhang M, Coutinho PM, Kenerley CM, Monte E, Baker SE, Grigoriev IV (2011) Comparative genome sequence analysis underscores mycoparasitism as the ancestral life style of *Trichoderma*. Genome Biol 12:R40. https://doi.org/10.1186/gb-2011-12-4-r40

Kües U, Liu Y (2000) Fruiting body production in basidiomycetes. Appl Microbiol Biotechnol 54:141–152. https://doi.org/10.1007/s002530000396

Kües U, Navarro-González M (2015) How do Agaricomycetes shape their fruiting bodies? 1. Morphological aspects of development. Fungal Biol Rev 29:63–97. https://doi.org/10.1016/J.FBR.2015.05.001

Lee E, Helt GA, Reese JT, Munoz-Torres MC, Childers CP, Buels RM, Stein L, Holmes IH, Elsik CG, Lewis SE (2013) Web Apollo: a web-based genomic annotation editing platform. Genome Biol 14:R93. https://doi.org/10.1186/gb-2013-14-8-r93

Loftus BJ, Fung E, Roncaglia P, Rowley D, Amedeo P, Bruno D, Vamathevan J, Miranda M, Anderson IJ, Fraser JA, Allen JE, Bosdet IE, Brent MR, Chiu R, Doering TL, Donlin MJ, D'Souza CA, Fox DS, Grinberg V, Fu J, Fukushima M, Haas BJ, Huang JC, Janbon G, Jones SJM, Koo HL, Krzywinski MI, Kwon-Chung JK, Lengeler KB, Maiti R, Marra MA, Marra RE, Mathewson CA, Mitchell TG, Pertea M, Riggs FR, Salzberg SL, Schein JE, Shvartsbeyn A, Shin H, Shumway M, Specht CA, Suh BB, Tenney A, Utterback TR, Wickes BL, Wortman JR, Wye NH, Kronstad JW, Lodge JK, Heitman J, Davis RW, Fraser CM, Hyman RW (2005) The genome of the basidiomycetous yeast and human pathogen *Cryptococcus neoformans*. Science 307:1321–1324. https://doi.org/10.1126/science.1103773

Lombard V, Golaconda Ramulu H, Drula E, Coutinho PM, Henrissat B (2014) The carbohydrate-active enzymes database (CAZy) in 2013. Nucleic Acids Res 42:D490–D495. https://doi.org/10.1093/nar/gkt1178

Lomsadze A, Burns PD, Borodovsky M (2014) Integration of mapped RNA-Seq reads into automatic training of eukaryotic gene finding algorithm.

Nucleic Acids Res 42:e119–e119. https://doi.org/10.1093/nar/gku557

Luo R, Liu B, Xie Y, Li Z, Huang W, Yuan J, He G, Chen Y, Pan Q, Liu Y, Tang J, Wu G, Zhang H, Shi Y, Liu Y, Yu C, Wang B, Lu Y, Han C, Cheung DW, Yiu S-M, Peng S, Xiaoqian Z, Liu G, Liao X, Li Y, Yang H, Wang J, Lam T-W, Wang J (2012) SOAPdenovo2: an empirically improved memory-efficient short-read de novo assembler. Gigascience 1:18. https://doi.org/10.1186/2047-217X-1-18

Ma L-J, van der Does HC, Borkovich KA, Coleman JJ, Daboussi M-J, Di Pietro A, Dufresne M, Freitag M, Grabherr M, Henrissat B, Houterman PM, Kang S, Shim W-B, Woloshuk C, Xie X, Xu J-R, Antoniw J, Baker SE, Bluhm BH, Breakspear A, Brown DW, Butchko RAE, Chapman S, Coulson R, Coutinho PM, Danchin EGJ, Diener A, Gale LR, Gardiner DM, Goff S, Hammond-Kosack KE, Hilburn K, Hua-Van A, Jonkers W, Kazan K, Kodira CD, Koehrsen M, Kumar L, Lee Y-H, Li L, Manners JM, Miranda-Saavedra D, Mukherjee M, Park G, Park J, Park S-Y, Proctor RH, Regev A, Ruiz-Roldan MC, Sain D, Sakthikumar S, Sykes S, Schwartz DC, Turgeon BG, Wapinski I, Yoder O, Young S, Zeng Q, Zhou S, Galagan J, Cuomo CA, Kistler HC, Rep M (2010) Comparative genomics reveals mobile pathogenicity chromosomes in *Fusarium*. Nature 464:367–373. https://doi.org/10.1038/nature08850

Mardis ER (2017) DNA sequencing technologies: 2006–2016. Nat Protoc 12:213–218. https://doi.org/10.1038/nprot.2016.182

Margulies M, Egholm M, Altman WE, Attiya S, Bader JS, Bemben LA, Berka J, Braverman MS, Chen Y-J, Chen Z, Dewell SB, Du L, Fierro JM, Gomes XV, Godwin BC, He W, Helgesen S, Ho CH, Irzyk GP, Jando SC, Alenquer MLI, Jarvie TP, Jirage KB, Kim J-B, Knight JR, Lanza JR, Leamon JH, Lefkowitz SM, Lei M, Li J, Lohman KL, Lu H, Makhijani VB, McDade KE, McKenna MP, Myers EW, Nickerson E, Nobile JR, Plant R, Puc BP, Ronan MT, Roth GT, Sarkis GJ, Simons JF, Simpson JW, Srinivasan M, Tartaro KR, Tomasz A, Vogt KA, Volkmer GA, Wang SH, Wang Y, Weiner MP, Yu P, Begley RF, Rothberg JM (2005) Genome sequencing in microfabricated high-density picolitre reactors. Nature 437:376–380. https://doi.org/10.1038/nature03959

Martin F, Aerts A, Ahrén D, Brun A, Danchin EGJ, Duchaussoy F, Gibon J, Kohler A, Lindquist E, Pereda V, Salamov A, Shapiro HJ, Wuyts J, Blaudez D, Buée M, Brokstein P, Canbäck B, Cohen D, Courty PE, Coutinho PM, Delaruelle C, Detter JC, Deveau A, DiFazio S, Duplessis S, Fraissinet-Tachet L, Lucic E, Frey-Klett P, Fourrey C, Feussner I, Gay G, Grimwood J, Hoegger PJ, Jain P, Kilaru S, Labbé J, Lin YC, Legué V, Le Tacon F, Marmeisse R, Melayah D, Montanini B, Muratet M, Nehls U, Niculita-Hirzel H, Oudot-Le Secq MP, Peter M, Quesneville H, Rajashekar B, Reich M, Rouhier N, Schmutz J, Yin T, Chalot M, Henrissat B, Kües U, Lucas S, Van de Peer Y, Podila GK, Polle A, Pukkila PJ, Richardson PM, Rouzé P, Sanders IR, Stajich JE, Tunlid A, Tuskan G, Grigoriev IV (2008) The genome of *Laccaria bicolor* provides insights into mycorrhizal symbiosis. Nature 452:88–92. https://doi.org/10.1038/nature06556

Martin F, Kohler A, Murat C, Balestrini R, Coutinho PM, Jaillon O, Montanini B, Morin E, Noel B, Percudani R, Porcel B, Rubini A, Amicucci A, Amselem J, Anthouard V, Arcioni S, Artiguenave F, Aury J-M, Ballario P, Bolchi A, Brenna A, Brun A, Buée M, Cantarel B, Chevalier G, Couloux A, Da Silva C, Denoeud F, Duplessis S, Ghignone S, Hilselberger B, Iotti M, Marçais B, Mello A, Miranda M, Pacioni G, Quesneville H, Riccioni C, Ruotolo R, Splivallo R, Stocchi V, Tisserant E, Viscomi AR, Zambonelli A, Zampieri E, Henrissat B, Lebrun M-H, Paolocci F, Bonfante P, Ottonello S, Wincker P (2010) Périgord black truffle genome uncovers evolutionary origins and mechanisms of symbiosis. Nature 464:1033–1038. https://doi.org/10.1038/nature08867

Martinez D, Berka RM, Henrissat B, Saloheimo M, Arvas M, Baker SE, Chapman J, Chertkov O, Coutinho PM, Cullen D, Danchin EGJ, Grigoriev IV, Harris P, Jackson M, Kubicek CP, Han CS, Ho I, Larrondo LF, de Leon AL, Magnuson JK, Merino S, Misra M, Nelson B, Putnam N, Robbertse B, Salamov AA, Schmoll M, Terry A, Thayer N, Westerholm-Parvinen A, Schoch CL, Yao J, Barabote R, Nelson MA, Detter C, Bruce D, Kuske CR, Xie G, Richardson P, Rokhsar DS, Lucas SM, Rubin EM, Dunn-Coleman N, Ward M, Brettin TS, Brettin TS (2008) Genome sequencing and analysis of the biomass-degrading fungus *Trichoderma reesei* (syn. Hypocrea jecorina). Nat Biotechnol 26:553–560. https://doi.org/10.1038/nbt1403

Martino E, Morin E, Grelet G-A, Kuo A, Kohler A, Daghino S, Barry KW, Cichocki N, Clum A, Dockter RB, Hainaut M, Kuo RC, LaButti K, Lindahl BD, Lindquist EA, Lipzen A, Khouja H-R, Magnuson J, Murat C, Ohm RA, Singer SW, Spatafora JW, Wang M, Veneault-Fourrey C, Henrissat B, Grigoriev IV, Martin FM, Perotto S (2018) Comparative genomics and transcriptomics depict ericoid mycorrhizal fungi as versatile saprotrophs and plant mutualists. New Phytol 217:1213–1229. https://doi.org/10.1111/nph.14974

Min B, Grigoriev IV, Choi I-G (2017) FunGAP: fungal genome annotation pipeline using evidence-based gene model evaluation. Bioinformatics 33:2936–2937. https://doi.org/10.1093/bioinformatics/btx353

Nützmann H-W, Scazzocchio C, Osbourn A (2018) Metabolic gene clusters in eukaryotes. Annu Rev

Genet 52:159–183. https://doi.org/10.1146/annurev-genet-120417-031237

Ohm RA, De Jong JF, Lugones LG, Aerts A, Kothe E, Stajich JE, De Vries RP, Record E, Levasseur A, Baker SE, Bartholomew KA, Coutinho PM, Erdmann S, Fowler TJ, Gathman AC, Lombard V, Henrissat B, Knabe N, Kües U, Lilly WW, Lindquist E, Lucas S, Magnuson JK, Piumi F, Raudaskoski M, Salamov A, Schmutz J, Schwarze FWMR, vanKuyk PA, Horton JS, Grigoriev IV, Wösten HAB (2010) Genome sequence of the model mushroom *Schizophyllum commune*. Nat Biotechnol 28:957–963

Ohm RA, de Jong JF, de Bekker C, Wösten HAB, Lugones LG (2011) Transcription factor genes of *Schizophyllum commune* involved in regulation of mushroom formation. Mol Microbiol 81:1433–1445. https://doi.org/10.1111/j.1365-2958.2011.07776.x

Ohm RA, Feau N, Henrissat B, Schoch CL, Horwitz BA, Barry KW, Condon BJ, Copeland AC, Dhillon B, Glaser F, Hesse CN, Kosti I, LaButti K, Lindquist EA, Lucas S, Salamov AA, Bradshaw RE, Ciuffetti L, Hamelin RC, Kema GHJ, Lawrence C, Scott JA, Spatafora JW, Turgeon BG, de Wit PJGM, Zhong S, Goodwin SB, Grigoriev IV (2012) Diverse lifestyles and strategies of plant pathogenesis encoded in the genomes of eighteen Dothideomycetes fungi. PLoS Pathog 8:e1003037. https://doi.org/10.1371/journal.ppat.1003037

Ohm RA, Aerts D, Wösten HAB, Lugones LG (2013) The blue light receptor complex WC-1/2 of *Schizophyllum commune* is involved in mushroom formation and protection against phototoxicity. Environ Microbiol 15:943–955. https://doi.org/10.1111/j.1462-2920.2012.02878.x

Ohm RA, Riley R, Salamov A, Min B, Choi I-G, Grigoriev IV (2014) Genomics of wood-degrading fungi. Fungal Genet Biol 72:82–90. https://doi.org/10.1016/J.FGB.2014.05.001

Parra G, Bradnam K, Korf I (2007) CEGMA: a pipeline to accurately annotate core genes in eukaryotic genomes. Bioinformatics 23:1061–1067. https://doi.org/10.1093/bioinformatics/btm071

Pel HJ, de Winde JH, Archer DB, Dyer PS, Hofmann G, Schaap PJ, Turner G, de Vries RP, Albang R, Albermann K, Andersen MR, Bendtsen JD, Benen JAE, van den Berg M, Breestraat S, Caddick MX, Contreras R, Cornell M, Coutinho PM, Danchin EGJ, Debets AJM, Dekker P, van Dijck PWM, van Dijk A, Dijkhuizen L, Driessen AJM, d'Enfert C, Geysens S, Goosen C, Groot GSP, de Groot PWJ, Guillemette T, Henrissat B, Herweijer M, van den Hombergh JPTW, van den Hondel CAMJJ, van der Heijden RTJM, van der Kaaij RM, Klis FM, Kools HJ, Kubicek CP, van Kuyk PA, Lauber J, Lu X, van der Maarel MJEC, Meulenberg R, Menke H, Mortimer MA, Nielsen J, Oliver SG, Olsthoorn M, Pal K, van Peij NNME, Ram AFJ, Rinas U, Roubos JA, Sagt CMJ, Schmoll M, Sun J, Ussery D, Varga J, Vervecken W, van de Vondervoort PJJ, Wedler H, Wösten HAB, Zeng A-P, van Ooyen AJJ, Visser J, Stam H (2007) Genome sequencing and analysis of the versatile cell factory *Aspergillus niger* CBS 513.88. Nat Biotechnol 25:221–231. https://doi.org/10.1038/nbt1282

Peter M, Kohler A, Ohm RA, Kuo A, Krützmann J, Morin E, Arend M, Barry KW, Binder M, Choi C, Clum A, Copeland A, Grisel N, Haridas S, Kipfer T, LaButti K, Lindquist E, Lipzen A, Maire R, Meier B, Mihaltcheva S, Molinier V, Murat C, Pöggeler S, Quandt CA, Sperisen C, Tritt A, Tisserant E, Crous PW, Henrissat B, Nehls U, Egli S, Spatafora JW, Grigoriev IV, Martin FM (2016) Ectomycorrhizal ecology is imprinted in the genome of the dominant symbiotic fungus *Cenococcum geophilum*. Nat Commun 7:12662. https://doi.org/10.1038/ncomms12662

Petersen TN, Brunak S, von Heijne G, Nielsen H (2011) SignalP 4.0: discriminating signal peptides from transmembrane regions. Nat Methods 8:785–786. https://doi.org/10.1038/nmeth.1701

Price AL, Jones NC, Pevzner PA (2005) De novo identification of repeat families in large genomes. Bioinformatics 21:i351–i358. https://doi.org/10.1093/bioinformatics/bti1018

Rawlings ND, Waller M, Barrett AJ, Bateman A (2014) *MEROPS*: the database of proteolytic enzymes, their substrates and inhibitors. Nucleic Acids Res 42:D503–D509. https://doi.org/10.1093/nar/gkt953

Rouxel T, Grandaubert J, Hane JK, Hoede C, van de Wouw AP, Couloux A, Dominguez V, Anthouard V, Bally P, Bourras S, Cozijnsen AJ, Ciuffetti LM, Degrave A, Dilmaghani A, Duret L, Fudal I, Goodwin SB, Gout L, Glaser N, Linglin J, Kema GHJ, Lapalu N, Lawrence CB, May K, Meyer M, Ollivier B, Poulain J, Schoch CL, Simon A, Spatafora JW, Stachowiak A, Turgeon BG, Tyler BM, Vincent D, Weissenbach J, Amselem J, Quesneville H, Oliver RP, Wincker P, Balesdent M-H, Howlett BJ (2011) Effector diversification within compartments of the *Leptosphaeria maculans* genome affected by repeat-induced point mutations. Nat Commun 2:202. https://doi.org/10.1038/ncomms1189

Sanger F, Air GM, Barrell BG, Brown NL, Coulson AR, Fiddes JC, Hutchison CA, Slocombe PM, Smith M (1977a) Nucleotide sequence of bacteriophage φX174 DNA. Nature 265:687–695. https://doi.org/10.1038/265687a0

Sanger F, Nicklen S, Coulson AR (1977b) DNA sequencing with chain-terminating inhibitors. Proc Natl Acad Sci USA 74:5463–5467. https://doi.org/10.1073/PNAS.74.12.5463

Shi T-Q, Liu G-N, Ji R-Y, Shi K, Song P, Ren L-J, Huang H, Ji X-J (2017) CRISPR/Cas9-based genome editing of the filamentous fungi: the state of the art.

Appl Microbiol Biotechnol 101:7435–7443. https://doi.org/10.1007/s00253-017-8497-9

Shi-Kunne X, Faino L, van den Berg GCM, Thomma BPHJ, Seidl MF (2018) Evolution within the fungal genus *Verticillium* is characterized by chromosomal rearrangement and gene loss. Environ Microbiol 20:1362–1373. https://doi.org/10.1111/1462-2920.14037

Simão FA, Waterhouse RM, Ioannidis P, Kriventseva EV, Zdobnov EM (2015) BUSCO: assessing genome assembly and annotation completeness with single-copy orthologs. Bioinformatics 31:3210–3212. https://doi.org/10.1093/bioinformatics/btv351

Simpson JT, Wong K, Jackman SD, Schein JE, Jones SJM, Birol I (2009) ABySS: a parallel assembler for short read sequence data. Genome Res 19:1117–1123. https://doi.org/10.1101/gr.089532.108

Sipos G, Prasanna AN, Walter MC, O'Connor E, Bálint B, Krizsán K, Kiss B, Hess J, Varga T, Slot J, Riley R, Bóka B, Rigling D, Barry K, Lee J, Mihaltcheva S, LaButti K, Lipzen A, Waldron R, Moloney NM, Sperisen C, Kredics L, Vágvölgyi C, Patrignani A, Fitzpatrick D, Nagy I, Doyle S, Anderson JB, Grigoriev IV, Güldener U, Münsterkötter M, Nagy LG (2017) Genome expansion and lineage-specific genetic innovations in the forest pathogenic fungi *Armillaria*. Nat Ecol Evol 1:1931–1941. https://doi.org/10.1038/s41559-017-0347-8

Smit AFA, Hubley R, Green P (2015) RepeatMasker Open-4.0 2013-2015. http://www.repeatmasker.org

Stanke M, Waack S (2003) Gene prediction with a hidden Markov model and a new intron submodel. Bioinformatics 19:ii215–ii225. https://doi.org/10.1093/bioinformatics/btg1080

Stein LD (2013) Using GBrowse 2.0 to visualize and share next-generation sequence data. Brief Bioinform 14:162–171. https://doi.org/10.1093/bib/bbt001

Sugano SS, Suzuki H, Shimokita E, Chiba H, Noji S, Osakabe Y, Osakabe K (2017) Genome editing in the mushroom-forming basidiomycete *Coprinopsis cinerea*, optimized by a high-throughput transformation system. Sci Rep 7:1260. https://doi.org/10.1038/s41598-017-00883-5

Terashima K, Yuki K, Muraguchi H, Akiyama M, Kamada T (2005) The dst1 gene involved in mushroom photomorphogenesis of *Coprinus cinereus* encodes a putative photoreceptor for blue light. Genetics 171:101–108. https://doi.org/10.1534/genetics.104.040048

Testa AC, Oliver RP, Hane JK (2016) OcculterCut: a comprehensive survey of AT-rich regions in fungal genomes. Genome Biol Evol 8:2044–2064. https://doi.org/10.1093/gbe/evw121

The C elegans Sequencing Consortium (1998) Genome sequence of the nematode *C. elegans*: a platform for investigating biology. Science 282:2012–2018. https://doi.org/10.1126/SCIENCE.282.5396.2012

Toome M, Ohm RA, Riley RW, James TY, Lazarus KL, Henrissat B, Albu S, Boyd A, Chow J, Clum A, Heller G, Lipzen A, Nolan M, Sandor L, Zvenigorodsky N, Grigoriev IV, Spatafora JW, Aime MC (2014) Genome sequencing provides insight into the reproductive biology, nutritional mode and ploidy of the fern pathogen *Mixia osmundae*. New Phytol 202:554–564. https://doi.org/10.1111/nph.12653

Varga T, Krizsán K, Földi C, Dima B, Sánchez-García M, Sánchez-Ramírez S, Szöllősi GJ, Szarkándi JG, Papp V, Albert L, Andreopoulos W, Angelini C, Antonín V, Barry KW, Bougher NL, Buchanan P, Buyck B, Bense V, Catcheside P, Chovatia M, Cooper J, Dämon W, Desjardin D, Finy P, Geml J, Haridas S, Hughes K, Justo A, Karasiński D, Kautmanova I, Kiss B, Kocsubé S, Kotiranta H, LaButti KM, Lechner BE, Liimatainen K, Lipzen A, Lukács Z, Mihaltcheva S, Morgado LN, Niskanen T, Noordeloos ME, Ohm RA, Ortiz-Santana B, Ovrebo C, Rácz N, Riley R, Savchenko A, Shiryaev A, Soop K, Spirin V, Szebenyi C, Tomšovský M, Tulloss RE, Uehling J, Grigoriev IV, Vágvölgyi C, Papp T, Martin FM, Miettinen O, Hibbett DS, Nagy LG (2019) Megaphylogeny resolves global patterns of mushroom evolution. Nat Ecol Evol 3:668–678. https://doi.org/10.1038/s41559-019-0834-1

Vesth TC, Nybo JL, Theobald S, Frisvad JC, Larsen TO, Nielsen KF, Hoof JB, Brandl J, Salamov A, Riley R, Gladden JM, Phatale P, Nielsen MT, Lyhne EK, Kogle ME, Strasser K, McDonnell E, Barry K, Clum A, Chen C, LaButti K, Haridas S, Nolan M, Sandor L, Kuo A, Lipzen A, Hainaut M, Drula E, Tsang A, Magnuson JK, Henrissat B, Wiebenga A, Simmons BA, Mäkelä MR, de Vries RP, Grigoriev IV, Mortensen UH, Baker SE, Andersen MR (2018) Investigation of inter- and intraspecies variation through genome sequencing of *Aspergillus* section Nigri. Nat Genet 50:1688–1695. https://doi.org/10.1038/s41588-018-0246-1

Vonk PJ, Escobar N, Wösten HAB, Lugones LG, Ohm RA (2019) High-throughput targeted gene deletion in the model mushroom *Schizophyllum commune* using pre-assembled Cas9 ribonucleoproteins. Sci Rep 9(1):7632. https://doi.org/10.1038/s41598-019-44133-2

Walker BJ, Abeel T, Shea T, Priest M, Abouelliel A, Sakthikumar S, Cuomo CA, Zeng Q, Wortman J, Young SK, Earl AM (2014) Pilon: an integrated tool for comprehensive microbial variant detection and genome assembly improvement. PLoS One 9:e112963. https://doi.org/10.1371/journal.pone.0112963

Wicker T, Sabot F, Hua-Van A, Bennetzen JL, Capy P, Chalhoub B, Flavell A, Leroy P, Morgante M,

Panaud O, Paux E, SanMiguel P, Schulman AH (2007) A unified classification system for eukaryotic transposable elements. Nat Rev Genet 8:973–982. https://doi.org/10.1038/nrg2165

Wösten HAB (2001) Hydrophobins: multipurpose proteins. Annu Rev Microbiol 55:625–646. https://doi.org/10.1146/annurev.micro.55.1.625

Yandell M, Ence D (2012) A beginner's guide to eukaryotic genome annotation. Nat Rev Genet 13:329–342. https://doi.org/10.1038/nrg3174

Zerbino DR, Birney E (2008) Velvet: algorithms for de novo short read assembly using de Bruijn graphs. Genome Res 18:821–829. https://doi.org/10.1101/gr.074492.107

Biotechnology

10 Filamentous Fungi as Hosts for Heterologous Production of Proteins and Secondary Metabolites in the Post-Genomic Era

Jakob K. H. Rendsvig[1], Malgorzata E. Futyma[1], Zofia D. Jarczynska[1], Uffe H. Mortensen[1]

CONTENTS

I. Introduction

Fungi are well-established producers of foods, food additives, industrial enzymes, and pharmaceutical drugs and contribute significantly to human health and economy. Production of industrial enzymes alone had an annual worth of 3.5 billion in 2015, which is a mere fraction of the general white biotechnology products estimated to reach 450 billion in 2020 (Meyer et al. 2016). The vast majority of fungal processes are performed with species that have not been genetically engineered. Hence, the fungi have either been domesticated through millennials of human use, or their performance and yields have been improved via classical genetic methods like mutagenesis. The importance of fungi has sparked several multi-species genome sequencing projects where one aim is to uncover novel fungal genes involved in enzyme secretion and secondary metabolite (SM) formation. The global fungal diversity currently encompasses 144,000 classified species and estimates reaching 3.8–6.0 million species (Taylor et al. 2014; Willis 2018), of which ~1500 species have been genome sequenced (de Vries et al. 2018; Grigoriev et al. 2014). As such, it is becoming increasingly clear that fungi represent a vast reservoir of potentially useful enzymes and secondary metabolites yet to be discovered. As part of the sequencing project of genus *Aspergillus*, Vesth et al. analyzed the genomes of 36 species of *Aspergillus* section *Nigri*, predicting 40,424 unique genes, including 17,903 carbohydrate-active enzymes and an average of 70 putative secondary metabolite gene clusters per species (near 2700 total), comprising 450 distinct compound classes (Vesth et al. 2018). With the increasing number of fungal genome sequences, many new genes and potentially interesting products will be uncovered in this post-genomic era.

However, many of these new products are likely made by natural fungal producers that are challenging to incorporate into industrial

[1] Technical University of Denmark, Kgs. Lyngby, Denmark; e-mail: jakren@dtu.dk; malfu@dtu.dk; zofja@bio.dtu.dk; um@bio.dtu.dk

Genetics and Biotechnology, 3rd Edition
The Mycota II
J.P. Benz, K. Schipper (Eds.)

processes, either because they are problematic to propagate efficiently in bioreactors, difficult to engineer, or opportunistic human pathogens. Based on the logical assumption that fungi by evolution likely are the best producers of fungal products, we therefore envision that there will be an increasing demand of transferring relevant genes and pathways from novel, exotic fungi to well-characterized fungal cell factories.

Filamentous fungi have been widely used for heterologous production of industrial enzymes, taking advantage of their large protein secretory capacity. The most commonly used fungal species for this purpose are *Trichoderma reesei* and members of the genus *Aspergillus*, e.g., *Aspergillus niger* and *Aspergillus oryzae*, and thus, most studies have been performed with these species. More recently, filamentous fungi have also been employed for heterologous production of SMs, typically as a part of an SM pathway elucidation strategy. For these studies, classical model fungi like *Aspergillus nidulans* and *Neurospora crassa* have been added to the repertoire of cell factories. The usefulness of fungal industrial workhorses and general model systems is also reflected by the development of substantial genetic toolboxes and methodologies for heterologous gene expression for these species.

The current trend for heterologous gene expression is to implement a synthetic biology approach. Hence, a host strain is transformed with a gene-expression cassette, generated from libraries of bio-blocks (see Sect. III), which are individually functional molecular units that can be combined by simple and seamless DNA fusion strategies (Fig. 1). According to this concept, we will first describe strategies to set up synthetic biology-based expression systems followed by an overview of popular bio-blocks that can be used for construction of expression cassettes. Next, we will provide two sections with examples where these concepts have been partly or entirely used to produce either enzymes or secondary metabolites in a heterologous fungal host.

II. Expression Systems

Heterologous expression is achieved by transforming the new fungal host with a suitable **gene-expression cassette**, which is integrated into a chromosome or an extrachromosomal vector and thereby maintained during growth.

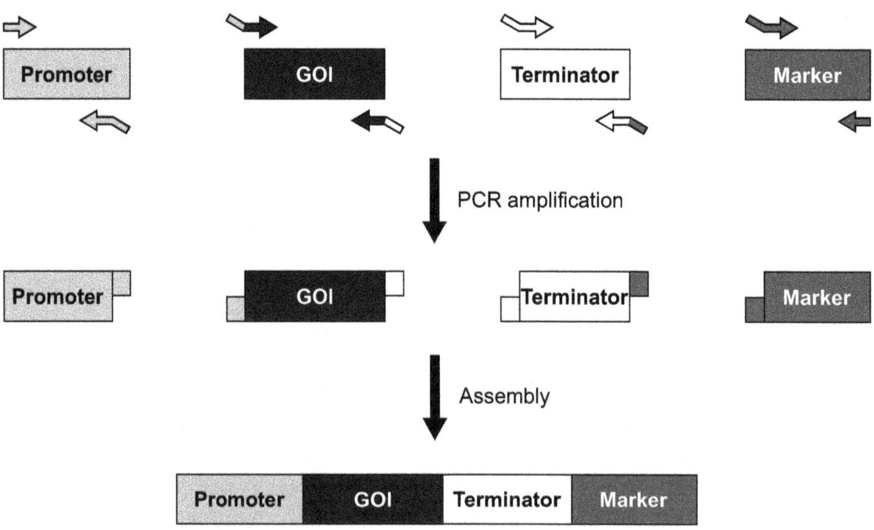

Fig. 1 Assembly of a basic gene-expression cassette. Bio-blocks are PCR amplified using primers (arrows) with tails containing sequences that are complementary to the tails of the adjacent bio-block (indicated by color); see main text for details. This allows matching overhangs (small boxes) to be formed ensuring that the bio-blocks are fused in the correct order. Tails for inserting the gene-expression cassette into a vector are not shown

Below, we review strategies for assembling gene-expression cassettes for heterologous enzyme or pathway expression, along with methods for introducing these into the fungal host and the benefits of these methods. The functional **bio-blocks** for fungal heterologous gene expression are presented in Sect. III.

A. Construction of Simple Gene-Expression Cassettes

In the simplest setups, heterologous production depends on the expression of a single **gene of interest** (GOI). In these cases, **gene-expression cassettes** (GEC) can be constructed by fusing the **GOI bio-block** to relevant components of a basic set of bio-blocks that includes promoters, terminators, and selectable markers (Fig. 1). If necessary, this set can be expanded with **additional bio-blocks for specialized purposes** such as sequences encoding secretion signals, epitope- and purification tags, fluorescent proteins etc., or bio-blocks that are designed to target the expression cassette for integration at a specific site in the genome by homologous recombination. Since many individual bio-blocks need to be combined in a single **cloning step**, it is important that they can be joined with high efficiency. Several reliable systems are available for this task, e.g., In-Fusion assembly (Zhu et al. 2007), Gibson assembly (Gibson et al. 2009), USER fusion (Bitinaite et al. 2007), and Golden Gate cloning (Engler and Marillonnet 2014). Often these systems are directly compatible with accompanying vectors dedicated to gene transfer into a desirable host and the construction work commonly facilitated by ligation and propagation in *Escherichia coli*. Alternatively, bio-blocks can be assembled by in vivo homologous recombination using, e.g., the yeast *Saccharomyces cerevisiae* as a host (Finnigan and Thorner 2015).

B. Introducing Gene-Expression Cassettes into Fungal Hosts

The gene-expression cassettes can be introduced into the fungal host according to different principles. Below, we briefly review how gene-expression cassettes can be maintained by being inserted into an extrachromosomal vector or by being integrated into a random or defined position on a chromosome (Fig. 2).

1. Plasmid-Based Expression Systems

Plasmid-based expression systems utilize self-replicating plasmids carrying autonomous replication sequences (Fig. 2a). In many filamentous fungi, this can be achieved via the **AMA1** (autonomous maintenance in *Aspergillus*) element. AMA1 was originally discovered in a lab strain of *A. nidulans* (Gems et al. 1991) in a search for **genetic elements enhancing trans-**

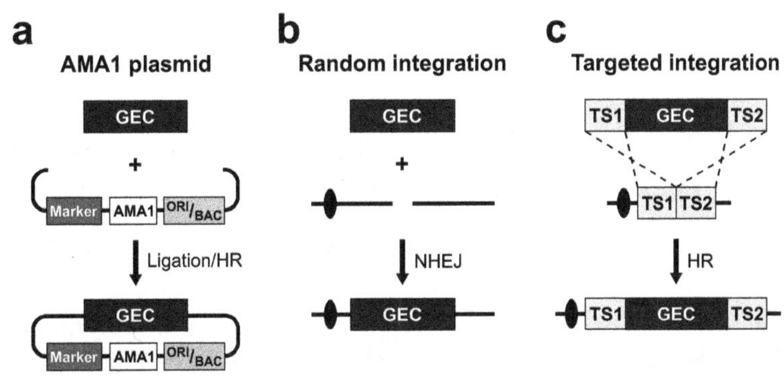

Fig. 2 Strategies for introducing gene-expression cassettes into fungi. The GEC can be (**a**) inserted into a self-replicating AMA1 plasmid via ligation or by in vivo HR or (**b**) integrated into the host genome by random or (**c**) targeted integration through NHEJ or HR, respectively

formation in *A. nidulans* (Johnstone et al. 1985). In subsequent studies, it has been shown that AMA1-based vectors increase fungal transformation efficiency in *A. nidulans* up to 2000 times in comparison to earlier developed ectopically integrating vectors (Aleksenko and Clutterbuck 1997). Moreover, AMA1 plasmids propagate in other *Aspergillus* species as well as in species belonging to other fungal genera, including *Penicillium* (Fierro et al. 1996), *Talaromyces* (Nielsen et al. 2017), *Trichoderma* (Kubodera et al. 2002), and *Rosellinia* (Shimizu et al. 2012).

Gene-expression cassettes can easily be incorporated into **extrachromosomal AMA1-based shuttle vectors**, using the cloning systems highlighted above. As an extreme case, fungal AMA1 plasmids have been combined with a bacterial artificial chromosome (Zhu et al. 1997) to form *fungal artificial chromosomes* (FACs). These vectors have been instrumental in the search for new secondary metabolite gene clusters as they allow up to 300 kbp of fungal DNA to be cloned into the FAC in *E. coli* and subsequently transferred into a new fungal host for product discovery (Bok et al. 2015; Clevenger et al. 2017). Importantly, AMA1-based plasmids are not very stable and their maintenance requires selection, thereby limiting their use to small-scale exploratory heterologous production experiments. However, we note that for some purposes, such as *cas9* expression in CRISPR experiments, plasmid instability is desirable as plasmid loss allows for transient gene expression, diminishing potential Cas9 off-target effects (Zhang et al. 2015).

2. Chromosome-Based Expression Systems

Integration of the expression cassette into the host genome provides a more stable expression, as compared to AMA1-based expression, and attenuates the need to maintain selection pressure. **Chromosomal integration** of foreign DNA occurs via one of the two DNA repair mechanisms (Krappmann 2007), either **non-homologous end joining** (NHEJ), where the DNA integrates into a random locus (Fig. 2b), or **homologous recombination** (HR), where the DNA integrates into a defined locus (Fig. 2c). Although both pathways are active in all fungi, foreign DNA typically integrates more efficiently via the NHEJ pathway in most filamentous fungal species (Meyer et al. 2007; Nødvig et al. 2015).

a) Random Chromosomal Integration of Gene-Expression Cassettes

Since NHEJ is the dominant pathway for chromosomal integration of DNA in most filamentous fungi, **random chromosomal integration** of gene-expression cassettes via NHEJ is therefore straightforward and can be efficiently performed in essentially all transformable fungal species. A drawback of the method is that the chromatin structures of the insertion sites are unpredictable and **expression levels may therefore vary** substantially between different transformants (Lubertozzi and Keasling 2006). On the other hand, if several gene-expression cassettes enter a nucleus, they may undergo recombination prior to genomic integration. As a result, transformants may contain multiple copies of the GOI inserted into a single locus. In these cases, the transformants will often be better producers of the product due to the many gene copies. However, since the cassettes will be organized as direct and/or inverted repeats, expression levels of such strains may be unstable as gene copies may be lost due to direct repeat recombination or formation of hairpin structures (Leach 1994; Petes 1988). Lastly, it is important to note that when gene-expression cassettes **integrate randomly** into the genome, they may **disrupt important genes**, or alter the expression levels of neighboring genes, causing undesired phenotypes. In summary, with this integration method, it is advisable to screen for transformants that produce high yields over many generations and at the same time do not display undesirable fitness defects.

b) Defined Chromosomal Integration of Gene-Expression Cassettes

The use of defined and well-characterized loci as **integration sites** for GECs allows for gene expression that is much **less prone to clonal**

differences. Hence, insertion of gene-expression cassettes into defined loci enables **comparative screening of expression levels** or enzyme activities based on different genetic elements employed in the cassette, such as promoter, secretion signal, and terminator sequences. Similarly, the effects achieved by changing the GOI sequence can be directly compared. In this way it is possible to address whether changes in the GOI codon composition increase yields or whether mutations infer changes in the heterologous protein that influences its activity, specificity, folding, and/or stability (Hansen et al. 2011b; Holm 2013).

Gene-expression cassettes can be **inserted into integration sites in the genome by HR**. To ensure high expression levels, it may be useful to position integration sites in intergenic regions located in **transcriptionally highly active sections** of a chromosome. With this method, the gene-expression cassettes must be flanked by up- and downstream targeting sequences matching the genomic expression site. Since fungi rarely prefer to integrate foreign DNA into its genome by the HR pathway, extensive screening for the desired transformant may be necessary. The screening workload can be reduced by using **bipartite gene-targeting substrates** that select for HR proficient protoplasts (Nielsen et al. 2006) or avoided by using **NHEJ-deficient strains** (Meyer et al. 2007). Alternatively, methods based on restriction enzymes or **CRISPR technology** can be employed (Ouedraogo et al. 2016; Zheng et al. 2017). Importantly with the latter technology, GOIs can be inserted in a marker-free manner and multiplexing is possible (Liu et al. 2015; Nødvig et al. 2018; Zhang et al. 2016a). In fact, CRISPR mediates very efficient gene targeting even in NHEJ-proficient strains (Nødvig et al. 2018), eliminating the risks of working with strains with a defective DNA repair pathway. We therefore envision that CRISPR-based methods will deliver the preferred tools for introducing gene-expression cassettes into defined chromosomal expression platforms.

C. Bio-Block-Based Multi-GOI Expression Strategies

In many cases, two or more genes are required for the synthesis of a desired product. For example, for heterologous production of SMs, several genes are commonly required to synthesize the product. Above, we described that it is possible to transfer large chromosomal fragments containing entire gene clusters from one fungus to another (Sect. II.B.1). However, in many cases the activation mechanism for the genes in the cluster is not known and may depend on unknown or even host-specific transcription factors (TF) (Keller 2019); see Chapter "A new avenue to drug discovery in Fungi". Moreover, not all genes contributing to the biosynthetic pathway of interest may be situated in the cluster (Schäpe et al. 2019). It may therefore be desirable to reconstruct the genes involved in formation of the heterologous compound using a setup that allows necessary GOIs to be equipped with known and well-characterized promoter and terminator sequences. Here we present **three strategies that allow for bio-block-based multi-GOI expression** (Fig. 3). Although multi-GOI expression cassettes can be assembled on AMA1-based vectors, we recommend using defined chromosomal expression sites for GOI insertion to gain genetic stability and to reduce dependency on selectable markers.

In the **first strategy** (Fig. 3a), individual expression cassettes for **different GOIs** are inserted into a number of **unique integration sites**. This strategy is advisable if all of the individual gene-expression cassettes contain an identical bio-block, such as a specific promoter or terminator. In this way, copy loss due to direct repeat recombination between the identical bio-blocks is avoided. With conventional gene-targeting methods, insertion of gene-expression cassettes into individual sites requires a number of selectable markers that matches the number of gene-expression cassettes required for establishing the biosynthetic pathway. Alternatively, the genes can be

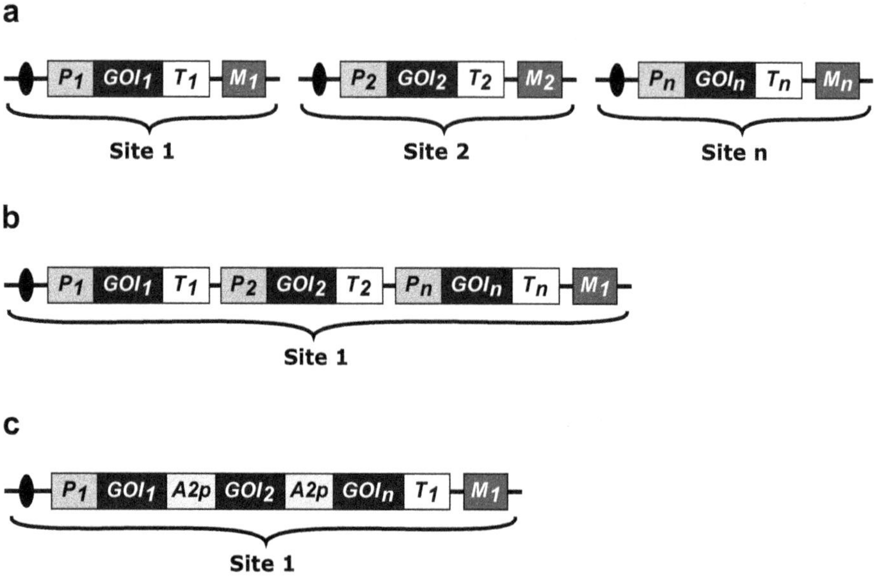

Fig. 3 Strategies for multi-GOI expression cassette assembly. Individual GECs composed of a promoter (P_x), a gene of interest (GOI_x), a terminator (T_x), and a selection marker (M_x). Individual GECs can be (**a**) integrated in different loci, or (**b**) assembled into a combined multi-GOI cassette for single locus integration, or (**c**) assembled as a single polycistronic GOI where coding sequences of individual polypeptides are separated by the sequence encoding the picornavirus 2A peptide (A2p)—see main text for details

inserted by iterative gene targeting using a recyclable marker (see Sect. III.B.5). Strain construction with these methods is cumbersome and restricts the method to biosynthetic pathways composed of a small number of genes. However, these limitations can be dramatically reduced by constructing the strains via CRISPR-based multiplexed marker-free gene insertions.

In the **second strategy** (Fig. 3b), all genes required to support a biosynthetic pathway are inserted into the same expression site. Arranging the **GOIs as a synthetic gene cluster** may be preferred if only few selectable markers are available. Note that if two markers are available, then an infinite number of consecutive gene-targeting events into the same locus can be performed. Hence, even very large synthetic gene clusters can be constructed at a defined expression site via multiple integration steps. In this case, the selectable marker used in a given integration step replaces the marker used in the previous integration step (see Sect. III.B.5). In addition, if the heterologous host has a sexual cycle, strains that contain a synthetic gene cluster in a single defined locus can be crossed to other strains without risking that the individual genes of the cluster segregates during meiosis. This feature can be used to combine the heterologous pathway with other beneficial traits harbored by other strains.

In the **third strategy** (Fig. 3c), the **GOIs are arranged as a polycistronic unit** where each open reading frame (ORF) is separated by a bio-block, encoding the **2A peptide** from the *Picornaviridae* virus family; see Table 3 (Schuetze and Meyer 2017). This strategy is based on the facts that; **firstly**, the ribosome fails to link glycine and proline residues during translation of the 2A spacer, thereby resulting in a break in the polypeptide chain. **Secondly**, this error does not result in release of the ribosome from the mRNA, thereby allowing for continued translation. Hence, in a simple manner, several proteins can be encoded from a single transcript generated from a single expression cassette. One drawback of this method may be that the final size of the polycistronic gene-expression cassette makes construction work difficult. Additionally, "cleavage" at the 2A spacer results in a partial 2A peptide sequence remaining at the C-terminus of the protein, which may interfere with protein function and therefore needs to be removed, e.g., by combinatory usage with other proteolytic sites (Hoefgen et al. 2018). Moreover, as "cleavage" at the 2A spacer is not 100%, the relative efficiency of protein production often depends on the position of the proteins ORF in the transcript (Schuetze and Meyer 2017).

Lastly, we note that for all strategies, two genes can be inserted simultaneously as a single cassette if bidirectional promoters (Wiemann et al. 2018; Rendsvig et al. 2019) are used to control gene expression of the heterologous gene pairs. This feature can be used to speed up strain construction.

III. Bio-Blocks

A **bio-block** is a DNA sequence that encompasses a **molecularly functional unit.** This includes **nucleotide sequences,** e.g., promoters and terminators, or sequences encoding a **protein,** e.g., the GOI and selection marker, or **shorter protein sequences** such as secretion signals and purification tags. Generation of the basic gene-expression cassette for production of a heterologous protein requires several bio-blocks, including the GOI, promoter, and terminator. Selection markers may also be included in the construct, offering versatile applications in different scenarios. In recent years, the use of -omics data has driven the discovery of natural promoters, which are active during the desired cultivation conditions for production. While the majority of strain engineering has been based on natural genetic elements for controlling gene expression, significant headway has been made in the development of **synthetic promoters** and terminators. In extension, even **full synthetic gene-expression systems** have been established, in which several individual genetic elements (natural or modified) are combined, to facilitate controlled gene expression and enhance productivity. Moreover, to enable secretion of the protein, a transport signal must be included as an additional bio-block (Sect. IV.A.1). For specialized purposes like protein visualization or purification, protein tags fulfilling such tasks are also included in the construct, thereby increasing the pool of bio-blocks.

A. Gene of Interest

Since the goal of heterologous expression is to produce products derived from one or more **genes of interest,** this element can also be considered the **most important bio-block.** As the GOIs coding potential in most cases should not be changed, it therefore constitutes the least flexible bio-block; however, its design requires **several considerations.** The GOI sequence may be derived from different sources. In the simplest scheme, if the donor strain is related to the new host, genomic DNA can typically serve as a template to make a functional PCR-derived GOI bio-block. However, if the donor strain is more distantly related, **intron splicing** and **codon bias** may compromise gene expression and translation, and we will briefly review the two latter issues below.

1. Introns

The presence of introns in a heterologous gene may result in reduced mRNA production due to **mis- and incomplete splicing** (He and Cox 2016; Zhao et al. 2013). In principle, the problem can be solved by employing cDNA as a PCR template for making the GOI bio-block or alternatively fuse PCR-derived exons by another round of PCR or by a one-step cloning method that allows for seamless assembly of multiple fragments (An et al. 2007). Alternatively, an entirely synthetic gene may be acquired from a commercial source. However, transcription, splicing, polyadenylation, and mRNA export are a coordinated process (Bentley 2014) and as a result the presence of introns may affect gene expression.

Indeed, **sequential elimination of three introns** in a gene encoding a protease from the thermophilic fungus *Malbranchea cinnamomea* cumulatively reduced production levels in *T. reesei* (Paloheimo et al. 2016). **Interestingly, the largest reduction was observed when the intron closest to the transcriptional start site was removed.** Similar results were obtained when a gene encoding an antifungal protein from *Aspergillus giganteus* was expressed in *Trichoderma viride* (Xu and Gong 2003). Moreover, the inclusion of an artificial intron was required for the heterologous production of GFP and mRFP in the basidiomycete *Armillaria mellea* (Ford et al. 2016).

Since it is not straightforward to predict the impact of retaining introns in a foreign gene on the final yield of mRNA, it may be useful to try different gene variants containing no, a few selected, or all introns, to optimize expression levels from the GOI.

2. Codon Optimization

Codon usage may vary dramatically between species (Iriarte et al. 2012) and, consequently, influence how well a heterologous mRNA is translated into a protein, reviewed in (Hanson and Coller 2018; Tanaka et al. 2014). For example, **codon composition** may change the chromatin structure of the GOI affecting **gene transcription efficiency** (Zhou et al. 2016), whereas rare codons in the beginning of the transcript may determine the efficiency of **translation initiation** (Pop et al. 2014). Moreover, if the mRNA contains abnormally high levels of rare codons, this may result in premature transcription termination and production of reduced length mRNA for the ORF (Zhou et al. 2018). Conversely, **rare codons** may be used in the native host as **translation pause sites** required for proper and timely folding of subdomains of the protein structure. If such codons are replaced by frequently used codons, overall folding may be compromised (Yu et al. 2015; Zhou et al. 2015).

Different algorithms exist in order to address the different challenges concerning optimal codon choices for heterologous protein production (Gould et al. 2014). Construction of a **codon-optimized gene** usually requires de novo synthesis of the entire gene, which can be done by fusing a set of overlapping oligonucleotides (Sect. III.E) (Hoover and Lubkowski 2002) or simply be acquired from a commercial source. Codon optimization may benefit heterologous protein production significantly, especially if the source of the foreign gene and the new host are **distantly related species**. However, it is important to stress that predicting the optimal codon composition for heterologous gene expression is a discipline still in its infancy. It is therefore often an advantage to test more than one gene variant during cell factory construction. This can be exemplified by the extracellular yield from *T. reesei* expressing a codon-optimized mammalian monoclonal immunoglobulin (IgG) antibody, which showed a 40-fold difference, dependent on which of two companies had performed the codon optimization (Lin et al. 2006).

B. Selection Markers

Selection markers are mainly used to facilitate identification of transformants containing the gene-expression cassette (Dave et al. 2015). Since targeted integration of the gene-expression cassette can be conducted in a marker-free manner via CRISPR technology, the presence of this bio-block in the cassette is no longer essential. Nevertheless, selective markers will likely remain a common feature in a gene-expression cassette since **strain identification** is simplified and more efficient with a marker and since the marker may also serve useful post-transformation roles. For example, a selectable marker can be used to quickly identify strains, or to promote breeding via sexual and parasexual cycles, or to maintain AMA1 plasmid-based expression. Moreover, to achieve high expression levels, it may be desirable to obtain a fungus with many copies of the expression cassette. Such strains can be selected by employing cassettes containing a **weak selection marker** as the strains with a high copy number will gain a fitness advantage at heavy selection pressure (Wernars et al. 1985). In some cases, it is necessary to eliminate the marker after transformation, for example, if a marker-free strain is desirable or if a marker needs to be recycled during iterative genetic engineering.

The many uses of selectable markers in strain development are reflected in the fact that several **different types of markers** have been developed, which can be divided into **four main categories**: resistance, visual, auxotrophic/nutritional, and counter-selectable markers. Importantly, the choice of selectable marker depends on the strain that needs to be engineered and on the purpose of the experiment. Below, we will briefly review the different categories of selectable markers and discuss in which situations a given marker can be advantageous.

1. Resistance Markers

Resistance markers encode proteins that **neutralize the toxic effect of antimicrobials** and thereby convey resistance as a **dominant trait.** Since the functionality of these markers (like some visual markers, see below) typically do not require any modifications of the host genome, they can be used with wild-type model fungi or with fungi where no or few genetic tools are available. Several antimicrobial/resistance-marker systems have been introduced for fungal genetic engineering (Table 1).

The **mechanism of these antimicrobials** typically involves interfering with protein translation, inhibition of metabolism, or induction of lethal DNA double-stranded breaks (DSBs). The resistance to the antimicrobials is achieved through protein-based mechanisms including 1:1 drug binding, drug turnover by chemical modification, and drug-resistant target enzymes; see Table 1.

The presence of an array of different marker systems is highly useful because it potentially supports that several cycles of genetic engineering steps can be performed. Perhaps more importantly, it also expands the number of species that can be engineered due to the fact that many fungal species possess an inherent resistance to certain antimicrobials due to cell wall impermeability, native efflux pumps, or catalytic activities (Garneau-Tsodikova and Labby 2016). Sometimes, **susceptibility depends on the media composition** (Roller and Covill 1999) exemplified by *Aspergillus* species, which show resistance toward hygromycin B and bleomycin at pH-values below five and in complex or hypertonic media (Punt and van den Hondel 1992). Hence, before using resistance markers, it is necessary to conduct susceptibility assays of the fungus to determine appropriate selective antimicrobials and have consistent properties of selective media batches. It should be noted that the use of antimicrobials may be undesirable, especially for large-scale cultivations, due to the risk of developing drug resistant strains, the price of the antimicrobial, and that some of the compounds are toxic to humans.

2. Visual Markers

Visual markers provide a **phenotypic trait** that can be **easily visualized** by conferring, e.g., a color change due to disruption of a host gene, or by heterologous expression of a color, fluorescent, or bioluminescent marker. Like with

Table 1 Commonly used fungal antimicrobials and their corresponding resistance genes

Antimicrobial	Gene	Source organism	Mechanism of action	References[a]
Hygromycins	*hph*	*Escherichia coli*	Translation inhibition	Cullen et al. (1987)
Bleomycins	*ble*	*Klebsiella pneumoniae*	DSB induction	Austin et al. (1990)
Kanamycins	*neo*	*Klebsiella pneumoniae*	Miscoding of RNA	Collis and Hall (1985)
Oligomycins	*oliC3*	*Aspergillus niger*	ATP synthase inhibition	Ward et al. (1988)
Pyrithiamine	*ptrA*	*Aspergillus oryzae*	Thiamine antagonist	Kubodera et al. (2002)
Phosphinothricin	*bar*	*Streptomyces hygroscopicus*	Glutamine synthetase inhibition	Avalos et al. (1989)
Nourseothricin	*nat1*	*Streptomyces noursei*	Miscoding of RNA	Krügel et al. (1993)

[a]References of resistance gene

antimicrobial markers, visual markers can be used for selection in wild-type strains, but without posing any risks in relation to compound toxicity or transference of resistance genes to other organisms.

Three types of heterologous visual markers are frequently used, all of which may be applied for spectrophotometric assays. **Firstly,** β-galactosidase LacZ (*lacZ*) (Lubertozzi and Keasling 2006) and β-glucuronidase GusA (*uidA*) of *E. coli* (Tada et al. 1991), respectively, convert the synthetic substrates X-gal and X-gluc to 5-bromo-4-chloro-3-hydroxyindole that undergoes dimerization forming a blue pigment. **Secondly,** luciferases convert substrates into bioluminescent products. **Thirdly,** fluorescent proteins derived from jellyfish *Aequorea victoria* GFP, which was adapted for fungal use by codon optimization and by eliminating a non-conventional/cryptic intron splice site (Lorang et al. 2001), or from RFP.

Visual markers do not offer any fitness advantage, and, if the goal is to detect transformants containing the expression cassette, selection requires **visual screening.** Selection by solely visual inspection can be achieved by inserting the cassette into conidial pigmentation genes (e.g., homologs of *A. nidulans wA* and *yA* genes), resulting in a distinct change in conidial pigmentation from the native black or green spores of *Aspergillus* species to white or yellow (Jørgensen et al. 2011; Nielsen et al. 2006). Similarly, the expression cassette can be inserted into *adeA* or *adeB* as this will produce easy detectable red colonies due to polymerization and oxidation of 4-amino-imidazole ribotide, the intermediate resulting from the disrupted purine biosynthetic pathway (Jin et al. 2004). Alternatively, the gene-expression cassette can be equipped with a heterologous marker that produces color, fluorescence, or bioluminescence to allow for visual selection. This is useful if the goal is to insert the cassette into an intergenic section of the genome by HR, thereby avoiding host gene disruption. In cases where correct transformants are rare, **fluorescent markers** set the stage for high-throughput detection schemes via **FACS analysis,** which can be employed if the fungus produces discrete entities like conidia (Bleichrodt and Read 2019; Vlaardingerbroek et al. 2015). Color, fluorescent, or bioluminescent markers can also be advantageously used if the goal is to insert the gene-expression cassette in multiple copies, e.g., via integration events catalyzed by the NHEJ pathway, as transformants with a high copy number can be selected for by the strength of the marker signal (Throndset et al. 2010).

When heterologous enzymes and fluorescent proteins are used as selective markers, it should be noted that some fungi produce enzymes that may catalyze the same reactions as those provided by the marker enzyme, or the fungi may display auto-fluorescence. It is therefore always necessary to test whether significant **native background** signals exist.

3. Nutritional Markers

Usage of nutritional markers relay on **auxotrophies** in the host organism, which thereby requires **supplementation with specific metabolites** to sustain growth. Introducing a functional copy of the given nutritional marker into the host complements the defective native gene function, allowing for growth without supplementation. In this way, nutritional markers offer selective pressure without the negative effects of antibiotics. On the other hand, prior to cell factory construction, the host needs to be mutagenized or genetically engineered to create the **relevant auxotrophic mutations.** If the starting point is a wild-type strain, this may require the use of mutagens, traditional genetic engineering using antimicrobial or visual markers, or CRISPR technology. Complementing homologous or heterologous genes can be used as markers, but often a heterologous marker is preferred as the sequence differences reduce the risk of generating false positives due to homologous recombination between the marker and the corresponding mutated locus.

Some **commonly used nutritional markers** include *argB* (Buxton et al. 1985), *trpC* (Goosen et al. 1989), *adeA and adeB* (Jin et al. 2004), *pyroA* (Osmani et al. 1999), and ***pyrG*** (Goosen et al. 1987), which are required for synthesis of arginine, tryptophan, adenine, pyridoxine, and uracil/uridine, respectively. Other frequently used markers are *niaD* required for nitrate assimilation (Unkles et al. 1989) and the ***amdS*** gene from *A. nidulans* allowing growth on acetamide as the sole nitrogen source (Kelly and Hynes 1985).

Like visual markers, the design of the nutritional marker can be used to select for strains containing multiple expression cassettes. For example, the complementary marker gene can be equipped with a poor promoter to ensure that only strains with many copies produce sufficient amounts of the missing enzyme.

Often several auxotrophies are introduced in the host to allow for several rounds of genetic engineering using a set of complementing nutritional markers. It is important to note that nutritional markers may change the metabolism of the host cell despite that the defects are compensated by addition of the missing metabolites to the growth medium. For physiological characterizations of the engineered strains, this may lead to undesired artefacts; and for cell factories, it may lead to suboptimal growth, or growth, which is restricted to specific media. It is therefore advisable to restore the functionality of those nutritional markers after genetic engineering. To facilitate this process, it has been recently shown that it is possible to functionally revert several mutated genes by multiplexing CRISPR technology (Nødvig et al. 2018).

4. Counter-Selectable Markers

Some nutritional markers can be **counter-selected** by using **analogs of their natural substrates that are converted into antimetabolites.** The two most frequently used counter-selectable wild-type markers are ***pyrG*** and ***amdS***, which are lethal in the presence of 5-fluoroorotic acid (5-FOA) and 5′-fluoroacetamide (FAA), respectively. For both *pyrG*, and to lesser extent *amdS*, a gene copy is naturally present in wild-type fungi. Hence, the endogenous gene needs to be disrupted before the marker can be applied in selection/counter-selection experiments. In *pyrG*$^+$ strains, 5-FOA is converted by the orotidine-5′-phosphate decarboxylase (PyrG) into 5-fluorouracil, which is further metabolized into toxic substances that interfere with RNA and DNA synthesis (Boeke et al. 1984; Longley et al. 2003). Similarly, in *amdS*$^+$ strains, FAA is converted by the acetamidase (AmdS) into the toxic compound fluoroacetate (Hynes and Pateman 1970), which forms a stable complex with Coenzyme A (fluoroacetyl-CoA), hence hindering normal functionality of the tricarboxylic acid cycle. Other counter-selectable markers include genes in the tryptophan pathway, most often *trpA* (Foureau et al. 2012), and genes in the lysine pathway, most often *lysB* (Alberti et al. 2003), which can be counter-selected in the presence of 5-fluoroanthranilic acid, 5-FAA, and α-aminoadipate, respectively. Similarly, expression of genes encoding herpes simplex virus 1 thymidine kinase can be counter-selected as the viral thymidine kinase, unlike the thymidine kinases of the hosts, activates nucleoside analogs like 5-fluoro-2′-deoxyuridine for toxic incorporation into nucleic acids (Lupton et al. 1991). Counter-selectable markers are highly useful as they facilitate **marker recycling** and iterative genetic engineering.

5. Marker Recycling for Iterative Engineering

In most fungi, the number of available markers is a limiting factor for multi-step genetic engineering strategies. If, e.g., an entire biosynthetic pathway needs to be implemented into a new host, **marker recycling** is therefore often a necessity. Specifically, if two markers are available for transformation of a host strain, an endless number of genes can in principle be inserted into the same site in the host genome by iterative **marker-swapping**. In this method, the marker gene, which was used for gene integration in the previous transformation step, is replaced by a new gene targeting construct that contains new GOIs and another marker (Nielsen et al. 2013).

A marker gene can also be eliminated from the genome by **spontaneous direct-repeat recombination** if the marker is flanked by sufficiently long direct repeats, typically 500–1000 bp (Nielsen et al. 2006). This method may be preferred if it is desirable to leave a genetically neutral DNA sequence (i.e. the direct repeat) as the only scar after genetic engineering, or if it is desirable to **integrate several genes** or sets of genes into several different loci in the genome. As spontaneous direct recombination events are rare, this method requires that marker loss is selectable, e.g., by using a **counter-selectable marker**; see above (Sect. III.B.4). Alternatively, marker loss can be achieved by **induced recombination** involving either site-specific recombinases like Cre and a marker, which is flanked by its target sequences, or gene deletion catalyzed by site-specific nucleases like I-SceI or a CRISPR nuclease (Ouedraogo et al. 2016; Zhang et al. 2013).

C. Promoters

Among the different bio-blocks, promoters have drawn most attention since they control gene transcription initiation. Promoters are typically divided in two groups: those that are **constitutively active**, i.e., promoters that are active under all/most circumstances, and those that are **inducible/repressible**. Often constitutive promoters are preferred for large-scale heterologous production as they are active throughout the fermentation process in inexpensive media. In cases where the product is toxic or unstable, it may be necessary to restrict production to a specific growth phase to maximize yields; and in these cases, it is necessary to use an inducible/repressible promoter. However, as addition of inducing or repressing agents comes with an additional cost, use of this type of promoters may be restricted to small scale production or exploratory studies.

The vast majority of heterologous gene-expression studies are based on **natural promoters** of which we have listed sets of frequently used promoters derived from different fungal genera (Table 2). Interestingly, many promoters involved in **basic metabolism** do not seem to be species specific. Hence, **promoter versatility** allows the same genetic element to be used in many different species, even species belonging to different genera. However, to our knowledge, no comparative studies have systematically analyzed how active a given promoter is in different species. Based on the fair assumption that a promoter is most active in the species from which it originates, and less active when applied in other fungi, it is advisable to equip the GOI with a promoter derived from the intended production host. Importantly, the efficiency of promoters, even constitutive promoters, often depends on the growth environment and growth phase of the host. This has motivated establishment of **synthetic promoters** that offer orthogonal setups that work **independently of the host metabolism.** Below we present examples of natural and synthetic promoters.

1. Natural Promoters

A wide selection of natural fungal promoters for species belonging to *Aspergillus*, *Penicillium*, and *Trichoderma* genera as well as others (Table 2) have been experimentally characterized as constitutive or inducible, along with their mode of induction and repression, as reviewed in Fitz et al. (2018), Fleissner and Dersch (2010), and Kluge et al. (2018). Typical **constitutive promoters** for heterologous expression in fungal cell factories are derived from genes involved in major metabolic pathways or cell functions that require high steady protein levels. Examples are the ***gpdA*** promoter (P*gpdA*) and the stronger ***tef1*** promoter (P*tef1*), which control production of glyceraldehyde-3-phosphate dehydrogenase acting in glycolysis and translation elongation factor 1α assisting in protein synthesis, respectively. Promoters catalyzing basic reactions in the cell are often functional if transferred to other species. In line with this view, P*gpdA* and P*tef1* from *A. nidulans* have been shown to work in other species of Aspergilli (Nødvig et al. 2015), in *Talaromyces atroroseus* (Nielsen et al. 2017), and the P*gpdA* of *A. nidulans* was even applied in the fungus *Metarhizium anisopliae* (Nakazato et al. 2006). **Inducible/repressible promoters** are also recruited from basic metabolic genes. This includes the commonly used alcohol and threonine inducible ***alcA*** promoter, which controls expression of alcohol dehydrogenase I

Table 2 Commonly used constitutive and inducible/repressible fungal promoters

Origin	Gene	Gene product	Inducer/repressor[a]	References
Aspergillus sp.	*gpdA*	Glyceraldehyde-3-phosphate dehydrogenase	–/–	Archer et al. (1990)
	tef1	Transcription elongation factor 1α	–/–	Kitamoto et al. (1998)
	pki1	Pyruvate kinase	–/–	de Graaff and Visser (1988)
	sodM	Manganese superoxide-dismutase	–/–	Ishida et al. (2004)
	glaA	Glucoamylase A	Starch, glucose/xylose	Boel et al. (1984)
	amyA/ amyB	α-amylase	Starch, maltose/ glucose	Gomi et al. (2000)
	alcA	Alcohol dehydrogenase I	Ethanol/glucose	Waring et al. (1989)
	niiA	Nitrite reductase	Nitrate/ammonium	Muller et al. (2002)
	thiA	Thiamine thiazole synthase	-/thiamine	Shoji et al. (2005)
	sorA/sorB	Sugar transporter/sorbitol dehydrogenase	Sorbitol/-	Oda et al. (2016)
Trichoderma reesei	*gpdA*	Glyceraldehyde-3-phosphate dehydrogenase	–/–	Li et al. (2012)
	tef1	Transcription elongation factor 1α	–/–	Nakari-Setälä and Penttilä (1995)
	pki1	Pyruvate kinase	–/–	Kurzatkowski et al. (1996)
	eno1	Enolase	–/–	Li et al. (2012)
	cDNA1	Hypothetical protein (Trire2:110879)	–/–	Nakari-Setälä and Penttilä (1995)
	hex1	Woronin body protein	–/–	Bergquist et al. (2004)
	cel6a	Cellobiohydrolase, CBH II	Cellulose, lactose/ glucose	Zeilinger et al. (1998)
	cel7a	Cellobiohydrolase, CBH I	Cellulose, lactose/ glucose	Ilmén et al. (1998)
	cel5a	Endoglucanase, EG2	Cellulose, lactose/ glucose	Miyauchi et al. (2013)
	cel12a	Endoglucanase, EG3	Cellulose, lactose/ glucose	Rahman et al. (2009)
	xyn1	Xylanase I	Xylan/glucose	Zeilinger et al. (1996)
Penicillium sp.	*pgkA*	3-phosphoglycerate kinase	–/–	Hoskins and Roberts (1994)
	gdhA	NADP-dependent glutamate dehydrogenase	–/–	Díez et al. (1999)
	paf	Antifungal protein	–/–	Marx et al. (1995)
	xylP	Endoxylanase	Xylose/glucose	Zadra et al. (2000)
	phoA	Acid phosphatase	-/phosphate	Haas et al. (1992)
	pbcC	Isopenicillin N synthase	Lactose, alkaline pH/n.d.	Barredo et al. (1989)
	gla1	Glucoamylase	Glucose/xylose	Bulakhov et al. (2017)
Neurospora crassa	*tub*	Beta-tubulin	–/–	Nakano et al. (1993)
	bli-3	Blue light-inducible gene	Blue light/darkness	Eberle and Russo (1994)
	qa-2	Catabolic 3-dehydroquinase	High/low quinic acid	Giles et al. (1985)
	tcu-1	Copper transporter	Copper depletion/ availability	Lamb et al. (2013)
	vvd	Blue light receptor	Light/darkness	Hurley et al. (2012)
	ccg-1	Clock-controlled gene 1	Light/glucose	McNally and Free (1988)
	gla1	Glucoamylase	Starch, maltose/ glucose	Stone et al. (1993)

[a]Constitutive promoters are labelled (–/–). For regulated promoters, the inducing/repressing compounds or conditions are indicated

involved in alcohol catabolism, and the thiamine repressible ***thiA*** promoter, controlling expression of thiamine thiazole synthase involved in thiamine synthesis. This type of promoters can typically also be expected to function in other species, such as the P*thiA* of *A. oryzae* shown to also be functional in *A. nidulans* (Shoji et al. 2005) and the *A. nidulans* P*alcA* applied in *A. fumigatus* (Romero et al. 2003). Some very **strong inducible promoters** depend on degradation of the feedstock, e.g., the starch-inducible promoter of *A. niger* ***glaA***, controlling expression of a secreted glucoamylase/1,4-alpha-glucosidase for starch hydrolysis. Similarly, a favorite promoter from *T. reesei* is the ***cel7a*** promoter controlling expression of secreted cellobiohydrolase I, which is induced on media containing cellulose or lactose. For promoters that normally drive expression of secreted biomass-degrading enzymes, the conditions inducing the gene expression typically involve co-induction of other endogenous genes of secreted biomass-degrading enzymes. Thereby, this may lead to production of these enzymes as undesired side-products. The activation mechanisms of such promoters are more specialized than for genes of the central metabolism and may not be directly transferrable from species to species. For example, activation of *T. reesei* P*cel7a* involves recruitment of a general transcription factor of cellulase-encoding genes, Xyr1 (Castro et al. 2016), in complex with a non-coding RNA, *HAX1* (Till et al. 2018; Table 2).

a) Approaches for Identification of Natural Promoters

Classical promoters may often be chosen due to their **availability and history** rather than due to considerations on whether their expression profile and strength fit the heterologous production process. Hence, in many cases the default choice is to employ the **strongest promoter** available as a basis for heterologous production. However, in some cases, maximum transcription levels are not desirable. For example, if the goal is to produce a complex secondary metabolite, the set of genes required to synthesize this compound may need to be expressed at different levels to achieve a balanced biosynthetic pathway. **Expanding the library** of fungal promoters to include members displaying a wide range of different expression strengths and other properties is therefore desirable. Fortunately, their discovery is accelerated by the availability of fully or partially sequenced fungal genomes. For example, by analyzing **genome-wide transcription profiles** obtained at different growth stages on different relevant growth media, it is possible to identify promoters with properties that are tailored to fit a given process. In the simplest scheme, one may apply small-scale targeted transcriptomic approaches, like **RT-qPCR** to determine the promoter strength under certain conditions (Li et al. 2012). Alternatively, large publically available datasets can be applied, such as the *A. niger* transcriptome microarrays covering several different growth conditions (Andersen et al. 2008; Breakspear and Momany 2007). Currently, **transcriptomic microarray** datasets covering 155 different cultivation conditions are available for *A. niger* (Schäpe et al. 2019). Using this approach to their advantage, Blumhoff and co-workers identified six novel constitutive *A. niger* promoters with varying expression strength (Blumhoff et al. 2013). The library contains promoters that are stronger and weaker than *A. niger* P*gpdA* displaying ten-fold differences in expression levels of native genes, and when applied to produce the β-glucuronidase (*uidA*) from *E. coli*, 1000-fold in specific activity yields were obtained. A similar set of *T. reesei* promoters displaying varying expression strengths has been derived from its glycolytic genes (Li et al. 2012), and in *Penicillium chrysogenum* both homologous and heterologous promoters have been evaluated for expression efficiency (Polli et al. 2016). It is important to note that although genome-wide transcription profiles provide mRNA levels for specific genes, they do not define the exact promoter sequences that control these genes. To this end, a global map of **transcription initiation sites** in *A. nidulans* was generated based on full-scale RNA-sequencing of gene transcripts obtained at six different growth conditions (Sibthorp et al. 2013), and this will facilitate extraction of new *A. nidulans* promoters as the map simplifies promoter sequence identification.

2. Synthetic Promoters and Gene-Expression Systems

Natural promoters are often influenced by the **state of metabolism** of the host, and this may be undesirable for heterologous production. Synthetic biology-based approaches are therefore adopted to generate artificial/**synthetic gene-expression systems** that ideally act **independent** of the host metabolism.

For example, Gressler et al. developed an expression system based on a fusion of the maltose-inducible *amyB* promoter from *A. oryzae* with the ORF encoding the transcriptional activator TerR from the *Aspergillus terreus* terrein gene cluster (Gressler et al. 2015). In addition, the system contains an expression cassette with a bidirectional promoter that binds TerR. Subsequently, the system was used to heterologously express two GOIs in *A. niger* by addition of maltose.

Another synthetic expression system was developed in *T. reesei* to facilitate cellulose degradation. Cre1 and Ace1; two general glucose stimulated repressors of genes involved in production of cellulolytic enzymes were fused to the activation domain of herpes simplex virus protein 16 (VP16), hence turning glucose repression into gene activation (Zhang et al. 2018).

More advanced orthogonal systems based on **synthetic transcriptional regulators** and matching **synthetic promoter sequences** have made gene expression much more controllable. For example, synthetic biology based methods have elegantly been applied to develop a constitutive expression system, which is almost equally functional in distantly related fungi including *A. niger*, *T. reesei*, and several yeasts (Rantasalo et al. 2018). Such synthetic tools **bypass the organism specificity** of conventional promoters and may become a valuable bio-block for multi-species studies. This system consists of a synthetic transcription factor (sTF) composed of the LexA repressor from *E. coli* fused to the activation domain from VP16. Expression of sTF is controlled by promoters with different arrays of LexA binding sites positioned upstream of a fungal core promoter, which was selected for functionality in several species. Low constitutive expression of the gene encoding the synthetic transcription factor is ensured via another universally active fungal core promoter. Importantly, promoter strength can be regulated by varying the number of LexA binding sites in the synthetic promoter. The system was recently combined with CRISPR-Cas9 multiplexing technology enabling triple targeted integration of the GOI cassette in three different loci in *T. reesei* (Rantasalo et al. 2019).

Orthogonal inducible expression platforms based on the Tet-on and Tet-off systems have been implemented in fungi (Fig. 4) (Meyer et al. 2011; Wanka et al. 2016). The **Tet-off** system is based on the synthetic transcription factor (tTA) a fusion of the *E. coli* repressor TetR with the activation domain of VP16 from herpes simplex virus, which binds to *tetO* operator elements in the absence of the tetracycline derivative doxycycline, Dox (Gossen and Bujard 1992). In contrast, the **Tet-on** system is based on rtTA-M2 (or other variants of rtTA) (Gossen et al. 1995; Urlinger et al. 2000), which are mutated versions of tTA that binds to *tetO* operator elements in the presence of Dox. A Dox responding synthetic promoter is made by positioning *tetO* elements upstream of a core promoter, which provides low or undetectable basal expression level in the absence of the synthetic transcription factors tTA and rtTA-M2 (Wanka et al. 2016). In the Tet-on system, which constitutively produces rtTA-M2, the promoter is activated by addition of Dox, whereas in the Tet-off system, which constitutively produces tTA, the promoter is activated in the absence of Dox. Note that the Tet-on system is often preferred over Tet-off due to its faster response time (Wanka et al. 2016).

D. Terminators

Terminators coordinate transcription termination and the extent of polyadenylation at the 3′-end of the new transcript, which in turn is important for its nuclear export and stability

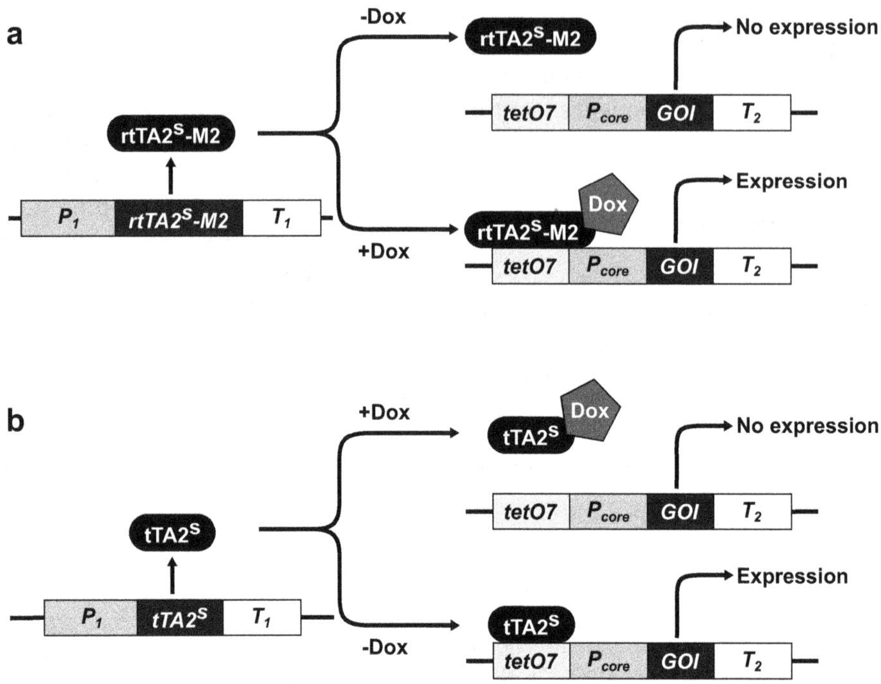

Fig. 4 Graphic representations of the Tet-on and Tet-off expression systems. (**a**) Tet-on system. (**b**) Tet-off system. In the Tet-on system, the Dox ligands acts as an activator of the synthetic TF rtTA2^{s}-M2 allowing it to bind to the *tetO7* sites of the promoter. In the Tet-off system, the Dox ligands repress binding of the synthetic TF tTA2^{s} to the *tetO7* sites of the promoter; see main text for details

(Bentley 2014). In fungi, gene-expression cassettes have typically employed terminators derived from ***tef1*** and ***trpC*** from *Aspergilli*, and it appears that terminators can be **functionally transferred** from one fungal species or genera to another. For example, *A. nidulans tef1* and *trpC* terminators have successfully been used in various *Aspergilli* (Nødvig et al. 2015) and *T. atroroseus* (Nielsen et al. 2017), while the terminators of *T. reesei cbhII* and *pdc* were applied in *A. niger* (Blumhoff et al. 2013) and the basidiomycete *Ganoderma lucidum* (Qin et al. 2017), respectively. Since functionality is often conserved during heterologous application of terminators, an alternative approach for gene expression uses the **GOI and its natural terminator** directly as a functional unit for assembly of the gene-expression cassette (Gressler et al. 2015; Li et al. 2018).

We note that no thorough comparative analyses of the impact of terminators on the overall protein production have been performed for filamentous fungi despite the prominent roles of terminators in the RNA life-cycle. However, in other expression systems, e.g., in yeasts, the choice of terminator has been demonstrated to significantly **influence production yields** (Curran et al. 2013; Morse et al. 2017), and it may therefore be useful to assemble expression cassettes with terminator variants if production yields are suboptimal. To this end, we also note that small functional synthetic terminator bio-blocks have been developed for *S. cerevisiae* to facilitate gene-expression cassette assembly and control of production yields (Curran et al. 2015).

E. Protein Tags and Linkers

In many cases, it is desirable to expand the sequence of a protein with **additional sequences** encoding domains that provide **new properties** to the protein to facilitate its secretion, purification, or visibility (Table 3). These new domains often need to be attached to the

Table 3 Protein-tag and functional-linker sequences

Purpose	Tag	Origin	Sequence[a]	References
Secretion signal	*glaA*-preprosequence	*A. niger*	24 AA	–
	cbhI-presequence	*T. reesei*	17 AA	–
	amyB-presequence	*A. oryzae*	21 AA	–
Affinity and purification	Poly-histidine	Synthetic	HHHHHH	Chaga et al. (1999)
	FLAG	Synthetic	DYKDDDDK	Hopp et al. (1988)
	c-myc	*Homo sapiens*	EQKLISEEDL	Kipriyanov et al. (1996)
	HA	Hemagglutinin antigen	YPYDVPDYA	Jackson et al. (1986)
	Hydrophobin I	*T. reesei*	97 AA	Linder et al. (2004)
Proteolytic site	Kex2 linker and cleavage site	*A. niger* glaA pro-seq	NVIS*KR**	Jalving et al. (2000)
	TEV cleavage site	*Tobacco etch virus*	*ENLYFQ*S*	Carrington and Dougherty (1988)
	Thrombin cleavage site	*Homo sapiens*	*LVPR*GS*	Sticha et al. (1997)
	A2 peptide	*Picornavirus*	20 AA	Ryan et al. (1991)
Visualization	Green fluorescent protein	*Aequorea victoria*	238 AA	Prasher et al. (1992)
	Fluorescent proteins	Various variants	–	Czymmek et al. (2005)
Nuclear localization	SV40-NLS	Simian virus 40	PKKKRKV	Ishii et al. (1996)

[a]Peptide motifs or sequence lengths of the tags are listed. For proteolytic sites, the protease recognition sequence is italics and the cleavage site is indicated by an asterisk

main protein via **linker sequences**, which may act solely as spacers (Chen et al. 2013), or contain a proteolytic site that allows for removal of the attached domain (in vivo or in vitro) when its function is no longer required. This adds to the complexity of the gene-expression cassette structure as additional bio-blocks need to be designed and incorporated into the cassette at the appropriate positions.

The design of these bio-blocks depends on the size of the new functional unit and strategies for their assembly are presented in Fig. 5. **Firstly**, bio-blocks encoding **large size addendums** (>30 amino acid residues) like fluorescent proteins can be made as individual bio-blocks by **PCR**. Similar to the assembly of the basic gene-expression cassette (in Fig. 1), proper incorporation of the new bio-block is ensured by sequences in the tails of the primers used to generate the individual bio-blocks, which can be enzymatically cut to provide overhangs that direct the correct fusion order of all components. **Secondly**, for **medium size bio-blocks** encoding sequences of 10–30 amino acid residues (e.g. secretion signals), the information can be incorporated into each of **two oligonucleotides** that are **annealed** in vitro to form the complete bio-block. Note that small sequence extensions may be added to the oligonucleotides to produce bio-blocks, which are directly equipped with short ssDNA overhangs for bio-block assembly. **Thirdly**, for **shorter sequences** (<10 amino acid residues), e.g. purification tags or proteolytic sites, the information may be incorporated into the **primer tails** of neighboring bio-blocks. For example, if the C-terminus of a protein needs to be extended with a poly-histidine tag for purification, the primer tail used to link the GOI to the next bio-block could be elongated with the sequence encoding the tag. The combinations in which these basic strategies can be employed to construct GECs are numerous, and as the prices decrease, it is likely that even large size protein tags will be synthesized de novo rather than by PCR amplification.

IV. Heterologous Protein Production in Filamentous Fungi

Filamentous fungi serve as favorite hosts for production of industrial enzymes due to their superior **protein secretory** potential. This

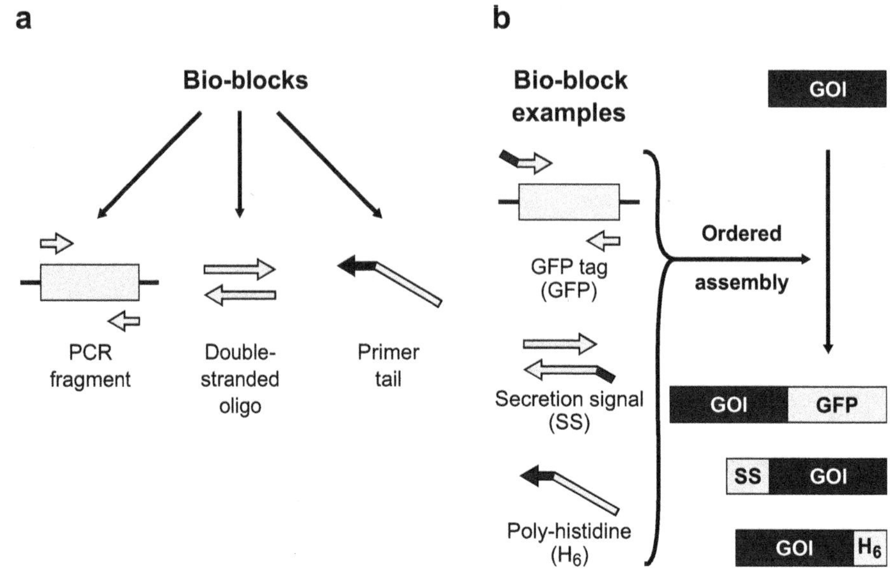

Fig. 5 Assembly strategies for employing protein tags. (**a**) Bio-blocks can be generated by three general strategies depending on their size: PCR amplification, annealing of two oligonucleotides, or incorporation into the tail of a primer used to PCR amplify another bio-block. (**b**) The resulting bio-blocks can be fused in a directed manner as illustrated in Fig. 1

potential has been further improved by classical mutagenesis yielding strains with superior protein secretion properties. This includes *A. niger* CBS 513.88 and *T. reesei* RUT-C30, which can reach titers of 30 and 100 g/L of **endogenous cellulolytic enzymes** by optimized cultivations, respectively (Cairns et al. 2018; Cherry and Fidantsef 2003). Detailed mechanistic insights into the secretory pathway combined with fully sequenced genomes and a diverse range of omics data set the stage for **rational strain engineering.** In this section, we will present examples of genetically modified strains, which serve as platforms for heterologous protein production. This includes strains with improved production and secretion physiology obtained by introducing defined mutations influencing processes ranging from the delivery of amino acid building blocks for protein synthesis to processes involved in exocytosis.

Specifically, we will treat two major topics: firstly, engineering the **secretory pathway** focusing on transport signals, the associated processes of protein glycosylation and folding, and vesicular trafficking and secondly, expression in **protease-deficient strains** as a method for reducing degradation of the heterologous protein.

A. Engineering the Secretory Pathway

The protein secretion pathway is complex and provides several processes that can be engineered to enhance production yields. Generally, protein secretion processes can be grouped into three major themes. Firstly, **secretion signals** that mediate translocation into the **endoplasmic reticulum** (ER). Secondly, **folding and glycosylation** of secretory proteins. Thirdly, the complex machinery, which transports proteins through the secretory pathway and leads to **secretion by exocytosis.** Below, we provide examples showing how each of these steps can be optimized by rational genetic engineering.

1. Transport Signals

A significant obstacle in optimizing secretion of a heterologous protein is choosing the right pre- and pro-sequence, as protein secretion is inefficient when suboptimal signals are applied.

Pre-sequences encode secretion signals that mediate protein translocation into the ER, while **pro-sequences** may facilitate the folding process and maintain proteins as inactive forms until its removal. One of the three general approaches is typically used to ensure secretion of a heterologous protein. In the **first approach**, secretion of the recombinant protein simply relies on the secretion signals of the native protein. In the **second approach**, the N-terminus of the recombinant protein is extended with either a pre- or pre-pro-sequence from a highly secreted protein of the production host (or from a closely related species) or **thirdly** N-terminal extension with a secreted carrier protein.

The first, and most simple, approach employs the **native secretion signal** of the heterologous protein. However, the fact that the sorting signal are from a different organism raises questions concerning their ability to **support efficient secretion in the host**. The observation that heterologous proteins originating from species closely related to the host often sort quite efficiently suggests that their secretion signals are functional.

For example, glucoamylase (*glaA*) from *A. niger* and rhamnogalacturonate lyase A (*rglA*) from *Aspergillus sojae* have been efficiently produced via their own secretion signals in *A. nidulans* (Schalén et al. 2016) and in *A. oryzae* (Yoshino-Yasuda et al. 2012), respectively.

Secretion signals may also be functional between more distantly related fungi, as observed with two basidiomycete laccases, Lcc1 of *Pycnoporus coccineus* and Lcc of *Pycnoporus sanguineus*, successfully secreted by the hosts *A. oryzae* and *A. nidulans* using the proteins native secretion signals, albeit with low yields (Hoshida et al. 2005; Li et al. 2018). More efficient secretion may be achieved by using **host endogenous signals** as they are expected to be more proficiently recognized and processed by the host. The second approach therefore employs the pre- or pre-pro-sequence of a highly secreted protein originating from the host, or a closely related species, to enhance secretion of the heterologous protein.

A successful example of this approach is production of basidiomycete laccase, Lac1, from *Pycnoporus cinnabarinus* in *A. niger* (Record et al. 2002). In this study, an 80-fold increase in extracellular activity was achieved by replacing the natural pre-sequence of Lac1 with the pre-pro-sequence of *A. niger glaA*.

In the final approach, a **carrier protein** is fused to the heterologous protein to improve production by alleviating post-translational bottlenecks and, in some cases, by increasing mRNA levels (Gouka et al. 1997). Comparative DNA microarrays of *A. oryzae* strains expressing heterologous bovine chymosin, with and without a carrier protein (AmyB), showed that inclusion of the carrier protein promoted induction of the UPR (see Sect. IV.A.2.b) and increased expression of genes encoding secretory chaperones and proteins involved in intracellular trafficking, thereby facilitating folding and secretion (Ohno et al. 2011). Examples of proteins that have been used as carriers are GlaA, AmyB, and CbhI from *A. niger*, *A. oryzae*, and *T. reesei*, respectively. To liberate the heterologous protein from the carrier protein in vivo, the two proteins are fused via a linker containing a **Kex2 cleavage site** (KR/RR) to enable **proteolytic separation** in the late-Golgi (Jin et al. 2007; Landowski et al. 2016).

Employing this approach, Gouka et al. used *A. niger* GlaA as carrier protein for heterologous production of human interleukin-6 in *Aspergillus awamori*, thereby improving the extracellular protein yield 100-fold (Gouka et al. 1997).

In an interesting variant of this approach, Jin et al. produced human lysozyme in *A. oryzae* using a gene cassette containing the *amyB* gene fused to tandem gene copies of *HLY* encoding lysosome, which were separated by sequences encoding Kex2 cleavage sites (Jin et al. 2007). However, releasing the heterologous protein from the carrier protein may be a bottleneck. For example, during production of human interferon (IFNα-2b) in *T. reesei* using Cbh1 as a carrier protein, 44% of the fusion protein was not cleaved by Kex2 (Landowski et al. 2016). More **efficient Kex2 cleavage** has been achieved by overexpressing *kex2* (Landowski et al. 2016) or by optimizing Kex2 cleavage sites (Lin et al. 2006; Nakajima et al. 2006;

Yang et al. 2013), e.g., by using the entire **pro-sequence of GlaA**, NVISKR, as a Kex2 cleavage site (Landowski et al. 2016).

It is important to stress that the choice of **optimal signal sequences** is not straightforward. For example, Rantasalo et al. optimized secretion of lipase B from *Candida antarctica* (CalB) in *T. reesei* (Rantasalo et al. 2019) by using natural secretion signals of the host proteins CbhI and CbhII and secretion signals from heterologous AmyA and GlaA from *A. awamori* and *A. niger*, respectively, and by using CbhI as a carrier protein. When the lipase yields obtained with the different strains were compared, the strain employing the secretion signal of AmyA performed best. On the other hand, studies on heterologous production of Cbh1 in *A. awamori* using different secretion signals showed no differences illustrating that gains may be dependent on the heterologous protein and/or the host (Adney et al. 2003; Chou et al. 2004).

2. Glycosylation and Folding

Glycosylation of the heterologous protein is important for the subsequent folding, and without validity of these properties, the protein activity may be impaired or abolished. In the heterologous production host, these processes may be executed in alternate fashion compared to the native producer, leading to reduced yields or non-functional proteins. Therefore, substantial efforts have been conducted to improve both processes by increasing or reducing expression of genes encoding proteins involved in the glycosylation and folding machinery of the production host to enhance protein production.

a) Glycosylation

Protein glycosylation has multifaceted impact on heterologous protein production. Firstly, it affects **protein folding and secretion** efficiencies; secondly, it influences the final properties of the enzyme including its **stability, solubility, and catalytic** parameters; and finally, it acts to **protect the enzymes** from proteases (Gupta and Shukla 2018; Li and d'Anjou 2009). In agreement with this view, mutation of the four N-glycosylation sites of *A. terreus* β-glucosidase resulted in reduced thermal stability and 15–35% decreased specific activity and catalytic rate as compared to the wild-type variant when heterologously produced by *T. reesei* (Wei et al. 2013). Hence, proper glycosylation is often required to achieve high production yields. For example, analysis of bovine chymosin produced by *A. niger* showed that an N-glycosylation site was poorly glycosylated by this host. Importantly, the extracellular activity yields could be increased by up to 100% in *A. niger* by **optimizing the glycosylation efficiency** of this site by site-directed mutagenesis (van den Brink et al. 2006). Similarly, production yields and quality may also be increased by engineering entirely **new glycosylation sites** into proteins. Using this approach, heterologous production activity yields of cellobiohydrolase Cel7A from *Penicillium funiculosum* in *A. awamori* were increased by 70% by engineering a novel N-glycosylation motif into the protein at position N194 (Adney et al. 2009). On the other hand, production yields and quality may be decreased if **non-native glycosylation sites** are unintentionally or **aberrantly glycosylated** by the glycosylation machinery of the heterologous host. Hence, a 70% increase of *T. reesei* Cel7A activity yields were achieved by removing a glycosylation site (N384), which contained a larger glycan structure when using *A. awamori* as production host, than in the native organism *T. reesei* (Adney et al. 2009). Fungi may also potentially be used as hosts for production of **therapeutic proteins**. The fact that different species add different sugar structures to proteins poses an additional production challenge, as therapeutic proteins containing aberrant sugar moieties may cause immune responses in patients. In yeasts, this challenge has been successfully addressed by **humanizing the glycosylation pathway** by genetic engineering (Gupta and Shukla 2018) indicating that production of therapeutic proteins may also be possible in filamentous fungi after dedicated strain engineering (Anyaogu and Mortensen 2015).

b) Chaperones and ER Stress

Overexpression of secretory proteins may result in **accumulation of misfolded proteins** in the ER, and this problem poses a severe bottleneck toward high yields of secreted proteins. Cells respond to misfolded proteins in the ER by triggering the *unfolded protein response* (**UPR**), which is sensed by the ER transmembrane protein Ire1. Normally, Ire1 forms an inactive complex with the chaperone BipA, but as BipA is increasingly recruited to assist in folding, Ire1 forms a nucleolytically active dimer that catalyzes splicing of the nonconventional intron in the ***hac1*** **mRNA** (Krishnan and Askew 2014). The processed *hac1* mRNA encodes mature **Hac1 transcription factor**, which in turn triggers transcription of UPR genes. This includes genes encoding ER-resident molecular chaperones like BipA, the heat-shock protein family of chaperones (Hsp104, Hsp70, Hsp90) and the protein disulfide-isomerase PdiA, thus **satisfying the increased demand for folding capacity** (Zubieta et al. 2018). Persistently misfolded proteins are deleterious to the cell and are degraded through the ERAD (*ER-associated degradation*) pathway that includes transport to the cytoplasm and ubiquitin-mediated degradation by the proteasome (Carvalho et al. 2011).

Several strategies have been pursued to optimize folding. These include defensive methods, like lowering the growth temperature or applying weaker promoters, but also potentially more awarding methods based on stimulating the folding potential via genetic engineering of the host organism folding machinery.

To promote heterologous protein production in *A. awamori*, the UPR pathway has been constitutively activated by overexpressing an intron-free variant of the *hac1* gene. Using this strain, production of *Trametes versicolor* laccase (Lcc1) was increased sevenfold and bovine chymosin 2.8-fold (Valkonen et al. 2003). Similarly, heterologous production of *A. niger* glucose oxidase (Gox) in a *T. reesei* strain overexpressing intron-free *hac1* increased production 1.8-fold (Wu et al. 2017). In comparison, overexpression of the *bipA* chaperone alone in *T. reesei* increased production of Gox by 1.5-fold.

Folding of proteins containing multiple disulfide bonds may be particularly challenging and may benefit from increased levels of the protein disulfide isomerase PdiA. In agreement with this, production of thaumatin, a sweet-tasting plant protein containing eight disulfide bonds, was increased twofold by using an *A. awamori* strain overexpressing *pdiA* (Moralejo et al. 2001). Since heterologous protein production appears to benefit from increased levels of host chaperones, it is tempting to speculate that more process-specific folding assistance could be achieved by co-expressing genes encoding one or more chaperones from the natural source of the protein of interest.

3. Vesicular Trafficking and Polarized Growth

Filamentous fungi propagate as hyphae and the general view is that most **protein secretion occurs at the hyphal tip** (Cairns et al. 2019). Secreted proteins are packed into vesicles, either by cargo-receptors or as the result of bulk flow, thereby mediating transport from the ER to the Golgi and from the Golgi to the plasma membrane (Barlowe and Miller 2013). Most vesicles accumulate at the Spitzenkörper at the apex of the hyphae before fusing to the membrane in a process that depends on the exocyst octamer (Ahmed et al. 2018; Riquelme and Sánchez-León 2014). Since proteins are synthesized throughout the hyphae, **vesicles are transported by an elaborate transport system** based on actin filaments and microtubules to ensure efficient transport to the hyphal apex (Steinberg et al. 2017). Many of the individual steps in the secretory pathway are understood in molecular detail, and a popular strategy aiming at enhancing secretory transport is to overexpress genes encoding proteins that are directly involved in transport. For example, **loading of cargo proteins** into vesicles and **targeting vesicles** to a destination membrane have been engineered to increase secretion as exemplified below.

To stimulate secretion of a heterologous protein, the vesicular transport was enhanced by overexpressing *rabD*, encoding a Rab GTPase involved in transport of exocytic post-Golgi vesicles (Pantazopoulou et al.

2014), the deletion of which reduces protein secretion (Punt et al. 2001). Specifically, a fusion of carrier glucoamylase to mRFP via a Kex2 cleavable linker (Kex2cl) increased production of mRFP in *A. nidulans* by 40% (Schalén et al. 2016) in the *rabD* overexpression background. Hoang and co-workers showed that Vip36, a putative lectin-type cargo receptor inferred to be involved in vesicular cargo loading of glycoproteins between ER and Golgi, facilitates secretion of the recombinant fusion proteins AmyB-Kex2cl-GFP and AmyB-Kex2cl-chymosin in *A. oryzae*, presumably by reducing protein retention in the ER (Hoang et al. 2015). Importantly, overexpression of *vip36* in strains producing the recombinant GFP fusion protein led to sevenfold increase in the levels of extracellular GFP (Hoang et al. 2015). Finally, docking of vesicles to a target membrane was stimulated by overexpressing *sncl* encoding a vSNARE, and this feature increased secretion of *A. niger* glucose oxidase in *T. reesei* by 2.2-fold (Wu et al. 2017).

Similarly, specific genes involved in the secretory pathway can be deleted to positively influence heterologous protein production. For example, the vacuolar protein sorting receptor Vps10 mediates transport of proteins from the Golgi to the vacuoles; and it has been observed that heterologous proteins designated for secretion partially accumulate in the vacuoles (Masai et al. 2003). This principle has been exploited in *A. oryzae* where a deletion of *vps10* improved production of recombinant human lysosome and bovine chymosin 2.2- and 3-fold, respectively (Yoon et al. 2010). Even the **micromorphology** of the mycelium can be manipulated to enhance protein secretion. In an elegant study, Fiedler et al. produced a **hyperbranching** *A. niger* strain by deleting the gene encoding the Rho G-protein **RacA**, a key protein in maintaining polarized growth (Kwon et al. 2011); and this feature increased glucoamylase production four-fold (Fiedler et al. 2018).

C. Building Blocks for Protein Synthesis

Many filamentous fungi secrete large amounts of enzymes and production of these proteins consume building blocks that could be used for the desired heterologous protein. Hence, additional building blocks for heterologous protein production can be obtained by deleting genes encoding the most abundant secreted proteins. This strategy has the further advantages that it reduces pressure on the secretory pathway machinery and that subsequent product purification is simplified. In its simplest scheme, this approach can be utilized by **deleting genes encoding major secreted products**, such as α-amylase A and B of *A. oryzae*, thereby providing a strain with no detectable α-amylase activity (Kitamoto et al. 2015). Toward the same goal, genes encoding **transcriptional activators** of major cellulolytic enzymes have been deleted in both *A. niger* and *T. reesei*.

In *A. niger*, deletion of *amyR* reduced the total amount of secreted protein 16.4-fold (Zhang et al. 2016b). Likewise, deletion of *xyr1* in *T. reesei* abolished expression of the genes *cbhI* and *cbhII* encoding the major cellobiohydrolases CBHI and CBHII, which may constitute up to 80% of the total amount of secreted protein (Bergquist et al. 2004; Stricker et al. 2006).

While these approaches diminished expression of endogenous enzymes, it may also be advantageous to modify the GOI expression cassette. To this end, for driving secretion of a heterologously produced protein, Rantasalo et al. exchanged usage of a carrier protein for a smaller secretion signal, thereby reducing the total secreted protein twofold and inferring a substantial increase in recombinant enzyme purity (Rantasalo et al. 2019).

Comparative transcriptomics approaches have been used to uncover **engineering targets** that enhance heterologous protein production. For example, processes related to **biosynthesis of amino acids** and **tRNAs** are upregulated in strains of *A. nidulans* overexpressing heterologous enzymes (Zubieta et al. 2018) and in *A. niger* CBS 513.88, a classic enzyme production strain (Andersen et al. 2011). Hence, building block availability may be a limiting factor that could be improved by genetic engineering, e.g., by increasing expression of genes encoding specific transporters or genes required for making aminoacryl-tRNAs.

D. Protease-Deficient Strains

The **saprophytic lifestyle** of filamentous fungi is key to their high secretion capacity. In addition to facilitating heterologous protein production, a consequence of this lifestyle is secretion of **large amounts of proteases**, which the fungi use for extracellular **biomass degradation**. For example, proteomic analysis of the *T. reesei* QM6a secretome under submerged cultivation showed pH-dependent secretion of 39 peptidases and proteinases, with more than 20 secreted simultaneously (Adav et al. 2011). Since these proteases need to digest a broad spectrum of substrates, chances are that the **heterologous protein is also degraded** leading to decreasing activity of the recombinant protein over time (Kamaruddin et al. 2018). In addition, intracellular proteases in the secretory pathway, such as Kex2, may also degrade recombinant proteins, which can be resolved by mutagenizing the cleavage site (Lin et al. 2006). The fact that addition of protease inhibitors, such as trypsin and aspartic protease inhibitor (Landowski et al. 2015), or use of strains with different protease profiles (Sun et al. 2016) can lead to significantly higher protein yields suggests that a strategy aiming at **producing protease-deficient strains** could be advantageous. However, whereas elimination of some proteases does not impact fitness, some may be involved in cell wall maintenance, and their absence often causes an altered undesired morphology. Hence, a gain in product yield may be lost in growth rate or overall productivity, and the relevant strain modifications vary from product to product even with the same host (Landowski et al. 2015).

In a simple case, Li et al. applied *A. nidulans* as host for heterologous production of a *P. sanguineus* laccase (Pslcc) and demonstrated that deletion of two genes encoding major proteolytic activities, proteases dipeptidyl-peptidase DppV and aspartic protease PepA, increased laccase production 13-fold (Li et al. 2018).

In a more elaborate study, Kitamoto et al. generated an *A. oryzae* strain with five protease gene deletions (*alp1*, *npI*, *npII*, *pepA*, *pepE*) thereby reducing extracellular protease activity to 1% (Kitamoto et al. 2015) and providing a production strain with minimal proteolysis.

In the post-genomic era, it is possible to make a **rational gene deletion strategy** based on proteomics where **abundant proteases** can be identified and linked to their genes. Using this strategy in *T. reesei*, nine target genes (*pep1*, *tsp1*, *slp1*, *gap1*, *gap2*, *pep4*, *pep3*, *pep5*, *amp2*) were identified and deleted to produce a strain where the majority of proteolytic product degradation is abolished (Landowski et al. 2015, 2016). The authors employed this strain for production of human interferon alpha-2b (IFNα-2b) and obtained 2.4 g/L of correctly processed IFNα-2b.

In contrast to deleting individual protease genes, similar effects may be achieved by RNAi silencing of individual genes (Kitamoto et al. 2015; Qin et al. 2012). In another approach, reduction of a set of proteases was achieved by deleting a general **transcription factor gene**, namely, *prtT* in *A. niger*, which reduced the levels of several secreted proteases, including PepA, PepB, and PepF. After producing *Glomerella cingulata* cutinase in this *A. niger* strain a 25-fold increase of residual activity was observed in culture filtrates after 2 weeks (Kamaruddin et al. 2018).

V. Heterologous Production of Secondary Metabolites in Filamentous Fungi

The genome sequencing projects have uncovered a vast repertoire of fungal SMs, which serves as an underexploited source of new **food additives** and **drug candidates**. This process has been facilitated by the fact that all genes required for production of a specific SM are typically organized in a **biosynthetic gene cluster** (BGC) (Keller 2019; Keller et al. 2005; Rokas et al. 2018). However, it is difficult to exploit the fungal SM potential for several reasons. Firstly, the vast majority of the fungal **SM BGCs are silent**; secondly, many producer species are difficult to propagate in bioreactors; thirdly, many compounds are produced in tiny amounts; fourthly, no genetic toolbox exists for new natural production hosts; and finally, the "generally recognized as safe" (GRAS) production status may be more difficult to achieve with a new species.

Heterologous expression of fungal BCGs using synthetic biology based approaches in well-characterized fungal cell factories provides avenues to speed up **SM discovery**, characterization, and production. On the other hand, **heterologous SM production** is challenged by several features including toxicity of some SMs, poor gene annotations, intron splicing differences, compartmentalization of pathways, and the requirement for **simultaneous expression of many SM genes**. Unlike heterologous production of industrial enzymes, heterologous production of SMs is still in its infancy, and most studies are aiming at product discovery and pathway elucidation rather than large-scale production. In this section, we will present examples on how heterologous SM-gene expression has contributed to expand our insights into SM biosynthesis, as well as challenges toward their production.

A. Challenges in Heterologous Secondary Metabolite Production

1. Product Toxicity

Many SMs are antimicrobials, and they may therefore impair the growth of, or even kill, the new heterologous producer strain. In case the task is to produce large amounts of a known SM, the first experiment should therefore be to test whether the host can tolerate the desired SM. If **SM toxicity** is a problem, it may be necessary to develop a **resistant strain.** If the resistance mechanism is known in the native producer species, it may be possible to **transfer the resistance mechanism** from the native host to the new producer strain.

For example, mycophenolic acid (MPA) from *Penicillium brevicompactum* kills *A. nidulans* by inhibiting its inosine-5′-monophosphate dehydrogenase (IMPDH). However, *A. nidulans* can be engineered to tolerate mycophenolic acid by inserting *mpaF* of the *P. brevicompactum mpa* BCG, which encodes an MPA-insensitive IMPDH, into its genome (Hansen et al. 2011a).

Toxicity can also be avoided or reduced by **introducing a pump** that exports the new compound. For example, the gliotoxin sensitivity of a sirodesmin transporter-deficient strain of *Leptosphaeria maculans* can be rescued by introduction of the transporter gene *gliA* from *A. fumigatus* (Gardiner et al. 2005). Similarly, *S. cerevisiae* expressing the *mlcE* efflux pump gene from the compactin BGC is protected against statins (Ley et al. 2015). A different mode of detoxification is based on **glycosylation of the SM**, a principle which is commonly used in plants (Sandermann Jr. 1992), but which has also been observed in fungi. For example, during yanuthone production by *A. niger*, a toxic intermediate is glycosylated (Holm et al. 2014), or during co-culturing, *Trichoderma* species uses glycosylation as a defense against deoxynivalenol, a toxin produced by *Fusarium graminearum* (Tian et al. 2016). For cell factory construction, this strategy has been applied for vanillin production in *S. cerevisiae* by expressing a gene encoding a glycosyltransferase from *Arabidopsis thaliana* (Brochado et al. 2010).

If the task of heterologous production is **product discovery**, rather than large-scale production, toxicity issues may be less important as only small amounts are sufficient for product identification. However, if weak growth of the new cell factory is observed, it may be advantageous to use **inducible promoters** to control expression of key genes in the pathway. For example, the inducible *alcA* promoter was fused to genes in the asperfuranone BGC to avoid toxicity issues during pathway elucidation (Chiang et al. 2013).

2. Genome Annotation and Intron Splicing

Heterologous production requires that the genes in the native producers are correctly annotated and for many sequenced genes, this may not be the case. In cases of doubt, it may be necessary to determine the **correct 5′-end** of the transcript by the rapid amplification of cDNA ends (RACE) technique (Frohman et al. 1988). Subsequently, **introns** and the **correct stop codon** can be identified by generating a complete cDNA of the gene by RT-PCR exploiting that the 3′-end of the transcript is polyadenylated. However, in cases where the GOI is silent, this strategy is not possible, and it may therefore be advisable to produce a set of gene constructs covering different combinations of start and stop codon possibilities.

Intron recognition and splicing differs between species, especially for phylogenetically distant organisms (Kupfer et al. 2004), and flawed splicing may reduce or even prevent formation of the desired protein as described in Sect. III.A.1. For example, introns are more abundant in basidiomycetes than in ascomycetes (Stajich et al. 2007). Moreover, introns from basidiomycetes may not be recognized properly in ascomycetes.

In agreement with this, a recent study using *A. oryzae* as a cell factory demonstrated that only half of the mRNA species produced by 30 terpene synthase genes from two basidiomycetes, *Clitopilus pseudo-pinsitus* and *Stereum hirsutum*, exhibited correct splicing patterns (Nagamine et al. 2019).

Even among species that are more closely related, e.g., within ascomycetes, transfer of genes between species may cause **splicing errors**. For example, heterologous expression of the 3-methylorcinaldehyde synthase gene from *Acremonium strictum* (Fisch et al. 2010), the avirulence gene ACE1 from *Magnaporthe oryzae* (Song et al. 2015), and the citrinin synthase gene from *Monascus ruber* (He and Cox 2016) in *A. oryzae* resulted in incorrect processing of introns.

Moreover, some mRNAs of filamentous fungi are **regulated by alternative splicing** (Kempken 2013; Zhao et al. 2013). Hence, populations of different splice variants were detected in the transcriptome (Wang et al. 2010) with consequences for protein levels (Chang et al. 2010). Indeed, new splice variants may be inactive or the correct variant may be produced only in small amounts (He and Cox 2016; Kempken and Windhofer 2004). Mapping mRNA splicing in the native producer, if possible, is therefore advisable; and if it is different in the new host, it may be necessary to produce different synthetic gene variants covering different splice variants to achieve heterologous production.

3. Compartmentalization of Secondary Metabolite Biosynthetic Pathways

Fungal secondary metabolite biosynthetic pathways are often compartmentalized into **different subcellular locations**, such as ER, vesicles, peroxisomes, cytoplasm, or vacuoles (Kistler and Broz 2015; Roze et al. 2011). The reason for this is either to **channel the pathway** into the compartment with relevant precursors to increase biosynthesis efficiency as in the case of penicillin in *A. nidulans* (Herr and Fischer 2014) or to **sequester the toxic product** or intermediates from the rest of the cell as in the case of aflatoxin synthesis in *Aspergillus parasiticus* (Chanda et al. 2009). Little is known about how SM biosynthetic pathways are compartmentalized in fungi, and it is unclear how often this occurs. However, in cases where compartmentalization is important for **product formation** or to **avoid toxicity effects**, it is likely essential that the pathway is organized in a similar manner in a new host. Otherwise, it may result in lack of production or even cell death. In the future, we expect that pathway compartmentalization will achieve significant attention in SM cell factory construction. As a standard **control experiment**, we advise to determine whether individual enzymes in the pathway localize to the same compartments in the natural producer and in the new host, e.g., by GFP tagging. In cases where the host sorts an enzyme incorrectly, it may be possible to ensure proper localization of the relevant enzyme by fusing it to transport signals or a carrier protein that is sorted to the correct compartment.

B. Secondary Metabolite Discovery via Different Gene-Expression Systems

Multiple gene-expression strategies and different fungal hosts have been used for SMs discovery and pathway elucidation by heterologous gene expression. ***A. oryzae***, unlike most other common fungal cell factories, **produces few SMs**. *A. oryzae* is therefore often chosen as a host as the subsequent detection of the new SM in the **metabolite profile is relatively simple**. However, with *A. oryzae* genetic engineering is complicated by the fact that its asexual spores contain several nuclei (Maruyama et al. 2001), which makes purification of transformed strains more laborious. In contrast, *A. nidulans* produces asexual spores that contain only a single nucleus (Yuill 1950), and transformants are therefore relatively easy to purify. Hence, if pathway elucidation requires more elaborate genetic engineering, species like *A. nidulans* may be a better choice of host. Moreover, in some cases, the more **complex host-chemistry** can advantageously contribute to **formation of novel synthetic** and potentially useful compounds. Alternatively, it may facilitate construction of a new synthetic pathway by delivering a missing activity in the desired pathway. Below we will provide successful examples of heterologous SM production using different hosts and expression systems, which may serve as an inspiration for construction of new fungal SM cell factories.

1. Heterologous Expression of Secondary Metabolite Synthase Genes

To investigate whether a new SM can be produced in a given cell factory, it may often be advantageous to **establish the first step** of a multi-enzyme pathway, which produces the **scaffold of the final compound**. Indeed, the first demonstration of a heterologously produced SM was 6-methylsalicylic acid (6-MSA) (Fujii et al. 1996), which serves as a precursor of several SMs including patulin and yanuthones (Beck et al. 1990; Holm et al. 2014; Petersen et al. 2015; Read and Vining 1968). Hence, it is of interest to develop **efficient systems for single gene transfer**. In *A. oryzae*, a plasmid was constructed to allow a gene to be inserted ectopically into the genome in multiple copies (Fujii et al. 1995).

This plasmid has been widely applied for heterologous gene expression, and it has been used to deliver basic SM scaffolds including production of the naphthopyrone YWA1 (Watanabe et al. 1998, 1999), alternapyrone (Fujii et al. 2005), ferrirhodin (Munawar et al. 2013), and astellifadiene (Matsuda et al. 2016).

In *A. nidulans*, where genetic engineering is easier, genes have often been inserted into a defined locus to facilitate strain characterization and to obtain better gene-expression control. To ensure high expression levels, it may be useful to position integration sites in intergenic regions located in transcriptionally highly active sections of a chromosome.

Using this approach, it has been demonstrated that MpaC from *P. brevicompactum* produces 5-methylorsellinic acid (5-MOA) as the first intermediate toward production of the immunosuppressant drug mycophenolic acid by expressing *mpaC* in *A. nidulans* (Hansen et al. 2011b).

The same setup was used to make a cell factory yielding 1.8 g/l of 6-MSA by expressing *yanA* from *A. niger* in *A. nidulans* (Knudsen 2015), which is sixfold higher as compared to ectopic integration in *A. oryzae* reported by Fujii et al. (1996). Using the *A. nidulans* 6-MSA cell factory, it was possible to produce ^{13}C-labelled 6-MSA, which was subsequently used to clarify the biosynthetic pathway for yanuthone D production in *A. niger* (Holm et al. 2014). The *wA* locus of *A. nidulans* has also been used as a gene-expression site. In this case, correct transformants can easily be identified as they produce white conidia. Using this approach, Chiang and co-workers discovered several new compounds by expressing *A. terreus* polyketide synthase genes in *A. nidulans* (Chiang et al. 2013).

Finally, we note that synthesis of some scaffolds requires the action of additional enzymes, such as a *trans*-enoyl reductase, a *trans*-acyltransferase, a *trans*-thioesterase, or another synthase. For example, tenellin synthetase from *Beauveria bassiana* requires *trans*-acting enoyl reductase for correct polyketide scaffold assembly (Halo et al. 2008); during lovastatin synthesis, both acyltransferase and thioesterase are necessary in order to release the polyketide products from the synthases LovF (Xie et al. 2009) and LovB (Xu et al. 2013), respectively, whereas for the synthesis of the first intermediate toward asperfuranone, two polyketide synthases are needed (Chiang et al. 2009). In these cases, a multi-gene insertion strategy is required; see Sect. II.C.

2. Reconstitution of Secondary Metabolite Pathways in a Heterologous Host

Establishing entire pathways is more challenging as several genes need to be functionally expressed. In the following section, we will describe various strategies that have been used to ensure expression of some or all genes from a BGC.

In an early study by Sakai et al., a 20 kb fragment of DNA containing the citrinin BGC from *Monascus purpureus* was inserted into an *E. coli-Aspergillus* shuttle cosmid and ectopically integrated into the genome of *A. oryzae* to produce 4 µg/L of citrinin (Sakai et al. 2008). This **yield was increased** approximately 400-fold to 1.48 mg/L when the **cluster-specific transcriptional regulator** encoding gene (*ctnA*) was **constitutively expressed** under control of the *A. nidulans trpC* promoter. Pathways may also be transferred using a vector set developed by Pahirulzaman et al., which allows up to 12 different genes to be ectopically integrated into the genome via three vectors containing different selectable markers (Pahirulzaman et al. 2012). Hence, pathways may be partly or entirely established in a fungus to produce key intermediates for pathway elucidation as well as to identify the final product. Using this system and *A. oryzae* as a host, He and Cox produced a detailed model of the citrinin pathway and generated a cell factory producing citrinin to final titers of 19.1 mg/L (He and Cox 2016).

Defined integration of entire BGCs into fungal chromosomal expression sites has also been used to establish functional SM pathways. One advantage of this method is that **reverse genetics** can be applied to **dissect the pathway** once the pathway has been functionally established in a new host. This principle has been exploited to successfully transfer all 12 biosynthetic genes required for geodin production from *A. terreus* to *A. nidulans*. Additionally, to activate the biosynthetic genes, the TF gene of the cluster (*gedR*) was equipped with a constitutive *A. nidulans* promoter prior to genomic integration. Subsequently, the efficient gene deletion toolbox of *A. nidulans* was used to identify *gedL* as the gene encoding the halogenase necessary for geodin production (Nielsen et al. 2013).

A recent screening-friendly method for linking BGC to metabolites uses FACs as vectors for gene transfer (see Sect. II.B.1) as they can accommodate DNA fragments containing even large BGCs. FAC-based BGC libraries from *A. terreus*, *Aspergillus wentii*, and *Aspergillus aculeatus* have been transferred and analyzed in *A. nidulans* resulting in the assignment of 17 compounds to BGCs (Bok et al. 2015; Clevenger et al. 2017).

3. Synthetic Pathway Setups for Secondary Metabolite Production

For some compounds of interest, it is not possible to establish a cell factory based on the natural pathway. This is the case if the **biosynthetic pathway is unknown** or it has only been **partly elucidated** or if enzymes are compartmentalized in the native producer in a manner that cannot be implemented in the new host. In these cases, it may be possible to make a cell factory based on a **semi-synthetic or synthetic pathway** with enzymes of known (or predicted) activities from various species. In this way, it may be possible to synthesize the desired compound, but in a manner, which is **different from the pathway in the native host**. For example, production of the pharmacologically relevant meroterpenoid **daurichromenic acid** (DCA) from the plant *Rhododendron dauricum* in *A. oryzae* represents a successful case of a functional semi-synthetic pathway (Okada et al. 2017). Using fungal expression vectors for ectopic genome integration, the production of

DCA was achieved by expressing two fungal genes from *Stachybotrys bisbyi* encoding a polyketide synthase (StbA) and a prenyltransferase (StbC), as well as a plant gene from the native producer encoding the DCA synthase. Similarly, production of **carminic acid**, an important food colorant produced by the scale insect *Dactylopius coccus*, has been achieved in *A. nidulans* via a semi-synthetic pathway (Frandsen et al. 2018). In this case, the five-step pathway was based on a single gene from the natural producer encoding a C-glucosyltransferase, a plant gene from *Aloe arborescens* encoding a type III octaketide synthase, two bacterial genes from *Streptomyces* sp. R1128 encoding a cyclase (ZhuI) and a aromatase (ZhuJ), and finally an unknown fungal gene from *A. nidulans*, which putatively encodes a monooxygenase. Importantly, unlike the DCA cell factory, all non-fungal genes were inserted into fungal expression cassettes and integrated into defined expression sites. These examples highlight the potential of puzzling together pathways by combining genes from different organisms. Similar strategies may likely gain an increasing role in future development fungal cell factories for production of known and novel small molecules.

VI. Concluding Remarks and Perspectives

This review demonstrates that heterologous production in fungi is already well established and that synthetic biology-based methods are increasingly used to create new cell factories, typically via **bio-block-based strategies.** In the future, we envision that this trend will gather momentum and that libraries of bio-blocks containing mutated genes, scrambled homologous genes, or synthetic genes encoding new combinations of functional domains will serve as common resources for developing cell factories that produce, e.g., industrial enzymes with new or improved properties or SMs that may target new diseases. In addition, increasingly **efficient high-throughput** CRISPR-based methods will allow for the generation of genome-wide mutant-strain libraries. Hence, rather than using a single strain for construction of a heterologous cell factory, mutant libraries may serve as the preferred starting point toward developing new cell factories. In combination, these strategies will create an enormous number of potential cell factories; and screening for the successful candidates will very likely constitute a bottleneck. The expanding **synthetic biology toolbox** may also contribute to address this challenge. For example, it may offer biosensors that can be used to screen for high-yielding strains. In agreement with this view, membrane-bound RFP was recently used as a biosensor in allowing for high-throughput FACS-based screening of *T. reesei* mutants to identify strains with improved enzyme secretion potential (Gao et al. 2018). Interestingly, **biosensors** that monitor the levels of specific intracellular metabolites are already available for *E. coli* and *S. cerevisiae* (Adeniran et al. 2015; Morris 2010; Rogers et al. 2016; Zhang and Keasling 2011), and it is likely that this type biosensors will be used in the future to identify **superior fungal cell factories** for production of specific SMs.

Novel methods for controlling gene function during fermentation are highly desirable. To this end, we envision that new improved **orthogonal TFs with programmable specificities**, e.g., by using catalytically dead Cas9 variants as TFs (Qi et al. 2013), will be developed. Such TFs have already been shown in other organisms to facilitate the activation of silent genes (Cheng et al. 2013; Perez-Pinera et al. 2013), to regulate metabolism of a heterologous cell factories for increased production yields (Deaner and Alper 2017; Jensen et al. 2017; Vanegas et al. 2017), to study genetic interactions (Du et al. 2017; Peters et al. 2016), or even to induce directed evolution (Hess et al. 2016; Ma et al. 2016). We also envision that synthetic biology tools will change the fungal production platforms dramatically. Specifically, highly efficient genetic engineering tools will allow for the development of cell factories with **minimal genomes**, where all undesirable cell functions that do not significantly impact fitness during fermentation have been eliminated. For this purpose, **multiplex gene editing** has already been implemented in fungi (Foster et al. 2018; Liu et al. 2015; Nødvig et al. 2018; Pohl et al. 2016; Shi et al. 2019). Deletion of a number of BGCs in *A. nidulans* with the aim of increasing the pool of available SM precursors and to simplify metabolite analysis (Chiang et al. 2013, 2016) already serve as a small scale example of this strategy. More ambitiously, one may envision that entirely **synthetic filamentous fungi**, inspired by the synthetic yeast 2.0 project (Pretorius and Boeke 2018), may be developed. Altogether, we conclude that the rapidly expanding synthetic biology toolbox in combination with the large number of fully sequenced genomes will set the stage for a very exciting future of fungal heterologous product production.

Acknowledgments We are grateful to Michael Lynge Nielsen for carefully reading and editing this book chapter. This work has been supported by Innovation Fund Denmark, grant no.: 6150-00031B.

References

Adav SS, Ravindran A, Chao LT, Tan L, Singh S, Sze SK (2011) Proteomic analysis of pH and strains dependent protein secretion of *Trichoderma reesei*. J Proteome Res 10:4579–4596

Adeniran A, Sherer M, Tyo KEJ (2015) Yeast-based biosensors: design and applications. FEMS Yeast Res 15:1–15

Adney WS, Chou YC, Decker SR, Ding SY, Baker JO, Kunkel G, Vinzant TB, Himmel ME (2003) Heterologous expression of *Trichoderma reesei* 1,4-β-D-glucan cellobiohydrolase (Cel 7A). In: Mansfield SD, Saddler JN (eds) Applications of enzymes to lignocellulosics, vol 855. American Chemical Society, Washington, DC, pp 403–437

Adney WS, Jeoh T, Beckham GT, Chou YC, Baker JO, Michener W, Brunecky R, Himmel ME (2009) Probing the role of N-linked glycans in the stability and activity of fungal cellobiohydrolases by mutational analysis. Cellulose 16:699–709

Ahmed SM, Nishida-Fukuda H, Li Y, McDonald WH, Gradinaru CC, Macara IG (2018) Exocyst dynamics during vesicle tethering and fusion. Nat Commun 9:1–17

Alberti A, Ferrero I, Lodi T (2003) *LYS2* gene and its mutation in *Kluyveromyces lactis*. Yeast 20:1171–1175

Aleksenko A, Clutterbuck AJ (1997) Autonomous plasmid replication in *Aspergillus nidulans*: AMA1 and MATE elements. Fungal Genet Biol 21:373–387

An X, Lu J, Huang J, Zhang B, Liu D, Zhang X, Chen J, Zhou Y, Tong Y (2007) Rapid assembly of multiple-exon cDNA directly from genomic DNA. PLoS One 2:1–7

Andersen MR, Vongsangnak W, Panagiotou G, Salazar MP, Lehmann L, Nielsen J (2008) A trispecies *Aspergillus* microarray: comparative transcriptomics of three *Aspergillus* species. Proc Natl Acad Sci USA 105:4387–4392

Andersen MR, Salazar MP, Schaap PJ, Van De Vondervoort PJI, Culley D, Thykaer J, Frisvad JC, Nielsen KF, Albang R, Albermann K, Berka RM, Braus GH, Braus-Stromeyer SA, Corrochano LM, Dai Z, Van Dijck PWM, Hofmann G, Lasure LL, Magnuson JK, Menke H, Meijer M, Meijer SL, Nielsen JB, Nielsen ML, Van Ooyen AJJ, Pel HJ, Poulsen L, Samson RA, Stam H, Tsang A, Van Den Brink JM, Atkins A, Aerts A, Shapiro H, Pangilinan J, Salamov A, Lou Y, Lindquist E, Lucas S, Grimwood J, Grigoriev IV, Kubicek CP, Martinez D, van Peij NNME, Roubos JA, Nielsen J, Baker SE (2011) Comparative genomics of citric-acid-producing *Aspergillus niger ATCC* 1015 versus enzyme-producing CBS 513.88. Genome Res 21:885–897

Anyaogu DC, Mortensen UH (2015) Manipulating the glycosylation pathway in bacterial and lower eukaryotes for production of therapeutic proteins. Curr Opin Biotechnol 36:122–128

Archer D, Jeenes D, MacKenzie D, Brightwell G, Lambert N, Lowe G, Radford S, Dobson C (1990) Hen egg white lysozyme expressed in, and secreted from, *Aspergillus niger* is correctly processed and folded. Biotechnology 8:741–745

Austin B, Hail RM, Tyler BM (1990) Optimized vectors and selection for transformation of *Neurospora crassa* and *Aspergillus nidulans* to bleomycin and phleomycin. Gene 93:157–162

Avalos J, Geever RF, Case ME (1989) Bialaphos resistance as a dominant selectable marker in *Neurospora crassa*. Curr Genet 16:369–372

Barlowe C, Miller EA (2013) Secretory protein biogenesis and traffic in the early secretory pathway. Genetics 193:383–410

Barredo JL, Cantoral JM, Alvarez E, Díez B, Martín JF (1989) Cloning, sequence analysis and transcriptional study of the isopenicillin N synthase of *Penicillium chrysogenum* AS-P-78. Mol Gen Genet 216:91–98

Beck J, Ripka S, Siegner A, Schiltz E, Schweizer E (1990) The multifunctional 6-methylsalicylic acid synthase gene of *Penicillium patulum*. Its gene structure relative to that of other polyketide synthases. Eur J Biochem 192:487–498

Bentley DL (2014) Coupling mRNA processing with transcription in time and space. Nat Rev Genet 15:163–175

Bergquist PL, Te'o VSJ, Gibbs MD, Curach NC, Nevalainen KMH (2004) Recombinant enzymes from thermophilic micro-organisms expressed in fungal hosts. Biochem Soc Trans 32:293–297

Bitinaite J, Rubino M, Varma KH, Schildkraut I, Vaisvila R, Vaiskunaite R (2007) USER friendly DNA engineering and cloning method by uracil excision. Nucleic Acids Res 35:1992–2002

Bleichrodt RJ, Read ND (2019) Flow cytometry and FACS applied to filamentous fungi. Fungal Biol Rev 33:1–15

Blumhoff M, Steiger MG, Marx H, Mattanovich D, Sauer M (2013) Six novel constitutive promoters for metabolic engineering of *Aspergillus niger*. Appl Microbiol Biotechnol 97:259–267

Boeke J, Lacroute F, Fink G (1984) A positive selection for mutants lacking orotidine-5′ phosphate decarboxylase activity in yeast: 5-fluoro-orotic acid resistance. Mol Gen Genet 197:346–346

Boel E, Hansen MT, Hjort I, Høegh I, Fiil NP (1984) Two different types of intervening sequences in the glucoamylase gene from *Aspergillus niger*. Eur Mol Biol Organ 3:1581–1585

Bok JW, Ye R, Clevenger KD, Mead D, Wagner M, Krerowicz A, Albright JC, Goering AW, Thomas PM, Kelleher NL, Keller NP, Wu CC (2015) Fungal artificial chromosomes for mining of the fungal secondary metabolome. BMC Genomics 16:1–10

Breakspear A, Momany M (2007) The first fifty microarray studies in filamentous fungi. Microbiology 153:7–15

Brochado AR, Matos C, Møller BL, Hansen J, Mortensen UH, Patil KR (2010) Improved vanillin production in baker's yeast through in silico design. Microb Cell Factories 9:1–15

Bulakhov AG, Volkov PV, Rozhkova AM, Gusakov AV, Nemashkalov VA, Satrutdinov AD, Sinitsyn AP (2017) Using an inducible promoter of a gene encoding *Penicillium verruculosum* glucoamylase for production of enzyme preparations with enhanced cellulase performance. PLoS One 12:1–13

Buxton FP, Gwynne DI, Davies RW (1985) Transformation of *Aspergillus niger* using the *argB* gene of *Aspergillus nidulans*. Gene 37:207–214

Cairns TC, Nai C, Meyer V (2018) How a fungus shapes biotechnology: 100 years of *Aspergillus niger* research. Fungal Biol Biotechnol 5:1–14

Cairns TC, Zheng X, Zheng P, Sun J, Meyer V (2019) Moulding the mould: understanding and reprogramming filamentous fungal growth and morphogenesis for next generation cell factories. Biotechnol Biofuels 12:1–18

Carrington JC, Dougherty WG (1988) A viral cleavage site cassette: identification of amino acid sequences required for tobacco etch virus polyprotein processing. Biochemistry 85:3391–3395

Carvalho NDSP, Arentshorst M, Kooistra R, Stam H, Sagt CM, Van Den Hondel CAMJJ, Ram AFJ (2011) Effects of a defective ERAD pathway on growth and heterologous protein production in *Aspergillus niger*. Appl Microbiol Biotechnol 89:357–373

Chaga G, Bochkariov DE, Jokhadze GG, Hopp J, Nelson P (1999) Natural poly-histidine affinity tag for purification of recombinant proteins on cobalt (II)-carboxymethylaspartate crosslinked agarose. J Chromatogr A 864:247–256

Chanda A, Roze LV, Kang S, Artymovich KA, Hicks GR, Raikhel NV, Calvo AM, Linz JE (2009) A key role for vesicles in fungal secondary metabolism. Proc Natl Acad Sci USA 106:19533–19538

Chang KY, Georgianna RD, Heber S, Payne GA, Muddiman DC (2010) Detection of alternative splice variants at the proteome level in *Aspergillus flavus*. J Proteome Res 9:1209–1217

Chen X, Zaro J, Shen WC (2013) Fusion protein linkers: property, design and functionality. Adv Drug Deliv Rev 65:1357–1369

Cheng AW, Wang H, Yang H, Shi L, Katz Y, Theunissen TW, Rangarajan S, Shivalila CS, Dadon DB, Jaenisch R (2013) Multiplexed activation of endogenous genes by CRISPR-on, an RNA-guided transcriptional activator system. Cell Res 23:1163–1171

Cherry JR, Fidantsef AL (2003) Directed evolution of industrial enzymes: an update. Curr Opin Biotechnol 14:438–443

Chiang YM, Szewczyk E, Davidson AD, Keller N, Oakley BR, Wang CCC (2009) A gene cluster containing two fungal polyketide synthases encodes the biosynthetic pathway for a polyketide, asperfuranone, in *Aspergillus nidulans*. J Am Chem Soc 131:2965–2970

Chiang Y-M, Oakley CE, Ahuja M, Entwistle R, Schultz A, Chang S-L, Sung CT, Wang CCC, Oakley BR (2013) An efficient system for heterologous expression of secondary metabolite genes in *Aspergillus nidulans*. J Am Chem Soc 135:7720–7731

Chiang YM, Ahuja M, Oakley CE, Entwistle R, Asokan A, Zutz C, Wang CCC, Oakley BR (2016) Development of genetic dereplication strains in *Aspergillus nidulans* results in the discovery of aspercryptin. Angew Chem Int Ed 55:1662–1665

Chou YC, Adney WS, Decker SR, Baker JO, Kunkel G, Templeton DW, Himmel ME (2004) Cloning and heterologous expression of the gene encoding a family 7 glycosyl hydrolase from *Penicillium funiculosum*. In: Saha BC, Hayashi K (eds) Lignocellulose biodegradation, vol 889. American Chemical Society, Washington, DC, pp 170–193

Clevenger KD, Bok JW, Ye R, Miley GP, Verdan MH, Velk T, Chen C, Yang K, Robey MT, Gao P, Lamprecht M, Thomas PM, Islam MN, Palmer JM, Wu CC, Keller NP, Kelleher NL (2017) A scalable platform to identify fungal secondary metabolites and their gene clusters. Nat Chem Biol 13:895–901

Collis CM, Hall RM (1985) Identification of a *Tn5* determinant conferring resistance to phleomycins, bleomycins, and tallysomycins. Plasmid 14:143–151

Cullen D, Leong SA, Wilson LJ, Henner DJ (1987) Transformation of *Aspergillus nidulans* with the hygromycin-resistance gene, *hph*. Gene 57:21–26

Curran KA, Karim AS, Gupta A, Alper HS (2013) Use of high capacity terminators in *Saccharomyces cerevisiae* to increase mRNA half-life and improve gene expression control for metabolic engineering applications. Metab Eng 19:88–97

Curran KA, Morse NJ, Markham KA, Wagman AM, Gupta A, Alper HS (2015) Short synthetic terminators for improved heterologous gene expression in yeast. ACS Synth Biol 4:824–832

Czymmek KJ, Bourett TM, Howard RJ (2005) Fluorescent protein probes in fungi. In: Savidge T, Pothulakis C (eds) Microbial imaging, vol 34. Academic, San Diego, pp 27–62

Dave K, Prabha VL, Ahuja M, Dave K, Tejaswini S, Punekar NS (2015) Expanding the repertoire of selectable markers for *Aspergillus* transformation. In: van den Berg MA, Maruthachalam K (eds) Genetic transformation systems in fungi, vol 1. Springer International Publishing, Cham, pp 141–153

de Graaff L, Visser J (1988) Structure of the *Aspergillus nidulans* pyruvate kinase gene. Curr Genet 14:553–560

de Vries RP, Grigoriev IV, Tsang A (2018) Introduction: overview of fungal genomics. In: de Vries RP, Tsang A, Grigoriev IV (eds) Fungal genomics. Methods in molecular biology, vol 1775. Humana Press, New York, NY

Deaner M, Alper HS (2017) Systematic testing of enzyme perturbation sensitivities via graded dCas9 modulation in *Saccharomyces cerevisiae*. Metab Eng 40:14–22

Díez B, Mellado E, Rodríguez M, Bernasconi E, Barredo JL (1999) The NADP-dependent glutamate dehydrogenase gene from *Penicillium chrysogenum* and the construction of expression vectors for filamentous fungi. Appl Microbiol Biotechnol 52:196–207

dos Santos Castro S, de Paula RG, Antoniêto ACC, Persinoti GF, Silva-Rocha R, Silva RN (2016) Understanding the role of the master regulator XYR1 in *Trichoderma reesei* by global transcriptional analysis. Front Microbiol 7:1–16

Du D, Roguev A, Gordon DE, Chen M, Chen SH, Shales M, Shen JP, Ideker T, Mali P, Qi LS, Krogan NJ (2017) Genetic interaction mapping in mammalian cells using CRISPR interference. Nat Methods 14:577–583

Eberle J, Russo V (1994) *Neurospora crassa* blue light-inducible gene *bli-3*. Biochem Mol Biol Int 34:737–744

Engler C, Marillonnet S (2014) Golden gate cloning. In: Valla S, Lale R (eds) DNA cloning and assembly methods, vol 1116. Springer Science Business Media, New York, pp 119–131

Fiedler MRM, Barthel L, Kubisch C, Nai C, Meyer V (2018) Construction of an improved *Aspergillus niger* platform for enhanced glucoamylase secretion. Microb Cell Factories 17:1–12

Fierro F, Kosalková K, Gutiérrez S, Martín JF (1996) Autonomously replicating plasmids carrying the AMA1 region in *Penicillium chrysogenum*. Curr Genet 29:482–489

Finnigan GC, Thorner J (2015) Complex in vivo ligation using homologous recombination and high-efficiency plasmid rescue from *Saccharomyces cerevisiae*. Bio-protocol 5:1–13

Fisch KM, Skellam E, Ivison D, Cox RJ, Bailey AM, Lazarus CM, Simpson TJ (2010) Catalytic role of the C-terminal domains of a fungal non-reducing polyketide synthase. Chem Commun 46:5331–5333

Fitz E, Wanka F, Seiboth B (2018) The promoter toolbox for recombinant gene expression in *Trichoderma reesei*. Front Bioeng Biotechnol 6:1–15

Fleissner A, Dersch P (2010) Expression and export: recombinant protein production systems for *Aspergillus*. Appl Microbiol Biotechnol 87:1255–1270

Ford KL, Baumgartner K, Henricot B, Bailey AM, Foster GD (2016) A native promoter and inclusion of an intron is necessary for efficient expression of GFP or mRFP in *Armillaria mellea*. Sci Rep 6:1–10

Foster AJ, Martin-Urdiroz M, Yan X, Wright HS, Soanes DM, Talbot NJ (2018) CRISPR-Cas9 ribonucleoprotein-mediated co-editing and counterselection in the rice blast fungus. Sci Rep 8:1–12

Foureau E, Clastre M, Millerioux Y, Simkin AJ, Cornet L, Dutilleul C, Besseau S, Marais E, Melin C, Guillard J, Crèche J, Giglioli-Guivarc'h N, Courdavault V, Papon N (2012) A TRP5/5-fluoroanthranilic acid counter-selection system for gene disruption in *Candida guilliermondii*. Curr Genet 58:245–254

Frandsen RJ, Khorsand-Jamal P, Kongstad KT, Nafisi M, Kannangara RM, Staerk D, Okkels FT, Binderup K, Madsen B, Lindberg Møller B, Thrane U, Mortensen UH (2018) Heterologous production of the widely used natural food colorant carminic acid in *Aspergillus nidulans*. Nat Sci Rep 8:1–10

Frohman MA, Dush MK, Martin GR (1988) Rapid production of full-length cDNAs from rare transcripts: amplification using a single gene-specific oligonucleotide primer. Proc Natl Acad Sci USA 85:8998–9002

Fujii T, Yamaoka H, Gomi K, Kitamoto K, Kumaga C (1995) Cloning and nucleotide sequence of the ribonuclease T1 gene (*rntA*) from Aspergillus oryzae and its expression in *Saccharomyces cerevisiae* and *Aspergillus oryzae*. Biosci Biotechnol Biochem 59:1869–1874

Fujii I, Ono Y, Tada H, Gomi K, Ebizuka Y, Sankawa U (1996) Cloning of the polyketide synthase gene *atX* from *Aspergillus terreus* and its identification as the 6-methylsalicylic acid synthase gene by heterologous expression. Mol Gen Genomics 253:1–10

Fujii I, Yoshida N, Shimomaki S, Oikawa H, Ebizuka Y (2005) An iterative type I polyketide synthase PKSN catalyzes synthesis of the decaketide alternapyrone with regio-specific octa-methylation. Chem Biol 12:1301–1309

Gao F, Hao Z, Sun X, Qin L, Zhao T, Liu W, Luo H, Yao B, Su X (2018) A versatile system for fast screening and isolation of *Trichoderma reesei* cellulase hyperproducers based on DsRed and fluorescence-assisted cell sorting. Biotechnol Biofuels 11:1–13

Gardiner DM, Jarvis RS, Howlett BJ (2005) The ABC transporter gene in the sirodesmin biosynthetic gene cluster of *Leptosphaeria maculans* is not essential for sirodesmin production but facilitates self-protection. Fungal Genet Biol 42:257–263

Garneau-Tsodikova S, Labby KJ (2016) Mechanisms of resistance to aminoglycoside antibiotics: overview and perspectives. Medchemcomm 7:11–27

Gems D, Johnstone IL, Clutterbuck AJ (1991) An autonomously replicating plasmid transforms *Aspergillus nidulans* at high frequency. Gene 98:61–67

Gibson DG, Young L, Chuang RY, Craig Venter J, Hutchison CA III, Smith HO (2009) Enzymatic assembly of DNA molecules up to several hundred kilobases. Nat Methods 6:343–347

Giles NH, Case ME, Baum J, Geever R, Huiet L, Patel V, Tyler B (1985) Gene organization and regulation in the qa (quinic acid) gene cluster of *Neurospora crassa*. Microbiol Rev 49:338–358

Gomi K, Akeno T, Minetoki T, Ozeki K, Kumagai C, Okazaki N, Iimura Y (2000) Molecular cloning and characterization of a transcriptional activator gene, *amyR*, involved in the amylolytic gene expression in *Aspergillus oryzae*. Biosci Biotechnol Biochem 64:816–827

Goosen T, Bloemheuvel G, Gysler C, de Bie DA, van den Broek HWJ, Swart K (1987) Transformation of *Aspergillus niger* using the homologous orotidine-5′-phosphate-decarboxylase gene. Curr Genet 11:499–503

Goosen T, van Engelenburg F, Debets F, Swart K, Bos K, van den Broek H (1989) Tryptophan auxotrophic mutants in *Aspergillus niger*: inactivation of the *trpC* gene by cotransformation mutagenesis. Mol Gen Genet 219:282–288

Gossen M, Bujard H (1992) Tight control of gene expression in mammalian cells by tetracycline-responsive promoters. Proc Natl Acad Sci USA 89:5547–5551

Gossen M, Freundlieb S, Bender G, Müller G, Hillen W, Bujardt H (1995) Transcriptional activation by tetracyclines in mammalian cells. Science 268:1766–1769

Gouka RJ, Punt PJ, Van Den Hondel CAMJJ (1997) Glucoamylase gene fusions alleviate limitations for protein production in *Aspergillus awamori* at the transcriptional and (post)translational levels. Appl Environ Microbiol 63:488–497

Gould N, Hendy O, Papamichail D (2014) Computational tools and algorithms for designing customized synthetic genes. Front Bioeng Biotechnol 2:1–14

Gressler M, Hortschansky P, Geib E, Brock M (2015) A new high-performance heterologous fungal expression system based on regulatory elements from the *Aspergillus terreus* terrein gene cluster. Front Microbiol 6:1–16

Grigoriev IV, Nikitin R, Haridas S, Kuo A, Ohm R, Otillar R, Riley R, Salamov A, Zhao X, Korzeniewski F, Smirnova T, Nordberg H, Dubchak I, Shabalov I (2014) MycoCosm portal: gearing up for 1000 fungal genomes. Nucleic Acids Res 42: D699–D704

Gupta SK, Shukla P (2018) Glycosylation control technologies for recombinant therapeutic proteins. Appl Microbiol Biotechnol 102:10457–10468

Haas H, Redl B, Friedlin E, Stöffler G (1992) Isolation and analysis of the *Penicillium chrysogenum phoA* gene encoding a secreted phosphate-repressible acid phosphatase. Gene 113:129–133

Halo LM, Marshall JW, Yakasai AA, Song Z, Butts CP, Crump MP, Heneghan M, Bailey AM, Simpson TJ, Lazarus CM, Cox RJ (2008) Authentic heterologous expression of the tenellin iterative polyketide synthase nonribosomal peptide synthetase requires coexpression with an enoyl reductase. Chembiochem 9:585–594

Hansen BG, Genee HJ, Kaas CS, Nielsen JB, Regueira TB, Mortensen UH, Frisvad JC, Patil KR (2011a) A new class of IMP dehydrogenase with a role in self-resistance of mycophenolic acid producing fungi. BMC Microbiol 11:1–7

Hansen BG, Salomonsen B, Nielsen MT, Nielsen JB, Hansen NB, Nielsen KF, Regueira TB, Nielsen J, Patil KR, Mortensen UH (2011b) Versatile enzyme expression and characterization system for *Aspergillus nidulans*, with the *Penicillium brevicompactum* polyketide synthase gene from the mycophenolic acid gene cluster as a test case. Appl Environ Microbiol 77:3044–3051

Hanson G, Coller J (2018) Codon optimality, bias and usage in translation and mRNA decay. Nat Rev Mol Cell Biol 19:20–30

He Y, Cox RJ (2016) The molecular steps of citrinin biosynthesis in fungi. Chem Sci 7:2119–2127

Herr A, Fischer R (2014) Improvement of *Aspergillus nidulans* penicillin production by targeting AcvA to peroxisomes. Metab Eng 25:131–139

Hess GT, Frésard L, Han K, Lee CH, Li A, Cimprich KA, Montgomery SB, Bassik MC (2016) Directed evolution using dCas9-targeted somatic hypermutation in mammalian cells. Nat Methods 13:1036–1042

Hoang HD, Maruyama JI, Kitamoto K (2015) Modulating endoplasmic reticulum-Golgi cargo receptors for improving secretion of carrier-fused heterologous proteins in the filamentous fungus *Aspergillus oryzae*. Appl Environ Microbiol 81:533–543

Hoefgen S, Lin J, Fricke J, Stroe MC, Mattern DJ, Kufs JE, Hortschansky P, Brakhage AA, Hoffmeister D, Valiante V (2018) Facile assembly and fluorescence-based screening method for heterologous expression of biosynthetic pathways in fungi. Metab Eng 48:44–51

Holm DMK (2013) Development and implementation of novel genetic tools for investigation of fungal secondary metabolism. PhD thesis, Technical University of Denmark

Holm DK, Petersen LM, Klitgaard A, Knudsen PB, Jarczynska ZD, Nielsen KF, Gotfredsen CH, Larsen TO, Mortensen UH (2014) Molecular and chemical characterization of the biosynthesis of the 6-MSA-derived meroterpenoid yanuthone D in *Aspergillus niger*. Chem Biol 21:519–529

Hoover DM, Lubkowski J (2002) DNAWorks: an automated method for designing oligonucleotides for PCR-based gene synthesis. Nucleic Acids Res 30:1–7

Hopp TP, Prickett KS, Price VL, Libby RT, March CJ, Cerretti DP, Urdal DL, Conlon PJ (1988) A short polypeptide marker sequence useful for recombinant protein identification and purification. Bio/Technology 6:1204–1210

Hoshida H, Fujita T, Murata K, Kubo K, Akada R (2005) Copper-dependent production of a *Pycnoporus coccineus* extracellular laccase in *Aspergillus oryzae* and *Saccharomyces cerevisiae*. Biosci Biotechnol Biochem 69:1090–1097

Hoskins I, Roberts C (1994) Expression of the 3-phosphoglycerate kinase gene (*pgkA*) of *Penicillium chrysogenum*. Mol Gen Genet 243:270–276

Hurley JM, Chen CH, Loros JJ, Dunlap JC (2012) Light-inducible system for tunable protein expression in *Neurospora crassa*. G3 2:1207–1212

Hynes MJ, Pateman JAJ (1970) The genetic analysis of regulation of amidase synthesis in *Aspergillus nidulans*. Mol Gen Genet 108:107–116

Ilmén M, Onnela ML, Klemsdal S, Keränen S, Penttilä M (1998) Functional analysis of the cellobiohydrolase I promoter of the filamentous fungus *Trichoderma reesei*. Mol Gen Genet 253:303–314

Iriarte A, Sanguinetti M, Fernández-Calero T, Naya H, Ramón A, Musto H (2012) Translational selection on codon usage in the genus *Aspergillus*. Gene 506:98–105

Ishida H, Hata Y, Kawato A, Abe Y, Kashiwagi Y (2004) Isolation of a novel promoter for efficient protein production in *Aspergillus oryzae*. Biosci Biotechnol Biochem 68:1849–1857

Ishii N, Minami N, Chen EY, Medina AL, Chico MM, Kasamatsu H (1996) Analysis of a nuclear localization signal of simian virus 40 major capsid protein Vp1. J Virol 70:1317–1322

Jackson DC, Tang XL, Brown LE, Murray JM, White DO, Tregear GW (1986) Antigenic determinants of influenza virus hemagglutinin XII. Virology 155:625–632

Jalving R, Van De Vondervoort PJI, Visser J, Schaap PJ (2000) Characterization of the kexin-like maturase of *Aspergillus niger*. Appl Environ Microbiol 66:363–368

Jensen ED, Ferreira R, Jakočiunas T, Arsovska D, Zhang J, Ding L, Smith JD, David F, Nielsen J, Jensen MK, Keasling JD (2017) Transcriptional reprogramming in yeast using dCas9 and combinatorial gRNA strategies. Microb Cell Factories 16:1–16

Jin FJ, Maruyama J, Juvvadi PR, Arioka M, Kitamoto K (2004) Adenine auxotrophic mutants of *Aspergillus oryzae*: development of a novel transformation system with triple auxotrophic hosts. Biosci Biotechnol Biochem 68:656–662

Jin FJ, Watanabe T, Juvvadi PR, Maruyama JI, Arioka M, Kitamoto K (2007) Double disruption of the proteinase genes, *tppA* and *pepE*, increases the production level of human lysozyme by *Aspergillus oryzae*. Appl Microbiol Biotechnol 76:1059–1068

Johnstone IL, Hughes SG, Clutterbuck AJ (1985) Cloning an *Aspergillus nidulans* developmental gene by transformation. EMBO J 4:1307–1311

Jørgensen TR, Park J, Arentshorst M, van Welzen AM, Lamers G, VanKuyk PA, Damveld RA, van den Hondel CAM, Nielsen KF, Frisvad JC, Ram AFJ (2011) The molecular and genetic basis of conidial pigmentation in *Aspergillus niger*. Fungal Genet Biol 48:544–553

Kamaruddin N, Storms R, Mahadi NM, Illias RM, Bakar FDA, Murad AMA (2018) Reduction of extracellular proteases increased activity and stability of heterologous protein in *Aspergillus niger*. Arab J Sci Eng 43:3327–3338

Keller NP (2019) Fungal secondary metabolism: regulation, function and drug discovery. Nat Rev Microbiol 17:167–180

Keller NP, Turner G, Bennett JW (2005) Fungal secondary metabolism—from biochemistry to genomics. Nat Rev Microbiol 3:937–947

Kelly JM, Hynes MJ (1985) Transformation of *Aspergillus niger* by the *amdS* gene of *Aspergillus nidulans*. EMBO J 4:475–479

Kempken F (2013) Alternative splicing in ascomycetes. Appl Microbiol Biotechnol 97:4235–4241

Kempken F, Windhofer F (2004) Alternative splicing of transcripts of the transposon Restless is maintained in the foreign host *Neurospora crassa* and can be modified by introducing mutations at the 5′ and 3′ splice sites. Curr Genet 46:59–65

Kipriyanov SM, Kupriyanova OA, Little M, Moldenhauer G (1996) Rapid detection of recombinant antibody fragments directed against cell-surface antigens by flow cytometry. J Immunol Methods 196:51–62

Kistler HC, Broz K (2015) Cellular compartmentalization of secondary metabolism. Front Microbiol 6:68

Kitamoto N, Matsui J, Kawai Y, Kato A, Yoshino S, Ohmiya K, Tsukagoshi N (1998) Utilization of the TEF1-α gene (*TEF1*) promoter for expression of polygalacturonase genes, *pgaA* and *pgaB*, in *Aspergillus oryzae*. Appl Microbiol Biotechnol 50:85–92

Kitamoto N, Ono N, Yoshino Yasuda S (2015) Construction of quintuple protease and double amylase gene deletant for heterologous protein production in *Aspergillus oryzae* KBN616. Food Sci Technol Res 21:297–307

Kluge J, Terfehr D, Kück U (2018) Inducible promoters and functional genomic approaches for the genetic engineering of filamentous fungi. Appl Microbiol Biotechnol 102:6357–6372

Knudsen PB (2015) Development of scalable high throughput fermentation approaches for physiological characterisation of yeast and filamentous fungi. PhD thesis, Technical University of Denmark (DTU)

Krappmann S (2007) Gene targeting in filamentous fungi: the benefits of impaired repair. Fungal Biol Rev 21:25–29

Krishnan K, Askew DS (2014) The fungal UPR: a regulatory hub for virulence traits in the mold pathogen *Aspergillus fumigatus*. Virulence 5:334–340

Krügel H, Fiedler G, Smith C, Baumberg S (1993) Sequence and transcriptional analysis of the nourseothricin acetyltransferase-encoding gene *nat1* from *Streptomyces noursei*. Gene 127:127–131

Kubodera T, Yamashita N, Nishimura A (2002) Transformation of *Aspergillus* sp. and *Trichoderma reesei* using the pyrithiamine resistance gene (*ptrA*) of *Aspergillus oryzae*. Biosci Biotechnol Biochem 66:404–406

Kupfer DM, Drabenstot SD, Buchanan KL, Hongshing L, Zhu H, Dyer DW, Roe BA, Murphy JW (2004) Introns and splicing elements of five diverse fungi. Eukaryot Cell 3:1088–1100

Kurzatkowski W, Törrönen A, Filipek J, Mach RL, Herzog P, Sowka S, Kubicek CP (1996) Glucose-induced secretion of *Trichoderma reesei* xylanases. Appl Environ Microbiol 62:2859–2865

Kwon MJ, Arentshorst M, Roos ED, Van Den Hondel CAMJJ, Meyer V, Ram AFJ (2011) Functional characterization of Rho GTPases in *Aspergillus niger* uncovers conserved and diverged roles of Rho proteins within filamentous fungi. Mol Microbiol 79:1151–1167

Lamb TM, Vickery J, Bell-Pedersen D (2013) Regulation of gene expression in *Neurospora crassa* with a copper responsive promoter. G3 3:2273–2280

Landowski CP, Huuskonen A, Wahl R, Westerholm-Parvinen A, Kanerva A, Hänninen AL, Salovuori N, Penttilä M, Natunen J, Ostermeier C, Helk B, Saarinen J, Saloheimo M (2015) Enabling low cost biopharmaceuticals: a systematic approach to delete proteases from a well-known protein production host *Trichoderma reesei*. PLoS One 10:1–28

Landowski CP, Mustalahti E, Wahl R, Croute L, Sivasiddarthan D, Westerholm-Parvinen A, Sommer B, Ostermeier C, Helk B, Saarinen J, Saloheimo M (2016) Enabling low cost biopharmaceuticals: high level interferon alpha-2b production in *Trichoderma reesei*. Microb Cell Factories 15:1–15

Leach DRF (1994) Long DNA palindromes, cruciform structures, genetic instability and secondary structure repair. BioEssays 16:893–900

Ley A, Coumou HC, Frandsen RJN (2015) Heterologous expression of MlcE in *Saccharomyces cerevisiae* provides resistance to natural and semi-synthetic statins. Metab Eng Commun 2:117–123

Li H, d'Anjou M (2009) Pharmacological significance of glycosylation in therapeutic proteins. Curr Opin Biotechnol 20:678–684

Li J, Wang J, Wang S, Xing M, Yu S, Liu G (2012) Achieving efficient protein expression in *Trichoderma reesei* by using strong constitutive promoters. Microb Cell Factories 11:1–10

Li W, Yu J, Li Z, Yin WB (2018) Rational design for fungal laccase production in the model host *Aspergillus nidulans*. Sci China Life Sci 61:1–11

Lin C, Baldwin T, Estell D, Fung J, Lin C, Pucci J, Razo J, Rider B, Throndset W, Trimbur D, Gastel F Van Ward M (2006) Develop systems for manufacturing 100,000,000 doses of an emergency pharmaceutical (e.g. vaccine or monoclonal antibody) within 2 months of product identification. Progress report, Genencor International, pp 1–21

Linder MB, Qiao M, Laumen F, Selber K, Hyytiä T, Nakari-Setälä T, Penttilä ME (2004) Efficient purification of recombinant proteins using hydrophobins as tags in surfactant-based two-phase systems. Biochemistry 43:11873–11882

Liu R, Chen L, Jiang Y, Zhou Z, Zou G (2015) Efficient genome editing in filamentous fungus *Trichoderma reesei* using the CRISPR/Cas9 system. Cell Discov 1:1–11

Longley DB, Harkin DP, Johnston PG (2003) 5-fluorouracil: mechanisms of action and clinical strategies. Nat Rev Cancer 3:330–338

Lorang JM, Tuori RP, Martinez JP, Sawyer TL, Redman RS, Rollins JA, Wolpert TJ, Johnson KB, Rodriguez RJ, Dickman MB, Ciuffetti LM (2001) Green fluorescent protein is lighting up fungal biology. Appl Environ Microbiol 67:1987–1994

Lubertozzi D, Keasling JD (2006) Marker and promoter effects on heterologous expression in *Aspergillus nidulans*. Appl Microbiol Biotechnol 72:1014–1023

Lupton SD, Brunton LL, Kalberg VA, Overell RW (1991) Dominant positive and negative selection using a hygromycin phosphotransferase-thymidine kinase fusion gene. Mol Cell Biol 11:3374–3378

Ma Y, Zhang J, Yin W, Zhang Z, Song Y, Chang X (2016) Targeted AID-mediated mutagenesis (TAM) enables efficient genomic diversification in mammalian cells. Nat Methods 13:1029–1035

Maruyama J, Nakajima H, Kitamoto K (2001) Visualization of nuclei in *Aspergillus oryzae* with EGFP and analysis of the number of nuclei in each conidium by FACS. Biosci Biotechnol Biochem 65:1504–1510

Marx F, Haas H, Reindl M, Stöffler G, Lottspeich F, Redl B (1995) Cloning, structural organization and regulation of expression of the *Penicillium chrysogenum paf* gene encoding an abundantly secreted protein with antifungal activity. Gene 167:167–171

Masai K, Marumaya J, Nakajima H, Kitamoto K (2003) In vivo visualization of the distribution of a secretory protein in *Aspergillus oryzae* hyphae using the RntA-EGFP fusion protein. Biosci Biotechnol Biochem 67:455–459

Matsuda Y, Mitsuhashi T, Lee S, Hoshino M, Mori T, Okada M, Zhang H, Hayashi F, Fujita M, Abe I (2016) Astellifadiene: structure determination by NMR spectroscopy and crystalline sponge method, and elucidation of its biosynthesis. Angew Chem 128:5879–5882

McNally MT, Free SJ (1988) Isolation and characterization of a *Neurospora* glucose-repressible gene. Curr Genet 14:545–551

Meyer V, Arentshorst M, El-Ghezal A, Drews AC, Kooistra R, van den Hondel CAMJJ, Ram AFJ (2007) Highly efficient gene targeting in the *Aspergillus niger kusA* mutant. J Biotechnol 128:770–775

Meyer V, Wanka F, van Gent J, Arentshorst M, van den Hondel CAMJJ, Ram AFJ (2011) Fungal gene expression on demand: an inducible, tunable, and metabolism-independent expression system for *Aspergillus niger*. Appl Environ Microbiol 77:2975–2983

Meyer V, Andersen MR, Brakhage AA, Braus GH, Caddick MX, Cairns TC, de Vries RP, Haarmann T, Hansen K, Hertz-Fowler C, Krappmann S, Mortensen UH, Peñalva MA, Ram AFJ, Head RM (2016) Current challenges of research on filamentous fungi in relation to human welfare and a sustainable bio-economy: a white paper. Fungal Biol Biotechnol 3:1–17

Miyauchi S, Te'o VS, Bergquist PL, Nevalainen KMH (2013) Expression of a bacterial xylanase in *Trichoderma reesei* under the *egl2* and *cbh2* glycosyl hydrolase gene promoters. New Biotechnol 30:523–530

Moralejo FJ, Watson AJ, Jeenes DJ, Archer DB, Martín JF (2001) A defined level of protein disulfide isomerase expression is required for optimal secretion of thaumatin by *Aspergillus awamori*. Mol Gen Genomics 266:246–253

Morris MC (2010) Fluorescent biosensors of intracellular targets from genetically encoded reporters to modular polypeptide probes. Cell Biochem Biophys 56:19–37

Morse NJ, Gopal MR, Wagner JM, Alper HS (2017) Yeast terminator function can be modulated and designed on the basis of predictions of nucleosome occupancy. ACS Synth Biol 6:2086–2095

Muller C, Hjort CM, Hansen K, Nielsen J (2002) Altering the expression of two chitin synthase genes differentially affects the growth and morphology of *Aspergillus oryzae*. Microbiology 148:4025–4033

Munawar A, Marshall JW, Cox RJ, Bailey AM, Lazarus CM (2013) Isolation and characterisation of a ferrirhodin synthetase gene from the sugarcane pathogen *Fusarium sacchari*. Chembiochem 14:388–394

Nagamine S, Liu C, Nishishita J, Kozaki T, Sogahata K, Sato Y, Minami A, Ozaki T, Schmidt-Dannert C, Maruyama J, Oikawa H (2019) Ascomycota *Aspergillus oryzae* is an efficient expression host for production of basidiomycota terpenes using genomic DNA sequences. Appl Environ Microbiol 85:1–21

Nakajima KI, Asakura T, Maruyama JI, Morita Y, Oike H, Shimizu-Ibuka A, Misaka T, Sorimachi H, Arai S, Kitamoto K, Abe K (2006) Extracellular production of neoculin, a sweet-tasting heterodimeric protein with taste-modifying activity, by *Aspergillus oryzae*. Appl Environ Microbiol 72:3716–3723

Nakano ET, Fox RD, Clements DE, Koo K, Stuart WD, Ivy JM (1993) Expression vectors for *Neurospora crassa* and expression of a bovine preprochymosin cDNA. Fungal Genet Rep 40:54–56

Nakari-Setälä T, Penttilä M (1995) Production of *Trichoderma reesei* cellulases glucose-containing media. Appl Environ Microbiol 61:3650–3655

Nakazato L, Dutra V, Broetto L, Staats CC, Vainstein MH, Schrank A (2006) Development of an expression vector for *Metarhizium anisopliae* based on the *tef-1α* homologous promoter. Appl Microbiol Biotechnol 72:521–528

Nielsen ML, Albertsen L, Lettier G, Nielsen JB, Mortensen UH (2006) Efficient PCR-based gene targeting with a recyclable marker for *Aspergillus nidulans*. Fungal Genet Biol 43:54–64

Nielsen MT, Nielsen JB, Anyaogu DC, Holm DK, Nielsen KF, Larsen TO, Mortensen UH (2013) Heterologous reconstitution of the intact geodin gene cluster in *Aspergillus nidulans* through a simple and versatile PCR based approach. PLoS One 8: e72871

Nielsen ML, Isbrandt T, Rasmussen KB, Thrane U, Hoof JB, Larsen TO, Mortensen UH (2017) Genes linked to production of secondary metabolites in *Talaromyces atroroseus* revealed using CRISPR-Cas9. PLoS One 12:1–9

Nødvig CS, Nielsen JB, Kogle ME, Mortensen UH (2015) A CRISPR-Cas9 system for genetic engineering of filamentous fungi. PLoS One 10:1–18

Nødvig CS, Hoof JB, Kogle ME, Jarczynska ZD, Lehmbeck J, Klitgaard DK, Mortensen UH (2018) Efficient oligo nucleotide mediated CRISPR-Cas9 gene editing in Aspergilli. Fungal Genet Biol 115:78–89

Oda K, Terado S, Toyoura R, Fukuda H, Kawauchi M, Iwashita K (2016) Development of a promoter shutoff system in *Aspergillus oryzae* using a sorbitol-sensitive promoter. Biosci Biotechnol Biochem 80:1792–1801

Ohno A, Maruyama JI, Nemoto T, Arioka M, Kitamoto K (2011) A carrier fusion significantly induces unfolded protein response in heterologous protein production by *Aspergillus oryzae*. Appl Microbiol Biotechnol 92:1197–1206

Okada M, Saito K, Wong CP, Li C, Wang D, Iijima M, Taura F, Kurosaki F, Awakawa T, Abe I (2017) Combinatorial biosynthesis of (+)-daurichromenic acid and its halogenated analogue. Org Lett 19:3183–3186

Osmani AH, May GS, Osmani SA (1999) The extremely conserved pyroA gene of *Aspergillus nidulans* is required for pyridoxine synthesis and is required indirectly for resistance to photosensitizers. J Biol Chem 274:23565–23569

Ouedraogo JP, Arentshorst M, Nikolaev I, Barends S, Ram AFJ (2016) I-SceI enzyme mediated integration (SEMI) for fast and efficient gene targeting in *Trichoderma reesei*. J Biotechnol 222:25–28

Pahirulzaman KAK, Williams K, Lazarus CM (2012) A toolkit for heterologous expression of metabolic pathways in *Aspergillus oryzae*. In: Hopwood D (ed) Methods in enzymology, vol 517. Academic, San Diego, pp 241–260

Paloheimo M, Haarmann T, Mäkinen S, Vehmaanperä J (2016) Production of industrial enzymes in *Trichoderma reesei*. In: Schmoll M, Dattenböck C (eds) Gene expression systems in fungi: advancements and applications, vol 1. Springer International Publishing, Cham, pp 23–57

Pantazopoulou A, Pinar M, Xiang X, Peñalva MA (2014) Maturation of late Golgi cisternae into RabERAB11 exocytic post-Golgi carriers visualized in vivo. Mol Biol Cell 25:2428–2443

Perez-Pinera P, Ousterout DG, Brunger JM, Farin AM, Glass KA, Guilak F, Crawford GE, Hartemink AJ, Gersbach CA (2013) Synergistic and tunable human gene activation by combinations of synthetic transcription factors. Nat Methods 10:239–244

Peters JM, Colavin A, Shi H, Czarny TL, Larson MH, Wong S, Hawkins JS, Lu CHS, Koo BM, Marta E, Shiver AL, Whitehead EH, Weissman JS, Brown ED, Qi LS, Huang KC, Gross CA (2016) A comprehensive, CRISPR-based functional analysis of essential genes in bacteria. Cell 165:1493–1506

Petersen LM, Holm DK, Gotfredsen CH, Mortensen UH, Larsen TO (2015) Investigation of a 6-MSA synthase gene cluster in *Aspergillus aculeatus* reveals 6-MSA-derived aculinic acid, aculins A-B and epi-aculin A. Chembiochem 16:2200–2204

Petes T (1988) Recombination between repeated genes in microorganisms. Annu Rev Genet 22:147–168

Pohl C, Kiel J, Driessen A, Bovenberg R, Nygård Y (2016) CRISPR/Cas9 based genome editing of *Penicillium chrysogenum*. ACS Synth Biol 5:754–764

Polli F, Meijrink B, Bovenberg RAL, Driessen AJM (2016) New promoters for strain engineering of *Penicillium chrysogenum*. Fungal Genet Biol 89:62–71

Pop C, Rouskin S, Ingolia NT, Han L, Phizicky EM, Weissman JS, Koller D (2014) Causal signals between codon bias, mRNA structure, and the efficiency of translation and elongation. Mol Syst Biol 10:1–15

Prasher DC, Eckenrode VK, Ward WW, Prendergast FG, Cormier MJ (1992) Primary structure of the *Aequorea victoria* green-fluorescent protein. Gene 111:229–233

Pretorius IS, Boeke JD (2018) Yeast 2.0-connecting the dots in the construction of the world's first functional synthetic eukaryotic genome. FEMS Yeast Res 18:1–15

Punt PJ, van den Hondel CAMJJ (1992) Transformation of filamentous fungi based on hygromycin b and phleomycin resistance markers. In: Wu R (ed) Methods in enzymology, vol 216. Academic, San Diego, pp 447–457

Punt PJ, Seiboth B, Weenink XO, Van Zeijl C, Lenders M, Konetschny C, Ram AFJ, Montijn R, Kubicek CP, Van Den Hondel CAMJJ (2001) Identification and characterization of a family of secretion-related small GTPase-encoding genes from the filamentous fungus *Aspergillus niger*: a putative SEC4 homologue is not essential for growth. Mol Microbiol 41:513–525

Qi LS, Larson MH, Gilbert LA, Doudna JA, Weissman JS, Arkin AP, Lim WA (2013) Repurposing CRISPR as an RNA-guided platform for sequence-specific control of gene expression. Cell 152:1173–1183

Qin LN, Cai FR, Dong XR, Huang ZB, Tao Y, Huang JZ, Dong ZY (2012) Improved production of heterologous lipase in *Trichoderma reesei* by RNAi mediated gene silencing of an endogenic highly expressed gene. Bioresour Technol 109:116–122

Qin H, Xiao H, Zou G, Zhou Z, Zhong J (2017) CRISPR-Cas9 assisted gene disruption in the higher fungus Ganoderma species. Process Biochem 56:57–61

Rahman Z, Shida Y, Furukawa T, Suzuki Y, Okada H, Ogasawara W, Morikawa Y (2009) Evaluation and characterization of *Trichoderma reesei* cellulase and xylanase promoters. Appl Microbiol Biotechnol 82:899–908

Rantasalo A, Landowski CP, Kuivanen J, Korppoo A, Reuter L, Koivistoinen O, Valkonen M, Penttilä M, Jäntti J, Mojzita D (2018) A universal gene expression system for fungi. Nucleic Acids Res 1:1–15

Rantasalo A, Vitikainen M, Paasikallio T, Jäntti J, Landowski CP, Mojzita D (2019) Novel genetic tools that enable highly pure protein production in *Trichoderma reesei*. Sci Rep 9:1–12

Read G, Vining LC (1968) The biosynthesis of terreic acid. Chem Commun 16:935–937

Record E, Punt PJ, Chamkha M, Labat M, Van Den Hondel CAMJJ, Asther M (2002) Expression of the *Pycnoporus cinnabarinus* laccase gene in *Aspergillus niger* and characterization of the recombinant enzyme. Eur J Biochem 269:602–609

Rendsvig JKH, Workman CT, Hoof JB (2019) Bidirectional histone-gene promoters in *Aspergillus*: characterization and application for multi-gene expression. Fungal Biol Biotechnol 6:1–14

Riquelme M, Sánchez-León E (2014) The Spitzenkörper: a choreographer of fungal growth and morphogenesis. Curr Opin Microbiol 20:27–33

Rogers JK, Taylor ND, Church GM (2016) Biosensor-based engineering of biosynthetic pathways. Curr Opin Biotechnol 42:84–91

Rokas A, Wisecaver JH, Lind AL (2018) The birth, evolution and death of metabolic gene clusters in fungi. Nat Rev Microbiol 16:731–744

Roller S, Covill N (1999) The antifungal properties of chitosan in laboratory media and apple juice. Int J Food Microbiol 47:67–77

Romero B, Turner G, Olivas I, Laborda F, Ramón De Lucas J (2003) The *Aspergillus nidulans alcA* promoter drives tightly regulated conditional gene expression in *Aspergillus fumigatus* permitting validation of essential genes in this human pathogen. Fungal Genet Biol 40:103–114

Roze LV, Chanda A, Linz JE (2011) Compartmentalization and molecular traffic in secondary metabolism: a new understanding of established cellular processes. Fungal Genet Biol 48:35–48

Ryan MD, King AMQ, Thomas GP (1991) Cleavage of FMDV polyprotein is mediated by residues located within a 19 amino acid sequence. J Gen Virol 72:2727–2732

Sakai K, Kinoshita H, Shimizu T, Nihira T (2008) Construction of a citrinin gene cluster expression system in heterologous *Aspergillus oryzae*. J Biosci Bioeng 106:466–472

Sandermann H Jr (1992) Plant metabolism of xenobiotics. Trends Biochem Sci 17:82–84

Schalén M, Anyaogu DC, Hoof JB, Workman M (2016) Effect of secretory pathway gene overexpression on secretion of a fluorescent reporter protein in *Aspergillus nidulans*. Fungal Biol Biotechnol 3:1–14

Schäpe P, Kwon MJ, Baumann B, Gutschmann B, Jung S, Lenz S, Nitsche B, Paege N, Schütze T, Cairns TC, Meyer V (2019) Updating genome annotation for the microbial cell factory *Aspergillus niger* using gene co-expression networks. Nucleic Acids Res 47:559–569

Schuetze T, Meyer V (2017) Polycistronic gene expression in *Aspergillus niger*. Microb Cell Factories 16:1–8

Shi TQ, Gao J, Wang WJ, Wang KF, Xu GQ, Huang H, Ji XJ (2019) CRISPR/Cas9-based genome editing in the filamentous fungus *Fusarium fujikuroi* and its application in strain engineering for gibberellic acid production. ACS Synth Biol 8:445–454

Shimizu T, Ito T, Kanematsu S (2012) Transient and multivariate system for transformation of a fungal plant pathogen, *Rosellinia necatrix*, using autonomously replicating vectors. Curr Genet 58:129–138

Shoji J, Maruyama J, Arioka M, Kitamoto K (2005) Development of *Aspergillus oryzae thiA* promoter as a tool for molecular biological studies. FEMS Microbiol Lett 244:41–46

Sibthorp C, Wu H, Cowley G, Wong PWH, Palaima P, Morozov IY, Weedall GD, Caddick MX (2013) Transcriptome analysis of the filamentous fungus *Aspergillus nidulans* directed to the global identification of promoters. BMC Genomics 14:1–18

Song Z, Bakeer W, Marshall JW, Yakasai AA, Khalid RM, Collemare J, Skellam E, Tharreau D, Lebrun M-H, Lazarus CM, Bailey AM, Simpson TJ, Cox RJ (2015) Heterologous expression of the avirulence gene *ACE1* from the fungal rice pathogen *Magnaporthe oryzae*. Chem Sci 6:4837–4845

Stajich JE, Dietrich FS, Roy SW (2007) Comparative genomic analysis of fungal genomes reveals intron-rich ancestors. Genome Biol 8:1–13

Steinberg G, Peñalva MA, Riquelme M, Wösten HA, Harris SD (2017) Cell biology of hyphal growth. Microbiol Spectr 5:231–265

Sticha KRK, Sieg CA, Bergstrom CP, Hanna PE, Wagner CR (1997) Overexpression and large-scale purification of recombinant hamster polymorphic arylamine N-acetyltransferase as a dihydrofolate reductase fusion protein. Protein Expr Purif 10:141–153

Stone PJ, Makoff AJ, Parish JH, Radford A (1993) Cloning and sequence analysis of the glucoamylase gene of *Neurospora crassa*. Curr Genet 24:205–211

Stricker AR, Grosstessner-Hain K, Würleitner E, Mach RL (2006) Xyr1 (xylanase regulator 1) regulates both the hydrolytic enzyme system and d-xylose metabolism in *Hypocrea jecorina*. Eukaryot Cell 5:2128–2137

Sun A, Peterson R, Te'o J, Nevalainen H (2016) Expression of the mammalian peptide hormone obestatin in *Trichoderma reesei*. New Biotechnol 33:99–106

Tada S, Gomi K, Kitamoto K, Takahashi K, Tamura G, Hara S (1991) Construction of a fusion gene comprising the Taka-amylase a promoter and the *Escherichia coli* β-glucuronidase gene and analysis of its expression in *Aspergillus oryzae*. Mol Gen Genet 229:301–306

Tanaka M, Tokuoka M, Gomi K (2014) Effects of codon optimization on the mRNA levels of heterologous genes in filamentous fungi. Appl Microbiol Biotechnol 98:3859–3867

Taylor DL, Hollingsworth TN, McFarland JW, Lennon NJ, Nusbaum C, Ruess RW (2014) A first comprehensive census of fungi in soil reveals both hyper-diversity and fine-scale niche partitioning. Ecol Monogr 84:3–20

Throndset W, Kim S, Bower B, Lantz S, Kelemen B, Pepsin M, Chow N, Mitchinson C, Ward M (2010) Flow cytometric sorting of the filamentous fungus *Trichoderma reesei* for improved strains. Enzym Microb Technol 47:335–341

Tian Y, Tan Y, Liu N, Yan Z, Liao Y, Chen J, De Saeger S, Yang H, Zhang Q, Wu A (2016) Detoxification of deoxynivalenol via glycosylation represents novel insights on antagonistic activities of *Trichoderma* when confronted with *Fusarium graminearum*. Toxins 8:1–15

Till P, Pucher ME, Mach RL, Mach-Aigner AR (2018) A long noncoding RNA promotes cellulase expression in *Trichoderma reesei*. Biotechnol Biofuels 11:1–16

Unkles SE, Campbell EI, Carrez D, Grieve C, Contreras R, Fiers W, van den Hondel CAMJJ, Kinghorn JR (1989) Transformation of *Aspergillus niger* with the homologous nitrate reductase gene. Gene 78:157–166

Urlinger S, Baron U, Thellmann M, Hasan MT, Bujard H, Hillen W (2000) Exploring the sequence space for tetracycline-dependent transcriptional activators: novel mutations yield expanded range and sensitivity. Proc Natl Acad Sci USA 97:7963–7968

Valkonen M, Ward M, Wang H, Penttilä M, Saloheimo M (2003) Improvement of foreign-protein production in *Aspergillus niger* var. *awamori* by constitutive induction of the unfolded-protein response. Appl Environ Microbiol 69:6979–6986

van den Brink HJM, Petersen SG, Rahbek-Nielsen H, Hellmuth K, Harboe M (2006) Increased production of chymosin by glycosylation. J Biotechnol 125:304–310

Vanegas KG, Lehka BJ, Mortensen UH (2017) SWITCH: a dynamic CRISPR tool for genome engineering and metabolic pathway control for cell factory construction in *Saccharomyces cerevisiae*. Microb Cell Factories 16:1–12

Vesth TC, Nybo JL, Theobald S, Frisvad JC, Larsen TO, Nielsen KF, Hoof JB, Brandl J, Salamov A, Riley R, Gladden JM, Phatale P, Nielsen MT, Lyhne EK, Kogle ME, Strasser K, McDonnell E, Barry K, Clum A, Chen C, LaButti K, Haridas S, Nolan M, Sandor L, Kuo A, Lipzen A, Hainaut M, Drula E, Tsang A, Magnuson JK, Henrissat B, Wiebenga A, Simmons BA, Mäkelä MR, de Vries RP, Grigoriev IV, Mortensen UH, Baker SE, Andersen MR (2018) Investigation of inter- and intraspecies variation through genome sequencing of *Aspergillus* section *Nigri*. Nat Genet 50:1688–1695

Vlaardingerbroek I, Beerens B, Shahi S, Rep M (2015) Fluorescence assisted selection of Transformants (FAST): using flow cytometry to select fungal transformants. Fungal Genet Biol 76:104–109

Wang B, Guo G, Wang C, Lin Y, Wang X, Zhao M, Guo Y, He M, Zhang Y, Pan L (2010) Survey of the transcriptome of *Aspergillus oryzae* via massively parallel mRNA sequencing. Nucleic Acids Res 38:5075–5087

Wanka F, Cairns T, Boecker S, Berens C, Happel A, Zheng X, Sun J, Krappmann S, Meyer V (2016) Tet-on, or Tet-off, that is the question: advanced conditional gene expression in *Aspergillus*. Fungal Genet Biol 89:72–83

Ward M, Wilson LJ, Carmona CL, Turner G (1988) The *oliC3* gene of *Aspergillus niger*: isolation, sequence and use as a selectable marker for transformation. Curr Genet 14:37–42

Waring RB, May GS, Morris NR (1989) Characterization of an inducible expression system in *Aspergillus nidulans* using *alcA* and tubulin coding genes. Gene 79:119–130

Watanabe A, Ono Y, Fujii I, Sankawa U, Mayorga ME, Timberlake WE, Ebizuka Y (1998) Product identification of polyketide synthase coded by *Aspergillus nidulans wA* gene. Tetrahedron Lett 39:7733–7736

Watanabe A, Fujii I, Sankawa U, Mayorga ME, Timberlake WE, Ebizuka Y (1999) Re-identification of *Aspergillus nidulans wA* gene to code for a polyketide synthase of naphthopyrone. Tetrahedron Lett 40:91–94

Wei W, Chen L, Zou G, Wang Q, Yan X, Zhang J, Wang C, Zhou Z (2013) N-glycosylation affects the proper folding, enzymatic characteristics and production of a fungal β-glucosidase. Biotechnol Bioeng 110:3075–3084

Wernars K, Goosen T, Wennekes LMJ, Visser J, Bos CJ, van den Broek HWJ, van Gorcom RFM, van den Hondel CAMJJ, Pouwels PH (1985) Gene amplification in *Aspergillus nidulans* by transformation with vectors containing the *amdS* gene. Curr Genet 9:361–368

Wiemann P, Soukup AA, Folz JS, Wang PM, Noack A, Keller NP (2018) CoIN:co-inducible nitrate expression system for secondary metabolites in *Aspergillus nidulans*. Fungal Biol Biotechnol 5:1–10

Willis KJ (ed) (2018) State of the world's fungi 2018. Report. Royal Botanic Gardens, Kew

Wu Y, Sun X, Xue X, Luo H, Yao B, Xie X, Su X (2017) Overexpressing key component genes of the secretion pathway for enhanced secretion of an *Aspergillus niger* glucose oxidase in *Trichoderma reesei*. Enzym Microb Technol 106:83–87

Xie X, Meehan MJ, Xu W, Dorrestein PC, Tang Y (2009) Acyltransferase mediated polyketide release from a fungal megasynthase. J Am Chem Soc 131:8388–8389

Xu J, Gong ZZ (2003) Intron requirement for AFP gene expression in *Trichoderma viride*. Microbiology 149:3093–3097

Xu W, Chooi Y-H, Choi JW, Li S, Vederas JC, Da Silva NA, Tang Y (2013) LovG: the thioesterase required for dihydromonacolin L release and lovastatin nonaketide synthase turnover in lovastatin biosynthesis. Angew Chem Int Ed 52:6472–6475

Yang S, Kuang Y, Li H, Liu Y, Hui X, Li P, Jiang Z, Zhou Y, Wang Y, Xu A, Li S, Liu P, Wu D (2013) Enhanced production of recombinant secretory proteins in *Pichia pastoris* by optimizing Kex2 P1' site. PLoS One 8:1–11

Yoon J, Aishan T, Maruyama JI, Kitamoto K (2010) Enhanced production and secretion of heterologous proteins by the filamentous fungus *Aspergillus oryzae* via disruption of vacuolar protein sorting receptor gene AoVps10. Appl Environ Microbiol 76:5718–5727

Yoshino-Yasuda S, Karita S, Kato M, Kitamoto N (2012) Sequence analysis and heterologous expression of rhamnogalacturonan lyase A gene (AsrglA) from Shoyu koji mold, *Aspergillus sojae* KBN1340. Food Sci Technol Res 18:901–909

Yu CH, Dang Y, Zhou Z, Wu C, Zhao F, Sachs MS, Liu Y (2015) Codon usage influences the local rate of translation elongation to regulate co-translational protein folding. Mol Cell 59:744–754

Yuill E (1950) The numbers of nuclei in conidia of Aspergilli. Trans Br Mycol Soc 33:324–331

Zadra I, Abt B, Parson W, Haas H (2000) xylP promoter-based expression system and its use for antisense downregulation of the *Penicillium chrysogenum* nitrogen regulator NRE. Appl Environ Microbiol 66:4810–4816

Zeilinger S, Mach RL, Schindler M, Herzog P, Kubicek CP (1996) Different inducibility of expression of the two xylanase genes xyn1 and xyn2 in *Trichoderma reesei*. J Biol Chem 271:25624–25629

Zeilinger S, Mach R, Kubicek C (1998) Two adjacent protein binding motifs in the cbh2 (Cellobiohydrolase II-encoding) promoter of the fungus *Hypocrea jecorina* (*Trichoderma reesei*) cooperate in the induction by cellulose. J Biol Chem 273:34463–34471

Zhang F, Keasling J (2011) Biosensors and their applications in microbial metabolic engineering. Trends Microbiol 19:323–329

Zhang DX, Lu HL, Liao X, St Leger RJ, Nuss DL (2013) Simple and efficient recycling of fungal selectable marker genes with the Cre-loxP recombination system via anastomosis. Fungal Genet Biol 61:1–9

Zhang XH, Tee LY, Wang XG, Huang QS, Yang SH (2015) Off-target effects in CRISPR/Cas9-mediated genome engineering. Mol Ther Nucleic Acids 4:1–8

Zhang C, Meng X, Wei X, Lu L (2016a) Highly efficient CRISPR mutagenesis by microhomology-mediated end joining in *Aspergillus fumigatus*. Fungal Genet Biol 86:47–57

Zhang H, Wang S, Zhang XX, Ji W, Song F, Zhao Y, Li J (2016b) The *amyR*-deletion strain of *Aspergillus niger* CICC2462 is a suitable host strain to express secreted protein with a low background. Microb Cell Factories 15:1–11

Zhang J, Zhang G, Wang W, Wei D (2018) Enhanced cellulase production in *Trichoderma reesei* RUT C30 via constitution of minimal transcriptional activators. Microb Cell Factories 17:1–14

Zhao C, Waalwijk C, de Wit PJGM, Tang D, van der Lee T (2013) RNA-Seq analysis reveals new gene models and alternative splicing in the fungal pathogen *Fusarium graminearum*. BMC Genomics 14:1–16

Zheng YM, Lin FL, Gao H, Zou G, Zhang JW, Wang GQ, Chen GD, Zhou ZH, Yao XS, Hu D (2017) Development of a versatile and conventional technique for gene disruption in filamentous fungi based on CRISPR-Cas9 technology. Sci Rep 7:1–10

Zhou M, Wang T, Fu J, Xiao G, Liu Y (2015) Nonoptimal codon usage influences protein structure in intrinsically disordered regions. Mol Microbiol 97:974–987

Zhou Z, Dang Y, Zhou M, Li L, Yu C, Fu J, Chen S, Liu Y (2016) Codon usage is an important determinant of gene expression levels largely through its effects on transcription. Proc Natl Acad Sci USA 113: E6117–E6125

Zhou Z, Dang Y, Zhou M, Yuan H, Liu Y (2018) Codon usage biases co-evolve with transcription termination machinery to suppress premature cleavage and polyadenylation. Elife 7:1–29

Zhu H, Choi S, Johnston AK, Wing RA, Dean RA (1997) A large-insert (130 kbp) bacterial artificial chromosome library of the rice blast fungus *Magnaporthe grisea*: genome analysis, contig assembly, and gene cloning. Fungal Genet Biol 21:337–347

Zhu B, Cai G, Hall EO, Freeman GJ (2007) In-Fusion™ assembly: seamless engineering of multidomain fusion proteins, modular vectors, and mutations. BioTechniques 43:354–359

Zubieta MP, Contesini FJ, Rubio MV, Gonçalves AEDSS, Gerhardt JA, Prade RA, Damasio ARDL (2018) Protein profile in *Aspergillus nidulans* recombinant strains overproducing heterologous enzymes. Microb Biotechnol 11:346–358

11 New Avenues Toward Drug Discovery in Fungi

Michal Flak[1,2], Mario K. C. Krespach[1,2], Annica J. Pschibul[1,2], Volker Schroeckh[1], Axel A. Brakhage[1,2]

CONTENTS

[1] Department of Molecular and Applied Microbiology, Leibniz Institute for Natural Product Research and Infection Biology (HKI), Jena, Germany; e-mail: michal.flak@leibniz-hki.de; mario.krespach@leibniz-hki.de; annica.pschibul@leibniz-hki.de; volker.schroeckh@leibniz-hki.de; axel.brakhage@leibniz-hki.de

[2] Institute of Microbiology, Friedrich Schiller University, Jena, Germany

I. Introduction

Fungi are a valuable source of medically important natural products (NPs), mostly secondary metabolites (SMs), low- to medium-molecular-mass organic compounds that are not directly involved in growth, development, or reproduction of the producing organism. Among them, the antibiotic penicillin, the antifungal griseofulvin, the anti-hypercholesterolemic agent lovastatin, the immunosuppressant cyclosporin, and the antitumor compound asperlin are prominent examples (Brakhage 2013).

Classical techniques to discover bioactive SMs often begin with microbial culture extracts, followed by iterative rounds of fractionation and bioassays for activity measurement (Brakhage and Schroeckh 2011). This strategy was highly successful, yielding a number of "key access antibiotics" on the WHO Model List of Essential Medicines (WHO 2017). On the other hand, metabolites, which are produced in a sufficient amount, facilitating isolation, were preferentially detected. Despite several thousand SMs having been isolated in this way, with a growing number of identified compounds (Hautbergue et al. 2018), re-identification of known molecules has become a bottleneck of NP discovery, hindering the discovery of novel ones (Bérdy 2005). The lack of new antibiotics introduced into the market, combined with the rise of antibiotic-resistant bacterial and fungal strains, forms the basis of a phenomenon called the "**antibiotic innovation gap**" (Payne et al. 2007). Due to this problematic development, new strategies to discover novel, unexploited antibiotics are urgently needed.

Genetics and Biotechnology, 3rd Edition
The Mycota II
J.P. Benz, K. Schipper (Eds.)

Fortunately, however, the biosynthetic potential of these organisms is still greatly unexplored, even underestimated. Mining the increasing number of published fungal genomes revealed that the vast majority of fungal SMs still await discovery. Most fungal genomes contain more **biosynthetic gene clusters** (BGCs) than the number of NPs known to be produced by the respective fungus (e.g., de Vries et al. 2017; Rutledge and Challis 2015; Sanchez et al. 2012; Bergmann et al. 2007). Two main reasons hinder a comprehensive mining of BGCs: (1) the overwhelming majority of microorganisms, including fungi, cannot be cultivated independently as axenic cultures in the laboratory, and (2) most SM gene clusters are silent under standard laboratory growth conditions. In the previous decade, a number of strategies have been developed to overcome these drawbacks (reviewed extensively in (Brakhage and Schroeckh 2011; Brakhage 2013). Strategies to stimulate expression of silent fungal BGCs include media and growth condition variation (one strain many compounds, OSMAC), manipulation of regulatory elements of BGCs (e.g., replacement of promoters) and of factors that affect SM production more generally (*laeA*, *mcrA*), use of chromatin modifiers (SAHA), heterologous expression of whole BGCs, and bioinformatic approaches utilizing genome, transcriptome, proteome, and metabolome datasets (Brakhage 2013; Mattern et al. 2015b; Hautbergue et al. 2018) (Fig. 1). A relatively new approach is the discovery of antimicrobials inspired by ecological interactions. Prominent successful examples are co-cultivations of fungi with other microorganisms, e.g., actinobacteria or the well-studied and astonishingly stable communities of leaf cutter ants associated with mutualistic bacteria and the food forage fungi, both discussed here (Haeder et al. 2009; Schroeckh et al. 2009).

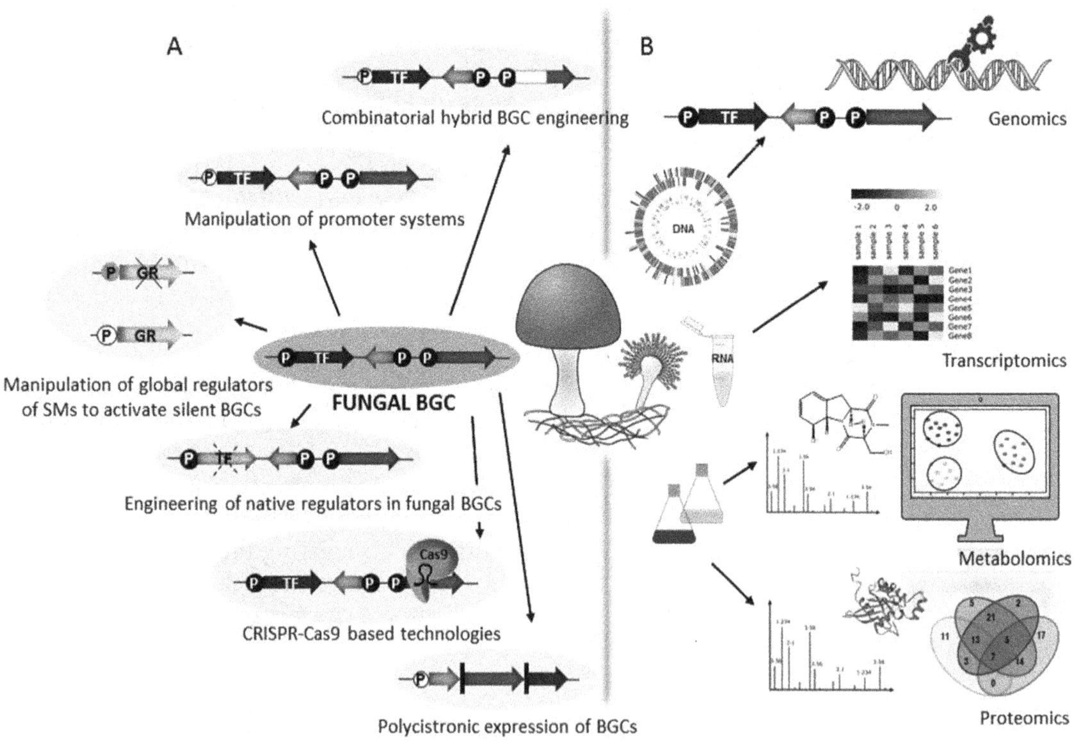

Fig. 1 Novel avenues toward the discovery of SMs in fungi. (**a**) Molecular methods and tools used in activation of biosynthetic gene clusters (BGCs) in fungi, discussed in Sect. II. (**b**) omics-based methods and various degrees of their integration in experimental approaches have proved effective in the search for new fungal-derived chemistries. Section IV provides an overview of these methodologies

Simulation of complex natural environmental conditions has been shown to allow growth of "unculturable" organisms [e.g., iChip device (Nichols et al. 2010)] and to mine and isolate their BGCs.

II. Molecular Biology Methods for Fungal Natural Product Discovery

A. Manipulation of Regulatory Elements

1. Global Regulators of Fungal Secondary Metabolism

Secondary metabolism in filamentous fungi, as well as in actinobacteria, is often associated with developmental maturity and a transition to a generative (sporulating) growth mode (Calvo et al. 2002; Wang 2003). It is therefore unsurprising that the master regulators that have a profound impact on the SM in filamentous fungi act pleiotropically and often have an effect on fungal development and/or stress response mechanisms (García-Estrada et al. 2018).

It is important to note that while the term 'global' is conventionally used to describe elements with a regulatory effect that is not limited to a single cluster, no regulator has been described so far that controls all, rather than a fraction, of SM pathways in a given organism.

Beginning with the observation that *veA*, a gene previously known for its role in the control of the light-dependent morphological switch between sexual and asexual development in *Aspergillus nidulans*, has an impact on the biosynthesis of sterigmatocystin and penicillin (Kato et al. 2003), researchers in the field have established a **direct link between morphology, primary metabolism, and the regulatory response in secondary metabolism.**

The velvet domain-containing family of proteins is conserved within the fungal kingdom (Bayram and Braus 2012). The members of this family, namely, *veA* and *velB*, interact with a methyltransferase domain-containing nuclear protein *laeA*, which was initially found to be essential for the production of sterigmatocystin (carcinogen), and penicillin (antibiotic) (Bok and Keller 2004). Since the discovery of *laeA*, this regulatory element has been characterized in other fungi, where its deletion led to the loss of production of SMs, such as gliotoxin in *Aspergillus fumigatus* (Keller et al. 2006), the antihypercholesterolemic lovastatin in *Aspergillus terreus*, as well as aflatoxin and aflatrem in *Aspergillus flavus* (Bok et al. 2006a). The *laeA* regulome investigation helped correctly identify the biosynthetic gene cluster of the antitumor compound terrequinone A, previously undescribed in *A. nidulans* (Bok et al. 2006b; Bouhired et al. 2007). Similarly, Hong and coworkers identified *Aspergillus fumisynnematus* to be a producer of cyclopiazonic acid upon overexpression of *laeA* (Hong et al. 2015), underlining the potential of manipulation of this global regulator for the purpose of discovery of novel metabolites even in strains with no readily available genome sequence. For a further overview of research focused on the *laeA* regulation of SM, the reader is directed to Yin and Keller (2011).

Another interesting discovery was that of a transcriptional regulator designated *mcrA* (*m*ulti*c*luster *r*egulator) in *Aspergillus nidulans* (Oakley et al. 2017). The authors used a mutagenic approach to find genes that *trans*-activated the backbone enzymes of biosynthetic gene clusters, which led to the discovery of a transcription factor with a global repressive effect on secondary metabolism. When deleted, the mutant strain not only shows a different profile of known SMs from the parental strain but also produces two new cichorine derivatives, along with trace amounts of felinone A, an antibiotic compound recently discovered in *Beauveria felina* (Du et al. 2014). It has also been shown that *mcrA* works independently of *laeA*, as its deletion stimulates SM biosynthesis regardless of the presence of *laeA*; furthermore, new masses were observed when *laeA* was overexpressed in a *mcrA* deletion background. Expectedly, the deletion of *mcrA* orthologues in other fungal strains, namely, *A. terreus* and *Penicillium canescens*, also led to altered metabolic profiles in the nascent mutants (Oakley et al. 2017). Given the high level of conservation of *mcrA* among ascomycetes, it is likely that the manipulation in other filamentous fungi will reveal many more compounds with bioactivity potential.

A recent resurrection of forward genetic screening approach has managed to identify mutations in three novel secondary metabolism-regulating elements in *A. nidulans* and *A. flavus*. Genomes of chemically mutagenized *A. nidulans* strains suffering from dysregulation of sterigmatocystin synthesis were sequenced to reveal *laeB*, *sntB*, and *hamI*, presumably acting alongside *laeA* as regulatory proteins required for the production of sterigmatocystin. This hypothesis is further supported by their conserved presence in the genomes of other aflatoxigenic fungi, as well as a radical decrease in sterigmatocystin production in the *A. flavus* Δ*laeB* strain (Pfannenstiel et al. 2017). The deletion of *laeB* in *A. nidulans* led to the discovery of two novel phthalide compounds (3- and 7-methoxyporriolide) and two dibenzo[1,4]dioxins (gibellulin C and D) (Lin et al. 2018), indicating that the deletion of the new regulatory proteins in other fungi may yet prove to be a fruitful approach to novel metabolite discovery.

The response of fungi to stress triggers an arsenal of regulatory machineries, often with a profound influence on the secondary metabolome of the fungus (reviewed by Brakhage 2013). This happens through signaling pathways, which often converge at master regulator nodes. The cell wall integrity (CWI) pathway is one such responsive signal cascade, formally ending with a terminal MAP kinase. The deletion of this enzyme often leads to an impaired growth phenotype, and the inhibition of this pathway has been shown to lead to decreased titers or complete abolition of SM production, e.g., penicillin in *A. nidulans* or gliotoxin in *A. fumigatus* (Valiante 2017 and references therein).

Other MAPK-regulated SMs include fumonisin B1 in *Fusarium verticillioides*, or ACT toxin in *Alternaria alternata* (Valiante 2017 and references therein).

Similarly, signals sensed by G-protein-coupled receptors embedded in the fungal cell membrane activate the cAMP/protein kinase A pathway, which relays developmental signals and has been shown to activate a silent SM cluster in *A. fumigatus*, revealing a novel compound, fumipyrrole (Macheleidt et al. 2015). Taken together, these findings suggest that the genetic manipulation of signaling proteins and regulators may provide insight not only into general biology but also cryptic secondary metabolism in fungi.

Stress signals inducing secondary metabolism are not required to be externally provided. Activation of the amino acid starvation pathway, particularly the Myb-like transcription factor BasR, in *A. nidulans* induced the expression of orsellinic acid gene cluster. Interestingly, activation of BasR also appears to constitute the fungal response to a biotic challenge presented in co-culture with *Streptomyces rapamycinicus* (Fischer et al. 2018).

2. Engineering Native Regulatory Elements in Fungal Biosynthetic Clusters

The plentiful amount of fungal genomic data produced and made available in the past decade has enabled mining for, and prediction of, biosynthetic clusters, often containing a cognate transcriptional regulatory protein. A seminal study of Brakhage and coworkers proved that a silent SM cluster could be awakened by controlling the expression of its native transcription factor using a promoter induced by the presence of ethanol (Bergmann et al. 2007). The *A. nidulans* aspyridone cluster was thus activated using the *alcA* promoter. The use of different regulatory elements, such as **promoter exchange** and **transcription factor overexpression**, with the purpose of discovery and production of natural products in filamentous fungi has since become a popular research strategy (Brakhage and Schroeckh 2011).

Unsurprisingly, genetic engineering of such basic regulatory elements has led to the discovery of a multitude of secondary metabolites across various species. A biosynthetic cluster's regulatory elements, namely, core enzyme promoter (P*eqxS*) and transcription factor (*eqxR*) are assumed to be responsible for the high native production of equisetin in *Fusarium heterosporum*. As strong transcriptional activators, these elements were exploited to resurrect an "extinct" pathway (i.e., a pathway from a producing organism that is no longer available) responsible for the production of pyrrolocin and its two derivatives, all with an antituberculosis activity (Kakule et al. 2015). In a similar approach, the *A. terreus* terrein cluster transcription factor *terR* combined with the transcriptional control of the native promoters, served as a basis for the development of a strong constitutive heterologous expression system (Gressler et al. 2015).

Further example of genetic manipulation of regulatory genes involves the overexpression of two pathways in *A. fumigatus*. (König et al. 2013) generated a strain overproducing fumicyclines, originally a product induced in a bacterial co-cultivation, by putting the cluster's associated transcription factor under the control of the inducible *icl* promoter. Similarly, fumipyrrole was found as the product of a protein kinase A/cAMP-signaling pathway-activated NRPS cluster (Macheleidt et al. 2015). The Tet-On promoter-facilitated overexpression of the cluster-specific transcription factor activated the biosynthesis of fumipyrrole even in absence of an active cAMP signaling cascade.

A. nidulans, similarly, produced asperfuranones as a result of replacement of the native promoter of one of the cluster's own transcription factors by the alcohol dehydrogenase promoter (Chiang et al. 2009).

Relying on the activating properties of many transcription factors in fungi, other studies opted for constitutive, rather than inducible, expression of transcription factors in order to awaken the entire cluster. A novel diterpene gene cluster was characterized after *gpdA*-induced expression of a newly discovered transcription factor *pbcR* in *A. nidulans*. While its overexpression generally attenuated secondary metabolite production, the seven genes adjacent to the cluster were shown to produce *ent*-pimara-8,14(15)-diene (Bromann et al. 2012).

Similarly, when the *gpdA* promoter was exchanged for the native promoter of *berA*, a transcription factor of a paraherquonin-like cluster encoded in *Neosartorya glabra* (Zhang et al. 2018), the pathway activation led to the discovery of four berkeleyacetal derivatives.

It needs to be noted that while the selected published examples suggest that the induction of regulatory proteins is sufficient for the activation of a silent secondary metabolite cluster, this is not always the case. Ahuja et al. (2012) systematically replaced the promoters of transcription factors of 18 non-reducing PKS gene clusters in *A. nidulans* with an inducible *alcA* promoter. Surprisingly, this approach only worked in three instances and yielded no or very low increases in the production of the remaining 15 respective SMs. Thus, additional considerations and genetic engineering methods are necessary and have indeed proved vital in the process of activating silent gene clusters in fungi. This necessity is elegantly illustrated by Grau et al. (2018). In their work, they exchanged the DNA-binding domain of a highly active TF associated with the asperfuranone gene cluster for the DNA-binding domain of a silent cluster-specific TF. This hybrid synthetic regulator was able to activate the *aln* cluster, producing the anti-inflammatory, antitumor polyketide asperlin, as well as a photoprotective carboxylic acid intermediate, potentially useful in cosmetics.

B. Heterologous Expression of Fungal Biosynthetic Gene Clusters

Expression of both, bacterial and fungal natural product biosynthetic pathways in heterologous hosts, has been established as a powerful technique for discovery, as well as production of secondary metabolites (Alberti et al. 2017). With the explosion of genomic data availability and remarkable advances in synthetic biology tools, **heterologous expression has become a viable alternative to total synthesis** in many cases. Numerous advantages of this approach, including state-of-the-art molecular tools for SM production enhancement, as well as the potential to refactor the natural pathways to synthesize novel product analogs, render this practice attractive in the field of fungal research (Billingsley et al. 2016). Heterologous expression is also considered to be the optimal tactic in the attempts to elucidate biosynthetic pathways in organisms exhibiting slow growth under laboratory conditions, which are not yet culturable, or in those that are not genetically amenable at present (Schmidt-Dannert 2015).

Multiple criteria must be considered and addressed in order to achieve a successful expression of a given pathway in a non-native host. Careful appraisal of potentially problematic sequence elements such as introns, along with codon optimization and the choice of a suitable host, may prove crucial (Billingsley et al. 2016). Generally, selecting a related organism naturally possessing the ability to produce secondary metabolites improves the chances of success in expression of an exogenous pathway (Mattern et al. 2015b). It is, however, important to acknowledge that even microorganisms devoid of secondary metabolism can be engineered for production of a wide array of bioactive secondary metabolites. This was the case with *Saccharomyces cerevisiae*, a biotechnology workhorse, producing the fungal polyketide lovastatin (Ley et al. 2015), or even the plant-derived antimalarial terpenoid artemisinin (Arsenault et al. 2008) or alkaloid hydrocodone (Galanie et al. 2015).

A recent work by Hillenmeyer and coworkers established a large-scale heterologous SM expression platform (HEx) in yeast (Harvey et al. 2018). The authors scanned all available fungal genomes to date to identify 3512 biosynthetic gene clusters from which 41 previously uncharacterized clusters were selected (from both ascomycetes and basidiomycetes), harboring unique PKS and UbiA-like terpene cyclase (UTC) core genes. The constituent genes from these clusters, placed under the control of the yeast P_{ADH2} promoter, were cloned into highly optimized *S. cerevisiae* strains using plasmid vectors. Twenty-two heterologous expression strains were reported to synthesize compounds absent in the control strains, 11 of these appeared to be completely novel. The structure of these new compounds was elucidated; however, none were tested for bioactive properties. Nevertheless, this effort underlines the power of heterologous expression of secondary metabolite gene clusters.

Among filamentous fungi, Aspergilli have proved to be particularly rewarding hosts for heterologous expression of biosynthetic gene clusters. *Aspergillus oryzae* has been used to heterologously synthesize high yields of antibacterial citrinin from *Monascus purpureus* (Sakai et al. 2008), an inhibitor of potassium pump with antibacterial effect paxilline from *Penicillium paxilli* (Tagami et al. 2014), or pleuromutilin, an antimicrobial compound from *Clitopilus passeckerianus* (Bailey et al. 2016).

A similar approach has been used for the discovery of a multitude of novel compounds, examples ranging from prenylated cyclic dipeptides produced in *A. nidulans* (Wunsch et al. 2015), to two novel ergot alkaloids being produced in *A. fumigatus* (Robinson and Panaccione 2014). Other examples of heterologous expression in filamentous fungi have been extensively reviewed in the past (Schümann and Hertweck 2006; Lubertozzi and Keasling 2009; Anyaogu and Mortensen 2015).

Of special note is the report of a **modification of the product of known pathways by the heterologous host**, raising the question whether compounds discovered using this approach with other, uncharacterized clusters, truly represent their *bona fide* products. The expression of the cytochalasin synthetase of *A. clavatus*, surprisingly, led to the production of a modified intermediate, niduclavine, in *A. nidulans* (Nielsen et al. 2016).

From a prognostic point of view, it is reasonable to assume that filamentous fungi (as well as their relatives from other clades) will play a consequential role in basic and applied research, as well as industrial production of pharmaceuticals and agrochemicals, be it in the role of heterologous expression hosts, ample sources for new compounds discovery, or as platforms for directed synthetic biology aimed at novel compound analogs with desired properties.

C. Molecular Engineering of Biosynthetic Clusters and Combinatorial Synthetic Biology

1. Manipulation of Promoter Systems

The vast economic relevance of fungi in the biotech industry, as well as complex basic and applied research endeavors over several decades have prompted a need for **tunable gene expression systems.** The ever-improving mechanistic understanding of gene regulation has enabled the development of promoter structures that facilitate the initiation of transcription of a gene of interest in a constitutive or inducible manner.

Features typical for **inducible promoters** include tight and controllable regulation, cost-optimal induction possibility, and a high expression of the gene downstream of the promoter (Kluge et al. 2018). Such systems developed over years have mostly been dependent on the metabolic activity of the induced organism and the carbon source present in the medium, e.g., the *xylP* promoter induced by xylose and repressed by glucose (Zadra et al. 2000) or the *alcA* promoter induced by ethanol and repressed by glucose (Waring et al. 1989). The latter was used in a pioneering attempt to activate a silent SM cluster in *A. nidulans* by putting the cognate cluster regulator, *apdR* under the control of the *alcA* promoter. This strategy led to the discovery of two novel compounds, aspyridones A and B (Bergmann et al. 2007).

Recently, the *cbhI* promoter, induced by various plant saccharides, was cloned to control the heterologous expression of two feruloyl esterase genes in *Trichoderma reesei* in order to liberate ferulate from wheat bran.

Inducible promoters that do not depend on the carbon source circumvent the obvious obstacles of manipulating growth conditions. Among these, the tetracycline-inducible promoter, originally based on the bacterial tetracycline resistance operon, was optimized for eukaryotes and has been in use for decades now (reviewed by Kluge et al. 2018). Tet-On and Tet-Off (active in presence and absence of a tetracycline core structure, respectively) promoters have been optimized for a number of filamentous fungi (Vogt et al. 2005; Meyer et al. 2011; Wanka et al. 2016; Janevska et al. 2017), ensuring the possibility of a fine-tuned expression regulation. Sun et al. (2016) conducted an elegant study in which the Tet-On promoter was used to overexpress the last uncharacterized PKS-NRPS hybrid gene in *A. terreus*. This approach, further validated by heterologous expression in *A. nidulans*, suggests that *pgnA* is the only gene necessary for the production of the metabolite.

Other particular examples of the use of the tetracycline-inducible promoter system are discussed throughout the present work.

Recently, a novel nitrate-inducible promoter system tailored for secondary metabolite expression, designated "CoIN" was developed in *A. nidulans* (Wiemann et al. 2018). Utilizing the bidirectional *niiA/niaD* promoter and supplementing the growth medium with nitrate, the authors were capable of expressing the β-carotene pathway from *Fusarium fujikuroi* in *A. nidulans*.

2. Combinatorial Hybrid Biosynthetic Cluster Engineering

A particularly interesting attitude to the discovery of novel fungal natural products is the rational genetic engineering of novel enzymes comprising the domains of biosynthetic genes from distinct clusters. These **chimeric enzymes** are then transferred into a tractable heterologous host for expression. Inspired by the **combinatorial biosynthesis** approaches long used in the research of actinobacteria (Weissman 2016), domain exchanges and module swapping have helped create novel "unnatural" products since as early as 1985, when mederrhodins, hybrid antibiotics from *Streptomyces* sp. were reported (Omura et al. 1986).

Another hybrid PKS product of mutational biosynthesis is doramectin, an antiparasitic molecule widely used in veterinary medicine today (Goudie et al. 1993).

The PKS and NRPS **biosynthetic genes in fungi share a similar modularity to those in bacteria**, facilitating their genetic shuffling and combination as building blocks into novel protein chimeras with desired catalytic properties (Nielsen et al. 2016). With the advancing structural information on fungal SM genes, it has become possible to predict the enzyme programming and product synthesis (Hertweck 2015).

An inviting type of biosynthetic enzymes presents itself in the fungal PKS-NRPS hybrids: multimodular enzymes, usually consisting of a type I iterative highly reducing PKS and a single NRPS module. These naturally hybrid enzymes are known to produce highly bioactive compounds, including cyclopiazonic acid (Seidler et al. 1989), pseurotin (Wiemann et al. 2013a), or fusarin C (Sondergaard et al. 2011). Given their inherently dual character, researchers have successfully attempted to combine the PKS and NRPS modules from different species, or even the same organism. Bassianin, an extinct metabolite (i.e., a previously reported metabolite whose producing strain is unavailable) from *Beauveria bassiana*, was brought back by domain and module swapping between two genes from the fungus, *tenS* and *dmbS*, responsible for tenellin and desmethylbassianin production, respectively (Heneghan et al. 2010; Fisch et al. 2011).

A hybrid compound incorporating a tryptophan residue into the aspyridone polyketide backbone was generated by an ApdA (aspyridone synthase) and CpaS (cyclopiazonate synthetase) chimeric enzyme (Xu et al. 2010).

A more comprehensive approach was employed, where PKS-NRPS compatibility was studied by constructing 34 distinct module swaps between five genes from four different organisms. Many expected chimeric compounds were detected and the approach yielded six novel metabolites (Kakule et al. 2014). A rather interesting set of results was provided by Nielsen et al. (2016), who used the cytochalasin PKS-NRPS hybrid, *ccsA*, along with a *trans*-acting enoyl reductase *ccsC*, to create chimeras by module swapping with *syn2*, an uncharacterized PKS-NRPS from *Magnaporthe oryzae*, leading to identification of novel products in *A. nidulans* as the heterologous host.

By reciprocally combining the PKS and NRPS modules of the respective genes, they created heterologous host mutants expressing functional fusion proteins producing niduchimaeralin A (chimeric *ccsA-syn2* PKS-NRPS), and a structurally similar niduchimaeralin B (chimeric *syn2-ccsA* PKS-NRPS). The authors speculate that the success of this study depended on the structural similarity of the intermediates, rather than protein compatibility, based on the ability of non-native condensation domain to recognize the polyketide.

This strategy to generate novel fungal chemistries is therefore potentially valuable but inherently limited by the structural affinities and substrate preferences of the nascent hybrid biosynthetic enzymes.

Engineering fungal SM enzymes is by no means limited to PKS-NRPS hybrid pathways. Previously, bacterial NRPSs have been manipulated to generate functional recombinant enzymes (e.g., Owen et al. 2016; Fischbach et al. 2007), while their fungal counterparts remained understudied. Three NRPS-like genes from *A. terreus* responsible for the production of aspulvinone E, butyrolactone IIa, and phenguignardic acid were engineered by Wang and coworkers, swapping individual adenylation (A), thiolation (T), and thioesterase (TE) domains (van Dijk et al. 2016). This approach resulted in the finding that NRPS-like genes can be engineered to yield functional hybrid enzymes that synthesize a novel metabolite.

Specifically, exchanging the A and T domain in butyrolactone IIa synthetase for phenguignardate A or A-T domain fragment results in the production of a nonhydroxylated metabolite phenylbutyrolactone, suggesting that the function of the TE domain is independent of the A domain.

D. Novel Techniques for Natural Product Discovery

1. CRISPR-Cas9 in Filamentous Fungi

The innovation in the field of molecular biology has overcome the restrictions imposed by the traditional molecular toolkit used in filamentous fungi: gene deletion, overexpression, or mutagenesis. The advent of CRISPR-Cas9 gene editing technology (Jinek et al. 2012), based on the bacterial adaptive immunity derived from *Streptococcus pyogenes*, and suitable for multiple concurrent gene-editing events, was promptly appropriated and installed in fungal hosts (Krappmann 2017; Weber et al. 2017).

The system was established in the model species *Neurospora crassa* (Matsu-ura et al. 2015), *A. nidulans* (Nødvig et al. 2015), the industrial workhorse, *Trichoderma reesei* (Liu et al. 2015), phytopathogenic *Ustilago maydis* (Schuster et al. 2016), and the human pathogens *Candida albicans* (Vyas et al. 2015) and *A. fumigatus* (Fuller et al. 2015).

This technology promises to accelerate genetic engineering by offering highly targeted, markerless editing of DNA. Deng et al. (2017) provides a comprehensive overview of the state of the art of CRISPR-Cas9 technology in filamentous fungi.

In *A. fumigatus* in particular, the CRISPS-Cas9 gene editing coupled with a split marker system has been used to reconstitute the functionality of a mutated trypacidin biosynthetic gene cluster (Weber et al. 2017). In this study, the authors utilized the CRISPR capability to edit single bases within the genome and "corrected" a point mutation that resulted in a premature stop codon in the PKS-coding *tynC* gene. To advance the efficiency of the method and prevent further alterations to the locus of interest, a spatially separated split marker was introduced that contained a gRNA recognition site and PAM identical to that of the *tynC* target site on each fragment. Thus, the selection marker was only active after cleavage by Cas9 endonuclease, which concurrently also introduced a double-strand break in *tynC*, allowing for a recombination event with provided donor DNA, effectively reconstituting the gene's functionality in a formerly non-producing strain. Trypacidin has been shown to be present in fungal spores, and it is hypothesized to have a protective effect against phagocytosis due to its amoebicidal activity (Mattern et al. 2015a).

CRISPR-Cas9 technology was also used to genetically engineer and explore the secondary metabolism of *Talaromyces atroroseus*, a fungus previously recalcitrant to genetic manipulation. Using the CRISPR-Cas9-enabled efficient gene targeting, the authors were able to delete the gene encoding a hybrid PKS-NRPS, identified as an orthologue of those producing nitrogen-containing tetramic acid derivatives in other filamentous fungi (Nielsen et al. 2017). This hybrid gene is responsible for the production of talaroconvolutin A and its medically relevant stereoisomer ZG-1494α.

The latter is an inhibitor of platelet-activating factor acetyltransferase (West et al. 1996), while talaroconvolutins have been evaluated for their antifungal activity (Suzuki et al. 2000).

2. Polycistronic Expression of Biosynthetic Pathways

Pioneered by the recent publication of the introduction of a construct containing the penicillin biosynthetic pathway genes encoded in a single **polycistronic mRNA** in a non-producing strain of *A. nidulans* (Unkles et al. 2014), the technique has sparked interest among fungal molecular biologists. This novel synthetic biology approach assembles multiple genes in a single mRNA and utilizes viral 2A peptide sequences to facilitate a co-translational cleavage of the nascent peptide chain by the ribosomes. That opens the possibility of introducing entire biosynthetic gene pathways in convenient heterologous hosts, circumventing promoter exchange limitations, the need for individual gene overexpression, or selection marker recycling.

This approach has since been validated and optimized for *A. niger*. Enniatin production native to *Fusarium oxysporum* was installed in the biotechnology workhorse by introducing ketoisovalerate reductase and enniatin synthetase genes, required for the SM synthesis, linked with 2A peptide sequences (Schuetze and Meyer 2017). Geib and Brock (2017) used an approach combining a double promoter exchange and polycistronic heterologous expression to optimize the expression levels of a reporter gene, subsequently verifying the capacity of the system using the aspulvinone E synthetase gene from *A. terreus*. The result was a finely tunable, strong gene expression platform.

Crossing the borders of secondary metabolism research, the 2A peptide-based system was also used to assemble a bicistronic construct of a *Penicillium funiculosum* cellobiohydrolase enzyme and a GFP, which was introduced into *T. reesei*, allowing the authors to monitor the expression of the enzyme (Subramanian et al. 2017).

Recently, Hoefgen et al. (2018) introduced an advanced method of polycistronic expression, introducing a **split fluorescent marker**, active only upon successful assembly and allowing for simple selection of positive transformants using fluorescence microscopy, along with a TEV protease to reduce the remaining C-terminus tag of the 2A peptides. This improved method was used to heterologously produce exceptionally high yields of the psychotropic alkaloid psilocybin in *A. nidulans*. The hallucinogenic compound has recently regained interest, showing a pharmaceutical potential as an anxiolytic in terminal-stage cancer patients and antidepressant (Griffiths et al. 2016).

It is worthwhile to note that this research is recent, and while no new fungal metabolites have been discovered yet, it is conceivable that the polycistronic expression of whole fungal pathways will, in the future, yield novel secondary metabolites with interesting bioactive potential.

3. Fungal Artificial Chromosomes

An elegant, albeit isolated technique incorporated the idea of bacterial and yeast artificial chromosomes and translated it to filamentous fungi, giving rise to so-called fungal artificial chromosomes (FACs). Constructed on a bacterial plasmid backbone harboring the fungal autonomously replicating fragment AMA1, these genetic elements are capable of replicating in both *E. coli* and *A. nidulans*. Randomly sheared segments of *A. terreus* genomic DNA were cloned into the backbone, creating the FACs. All 56 annotated biosynthetic gene clusters from *A. terreus* were introduced into FACs in this fashion and successfully shuttled into *A. nidulans* as a heterologous host (Bok et al. 2015). To validate the system, the authors analyzed a host mutant carrying the *A. terreus* astechrome cluster and discovered terrezine D, a precursor molecule in astechrome biosynthesis machinery.

The utilization of unbiased, randomly sheared gDNA is specifically of note. This renders the FACs methodology particularly applicable in the case of unsequenced microbes and complex microbial metagenomes and highlights its enabling potential to identify SM gene clusters and their corresponding products.

III. Fungal Interactions as a Source of New Compounds

A. Induction of Natural Product Formation by Co-cultivation

At the beginning of natural product research, isolates of fungi and other microorganisms were taken from distinct habitats, such as soil, grown in monoculture, and fermented to find metabolites showing desirable activity. This approach was highly successful and yielded an abundance of potent pharmaceuticals, which are still being used to date. It is widely accepted that fungi, as well as all other microorganisms, do not live in isolation in their natural habitat (Fig. 2). Therefore, **organismal interactions** are highly likely to occur and secondary metabolites are mainly thought to serve as arsenal in the defense of an ecological niche (Netzker et al. 2018). An abundance of **co-cultivation experiments** has been carried out in order to obtain novel antibiotics to refill the antibiotic pipeline, as well as to gain insight into the biology of organismal interactions. In the next sections we aim to give an overview of examples in which natural product formation was induced

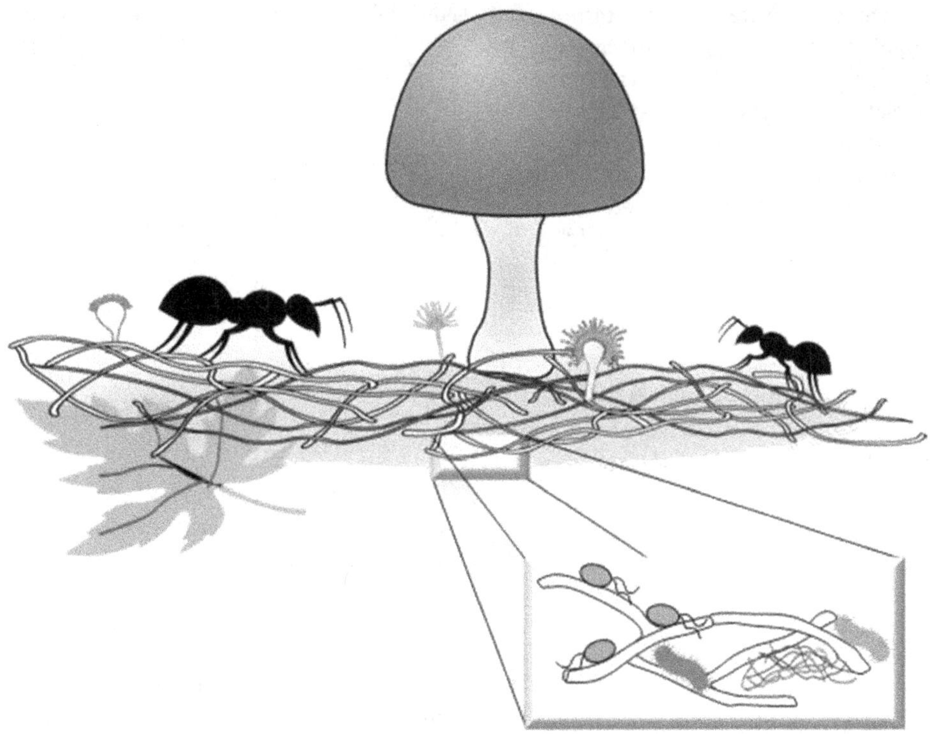

Fig. 2 Secondary metabolites fulfill a diverse array of functions in the natural habitat of the producing organisms. As discussed in Sect. III, fungal secondary metabolites are instrumental in mediating the communication between various fungal species, their bacterial symbionts, as well as the ecological interactions between fungi and various invertebrates and plants

by fungal co-cultivation with bacteria and other fungi.

1. Induction of Silent Natural Product Gene Clusters by Co-cultivation of Fungi with Bacteria

As aforementioned, fungi do not live in isolation, but encounter various neighboring microorganisms, to whose presence they have to react. It is conceivable that upon co-cultivation of fungi with other microorganisms in the laboratory, each interaction partner might produce compounds to communicate or defend itself.

Early co-cultivation approaches proved that fungi could display specific reactions to co-cultivated bacteria. Following the assumption that fungi coexist with actinobacteria in the soil, these organisms were among the first to be used in controlled co-cultivation experiments. The filamentous ascomycete *Aspergillus nidulans* reacts to the close physical contact with the soil bacterium *Streptomyces rapamycinicus* by producing orsellinic acid, lecanoric acid, and the cathepsin K inhibitors F-9775A and B (Schroeckh et al. 2009). Interestingly, the same bacterium induces the fumicycline biosynthetic gene cluster in the human pathogenic *A. fumigatus*, leading to the production of fumicyclines A and B (König et al. 2013). However, *S. rapamycinicus* is not the only bacterium, which elicits a specific reaction in *A. fumigatus*.

In co-culture with *Streptomyces peucetius*, the fungus produces two formyl xanthocillin analogues, fumiformamide and N,N′-((1Z,3Z)-1,4-bis(4-methoxyphenyl) buta-1,3-diene-2,3-diyl)diformamide, the latter showing significant cytotoxic activity against tumor cells (Zuck et al. 2011).

There are also quite recent examples of novel fungal natural products discovered by studying fungal-bacterial interactions. A marine isolate of *A. fumigatus* produced the undiscovered luteoride D, along with pseurotin G, 11-*O*-methylpseurotin, and terezine D, when co-cultured with two desert-derived strains of *Streptomyces leeuwenhoekii* (Wakefield et al. 2017).

Expectedly, the effect of activation of silent metabolic clusters in fungi is not limited to streptomycetes. Abdelwahab et al. (2018) described the production of aspvanicin A and its epimer aspvanicin B in co-culture of *Aspergillus versicolor* with *Bacillus subtilis.*

Aspvanicin B shows anti-proliferative activity towards mouse lymphoma cell lines.

Similarly, *Aspergillus* is not the only fungal genus useful for co-cultivation approaches. In the co-culture of an endophytic fungus *Trichoderma* sp. with *Acinetobacter johnsonii,* two previously undescribed sesquiterpenes, namely, microsphaeropsisin B and C, and two new de-O-methyllasiodiplodins were detected (Zhang et al. 2017).

In addition to the novel compounds, these co-cultivations yielded 12 known products.

Basidiomycetes are also capable of adjusting their secondary metabolism in interaction with bacteria. The brown rot fungus *Serpula lacrymans* shows production of pigmented non-ribosomal peptides when in contact with one of at least 13 different bacteria, but not in monoculture. Among these colored compounds, variegatic acid and xerocomic acid decreased swarming and biofilm formation of *B. subtilis* by a potentially novel, nontoxic mode of action (Tauber et al. 2016, 2018). *Streptomyces* AcH 505 induced the formation of a 5-formylsalicylic acid (5-FSA) in the plant pathogenic basidiomycete *Heterobasidion abietinum.*

The addition of this compound to *H. abietinum* infecting Norway spruce seedlings led to increased detrimental effects to the plant and an elevation of fungal biomass. However, 5-FSA alone added to the plant did not inhibit plant growth (Keilhofer et al. 2018). This leads to the assumption that the streptomycete helps the fungus to infect the host plant by production of a natural product.

Taken together, bacterial-fungal co-cultivation has been highly successful in discovery of novel natural products and might also be used to shed light on their role in the natural habitats of the producing organisms.

2. Induction of Silent Secondary Metabolite Gene Clusters by Co-cultivation of Fungi with Other Fungi

In accordance with the idea that co-cultivation of fungi with bacteria might result in the induction of otherwise silent natural product biosynthetic gene clusters, there are successful approaches of co-cultivation of one fungal species with another that yield novel secondary metabolites with antifungal properties. Three novel lipoaminopeptides, the acremostatins A, B, and C, were discovered when the endophytic ascomycete *Acremonium* sp. and the mycoparasite *Mycogone rosea* were cultivated together (Degenkolb et al. 2002).

These SMs bear structural resemblance to leucinostatins, which show activity against the major potato pathogen *Phytophtora infestans* (Wang et al. 2016a).

Antifungal natural products are also formed when isolates of *Phomopsis* sp. and *Alternaria* sp. are co-cultured. Li et al. (2014) reported the production of a new cyclic tetrapeptide in mixed fermentation. Interestingly, the newly discovered natural product shows promising activity against several plant pathogenic fungi (Li et al. 2014). The co-cultivation of the two mollusk-derived fungi, *Chaunopycnis* sp. and *Trichoderma hamatum*, is an interesting showcase for multi-level fungal interactions. *Chaunopycnis* sp. was shown to not only produce the novel natural product chaunopyran A but also biotransform pyridoxatin produced by *Trichoderma hamatum* to methylpyridoxatin, deactivating the molecule's antifungal properties (Shang et al. 2017).

Novel SMs from fungal-fungal interaction are not limited to antifungal activity: there are also reports about production of antibacterial compounds. The citrifelins A and B, citrinine derivatives with a unique tetracyclic backbone, were isolated from the culture broth of a co-culture of marine-derived isolates of *Penicillium citrinum* and *Beauveria feline* (Meng et al. 2015). Another example for metabolites identified in co-cultures involving fungi of the genus *Penicillium* are the berkeleylactones.

P. fuscum co-cultured with *P. camembertii/clavigerum* yielded eight berkeleylactone derivatives.

Interestingly, these show a very high antimicrobial activity against Gram-positive bacteria such as MRSA, as well as the opportunistic fungal pathogen *Candida albicans*. However, unlike other macrolide antibiotics resembling berkeleylactones, they do not interfere with protein biosynthesis, indicating a novel mode of action (Stierle et al. 2017).

Furthermore, the antibacterial alkaloid aspergicin was produced, when two epiphytic *Aspergillus* sp. strains were co-cultured (Zhu et al. 2011, 2017).

Another interesting instance of fungal co-cultivation is that of two developmental stages of *Aspergillus alliaceus* with one another. In this fungus, the vegetative, asexual anamorph can be separated from the sclerotia-forming morph that induced sexual development. Both morphotypes produce distinct natural products when grown in monoculture. Ochratoxin, a potent mycotoxin, is being produced mainly by the sexual teleomorph, while nalgiovensin, an anthraquinone pigment, was produced by the asexual stage. However, the natural product pattern changed remarkably, when the asexual and the sexual morph were co-cultivated. The chlorinated nalgiovensin derivative, nalgiolaxin, increased in abundance, and weakly cytotoxic allianthrones A, B and C, novel natural products, were produced (Mandelare et al. 2018). This example illustrates that not only different species of fungi, but different developmental stages of the same strain, can interact with each other and lead to different natural product formation.

3. Induction of Fungal Volatile Compounds and Their Role in Co-cultivation

Volatile organic compounds (VOCs) are generally considered to be small ($\leq$20 C atoms), possessing low molecular mass, high vapor pressure, and a low boiling point. These characteristics enable easy transfer of VOCs in soil and aboveground, which suggests that these compounds serve as long-distance messengers (Schmidt and Podmore 2015). Extensive research was conducted investigating the role of VOCs in communication of microorganisms with their surroundings (Piechulla & Degenhardt 2014). As described below, there is now growing evidence that VOCs also play an important **role in communication, signaling, and growth inhibition** of fungi, impacting their physiological state, as well as secondary metabolism. Evans et al. (2008) investigated the production and influence of VOCs when various fungi where challenged with each other. The VOC profiles were species-specific and showed variation over time. Notably, the VOC pattern generally changed when the fungi were in contact with other fungi. Among a number of other VOCs, a potentially antifungal quinolinium-like compound was formed when *Hypholoma fasciculare* was either self-paired or co-cultivated with *Trametes versicolor*.

A similar approach was pursued when the gas phase of the plant pathogens *Eutypa lata* and *Botryosphaeria obtusa* were connected. The produced VOCs were species-specific, variable over time, and subject to change when fungal partners were in contact. One of the detected VOCs, antifungal 2-nonanone, gradually increased in abundance during co-cultivation. However, both fungi, i.e. including the producer strain, were susceptible to the antifungal activity of 2-nonanone (Azzollini et al. 2018).

There is also a specific volatome response between two peanut pathogens *Aspergillus flavus* and *Ralstonia solanacearum*. *A. flavus* VOCs reduced the production of exopolysaccharides, *R. solanacearum*'s major virulence factor. In contrast, *A. flavus* decreased spore production but increased aflatoxin production when exposed to *R. solanacearum* VOCs (Spraker et al. 2014). This highlights not only the diverse patterns of VOCs but also the very specific "scent" fungi and their interactions seem to have.

Even though no novel fungal SMs have been identified thus far through methods involving VOCs, it is conceivable that this approach will become a valid strategy for silent SM gene cluster induction in the future.

4. Induction of Silent Natural Product Gene Clusters by Co-cultivation of Fungi with Non-microorganisms

Co-cultivation approaches need not be limited to confrontations of microorganisms. This has been elegantly demonstrated in a co-cultivation experiment of *Trichoderma harzianum* and callus of the Madagascar periwinkle, *Catharanthus roseus*, leading to the discovery of a novel fungal metabolite, trichosetin. This previously undescribed tetramic acid exhibits remarkable anti-Gram-positive activity also against MRSA strains (Marfori et al. 2002).

B. Discovery of Fungal Natural Products Inspired by Ecological Interactions

1. Natural Products from Fungi in Association with Marine Organisms

Finding novel metabolites from fungi can be achieved by exploring previously **underinvestigated habitats and ecosystems.** The sea is one of these ecosystems, gaining increasing attention in screening for potential pharmaceutically relevant compounds; indeed, a number of antibacterial natural products have been discovered this way [recently reviewed in (Blunt et al. 2018)]. An early report describes the production of the chlorinated benzophenone antibiotic pestalone with potent anti-Gram-positive activity by a fungus of the genus *Pestalotia.* This fungus was originally isolated from the surface of the marine alga *Rosenvingea* sp. and only produced pestalone when co-cultured with an unidentified unicellular marine bacterium (Cueto et al. 2001).

Similarly, three novel chlorinated benzophenones, as well as four novel natural products, pestaloisocoumarins A and B, isopolisin B and pestalotiol A were isolated from the sponge-derived fungus *Pestalotiopsis heterocornis.* The chlorinated benzophenones as well as pestaloisocoumarins A and B exhibited potent anti-Gram-positive activity (Lei et al. 2017).

GKK1032B and secalonic acid A were isolated from the marine sponge-associated fungus *Penicillium erubescens* and showed activity against Gram-positive bacteria and in the case of secalonic acid A also against MRSA (Kumla et al. 2018). A marine *A. fumigatus* strain, found in association with a sponge, was able to produce seven novel antibacterial helvolic acid derivatives with potent activity against the human and fish pathogenic bacterium *Streptococcus agalactiae* (Kong et al. 2018). Not only antibacterial natural products were discovered by sampling fungi from marine backgrounds. An *A. fumigatus* isolate from the gastrointestinal tract of the marine fish *Pseudolabrus japonicus* produced seven cytotoxic compounds, the fumiquinazolines A-G, when cultured in artificial sea water for 3 weeks (Takahashi et al. 1995). Liu et al. (2018) discovered three cyclic peptides from the sponge-derived *Aspergillus violaceofuscus.*

Among these compounds, a cyclic tetrapeptide, as well as a diketopiperazine dimer exhibited anti-inflammatory activity against interleukin-10 expression of lipopolysaccharide-induced THP-1 cells.

Another isolate from a marine sponge is the ascomycete *Truncatella angustata*, which produced 22 isoprenylated cyclohexanols on solid medium. The novel derivative truncateol O inhibited HIV-1 and H_1N_1 viruses, while truncateol P exhibited activity against HIV-1 (Zhao et al. 2018). Marine habitats and fungi in association with marine organisms like algae and sponges can be regarded as a valuable source of novel natural products with various pharmaceutically relevant features.

2. Natural Products from Fungi in Association with Arthropods

Leaf cutter ants in association with mutualistic bacteria and their forage fungi are well studied and remarkably stable communities. For 50–60 million years, these communities have developed to a **commensal relationship** with the ultimate goal of defense of the community against attacking pathogens. In this multipartner system, the basidiomycete *Leucoagaricus gongylophorus* serves as a food source to the attine ants *Acromyrmex octospinosus*. The ants culture the food fungus by feeding it gathered plant material, as well as by grooming and weeding the fungal garden (Currie 2001; Haeder et al. 2009). Pathogenic fungi of the genus *Escovopsis* disturb the delicate symbiosis by overgrowing the food fungus, leading to a colony collapse. Ant colony-associated *Streptomyces* sp. Ao10, in turn, produces candicidin, which can inhibit the pathogenic fungus, while showing no effect on *L. gongylophorus* (Haeder et al. 2009). There is evidence that the pathogenic fungus also features a distinct natural product arsenal, suggested to serve as virulence factors to promote the infection of the ant colony. *Escovopsis weberi* has been found to produce cycloarthropsone and emodin, polyketides that inhibit the growth of *L. gongylophorus* and, in the case of emodin, also the associated streptomycete symbionts.

Furthermore, the pathogen also produced a novel shearinine derivative, shearinine L, which shows repellent activity towards the ant (Dhodary et al. 2018). MALDI-MS imaging revealed that in interaction assays of the ant-associated bacteria *Streptomyces* CBR38 and *Streptomyces* CBR59 with *Escovopsis* TZ49, similar compounds are involved. *Escovopsis* TZ49 produced the shearinines D, F, and J, while the bacteria produced a number of natural products, including antifungal compounds (Boya et al. 2017).

This network was further expanded by the identification of two more natural products playing a role in the interaction of leaf cutter ants, bacteria and fungi. *E. weberi* produced melinacidin IV, which inhibits ant-mutualistic *Pseudonocardia* bacteria (Heine et al. 2018). In addition, it produces shearinine D, which not only inhibits mutualistic bacteria but also accumulates in ant workers and alters their behavior, causing them to fail to remove infected material from the ant colony. This portrays a complex, multilayered system of natural warfare between closely associated fungi, bacteria, and higher eukaryotes, in which secondary metabolites play a leading strategic role.

IV. Natural Product Discovery in Fungi in the Age of Omics

To combat the shortcomings associated with the classical approaches toward discovery of novel microbe-derived bioactive compounds based on targeted purification, the postgenomic era offers its own set of valuable tools. The use of omics-based techniques aimed at secondary metabolism in fungi provides a route toward identification of the involved genetic elements, direct detection of natural products, as well as insights into their role and regulation in native environments.

A vast **spectrum of omics strategies** has been developed to advance the endeavors of secondary metabolism research: genome-based methods and bioinformatic algorithms to identify the metabolic potential, high-throughput metabolome studies used to complement and verify the in silico predictions, along with transcriptomic and proteomic tools that form a link between the genomes and the nascent metabolomes (Fig. 1b). As reviewed in the following passage, the application and integration of these strategies in fungal research have proved fruitful in discovery of natural products.

A. Genomics

Computational methods are routinely used to identify biosynthetic gene clusters in the wealth of genomic data made available in public repositories thanks to the advancements in high-throughput sequencing. Systematic analysis of sequencing data has consistently revealed that most fungal genomes contain more BGCs than there are known NPs produced by the respective organisms (e.g., Macheleidt et al. 2016; Rutledge and Challis 2015; Sanchez et al. 2012), representing an underexploited reservoir of novel chemistries. As such, the genomic approaches used in natural product discovery are a valuable toolkit for capturing the general biosynthetic potential of fungi.

Targeted genome mining approaches routinely used to identify BGCs are based primarily on domain-dependent protocols, which identify conserved elements within the constituent genes of a cluster. The antiSMASH framework (Medema et al. 2011), with its latest version, antiSMASH 4.0 (Blin et al. 2017), represents one of the most comprehensive algorithms, capable of recognizing 44 different classes of biosynthetic enzymes. Others, like SMURF (Khaldi et al. 2010), ClustScan (Starcevic et al. 2008), or PRISM (Skinnider et al. 2015), recognize core enzymes typical for fungal secondary metabolite clusters, including PKSs, NRPSs, DMATSs, and/or hybrid PKS-NRPS entities.

For a recent overview of software tools and databases for analysis of biosynthetic gene clusters in genomic data, the reader is directed to a review by Weber (2014) and references therein.

In contrast to the motif-dependent tools, protocols have been designed to identify clusters lacking a core enzyme. MIPS-CG, an algorithm developed by Takeda et al. (2014), utilizes comparative genomics of two species, recognizing gene order within a nonsyntenic genomic background, rather than conserved motifs. Similarly, comparative **phylogenomics** uses the distribution and evolution of gene clusters to predict the roles of genes, potentially linking genetic elements and SMs (Brown and Proctor 2016), identifying novel clusters from phylogenetic alignment outliers (Kang 2017), or revealing an evolutionary plasticity in SM profiles (Mattern et al. 2017). **Comparative genomics** approaches are also highly suited to the identification of BGCs in a scenario where the product is known, but the encoding cluster remains to be discovered. Using the product structure, information can be gained about the types of enzymes involved in its production. This strategy led to the identification of the gene cluster of griseofulvin, an antifungal polyketide produced by several *Penicillium* species. Comparing the genome of the producing *P. aethiopicum* with that of *P. chrysogenum*, a nonproducer, facilitated the elimination of orthologous clusters. From the remaining ones, only one contained methyltransferases and a halogenase necessary for the formation of griseofulvin. The deletion of the PKS confirmed the prediction (Chooi et al. 2010).

The SM cluster producing echinocandin B1, a novel antifungal from *Emericella rugulosa*, was identified in a similar fashion (Cacho et al. 2012).

While mining of genome data using bioinformatic tools is valuable due to its speed of processing and high throughput potential (e.g., Cacho et al. 2015), it should be noted that this approach is invariably predictive and the derived hypothesis needs to be validated experimentally. That links the computational prediction with wet lab techniques, based either on the OSMAC strategy and a relatively inefficient screening of modified physicochemical cultivation conditions [e.g., isolation of new antibiotics from *Aspergillus parasiticus* (Bracarense and Takahashi 2014)], or one of the molecular techniques discussed in the present work.

Traditionally, an SM cluster identified by genomic approaches is validated by gene knockout or overexpression to establish a link between the genetic prediction and SM production. Deletion of two genes in *A. nidulans* led to the identification of the four emericellamides C-F evident from the comparison between the metabolic profiles of the wild type and deletion strains (Chiang et al. 2009). Six azaphilones, azanigerones A-F, were also discovered in the same manner in *A. niger* and further validated by overexpression and gene deletion mutants (Zabala et al. 2012). Alternatively, a cluster can be heterologously expressed, as was the case of didymellamide B, whose cluster was identified in *A. solani* and transferred to *A. oryzae* for expression (Ugai et al. 2016).

1. Metagenomics

The sequencing and analysis of DNA is not limited to organisms culturable in laboratory conditions. In the context of metagenomics, **DNA from environmental samples** can be purified and even enriched to select for a particular group of organisms, e.g., eukaryotes or fungi (Chávez et al. 2015). The subsequent treatment branches out and may involve the complex, labor-intensive process of sequencing, assembly, annotation, and bioinformatic search for biosynthetic clusters. Alternatively, a **metagenomic library construction** may follow to produce vectors carrying the isolated DNA. This approach is well established in bacteria [e.g., Iqbal et al. (2016)] and has led to the discovery of numerous bioactive compounds, including the recent malacidins (Hover et al. 2018). In fungi, this has been elegantly demonstrated in the construction of fungal artificial chromosomes (Bok et al. 2015), as discussed above. Following suit, Robey et al. (2018) used the same technology to identify a novel biosynthetic mechanism from an *Aspergillus aculeatus* NRPS. Metagenomic mining is a prospective tool for the identification of potentially valuable natural products from microbes that are unculturable under laboratory conditions (Milshteyn et al. 2014). Chávez et al. (2015) underline the importance of extremophilic ascomycetes as an untapped source. Similarly, symbiotic and endophytic fungi are garnering attention as a poorly characterized biological system with biosynthetic capacity (Borges et al. 2009).

B. Transcriptomics

The utility of transcriptomics in the identification of active BGCs, as an intermediate approach between genomics and metabolomics, is evident. The **comparative analysis of expression profiles** between samples (e.g., treated/untreated, induced/uninduced, wild type/mutant) yields information about pathway expression, thus supporting other omics approaches, and helping to identify (co-) regulation patterns.

In an aforementioned example, Schroeckh et al. (2009) used a microarray-based transcriptomics approach to analyze the changes induced in *A. nidulans* upon co-cultivation with *S. rapamycinicus*. Within the large set of differentially expressed genes, genes from several SM clusters were found whose expression was induced by contact with the bacterium. One of these was the previously unidentified cluster encoding orsellinic acid biosynthetic genes.

Similarly, an uncharacterized NRPS cluster with an unidentified product was linked to virulence and late stage of infection in *Fusarium graminearum* (Zhang et al. 2012) using microarray transcriptional profiling.

Transcriptional regulation at a global scale can be revealed via transcriptome analyses. The comparison of expression profiles between samples was used to identify regulators linked to SM silencing in bacteria (Amos et al. 2017); it is conceivable that a similar approach may be used in identifying regulatory elements in fungi. Furthermore, an apparent co-regulation of genes uncharacteristically scattered over multiple chromosomes (rendering them unidentifiable by conventional genomic predictions) revealed a regulatory crosstalk between gene clusters. In *A. nidulans*, an activator gene from a putative NRPS cluster on chromosome II induced the expression of a silent PKS cluster on chromosome VIII, leading to the identification of the polyketide asperfuranone (Bergmann et al. 2010).

The production of nidulanin A, discovered in the same organism more recently, also requires two biosynthetic genes, an NRPS and a prenyltransferase, physically distant in their genomic context (Andersen et al. 2013).

MIDDAS-M, a bioinformatics tool designed to detect BGCs, including those lacking a canonical backbone enzyme, incorporates an algorithm that can be used to analyze co-regulatory phenomena in transcriptome data by generating virtual clusters and scoring expression differences. This method was applied to identify the gene cluster responsible for ustiloxin B in *A. flavus* (Takeda et al. 2014).

While ustiloxin B is a nonribosomal peptide-like compound, no predicted NRPS genes could be linked to its biosynthesis. The MIDDAS-M analysis of the transcriptome data from *A. flavus* grown under different conditions showed three differentially expressed clusters. Finally, the ustiloxin B cluster was identified and verified by gene deletion.

C. Proteomics

To complement the genomic and metabolomic strategies, proteomic studies offer a snapshot of an organism's secondary metabolism from the perspective of active biosynthetic enzymes. This approach is, naturally, based on an assumption that the changes in enzyme types and quantities are projected in modifications of the metabolome. Indeed, the expression levels of biosynthetic proteins are often directly correlated to the levels of NP produced (Vödisch et al. 2011; Gubbens et al. 2014).

Furthermore, the comprehensive nature of a proteome study allows for the detection of **posttranslational modifications** that may occur to identify active enzymes. Du and van Wezel (2018) provide an in-depth review of activity-based protein profiling (ABPP), a cost-efficient set of methods providing a snapshot of active biosynthetic potential in a complex biological sample. Combining ABPP with probes for conserved features of biosynthetic enzymes, such as carrier protein (CP) and thioesterase domains, and coupling to mass spectrometry was used to explore and quantify PKS and NRPS enzymes [the OASIS (Orthogonal Active Site Identification System) by Meier et al. (2009)]. It is important to note that these methods are limited in their range by the availability of target domains in the sample, potentiating bias in the final results. Thus, future developments of novel probes for modular enzymes could enable a wide implementation of probe-based methods in secondary metabolism research in fungi, as well as other phyla (Meier and Burkart 2011).

Beyond discovery of biosynthetic enzymes, proteomics is also useful for characterizing peptidic NPs and determining the mode of action of bioactive NPs. Comparing the whole proteome of an NP-treated organism to an untreated control provides a global overview of changed pathways. Alternatively, a metabolite can be immobilized and used as bait to identify proteins with a sufficient binding affinity, thus identifying potential target proteins that can be further characterized by proteomics (Du and van Wezel 2018).

The effects of gliotoxin on *A. fumigatus* were elucidated in this manner: the toxin was found to trigger oxidative stress by induction of a superoxide dismutase (Carberry et al. 2012), but also attenuated the effects of H_2O_2-induced stress on the fungus (Owens et al. 2014).

Even a modified version of genome mining is possible from a proteomics starting point: the platform PrISM [Proteomic Investigation of Secondary Metabolites (Bumpus et al. 2009)] allows for targeted detection of NRPS- and PKS-derived peptides as well as the corresponding BGC in unsequenced genomes. The method makes use of the labile phosphodiester linkage of the phosphopantetheinyl cofactor to the thiolation domain of NRPS and PKS enzymes.

This bond easily breaks during mass spectrometry, releasing a cofactor fragment, thus creating a characteristic mass difference that allows the identification of peptides from the thiolation domain.

Peptide sequences can then be used as templates to create primers and amplify the corresponding gene from the genome without any previous sequence information. Sequencing of the gene, domain predictions, and alignment to known genes can then be used to postulate a hypothesis regarding the enzyme product, leading to its targeted detection.

The complementation of proteomic methods with other approaches, e.g., **proteogenomics** (an integrated methodology combining genomics, transcriptomics, and proteomics) (Albright et al. 2014), along with the advancing possibilities of exploiting **metaproteomics** from microbial communities (Wang et al. 2016b), promises deep insight into, and large expansion of our knowledge of fungal biosynthetic pathways.

D. Metabolomics

In line with the other omics methods, metabolomics provides a **temporally limited snapshot of an organism's chemical potential,** pooling all metabolites available in the sample, or enriching for secondary **metabolome fingerprint,** using an adequate extraction method. Experiments aiming at the discovery of novel NPs thus usually involve an untargeted screening of organic extracts of whole cultures, followed by a chromatographic separation and mass spectrometry for compound analysis. Lately, a shift has been observed toward in situ analysis strategies, including mass spectrometry imaging (MSI), circumventing the need for sample preparation. As the analytical methods are beyond the scope of the present work, the reader is directed to the excellent reviews by Cox et al. (2014), Wolfender et al. (2015), and Netzker et al. (2018).

Dereplication (the process of identification of known compounds in a mixture) of experimental data represents a key step in metabolomics—indeed, rediscovery of SMs has hampered research in this field for years (see above).

For illustration, an OSMAC-coupled de-replication approach was used to identify three new penitremone derivatives from a marine-derived strain of *Penicillium canescens* amidst a large number of already known compounds (Vansteelandt et al. 2012).

Multiple public data repositories are available, including METLIN (Zhu et al. 2013), GNPS (Wang et al. 2016c), or MassBank (Horai et al. 2010), where researchers are encouraged to deposit their spectrometric data to aid dereplication efforts and acceleration of the process of novel compound discovery. The software tools used in database interrogation usually also provide the option of performing multivariate statistical analysis, e.g., principal component analysis and partial least squares (Cox et al. 2014; Covington et al. 2017), as well as hierarchical clustering and pathway enrichment analysis for further exploration of the data [for an overview of software, see Liu and Locasale (2017)].

Molecular networking based on metabolomic data, i.e., creating clusters of molecule families according to correlations between fragment ions was used to identify new analogs of fungisporin and nidulanin A in *A. nidulans* (Klitgaard et al. 2015). Metabolomics research targeting secondary metabolites has often seen appropriation of methods originally designed for different research purposes [e.g., MS imaging, discussed in Netzker et al. (2018)]. This also holds true for computational strategies assisting with the analysis of complex data. A commonly used tool for mapping of compounds into metabolic pathways and networks is Cytoscape, with the metabolism plugin Metscape, originally devised for human metabolome experiments (Gao et al. 2010). Other platforms include the KEGG Atlas (Okuda et al. 2008) or MetaMapR (Grapov et al. 2015), a software which can use information from KEGG and PubChem databases to find associations between metabolites which have been detected in a metabolomics approach, but whose biochemistry and structure are not yet known, and integrate this information with genomic or proteomic data into metabolic networks.

E. Integration of Omics Data

Due to **inherent limitations** and properties of each of the omics methods, it is possible to introduce bias and/or obtain an imperfect, partial understanding of a biological system at hand (Yeger-Lotem et al. 2009). Following the trend evident in other areas, fungal SM research has also started moving toward the adoption of an **integrative approach to novel compound discovery, as well as mechanistic characterization of biosynthetic processes and the roles of SMs** (Macheleidt et al. 2016; Hautbergue et al. 2018).

An excellent example of the power of a multi-omics strategy is the study of the rice pathogen *Fusarium fujikuroi*, which amalgamated genome mining, transcriptome and proteome studies, chromatin immunoprecipitation, and metabolic profiling to reveal the complexities in the regulation of the secondary metabolism in the fungus and identify two novel compounds (Wiemann et al. 2013b). Similarly, a large-scale genomic comparative functional analysis of the biosynthetic potential of *Penicillia*, coupled with phylogenetic analysis and a metabolomics approach, led to the identification of novel antimicrobial yanuthone derivatives and underlines the importance and potential of this fungal clade as a bioactive compound source (Nielsen et al. 2017). Acharya et al. (2019) selected a combined omics approach for an in-depth characterization of the interaction between a *Rhodococcus* sp. strain and a strain of *Micromonospora* sp. producing a novel glycosylated anthracycline, keyicin, with antimicrobial properties.

Using genomics, transcriptomics, and proteomics technologies, the authors were able to elucidate regulatory pathways that affect keycin production and correlate changes appearing in the protein and transcript profiles with changes consistent with quorum sensing signaling.

The intrinsic **complementarity of individual omics** datasets has been recognized, and various tools have been developed to aid the integration that will eventually improve efficiency and reduce complexity of sample analysis, as well as accelerate the discovery of novel compounds in fungi. XCMS Online is a data-processing platform originally developed for metabolomics studies, which has been expanded to integrate genomic, proteomic and transcriptomic data. It provides metabolic pathway analysis and cross-referencing with other types of omics data for a comprehensive characterization of the sample (Forsberg et al. 2018).

Andersen et al. (2013) developed a method for identification of new BGCs that combines transcriptomic and metabolomic data with genomic analysis. Clusters were predicted from genome data using the SMURF algorithm and scored based on similarity of the expression profiles. These predictions correlated well with published data and the algorithm could be extended to include biosynthetic superclusters spread over several chromosomes; the aforementioned nidulanin A was discovered using this integrative strategy (see above).

It is, however, worthwhile to note that the integration of omics methods produces big data volumes, compelling further development of bioinformatics tools for processing, as well as a consolidated effort among researchers to ensure standardization and improve accessibility and comparability of individual datasets in fungal research.

As has been illustrated, each type of omics data provides unique information about a biological system. In fungi, the inclusion of several complementary techniques may significantly increase the NP discovery power of the approach. Thus, omics-based methods represent a promising strategy for the identification of new leads for drug discovery and development.

V. Conclusion

The latest report on fungal diversity estimates that there are 2.2 to 3.8 million species in this ancient clade, of which 120,000 (i.e. *ca.* 8%) have been described (Hawksworth and Lücking 2017). Considering the majority of fungi yet to be identified in underexplored environments (e.g., lichens, tropical diversity hotspots, extreme environments), as well as the wealth of NPs discovered to date, it is justifiable to predict that a vast number of fungal SMs will be described in future. The emergence of novel techniques, as well as advancements in established protocols, in the post-omics era promises a bright future for NP research.

Still, there persists a fundamental disconnect between the ample natural products being discovered at present and the ever-growing call for novel antimicrobial compounds used as drugs in a pharmaceutical context. The contrast is evident when considering the drying out antibiotics pipeline (Cooper and Shlaes 2011) and the overwhelming number of claims regarding the unexplored potential of microbes as a largely untapped source of natural products, many of them drug-like. As evidenced by this review, fungi are not exempt. Moreover, naturally derived SMs are intrinsically biologically relevant, thus occupying a much larger drug-like chemical space than synthetic compounds (Harvey et al. 2015). Logically, NPs, including those from fungi, should hold the key to tackling the problem of missing anti-infective drugs and increasing antimicrobial resistance.

Despite the apparent simplicity of such claims, the issue is fundamentally complex and multilayered. Initially, a considerable effort is needed to subject a newly isolated compound to relevant bioactivity testing—a task to which many research groups are ill-equipped. Subsequently, there is a long (and financially demanding) way to go between the detection of a biological activity of a compound and its potential introduction to the market as a drug. This difficult endeavor is further hindered by the reluctance of large pharmaceutical companies to engage and invest considerable capital in antimicrobial products, which have a short life span and a low, if not negative, net value (Cooper and Shlaes 2011).

Integrative approaches inside the lab, as well as in collaborative efforts between academic researchers, the industry, and regulatory bodies, will be crucial in overcoming the threat of antimicrobial resistance and the introduction of novel antibiotics into clinical practice. With their predicted ecological diversity and secondary metabolic capacity, fungi will undoubtedly prove to be an influential source of lead molecules.

References

Abdelwahab MF, Kurtán T, Mándi A et al (2018) Induced secondary metabolites from the endophytic fungus *Aspergillus versicolor* through bacterial co-culture and OSMAC approaches. Tetrahedron Lett 59:2647–2652. https://doi.org/10.1016/j.tetlet.2018.05.067

Acharya D, Miller I, Cui Y et al (2019) omics technologies to understand activation of a biosynthetic gene cluster in *Micromonospora* sp. WMMB235: deciphering keyicin biosynthesis. ACS Chem Biol 14:1260–1270. https://doi.org/10.1021/acschembio.9b00223

Ahuja M, Chiang YM, Chang SL et al (2012) Illuminating the diversity of aromatic polyketide synthases in *Aspergillus nidulans*. J Am Chem Soc 134:8212–8221. https://doi.org/10.1021/ja3016395

Alberti F, Foster GD, Bailey AM (2017) Natural products from filamentous fungi and production by heterologous expression. Appl Microbiol Biotechnol 101:493–500. https://doi.org/10.1007/s00253-016-8034-2

Albright JC, Goering AW, Doroghazi JR et al (2014) Strain-specific proteogenomics accelerates the discovery of natural products via their biosynthetic pathways. J Ind Microbiol Biotechnol 41:451–459. https://doi.org/10.1007/s10295-013-1373-4

Amos GCA, Awakawa T, Tuttle RN et al (2017) Comparative transcriptomics as a guide to natural product discovery and biosynthetic gene cluster functionality. Proc Natl Acad Sci U S A 114:E11121–E11130. https://doi.org/10.1073/pnas.1714381115

Andersen MR, Nielsen JB, Klitgaard A et al (2013) Accurate prediction of secondary metabolite gene clusters in filamentous fungi. Proc Natl Acad Sci U S A 110:E99–E107. https://doi.org/10.1073/pnas.1205532110

Anyaogu DC, Mortensen UH (2015) Heterologous production of fungal secondary metabolites in *Aspergilli*. Front Microbiol 6:77. https://doi.org/10.3389/fmicb.2015.00077

Arsenault P, Wobbe K, Weathers P (2008) Recent advances in artemisinin production through heterologous expression. Curr Med Chem 15:2886–2896. https://doi.org/10.2174/092986708786242813

Azzollini A, Boggia L, Boccard J et al (2018) Dynamics of metabolite induction in fungal co-cultures by metabolomics at both volatile and non-volatile levels. Front Microbiol 9:72. https://doi.org/10.3389/fmicb.2018.00072

Bailey AM, Alberti F, Kilaru S et al (2016) Identification and manipulation of the pleuromutilin gene cluster from *Clitopilus passeckerianus* for increased rapid antibiotic production. Sci Rep 6:25202. https://doi.org/10.1038/srep25202

Bayram Ö, Braus GH (2012) Coordination of secondary metabolism and development in fungi: the velvet family of regulatory proteins. FEMS Microbiol Rev 36:1–24. https://doi.org/10.1111/j.1574-6976.2011.00285.x

Bérdy J (2005) Bioactive microbial metabolites. J Antibiot (Tokyo) 58:1–26. https://doi.org/10.1038/ja.2005.1

Bergmann S, Schümann J, Scherlach K et al (2007) Genomics-driven discovery of PKS-NRPS hybrid metabolites from *Aspergillus nidulans*. Nat Chem Biol 3:213–217. https://doi.org/10.1038/nchembio869

Bergmann S, Funk AN, Scherlach K et al (2010) Activation of a silent fungal polyketide biosynthesis pathway through regulatory cross talk with a cryptic nonribosomal peptide synthetase gene cluster. Appl Environ Microbiol 76:8143–8149. https://doi.org/10.1128/AEM.00683-10

Billingsley JM, DeNicola AB, Tang Y (2016) Technology development for natural product biosynthesis in *Saccharomyces cerevisiae*. Curr Opin Biotechnol 42:74–83. https://doi.org/10.1016/j.copbio.2016.02.033

Blin K, Wolf T, Chevrette MG et al (2017) antiSMASH 4.0—improvements in chemistry prediction and gene cluster boundary identification. Nucleic Acids Res 45:W36–W41. https://doi.org/10.1093/nar/gkx319

Blunt JW, Carroll AR, Copp BR et al (2018) Marine natural products. Nat Prod Rep 35:8–53. https://doi.org/10.1039/C7NP00052A

Bok JW, Keller NP (2004) LaeA, a regulator of secondary metabolism in *Aspergillus* spp. Eukaryot Cell 3:527–535. https://doi.org/10.1128/EC.3.2.527-535.2004

Bok JW, Hoffmeister D, Maggio-Hall LA et al (2006a) Genomic mining for *Aspergillus* natural products. Chem Biol 13:31–37. https://doi.org/10.1016/j.chembiol.2005.10.008

Bok JW, Noordermeer D, Kale SP, Keller NP (2006b) Secondary metabolic gene cluster silencing in *Aspergillus nidulans*. Mol Microbiol 61:1636–1645. https://doi.org/10.1111/j.1365-2958.2006.05330.x

Bok JW, Ye R, Clevenger KD et al (2015) Fungal artificial chromosomes for mining of the fungal secondary metabolome. BMC Genomics 16:343. https://doi.org/10.1186/s12864-015-1561-x

Bouhired S, Weber M, Kempf-Sontag A et al (2007) Accurate prediction of the *Aspergillus nidulans* terrequinone gene cluster boundaries using the transcriptional regulator LaeA. Fungal Genet Biol 44:1134–1145. https://doi.org/10.1016/j.fgb.2006.12.010

Boya CAP, Fernández-Marín H, Mejía LC et al (2017) Imaging mass spectrometry and MS/MS molecular networking reveals chemical interactions among cuticular bacteria and pathogenic fungi associated with fungus-growing ants. Sci Rep 7:5604. https://doi.org/10.1038/s41598-017-05515-6

Bracarense AAP, Takahashi JA (2014) Modulation of antimicrobial metabolites production by the fungus *Aspergillus parasiticus*. Braz J Microbiol 45:313–321. https://doi.org/10.1590/S1517-83822014000100045

Brakhage AA (2013) Regulation of fungal secondary metabolism. Nat Rev Microbiol 11:21–32. https://doi.org/10.1038/nrmicro2916

Brakhage AA, Schroeckh V (2011) Fungal secondary metabolites—strategies to activate silent gene clusters. Fungal Genet Biol 48:15–22. https://doi.org/10.1016/j.fgb.2010.04.004

Bromann K, Toivari M, Viljanen K et al (2012) Identification and characterization of a novel diterpene gene cluster in *Aspergillus nidulans*. PLoS One 7:e35450. https://doi.org/10.1371/journal.pone.0035450

Brown DW, Proctor RH (2016) Insights into natural products biosynthesis from analysis of 490 polyketide synthases from *Fusarium*. Fungal Genet Biol 89:37–51. https://doi.org/10.1016/j.fgb.2016.01.008

Bumpus SB, Evans BS, Thomas PM et al (2009) A proteomics approach to discovering natural products and their biosynthetic pathways. Nat Biotechnol 27:951–956. https://doi.org/10.1038/nbt.1565

Cacho RA, Jiang W, Chooi YH et al (2012) Identification and characterization of the echinocandin B biosynthetic gene cluster from *Emericella rugulosa* NRRL 11440. J Am Chem Soc 134:16781–16790. https://doi.org/10.1021/ja307220z

Cacho RA, Tang Y, Chooi YH (2015) Next-generation sequencing approach for connecting secondary metabolites to biosynthetic gene clusters in fungi. Front Microbiol 5:774. https://doi.org/10.3389/fmicb.2014.00774

Calvo AM, Wilson RA, Bok JW, Keller NP (2002) Relationship between secondary metabolism and fungal development. Microbiol Mol Biol Rev 66:447–459. https://doi.org/10.1128/MMBR.66.3.447-459.2002

Carberry S, Molloy E, Hammel S et al (2012) Gliotoxin effects on fungal growth: mechanisms and exploitation. Fungal Genet Biol 49:302–312. https://doi.org/10.1016/j.fgb.2012.02.003

Chávez R, Fierro F, García-Rico RO, Vaca I (2015) Filamentous fungi from extreme environments as a promising source of novel bioactive secondary metabolites. Front Microbiol 6:903. https://doi.org/10.3389/fmicb.2015.00903

Chiang Y-M, Szewczyk E, Davidson AD et al (2009) A gene cluster containing two fungal polyketide synthases encodes the biosynthetic pathway for a polyketide, asperfuranone, in *Aspergillus nidulans*. J Am Chem Soc 131:2965–2970. https://doi.org/10.1021/ja8088185

Chooi YH, Cacho R, Tang Y (2010) Identification of the viridicatumtoxin and griseofulvin gene clusters from *Penicillium aethiopicum*. Chem Biol 17:483–494. https://doi.org/10.1016/j.chembiol.2010.03.015

Cooper MA, Shlaes D (2011) Fix the antibiotics pipeline. Nature 472:32–32. https://doi.org/10.1038/472032a

Covington BC, McLean JA, Bachmann BO (2017) Comparative mass spectrometry-based metabolomics strategies for the investigation of microbial secondary metabolites. Nat Prod Rep 34:6–24. https://doi.org/10.1039/C6NP00048G

Cox DG, Oh J, Keasling A et al (2014) The utility of metabolomics in natural product and biomarker characterization. Biochim Biophys Acta Gen Subj 1840:3460–3474. https://doi.org/10.1016/j.bbagen.2014.08.007

Cueto M, Jensen PR, Kauffman C et al (2001) Pestalone, a new antibiotic produced by a marine fungus in response to bacterial challenge. J Nat Prod 64:1444–1446

Currie CR (2001) A community of ants, fungi, and bacteria: a multilateral approach to studying symbiosis. Annu Rev Microbiol 55:357–380. https://doi.org/10.1146/annurev.micro.55.1.357

de Borges W, Borges K, Bonato P et al (2009) Endophytic fungi: natural products, enzymes and biotransformation reactions. Curr Org Chem 13:1137–1163. https://doi.org/10.2174/138527209788921783

de Vries RP, Riley R, Wiebenga A et al (2017) Comparative genomics reveals high biological diversity and specific adaptations in the industrially and medically important fungal genus *Aspergillus*. Genome Biol 18:28. https://doi.org/10.1186/s13059-017-1151-0

Degenkolb T, Heinze S, Schlegel B et al (2002) Formation of new lipoaminopeptides, acremostatins A, B, and C, by co-cultivation of *Acremonium* sp. Tbp-5 and *Mycogone rosea* DSM 12973. Biosci Biotechnol Biochem 66:883–886

Deng H, Gao R, Liao X, Cai Y (2017) CRISPR system in filamentous fungi: current achievements and future directions. Gene 627:212–221. https://doi.org/10.1016/j.gene.2017.06.019

Dhodary B, Schilg M, Wirth R, Spiteller D (2018) Secondary metabolites from *Escovopsis weberi* and their role in attacking the garden fungus of leaf-cutting ants. Chem A Eur J 24:4445–4452. https://doi.org/10.1002/chem.201706071

Du C, van Wezel GP (2018) Mining for microbial gems: integrating proteomics in the postgenomic natural product discovery pipeline. Proteomics 18:1700332. https://doi.org/10.1002/pmic.201700332

Du FY, Li XM, Zhang P et al (2014) Cyclodepsipeptides and other O-containing heterocyclic metabolites from *Beauveria felina* EN-135, a marine-derived entomopathogenic fungus. Mar Drugs 12:2816–2826. https://doi.org/10.3390/md12052816

Evans JA, Eyre CA, Rogers HJ et al (2008) Changes in volatile production during interspecific interactions between four wood rotting fungi growing in artificial media. Fungal Ecol 1:57–68. https://doi.org/10.1016/j.funeco.2008.06.001

Fisch KM, Bakeer W, Yakasai AA et al (2011) Rational domain swaps decipher programming in fungal highly reducing polyketide synthases and resurrect an extinct metabolite. J Am Chem Soc 133:16635–16641. https://doi.org/10.1021/ja206914q

Fischbach MA, Lai JR, Roche ED et al (2007) Directed evolution can rapidly improve the activity of chimeric assembly-line enzymes. Proc Natl Acad Sci USA 104:11951–11956. https://doi.org/10.1073/pnas.0705348104

Fischer J, Müller SY, Netzker T et al (2018) Chromatin mapping identifies BasR, a key regulator of bacteria-triggered production of fungal secondary metabolites. eLife 7:e40969. https://doi.org/10.7554/eLife.40969

Forsberg EM, Huan T, Rinehart D et al (2018) Data processing, multi-omic pathway mapping, and metabolite activity analysis using XCMS Online. Nat Protoc 13:633–651. https://doi.org/10.1038/nprot.2017.151

Fuller KK, Chen S, Loros JJ, Dunlap JC (2015) Development of the CRISPR/Cas9 system for targeted gene disruption in *Aspergillus fumigatus*. Eukaryot Cell 14:1073–1080. https://doi.org/10.1128/EC.00107-15

Galanie S, Thodey K, Trenchard IJ et al (2015) Complete biosynthesis of opioids in yeast. Science 349 (6252):1095–1100. https://doi.org/10.1126/science.aac9373

Gao J, Tarcea VG, Karnovsky A et al (2010) Metscape: a Cytoscape plug-in for visualizing and interpreting metabolomic data in the context of human metabolic networks. Bioinformatics 26:971–973. https://doi.org/10.1093/bioinformatics/btq048

García-Estrada C, Domínguez-Santos R, Kosalková K, Martín JF (2018) Transcription factors controlling primary and secondary metabolism in filamentous fungi: the β-Lactam paradigm. Fermentation 4:47. https://doi.org/10.3390/fermentation4020047

Geib E, Brock M (2017) ATNT: an enhanced system for expression of polycistronic secondary metabolite gene clusters in *Aspergillus niger*. Fungal Biol Biotechnol 4:13. https://doi.org/10.1186/s40694-017-0042-1

Goudie AC, Evans NA, Gration KAF et al (1993) Doramectin—a potent novel endectocide. Vet Parasitol 49:5–15. https://doi.org/10.1016/0304-4017(93)90218-C

Grapov D, Wanichthanarak K, Fiehn O (2015) MetaMapR: pathway independent metabolomic network analysis incorporating unknowns. Bioinformatics 31:2757–2760. https://doi.org/10.1093/bioinformatics/btv194

Grau MF, Entwistle R, Chiang Y-M et al (2018) Hybrid transcription factor engineering activates the silent secondary metabolite gene cluster for (+)-asperlin in *Aspergillus nidulans*. ACS Chem Biol 13:3193–3205. https://doi.org/10.1021/acschembio.8b00679D

Gressler M, Hortschansky P, Geib E, Brock M (2015) A new high-performance heterologous fungal expression system based on regulatory elements from the *Aspergillus terreus* terrein gene cluster. Front Microbiol 6:184. https://doi.org/10.3389/fmicb.2015.00184

Griffiths RR, Johnson MW, Carducci MA et al (2016) Psilocybin produces substantial and sustained decreases in depression and anxiety in patients with life-threatening cancer: a randomized double-blind trial. J Psychopharmacol 30:1181–1197. https://doi.org/10.1177/0269881116675513

Gubbens J, Zhu H, Girard G et al (2014) Natural product proteomining, a quantitative proteomics platform, allows rapid discovery of biosynthetic gene clusters for different classes of natural products. Chem Biol 21:707–718. https://doi.org/10.1016/j.chembiol.2014.03.011

Haeder S, Wirth R, Herz H, Spiteller D (2009) Candicidin-producing *Streptomyces* support leaf-cutting ants to protect their fungus garden against the pathogenic fungus *Escovopsis*. Proc Natl Acad Sci USA 106:4742–4746. https://doi.org/10.1073/pnas.0812082106

Harvey AL, Edrada-Ebel R, Quinn RJ (2015) The re-emergence of natural products for drug discovery in the genomics era. Nat Rev Drug Discov 14:111–129. https://doi.org/10.1038/nrd4510

Harvey CJB, Tang M, Schlecht U et al (2018) HEx: a heterologous expression platform for the discovery of fungal natural products. Sci Adv 4:eaar5459. https://doi.org/10.1126/sciadv.aar5459

Hautbergue T, Jamin EL, Debrauwer L et al (2018) From genomics to metabolomics, moving toward an integrated strategy for the discovery of fungal secondary metabolites. Nat Prod Rep 35:147–173. https://doi.org/10.1039/C7NP00032D

Hawksworth DL, Lücking R (2017) Fungal diversity revisited: 2.2 to 3.8 million species. In: The fungal kingdom. American Society of Microbiology, Washington, DC, pp 79–95

Heine D, Holmes NA, Worsley SF et al (2018) Chemical warfare between leafcutter ant symbionts and a co-evolved pathogen. Nat Commun 9:2208. https://doi.org/10.1038/s41467-018-04520-1

Heneghan MN, Yakasai AA, Halo LM et al (2010) First heterologous reconstruction of a complete functional fungal biosynthetic multigene cluster. ChemBioChem 11:1508–1512. https://doi.org/10.1002/cbic.201000259

Hertweck C (2015) Decoding and reprogramming complex polyketide assembly lines: prospects for synthetic biology. Trends Biochem Sci 40:189–199. https://doi.org/10.1016/j.tibs.2015.02.001

Hoefgen S, Lin J, Fricke J et al (2018) Facile assembly and fluorescence-based screening method for heterologous expression of biosynthetic pathways in fungi. Metab Eng 48:44–51. https://doi.org/10.1016/j.ymben.2018.05.014

Hong EJ, Kim NK, Lee D et al (2015) Overexpression of the *laeA* gene leads to increased production of cyclopiazonic acid in *Aspergillus fumisynnematus*. Fungal Biol 119:973–983. https://doi.org/10.1016/j.funbio.2015.06.006

Horai H, Arita M, Kanaya S et al (2010) MassBank: a public repository for sharing mass spectral data for life sciences. J Mass Spectrom 45:703–714. https://doi.org/10.1002/jms.1777

Hover BM, Kim SH, Katz M et al (2018) Culture-independent discovery of the malacidins as calcium-dependent antibiotics with activity against multidrug-resistant Gram-positive pathogens. Nat Microbiol 3:415–422. https://doi.org/10.1038/s41564-018-0110-1

Iqbal HA, Low-Beinart L, Obiajulu JU, Brady SF (2016) Natural product discovery through improved functional metagenomics in *Streptomyces*. J Am Chem Soc 138:9341–9344. https://doi.org/10.1021/jacs.6b02921

Janevska S, Arndt B, Baumann L et al (2017) Establishment of the inducible Tet-On system for the activation of the silent trichosetin gene cluster in *Fusarium fujikuroi*. Toxins (Basel) 9:126. https://doi.org/10.3390/toxins9040126

Jinek M, Chylinski K, Fonfara I et al (2012) A programmable dual-RNA-guided DNA endonuclease in adaptive bacterial immunity. Science 337:816–821. https://doi.org/10.1126/science.1225829

Kakule TB, Zhenjian L, Schmidt EW (2014) Combinatorialization of fungal polyketide synthase–peptide synthetase hybrid proteins. J Am Chem Soc 136:17882–17890. https://doi.org/10.1021/ja511087p

Kakule TB, Jadulco RC, Koch M et al (2015) Native promoter strategy for high-yielding synthesis and engineering of fungal secondary metabolites. ACS Synth Biol 4:625–633. https://doi.org/10.1021/sb500296p

Kang HS (2017) Phylogeny-guided (meta)genome mining approach for the targeted discovery of new microbial natural products. J Ind Microbiol Biotechnol 44:285–293. https://doi.org/10.1007/s10295-016-1874-z

Kato N, Brooks W, Calvo AM (2003) The expression of sterigmatocystin and penicillin genes in *Aspergillus nidulans* is controlled by *veA*, a gene required for sexual development. Eukaryot Cell 2:1178–1186. https://doi.org/10.1128/EC.2.6.1178-1186.2003

Keilhofer N, Nachtigall J, Kulik A et al (2018) *Streptomyces* AcH 505 triggers production of a salicylic acid analogue in the fungal pathogen *Heterobasidion abietinum* that enhances infection of Norway spruce seedlings. Antonie Van Leeuwenhoek 111:691–704. https://doi.org/10.1007/s10482-018-1017-9

Keller N, Bok J, Chung D et al (2006) LaeA, a global regulator of *Aspergillus* toxins. Med Mycol 44:83–85. https://doi.org/10.1080/13693780600835773

Khaldi N, Seifuddin FT, Turner G et al (2010) SMURF: genomic mapping of fungal secondary metabolite clusters. Fungal Genet Biol 47:736–741. https://doi.org/10.1016/j.fgb.2010.06.003

Klitgaard A, Nielsen JB, Frandsen RJN et al (2015) Combining stable isotope labeling and molecular networking for biosynthetic pathway characterization. Anal Chem 87:6520–6526. https://doi.org/10.1021/acs.analchem.5b01934

Kluge J, Terfehr D, Kück U (2018) Inducible promoters and functional genomic approaches for the genetic engineering of filamentous fungi. Appl Microbiol Biotechnol 102:6357–6372. https://doi.org/10.1007/s00253-018-9115-1

Kong FD, Huang XL, Ma QY et al (2018) Helvolic acid derivatives with antibacterial activities against *Streptococcus agalactiae* from the marine-derived fungus *Aspergillus fumigatus* HNMF0047. J Nat Prod 81:1869–1876. https://doi.org/10.1021/acs.jnatprod.8b00382

König CC, Scherlach K, Schroeckh V et al (2013) Bacterium induces cryptic meroterpenoid pathway in the pathogenic fungus *Aspergillus fumigatus*. ChemBioChem 14:938–942. https://doi.org/10.1002/cbic.201300070

Krappmann S (2017) CRISPR-Cas9, the new kid on the block of fungal molecular biology. Med Mycol 55:16–23. https://doi.org/10.1093/mmy/myw097

Kumla D, Pereira J, Dethoup T et al (2018) Chromone derivatives and other constituents from cultures of the marine sponge-associated fungus *Penicillium erubescens* KUFA0220 and their antibacterial activity. Mar Drugs 16:289. https://doi.org/10.3390/md16080289

Lei H, Lin X, Han L et al (2017) New metabolites and bioactive chlorinated benzophenone derivatives produced by a marine-derived fungus *Pestalotiopsis heterocornis*. Mar Drugs 15:69. https://doi.org/10.3390/md15030069

Ley A, Coumou HC, Frandsen RJN (2015) Heterologous expression of MlcE in *Saccharomyces cerevisiae* provides resistance to natural and semi-synthetic statins. Metab Eng Commun 2:117–123. https://doi.org/10.1016/j.meteno.2015.09.003

Li C, Wang J, Luo C et al (2014) A new cyclopeptide with antifungal activity from the co-culture broth of two marine mangrove fungi. Nat Prod Res 28:616–621. https://doi.org/10.1080/14786419.2014.887074

Lin H, Lyu H, Zhou S et al (2018) Deletion of a global regulator LaeB leads to the discovery of novel polyketides in *Aspergillus nidulans*. Org Biomol Chem 16:4973–4976. https://doi.org/10.1039/C8OB01326H

Liu X, Locasale JW (2017) Metabolomics: a primer. Trends Biochem Sci 42:274–284. https://doi.org/10.1016/j.tibs.2017.01.004

Liu R, Chen L, Jiang Y et al (2015) Efficient genome editing in filamentous fungus *Trichoderma reesei* using the CRISPR/Cas9 system. Cell Discov 1:15007. https://doi.org/10.1038/celldisc.2015.7

Liu J, Gu B, Yang L et al (2018) New anti-inflammatory cyclopeptides from a sponge-derived fungus *Aspergillus violaceofuscus*. Front Chem 6:226. https://doi.org/10.3389/fchem.2018.00226

Lubertozzi D, Keasling JD (2009) Developing *Aspergillus* as a host for heterologous expression. Biotechnol Adv 27:53–75. https://doi.org/10.1016/j.biotechadv.2008.09.001

Macheleidt J, Scherlach K, Neuwirth T et al (2015) Transcriptome analysis of cyclic AMP-dependent protein kinase A-regulated genes reveals the production of the novel natural compound fumipyrrole by *Aspergillus fumigatus*. Mol Microbiol 96:148–162. https://doi.org/10.1111/mmi.12926

Macheleidt J, Mattern DJ, Fischer J et al (2016) Regulation and role of fungal secondary metabolites. Annu Rev Genet 50:371–392. https://doi.org/10.1146/annurev-genet-120215-035203

Mandelare PE, Adpressa DA, Kaweesa EN et al (2018) Coculture of two developmental stages of a marine-derived *Aspergillus alliaceus* results in the production of the cytotoxic bianthrone allianthrone A. J Nat Prod 81:1014–1022. https://doi.org/10.1021/acs.jnatprod.8b00024

Marfori EC, Kajiyama S, Fukusaki E, Kobayashi A (2002) Trichosetin, a novel tetramic acid antibiotic produced in dual culture of *Trichoderma harzianum* and *Catharanthus roseus* Callus. Z Naturforsch C 57:465–470

Matsu-ura T, Baek M, Kwon J, Hong C (2015) Efficient gene editing in *Neurospora crassa* with CRISPR technology. Fungal Biol Biotechnol 2:4. https://doi.org/10.1186/s40694-015-0015-1

Mattern DJ, Schoeler H, Weber J et al (2015a) Identification of the antiphagocytic trypacidin gene cluster in the human-pathogenic fungus *Aspergillus fumigatus*. Appl Microbiol Biotechnol 99:10151–10161. https://doi.org/10.1007/s00253-015-6898-1

Mattern DJ, Valiante V, Unkles SE, Brakhage AA (2015b) Synthetic biology of fungal natural products. Front Microbiol 6:775. https://doi.org/10.3389/fmicb.2015.00775

Mattern DJ, Valiante V, Horn F et al (2017) Rewiring of the austinoid biosynthetic pathway in filamentous fungi. ACS Chem Biol 12:2927–2933. https://doi.org/10.1021/acschembio.7b00814

Medema MH, Blin K, Cimermancic P et al (2011) antiSMASH: rapid identification, annotation and analysis of secondary metabolite biosynthesis gene clusters in bacterial and fungal genome sequences. Nucleic Acids Res 39:W339–W346. https://doi.org/10.1093/nar/gkr466

Meier JL, Burkart MD (2011) Proteomic analysis of polyketide and nonribosomal peptide biosynthesis. Curr Opin Chem Biol 15:48–56. https://doi.org/10.1016/j.cbpa.2010.10.021

Meier JL, Niessen S, Hoover HS et al (2009) An Orthogonal Active Site Identification System (OASIS) for proteomic profiling of natural product biosynthesis. ACS Chem Biol 4:948–957. https://doi.org/10.1021/cb9002128

Meng LH, Liu Y, Li XM et al (2015) Citrifelins A and B, citrinin adducts with a tetracyclic framework from cocultures of marine-derived isolates of *Penicillium citrinum* and *Beauveria felina*. J Nat Prod 78:2301–2305. https://doi.org/10.1021/acs.jnatprod.5b00450

Meyer V, Wanka F, van Gent J et al (2011) Fungal gene expression on demand: an inducible, tunable, and metabolism-independent expression system for *Aspergillus niger*. Appl Environ Microbiol 77:2975–2983. https://doi.org/10.1128/AEM.02740-10

Milshteyn A, Schneider JS, Brady SF (2014) Mining the metabiome: identifying novel natural products from microbial communities. Chem Biol 21:1211–1223. https://doi.org/10.1016/j.chembiol.2014.08.006

Netzker T, Flak M, Krespach MK et al (2018) Microbial interactions trigger the production of antibiotics. Curr Opin Microbiol 45:117–123. https://doi.org/10.1016/j.mib.2018.04.002

Nichols D, Cahoon N, Trakhtenberg EM et al (2010) Use of iChip for high-throughput in situ cultivation of 'uncultivable' microbial species. Appl Environ Microbiol 76:2445–2450. https://doi.org/10.1128/AEM.01754-09

Nielsen ML, Isbrandt T, Petersen LM et al (2016) Linker flexibility facilitates module exchange in fungal hybrid PKS-NRPS engineering. PLoS One 11:e0161199. https://doi.org/10.1371/journal.pone.0161199

Nielsen ML, Isbrandt T, Rasmussen KB et al (2017) Genes linked to production of secondary metabolites in *Talaromyces atroroseus* revealed using CRISPR-Cas9. PLoS One 12:e0169712. https://doi.org/10.1371/journal.pone.0169712

Nødvig CS, Nielsen JB, Kogle ME, Mortensen UH (2015) A CRISPR-Cas9 system for genetic engineering of filamentous fungi. PLoS One 10:0133085. https://doi.org/10.1371/journal.pone.0133085

Oakley CE, Ahuja M, Sun WW et al (2017) Discovery of McrA, a master regulator of *Aspergillus* secondary metabolism. Mol Microbiol 103:347–365. https://doi.org/10.1111/mmi.13562

Okuda S, Yamada T, Hamajima M et al (2008) KEGG Atlas mapping for global analysis of metabolic pathways. Nucleic Acids Res 36:W423–W426. https://doi.org/10.1093/nar/gkn282

Omura S, Ikeda H, Malpartida F et al (1986) Production of new hybrid antibiotics, mederrhodins A and B, by a genetically engineered strain. Antimicrob Agents Chemother 29:13–19. https://doi.org/10.1128/AAC.29.1.13

Owen JG, Calcott MJ, Robins KJ, Ackerley DF (2016) Generating functional recombinant NRPS enzymes in the laboratory setting via peptidyl carrier protein engineering. Cell Chem Biol 23:1395–1406. https://doi.org/10.1016/j.chembiol.2016.09.014

Owens RA, Hammel S, Sheridan KJ et al (2014) A proteomic approach to investigating gene cluster expression and secondary metabolite functionality in *Aspergillus fumigatus*. PLoS One 9:e106942. https://doi.org/10.1371/journal.pone.0106942

Payne DJ, Gwynn MN, Holmes DJ, Pompliano DL (2007) Drugs for bad bugs: confronting the challenges of antibacterial discovery. Nat Rev Drug Discov 6:29–40. https://doi.org/10.1038/nrd2201

Pfannenstiel BT, Zhao X, Wortman J et al (2017) Revitalization of a forward genetic screen identifies three new regulators of fungal secondary metabolism in the genus *Aspergillus*. MBio 8:e01246–e01217. https://doi.org/10.1128/mBio.01246-17

Piechulla B, Degenhardt J (2014) The emerging importance of microbial volatile organic compounds. Plant Cell Environ 37:811–812. https://doi.org/10.1111/pce.12254

Robey MT, Ye R, Bok JW et al (2018) Identification of the first diketomorpholine biosynthetic pathway using FAC-MS technology. ACS Chem Biol 13:1142–1147. https://doi.org/10.1021/acschembio.8b00024

Robinson SL, Panaccione DG (2014) Heterologous expression of lysergic acid and novel ergot alkaloids in *Aspergillus fumigatus*. Appl Environ Microbiol 80:6465–6472. https://doi.org/10.1128/AEM.02137-14

Rutledge PJ, Challis GL (2015) Discovery of microbial natural products by activation of silent biosynthetic gene clusters. Nat Rev Microbiol 13:509–523. https://doi.org/10.1038/nrmicro3496

Sakai K, Kinoshita H, Shimizu T, Nihira T (2008) Construction of a citrinin gene cluster expression system in heterologous *Aspergillus oryzae*. J Biosci Bioeng 106:466–472. https://doi.org/10.1263/jbb.106.466

Sanchez JF, Somoza AD, Keller NP, Wang CCC (2012) Advances in *Aspergillus* secondary metabolite research in the post-genomic era. Nat Prod Rep 29:351. https://doi.org/10.1039/c2np00084a

Schmidt K, Podmore I (2015) Current challenges in volatile organic compounds analysis as potential biomarkers of cancer. J Biomarkers 2015:1–16. https://doi.org/10.1155/2015/981458

Schmidt-Dannert C (2015) NextGen microbial natural products discovery. Microb Biotechnol 8:26–28. https://doi.org/10.1111/1751-7915.12184

Schroeckh V, Scherlach K, Nützmann HW et al (2009) Intimate bacterial-fungal interaction triggers biosynthesis of archetypal polyketides in *Aspergillus nidulans*. Proc Natl Acad Sci U S A 106:14558–14563. https://doi.org/10.1073/pnas.0901870106

Schuetze T, Meyer V (2017) Polycistronic gene expression in *Aspergillus niger*. Microb Cell Factories 16:162. https://doi.org/10.1186/s12934-017-0780-z

Schümann J, Hertweck C (2006) Advances in cloning, functional analysis and heterologous expression of fungal polyketide synthase genes. J Biotechnol 124:690–703. https://doi.org/10.1016/j.jbiotec.2006.03.046

Schuster M, Schweizer G, Reissmann S, Kahmann R (2016) Genome editing in *Ustilago maydis* using the CRISPR–Cas system. Fungal Genet Biol 89:3–9. https://doi.org/10.1016/j.fgb.2015.09.001

Seidler NW, Jona I, Vegh M, Martonosi A (1989) Cyclopiazonic acid is a specific inhibitor of the Ca2+-ATPase of sarcoplasmic reticulum. J Biol Chem 264:17816–17823

Shang Z, Salim AA, Capon RJ (2017) Chaunopyran A: co-cultivation of marine mollusk-derived fungi activates a rare class of 2-alkenyl-tetrahydropyran. J Nat Prod 80:1167–1172. https://doi.org/10.1021/acs.jnatprod.7b00144

Skinnider MA, Dejong CA, Rees PN et al (2015) Genomes to natural products PRediction Informatics for Secondary Metabolomes (PRISM). Nucleic Acids Res. https://doi.org/10.1093/nar/gkv1012

Sondergaard TE, Hansen FT, Purup S et al (2011) Fusarin C acts like an estrogenic agonist and stimulates breast cancer cells in vitro. Toxicol Lett 205:116–121. https://doi.org/10.1016/j.toxlet.2011.05.1029

Spraker JE, Jewell K, Roze LV et al (2014) A volatile relationship: profiling an inter-kingdom dialogue between two plant pathogens, *Ralstonia solanacearum* and *Aspergillus flavus*. J Chem Ecol 40:502–513. https://doi.org/10.1007/s10886-014-0432-2

Starcevic A, Zucko J, Simunkovic J et al (2008) ClustScan: an integrated program package for the semi-automatic annotation of modular biosynthetic gene clusters and in silico prediction of novel chemical structures. Nucleic Acids Res 36:6882–6892. https://doi.org/10.1093/nar/gkn685

Stierle AA, Stierle DB, Decato D et al (2017) The berkeleylactones, antibiotic macrolides from fungal coculture. J Nat Prod 80:1150–1160. https://doi.org/10.1021/acs.jnatprod.7b00133

Subramanian V, Schuster LA, Moore KT et al (2017) A versatile 2A peptide-based bicistronic protein expressing platform for the industrial cellulase producing fungus, *Trichoderma reesei*. Biotechnol Biofuels 10:34. https://doi.org/10.1186/s13068-017-0710-7

Sun WW, Guo C-J, Wang CCC (2016) Characterization of the product of a nonribosomal peptide synthetase-like (NRPS-like) gene using the doxycycline dependent Tet-on system in *Aspergillus terreus*. Fungal Genet Biol 89:84–88. https://doi.org/10.1016/j.fgb.2016.01.014

Suzuki S, Hosoe T, Nozawa K et al (2000) Antifungal substances against pathogenic fungi, talaroconvolutins, from *Talaromyces convolutus*. J Nat Prod 63:768–772. https://doi.org/10.1021/np990371x

Tagami K, Minami A, Fujii R et al (2014) Rapid reconstitution of biosynthetic machinery for fungal metabolites in *Aspergillus oryzae*: total biosynthesis of aflatrem. Chembiochem 15:2076–2080. https://doi.org/10.1002/cbic.201402195

Takahashi C, Matsushita T, Doi M et al (1995) Fumiquinazolines A–G, novel metabolites of a fungus separated from a *Pseudolabrus* marine fish. J Chem Soc Perkin Trans 1 18:2345–2353. https://doi.org/10.1039/P19950002345

Takeda I, Umemura M, Koike H et al (2014) Motif-independent prediction of a secondary metabolism gene cluster using comparative genomics: application to sequenced genomes of Aspergillus and ten other filamentous fungal species. DNA Res 21:447–457. https://doi.org/10.1093/dnares/dsu010

Tauber JP, Schroeckh V, Shelest E et al (2016) Bacteria induce pigment formation in the basidiomycete *Serpula lacrymans*. Environ Microbiol 18:5218–5227. https://doi.org/10.1111/1462-2920.13558

Tauber JP, Gallegos-Monterrosa R, Kovács ÁT et al (2018) Dissimilar pigment regulation in *Serpula lacrymans* and *Paxillus involutus* during inter-kingdom interactions. Microbiology 164:65–77. https://doi.org/10.1099/mic.0.000582

Ugai T, Minami A, Gomi K, Oikawa H (2016) Genome mining approach for harnessing the cryptic gene cluster in *Alternaria solani*: production of PKS–NRPS hybrid metabolite, didymellamide B. Tetrahedron Lett 57:2793–2796. https://doi.org/10.1016/j.tetlet.2016.05.043

Unkles SE, Valiante V, Mattern DJ, Brakhage AA (2014) Synthetic biology tools for bioprospecting of natural products in eukaryotes. Chem Biol 21:502–508. https://doi.org/10.1016/j.chembiol.2014.02.010

Valiante V (2017) The cell wall integrity signaling pathway and its involvement in secondary metabolite production. J Fungi 3:68. https://doi.org/10.3390/jof3040068

van Dijk JWA, Guo CJ, Wang CCC (2016) Engineering fungal nonribosomal peptide synthetase-like enzymes by heterologous expression and domain swapping. Org Lett 18:6236–6239. https://doi.org/10.1021/acs.orglett.6b02821

Vansteelandt M, Kerzaon I, Blanchet E et al (2012) Patulin and secondary metabolite production by marine-derived *Penicillium* strains. Fungal Biol 116:954–961. https://doi.org/10.1016/j.funbio.2012.06.005

Vödisch M, Scherlach K, Winkler R et al (2011) Analysis of the *Aspergillus fumigatus* proteome reveals metabolic changes and the activation of the pseurotin a biosynthesis gene cluster in response to hypoxia. J Proteome Res 10:2508–2524. https://doi.org/10.1021/pr1012812

Vogt K, Bhabhra R, Rhodes JC, Askew DS (2005) Doxycycline-regulated gene expression in the opportunistic fungal pathogen *Aspergillus fumigatus*. BMC Microbiol 5:1. https://doi.org/10.1186/1471-2180-5-1

Vyas VK, Barrasa MI, Fink GR (2015) A *Candida albicans* CRISPR system permits genetic engineering of essential genes and gene families. Sci Adv 1: e1500248. https://doi.org/10.1126/sciadv.1500248

Wakefield J, Hassan HM, Jaspars M et al (2017) Dual induction of new microbial secondary metabolites by fungal bacterial co-cultivation. Front Microbiol 8:1284. https://doi.org/10.3389/fmicb.2017.01284

Wang L (2003) Control of growth, secondary metabolism and sporulation in *Streptomyces venezuelae* ISP5230 by jadW1, a member of the afsA family of gamma-butyrolactone regulatory genes. Microbiology 149:1991–2004. https://doi.org/10.1099/mic.0.26209-0

Wang DZ, Kong LF, Li YY, Xie ZX (2016a) Environmental microbial community proteomics: status, challenges and perspectives. Int J Mol Sci 17:1275. https://doi.org/10.3390/ijms17081275

Wang G, Liu Z, Lin R et al (2016b) Biosynthesis of antibiotic leucinostatins in bio-control fungus *Purpureocillium lilacinum* and their inhibition on *Phytophthora* revealed by genome mining. PLoS Pathog 12:e1005685. https://doi.org/10.1371/journal.ppat.1005685

Wang M, Carver JJ, Phelan VV et al (2016c) Sharing and community curation of mass spectrometry data with Global Natural Products Social Molecular Networking. Nat Biotechnol 34:828–837. https://doi.org/10.1038/nbt.3597

Wanka F, Cairns T, Boecker S et al (2016) Tet-on, or Tet-off, that is the question: advanced conditional gene expression in *Aspergillus*. Fungal Genet Biol 89:72–83. https://doi.org/10.1016/j.fgb.2015.11.003

Waring RB, May GS, Morris NR (1989) Characterization of an inducible expression system in *Aspergillus nidulans* using alcA and tubulin coding genes. Gene 79:119–130. https://doi.org/10.1016/0378-1119(89)90097-8

Weber T (2014) In silico tools for the analysis of antibiotic biosynthetic pathways. Int J Med Microbiol 304:230–235. https://doi.org/10.1016/j.ijmm.2014.02.001

Weber J, Valiante V, Nødvig CS et al (2017) Functional reconstitution of a fungal natural product gene cluster by advanced genome editing. ACS Synth Biol 6:62–68. https://doi.org/10.1021/acssynbio.6b00203

Weissman KJ (2016) Genetic engineering of modular PKSs: from combinatorial biosynthesis to synthetic biology. Nat Prod Rep 33:203–230. https://doi.org/10.1039/C5NP00109A

West RR, van Ness J, Varming AM et al (1996) ZG-1494α, a novel platelet-activating factor acetyltransferase inhibitor from *Penicilium rubrum*, isolation, structure elucidation and biological activity. J Antibiot (Tokyo) 49:967–973. https://doi.org/10.7164/antibiotics.49.967

WHO (2017) 20th edition WHO model list of essential medicines (March 2017)

Wiemann P, Guo CJ, Palmer JM et al (2013a) Prototype of an intertwined secondary-metabolite supercluster. Proc Natl Acad Sci U S A 110:17065–17070. https://doi.org/10.1073/pnas.1313258110

Wiemann P, Sieber CMK, von Bargen KW et al (2013b) Deciphering the cryptic genome: genome-wide analyses of the rice pathogen *Fusarium fujikuroi* reveal complex regulation of secondary metabolism and novel metabolites. PLoS Pathog 9: e1003475. https://doi.org/10.1371/journal.ppat.1003475

Wiemann P, Soukup AA, Folz JS et al (2018) CoIN: co-inducible nitrate expression system for secondary metabolites in *Aspergillus nidulans*. Fungal Biol Biotechnol 5:6. https://doi.org/10.1186/s40694-018-0049-2

Wolfender JL, Marti G, Thomas A, Bertrand S (2015) Current approaches and challenges for the metabolite profiling of complex natural extracts. J Chromatogr A 1382:136–164. https://doi.org/10.1016/j.chroma.2014.10.091

Wunsch C, Mundt K, Li SM (2015) Targeted production of secondary metabolites by coexpression of non-ribosomal peptide synthetase and prenyltransferase genes in *Aspergillus*. Appl Microbiol Biotechnol 99:4213–4223. https://doi.org/10.1007/s00253-015-6490-8

Xu W, Cai X, Jung ME, Tang Y (2010) Analysis of intact and dissected fungal polyketide synthase-nonribosomal peptide synthetase in vitro and in *Saccharomyces cerevisiae*. J Am Chem Soc 132:13604–13607. https://doi.org/10.1021/ja107084d

Yeger-Lotem E, Riva L, Su LJ et al (2009) Bridging high-throughput genetic and transcriptional data reveals cellular responses to alpha-synuclein toxicity. Nat Genet 41:316–323. https://doi.org/10.1038/ng.337

Yin W, Keller NP (2011) Transcriptional regulatory elements in fungal secondary metabolism. J Microbiol 49:329–339. https://doi.org/10.1007/s12275-011-1009-1

Zabala AO, Xu W, Chooi Y-H, Tang Y (2012) Characterization of a silent azaphilone gene cluster from *Aspergillus niger* ATCC 1015 reveals a hydroxylation-mediated pyran-ring formation. Chem Biol 19:1049–1059. https://doi.org/10.1016/j.chembiol.2012.07.004

Zadra I, Abt B, Parson W, Haas H (2000) *xylP* promoter-based expression system and its use for antisense downregulation of the *Penicillium chrysogenum* nitrogen regulator NRE. Appl Environ Microbiol 66:4810–4816. https://doi.org/10.1128/AEM.66.11.4810-4816.2000

Zhang XW, Jia LJ, Zhang Y et al (2012) In planta stage-specific fungal gene profiling elucidates the molecular strategies of *Fusarium graminearum* growing inside wheat coleoptiles. Plant Cell 24:5159–5176. https://doi.org/10.1105/tpc.112.105957

Zhang L, Niaz S, Khan D et al (2017) Induction of diverse bioactive secondary metabolites from the mangrove endophytic fungus *Trichoderma* sp. (strain 307) by co-cultivation with *Acinetobacter johnsonii* (strain B2). Mar Drugs 15:35. https://doi.org/10.3390/md15020035

Zhang T, Wan J, Zhan Z et al (2018) Activation of an unconventional meroterpenoid gene cluster in *Neosartorya glabra* leads to the production of new berkeleyacetals. Acta Pharm Sin B 8:478–487. https://doi.org/10.1016/j.apsb.2017.12.005

Zhao Y, Liu D, Proksch P et al (2018) Truncateols O-V, further isoprenylated cyclohexanols from the sponge-associated fungus *Truncatella angustata* with antiviral activities. Phytochemistry 155:61–68. https://doi.org/10.1016/j.phytochem.2018.07.017

Zhu F, Chen G, Chen X et al (2011) Aspergicin, a new antibacterial alkaloid produced by mixed fermentation of two marine-derived mangrove epiphytic fungi. Chem Nat Compd 47:767–769. https://doi.org/10.1007/s10600-011-0053-8

Zhu ZJ, Schultz AW, Wang J et al (2013) Liquid chromatography quadrupole time-of-flight mass spectrometry characterization of metabolites guided by the METLIN database. Nat Protoc 8:451–460. https://doi.org/10.1038/nprot.2013.004

Zhu F, Li JS, Xie WC et al (2017) Structure revision of aspergicin by the crystal structure of aspergicine, a co-occurring isomer produced by co-culture of two mangrove epiphytic fungi. Nat Prod Res 31:2268–2272. https://doi.org/10.1080/14786419.2017.1300800

Zuck KM, Shipley S, Newman DJ (2011) Induced production of N-formyl alkaloids from *Aspergillus fumigatus* by co-culture with *Streptomyces peucetius*. J Nat Prod 74:1653–1657. https://doi.org/10.1021/np200255f

12 Exploiting Fungal Photobiology as a Source of Novel Bio-blocks for Optogenetic Systems

Vicente Rojas[1,2], Francisco Salinas[1,3], Leonardo Guzman-Zamora[1,2], Andrés Romero[1,2], Verónica Delgado[1,2], Luis F. Larrondo[1,2]

CONTENTS

I. Introduction

Living organisms are constantly exposed to environmental perturbations, which vary in time and space and lead to changes at the cellular level. In that context, **light** is a powerful environmental stimulus that drives, and also shapes, development, morphogenesis, and physiology in a wide range of organisms including plants, fungi, and bacteria. In these organisms, light can be perceived by several light-sensitive proteins that are known as **photoreceptors**, which are well spread in the tree of life, being involved in a variety of biological responses, such as phototropism in fungi and bacteria, flowering in plants and behavior in animals (Banerjee and Batschauer 2004; Navara and Nelson 2007; Briggs 2014). Although photoreceptors differ in their biochemical properties, in most cases it is possible to recognize specific domains that coordinate a chromophore as a cofactor. These molecules absorb photons of a specific wavelength, translating such difference in energy as a conformational change in the surrounding polypeptide, affecting the structure of the protein (Shcherbakova et al. 2015). Thereby, the activated state of the photoreceptor is reached, triggering a molecular signal which then modulates a light-regulated biological process (Lorrain et al. 2006).

In recent years, key questions in biology are being addressed by a discipline known as **synthetic biology**. This new field has promoted a revolution in the way that biological processes are studied. Via the rational generation of novel genetic circuits and artificial devices, it is possible to obtain biological systems more or less complex than natural ones, permitting the analysis of relevant cellular processes, while reprogramming their function in a predictable and robust manner (Benner and Sismour 2005; Andrianantoandro et al. 2006; Purnick and Weiss 2009; Bashor et al. 2010). One desirable characteristic in synthetic systems is the ability to turn off and on genetic programs at will. While this could be done using a classic chemical inducer, the latter presents several problems such as once added it may be hard to remove or

[1] Millennium Institute for Integrative Biology (iBio), Santiago, Chile; e-mail: llarrondo@bio.puc.cl
[2] Facultad de Ciencias Biológicas, Departamento de Genética Molecular y Microbiología, Pontificia Universidad Católica de Chile, Santiago, Chile
[3] Facultad de Ciencias, Instituto de Bioquímica y Microbiología, Universidad Austral de Chile (UACH), Valdivia, Chile

Genetics and Biotechnology, 3rd Edition
The Mycota II
J.P. Benz, K. Schipper (Eds.)

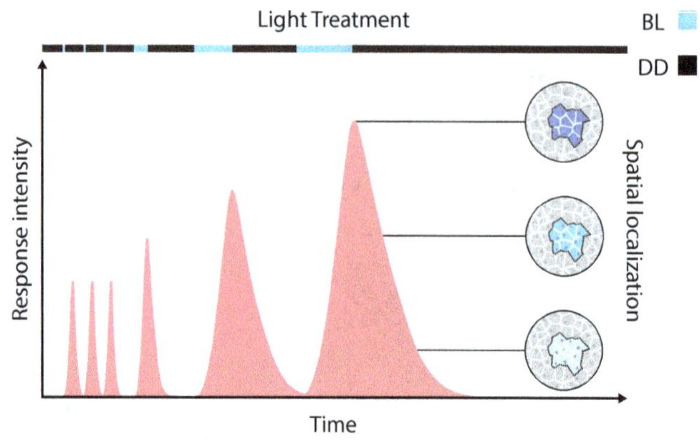

Fig. 1 Spatio-temporal and dynamic resolution of optogenetic systems. The diagram depicts light input and response properties in optogenetic systems, highlighting that variation in the length of the light pulses changes the intensity of the responses. Moreover, the latter can be localized spatially by giving an input in a certain area, which can quickly revert in the area of interest as the light stimulus is turned off. *BL* blue light, *DD* constant darkness

it can be metabolized, etc. On the other hand, light presents itself as a prominent tool to assess complex and dynamic biological phenomena, and thus, different synthetic approaches have used proteins containing light-sensitive domains as genetically encoded building blocks to manipulate cellular processes by luminous stimuli (Möglich and Moffat 2010; Schmidt and Cho 2015). Indeed, light is an ideal modulator of specific biological phenomena because it possesses various advantages compared to traditional induction strategies based on chemical inducers. First, light can be easily controlled by manageable sources, allowing a remarkably tunable induction. In this way, light induction constitutes an interesting alternative to increase resolution at the temporal level, as well as at the spatial scale (Fig. 1) (Drepper et al. 2011). Furthermore, light can be readily delivered to cultures, overcoming the problems associated with the addition of chemical inducers (i.e., uptake time or diffusion of the molecules). In addition, the capacity to turn on/off lights by an external tunable switch allows reversible induction, reducing potential phototoxicity that may be caused by prolonged illumination (Robertson et al. 2013; Canessa et al. 2013). Moreover, and importantly, light acts as an orthogonal input when used to interrogate biological processes in a blind organism, where the endogenous signaling pathways are minimally affected (Gautier et al. 2014). Finally, the significant advances in optic technologies are allowing the implementation of fully automated illumination systems, where essential variables such as duration, intensity, and location of light stimuli can be adjusted with exquisite accuracy (Gerhardt et al. 2016; Brenker et al. 2016; Rullan et al. 2018).

Historically, the development of synthetic systems based on light induction gave rise to an exciting area that is defined as "**Optogenetics**" (Deisseroth 2011). In the last decades, particularly in the past 5 years, an increasing number of light-controlled genetic devices and toolkits have been reported, in which photoreceptors responding to different wavelengths have been successfully implemented in distinct biological platforms (Gautier et al. 2014; Mühlhäuser et al. 2017; Salinas et al. 2017; Liu and Tucker 2017; An-adirekkun et al. 2019). In this review, we summarize the current light-sensitive tools and their applications, focusing our attention on systems with photoresponsive modules coming from fungi, and discussing how such parts, or bio-blocks, can expand the existing optogenetic toolbox.

II. Overview of Optogenetic Systems and Their Applications

As mentioned above, optogenetics provides a great way to probe complex biological phenomena. Specifically, the implementation of light-responsive proteins has the potential to allow deciphering the dynamics of cellular processes with great detail as a result of its striking spatiotemporal resolution (Fig. 1) and minimal invasiveness. In that context, different optogenetic systems have been reported to control gene expression, DNA recombination, subcellular protein localization, protein activity, cell morphology, cell signaling, and protein degradation (Zhang and Cui 2015; Shcherbakova et al. 2015).

A. The Early Days of Optogenetics

A clear example of a modular system in the early days of optogenetics can be tracked down to 2005 when the **light-sensitive microbial protein Channelrhodopsin-2 (ChR2)** was used to modulate mammalian neuronal activity (Boyden et al. 2005). ChR2 is a cation channel from the green algae *Chlamydomonas reinhardtii*, containing seven α-helix transmembrane domains and passing to an open state by isomerization with the retinal chromophore in response to blue light (~460 nm) (Nagel et al. 2003). Using a lentiviral transfection system, ChR2 was expressed in neurons achieving the control of synaptic transmission with an outstanding temporal resolution in the order of milliseconds (Boyden et al. 2005). After this successful optogenetic application, more light-sensitive ion channels (type-I opsins) were discovered and tested to excite/inhibit the generation of action potentials. For instance, **archaeal halorhodopsin** has been utilized to repress neuronal firing by yellow light-induced transport of chloride anions (Han and Boyden 2007). The use of photoactivated ion channels in the neuroscience field led to great knowledge advances in neuronal connectivity, being also employed in the study of cell signaling processes in other relevant mammalian tissues. In this sense, ChR2 has been used to control Ca^{2+} fluxes in muscular cells, where this ion is one of the most important secondary messengers in signal transduction processes (Carafoli and Krebs 2016; Sebille et al. 2017; Mao et al. 2019). Similarly, the light-induced opening of ChR2 permitted to study the maturation and contraction capacity of muscle cell lines, overcoming the problems of traditional electric stimulation (Bruegmann et al. 2010; Beiert et al. 2014). On the other hand, **type II opsins** have been used to manipulate cell signaling at the plasma membrane level, since they are G-protein-coupled receptors (Mühlhäuser et al. 2017). Depending on the nature of the G protein, the receptor is associated with different signaling pathways, being able to activate or inhibit signal propagation. In that context, blue light has been used to increase the concentration of inositol triphosphate (IP3), by activation of phospholipase C (PLC) in HeLa cells expressing **melanopsin** (Melyan et al. 2005).

However, the beginnings of optogenetics as the direct means of control of any biological process by light dates back to 2002. In that year, a plant photoreceptor was inadvertently used to implement a light-inducible expression system in the budding yeast *Saccharomyces cerevisiae*. Indeed, such experiments were not intended to create an optogenetic device, but instead to explore the interaction of two particular plant components. In *Arabidopsis thaliana*, **phytochrome B (PhyB)** senses red light (~ 660 nm) through its chromophore phytochromobilin, passing to the active state and binding the Protein Interacting Factor (PIF3 or PIF6) (Fig. 2a), which in turn activates the expression of multiple genes (Fairchild and Quail 1998; Quail 2010). Importantly, phytochrome B returns to its basal inactive state by far-red illumination (~740 nm), configuring a system with on and off states, and generating the first description of a light-controlled gene expression system (Shimizu-Sato et al. 2002). In this seminal work, the light-inducible system was based on the architecture of a yeast two-hybrid assay, where GAL4 DNA-binding domain (DBD) and GAL4 transactivation domain (AD) were fused to PhyB and PIF3, respectively (Shimizu-Sato et al. 2002). Thus, a

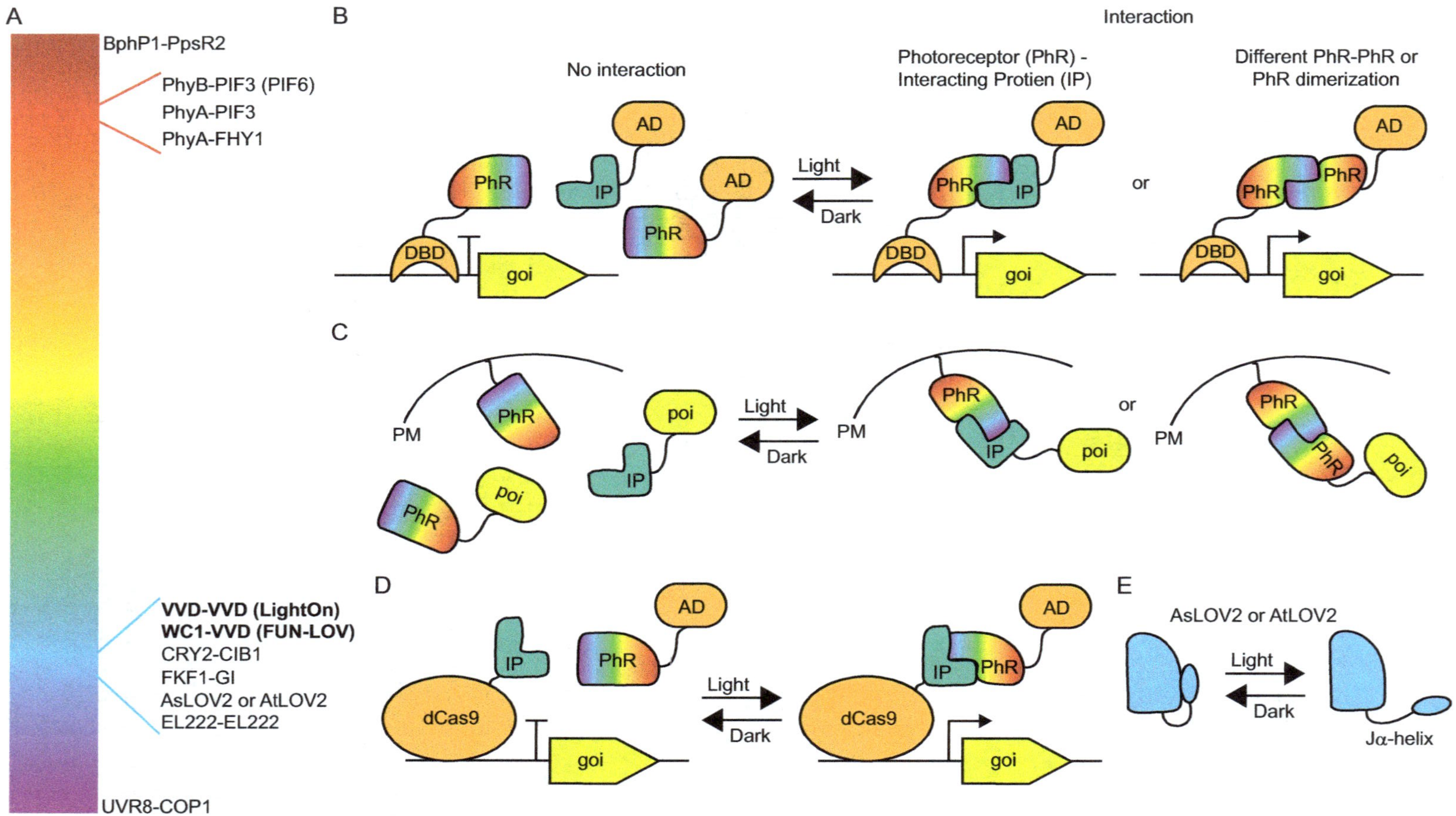

Fig. 2 Overview of different optogenetic systems. (**a**) Summary of different optogenetic systems described so far capable of responding to particular wavelengths. Defined photoreceptors and their interacting proteins are shown, considering also the systems based on photoreceptor self-dimerization. The optogenetic switches containing fungal photoreceptors are highlighted in bold ('LightOn' and 'FUN-LOV'). (**b**) General architecture of optogenetic systems used for light-controlled gene expression. The DNA-binding domain (DBD) is linked to the photoreceptor (PhR) and the Interacting Protein (IP) is tied to the transactivation domain (AD). Upon light stimulation, the interaction between PhR-IP or PhR-PhR (PhR dimerization), reconstitutes a chimeric transcription factor that activates expression of the gene of interest (*goi*). (**c**) Optogenetic control of subcellular protein localization. The PhR is linked to plasma membrane (PM) and the IP or PhR are tied to the protein of interest (poi). Upon light stimulation, the interaction between PhR-IP or PhR-PhR sequestrates the poi at the plasma membrane. (**d**) Optogenetic systems based on CRISPR-Cas9 technology for light-controlled gene expression. The dCas9 protein is linked to the IP protein and the PhR is tied to the AD, activating gene expression upon light stimulation. (**e**) Light induced conformational change in the AsLOV2 or AtLOV2 domains. The light activated displacement of the Jα-helix is shown

chimeric functional transcription factor is reconstituted by the red light-dependent PhyB/PIF3 interaction (Fig. 2a, b), inducing the expression of a *lacZ* reporter gene under the *gal1* promoter control (Shimizu-Sato et al. 2002). Similarly, optogenetic systems based on PhyB/PIF6 interaction have been implemented in different cell hosts such as yeast (Pathak et al. 2014), cell lines (Müller et al. 2013a, b) and plants (Müller et al. 2014). On the other hand, there are more phytochrome isoforms in *A. thaliana* which are potentially useful for building novel optogenetic switches. For instance, the interaction between **phytochrome A (PhyA)** and PIF3 was evaluated in response to red light (Fig. 2a, b), observing high and comparable levels of lacZ gene expression as in the PhyB/PIF3 system (Shimizu-Sato et al. 2002). Similarly, PhyA interacts with the far-red elongated hypocotyl 1 (FHY1) protein in response to red light (Fig. 2a), an interaction that has been also exploited for implementing an optogenetic system in yeast (Sorokina et al. 2009). The rest of *A. thaliana* phytochromes (PhyC, PhyD, and PhyE) have not been utilized to generate synthetic light-inducible devices yet. The red light-responsive phytochromes can also be found in cyanobacteria, where the cofactor chromophore corresponds to phycocyanobilin (PCB). Interestingly, the latter can replace phytochromobilin in optogenetic approaches that use plant phytochromes (Müller et al. 2013a, b). However, both chromophores must be supplemented in mammals because they are absent in these eukaryotic organisms (Müller et al. 2013a, b). Similarly, an alternative option to implement synthetic optogenetic systems is the use of bacterial phytochrome-like photoreceptors. For instance, CPH1 from *Synechocystis* sp. is capable of sensing red light, but unlike plant phytochromes, it forms homodimers in response to the stimulus (Hughes et al. 1997).

B. Blue Light Photosensitive Modules in Optogenetic Devices

Besides ChR2, there is a battery of natural flavoproteins that can perceive blue light and use flavin derivatives as a chromophore. In that context, the **blue light photoreceptor cryptochrome 2 (CRY2)** from *A. thaliana* provides a versatile optogenetic tool that has been used/cited in circa 160 articles to date according to OptoBase, an online database for molecular optogenetics (Kolar et al. 2018). The CRY2 protein senses blue light by its Photolyase Homology Region (PHR), which specifically binds flavin adenine dinucleotide (FAD) as a cofactor (Essen 2006). Upon illumination, CRY2 is capable of interacting with the amino-terminus of the cryptochrome-interacting basic-helix-loop-helix 1 (CIB1) protein (Fig. 2a, b) (Liu et al. 2008). Several groups have developed optogenetic approaches based on CRY2/CIB1 interaction to control diverse cellular processes. For instance, an optogenetic system of this type was implemented in mammalian cell lines to manipulate gene expression, subcellular protein localization, and DNA recombination (Kennedy et al. 2010). First, the authors based their approach on light-induced reconstitution of a chimeric GAL4 transcription factor by genetic fusions, using the *lacZ* reporter gene controlled by the *gal1* promoter to measure expression levels in response to blue light. Additionally, truncated versions of both proteins (CRY2-PHR domain and CIB1 N-terminus) were evaluated, observing higher expression levels than those using the full-length proteins. Then, the recruitment of a fluorescent protein to the plasma membrane was tested by this photoactivated interaction (Fig. 2c). In order to do this, a chimeric CIB1-GFP was attached to this specific cell location using a CAAX motif. Conversely, a CRY2-mCherry fusion was uniformly found in the cytoplasm before illumination, but upon a blue light treatment, *mcherry* fluorescence was observed at the plasma membrane level in a dose-dependent manner (Kennedy et al. 2010). Moreover, the optogenetic heterodimerization CYR2/CIB1 system successfully reconstituted the functionality of a split protein such as the Cre recombinase (Kennedy et al. 2010). The Cre/loxP recombination system, derived from bacteriophage P1, can perform diverse changes at the genome level (i.e., deletion, inversion and transplacement between strands) depending on the orientation of *loxP* sequences (McLellan et al. 2017). As proof of principle, Cre recombinase was divided into two parts that were fused to CRY2 and CIB1, and an *egfp* reporter was used to inform the recombination level in response to blue light. A 15-min treatment was enough to observe detectable levels of fluorescence, which increased with longer exposures to light (Kennedy et al. 2010). In yeast, the light-dependent heterodimerization between CRY2 and CIB1 has been also used to control gene expression (Hughes et al. 2012). Similarly, a chimeric transcription factor using non-yeast domains such as LexA-DBD and VP16-AD was designed (Fig. 2b), displaying high levels of reporter gene expression in response to blue light (Hughes et al. 2012). Furthermore, CRY2/CIB1 photoactivated interaction was used to control *sic1* gene expression, which encodes for a regulator of the yeast cell cycle whose degradation allows transition between G1 and S phases (Verma et al. 1997; Yang et al. 2013). Thus, yeast cells grew normally in darkness, but the optogenetic expression of a hyperstable Sic1p mutant led to the accumulation of this protein and cell cycle arrest, observing dramatic cell morphology changes in response to 4 h of blue light (Hughes et al. 2012). Recently, a similar approach was developed in yeast using the CRY2/CIB1 interaction. In order to avoid the interference that can exist by using endogenous building blocks, the DBD of the Zif268 mammalian transcription factor was fused to the PHR domain of CRY2 (An-Adirekkun et al. 2019). Furthermore, a synthetic promoter was generated by placing the 9-bp specific binding element of Zif68 in a *gal1* minimum promoter that lacks the GAL4-binding upstream activating sequence (UAS), which was previously demonstrated to work orthogonally in yeast (McIsaac et al. 2013). Characterization of the system included a variation of the number of Zif68-binding UAS element repetitions in the promoter as well as different orientations, changes in light intensity, and several illumination treatments. Various reporter genes and expression levels of the components were also tested, obtaining a wide range of responses that even surpassed the expression levels reached by native yeast promoters (An-Adirekkun et al. 2019). On the other hand, a two-component optogenetic expression system based on CRISPR/Cas9 technology was developed and implemented in mammalian cells (Fig. 2d). To this end, a catalytically inactive "dead" mutant of Cas9 (dCas9) was fused to CIB1 and different transactivation domains were fused to the CRY2-PHR domain. In response to blue light, a transcription system was thus reconstituted activating gene expression depending on the guide RNA (gRNA) targeting a specific promoter (Fig. 2d) (Nihongaki et al. 2015a, b). Using a luciferase reporter and three different gRNAs that recognize a *gal1* promoter, this system showed high levels of bioluminescence in response to illumination. This photoactivable transcription system also allowed the spatial expression of an mCherry fluorescent protein (Nihongaki et al. 2015a, b). Importantly, synthetic expression systems based on CRISPR/Cas9 allow easier activation of endogenous genes because it is not necessary to modify the promoters of the genes of interest. For this reason, the authors probed their system to induce transcription of several genes in mammalian cells, observing high transcript levels of *ascl1*, *myod1*, *nanog*, and *il1rn* in response to light using multiple gRNAs, showing the feasibility of simultaneously controlling the transcription of multiple genes (Nihongaki et al. 2015a, b). Considering the molecular size of the Cas9 protein, the CRISPR/Cas9 photoactivatable transcription system was optimized. For this, CIB1 was fused to the amino- and carboxyl-terminus of the dCas9 protein, developing a new system called **Light-Activated CRISPR/Cas9 Effector ("LACE")**. This approach increased the expression of endogenous genes in mammalian cell lines, with a precise spatiotemporal resolution (Polstein and Gersbach 2015). Besides its light-dependent heterodimerization with CIB1, CRY2 can oligomerize upon illumination (Bugaj et al. 2013). Thereby, several protein clustering approaches have been developed using this property, such as the **"CLICR" system** which fused CRY2 to an adaptor that can bind specific proteins (Bugaj et al. 2015). In darkness, CRY2 monomers bind their targets with low affinity; however, light-induced

CRY2 oligomerization achieved high-affinity clustering of these proteins (Bugaj et al. 2015). In the same context, **"CYR2olig"** is a point mutant protein (E490G) that shows increased levels of light-mediated clustering (Taslimi et al. 2014). Using a similar logic, the **"LARIAT" system** was developed to inactivate proteins by its clustering upon CRY2/CIB1 interaction. Thus, the fusion of CIB1 to a multimeric protein and CRY2 to the protein of interest permitted the sequestration of this protein in response to light (Lee et al. 2014).

An alternative set of blue light photoreceptors are the ones containing **light-oxygen-voltage (LOV) domains** (Pudasaini et al. 2015). These domains belong to the Per-Arnt-Sim (PAS) protein family, which are found in organisms across different kingdoms (Glantz et al. 2016). In general, LOV-containing proteins bind flavin derivatives (FMN or FAD) as the chromophore. In response to blue light, a thiol photo-adduct between the flavin chromophore and a conserved cysteine (Cys) residue in the LOV domain is formed, generating a conformational change in the protein (Pudasaini et al. 2015; Glantz et al. 2016). This cysteinyl-flavin interaction is spontaneously hydrolyzed in darkness, however, the time for photo-adduct reversion to basal inactive state depending on the photoreceptors being analyzed (Pudasaini et al. 2015; Glantz et al. 2016). For instance, **Flavin-binding kelch domain F-box protein (FKF1)** from *A. thaliana* is a LOV-containing protein that binds its natural partner GIGANTEA (GI) in response to blue light (Fig. 2a, b). This light-induced association occurs in a few minutes and is stable for several hours (Sawa et al. 2007). In this way, optogenetic systems based on FKF1/GI interaction have been developed for light-controlled gene expression and subcellular protein localization (Fig. 2b, c) (Yazawa et al. 2009), but its slow photocycle limits temporal resolution. The **blue light photoreceptor AsLOV2** from the plant *Avena sativa* overcomes the photocycle limits, showing fast kinetics between inactive and active states (Swartz et al. 2001). The LOV2 domain of this photoreceptor is flanked by two helixes (A'α and Jα) that are unfolded by blue light stimulation and return to the basal condition in a few seconds each (Fig. 2e) (Harper et al. 2003). In this way, AsLOV2 light-dependent unfolding allows implementing one-component optogenetic systems by caging. Caging approaches allow the temporal hindrance of binding/active sites of a specific protein, depending on how this protein is fused to AsLOV2. Thus, the caging interaction can be modified by blue light, exposing the functional domain of a protein and providing fine control at the post-translational level. The AsLOV2 domain was also used to implement an exciting approach aimed to diminish the variability of gene expression levels. By fusing this LOV domain to different components of the Tet system, gene circuits based on negative feedback allowed the reduction of transcriptional noise in human cell lines (Guinn and Balázsi 2019). In addition, the AsLOV2 domain has been adapted to implement two-component systems mediated by light-dependent interaction. For instance, the **"TULIP" system** exploits the caging mechanism in a fusion protein between AsLOV2 and a specific peptide where, in response to blue light, this epitope is exposed and interacts with a synthetic PSD95-Dlg1-zo1 (PDZ) domain (Strickland et al. 2012). Similarly, the **phototropin 1 (PHOT1) from *A. thaliana* has a LOV domain (AtLOV2)** that shares the basic mechanism of photoperception with AsLOV2 (Christie et al. 1999). In that context, Renicke et al. (2013) utilized a degradation sequence sterically blocked by the Jα-helix of the AtLOV2, which in turn was linked to a specific protein of interest. Thus, upon blue light stimulation, the displacement of Jα-helix exposed the degradation signal, changing the stability of relevant proteins in the yeast cell cycle (Renicke et al. 2013). On the other hand, LOV-containing photoreceptors can also be found in bacteria. For instance, the **transcription factor EL222** from *Erythrobacter litoralis* has been also used to implement synthetic light-inducible systems (Fig. 2a, b). Besides a helix-turn-helix (HTH) DBD, this protein contains a LOV module that allows transcriptional activation upon blue light stimulation (Zoltowski et al. 2013). In the dark, both domains interact and a critical dimerization sequence is blocked, avoiding the binding of EL222 to specific promoter regions of the genome. Under blue light conditions, the formation of a photo-adduct disrupts the interaction between LOV and HTH, leading to the protein homodimerization and light-dependent gene expression (Zoltowski et al. 2013). Considering these EL222 features, a one-component gene expression system was developed and tested in different mammalian cell lines, obtaining a wide range of luciferase expression levels in response to blue light stimulation. Moreover, the system showed fast kinetics of activation/deactivation and low basal activity in the dark (Motta-Mena et al. 2014). Notably, this approach was recently implemented in *S. cerevisiae*. In this work, the EL222 optogenetic system was used to directly control the expression of a *gfp* reporter gene under the synthetic C120 promoter. The **circuit, named OptoEXP**, was evaluated upon different light treatments, displaying fluorescence levels that are similar to those obtained with an *adh1* constitutive promoter (Zhao et al. 2018). On the other hand, the **OptoINVRT circuit** allows light-dependent gene repression. In this case, a constitutively activated GAL4 transcription factor allowed the constant expression of the fluorescent reporter controlled by *gal1* promoter in darkness by, while the transcription levels were switched off by the light-dependent production of the GAL80 repressor (Zhao et al. 2018). Thereby, depending on the gene controlled by the synthetic EL222 system, it was possible to activate or repress different genes in response to light using OptoEXP or OptoINVRT circuits, respectively. In this way, the biosynthesis of a metabolite of interest could be inhibited during the growth phase, whereas it could be produced (at will) when enough cell biomass has been accumulated in a yeast fermentation process. Using these **optogenetic metabolic valves**, it was possible to obtain high yields of isobutanol and 2-methyl-1-butanol production after a light/dark transition by controlling an essential mitochondrial pathway (Fig. 3a). This work is a remarkable case of control of multiple genes and represents one of the first examples of successful applications of optogenetics to metabolic engineering of molecules of interest in yeast (Zhao et al. 2018). Other blue light sensing domains will be covered in the upcoming section.

C. Beyond Red and Blue Light

The so far mentioned optogenetic systems respond to wavelengths that are part of classic white light (~400–700 nm). However, the sunlight also includes ultraviolet light (UV, ~100–400 nm) and near-infrared light (NIR, >700 nm). Since UV causes DNA damage and generation of reactive oxygen species (ROS), sessile organisms such as plants had to develop tolerance and acclimation mechanisms (Yin and Ulm 2017; Jenkins 2017). In *A. thaliana*, the **photoreceptor UV-B resistance 8 (UVR8)** forms inactive homodimers in darkness conditions by an interaction between charged amino acids. In response to UV-B light (~280–315 nm), a conformational change breaks these intermolecular interactions and UVR8 behaves as a monomer (Christie et al. 2012). In this state, the photoreceptor can interact with the constitutive photomorphogenic 1 (COP1) protein (Fig. 2a), triggering the UV-protection signaling pathways (Oravecz et al. 2006; Favory et al. 2009). Interestingly, UVR8 does not require a chromophore, and its photoperception is based on the intrinsic amino acids of the protein, where tryptophan (Trp233 and Trp285) residues are responsible for UV-B light sensing (Rizzini et al. 2011). Although optogenetic systems based on UVR8/COP1 interaction have been implemented, they have been less exploited due to the cellular damages caused by high-energy UV radiation. On the other hand, it has been reported that certain bacterial species have the capacity to respond to NIR. For instance, the **BphP1 protein** from *Rhodopseudomonas palustris* can perceive NIR, and upon light stimulation interacts with the PpsR2 protein (Fig. 2a, b) (Kojadinovic et al. 2008). In comparison with shorter wavelengths, NIR causes less phototoxicity at constant exposures and is more effective to penetrate into mammalian tissues (Weissleder and Ntziachristos 2003). Moreover, this interaction does not require a chromophore of bilin nature, such as PCB, utilizing instead biliverdin as cofactor. In that context, an optoinducible system based on BphP1/PpsR2 interaction was implemented in mammalian cell lines, allowing the control of cell signaling, gene expression and subcellular protein localization (Kaberniuk et al. 2016).

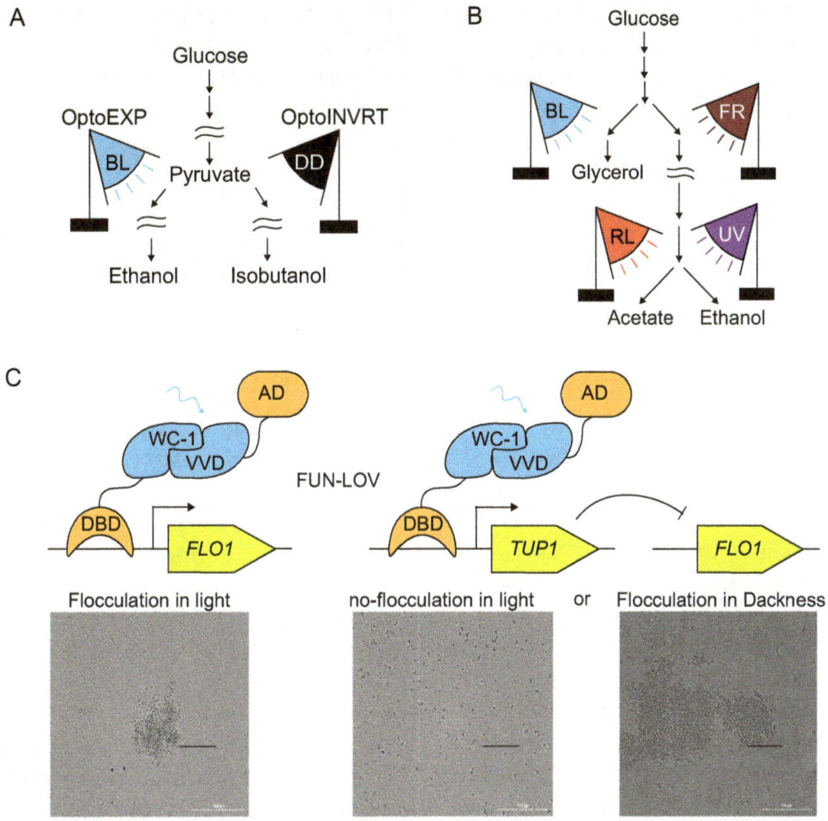

Fig. 3 Potential applications of optogenetic systems. (**a**) The OptoEXP and OptoINVRT circuits are optogenetic switches based on EL222 photoreceptor, which were assembled as molecular valves in a metabolic pathway (Zhao et al. 2018). Thus, ethanol production is triggered by blue light (BL) and isobutanol production is activated by the absence of light (darkness condition, DD). (**b**) A potential multichromatic control of a yeast metabolic pathway would generate different output metabolites: under blue light (BL) glycerol could be produced, whereas a combination of far-red (FR) and UV light (UV) could lead to ethanol, or instead acetate by the combination of far-red (FR) and red light. (**c**) In the 'FUN-LOV' optogenetic switch (Salinas et al. 2018), depending of the target gene, the phenotypic outcome (flocculation) is triggered by blue-light or, when the switch is wired in a different way, in the absence of light (flocculation in darkness). In the latter case the light-activated *tup1* expression represses the *flo1* gene, generating flocculation in the dark. In the bright-field microscopy images, the scale bar represents 100 μM

Obviously, the abovementioned examples do not encompass all optogenetic systems described to date. However, they constitute representative illustrations of synthetic devices responding to light of different wavelengths. Altogether, these diverse optogenetic systems have the capacity to control gene expression, subcellular protein localization, protein activity, cell morphology, cell signaling, and protein degradation in multiple biological platforms. However, **a major challenge still to be addressed is the combination of multiple optogenetic systems for the control of several processes by simultaneously responding to different wavelengths** (i.e., spectral multiplexing) (Fig. 3b). The latter is particularly interesting to manipulate pivotal nodes in a metabolic pathway, allowing to direct carbon flux and other important substrates depending on the light types used to stimulate the cell cultures (Fig. 3b). In this way, it might be possible to avoid metabolic burden in the production of high-value compounds. Moreover, an exciting advantage implies having a multi-chromatic system to achieve a simultaneous activation of various light-regulated tasks using white light, which covers a wide spectrum of wavelengths. Furthermore, **a second challenge in optogenetics is to expand the repertory of available photoreceptors, describing novel bio-blocks for these synthetic systems**, and where fungal photobiology may play an important role as a source of new opto-modules.

III. Exploiting Fungal Photobiology as a Source of Novel Bio-blocks for Optogenetic Systems

In the former section, we have focused our attention on optogenetic systems that use light-responsive domains belonging to bacteria, archaea, plants, and animals. However, what about **the fungal kingdom**? As mentioned earlier, light is a strong signal that provides important environmental information affecting several cellular processes. In fungi, these processes include developmental decisions, utilization of complex substrates, pigmentation, phototropism, and virulence (Schumacher 2017; Fuller et al. 2018; Schmoll 2018; Corrochano 2019). Specifically, the majority of the studies involving light responses in fungi have been performed in filamentous species, with some of them having a plethora of photoreceptors that contain light-sensing modules, including opsins, phytochromes, cryptochromes, and LOV-proteins (Corrochano 2007, 2019; Idnurm et al. 2010). Interestingly, however, genome analyses indicate **a complete lack of photoreceptor-encoding genes in the unicellular yeast *S. cerevisiae***, which has therefore already been used as a platform to test multiple optogenetic systems, since light can be considered an orthogonal stimulus in this organism (Goffeau et al. 1996).

A. Utilizing Light-Responsive Domains from Fungal Species

There are only few examples of light-inducible genetic devices using fungal photoreceptors, which could be due, in part, to the fact that most of them have not been fully characterized yet. In that context, studies carried out in the filamentous fungus *Neurospora crassa* have established clear examples of light-controlled mechanisms in fungi. This has led to the translation of information about *N. crassa* photobiology into applied optogenetic systems. For instance, Wang et al. (2012) reported an optogenetic system using the **VIVID protein (VVD)** from *N. crassa*. This protein possesses only 186 amino acids, being the smallest LOV-domain containing photoreceptor (Schwerdtfeger and Linden 2003). The authors took advantage of VVD homodimerization in response to blue light to develop a synthetic gene expression system in mammalian cells (Fig. 2a). They fused VVD to the p65-AD and the GAL4 DBD, the latter without its dimerization sequence, generating a chimeric transcription factor called GAVP. Thus, the blue light-dependent interaction VVD/VVD by their LOV domains was the unique possibility to form the functional dimer with the ability to turn on the reporter gene expression (Wang et al. 2012). To probe this synthetic design, the authors transfected GAVP into mammalian cell lines containing the luciferase reporter gene (*luc*) under the control of the *gal1* promoter. Although *luc* expression was high in blue light conditions, the background activity in darkness had to be reduced by point mutations in VVD. This optimized version was named GAVPO, and it was used to implement the one-component **"LightOn" optogenetic system** (Wang et al. 2012). The latter showed fast dose-dependent responses depending on the number of light pulses, length of the pulse, and light irradiance. Furthermore, the spatial resolution of the "LightOn" system was validated in mice, where liver and kidney were specifically irradiated by blue light and expression of *mcherry* and *lacZ* reporters confirmed the feasibility to control biological processes with accuracy in a complex free-moving animal (Wang et al. 2012). Importantly, decreased blood glucose levels are observed in diabetic mice by light-induced production of insulin (Wang et al. 2012). Subsequently, various parameters of the "LightOn" system were modified to optimize the light/dark fold-induction. Overall, the expression levels were increased and the background activity reduced by optimizing the promoter region including changes of the number of UAS, the length of the spacer between these sequences, and the core promoter. In addition, the concentration ratio between the GAVPO transcription factor and the luciferase reporter was also optimized (Ma et al. 2013).

Based on VVD blue light-dependent homodimerization, **an optimized system was developed receiving the name "Magnets."** This is an optogenetic system where two Neurospora VVD molecules are fused to complementary protein domains, avoiding the generation of non-functional homodimers. To achieve this, point mutations were introduced in the interface of the VVD LOV domain that is exposed to the solvent. Specifically, neutral residues such as isoleucine 52 (Ile52) and methionine 55 (Met55) were exchanged by charged amino acids, generating a positive component (pMag) and a negative component (nMag). Thus, the formation of functional heterodimers is optimized in response to blue light, minimizing homodimers by electrostatic repulsion. "Magnets" were evaluated by light-induced reconstitution of a split luciferase protein, showing high bioluminescence levels with the pMag/nMag pair (Kawano et al. 2015). Afterward, the same group reported the use of the "Magnets" approach to Cre recombinase reconstitution by blue light treatment in mammalian cells (Kawano et al. 2016). Using a *luc* reporter as outcome of the Cre-mediated recombination, the photoactivatable Cre was extensively verified using different *loxP* variants, cell lines, light intensities, and light exposures (Kawano et al. 2016). Interestingly, this optogenetic system showed better induction levels than a similar system based on CRY2/CIB1 interaction (Kawano et al. 2016). Overall, high levels of DNA recombination were obtained by this system, even showing activation in mice livers (Kawano et al. 2016). In the same context, "Magnets" dimerization domains were used to control the reconstitution of a split Cas9 protein in mammalian cells (Nihongaki et al. 2015a, b). In this way, a specific editing of genetic material was light-induced by activation of endogenous repair pathways. The activity of this optogenetic CRISPR/Cas9 system is rapidly turned off in darkness and showed spatial activation defined by certain illumination patterns (Nihongaki et al. 2015a, b). Likewise, the "Magnets" approach was also utilized to develop a light-inducible transcription system through the reassembly of an orthogonal RNA polymerase (RNAP) in *E. coli*. To this end, the RNAP from the T7 virus was divided and each part fused to pMag and nMag, generating different variants of this optogenetic system and showing high induction of a mCherry reporter (Baumschlager et al. 2017).

Interestingly, VVD photocycle mutants could be also applied in optogenetics. By mutating specific VVD residues, different variants of the protein that display altered photocycle lengths have been generated (Dasgupta et al. 2015). For instance, the mutagenesis of Ile74 and Ile85 by valine (Val) residues led to a version with faster reversion between activated and inactivated states, impairing the biological function of VVD as shown by the incapacity of the strain to attenuate gene expression and to respond to increasing light intensities (Dasgupta et al. 2015). On the other hand, VVD was also modified (Met135Ile/Met165Ile) to generate a protein in which the stability of the photo-adduct was 10 times longer (Dasgupta et al. 2015). In that context, one of the main advantages of optogenetic tools is the high temporal resolution that can be reached, permitting the assessment of transient biological phenomena. For this reason, the light-inducible systems reported so far have preferred photoreceptors with fast-cycling kinetics, discarding the massive use of proteins such as FKF1. Nevertheless, the development of slow-cycling versions could be useful to induce cellular processes that are needed to be permanent overtime with only a brief illumination input.

Despite its capacity of light-induced homodimerization, VVD can also heterodimerize with the **transcription factor White Collar 1 (WC-1)**. In *N. crassa*, WC-1 interacts with the **WC-2 transcription factor** by direct protein-protein interaction via a different type of PAS domain, forming **the White-Collar Complex (WCC)** that commands circadian gene expression (Linden and Macino 1997). Additionally, WC-1 is a blue light photoreceptor containing a

LOV domain (He et al. 2002; Froehlich et al. 2002), which under blue light illumination passes to an active form allowing interaction with another WC-1 protein through a LOV-LOV interaction, affecting the WCC stoichiometry (Cheng et al. 2003). Thus, the sequence specificity of WCC changes and a photoactivated multimer binds to specific elements found in promoters of several light-activated genes (He and Liu 2005; Chen et al. 2009). One of these genes is *vvd*, and, therefore, after prolonged illumination, VVD is accumulated and becomes light-activated, breaking the WCC-WCC interactions by competitive LOV-LOV binding between VVD and WC-1. This latter interaction leads to the attenuation of light-inducible transcription in *N. crassa* in a process known as photoadaptation (Chen et al. 2010; Malzahn et al. 2010; Hunt et al. 2010; Gin et al. 2013). Thereby, the natural WC-1/VVD interaction is an interesting combination of fungal photoreceptors to generate a novel light-inducible approach. Recently, our group used this feature from *N. crassa* photobiology to implement **an optogenetic system in yeast** that we termed **"FUN-LOV"** (Fig. 2a). We showed that "FUN-LOV" allows high expression of *luc* upon constant light illumination, showing almost null residual activity in the dark (Salinas et al. 2018). Thus, "FUN-LOV" yielded high levels of the reporter under constant white or blue lights, achieving superior results over those obtained with the classical galactose induction strategy. Importantly, we achieved induction levels of ~1300-fold when comparing light versus dark conditions, and even over 2500-fold when analyzing luciferase levels from extracts of cells grown in the light versus dark. Moreover, the "FUN-LOV" system showed a very dynamic on/off gene expression pattern in response to light/dark cycles, reaching strong induction during lights on and quick decrease of the signal as lights were turned off. We also tested if longer exposure to light pulses would increase expression of the reporter gene, observing a rise in the signal as the duration of exposure increases. On the other hand, despite that a doubling in light intensity (from 20 to 40 μmol m^{-2} s^{-1}) did not produce an increase in luciferase levels (Salinas et al. 2018), we have observed a slower induction kinetic and diminished maximal levels in low light intensity such as 1 μmol m^{-2} s^{-1} (Delgado, unpublished results). These results confirm that "FUN-LOV" not only allows reaching high levels of expression and displaying low background signals but that it is also tunable by modifying the intensity and the duration of the light stimulus. We also assessed the feasibility of the "FUN-LOV" system to control the expression of a heterologous protein such as limonene synthase from *Cannabis sativa*, observing a strong production of the enzyme in constant light conditions by western blot assays (Salinas et al. 2018). Importantly, as proof of principle, we evaluated the performance of this synthetic optogenetic switch to induce a biotechnologically relevant phenotype such as flocculation (Fig. 3c). Depending on the genetic target of "FUN-LOV," we controlled the formation of yeast cell aggregates by light or darkness, demonstrating the versatility of our system (Salinas et al. 2018).

B. Optogenetic Approaches Directly Implemented in Filamentous Fungi

In past years, different groups have attempted to harness the power of light to control gene expression directly in filamentous fungi, which are already capable of sensing light. An example of this was utilizing the promoter of the *vvd* gene from *N. crassa*, to control the expression of autologous and heterologous proteins in this organism. The expression of *vvd* is quite high, reaching augmented expression within minutes of turning on the lights, while levels are extremely low in constant darkness (Cesbron et al. 2013). Thus, the *vvd* promoter was set to control *gfp* reporter gene expression and transformed ectopically both in wild-type (WT) and Δ*vvd* backgrounds, the latter to eliminate photoadaptation (for continuous high expression in constant light). As expected, the system revealed tight control as GFP levels were almost undetectable when the fungus was kept in constant darkness, whereas within 1 h of light exposure, GFP was high and its levels increased linearly for almost 2 days in constant light. Regarding *gfp* transcript levels, they were ~50-

fold induced after 1 h of light, staying as high as 80-fold throughout the 48 h of the time course (Hurley et al. 2012). Importantly, rapid and strong increase in *vvd* expression has been reported by the use of destabilized *luc* as gene reporter (Cesbron et al. 2013). Thus, we have observed that 1 s of intense light can create a strong wave of induction that is detectable within a couple of minutes, that can last for over 1 h after the light input disappears (Fig. 4). In addition, the *vvd* promoter is quite sensitive to different amounts of light, allowing to cover a wide range of transcriptional strengths in response to light of different intensities (Cesbron et al. 2013). Interestingly, the *vvd* promoter can respond to successive stimuli, while other light responsive promoters, such as *frequency* (*frq*), are unresponsive (refractory) to new stimuli for a given amount of time (Cesbron et al. 2013). In an attempt to provide an example of the applicability of this endogenous optogenetic system in *N. crassa*, the autologous gene encoding the GH5-1 cellulase was expressed under the control of the *vvd* promoter, showing that high levels of expression can be achieved. This served as a proof of concept that biotechnologically relevant genes can be subjected to optogenetic control by the native light sensitive machinery (Hurley et al. 2012; Fuller et al. 2018). Importantly, *N. crassa* contains hundreds of genes that react to blue light as a direct response to light-activated WCC, displaying a great variety of dynamics and strengths (Chen et al. 2009). Therefore, different promoters could be utilized, to optogenetically control a gene of interest, providing broad expression dynamics upon the same light regime. In principle, similar strategies could be adopted in other fungi where genome-wide responses to light have been described and,

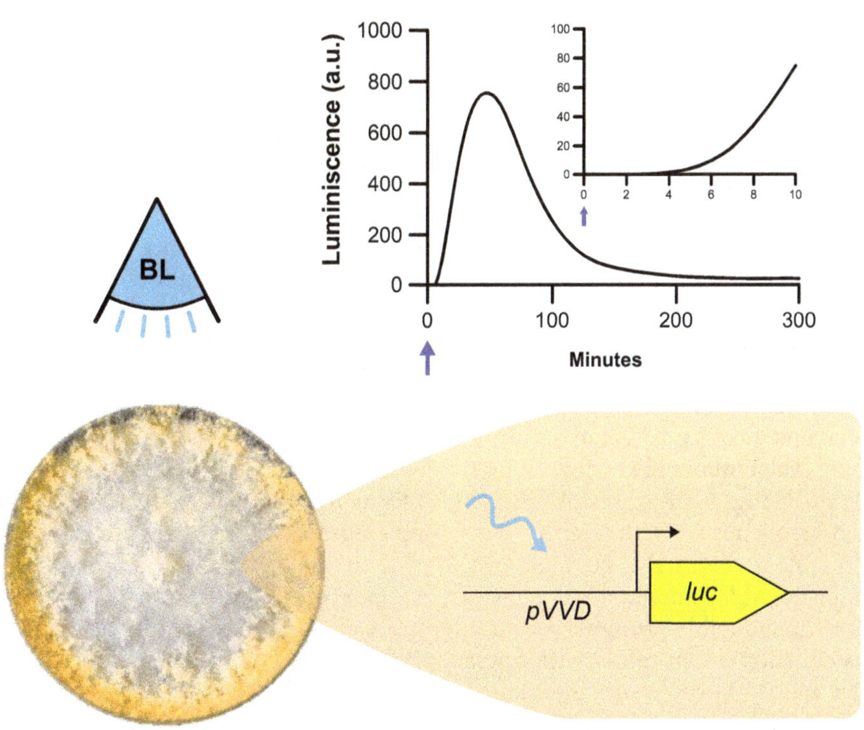

Fig. 4 Dynamics of *vvd* expression in response to a discrete light pulse. A *N. crassa* strain containing a destabilized *luciferase* reporter under the control of *vvd* promoter was grown under constant darkness for 24 h and then a 1 s light pulse, delivered with a blue laser (blue arrow), was applied and luciferase activity was quantified for 300 min (first 10 min in inset). As it can be observed, the response is strong and prompt, lasting for over 120 min (results obtained at the Larrondo Lab)

therefore, a list of promoters can be drawn from genes with strong induction and kinetic responses. A recent review summarizes such studies, highlighting the top 10 most light-induced candidates in several fungi (Fuller et al. 2018): while in many organisms *vvd* homologs are part of the list, other genes such as *con-6* or *grg-1* appear as interesting candidates that could drive strong expression upon light in a WCC-dependent manner.

The industrial workhorse *Trichoderma reesei* also has been used as a biological platform to control the expression of genes of interest by an artificial optogenetic system following the logic of the previously mentioned "LightOn" approach (Wang et al. 2014). The switch allowed to control the expression of different reporter genes such as *dsred2* and *gus*, obtaining positive signals after 36 h of light stimulation. Subsequently, the system was improved by using a light-switchable bidirectional promoter version based on a "light-on" and "light-off" divergent promoter (Zhang et al. 2016a, b). The optimization also included testing different transactivation domains (selecting VP16) as well as different flexible linkers between VVD and the latter. The bidirectional expression system evaluated the expression of an alkaline cellulase encoding gene and *rfp*, achieving 70-fold of expression comparing light versus dark conditions. Both the duration of light stimuli and light intensities were important in dosing expression levels. Nevertheless, at high irradiance (above 137 W) a negative effect was observed, which could be related to cellular stress and toxicity associated with an excess of light. The authors also tried different work cycles, providing light pulses of different duration and frequency, concluding that continuous light is not necessary to get high levels of expression. In order to obtain a light-off promoter, a light-inducible promoter was used along with a constitutive promoter with operators between the TATA-box and the gene of interest, allowing the expression of two different reporters in a light/dark dependent manner, respectively (Zhang et al. 2016a, b). Another approach developed by this group is a one-step method that combines self-excisable marker rescue and the control of the non-homologous end joining (NHEJ) pathway in such industrial eukaryotic microorganisms. Using *Agrobacterium tumefaciens*-mediated transformation and a Cre-LoxP system induced by light, this team achieved simultaneous excision of a selectable marker gene and the *cre* gene by Cre recombinase, in *T. reesei*, *N. crassa* and *Aspergillus niger* (Zhang et al. 2016a, b).

In the context of harnessing potential applications of light in basidiomycetes, recent studies in *Pleurotus ostreatus* have shown that the promoter of a lectin encoding gene could be used in this organism to provide light-controlled gene expression. Indeed, by putting *egfp* under its control, the authors observed an increase of fluorescence signal in the presence of light compared to dark, a result that was also supported by measurement of *gfp* mRNA levels (Yin et al. 2019). Interestingly, such an effect occurred only in the presence of white light, but not blue light, a result that is difficult to interpret based on the little information associated with the promoter under study (Yin et al. 2019).

C. Expanding Optogenetic Toolbox by Unexplored Fungal Photoreceptors

Curiously, only VVD/VVD and VVD/WC-1 interactions from *N. crassa* have been used to implement fungal light-controlled systems so far. In fact, WC-1 homodimerization via LOV domains in response to blue light has not been tested for optogenetic applications yet. This only being an example, **the kingdom mycota could be considered an underexploited and rather unexplored source of novel building blocks for optogenetic systems.** In fact, photoreceptors of different nature have been identified in many fungal species, including other ascomycetes as well as basidiomycetes, zygomycetes, and chytrids (Herrera-Estrella and Horwitz 2007; Corrochano 2007, 2019; Idnurm et al. 2010; Fuller et al. 2016). **Even in *N. crassa* more is to be uncovered**: for instance, this fungus also expresses a cryptochrome (*cry*) (Galagan et al. 2003) and an opsin (*nop-1*) (Bieszke et al. 1999a, b), whose roles are still rather obscure. Despite the fact that both proteins can bind their chromophores (FAD and retinal, respectively) (Bieszke et al. 1999a, b), light-dependent gene expression is not altered in the knockout strains grown under vegetative conditions (Chen et al. 2009; Froehlich et al. 2010). Moreover, *N. crassa* possesses two genes (*phy-1* and *phy-2*) with similarities to red light-responsive phytochromes (Galagan et al. 2003). Although further studies must be performed to confirm PHY-1 and PHY-2-proteins as bona fide active photoreceptors, it has also been reported that the amplitude of blue light induction for some genes such as *con-10* are affected in a *phy-2* null mutant (Olmedo et al. 2010), suggesting the existence of a connection between blue and red light photoreceptors in *N. crassa*.

An interplay between both wavelengths is better understood by pioneering studies performed in ***Aspergillus nidulans*** (Idnurm and Heitman 2005). In this saprophytic fungus, sexual/asexual development is tightly controlled by light. Thus, red light is sensed by the **phytochrome FphA**, inducing conidiation and inhibiting the formation of fruiting bodies (Blumenstein et al. 2005). Despite FphA being able to coordinate biliverdin in vitro, the endogenous chromophore is still a mystery. On the other hand, **the WC-1 and WC-2 orthologs LreA and LreB** interact to form the WCC-like complex involved in blue light perception, which increases the generation of conidiospores but also activates the formation of reproductive structures. In order to do this, FphA binds to the LreA/LreB heterodimer in the nucleus (Purschwitz et al. 2008), forming a larger complex involved in balancing the action of both wavelengths. Notably, the repressor role of red light-activated FphA is mediated by its target protein VeA, which also interacts with several proteins that form the Velvet complex (Bayram et al. 2010). The latter includes LaeA, a protein with a major role in the production of secondary metabolites. For this reason, VeA is considered a bridge in the regulation of relevant biological processes such as life cycle and secondary metabolism by light (Bayram et al. 2010). Interestingly, the phenotype of a *veA* knockout strain is partially restored by complementation with a *N. crassa* ortholog (Bayram et al. 2008).

Light responses have also been studied in the fungus ***Botrytis cinerea***, which is a relevant pathogen in the agroindustry due to its capacity to infect many plant species (Dean et al. 2012). In that context, a bioinformatic analysis revealed that its genome encodes for 11 photoreceptors (Schumacher 2017), but only few of these have been studied in detail. As it could be expected, photoreception in *B. cinerea* is based on functional WC-1 and WC-2 proteins (BcWCL1 and BcWCL2, respectively). Just as in *N. crassa*, these orthologs interact in the nuclei (Schumacher 2012) forming **the *B. cinerea* White Collar Complex (BcWCC)**. The BcWCC is key in regulating the circadian clock of this phytopathogen, providing daily rhythms of its virulence on plant hosts (Hevia et al. 2015). Yet unexpectedly, while some genes loose regulation by light in the absence of BcWCL1, others appear to remain light-responsive, which contrasts the essential role that WC-1 plays in *N. crassa* (Canessa et al. 2013). Additionally, the photochemical properties of a ***B. cinerea* VVD homolog (BcVVD1)** have been studied, being *bcvvd1* one of the genes regulated by light in a BcWCC-dependent manner (Canessa et al. 2013). However, unlike the *N. crassa* protein, BcVVD1 does not show homodimerization in response to blue light (Foley et al. 2018). Moreover, the existence of a photoadaptation process mediated by BcVVD1/BcWCL-1 interaction is not clear yet in *B. cinerea*. Recently, another LOV-containing photoreceptor with unique features has been described in *B. cinerea*. The **BcLOV4 protein** contains a polybasic amphipathic helix between the LOV domain and the carboxyl-terminus of the protein, which is rapidly recruited to the plasma membrane upon illumination and interacts with anionic phospholipids in an unspecific way (Glantz et al. 2018). Notably, this work presented the first natural photoreceptor with the capacity to associate with membrane components by light (Glantz et al. 2018).

In contrast to *B. cinerea*, photoadaptation is well-studied in ***T. reesei***, in which it occurs in a similar way to *N. crassa*. Several studies have shown that light modifies the physiology of *T. reesei*, including its striking ability to degrade cellulose (Schmoll 2018). Specifically, the expression of a cellobiohydrolase encoding gene (*cbh-1*) is a light-regulated process (Gyalai-Korpos et al. 2010). In the same way, as occurs in the fungi mentioned above, **the orthologs of WC-1 and WC-2 (BLR1 and BLR2)** are key to command the light responses in *T. reesei*, by forming a WCC-like complex (Castellanos et al. 2010). One of the genes regulated by the light-dependent activity of the BLR1/BLR2 complex is the *vvd* ortholog ***env1***, that encodes a protein involved in photoadaptation to constant illumination as well as other signaling pathways such as growth, reproduction, nutrient metabolism, response to stresses, and biosynthesis of specific molecules (Schmoll et al. 2005; Seibel et al. 2012). Moreover, a variety of photoreceptors have been identified in the *T. reesei* genome, including a cryptochrome, a phytochrome, and an opsin (Martinez et al. 2008).

Nonetheless, fungal photobiology is not exclusive for ascomycete species. Bioinformatic analyses of available basidiomycete, zygomycete, and chytrid genomes have found several orthologs of described fungal photoreceptors (Idnurm et al. 2010). In fact, WCC-, cryptochrome-, opsin-, and phytochrome-encoding genes have been identified in **basidiomycetes such as *Coprinopsis cinerea*, *Cryptococcus neoformans*, *Ustilago maydis*, and *Puccinia graminis*** (Salichos and Rokas 2010; Corrochano 2019). Furthermore, cellular processes, such as phototropism of fruiting bodies, development of sexual structures, and activation of carotenogenesis have been related to light regulation in **the zygomycete fungi *Phycomyces blakesleeanus*, *Mucor circinelloides*, and *Rhizopus delemar*** (Corrochano and Garre 2010; Corrochano 2019). Notably, unlike ascomycete and basidiomycete species, each one of these zygomycetes has multiple homolog genes to the WC-1 photoreceptor and its partner WC-2 (Corrochano and Garre 2010). Even **the chytridiomycete *Spizellomyces punctatus*** possesses WCC components and a phytochrome (Corrochano 2019).

As it has become clear throughout this chapter, many of the reported optogenetic experiences have been obtained in yeast, utilizing a wide range of optogenetic modules coming from different organisms, with just few experiences using fungal bio-blocks such as VVD and WC-1 LOV domains from *N. crassa* (Fig. 5). Although diverse (and even antagonistic) biological processes depending on WCC proteins appear to be well-spread in fungi, a conserved role of this complex in photoreception processes is nevertheless clear. In contrast, the presence of VVD orthologs seems to be infrequent, suggesting that it may not be a vital element of the machinery involved in light-responses across different fungal phyla. This latter opens the exciting possibility of using the LOV domain of Neurospora VVD as well as a cognate promoter to control biological processes by blue light in those fungi that do not possess it naturally. On the other hand, genes encoding cryptochromes, opsin, and phytochromes also have been reported in several fungal organisms. **Therefore, a wide repertoire of fungal photoreceptors is available for the implementation of new optogenetic tools, expanding the current toolbox of this discipline.**

The study of light-controlled processes requires, yet, the ability to properly control experimental light conditions. Thus, whether conducting studies aimed at further exploring the photobiology of fungal species (i.e., *N. crassa*, *A. nidulans*, *T. reesei*, *B. cinerea*, etc.), or implementing optogenetic switches in yeast or other fungi, it is important to have a proper setup that allows tight control of full darkness or defined light stimuli. Thus, a customized darkroom is recommended, containing incubators that can set defined photocycles and light intensities, equipped with safety-red lights (when manipulating blue light optogenetic systems), and conditions securing manipulation of samples in full darkness when needed (Fig. 6).

IV. Concluding Remarks

In past years, synthetic biologists have begun to interrogate different cell processes by a new set of light-controlled tools. Light acts as an orthogonal, non-toxic, and inexpensive modu-

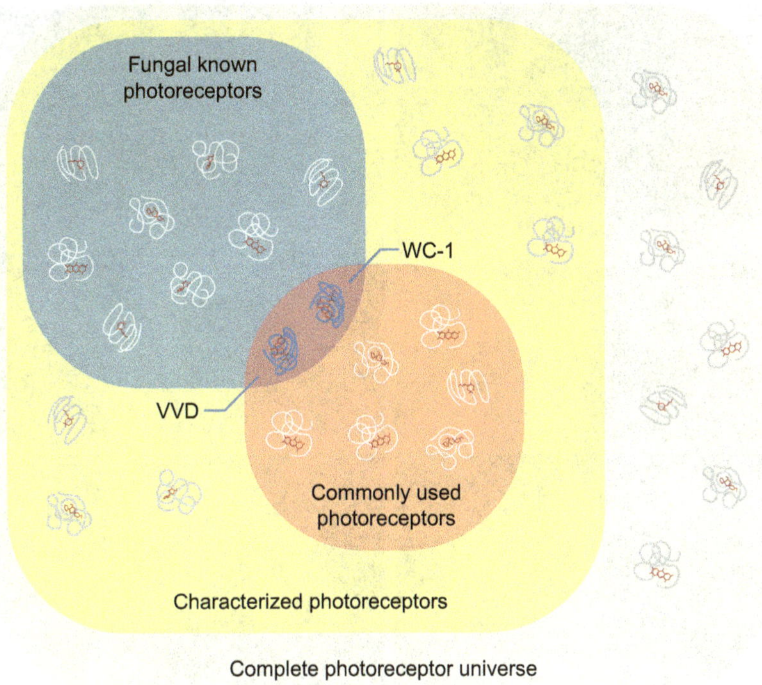

Fig. 5 Photoreceptor universe compared to fungal known and exploited photoreceptors. The diagram shows that despite the vastly known photoreceptors present in fungi, only a few ones are used, specifically the LOV-containing VVD and WC-1 proteins from *Neurospora crassa*, in the implementation of optogenetic switches. Thus, there is a great opportunity to expand the existing optogenetic toolkit, developing novel switches with interesting properties

lator of biological phenomena that is easily delivered by external illumination technologies. In this way, light-controlled approaches improve spatiotemporal resolution in relation to traditional chemical inducers. However, light has some issues that must be solved to establish it as the main modulator of biological processes in scientific investigation. For instance, phototoxicity can be a relevant point that depends on light intensity and duration of its exposure. In addition, the penetrance of light is crucial when it is applied in high-density cell cultures or opaque tissues.

Considering the great advantages and little disadvantages of light as inducer, a broad number of optogenetic tools have been developed in several biological chassis and responding to different wavelengths. Based on natural and engineered photoreceptors, it is now possible to study and manipulate different cell processes with great precision. To date, over 500 articles/reviews related to optogenetics research have been published, where 318 use mammalian hosts, while only 56 cover fungi. In that context, the number of articles mentioning LOV domains reaches 128, in which there are just 28 publications related to VVD and/or WC-1 (https://www.optobase.org). Considering the kingdom mycota comprises some 100,000 described species, with a diversity estimated at 0.8 million to 5.1 million species (Blackwell 2011), this poses an open invitation to the scientific community to consider the rather unexplored photobiology of fungi in the development of new optogenetic devices to address pending biological questions. In that context, we expect that advances in optics technology and discovery of light-sensitive domains with novel biochemical properties can enrich the optogenetic toolkit and the applications of this field in basic and applied research.

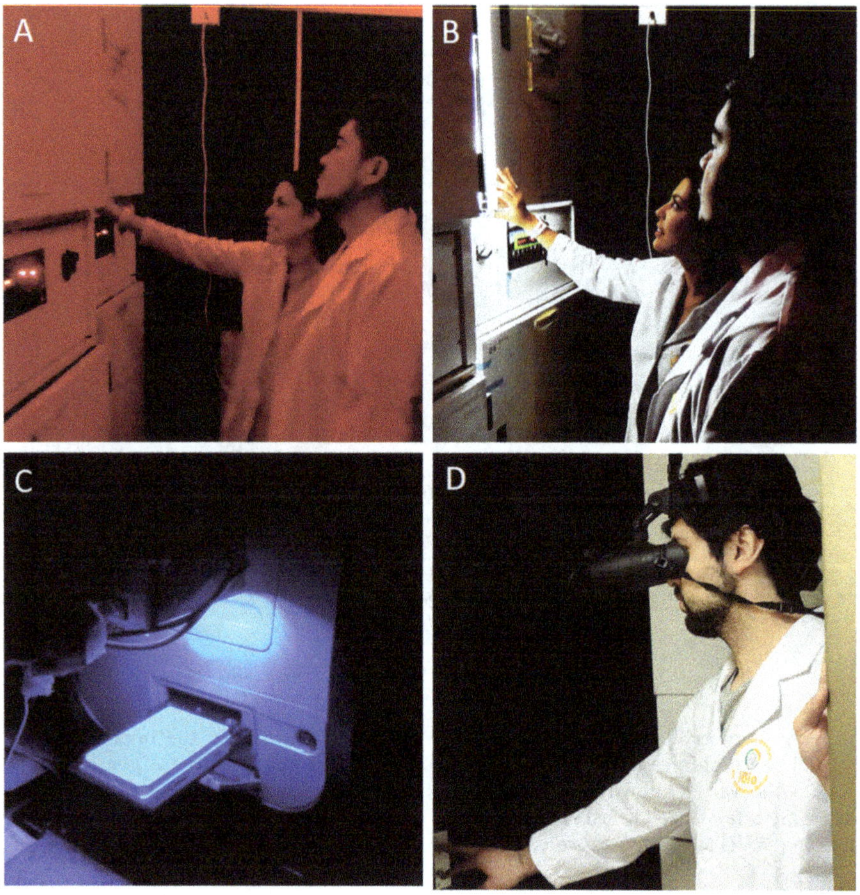

Fig. 6 Conducting optogenetic experiments in a darkroom. (**a**) When utilizing a darkroom one can control full darkness or provide stimulation with defined light intensities and wavelengths. In experiments utilizing *N. crassa* or yeast with LOV optogenetic switches, safety red lights can be used as "darkness" to manipulate these organisms, as they don't respond to such wavelengths. In order to induce responses, one can utilize incubators implemented with white light (**b**), or incubators (or plate-reader systems) where cultures are exposed to blue light for a defined amount of time (**c**). In cases where handling samples in complete darkness is required (avoiding even safety red lights) night vision goggles can be utilized (**d**)

Acknowledgments Research in our labs is funded by iBio Iniciativa Cientifica Milenio-MINECON, CONICYT/FONDEQUIP EQM130158, CONICYT/FONDECYT 1171151, 11170158, and the International Research Scholar program of the Howard Hughes Medical Institute.

References

An-Adirekkun JM, Stewart CJ, Geller SH et al (2019) A yeast optogenetic toolkit (yOTK) for gene expression control in *Saccharomyces cerevisiae*. Biotechnol Bioeng 117(3):886–893. https://doi.org/10.1002/bit.27234

Andrianantoandro E, Basu S, Karig DK, Weiss R (2006) Synthetic biology: new engineering rules for an emerging discipline. Mol Syst Biol 2:2006.0028

Banerjee R, Batschauer A (2004) Plant blue-light receptors. Planta 220:498–502. https://doi.org/10.1007/s00425-004-1418-z

Bashor CJ, Horwitz AA, Peisajovich SG, Lim WA (2010) Rewiring cells: synthetic biology as a tool to interrogate the organizational principles of living systems. Annu Rev Biophys 39:515–537

Baumschlager A, Aoki SK, Khammash M (2017) Dynamic blue light-inducible T7 RNA polymerases (Opto-T7RNAPs) for precise spatiotemporal gene expression control. ACS Synth Biol 6:2157–2167

Bayram O, Krappmann S, Seiler S et al (2008) *Neurospora crassa ve-1* affects asexual conidiation. Fungal Genet Biol 45:127–138

Bayram O, Braus GH, Fischer R, Rodriguez-Romero J (2010) Spotlight on *Aspergillus nidulans* photosensory systems. Fungal Genet Biol 47:900–908

Beiert T, Bruegmann T, Sasse P (2014) Optogenetic activation of Gq signalling modulates pacemaker activity of cardiomyocytes. Cardiovasc Res 102:507–516

Benner SA, Sismour AM (2005) Synthetic biology. Nat Rev Genet 6:533–543

Bieszke JA, Braun EL, Bean LE et al (1999a) The *nop-1* gene of *Neurospora crassa* encodes a seven transmembrane helix retinal-binding protein homologous to archaeal rhodopsins. Proc Natl Acad Sci U S A 96:8034–8039

Bieszke JA, Spudich EN, Scott KL et al (1999b) A eukaryotic protein, NOP-1, binds retinal to form an archaeal rhodopsin-like photochemically reactive pigment. Biochemistry 38:14138–14145

Blackwell M (2011) The Fungi: 1 2, 3 … 5.1 million species? Am J Bot 98:426–438. https://doi.org/10.3732/ajb.1000298

Blumenstein A, Vienken K, Tasler R et al (2005) The *Aspergillus nidulans* phytochrome FphA represses sexual development in red light. Curr Biol 15:1833–1838

Boyden ES, Zhang F, Bamberg E et al (2005) Millisecond-timescale, genetically targeted optical control of neural activity. Nat Neurosci 8:1263–1268

Brenker K, Osthof K, Yang J, Reth M (2016) LED thermo flow—combining optogenetics with flow cytometry. J Vis Exp. https://doi.org/10.3791/54707

Briggs WR (2014) Phototropism: some history some puzzles, and a look ahead. Plant Physiol 164:13–23. https://doi.org/10.1104/pp.113.230573

Bruegmann T, Malan D, Hesse M et al (2010) Optogenetic control of heart muscle in vitro and in vivo. Nat Methods 7:897–900

Bugaj LJ, Choksi AT, Mesuda CK et al (2013) Optogenetic protein clustering and signaling activation in mammalian cells. Nat Methods 10:249–252

Bugaj LJ, Spelke DP, Mesuda CK et al (2015) Regulation of endogenous transmembrane receptors through optogenetic Cry2 clustering. Nat Commun 6:6898

Canessa P, Schumacher J, Hevia MA et al (2013) Assessing the effects of light on differentiation and virulence of the plant pathogen *Botrytis cinerea*: characterization of the White Collar Complex. PLoS One 8:e84223

Carafoli E, Krebs J (2016) Why calcium? How calcium became the best communicator. J Biol Chem 291:20849–20857

Castellanos F, Schmoll M, Martínez P et al (2010) Crucial factors of the light perception machinery and their impact on growth and cellulase gene transcription in *Trichoderma reesei*. Fungal Genet Biol 47:468–476

Cesbron F, Brunner M, Diernfellner AC (2013) Light-dependent and circadian transcription dynamics in vivo recorded with a destabilized luciferase reporter in Neurospora. PLoS One 8:e83660

Chen CH, Ringelberg CS, Gross RH et al (2009) Genome-wide analysis of light-inducible responses reveals hierarchical light signalling in Neurospora. EMBO J 28:1029–1042

Chen CH, DeMay BS, Gladfelter AS et al (2010) Physical interaction between VIVID and white collar complex regulates photoadaptation in Neurospora. Proc Natl Acad Sci U S A 107:16715–16720

Cheng P, Yang Y, Wang L et al (2003) WHITE COLLAR-1, a multifunctional neurospora protein involved in the circadian feedback loops, light sensing, and transcription repression of wc-2. J Biol Chem 278:3801–3808

Christie JM, Salomon M, Nozue K et al (1999) LOV (light, oxygen, or voltage) domains of the blue-light photoreceptor phototropin (nph1): binding sites for the chromophore flavin mononucleotide. Proc Natl Acad Sci U S A 96:8779–8783

Christie JM, Arvai AS, Baxter KJ et al (2012) Plant UVR8 photoreceptor senses UV-B by tryptophan-mediated disruption of cross-dimer salt bridges. Science 335:1492–1496

Corrochano LM (2007) Fungal photoreceptors: sensory molecules for fungal development and behaviour. Photochem Photobiol Sci 6:725–736

Corrochano LM (2019) Light in the fungal world: from photoreception to gene transcription and beyond. Annu Rev Genet 53:149–170

Corrochano LM, Garre V (2010) Photobiology in the Zygomycota: multiple photoreceptor genes for complex responses to light. Fungal Genet Biol 47:893–899

Dasgupta A, Chen CH, Lee C et al (2015) Biological significance of photoreceptor photocycle length: VIVID photocycle governs the dynamic VIVID-white collar complex pool mediating photoadaptation and response to changes in light intensity. PLoS Genet 11:e1005215

Dean R, Van KJA, Pretorius ZA et al (2012) The Top 10 fungal pathogens in molecular plant pathology. Mol Plant Pathol 13:414–430

Deisseroth K (2011) Optogenetics. Nat Methods 8:26–29

Drepper T, Krauss U, Meyer zu BS et al (2011) Lights on and action! Controlling microbial gene expression by light. Appl Microbiol Biotechnol 90:23–40

Essen LO (2006) Photolyases and cryptochromes: common mechanisms of DNA repair and light-driven signaling? Curr Opin Struct Biol 16:51–59

Fairchild CD, Quail PH (1998) The phytochromes: photosensory perception and signal transduction. Symp Soc Exp Biol 51:85–92

Favory JJ, Stec A, Gruber H et al (2009) Interaction of COP1 and UVR8 regulates UV-B-induced photomorphogenesis and stress acclimation in Arabidopsis. EMBO J 28:591–601

Foley BJ, Stutts H, Schmitt SL et al (2018) Characterization of a vivid homolog in *Botrytis cinerea*. Photochem Photobiol 94:985–993

Froehlich AC, Liu Y, Loros JJ, Dunlap JC (2002) White Collar-1, a circadian blue light photoreceptor, binding to the frequency promoter. Science 297:815–819

Froehlich AC, Chen CH, Belden WJ et al (2010) Genetic and molecular characterization of a cryptochrome from the filamentous fungus *Neurospora crassa*. Eukaryot Cell 9:738–750

Fuller KK, Dunlap JC, Loros JJ (2016) Fungal light sensing at the bench and beyond. Adv Genet 96:1–51

Fuller KK, Dunlap JC, Loros JJ (2018) Light-regulated promoters for tunable, temporal, and affordable control of fungal gene expression. Appl Microbiol Biotechnol 102:3849–3863

Galagan JE, Calvo SE, Borkovich KA et al (2003) The genome sequence of the filamentous fungus *Neurospora crassa*. Nature 422:859–868

Gautier A, Gauron C, Volovitch M et al (2014) How to control proteins with light in living systems. Nat Chem Biol 10:533–541

Gerhardt KP, Olson EJ, Castillo-Hair SM et al (2016) An open-hardware platform for optogenetics and photobiology. Sci Rep 6:35363

Gin E, Diernfellner AC, Brunner M, Höfer T (2013) The Neurospora photoreceptor VIVID exerts negative and positive control on light sensing to achieve adaptation. Mol Syst Biol 9:667

Glantz ST, Carpenter EJ, Melkonian M et al (2016) Functional and topological diversity of LOV domain photoreceptors. Proc Natl Acad Sci U S A 113:E1442–E1451

Glantz ST, Berlew EE, Jaber Z et al (2018) Directly light-regulated binding of RGS-LOV photoreceptors to anionic membrane phospholipids. Proc Natl Acad Sci U S A 115:E7720–E7727

Goffeau A, Barrell BG, Bussey H et al (1996) Life with 6000 genes. Science 274(546):563–567

Guinn MT, Balázsi G (2019) Noise-reducing optogenetic negative-feedback gene circuits in human cells. Nucleic Acids Res 47:7703–7714

Gyalai-Korpos M, Nagy G, Mareczky Z et al (2010) Relevance of the light signaling machinery for cellulase expression in *Trichoderma reesei* (*Hypocrea jecorina*). BMC Res Notes 3:330

Han X, Boyden ES (2007) Multiple-color optical activation, silencing, and desynchronization of neural activity, with single-spike temporal resolution. PLoS One 2:e299

Harper SM, Neil LC, Gardner KH (2003) Structural basis of a phototropin light switch. Science 301:1541–1544

He Q, Liu Y (2005) Molecular mechanism of light responses in Neurospora: from light-induced transcription to photoadaptation. Genes Dev 19:2888–2899

He Q, Cheng P, Yang Y et al (2002) White collar-1, a DNA binding transcription factor and a light sensor. Science 297:840–843

Herrera-Estrella A, Horwitz BA (2007) Looking through the eyes of fungi: molecular genetics of photoreception. Mol Microbiol 64:5–15

Hevia MA, Canessa P, Müller-Esparza H, Larrondo LF (2015) A circadian oscillator in the fungus *Botrytis cinerea* regulates virulence when infecting *Arabidopsis thaliana*. Proc Natl Acad Sci U S A 112:8744–8749

Hughes J, Lamparter T, Mittmann F et al (1997) A prokaryotic phytochrome. Nature 386:663

Hughes RM, Bolger S, Tapadia H, Tucker CL (2012) Light-mediated control of DNA transcription in yeast. Methods 58:385–391

Hunt SM, Thompson S, Elvin M, Heintzen C (2010) VIVID interacts with the WHITE COLLAR complex and FREQUENCY-interacting RNA helicase to alter light and clock responses in Neurospora. Proc Natl Acad Sci U S A 107:16709–16714

Hurley JM, Chen CH, Loros JJ, Dunlap JC (2012) Light-inducible system for tunable protein expression in *Neurospora crassa*. G3 (Bethesda) 2:1207–1212

Idnurm A, Heitman J (2005) Photosensing fungi: phytochrome in the spotlight. Curr Biol 15:R829–R832

Idnurm A, Verma S, Corrochano LM (2010) A glimpse into the basis of vision in the kingdom Mycota. Fungal Genet Biol 47:881–892

Jenkins GI (2017) Photomorphogenic responses to ultraviolet-B light. Plant Cell Environ 40:2544–2557

Kaberniuk AA, Shemetov AA, Verkhusha VV (2016) A bacterial phytochrome-based optogenetic system controllable with near-infrared light. Nat Methods 13:591–597

Kawano F, Suzuki H, Furuya A, Sato M (2015) Engineered pairs of distinct photoswitches for optogenetic control of cellular proteins. Nat Commun 6:6256

Kawano F, Okazaki R, Yazawa M, Sato M (2016) A photoactivatable Cre-loxP recombination system for optogenetic genome engineering. Nat Chem Biol 12:1059–1064

Kennedy MJ, Hughes RM, Peteya LA et al (2010) Rapid blue-light-mediated induction of protein interactions in living cells. Nat Methods 7:973–975

Kojadinovic M, Laugraud A, Vuillet L et al (2008) Dual role for a bacteriophytochrome in the bioenergetic control of *Rhodopseudomonas palustris*: enhancement of photosystem synthesis and limitation of respiration. Biochim Biophys Acta 1777:163–172

Kolar K, Knobloch C, Stork H et al (2018) OptoBase: a web platform for molecular optogenetics. ACS Synth Biol 7:1825–1828

Lee S, Park H, Kyung T et al (2014) Reversible protein inactivation by optogenetic trapping in cells. Nat Methods 11:633–636

Linden H, Macino G (1997) White collar 2, a partner in blue-light signal transduction, controlling expression of light-regulated genes in *Neurospora crassa*. EMBO J 16:98–109

Liu Q, Tucker CL (2017) Engineering genetically-encoded tools for optogenetic control of protein activity. Curr Opin Chem Biol 40:17–23

Liu H, Yu X, Li K et al (2008) Photoexcited CRY2 interacts with CIB1 to regulate transcription and floral initiation in Arabidopsis. Science 322:1535–1539

Lorrain S, Genoud T, Fankhauser C (2006) Let there be light in the nucleus! Curr Opin Plant Biol 9:509–514

Ma Z, Du Z, Chen X et al (2013) Fine tuning the LightOn light-switchable transgene expression system. Biochem Biophys Res Commun 440:419–423

Malzahn E, Ciprianidis S, Káldi K et al (2010) Photoadaptation in Neurospora by competitive interaction of activating and inhibitory LOV domains. Cell 142:762–772

Mao D, Li N, Xiong Z et al (2019) Single-cell optogenetic control of calcium signaling with a high-density micro-LED array. iScience 21:403–412

Martinez D, Berka RM, Henrissat B et al (2008) Genome sequencing and analysis of the biomass-degrading fungus *Trichoderma reesei* (syn. *Hypocrea jecorina*). Nat Biotechnol 26:553–560

McIsaac RS, Oakes BL, Wang X et al (2013) Synthetic gene expression perturbation systems with rapid, tunable, single-gene specificity in yeast. Nucleic Acids Res 41:e57

McLellan MA, Rosenthal NA, Pinto AR (2017) Cre-loxP-mediated recombination: general principles and experimental considerations. Curr Protoc Mouse Biol 7:1–12

Melyan Z, Tarttelin EE, Bellingham J et al (2005) Addition of human melanopsin renders mammalian cells photoresponsive. Nature 433:741–745

Möglich A, Moffat K (2010) Engineered photoreceptors as novel optogenetic tools. Photochem Photobiol Sci 9:1286–1300

Motta-Mena LB, Reade A, Mallory MJ et al (2014) An optogenetic gene expression system with rapid activation and deactivation kinetics. Nat Chem Biol 10:196–202

Mühlhäuser WW, Fischer A, Weber W, Radziwill G (2017) Optogenetics—bringing light into the darkness of mammalian signal transduction. Biochim Biophys Acta Mol Cell Res 1864:280–292

Müller K, Engesser R, Metzger S et al (2013a) A red/far-red light-responsive bi-stable toggle switch to control gene expression in mammalian cells. Nucleic Acids Res 41:e77

Müller K, Engesser R, Timmer J et al (2013b) Synthesis of phycocyanobilin in mammalian cells. Chem Commun (Camb) 49:8970–8972

Müller K, Zurbriggen MD, Weber W (2014) Control of gene expression using a red- and far-red light-responsive bi-stable toggle switch. Nat Protoc 9:622–632

Nagel G, Szellas T, Huhn W et al (2003) Channelrhodopsin-2, a directly light-gated cation-selective membrane channel. Proc Natl Acad Sci U S A 100:13940–13945

Navara KJ, Nelson RJ (2007) The dark side of light at night: physiological epidemiological, and ecological consequences. J Pineal Res 43:215–224. https://doi.org/10.1111/j.1600-079x.2007.00473.x

Nihongaki Y, Kawano F, Nakajima T, Sato M (2015a) Photoactivatable CRISPR-Cas9 for optogenetic genome editing. Nat Biotechnol 33:755–760

Nihongaki Y, Yamamoto S, Kawano F et al (2015b) CRISPR-Cas9-based photoactivatable transcription system. Chem Biol 22:169–174

Olmedo M, Ruger-Herreros C, Luque EM, Corrochano LM (2010) A complex photoreceptor system mediates the regulation by light of the conidiation genes con-10 and con-6 in *Neurospora crassa*. Fungal Genet Biol 47:352–363

Oravecz A, Baumann A, Máté Z et al (2006) CONSTITUTIVELY PHOTOMORPHOGENIC1 is required for the UV-B response in Arabidopsis. Plant Cell 18:1975–1990

Pathak GP, Strickland D, Vrana JD, Tucker CL (2014) Benchmarking of optical dimerizer systems. ACS Synth Biol 3:832–838

Polstein LR, Gersbach CA (2015) A light-inducible CRISPR-Cas9 system for control of endogenous gene activation. Nat Chem Biol 11:198–200

Pudasaini A, El-Arab KK, Zoltowski BD (2015) LOV-based optogenetic devices: light-driven modules to impart photoregulated control of cellular signaling. Front Mol Biosci 2:18

Purnick PE, Weiss R (2009) The second wave of synthetic biology: from modules to systems. Nat Rev Mol Cell Biol 10:410–422

Purschwitz J, Müller S, Kastner C et al (2008) Functional and physical interaction of blue- and red-light sensors in *Aspergillus nidulans*. Curr Biol 18:255–259

Quail PH (2010) Phytochromes. Curr Biol 20:R504–R507

Renicke C, Schuster D, Usherenko S et al (2013) A LOV2 domain-based optogenetic tool to control protein degradation and cellular function. Chem Biol 20:619–626

Rizzini L, Favory JJ, Cloix C et al (2011) Perception of UV-B by the Arabidopsis UVR8 protein. Science 332:103–106

Robertson JB, Davis CR, Johnson CH (2013) Visible light alters yeast metabolic rhythms by inhibiting respiration. Proc Natl Acad Sci U S A 110:21130–21135

Rullan M, Benzinger D, Schmidt GW et al (2018) An optogenetic platform for real-time, single-cell interrogation of stochastic transcriptional regulation. Mol Cell 70:745–756.e6

Salichos L, Rokas A (2010) The diversity and evolution of circadian clock proteins in fungi. Mycologia 102:269–278

Salinas F, Rojas V, Delgado V et al (2017) Optogenetic switches for light-controlled gene expression in yeast. Appl Microbiol Biotechnol 101:2629–2640

Salinas F, Rojas V, Delgado V et al (2018) Fungal light-oxygen-voltage domains for optogenetic control of gene expression and flocculation in yeast. MBio 9: e00626-18

Sawa M, Nusinow DA, Kay SA, Imaizumi T (2007) FKF1 and GIGANTEA complex formation is required for day-length measurement in Arabidopsis. Science 318:261–265

Schmidt D, Cho YK (2015) Natural photoreceptors and their application to synthetic biology. Trends Biotechnol 33:80–91

Schmoll M (2018) Regulation of plant cell wall degradation by light in Trichoderma. Fungal Biol Biotechnol 5:10

Schmoll M, Franchi L, Kubicek CP (2005) Envoy, a PAS/LOV domain protein of *Hypocrea jecorina* (Anamorph *Trichoderma reesei*), modulates cellulase gene transcription in response to light. Eukaryot Cell 4:1998–2007

Schumacher J (2012) Tools for *Botrytis cinerea*: new expression vectors make the gray mold fungus more accessible to cell biology approaches. Fungal Genet Biol 49:483–497

Schumacher J (2017) How light affects the life of Botrytis. Fungal Genet Biol 106:26–41

Schwerdtfeger C, Linden H (2003) VIVID is a flavoprotein and serves as a fungal blue light photoreceptor for photoadaptation. EMBO J 22:4846–4855

Sebille S, Ayad O, Chapotte-Baldacci CA et al (2017) Optogenetic approach for targeted activation of global calcium transients in differentiated C2C12 myotubes. Sci Rep 7:11108

Seibel C, Tisch D, Kubicek CP, Schmoll M (2012) ENVOY is a major determinant in regulation of sexual development in *Hypocrea jecorina* (*Trichoderma reesei*). Eukaryot Cell 11:885–895

Shcherbakova DM, Shemetov AA, Kaberniuk AA, Verkhusha VV (2015) Natural photoreceptors as a source of fluorescent proteins, biosensors, and optogenetic tools. Annu Rev Biochem 84:519–550

Shimizu-Sato S, Huq E, Tepperman JM, Quail PH (2002) A light-switchable gene promoter system. Nat Biotechnol 20:1041–1044

Sorokina O, Kapus A, Terecskei K et al (2009) A switchable light-input, light-output system modelled and constructed in yeast. J Biol Eng 3:15

Strickland D, Lin Y, Wagner E et al (2012) TULIPs: tunable, light-controlled interacting protein tags for cell biology. Nat Methods 9:379–384

Swartz TE, Corchnoy SB, Christie JM et al (2001) The photocycle of a flavin-binding domain of the blue light photoreceptor phototropin. J Biol Chem 276:36493–36500

Taslimi A, Vrana JD, Chen D et al (2014) An optimized optogenetic clustering tool for probing protein interaction and function. Nat Commun 5:4925

Verma R, Annan RS, Huddleston MJ et al (1997) Phosphorylation of Sic1p by G1 Cdk required for its degradation and entry into S phase. Science 278:455–460

Wang X, Chen X, Yang Y (2012) Spatiotemporal control of gene expression by a light-switchable transgene system. Nat Methods 9:266–269

Wang W, Shi XY, Wei DZ (2014) Light-mediated control of gene expression in filamentous fungus *Trichoderma reesei*. J Microbiol Methods 103:37–39

Weissleder R, Ntziachristos V (2003) Shedding light onto live molecular targets. Nat Med 9:123–128

Yang X, Lau KY, Sevim V, Tang C (2013) Design principles of the yeast G1/S switch. PLoS Biol 11:e1001673

Yazawa M, Sadaghiani AM, Hsueh B, Dolmetsch RE (2009) Induction of protein-protein interactions in live cells using light. Nat Biotechnol 27:941–945

Yin R, Ulm R (2017) How plants cope with UV-B: from perception to response. Curr Opin Plant Biol 37:42–48

Yin C, Fan X, Ma K et al (2019) Identification and characterization of a novel light-induced promoter for recombinant protein production in *Pleurotus ostreatus*. J Microbiol 58(1):39–45

Zhang K, Cui B (2015) Optogenetic control of intracellular signaling pathways. Trends Biotechnol 33:92–100

Zhang G, Liu P, Wei W et al (2016a) A light-switchable bidirectional expression system in filamentous fungus *Trichoderma reesei*. J Biotechnol 240:85–93

Zhang L, Zhao X, Zhang G et al (2016b) Light-inducible genetic engineering and control of non-homologous end-joining in industrial eukaryotic microorganisms: LML 3.0 and OFN 1.0. Sci Rep 6:20761

Zhao EM, Zhang Y, Mehl J et al (2018) Optogenetic regulation of engineered cellular metabolism for microbial chemical production. Nature 555:683–687

Zoltowski BD, Motta-Mena LB, Gardner KH (2013) Blue light-induced dimerization of a bacterial LOV-HTH DNA-binding protein. Biochemistry 52:6653–6661

13 Yeast Cell Factories

Bernhard Schmelzer[1,2], Martin Altvater[1,2], Brigitte Gasser[1,2,3], Michael Sauer[1,2,4], Diethard Mattanovich[1,2]

CONTENTS

I. Introduction

Yeasts are unicellular fungi which have evolved multiple times mainly in the phylum Ascomycota and more rarely among Basidiomycota. While most yeasts of both phyla reproduce asexually by asymmetric cell division or budding, the yeasts of the ascomycete subphylum Saccharomycotina are often referred to as the budding yeasts (Shen et al. 2018). Among them, *Saccharomyces cerevisiae* occupies an exceptional role as the species that is most intensely used by humankind and best studied among all yeasts. In literature it is often simply named "yeast" or "budding yeast," while the plural form "yeasts" encompasses all yeast species. To avoid confusion, we refer here to "yeasts" when we mean any yeast species and to "Baker's yeast" or "*S. cerevisiae*" when this species is addressed.

Yeasts have been employed for **food processing** and **production of alcoholic beverages** since millennia, marking an early inventive step of biotechnology even long before the biological nature behind was known. With the industrialization of baking and brewing in the nineteenth century, the prototypes of modern biotechnological processes were developed, among them the "Zulaufverfahren" for efficient production of baker's yeast (International Yeast Co. Ltd. 1933). Termed "Fed-batch process" today, this process type of substrate limited feed regime is used to avoid overflow metabolism in most industrial bioprocesses. It was found soon thereafter that yeast is able to overproduce other metabolites besides ethanol, such as glycerol (Klein et al. 2017). These early

[1] Department of Biotechnology, University of Natural Resources and Life Sciences Vienna, Vienna, Austria; e-mail: Bernhard.Schmelzer@boku.ac.at; Martin.Altvater@boku.ac.at; Brigitte.Gasser@boku.ac.at; Michael.Sauer@boku.ac.at; Diethard.Mattanovich@boku.ac.at
[2] Austrian Centre of Industrial Biotechnology (ACIB GmbH), Vienna, Austria
[3] CD Laboratory for Growth-Decoupled Protein Production in Yeast, University of Natural Resources and Life Sciences, Vienna, Austria
[4] CD Laboratory for Biotechnology of Glycerol, University of Natural Resources and Life Sciences, Vienna, Austria

Genetics and Biotechnology, 3rd Edition
The Mycota II
J.P. Benz, K. Schipper (Eds.)

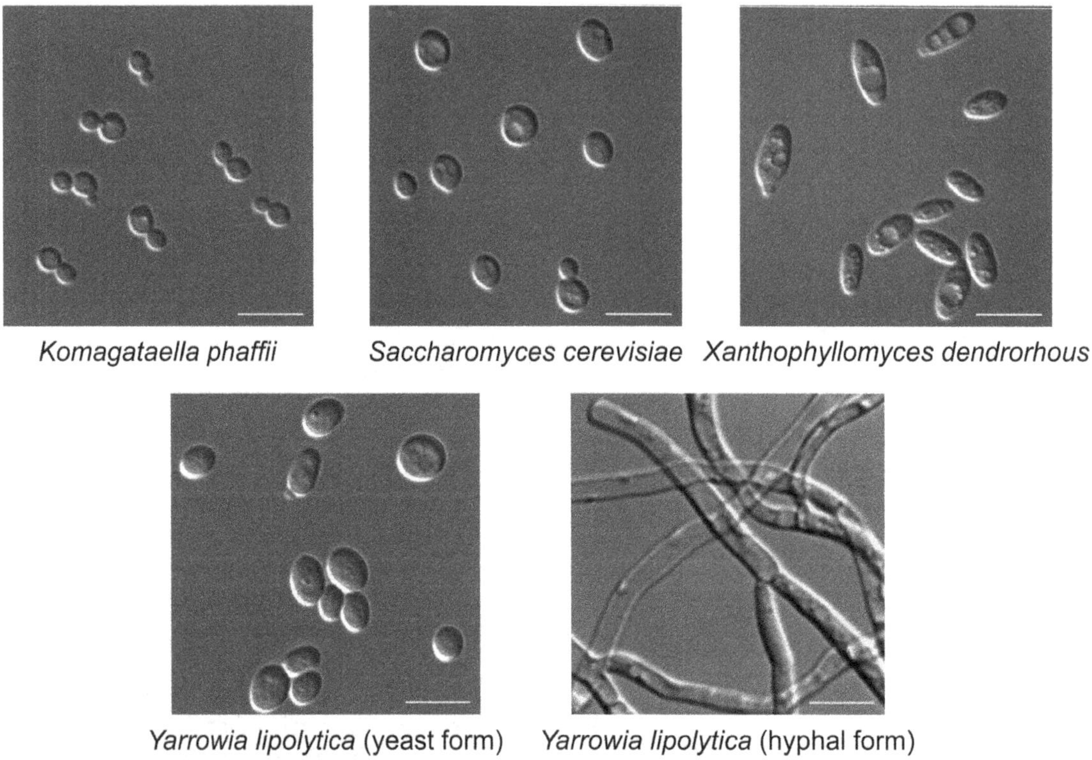

Fig. 1 Differential interference contrast micrographs of selected biotechnologically relevant yeasts. Scale bar: 10 μm. Budding yeast species occur in different sizes and shapes, and some species may even adopt pseudohyphal morphology in certain conditions, as shown for the case of *Yarrowia lipolytica*. Images by Lina Heistinger (BOKU)

developments of yeast biotechnology, using *Saccharomyces cerevisiae* and its sibling species, laid the basis for a broad development of bioproductions based on a range of species, mostly of ascomycete yeasts.

Besides its value for food industry, ***S. cerevisiae* is still the dominant yeast for biotechnology applications.** This is partly due to its outstanding ability to produce ethanol, applied to manufacture biofuel from corn and sugar cane, and partly based on its traditional use as a platform for metabolic and cell engineering, e.g., as a host for production of recombinant proteins or secondary metabolites. However, an increasing variety of ascomycete and basidiomycete yeasts are nowadays also employed for biotechnological production, such as *Yarrowia lipolytica*, *Kluyveromyces lactis*, *Pichia pastoris* (syn *Komagataella* spp.), and *Xanthophyllomyces dendrorhous*, to name a few prominent examples. Although yeasts are defined as unicellular fungi, they vary significantly in their size and morphology and may even attain pseudohyphal structures (Fig. 1).

Among the ascomycete yeasts, budding yeasts have evolved from a postulated common ancestor about 400 million years ago (mya), with a large expansion of metabolic diversity from approximately 150 mya onward. Evolutionary sequence diversity of budding yeasts equals that of plants and animals (Fig. 2, Shen et al. 2018). This huge diversity is reflected in a **broad variety of physiological traits**, providing a largely untapped resource of metabolic pathways and resistances to specific process conditions, waiting to be utilized in biotechnology.

S. cerevisiae was the first eukaryotic organism having its genome sequenced (Goffeau et al. 1996). This milestone enabled efficient **metabolic engineering** of *S. cerevisiae* for pro-

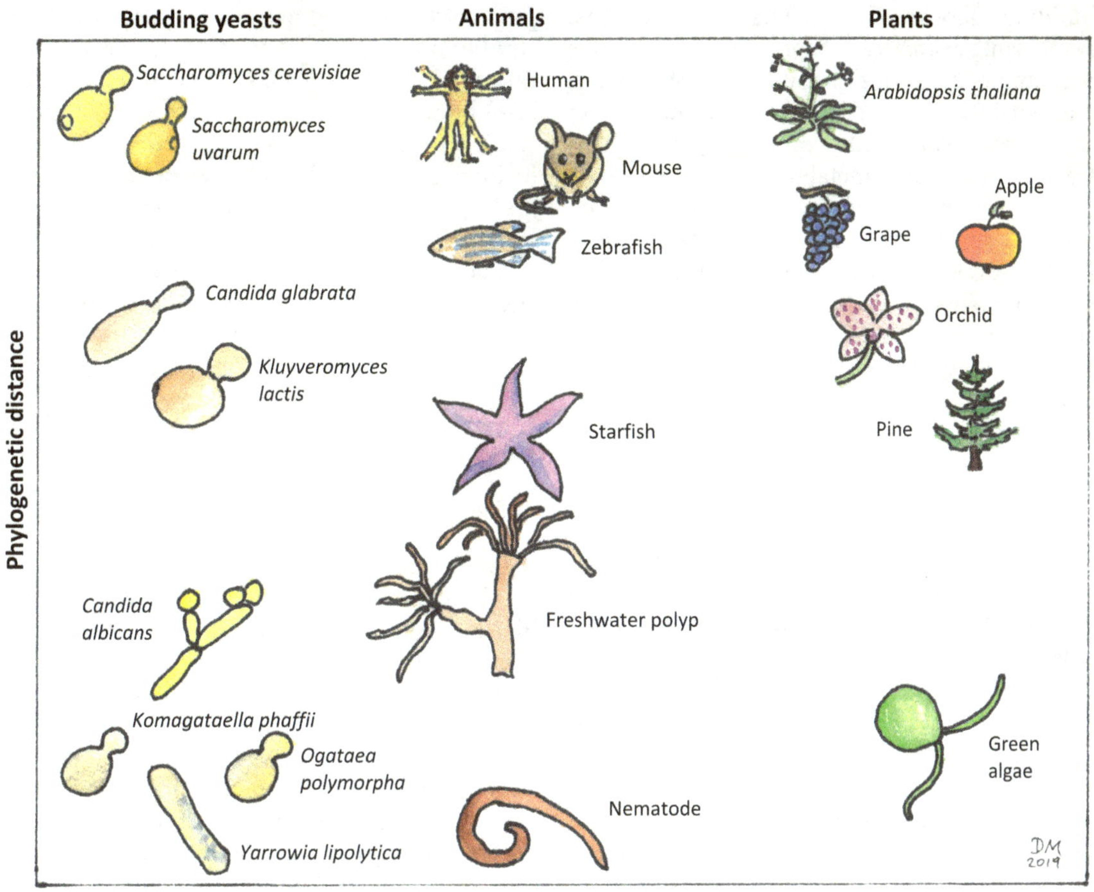

Fig. 2 Phylogenetic distance of the budding yeasts compared to animals and plants. The distances, estimated by genome sequence diversities of yeast species discussed in this chapter together with some pathogenic yeasts, are illustrated based on data of Shen et al. (2018). The concept of this image follows illustrations by Shen et al. (2018) and Bernard Dujon (unpublished)

duction of a plethora of metabolites, and it set the ground for the development of **synthetic biology tools** for baker's yeast. While genome sequencing of yet uncharacterized yeast species is easy and rather cheap today, baker's yeast has a lead in the availability of tools over other yeasts and is therefore (still) the preferred eukaryotic platform for synthetic biology applications (Chen et al. 2018; Besada-Lombana et al. 2018).

In the following, we describe the current industrial use of yeasts for production of chemicals and heterologous proteins, with a focus on the major industrial products and current developments that may change bioproduction of building blocks for chemical industry in the near future.

II. Yeasts for Production of Primary Metabolites

Primary metabolites play an important role in the central metabolism and are directly involved in the growth and reproduction of a living cell. Through the pathways of the central carbon metabolism, organisms convert organic nutrients into metabolites that contribute to energy generation for biosynthesis or serve as building blocks for cellular components. Important examples of primary metabolites are amino acids for biosynthesis or ethanol and lactic acid as products of fermentation processes in many different cell types. In contrast to secondary metabolites (see Sect. III) primary metabolites do not have direct pharmacological effects in humans. **Biotechnologically produced primary metabolites rather play important roles in everyday human life,** serving as preservative or flavoring components in food, beverages, cosmetics and pharmaceuticals, as biofuels, cleaning agents, precursor for biopolymers, dyes, etc. Due to its robustness and well-established metabolic engineering tools, yeast is often the preferred production host for primary metabolites in today's industrial biotechnology. Below you find several examples of industrially produced primary metabolites in different kinds of yeast species.

A. Ethanol

With over 80 million tons globally produced in 2017 (Renewable Fuels Association 2017), **ethanol is one of the most important products** in industrial biotechnology. Bioethanol is used as renewable fuel and widely applied as additive in gasoline (e.g., E10: 10% ethanol + 90% gasoline). Currently, most of the industrial ethanol is produced by fermenting substrates such as hydrolyzed corn starch or cane sugar ("first generation" biofuel). Due to its rapid substrate consumption and ethanol fermentation ability, and its robustness against harsh conditions, *Saccharomyces cerevisiae* is ideal for industrial ethanol production (Nielsen et al. 2013). Over the last two decades, a lot of research effort was aimed at using cheap and abundant forestry and agricultural substrates ("second generation" biofuel) with the goal of reducing the carbon footprint of ethanol production and avoiding the ethical dilemma of using farmland for biofuels and not for food (Rude and Schirmer 2009). Such non-food substrates primarily include **lignocellulosic plant biomass derived from agricultural waste** such as straw, fruit pulp, leaves, unused parts of food crops, wood chips, and others. Lignocellulose consists of carbohydrate polymers (cellulose, hemicellulose) bound to cross-linked phenolic polymers called lignin. By a series of pre-treatment steps including heat, acid, and enzymatic methods, the sugar monomers trapped inside the lignocellulose are extracted and used as substrate in ethanol fermentations. Rational metabolic engineering and adaptive laboratory evolution strategies further improved the performance of *S. cerevisiae* in bioethanol production (Jansen et al. 2017).

B. Butanol

Butanol isomers (1-butanol, 2-butanol, isobutanol, 2-methyl-2-propanol) are very **attractive biofuel alternatives** with several advantages over ethanol, which is currently the most important biofuel. These advantages include higher energy density, better blending ability, lower moisture adsorption, and lower corrosiveness (Dürre 2007). Furthermore, butanol is an important chemical building block for paints, coatings, plastic polymers, lubricants, fragrances, etc. In the first half of the twentieth century, butanol was mainly produced by the acetone-butanol-ethanol (ABE) fermentation of *Clostridia* species. However, high purification and raw material costs, growth inhibition by the products, low butanol recovery yield, and the availability of cheap oil led the butanol production away from the bioprocess to the more economical petrochemical industry. In the recent decades, modern metabolic engineering and advanced fermentation technologies helped to revive the fermentation route. High butanol production was achieved by metabolic engineering and fermentation of *Escherichia coli* (Dellomonaco et al. 2011; Shen et al. 2011). Due to its butanol tolerance and well-established genetic tools, *S. cerevisiae* shows great potential as butanol production host. By extensive metabolic engineering of *S. cerevisiae*, butanol titers and yields could be improved significantly (Generoso et al. 2015; Buijs et al. 2013). Further research and process optimization are on the way to make budding yeast the ideal microbial cell factory for the production of biobutanol.

C. Lactic Acid

Lactic acid is a versatile organic acid used in food, cosmetic, pharmaceutical, and plastic industries, in the latter serving as **building block for polylactic acid (PLA), a biodegradable and biocompatible polymer**. The increasing global lactic acid market is projected to reach 1.6 million tons in 2024 with a revenue of US$3.7 billion (Global Industry Analysts, Inc. 2017). Currently, lactic acid bacteria and different yeast genera are most widely used for the industrial production of lactic acid, although, since the advent of modern metabolic engineering tools, yeast-based production processes became increasingly important (Sauer et al. 2010). Besides Baker's yeast, also other yeast species are currently being developed as host for lactic acid production. While many bacteria require media supplementation with peptides, amino acids, nucleotides, or vitamins, most yeasts can grow on simple mineral medium (maybe with the addition of few vitamins). Therefore, the nutrient and downstream purification costs are significantly lower as compared to processes using bacteria. Moreover, yeast cells are more tolerant to low pH reducing the addition of neutralizing agents during the fermentation process. In addition, the low pH of the process greatly decreases the risk of bacterial contaminations or bacteriophage infections. However, yeast cells do not produce lactic acid naturally; thus, metabolic engineering is required. Converting pyruvate to lactic acid can be achieved by introduction of a lactate dehydrogenase (e.g., from lactic acid bacteria). To avoid any carbon loss, the competing ethanol fermentation pathway must be shut down, e.g., by deleting the genomic pyruvate decarboxylase genes. Furthermore, adaptive laboratory evolution, random mutagenesis, and process engineering are applied in academic and industrial biotechnology to further optimize lactic acid production in yeast (Miller et al. 2011; Adachi et al. 1998; Baek et al. 2016; Rajgarhia et al. 2007; Porro et al. 1999).

D. Citrate

Citric acid is one of the most important organic acids in biotechnology with over 2 million tons produced per year. Around 75% is used as acidity regulator and flavor enhancer in beverages and food products. Furthermore, due to its broad range of applications, it is widely applied in the chemical, cosmetic, pharmaceutical, and agricultural industries (IHS Markit 2015). Naturally, citric acid is found in high concentrations in a variety of fruits and vegetables. Moreover, it is an important metabolic intermediate in the TCA cycle and thus present in all aerobic organisms. Industrial production of citric acid began at the end of the nineteenth century by extraction from concentrated lemon juice. In the early twentieth century, the filamentous fungus *Aspergillus niger* was identified as natural citric acid producer (Currie 1917; Cavallo et al. 2017). From then on, fermentation of *A. niger* was developed and optimized and is, to date, the preferred production route of numerous companies around the world (Ciriminna et al. 2017). To further lower the production costs, research in the last two decades focused on identifying and engineering other citric acid producers, such as the yeast *Yarrowia lipolytica*. This yeast species has many advantages over filamentous fungi: it can metabolize a wide range of substrates and it is resistant to high substrate concentrations and contaminants such as metal ions (Cavallo et al. 2017). One low-cost substrate that can be utilized by *Y. lipolytica* is glycerol, an inexpensive by-product of the prospering biodiesel economy. Recent studies proved that optimizing strain development, fermentation processes and the utilization of waste products as substrates can bring environmental and economical benefits in the near future (Hu et al. 2019; Egermeier et al. 2017; Khanna et al. 2012; Almeida et al. 2012).

E. Succinic Acid

Succinic acid is another very important molecule produced by yeast with a major role as **precursor of many valuable bio-based chemicals**. These include biodegradable polyesters, surfactants, additives in food and agriculture, pharmaceutical products, dyes, etc. Although succinic acid is still produced by petrochemical processes in large part, environmental reasons and advancements in fermentation and metabolic engineering technologies make bio-based succinic acid production routes a competitive and future-oriented alternative (Pinazo et al. 2015). As intermediate in the tricarboxylic acid (TCA) cycle, succinic acid is synthesized in almost all microbes, plants, and animals. The ability of growing at low pH and the availability of genetic information, metabolic engineering, and -omics tools make yeast an important production host. By using rational metabolic engineering or systems biology approaches mainly focused on the TCA cycle and the succinic acid export, strains and fermentation processes were generated proving competitiveness with the current petrochemical production (Ito et al. 2014; Otero et al. 2013; Raab et al. 2010). As an example for strain engineering, a part of the metabolic engineering strategy of DSM/Roquette (former Reverdia) using the succinic acid-tolerant yeast *S. cerevisiae* is outlined in Fig. 3.

III. Yeasts for Production of Secondary Metabolites

Secondary metabolites are usually defined as organic compounds that are not directly connected to the growth of the organism. As such this is a negative definition and so essentially every metabolic product qualifies as secondary metabolite as long as it is not directly associated to growth. However, usually only those metabolites are referred to as secondary metabolites, which are somehow connected to specific functions like protection, competition, or interactions with other living organisms. Often their effect is very specific and their production is restricted to a few species. Unlike primary metabolites, the absence of secondary metabolites does not result in the demise of the organism. The specificity of these molecules in various contexts makes them interesting for many applications, in particular for pharmaceuticals or cosmetics but also as food or feed additives, antibiotics, herbicides or fungicides, and many more. When microbial cells are the source of the desired metabolites, they are often directly useful for mass production of the desired compounds—for example, fungi for penicillin production. However, often the metabolites are plant derived, and large-scale production is very expensive or even impossible. In such cases metabolic engineering of microbial cells offers a promising alternative. Since the pathways often involve unusual enzymes and complicated reactions, yeasts are the preferred host cells for such metabolic engineering endeavors.

A. Carotenoids

Carotenoids are among the few **secondary metabolites** that are **naturally produced by certain yeast species**. They are pigments with antioxidant properties and chemically part of the terpenoid family of compounds, consisting usually of eight isoprene units (C_{40}). Carotenoids are natural colorants with many pharmaceutical functions and nutraceutical applications, which make them very valuable. The most prominent producer of the carotenoid **astaxanthin** is the heterobasidiomycetous yeast *Xanthophyllomyces dendrorhous* (teleomorph of *Phaffia rhodozyma*). This yeast has been isolated in the late 1960s from tree exudates in such distant points as Japan and Alaska (Barredo et al. 2017). Another carotenoid naturally produced by yeast is **beta-carotene**. The most prominent genus accumulating beta-carotene is *Rhodotorula* found in a wide variety of habitats all over the world (Tang et al. 2019).

X. dendrorhous naturally accumulates astaxanthin in high amounts. With distinct strains grams per liter of astaxanthin can be accumulated under appropriate culture conditions. Since this yeast is a natural producer,

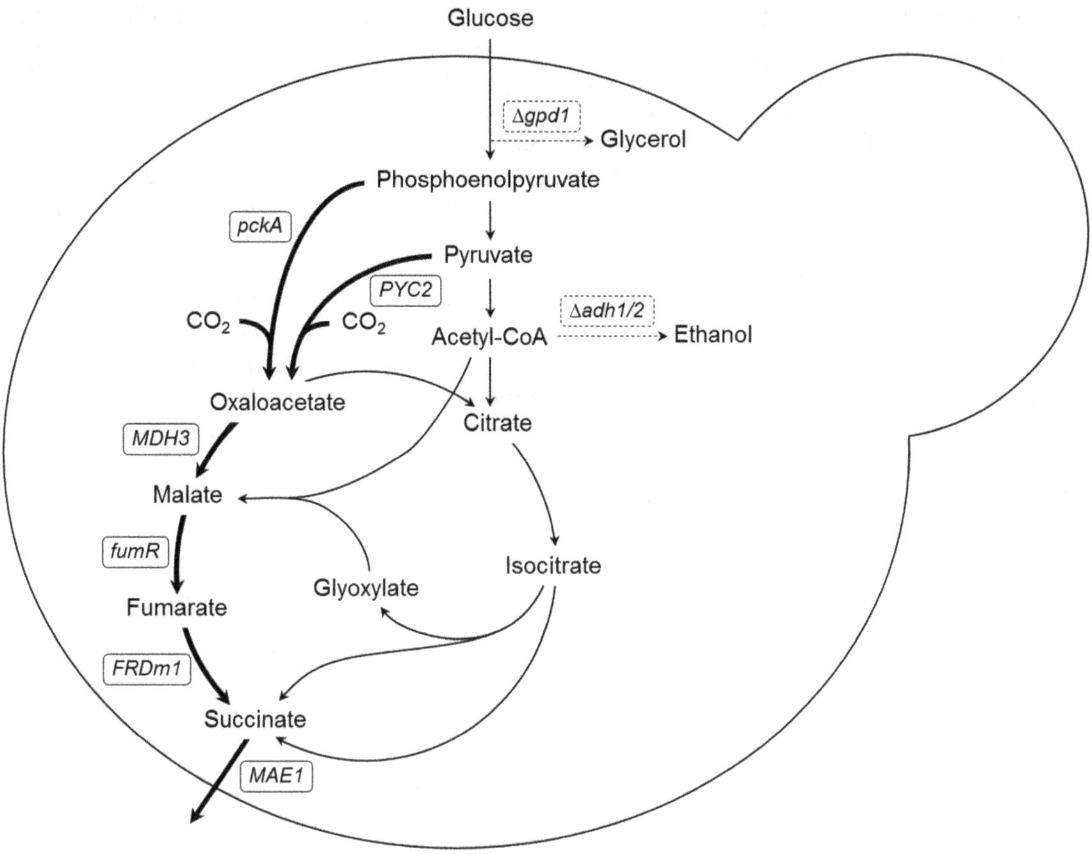

Fig. 3 Metabolic engineering for enhanced succinic acid production in *Saccharomyces cerevisiae*. Succinic acid formation can be achieved through three different branches: The oxidative tricarboxylic acid (TCA) cycle, the oxidative glyoxylate branch or the reductive pathway of the TCA. To enhance the flux from phosphoenolpyruvate or pyruvate through the energetically unfavorable reductive branch to succinate, following genes have been overexpressed in the production strain of DSM/Roquette (former Reverdia, bold arrows): Phosphoenolpyruvate carboxykinase *pckA* from *Mannheimia succiniciproducens*, pyruvate carboxylase *PYC2* and malate dehydrogenase *MDH3* from *S. cerevisiae*, fumarase *fumR* from *Rhizopus oryzae* and fumarate dehydrogenase *FRDm1* from *Trypanosoma brucei*. Transport of succinate across the cell membrane is increased by overexpressing *MAE1* from *S. cerevisiae*, encoding a permease for malate and other C4 dicarboxylic acids, such as succinate. Furthermore, to avoid carbon loss to glycerol or ethanol formation pathways, endogenous glycerol-3-phosphate dehydrogenase *gpd1* and alcohol dehydrogenases *adh1* and *adh2* were deleted (dashed arrows; Ahn et al. 2016; Van De Graaf et al. 2015)

traditional methods like mutagenesis and screening as well as bioprocess engineering have been major approaches to increase productivity. The question if accumulation of astaxanthin is growth related or not was highly debated in the literature and still seems not really easy and straightforward to answer. It turns out, however, that the optimal production process is divided into two phases—a cell growth phase with low carbon-to-nitrogen ratio, allowing for rapid growth and moderate astaxanthin production, followed by a maturation phase with a high carbon-to-nitrogen ratio, limiting growth, but leading to the accumulation of high amounts of the desired compound (Schmidt et al. 2011).

The enzymatic steps of the carotenoid biosynthetic pathway have been elucidated long time ago. The products are commercially valuable and they are colorful—making engineering fun and success easy to recognize—so it is no wonder that the carotenoid biosynthesis has long been a playground for the field of metabolic engineering. The first reports about successful engineering of baker's yeast for betacarotene and lycopene production date back to the beginning of the 1990s. Many studies have since then been published transferring the synthesis of various carotenoids into many yeasts species, among them *Saccharomyces cerevisiae*, *Pichia pastoris* (syn *Komagataella* spp), and *Yarrowia lipolytica.* While most of these studies show a proof-of-principle, the titers and productivities reached with natural producers have not been exceeded (Mata-Gómez et al. 2014). Very successful was a recent attempt to reroute the pathway from astaxanthin to zeaxanthin production in *X. dendrorhous* (Breitenbach et al. 2019).

B. Omega-3 Fatty Acids

The first example of a commercial product to be produced by metabolically engineered yeasts was the omega-3 long-chain **polyunsaturated fatty acid eicosapentaenoic acid** (**EPA**; Xue et al. 2013). EPA and **docosahexaenoic acid** (**DHA**) are natural products of high commercial value. They are essential for human health and have therefore broad applications as **nutritional supplements,** but they are also of great importance as animal feed, particularly in aquaculture. In nature EPA and DHA are mainly produced by marine microorganisms and a few plants. They then accumulate in other organisms through the food chain. The demand for EPA and DHA exceeds the natural reserves, and increasing pollution of the sea and concomitant accumulation of pollutants with the desired acids in fish oil, for example, led to the strong request for other sources. First attempts to produce EPA in baker's yeast were not successful due to very low yields. The successful approach started from *Y. lipolytica*, which is an **oleaginous yeast**, naturally accumulating high amounts of lipids under conditions of carbon excess. Expression of the enzymes required to modify the natural fatty acids to EPA led to a moderately productive strain. Interestingly, the final step to success happened by chance: when the strain was constructed by random integration of the expression cassettes, the authors found one "jackpot" clone with decisively increased EPA accumulation. They then found out, that a pex gene had been interrupted, which led to compromised peroxisome morphology and abolished betaoxidation. Reconstruction of the same modification in another background confirmed this finding. The final strain produces up to 30% of the cell dry weight as lipids, of which 56.6% are EPA. This strain is used for **two commercial products—one nutritional supplement for humans and one for the feeding of cultured salmon** (Xue et al. 2013). The first production approach was a batch culture. In a two-stage continuous fermentation, productivity could be improved by 80% and EPA concentration in the reactor by 40% without compromising the yield or EPA concentration in the biomass (Xie et al. 2017).

Fatty acids and derivatives are generally an important product of many yeast species. Further examples and metabolic engineering approaches to increase yield and modify the accumulated products are highlighted in chapter "Engineering *Saccharomyces cerevisiae* for production of fatty acids and their derivatives" (by Oreb and Boles).

C. Artemisinic Acid

Another success story for **metabolic engineering of yeast** is the market entrance of semisynthetic artemisinin in 2013. Artemisinin is a plant-derived **antimalarial drug**, which has been used by traditional Chinese medicine for a long time. After a recommendation of this drug by the World Health Organization, production of the drug by isolation from plants in the required amounts became difficult—so other production routes had to be explored.

Extensive metabolic engineering of baker's yeast allowed for the production of artemisinic acid, which can be chemically converted into artemisinin. Three modules had to be optimized for efficient production of the desired acid: firstly, to provide the isoprenoid precursors efficiently, the enzymes of the mevalonate pathway needed to be overexpressed. Secondly, expression of the plant-derived amorphadiene synthase had to be optimized, and, finally, cytochromes and dehydrogenases involved in the oxidation of amorphadiene to artemisinic acid had to be optimized. With these modifications 25 g/L of the desired compound can be accumulated, which led to the industrialization of the process (Kung et al. 2018). The time from initial inception of the project to market entrance was about 14 years, this included an increase of the amorphadiene titer greater than ten million-fold from the first proof-of-concept to industry-readiness (Hale et al. 2007).

D. Opioids

Sixteen different opiate alkaloids are currently approved by the US Food and Drug Administration (USFDA) for medical use. Currently, the sole commercial source of opiate alkaloids is opium poppy cultivation, which leads to fluctuations in supply and quality of the pharmaceutically important substances. The core structure of opioids is pentacyclic with multiple chiral centers, which makes chemical synthesis on commercial scale impossible. This prompted efforts to reconstitute the opiate alkaloid production in microbial systems. Expression of 21 enzymes in baker's yeast allowed for the first time the **production of thebaine** from sugar in 2015 (Galanie et al. 2015). Thebaine is an opioid structure, from which various pharmaceutically important opiates can be chemically or microbially synthesized. This work was the result of a decade of work and provided the proof-of-principle that **heterologous microbial opiate production** is possible. This work is not only important from an industrial point of view, but it is a prime example for how such an **applied aim helps to solve groundbreaking questions about biology** itself. Many crucial, hitherto unknown enzymes have been identified. The key enzyme of the pathway catalyzes the epimerization of S-reticuline to R-reticuline. It was identified in the course of such projects from three groups at the same time. Only its identification allowed the proof-of-principle of opiate production yeasts. However, the insights go much deeper: the final step of thebaine synthesis was long time thought to occur spontaneously in nature. Recently, this hypothesis could be falsified and an enzyme for this reaction has been identified. The initially produced amounts of thebaine were in a scale of µg/L, which differs about six orders of magnitude from the commercially useful concentration of 5 g/L (Galanie et al. 2015). Expression of this enzyme catalyzing the last step of thebaine synthesis together with the remaining pathway dramatically increases thebaine production to about 700 µg/L (Chen et al. 2018).

The construction of the first proof-of-concept strain is quite interesting also from a **synthetic biology** point of view: the pathway has been split into seven functional modules (Höhne and Kabisch 2016). One module ensures sufficient supply of the primary metabolites, four modules comprise the main pathway of the opiate production—split at branching points in the biosynthetic pathway. One module encodes the enzymes for tetrahydrobiopeterin production—a required enzymatic co-factor, which is naturally not present in yeast, and finally, one module comprises enzymes catalyzing rate-limiting steps, which have been identified by metabolic flux analyses. Starting from these modules other related alkaloids are accessible. Li et al., for example, report the successful production of noscapine—an antitussive, with anticancer properties, in baker's yeast. Expression of 31 enzymes was required to obtain this substance, of which 25 are heterologous from plants, mammals, and bacteria (Li et al. 2018).

In the meantime many other alkaloids have been successfully synthesized in yeasts. Interestingly, there are only very few examples of other production hosts than baker's yeast. For the assembly of very complicated heterologous pathways, this traditional workhorse seems still to be the preferred choice. It can be noted also that most of the endeavors are at a proof-of-principle level. It remains to be seen which of the substances ever reach the markets as microbial products.

E. Aromatic Compounds

A wide variety of commercially relevant chemicals can be produced starting from aromatic amino acids. The choice of these chemicals ranges from **small building block molecules to complicated plant secondary metabolites** such as flavonoids, stilbenes, or the abovementioned opioids (Suástegui and Shao 2016). Most of these substances are produced in low concentrations in their natural sources, because many are toxic and the metabolic pathways generally complicated, often involving cytochrome P450 enzymes. This makes microbial production, where high concentrations of the product are required for economic reasons, challenging. Also in this case, most of the metabolic engineering endeavors have to be been done so far with *S. cerevisiae* as cell factory. One reason for this—apart of the advantageous availability of many tools for this model organism—is that cytochrome P450 enzymes can be functionally expressed quite well in baker's yeast. There is, however, an intrinsic disadvantage of this microbial host. The shikimate pathway—which is the point of origin for the aromatic amino acid biosynthesis—combines two precursor molecules from the primary metabolism: erythrose-4-phosphate (E4P) and phosphoenolpyruvate (PEP). PEP derives directly from glycolysis, while E4P derives from the pentose phosphate pathway (PPP). In the Crabtree-positive yeast *S. cerevisiae*, glycolysis is extremely active, while the carbon flux via the PPP is at least one order of magnitude lower. This means that the required precursors are entirely unbalanced, which leads to very low yields. Rewiring the entire central carbon metabolism is therefore a prerequisite for efficient aromatics production. Other yeasts, particularly Crabtree-negative yeasts, such as *P. pastoris* (syn. *Komagataella* spp.) or *Scheffersomyces stipitis* might be better candidates for such endeavors. The next challenge is the deregulation of the amino acid biosynthesis itself. Aromatic amino acids are the most "cost intensive" amino acids for the cellular metabolism. Consequently, their biosynthesis is strictly regulated on many levels—a hurdle which needs to be overcome. When all of this is achieved, the more or less complicated further pathway has to be clicked in, and finally all toxicity issues have to be addressed. This explains why most of the endeavors so far did not exceed the proof-of-principle level.

Among the few substances which can be produced in comparatively high amounts is resveratrol, a plant stilbene, which got a lot of attention due to alleged health benefits. Following the strategy outlined above, that is starting with a complete rewiring of the central metabolism, about 800 mg/L of this compound has been produced with baker's yeast (Li et al. 2016).

IV. Yeasts for Production of Recombinant Proteins

The market for recombinant proteins comprises of **protein therapeutics, diagnostic proteins, and industrial enzymes**. The production of recombinant protein therapeutics generated sales of more than US$180 billion in 2017 with predicted global annual growth rates of 6–8% (Walsh 2018). In the period of 2014 to July 2018, 129 biopharmaceutically active agents entered the marked and a total number of 316 distinct biopharmaceutical ingredients are available in licensed products. In 2017, the Top 10 biopharma products generated sales of more than US$80 billion. In addition to that, the market for industrial enzymes, such as amylases, lipases, cellulases, proteases, carbohydrases, etc., yielded revenues of US$4.5 billion worldwide in 2012 (Dewan 2014) and is expected to grow to more than US$5.5 billion by 2022 (Market Research Engine 2019).

Approximately 20–30% of biopharmaceutical products and industrial enzymes are produced in yeasts (Langer et al. 2018). In contrast to primary and secondary metabolite production, which is largely dominated by *Saccharomyces cerevisiae*, the range of yeast hosts used for recombinant protein production is more diverse (Gündüz Ergün et al. 2019; Baghban et al. 2019; Vieira Gomes et al. 2018). Yeast cell factories for protein production combine the strengths of microbial and higher eukaryotic expression systems, providing a set of advantages in comparison to bacterial and mammalian production hosts. Among them are the fast growth to high cell densities in cheap media, fully sequenced and stable genomes, the easy genetic manipulation with developed toolbox sets, the ability of post-translational modifications and protein secretion, as well as the microbiological safety (GRAS status of some yeasts themselves and their products, and the absence of pyrogens and adventitious pathogens) (Roohvand et al. 2017). The lower specific productivity compared to mammalian hosts is usually compensated by high cell densities (up to 150 g/L dry cell weight), which on the other side also demands a higher fraction of energy and resources directed into biomass formation (Maccani et al. 2014).

Due to the dominance of monoclonal antibodies (which represent 52% of new approvals in the last 3 years), products derived from mammalian expression systems have more than 65% share in sales in the biopharmaceutical markets (Walsh 2018). However, if comparing product amounts, of the 26.4 metric tons pure active protein, 68% are produced in microbial systems (Walsh 2014), and **microbially produced insulin** is the most profitable non-antibody-based product on the market. Yeasts are primarily used for production of **secreted proteins** which enables easy purification of the product from the supernatant. Among them mainly hormones, growth factors, antibody fragments, and enzymes are currently produced on an industrial scale. Additionally, yeasts are excellent **producers of vaccines** such as for Hepatitis B (containing recombinant hepatitis B surface antigen) and Human papilloma virus (HPV) (Walsh 2018). While in the early 2000s attempts to produce full length monoclonal antibodies in yeast were initiating extensive glycoengineering efforts, the recent years saw the re-advent of yeast hosts for novel (mostly non-glycosylated) targets such as single chain variable antibody fragments (scFv), camelid single domain antibodies (vHH), and other binding scaffolds and antibody mimetics (Liu and Huang 2018; Spadiut et al. 2014).

Glycosylation is an important topic, especially for biopharmaceutical products. Yeasts, as eukaryotic organisms, are generally able to perform these **post-translational modifications**, but their N-glycan structure is different to the complex mammalian patterns. Yeasts synthesize mannose-rich structures (depending on the species, these contain from 6 up to 200 mannose residues) that are lacking galactose and sialic acid residues, which are present in human glycans (Gemmill and Trimble 1999). This might have two effects: first, the three dimensional structure of the proteins can be impaired, which can hamper or disable their function. The second issue is immunogenicity: the human immune system may recognize these patterns as "unfamiliar" and elicited immune reactions can lead to severe shocks. Interestingly, the terminal glycosidic bonds seem to evoke the immune reaction. While in *S. cerevisiae*, the terminal mannose residues have alpha-1,3 connections (Romero et al. 1999; Nakajima and Ballou 1975), other yeasts such as *Pichia pastoris* and *Hansenula polymorpha* (syn. *Ogataea polymorpha*) lack the enzymes for this reaction and produce terminal alpha-1,2-glycosidic bonds instead, which were shown to be less immunogenic to humans (Song et al. 2007; Kim et al. 2004). During the last decade, major cell engineering efforts were directed toward **glycoengineering of different yeast species**. After blocking the initial steps of mannose chain elongation, N-glycans of the GlcNAc2Man3 or GlcNacMan5 type could be obtained in *S. cerevisiae*, *P. pastoris*, *Yarrowia lipolytica*, and *H. polymorpha* thereby avoiding high-mannose type N-glycans [reviewed by Anyaogu and Mortensen (2015) and De Wachter et al. (2018)]. For *P. pastoris*, more advanced cell engineering was pursued involving 17 gene deletions or overexpressions resulting in glycoengineered strains that do not only prevent high-mannosylation but even allowed mammalian-like N-glycans containing terminal galactose and sialic acid (Hamilton and Gerngross 2007). These made the yeast-based production of highly glycosylated human proteins such as full length IgGs and erythropoietin (EPO) possible (Ye et al. 2011; Hamilton et al. 2006).

Although the yeast quota of the recombinant protein market is dominated by products of the "conventional" yeast *S. cerevisiae*, there are many other—so-called non-conventional—yeast species that provide interesting features for heterologous protein production (Meehl and Stadheim 2014). In this chapter, we summarize the characteristics and application of different yeast species and their potential for protein production with a focus on the industrial and biopharmaceutical use.

A. *Saccharomyces cerevisiae*

Baker's yeast *S. cerevisiae* is one of the most prominent and important model organisms for cell and molecular biology and consequently also highly used for recombinant protein production (Duina et al. 2014). Thus, most marketed yeast-derived therapeutic proteins are produced in *S. cerevisiae*, as the production processes are well established and authorities-approved. As a respiro-fermentative (so-called Crabtree positive) yeast *S. cerevisiae* predominantely ferments a surplus of glucose to ethanol even in the presence of oxygen. Therefore special requirements are necessary for the cultivation in the bioprocess, namely, a glucose limited feed, mostly as a fed batch cultivation.

Efficient production of insulin (see Sect. IV. A.1) and related biopharmaceuticals has been a major driving force in the last decades of the previous century to improve *S. cerevisiae* as a protein production host, including efficient promoter and stable vector systems like the promoter of alcohol dehydrogenase *ADH2*, the galactose-inducible GAL promoter or the constitutive heterologous *TPI1*-vector system (Stepién et al. 1983; Hitzeman et al. 1981; Liu et al. 2012), improvement of secretion leaders (Kjeldsen et al. 1997; Kjeldsen 2000), and optimization of bioreactor cultivation (Diers et al. 1991; Mendoza-Vega et al. 1994; Vasavada 1995). Pioneering work in *S. cerevisiae* also elucidated the function of genes in the secretory pathway [reviewed by Feyder et al. (2015)] and led to the identification of cellular bottlenecks during protein secretion (Mori 2015) as well as strategies to overcome them [reviewed by Hou et al. (2012) and Zahrl et al. (2019)]. Other products produced by *S. cerevisiae* include the hormones glucagon, human growth hormone, interferon, and subunit vaccines such as Hepatitis B surface antigen and HPV vaccines (Wang et al. 2017). Nowadays strategies for improving protein production with *S. cerevisiae* are mainly driven by the high demand of cellulosic and lignocellulosic enzymes needed for the production of second-generation biofuels (Tang et al. 2017; Xu et al. 2014; Kitagawa et al. 2011; Van Zyl et al. 2016) aiming for high-level secretion or surface display of these enzymes.

1. Insulin: The First Marketed Recombinant Protein

Insulin was the first fully sequenced protein (by Sanger in 1955) and also the first licensed drug produced using recombinant DNA technology in 1982. Today recombinant human insulin is mainly produced either by *Escherichia coli* (insoluble production as inclusion bodies and refolding) or by the yeast *S. cerevisiae* [secretion of soluble insulin precursor, Baeshen et al. (2014)]. In insulin precursor (IP), the naturally 35 amino acids long C-peptide is exchanged for a mini-C-peptide comprising of Ala-Ala-Lys, which links the insulin B-chain (29 amino acids) to the A-chain (21 amino acids). Secreted and purified IP is then converted to human insulin by tryptic transpeptidation (Kjeldsen 2000). The global insulin market (insulin and insulin analogs) was US$22 billion in 2017 (Walsh 2018), with an annual sales increase of more than 10% due to the increasing worldwide prevalence of diabetes (Wirtz 2016). **Yeast-based production of insulin** was first achieved in 1986 and is now accounting for more than half of market share. Novo Nordisk sells *S. cerevisiae* derived insulin to more than 100 countries worldwide and has 44% value share and 52% volume share. Additionally the Indian biomanufacturers Biocon and Wockhardt produce insulin and insulin-derivatives such as glargine in yeast hosts. While Novo Nordisk started in 1991 to produce human insulin in the *S. cerevisiae* expression system, Biocon was the first company to produce human insulin in *P. pastoris* in 2003 (Meehl and Stadheim 2014). In 2018 the glargine biosimilar "Semglee" produced in *P. pastoris* was approved by the EU authorities.

B. *Pichia pastoris*

Among the section of **methylotrophic yeasts,** one of the most prominent representatives *Pichia pastoris* (*Komagataella* spp.) is a widespread host for protein production. As these yeasts can utilize methanol as the sole carbon source, they were first used for single cell protein production for animal feed by Philips Petroleum (later ConocoPhilips), since the price for methanol derived from oil refining was very cheap. After the rise of the feedstock prices during the late 1970s, single cell protein production was not profitable any longer and *P. pastoris* was "set free" (Wegner 1990) to the scientific community for research purposes and further developed as a protein production system in the 1980s. The most important reason for its relevance in protein production is the ability to utilize methanol as sole carbon source. The genes that are responsible for this metabolic pathway provide a set of strong inducible and tightly regulated promoters (alcohol oxidase AOX promoter system) that are frequently utilized for overexpression of recombinant proteins. Besides this, in the recent years some alternatives to the AOX inducible promoter systems (rhamnose-inducible, glucose-limit induced P_{GTH1} etc.) were developed to circumvent fire-hazard and toxicity issues when using methanol in industrial scale (Ahmad et al. 2014; Yang and Zhang 2018). Also the availability of commercial cloning kits that provide beginners of this host a full guide from cloning to protein production helped to propagate the application of *P. pastoris.* A unique feature of *P. pastoris* is the GlycoFi-technology owned by Merck & Co., a cell engineering technology that enables manufacturers to produce human-like biantennary N-glycosylated proteins, especially necessary for antibody production. For bioprocess optimization, *P. pastoris* has some important features. It can grow to very high cell densities up to 100 g/L and more (Shay et al. 1987), and it secretes only a comparably small number of host cell proteins (Huang et al. 2011) into the culture supernatant. Therefore downstream processing is facilitated.

The most important industrial biopharmaceutical produced in *P. pastoris* is Insulin glargine (Basalog by Biocon Ltd.—India). Also Hepatitis B vaccine and Interferon alpha 2b are produced in this system (Shantha Biotechnics—India). In 2018, Semglee, a biosimilar to insulin glargine Lantus, produced in *P. pastoris* was approved by the EU authorities for the treatment of Diabetes mellitus. Also Kalbitor (INN: ecallantide), a plasma kallikrein inhibitor for the treatment of hereditary angioedema, is produced in this system. It was the first ever *P. pastoris*-produced drug approved for human treatment in 2009. A remarkable fact is that despite the technical development by GlycoFi which already enabled the production of glycoproteins with humanized N-glycosylation structures, not a single *P. pastoris*-produced antibody is on the biopharmaceutical market. However, some exciting studies were published in the last years, where nanobodies (single domain antibodies lacking N-glycosylation) were produced in high product titers [>10 g/L, Schotte et al. (2016)]. There are also some *P. pastoris*-derived products in the sector of industrial enzymes that play an important part: Quantum is an animal feed additive containing recombinant phytase approved by the FDA Center of Veterinary Medicine (CVM) in 2008. Furthermore, carboxypeptidase, lipase, and phospholipase C are also produced in *P. pastoris.*

C. *Hansenula polymorpha*

Another important cell factory system for the production of recombinant proteins among the group of methylotrophic yeasts is *Hansenula polymorpha* (syn. *Ogataea polymorpha*). Apart from its methylotrophy, it has some other specifically interesting characteristics. First it is able to assimilate nitrate (Siverio 2002), which makes it one of just a few yeast species with this ability. Furthermore, it shows an extraordinary thermotolerance up to 50 °C (Wijeyaratne et al. 1986) which is quite unusual for yeasts. The ability to induce methanol pathway promoters such as P_{MOX} in carbon source (e.g., glucose) limit without the presence of methanol (in contrast to other methylotrophic yeasts) is a unique feature of the *H. polymorpha* expression system that makes it safe to use in industrial bioprocesses. Recent advances regarding genetic tools and bioprocess engineering of the *H. polymorpha* platform are summarized by Manfrão-Netto et al. (2019), while Yoo et al. (2019) provides guidelines on vectors and strains that are suitable for production of secretory recombinant proteins.

The most prominent commercial product made by *H. polymorpha* is HEPLISAV-B, a recombinant hepatitis B vaccine launched by Dynavax Technologies (USA) in 2017. Hexacima, a multi-component vaccine containing recombinant Hepatitis B antigens is also manufactured in *H. polymorpha*. Wockhardt Ltd. (India) launched Asia's first human recombinant insulin (Wosulin) in 2003, which is produced in *H. polymorpha*.

D. *Yarrowia lipolytica*

Yarrowia lipolytica is a yeast species with the special ability to utilize higher hydrocarbons, especially long-chain fatty acids. It also shows a strong morphological dimorphism, as it can grow from ovoid to hyphae form depending on the present environmental conditions. While the morphological status has a yet unresolved impact on the ability of *Y. lipolytica* to produce metabolic products such as citric acid and polyols (Timoumi et al. 2018), some researchers hypothesized that the mechanisms involved in dimorphic transition positively influence the protein secretion behavior of this yeast (Celińska and Nicaud 2019). On the other hand, dimorphism poses a challenge for establishing efficient bioprocesses for recombinant secretory protein production (Vandermies and Fickers 2019). Based on its natural habitats (cheese, oil etc.), it secretes large amounts of lipases and proteases for the retrieval of nutrients for growth. Many *Y. lipolytica*-derived products (citric acid, erythritol, eicosapentaenoic acid) received GRAS status in the last decade (Groenewald et al. 2014). Therefore, it can unhesitatingly be utilized in food and feed products. For expression of recombinant genes either the constitutive TEF promoter or inducible hybrid promoters are used (reviewed, e.g., by Madzak 2018). The most important recombinant products are mainly industrial enzymes for applications in food and feed supplements, cosmetics, detergents, textiles and paper industries, biofuels, and waste stream depollution (Madzak 2015).

E. *Kluyveromyces lactis*

Kluyveromyces lactis, formerly classified as *Saccharomyces lactis*, is known for its **ability to assimilate milk sugar (lactose)**. It can be isolated from milk, butter, yoghurt, cheese, and other fermented dairy products. Since *K. lactis* is present in several dairy products, it is generally recognized as safe (GRAS) by the FDA authorities playing an important role for the production of food and feed ingredients. Due to the production of the pigment pulcherrimin, the color of the colonies appears cream to pink.

For the expression of heterologous genes, the availability of a strong and tightly regulated promoter is of major importance. The major promoter used in *K. lactis* is the promoter associated with the beta-galactosidase *LAC4* gene that is well repressed by low concentrations of glucose (Swinkels et al. 1993). An overview about genetic tools for recombinant protein production as well as bioprocessing strategies was given by Spohner et al. (2016). In terms of industrial applications, the ability to grow on inexpensive substrates, such as cheese whey, is a basis of economic competitiveness (Ornelas et al. 2007). The naturally secreted beta-galactosidase (lactase) is an important industrial product of *K. lactis* cultivation, since it is used for the production of lactose-free milk products to a large extent. Furthermore, the expression of heterologous chymosins plays a major role for cheese production that nowadays replaced the use of rennet from slaughtered animals (van Ooyen et al. 2006).

V. Conclusion and Outlook

Industrial application of yeasts has not only shaped modern biotechnology in many ways but is also contributing to our environment and human health and well-being in many ways. As outlined above yeast is employed to produce bioethanol as a biogenic substitute to transport fuels, natural colorants, and fragrances as well as biopharmaceutical proteins and industrial enzymes. The detailed knowledge on ethanol production has guided the development of strains producing different organic acids as well as higher alcohols which are likely to become chemical building blocks and energy carrier derived from biogenic resources in the near future. Yeast derived secondary metabolites like artemisinin or opioids are close to being available for patients, constituting a novel class of complex biochemicals not naturally produced in yeast.

Two major developments are shaping the future of yeast biotechnology:

1. **Development of so-called non-conventional yeasts** (i.e., non-*Saccharomyces* yeasts) into production platforms at par with *S. cerevisiae* has gained enormous momentum by modern genomics and systems biology, speeding up research on physiology of these yeasts vastly.
2. **Synthetic biology is accelerating development of new production strains** by providing standardized methods and tools to assemble genetic elements and engineer genomes. The advent of a fully synthetic yeast genome (Schindler et al. 2018) adds a complementary perspective to this development which may impact yeast biotechnology in many ways in future (Pretorius and Boeke 2018).

The availability of fully synthetic yeast chromosomes (and full genomes) will likely shape yeast biotechnology in an unprecedented way, once the newly developed technologies for genome design, synthesis, and assembly have developed toward routine, and can be fully combined with synthetic biology standards and tools toward the **industrial application of yeasts for production of medicines, consumer goods and energy**. By this, yeasts will likely play a major role in a future bioeconomy with the aim to master some of our world's great challenges.

Acknowledgments Research on yeast biotechnology in our labs is supported by the Federal Ministry for Digital and Economic Affairs (BMDW); the Federal Ministry for Transport, Innovation and Technology (BMVIT); the Styrian Business Promotion Agency SFG; the Standortagentur Tirol; Government of Lower Austria; and ZIT—Technology Agency of the City of Vienna through the COMET-Funding Program managed by the Austrian Research Promotion Agency FFG. Further support by the BMDW and the National Foundation for Research, Technology and Development through the Christian Doppler Research Association is acknowledged.

References

Adachi E, Torigoe M, Sugiyama M, Nikawa J-I, Shimizu K (1998) Modification of metabolic pathways of *Saccharomyces cerevisiae* by the expression of lactate dehydrogenase and deletion of pyruvate decarboxylase genes for the lactic acid fermentation at low pH value. J Ferment Bioeng 86(3):284–289

Ahmad M, Hirz M, Pichler H, Schwab H (2014) Protein expression in *Pichia pastoris*: recent achievements and perspectives for heterologous protein production. Appl Microbiol Biotechnol 98(12):5301–5317

Ahn JH, Jang YS, Lee SY (2016) Production of succinic acid by metabolically engineered microorganisms. Curr Opin Biotechnol 42:54–66

Almeida JR, Fávaro LC, Quirino BF (2012) Biodiesel biorefinery: opportunities and challenges for microbial production of fuels and chemicals from glycerol waste. Biotechnol Biofuels 5(1):48

Anyaogu DC, Mortensen UH (2015) Manipulating the glycosylation pathway in bacterial and lower eukaryotes for production of therapeutic proteins. Curr Opin Biotechnol 36:122–128

Baek SH, Kwon EY, Kim YH, Hahn JS (2016) Metabolic engineering and adaptive evolution for efficient production of D-lactic acid in *Saccharomyces cerevisiae*. Appl Microbiol Biotechnol 100(6):2737–2748

Baeshen NA, Baeshen MN, Sheikh A, Bora RS, Ahmed MM, Ramadan HA, Saini KS, Redwan EM (2014) Cell factories for insulin production. Microb Cell Factories 13:141

Baghban R, Farajnia S, Rajabibazl M, Ghasemi Y, Mafi A, Hoseinpoor R, Rahbarnia L, Aria M (2019) Yeast expression systems: overview and recent advances. Mol Biotechnol 61(5):365–384

Barredo JL, García-Estrada C, Kosalkova K, Barreiro C (2017) Biosynthesis of astaxanthin as a main carotenoid in the Heterobasidiomycetous Yeast *Xanthophyllomyces dendrorhous*. J Fungi (Basel) 3(3):44

Besada-Lombana PB, McTaggart TL, Da Silva NA (2018) Molecular tools for pathway engineering in *Saccharomyces cerevisiae*. Curr Opin Biotechnol 53:39–49

Breitenbach J, Pollmann H, Sandmann G (2019) Genetic modification of the carotenoid pathway in the red yeast *Xanthophyllomyces dendrorhous*: engineering of a high-yield zeaxanthin strain. J Biotechnol 289:112–117

Buijs NA, Siewers V, Nielsen J (2013) Advanced biofuel production by the yeast *Saccharomyces cerevisiae*. Curr Opin Chem Biol 17(3):480–488

Cavallo E, Charreau H, Cerrutti P, Foresti ML (2017) *Yarrowia lipolytica*: a model yeast for citric acid production. FEMS Yeast Res 17(8):fox084

Celińska E, Nicaud JM (2019) Filamentous fungi-like secretory pathway strayed in a yeast system: peculiarities of *Yarrowia lipolytica* secretory pathway underlying its extraordinary performance. Appl Microbiol Biotechnol 103(1):39–52

Chen X, Hagel JM, Chang L, Tucker JE, Shiigi SA, Yelpaala Y, Chen HY, Estrada R, Colbeck J, Enquist-Newman M, Ibáñez AB, Cottarel G, Vidanes GM, Facchini PJ (2018) A pathogenesis-related 10 protein catalyzes the final step in thebaine biosynthesis. Nat Chem Biol 14(7):738–743

Ciriminna R, Meneguzzo F, Delisi R, Pagliaro M (2017) Citric acid: emerging applications of key biotechnology industrial product. Chem Cent J 11(1):22

Currie JN (1917) The citric acid fermentation of *Aspergillus niger*. J Biol Chem 31(1):15–37

De Wachter C, Van Landuyt L, Callewaert N (2018) Engineering of yeast glycoprotein expression. Adv Biochem Eng Biotechnol. https://doi.org/10.1007/10_2018_69

Dellomonaco C, Clomburg JM, Miller EN, Gonzalez R (2011) Engineered reversal of the β-oxidation cycle for the synthesis of fuels and chemicals. Nature 476(7360):355–359

Dewan SS (2014) Global markets for enzymes in industrial applications. BIO0300J. BBC Research, Wellesley, MA

Diers IV, Rasmussen E, Larsen PH, Kjaersig IL (1991) Yeast fermentation processes for insulin production. Bioprocess Technol 13:166–176

Duina AA, Miller ME, Keeney JB (2014) Budding yeast for budding geneticists: a primer on the *Saccharomyces cerevisiae* model system. Genetics 197 (1):33–48

Dürre P (2007) Biobutanol: an attractive biofuel. Biotechnol J 2(12):1525–1534

Egermeier M, Russmayer H, Sauer M, Marx H (2017) Metabolic flexibility of *Yarrowia lipolytica* growing on glycerol. Front Microbiol 8:49

Feyder S, De Craene JO, Bär S, Bertazzi DL, Friant S (2015) Membrane trafficking in the yeast *Saccharomyces cerevisiae* model. Int J Mol Sci 16(1):1509–1525

Galanie S, Thodey K, Trenchard IJ, Filsinger Interrante M, Smolke CD (2015) Complete biosynthesis of opioids in yeast. Science 349(6252):1095–1100

Gemmill TR, Trimble RB (1999) Overview of N- and O-linked oligosaccharide structures found in various yeast species. Biochim Biophys Acta 1426(2):227–237

Generoso WC, Schadeweg V, Oreb M, Boles E (2015) Metabolic engineering of *Saccharomyces cerevisiae* for production of butanol isomers. Curr Opin Biotechnol 33:1–7

Global Industry Analysts, Inc. (2017) The global market for lactic acid is projected to reach 1.6 million metric tons by 2024. https://strategyr.blogspot.com/2017/06/the-global-market-for-lactic-acid-is.html. Cited 20 Mar 2019

Goffeau A, Barrell BG, Bussey H, Davis RW, Dujon B, Feldmann H, Galibert F, Hoheisel JD, Jacq C, Johnston M, Louis EJ, Mewes HW, Murakami Y, Philippsen P, Tettelin H, Oliver SG (1996) Life with 6000 genes. Science 274 (5287):546, 563–547

Groenewald M, Boekhout T, Neuvéglise C, Gaillardin C, van Dijck PW, Wyss M (2014) *Yarrowia lipolytica:* safety assessment of an oleaginous yeast with a great industrial potential. Crit Rev Microbiol 40 (3):187–206

Gündüz Ergün B, Hüccetoğulları D, Öztürk S, Çelik E, Çalık P (2019) Established and upcoming yeast expression systems. Methods Mol Biol 1923:1–74

Hale V, Keasling JD, Renninger N, Diagana TT (2007) Microbially derived artemisinin: a biotechnology solution to the global problem of access to affordable antimalarial drugs. Am J Trop Med Hyg 77(6 Suppl):198–202

Hamilton SR, Gerngross TU (2007) Glycosylation engineering in yeast: the advent of fully humanized yeast. Curr Opin Biotechnol 18(5):387–392

Hamilton SR, Davidson RC, Sethuraman N, Nett JH, Jiang Y, Rios S, Bobrowicz P, Stadheim TA, Li H, Choi BK, Hopkins D, Wischnewski H, Roser J, Mitchell T, Strawbridge RR, Hoopes J, Wildt S, Gerngross TU (2006) Humanization of yeast to produce complex terminally sialylated glycoproteins. Science 313(5792):1441–1443

Hitzeman RA, Hagie FE, Levine HL, Goeddel DV, Ammerer G, Hall BD (1981) Expression of a human gene for interferon in yeast. Nature 293 (5835):717–722

Höhne M, Kabisch J (2016) Brewing painkillers: a yeast cell factory for the production of opioids from sugar. Angew Chem Int Ed Engl 55(4):1248–1250

Hou J, Tyo KE, Liu Z, Petranovic D, Nielsen J (2012) Metabolic engineering of recombinant protein secretion by *Saccharomyces cerevisiae*. FEMS Yeast Res 12(5):491–510

Hu W, Li WJ, Yang HQ, Chen JH (2019) Current strategies and future prospects for enhancing microbial production of citric acid. Appl Microbiol Biotechnol 103(1):201–209

Huang CJ, Damasceno LM, Anderson KA, Zhang S, Old LJ, Batt CA (2011) A proteomic analysis of the *Pichia pastoris* secretome in methanol-induced cultures. Appl Microbiol Biotechnol 90(1):235–247

IHS Markit (2015) Citric acid-chemical economics handbook. IHS Markit, London

International Yeast Co. Ltd. (1933) Verfahren zur Herstellung von Hefe nach dem Zulaufverfahren. German patent DE583760 C

Ito Y, Hirasawa T, Shimizu H (2014) Metabolic engineering of *Saccharomyces cerevisiae* to improve succinic acid production based on metabolic profiling. Biosci Biotechnol Biochem 78(1):151–159

Jansen MLA, Bracher JM, Papapetridis I, Verhoeven MD, de Bruijn H, de Waal PP, van Maris AJA, Klaassen P, Pronk JT (2017) *Saccharomyces cerevisiae* strains for second-generation ethanol production: from academic exploration to industrial implementation. FEMS Yeast Res 17(5):fox044

Khanna S, Goyal A, Moholkar VS (2012) Microbial conversion of glycerol: present status and future prospects. Crit Rev Biotechnol 32(3):235–262

Kim MW, Rhee SK, Kim JY, Shimma Y, Chiba Y, Jigami Y, Kang HA (2004) Characterization of N-linked oligosaccharides assembled on secretory recombinant glucose oxidase and cell wall mannoproteins from the methylotrophic yeast *Hansenula polymorpha*. Glycobiology 14(3):243–251

Kitagawa T, Kohda K, Tokuhiro K, Hoshida H, Akada R, Takahashi H, Imaeda T (2011) Identification of genes that enhance cellulase protein production in yeast. J Biotechnol 151(2):194–203

Kjeldsen T (2000) Yeast secretory expression of insulin precursors. Appl Microbiol Biotechnol 54(3):277–286

Kjeldsen T, Pettersson AF, Hach M, Diers I, Havelund S, Hansen PH, Andersen AS (1997) Synthetic leaders with potential BiP binding mediate high-yield secretion of correctly folded insulin precursors from *Saccharomyces cerevisiae*. Protein Expr Purif 9(3):331–336

Klein M, Swinnen S, Thevelein JM, Nevoigt E (2017) Glycerol metabolism and transport in yeast and fungi: established knowledge and ambiguities. Environ Microbiol 19(3):878–893

Kung SH, Lund S, Murarka A, McPhee D, Paddon CJ (2018) Approaches and recent developments for the commercial production of semi-synthetic artemisinin. Front Plant Sci 9:87

Langer ES, Rader RA, Gillespie DE (eds) (2018) Fifteenth annual report and survey of biopharmaceutical manufacturing capacity and production. BioPlan Associates, Rockville, MD. isbn:978-1-934106-33-4

Li M, Schneider K, Kristensen M, Borodina I, Nielsen J (2016) Engineering yeast for high-level production of stilbenoid antioxidants. Sci Rep 6:36827

Li Y, Li S, Thodey K, Trenchard I, Cravens A, Smolke CD (2018) Complete biosynthesis of noscapine and halogenated alkaloids in yeast. Proc Natl Acad Sci U S A 115(17):E3922–E3931

Liu Y, Huang H (2018) Expression of single-domain antibody in different systems. Appl Microbiol Biotechnol 102(2):539–551

Liu Z, Tyo KE, Martínez JL, Petranovic D, Nielsen J (2012) Different expression systems for production of recombinant proteins in *Saccharomyces cerevisiae*. Biotechnol Bioeng 109(5):1259–1268

Maccani A, Landes N, Stadlmayr G, Maresch D, Leitner C, Maurer M, Gasser B, Ernst W, Kunert R (2014) Mattanovich D (2014) *Pichia pastoris* secretes recombinant proteins less efficiently than Chinese hamster ovary cells but allows higher space-time yields for less complex proteins. Biotechnol J 9 (4):526–537

Madzak C (2015) *Yarrowia lipolytica*: recent achievements in heterologous protein expression and pathway engineering. Appl Microbiol Biotechnol 99(11):4559–4577

Madzak C (2018) Engineering *Yarrowia lipolytica* for use in biotechnological applications: a review of major achievements and recent innovations. Mol Biotechnol 60(8):621–635

Manfrão-Netto JHC, Gomes AMV, Parachin NS (2019) Advances in using *Hansenula polymorpha* as chassis for recombinant protein production. Front Bioeng Biotechnol 7:94

Market Research Engine (2019) Specialty enzymes market expected to be worth US$ 6.50 billion by 2024. https://www.marketwatch.com/press-release/specialty-enzymes-market-expected-to-be-worth-us-55-billion-by-2022-2018-08-28. Cited 20 Mar 2019

Mata-Gómez LC, Montañez JC, Méndez-Zavala A, Aguilar CN (2014) Biotechnological production of carotenoids by yeasts: an overview. Microb Cell Factories 13:12

Meehl MA, Stadheim TA (2014) Biopharmaceutical discovery and production in yeast. Curr Opin Biotechnol 30:120–127

Mendoza-Vega O, Sabatié J, Brown SW (1994) Industrial production of heterologous proteins by fed-batch cultures of the yeast *Saccharomyces cerevisiae*. FEMS Microbiol Rev 15(4):369–410

Miller C, Fosmer A, Rush B, McMullin T, Beacom D, Suominen P (2011) Industrial production of lactic acid. In: Moo-Young M (ed) Comprehensive biotechnology, 2nd edn. Academic Press, Burlington, pp 179–188

Mori K (2015) The unfolded protein response: the dawn of a new field. Proc Jpn Acad Ser B Phys Biol Sci 91 (9):469–480

Nakajima T, Ballou CE (1975) Yeast manno-protein biosynthesis: solubilization and selective assay of four mannosyltransferases. Proc Natl Acad Sci U S A 72(10):3912–3916

Nielsen J, Larsson C, van Maris A, Pronk J (2013) Metabolic engineering of yeast for production of fuels and chemicals. Curr Opin Biotechnol 24 (3):398–404

Ornelas AP, Silveira WB, Sampaio FC, Passos FML (2007) The activity of β-galactosidase and lactose metabolism in *Kluyveromyces lactis* cultured in cheese whey as a function of growth rate. J Appl Microbiol 104:1008–1013

Otero JM, Cimini D, Patil KR, Poulsen SG, Olsson L, Nielsen J (2013) Industrial systems biology of *Saccharomyces cerevisiae* enables novel succinic acid cell factory. PLoS One 8(1):e54144

Pinazo JM, Domine ME, Parvulescu V, Petru F (2015) Sustainability metrics for succinic acid production: a comparison between biomass-based and petrochemical routes. Catal Today 239:17–24

Porro D, Bianchi MM, Brambilla L, Menghini R, Bolzani D, Carrera V, Lievense J, Liu CL, Ranzi BM, Frontali L, Alberghina L (1999) Replacement of a metabolic pathway for large-scale production of lactic acid from engineered yeasts. Appl Environ Microbiol 65(9):4211–4215

Pretorius IS, Boeke JD (2018) Yeast 2.0-connecting the dots in the construction of the world's first functional synthetic eukaryotic genome. FEMS Yeast Res 18(4):foy032

Raab AM, Gebhardt G, Bolotina N, Weuster-Botz D, Lang C (2010) Metabolic engineering of *Saccharomyces cerevisiae* for the biotechnological production of succinic acid. Metab Eng 12(6):518–525

Rajgarhia V, Asleson Dundon C, Olson S, Suominen P, Hause B (2007) Methods and materials for the production of D-lactic acid in yeast. EP Patent 1513923 B1

Renewable Fuels Association (2017) World fuel ethanol production. https://ethanolrfaorg/resources/industry/statistics/#1537559649968-e206480c-7160. Cited 20 Mar 2019

Romero PA, Lussier M, Véronneau S, Sdicu AM, Herscovics A, Bussey H (1999) Mnt2p and Mnt3p of *Saccharomyces cerevisiae* are members of the Mnn1p family of alpha-1,3-mannosyltransferases responsible for adding the terminal mannose residues of O-linked oligosaccharides. Glycobiology 9 (10):1045–1051

Roohvand F, Shokri M, Abdollahpour-Alitappeh M, Ehsani P (2017) Biomedical applications of yeast-a patent view, part one: yeasts as workhorses for the production of therapeutics and vaccines. Expert Opin Ther Pat 27(8):929–951

Rude MA, Schirmer A (2009) New microbial fuels: a biotech perspective. Curr Opin Microbiol 12 (3):274–281

Sauer M, Porro D, Mattanovich D, Branduardi P (2010) 16 years research on lactic acid production with yeast—ready for the market? Biotechnol Genet Eng Rev 27:229–256

Schindler D, Dai J, Cai Y (2018) Synthetic genomics: a new venture to dissect genome fundamentals and engineer new functions. Curr Opin Chem Biol 46:56–62

Schmidt I, Schewe H, Gassel S, Jin C, Buckingham J, Hümbelin M, Sandmann G, Schrader J (2011) Biotechnological production of astaxanthin with *Phaffia rhodozyma/Xanthophyllomyces dendrorhous*. Appl Microbiol Biotechnol 89(3):555–571

Schotte P, Dewerte I, De Groeve M, De Keyser S, De Brabandere V, Stanssens P (2016) *Pichia pastoris* Mut(S) strains are prone to misincorporation of O-methyl-L-homoserine at methionine residues when methanol is used as the sole carbon source. Microb Cell Factories 15:98

Shay LK, Hunt HR, Wegner GH (1987) High-productivity fermentation process for cultivating industrial microorganisms. J Ind Microbiol 2 (2):79–85

Shen CR, Lan EI, Dekishima Y, Baez A, Cho KM, Liao JC (2011) Driving forces enable high-titer anaerobic 1-butanol synthesis in *Escherichia coli*. Appl Environ Microbiol 77(9):2905–2915

Shen XX, Opulente DA, Kominek J, Zhou X, Steenwyk JL, Buh KV, Haase MAB, Wisecaver JH, Wang M, Doering DT, Boudouris JT, Schneider RM, Langdon QK, Ohkuma M, Endoh R, Takashima M, Manabe RI, Čadež N, Libkind D, Rosa CA, DeVirgilio J, Hulfachor AB, Groenewald M, Kurtzman CP, Hittinger CT, Rokas A (2018) Tempo and mode of genome evolution in the budding yeast subphylum. Cell 175(6):1533–1545.e1520

Siverio JM (2002) Assimilation of nitrate by yeasts. FEMS Microbiol Rev 26(3):277–284

Song Y, Choi MH, Park JN, Kim MW, Kim EJ, Kang HA, Kim JY (2007) Engineering of the yeast *Yarrowia lipolytica* for the production of glycoproteins lacking the outer-chain mannose residues of N-glycans. Appl Environ Microbiol 73(14):4446–4454

Spadiut O, Capone S, Krainer F, Glieder A, Herwig C (2014) Microbials for the production of monoclonal antibodies and antibody fragments. Trends Biotechnol 32(1):54–60

Spohner SC, Schaum V, Quitmann H, Czermak P (2016) *Kluyveromyces lactis*: an emerging tool in biotechnology. J Biotechnol 222:104–116

Stepién PP, Brousseau R, Wu R, Narang S, Thomas DY (1983) Synthesis of a human insulin gene. VI Expression of the synthetic proinsulin gene in yeast. Gene 24(2–3):289–297

Suástegui M, Shao Z (2016) Yeast factories for the production of aromatic compounds: from building blocks to plant secondary metabolites. J Ind Microbiol Biotechnol 43(11):1611–1624

Swinkels BW, van Ooyen AJJ, Bonekamp FJ (1993) The yeast *Kluyveromyces lactis* as an efficient host for heterologous gene expression. Antonie Van Leeuwenhoek 64:187–201

Tang H, Song M, He Y, Wang J, Wang S, Shen Y, Hou J, Bao X (2017) Engineering vesicle trafficking improves the extracellular activity and surface display efficiency of cellulases in. Biotechnol Biofuels 10:53

Tang W, Wang Y, Zhang J, Cai Y, He Z (2019) Biosynthetic pathway of carotenoids in *Rhodotorula* and strategies for enhanced their production. J Microbiol Biotechnol 29:507–517

Timoumi A, Guillouet SE, Molina-Jouve C, Fillaudeau L, Gorret N (2018) Impacts of environmental conditions on product formation and morphology of *Yarrowia lipolytica*. Appl Microbiol Biotechnol 102(9):3831–3848

van de Graaf MJ, Valianpoer F, Fiey G, Delattre L, Schulten EAM (2015) Process for the crystallization of succinic acid. US patent US20150057425A1

van Ooyen AJJ, Dekker P, Huang M, Olsthoorn MMA, Jacobs DI, Colussi PA, Taron CH (2006) Heterologous protein production in the yeast *Kluyveromyces lactis*. FEMS Yeast Res 6:381–392

van Zyl JH, Den Haan R, Van Zyl WH (2016) Overexpression of native *Saccharomyces cerevisiae* ER-to-Golgi SNARE genes increased heterologous cellulase secretion. Appl Microbiol Biotechnol 100 (1):505–518

Vandermies M, Fickers P (2019) Bioreactor-scale strategies for the production of recombinant protein in the yeast *Yarrowia lipolytica*. Microorganisms 7 (2):E40

Vasavada A (1995) Improving productivity of heterologous proteins in recombinant *Saccharomyces cerevisiae* fermentations. Adv Appl Microbiol 41:25–54

Vieira Gomes AM, Souza Carmo T, Silva Carvalho L, Mendonça Bahia F, Parachin NS (2018) Comparison of yeasts as hosts for recombinant protein production. Microorganisms 6(2):E38

Walsh G (2014) Biopharmaceutical benchmarks 2014. Nat Biotechnol 32(10):992–1000

Walsh G (2018) Biopharmaceutical benchmarks 2018. Nat Biotechnol 36(12):1136–1145

Wang G, Huang M, Nielsen J (2017) Exploring the potential of *Saccharomyces cerevisiae* for biopharmaceutical protein production. Curr Opin Biotechnol 48:77–84

Wegner GH (1990) Emerging applications of the methylotrophic yeasts. FEMS Microbiol Rev 7(3-4):279–283

Wijeyaratne SC, Ohta K, Chavanich S, Mahamontri V, Nilubol N, Hayashida S (1986) Lipid composition of a thermotolerant yeast, *Hansenula polymorpha*. Agric Biol Chem 50(4):827–832

Wirtz VJ (2016) Insulin market profile. Health Action International, Amsterdam

Xie D, Miller E, Sharpe P, Jackson E, Zhu Q (2017) Omega-3 production by fermentation of *Yarrowia lipolytica*: from fed-batch to continuous. Biotechnol Bioeng 114(4):798–812

Xu L, Shen Y, Hou J, Peng B, Tang H, Bao X (2014) Secretory pathway engineering enhances secretion of cellobiohydrolase I from *Trichoderma reesei* in *Saccharomyces cerevisiae*. J Biosci Bioeng 117 (1):45–52

Xue Z, Sharpe PL, Hong SP, Yadav NS, Xie D, Short DR, Damude HG, Rupert RA, Seip JE, Wang J, Pollak DW, Bostick MW, Bosak MD, Macool DJ, Hollerbach DH, Zhang H, Arcilla DM, Bledsoe SA, Croker K, McCord EF, Tyreus BD, Jackson EN, Zhu Q (2013) Production of omega-3 eicosapentaenoic acid by metabolic engineering of *Yarrowia lipolytica*. Nat Biotechnol 31(8):734–740

Yang Z, Zhang Z (2018) Engineering strategies for enhanced production of protein and bio-products in *Pichia pastoris*: a review. Biotechnol Adv 36 (1):182–195

Ye J, Ly J, Watts K, Hsu A, Walker A, McLaughlin K, Berdichevsky M, Prinz B, Sean Kersey D, d'Anjou M, Pollard D, Potgieter T (2011) Optimization of a glycoengineered *Pichia pastoris* cultivation process for commercial antibody production. Biotechnol Prog 27(6):1744–1750

Yoo SJ, Moon HY, Kang HA (2019) Screening and selection of production strains: secretory protein expression and analysis in *Hansenula polymorpha*. Methods Mol Biol 1923:133–151

Zahrl RJ, Gasser B, Mattanovich D, Ferrer P (2019) Detection and elimination of cellular bottlenecks in protein-producing yeasts. Methods Mol Biol 1923:75–95

14 Engineering *Saccharomyces cerevisiae* for Production of Fatty Acids and Their Derivatives

Leonie Baumann[1], Florian Wernig[1], Sandra Born[1], Mislav Oreb[1]

CONTENTS

[1] Faculty of Biological Sciences, Institute of Molecular Biosciences, Goethe-University Frankfurt, Frankfurt, Germany; e-mail: M.Oreb@bio.uni-frankfurt.de

Abbreviations

AcCoA	Acetyl-CoA
AcP	Acetyl-phosphate
ALE	Adaptive laboratory evolution
CoA	Coenzyme A
FA	Fatty acid(s)
FACS	Fluorescence-activated cell sorting
FAEE	Fatty acid ethyl ester(s)
GAP	Glyceraldehyde-3-phosphate
gTME	Global transcription machinery engineering
MalCoA	Malonyl-CoA
PPP	Pentose phosphate pathway
SE	Steryl ester
TAG	Triacylglycerol
TCA	Tricarboxylic acid
TE	Thioesterase
X5P	Xylulose 5-phosphate

I. Introduction

Fatty acids (FAs) and their derivatives, such as fatty alcohols, dicarboxylic acids, FA esters, alkanes, and alkenes, are frequently summarized under the generic term "**oleochemicals.**" The diversity of physicochemical properties of oleochemicals is determined by their functional groups and by the length of the aliphatic chains. This is reflected by a variety of products containing these classes of compounds, including fuels, lubricants, surfactants, detergents, cosmetics, food additives, and pharmaceuticals. Consequently, the production volume of different oleochemicals is measured in millions of

Genetics and Biotechnology, 3rd Edition
The Mycota II
J.P. Benz, K. Schipper (Eds.)

tons per year (Rupilius and Ahma 2006), and the demand is steadily increasing. They are currently produced from petroleum (in which case the term "petrochemicals" is also used) or biomass-derived (mainly plant) fats. While the concerns associated with the exploitation of fossil resources have been known for a long time, there is an increasing awareness of problems caused by mass cultivation of oily plants (e.g., palms), for instance, excessive land use and water consumption (Schmidt 2015). The biorefinery concept, in which **engineered microbes** are used to convert various biomass feedstocks into desired products (Nielsen et al. 2013), offers a potentially more effective exploitation of resources and thereby reduces the environmental impact compared to extraction from plants. Indeed, significant advances in production of oleochemicals with different chain lengths and functionalities have been achieved in diverse microbial hosts (for review, see Yu et al. 2014; d'Espaux et al. 2015; Pfleger et al. 2015; Sarria et al. 2017; Marella et al. 2018; Xue et al. 2018; Zhang et al. 2018). Since rapid progress is made in the field, we feel that a critical re-evaluation of the literature is justified. Here, we will focus on strategies developed in *S. cerevisiae*, one of the best characterized and most popular biotechnological host organisms (see also Chap. 13 by Schmelzer et al.), in which highest yields of microbially produced FA have been recently reported (Yu et al. 2018). Different approaches for engineering precursor and cofactor supply, chain length control, elimination of by-product formation, and overcoming product toxicity as well as the development of biosensors for accelerating the screening for best performing strains are summarized and evaluated. Throughout the manuscript, we consciously do not collect all data on achieved titers and yields for different FA-derived products, as another recent review has provided a very detailed summary (Fernandez-Moya and Da Silva 2017). Strategies that have been developed in other host organisms (e.g., *Escherichia coli*) are occasionally discussed, if they appear suitable to expand the pathways already established in *S. cerevisiae*. Lastly, we give an overview of the production of FA-derived products in *S. cerevisiae* as well as methods for optimizing fermentation conditions, enabling the scale-up of current lab-scale fermentation processes.

II. Providing Carbon, Redox Power, and Energy for Fatty Acid Synthesis: The Options for Precursor Supply Routes

S. cerevisiae, an ethanologenic yeast, only requires relatively small amounts of FA as building blocks of membrane lipids. The FA synthesis naturally occurs in the cytosol by the FA synthase (FAS) that uses **acetyl-CoA (AcCoA)** and its derivative **malonyl-CoA (MalCoA)** as precursor molecules. AcCoA metabolism in yeast is compartmentalized, whereby the major part is synthesized inside mitochondria via the pyruvate dehydrogenase (PDH) complex (Krivoruchko et al. 2015) and the mitochondrial AcCoA cannot be exported to the cytosol (van Rossum et al. 2016a, c). The native pathway for the synthesis of cytosolic AcCoA in *S. cerevisiae* is referred to as **pyruvate dehydrogenase (PDH) bypass,** and it diverts only a minor part of the pyruvate produced from glucose (Pronk et al. 1996). Following decarboxylation of pyruvate by pyruvate decarboxylases, acetaldehyde is oxidized to acetate by **acetaldehyde dehydrogenases (ALDs)**, and acetate is subsequently ligated to CoA by the **acetyl-CoA synthetases (ACS)**. This reaction is thermodynamically driven by hydrolysis of ATP to AMP and inorganic pyrophosphate, which is subsequently hydrolyzed by the inorganic pyrophosphatase. Thus, each AcCoA molecule is synthesized at the expense of two ATP equivalents by the PDH bypass. This reaction scheme implies that the glycolytic ATP supply (net two ATP moles per mol glucose) is insufficient, because energy equivalents are also needed for cell proliferation and maintenance. Thus, a significant proportion of the available (sugar) substrate must be diverted to the tricarboxylic acid (TCA) cycle and respiration for the supply of energy equivalents. This reduces the attainable product yield (van Rossum et al. 2016b) and, considering that most FA and derived products have a relatively low commercial value, makes their production through the PDH bypass uneconomical.

AcCoA is not only a building block of FA but also of many compound classes that are of interest in biotechnology, such as isoprenoids (e.g., artemisinic acid; Paddon et al. 2013), polyesters (e.g., polyhydroxybutyric acid; Kocharin et al. 2012), polyketides (e.g., 6-methylsalicylic acid; Wattanachaisaereekul et al. 2008), or flavonoids (e.g., naringenin; Koopman et al. 2012). To avoid constraints such as limited metabolic capacity of organelles, availability of cofactors, and transport of products across organellar membranes, heterologous pathways for production of these compounds are preferably expressed in the cytosol. Therefore, significant effort has been devoted to improving the cytosolic AcCoA supply in *S. cerevisiae*. An excellent review article (van Rossum et al. 2016b) provided a systematic analysis of possible pathways to optimize the supply of cytosolic AcCoA under consideration of reaction stoichiometry, energy conservation, and maximally attainable yields for four model AcCoA-derived compounds including FA. Here, we will revisit the current literature that exploited some of those different possibilities (summarized in Fig. 1) to produce FA and other AcCoA-derived products. Since the provision of **reducing equivalents**—either NADH or NADPH—is equally important for FA production and intrinsically dependent on the choice of the AcCoA synthesis route, the redox cofactor supply will be discussed throughout this chapter. The synthesis of FA can be engineered to occur either via FAS that strictly requires NADPH or by reversing β-oxidation, a FA degradation pathway, which allows for more flexibility regarding the **cofactor dependence** (see Sect. III.B). Therefore, the choice of the appropriate AcCoA pathway also depends on the synthesis mode of FA.

A. Engineering the Pyruvate Dehydrogenase Bypass

In most proof-of-concept studies, the production of AcCoA-derived products in *S. cerevisiae* relied on the native PDH bypass or engineered variants of it, without introducing heterologous AcCoA routes. Even when various alternative AcCoA pathways were introduced, the native PDH bypass was present in the background of all engineered strains, with one exception (Meadows et al. 2016) known to us (see Sect. II.C).

An increased flux through the PDH bypass can be achieved by redirecting acetaldehyde away from reduction to ethanol by alcohol dehydrogenases (ADHs) toward oxidation to acetate by ALDs. In a first study (aiming to increase the production of the isoprenoid amorphadiene), it could be shown that overexpressing the acetaldehyde dehydrogenase ***ALD6*** alone substantially increased the accumulation of acetate (Shiba et al. 2007). Since Ald6 is NADP-specific, this step is also important to provide NADPH required by FAS. Next, the authors tested different ACS variants, whereby only the expression of a heterologous enzyme from *Salmonella enterica*, mutated to prevent inactivation by acetylation (**SeACS^{L641P}**), led to a substantial increase in amorphadiene production (Shiba et al. 2007). In the same study, the overexpression of endogenous enzymes had no (for Acs2) or little (for Acs1) effect on amorphadiene titers, which was attributed to negative regulatory mechanisms at the post-translational level, possibly acetylation. In subsequent studies, a positive effect of combined Ald6/SeACS^{L641P} overexpression on production of several AcCoA-derived products, including FA derivatives such as FA ethyl esters (FAEEs) and hexadecanol, was confirmed by different groups (Chen et al. 2013b; Krivoruchko et al. 2013; Jong et al. 2014; Lian et al. 2014; Feng et al. 2015).

As a strategy to utilize the major fermentation product ethanol for AcCoA synthesis, the ADH isoform II (Adh2), which is the main enzyme responsible for oxidation of ethanol to acetaldehyde, was co-expressed with Ald6/SeACS^{L641P} in some variants (Chen et al. 2013b; Krivoruchko et al. 2013; Jong et al. 2014; Feng et al. 2015). Unfortunately, the effect of Adh2 overexpression alone remained unclear, since it was overexpressed only in combination with other enzymes (e.g., acetoacetyl-CoA thiolase Erg10) that affect the product yield (Kocharin et al. 2012; Chen et al. 2013b; Krivoruchko et al. 2013) or together with Ald6 and SeACS^{L641P}, without a direct comparison to a control not overexpressing Adh2 (Jong et al. 2014; Feng et al. 2015). Considering that the interconversion of acetaldehyde and ethanol is rather driven by the chemical equilibrium, it remains yet to be demonstrated that Adh2 overexpression indeed diverts the flux toward AcCoA in engineered strains in the presence of glucose.

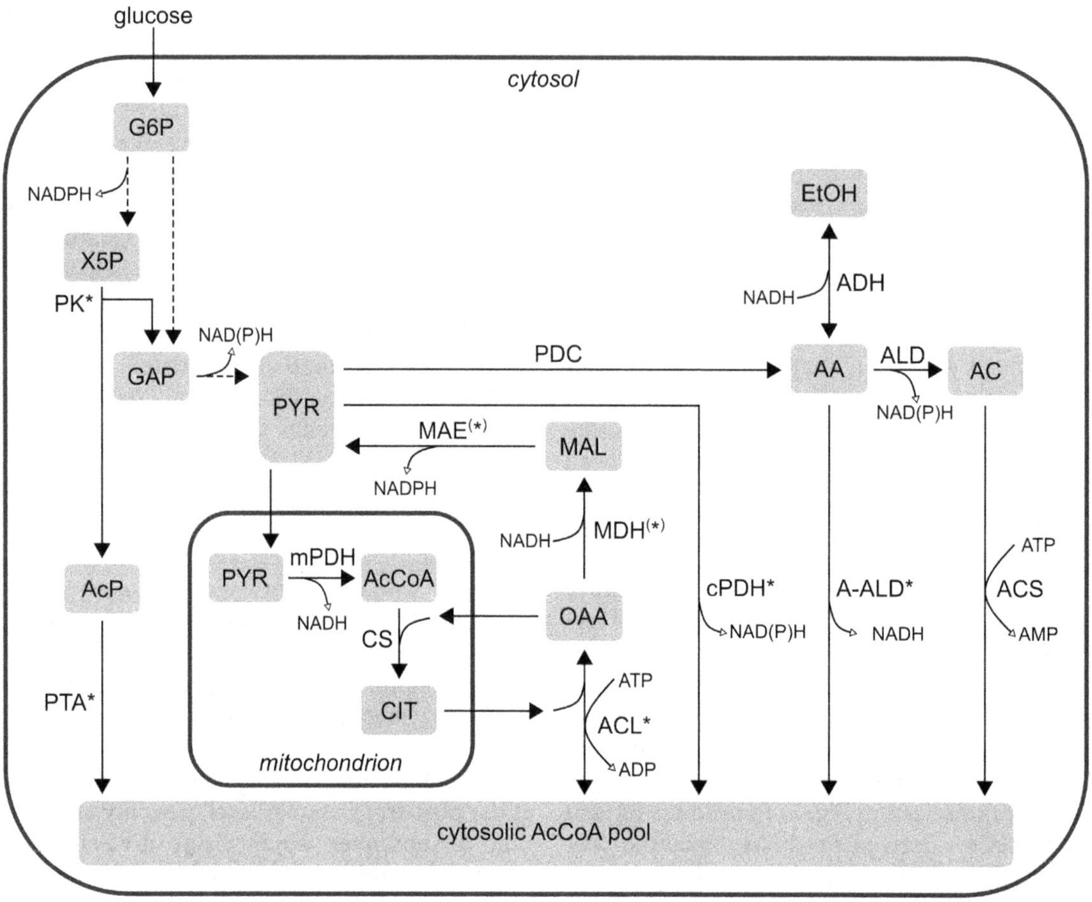

Fig. 1 A simplified scheme of precursor supply pathways for production of acetyl-CoA (AcCoA) derivatives in the cytosol of *S. cerevisiae*. Only the relevant pathway intermediates are shown in gray boxes: glucose-6-phosphate (G6P), xylulose 5-phosphate (X5P), glyceraldehyde-3-phosphate (GAP), pyruvate (PYR), acetaldehyde (AA), ethanol (EtOH), acetate (AC), acetyl-phosphate (AcP), citrate (CIT), oxaloacetate (OAA), and malate (MAL). Solid lines depict single-step reactions, whereas dashed lines stand for multiple enzymatic steps. Reversibility of reactions is indicated by double-headed arrows only if relevant in the context of AcCoA supply. For reactions that can be catalyzed by enzymes with different specificities, the generic notation "NAD(P)H" is used; otherwise, the specific cofactor is denoted. For clarity, only the reduced form of the redox cofactors is shown. The reaction stoichiometries are not considered, and energy equivalents are shown only for the AcCoA supply reactions. Enzyme names are abbreviated as follows: *PK* phosphoketolase, *PTA* phosphotransacetylase, *PDC* pyruvate decarboxylase, *ALD* aldehyde dehydrogenase, *mPDH* mitochondrial pyruvate dehydrogenase, *cPDH* cytosolic pyruvate dehydrogenase, *CS* citrate synthase, *ACL* ATP citrate lyase, *MDH* malate dehydrogenase, *MAE* malic enzyme, *A-ALD* acetylating aldehyde dehydrogenase, *ACS* acetyl-CoA synthetase; heterologous enzymes are marked with an asterisk * and endogenous enzymes that are expressed in a different compartment with an asterisk in brackets (*). Transporters are not shown. Influx/efflux of (pyro)phosphate, carbon dioxide, and coenzyme A (CoA) moieties is omitted for clarity

A logical approach to boost the production of cytosolic AcCoA via the PDH bypass is the **elimination of ethanol production** by deleting the ADH genes. Li et al. (2014) first demonstrated that the deletion of *ADH1* is beneficial for FA production. Lian et al. (2014) went a step further and deleted the isoforms *ADH1* and *ADH4* with a concomitant deletion of glycerol-3-phosphate dehydrogenases *GPD1* and *GPD2* to prevent accumulation of glycerol, a trait previously observed in ADH-deficient strains (Smidt et al. 2012). This elimination of competing products already increased the flux toward AcCoA, measured as an increase in n-butanol production through a simultaneously expressed heterologous pathway (Lian et al. 2014). Schadeweg and Boles (2016b) could later show that a deletion of ADH isoforms 1–5, even without a GPD deletion, substantially increases the production of n-butanol through a similarly designed pathway. When Ald6/SeACS^{L641P} were overexpressed in an ADH-/GPD-deficient strain, the production of n-butanol unexpectedly decreased due to a strong accumulation of acetate (Lian et al. 2014). Conversely, overexpression of SeACS^{L641P} alone improved the productivity in the same strain background (Lian et al. 2014; Lian and Zhao 2015). Taken together, the analyses performed in ADH positive (Shiba et al. 2007) and negative (Lian et al. 2014; Lian and Zhao 2015) strains suggest that the **balance of ALD and ACS activity** is a critical and context-dependent variable in optimizing the flux through the cytosolic PDH bypass due to the rate-limiting role of ACS and the accumulation of toxic acetate levels.

B. ATP-Independent Pyruvate-to-Acetyl-CoA Routes

The high energy requirement makes the PDH bypass a rather inefficient precursor supply pathway for manufacturing AcCoA-derived products. Therefore, several alternative ATP-independent AcCoA yielding pathways were tested in *S. cerevisiae* for different products. One possibility to convert acetaldehyde to AcCoA is via **acetylating aldehyde dehydrogenases (A-ALD)**. Their functionality in yeast was demonstrated by the ability to complement the growth defect of an ACS-deficient strain (Kozak et al. 2014b) and to replace endogenous acetaldehyde dehydrogenases (Kozak et al. 2016). Schadeweg and Boles (2016b) demonstrated increased n-butanol production via reverse β-oxidation when they overexpressed A-ALD from *E. coli*, mutated to favor the (non-physiological) reaction direction from acetaldehyde to AcCoA (adhE$^{A267T/E568K}$; Membrillo-Hernandez et al. 2000) in an ADH-deficient strain. However, a positive effect of A-ALD overexpression could only be seen when the supply of **coenzyme A (CoA)** was concomitantly increased by overexpression of a heterologous pantothenate kinase (coaA from *E. coli*) and pantothenate feeding, demonstrating that the availability of not only acetyl moieties but also of the coenzyme is a factor limiting the synthesis of AcCoA (Schadeweg and Boles 2016b).

This notion likely applies to all AcCoA pathways, as the overexpression of the pantothenate kinase and pantothenate supplementation was also beneficial in combination with an engineered PDH bypass (Ald6/SeACS^{L641P} overexpression) in a strain constructed for production of the flavonoid naringenin (Liu et al. 2017). The endogenous pantothenate supply can be improved by overexpressing the polyamine oxidase Fms1 (Schadeweg and Boles 2016a), which catalyzes the limiting step of its biosynthesis.

A-ALD pathway is a very promising alternative to the PDH bypass to produce AcCoA derivatives, owing to its lower energy requirement. However, the theoretically attainable yield for FA produced by FAS is lower when A-ALD is used compared to PDH bypass (van Rossum et al. 2016b) due to cofactor incompatibility (the A-ALD pathway yields NADH, while FAS requires NADPH). Thus, only if the FAs are produced via reverse β-oxidation, which can be engineered to utilize exclusively NADH, the implementation of the A-ALD pathway appears feasible. The same is true for other alternative pathways that convert pyruvate to AcCoA in an ATP-independent manner and thereby yield NADH, namely, (1) pyruvate formate lyase (PFL) combined with formate dehydrogenase, and (2) **cytosolic pyruvate dehydrogenase complex (cPDH)** (van Rossum et al. 2016b). Although the functionality of PFL in yeast cytosol has been demonstrated (Waks and Silver 2009; Kozak et al. 2014b), it has not been used for high-level production of FA or other AcCoA derivatives so far, due to its complex biochemical properties, including strong oxygen sensitivity (Knappe et al. 1969). In contrast, a significant improvement of n-butanol production was reported when a bacterial PDH or the native PDH components lacking the mitochondrial targeting sequence were overexpressed in yeast cytosol (Lian et al. 2014).

It has to be noted, however, that the cytosolic PDH activity was not directly demonstrated in this study. In another study (Kozak et al. 2014a), it has been shown that the activity of PDH in the yeast cytosol is strictly dependent on supplementation of lipoic acid, a cofactor that is normally synthesized inside mitochondria and must be covalently linked to the E2 subunit of PDH. Moreover, overexpression of enzymes involved in lipoylation of E2 in the cytosol is crucial (Kozak et al. 2014a). Since Lian et al. (2014) did neither supplement lipoic acid nor express the components of the lipoylation machinery, the mechanism by which cPDH led to an improvement of n-butanol production in their work remains enigmatic.

Of particular interest for FA synthesis via the (NADPH-dependent) FAS would be an implementation of NADPH-yielding cPDH variants. Indeed, a mutated ($NADP^+$-accepting) PDH was developed (Bocanegra et al. 1993) and implemented to produce the polyketide triacetic acid lactone in yeast cytosol. Although the cytosolic PDH activity was not directly measured, this strategy led to a significantly increased NADPH/NADP ratio and AcCoA levels in whole cell extracts as well as to improved product titers in fermentations (Cardenas and Da Silva 2016). Since this approach has the same cofactor yield (one NADH and one NADPH per molecule AcCoA) but a better energetic balance compared to the PDH bypass, it could potentially improve the production of FA. However, the study did not address the abovementioned requirement for lipoylation, and this aspect will therefore require further investigation before this approach can be employed for FA production.

C. Phosphoketolase Pathway

Among all individually considered alternative AcCoA pathways, the phosphoketolase/transacetylase variant was predicted to enable the highest yield of FAS-derived FA (87% of the theoretical maximum), when the carbon flux is rerouted from glycolysis to the oxidative **pentose phosphate pathway (PPP)** to increase the NADPH pool (van Rossum et al. 2016b). In this scheme, glucose-6-phosphate is converted to xylulose 5-phosphate (X5P) through a consecutive action of the endogenous enzymes glucose-6-phosphate dehydrogenase, 6-phosphogluconolactonase, 6-phosphoglu conate dehydrogenase, and D-ribulose-5-phosphate 3-epimerase, whereby the first and the third reaction yield one NADPH molecule each (for a review of the PPP, see Stincone et al. 2015). X5P is then converted to **acetyl-phosphate (AcP)** and **glyceraldehyde-3-phosphate (GAP)** by a heterologous **phosphoketolase (PK)**. Subsequently, AcP is directly converted to AcCoA by a heterologous **phosphotransacetylase (PTA)**. The PK/PTA pathway was expressed in a strain engineered to produce FAEE and indeed led to a significant increase of product titers compared to the reference strain containing only the native PDH bypass (Jong et al. 2014).

Another variant to convert PKA-derived AcP to AcCoA involves the acetate kinase (ACK), which yields acetate and ATP by transferring the phosphate residue from AcP to ADP. Although ATP is produced in this reaction, the production of AcCoA via this route is energetically less favorable than via PTA since two ATP have to be invested into the subsequent activation of acetate by ACS (see above). The implementation of this strategy led to a significant improvement of polyhydroxybutyrate (Kocharin et al. 2013) and, to a lower extent than with the PK/PTA variant, FAEE production (Jong et al. 2014).

In one recent study (Meadows et al. 2016), the PK/PTA pathway was combined with the A-ALD to produce the isoprenoid farnesene, as the **combinatorial configuration** was calculated to lead to the highest possible product yields (van Rossum et al. 2016b). Interestingly, the authors found that AcP produced by PK is partly dephosphorylated by endogenous (promiscuous) glycerol-3-phosphate phosphatases and their activity needs to be reduced to favor the PTA reaction. This study stands out as the only one in which the endogenous PDH bypass was fully substituted (by *ald6 acs1 acs2* deletion) with heterologous AcCoA synthesis routes

(A-ALD and PK/PTA). Indeed, this approach led to the highest yields ever reported of an isoprenoid produced in *S. cerevisiae* and impressively demonstrated how an "ideal" combination of precursor supply pathways based on stoichiometric analyses can be designed to produce AcCoA derivatives at an industrially competitive level.

D. Citrate-Oxaloacetate Shuttle

The major breakthroughs regarding production of oleochemicals in *S. cerevisiae* have been recently achieved by implementing the citrate-oxaloacetate shuttle. If NADPH supply is not concomitantly engineered, this pathway, like the ATP-independent pyruvate to AcCoA routes, has the lowest expected yield for FA produced via FAS (74% of the theoretical maximum; van Rossum et al. 2016b) due to the imbalance of NADH production and NADPH requirement of FAS. Nevertheless, many oleaginous yeasts, such as *Yarrowia lipolytica*, which produce large amounts of lipids in the cytosol, rely on this system to provide cytosolic AcCoA (Vorapreeda et al. 2012). The term "shuttle" refers to the fact that AcCoA is formed from pyruvate inside mitochondria via the native PDH complex. Subsequently, the acetate moiety is transferred to oxaloacetate by **citrate synthase (CS)**. The resulting citrate either can enter the TCA cycle inside the mitochondria or be exported to the cytosol to serve as a shuttle for acetyl moieties. A cytosolic **ATP citrate lyase (ACL)**, an enzyme that is not present in *S. cerevisiae*, is required to form AcCoA and oxaloacetate under the expense of one ATP molecule. The cycle is closed by transporting oxaloacetate back into the mitochondria through a citrate-oxaloacetate antiporter. In a first attempt, the overexpression of a murine ACL in *S. cerevisiae* led to a slightly increased titer of total FA (Tang et al. 2013). The effect was mostly pronounced during the stationary phase, probably reflecting the redirection of the metabolism toward respiration of the accumulated ethanol.

In the same study, isocitrate dehydrogenase genes *IDH1* and *IDH2* were deleted to promote the accumulation of citrate; somewhat surprisingly, this intervention did not affect the total content but only the saturation profile of FA, suggesting that ACL activity and/or other factors were rate limiting. Later studies, performed in strains engineered for n-butanol (Lian et al. 2014), hexadecanol (Feng et al. 2015), or mevalonate (Rodriguez et al. 2016) production, showed that the choice of the heterologous enzyme has a certain impact on the pathway efficiency.

In most recent studies that reported the highest titers of FA or their derivatives (Zhou et al. 2016b; Yu et al. 2018), the citrate-oxaloacetate shuttle was further optimized and combined with strategies that increase the **NADPH supply**. The overexpression of ACL alone in a strain that was already engineered to block FA degradation led to a moderate improvement (up to 50%) of free FA titers, which is consistent with previous studies (Tang et al. 2013; Lian et al. 2014). A further improvement was achieved by introducing a **transhydrogenase**-like reaction sequence to increase NADPH supply, exemplifying the importance of redox cofactors as a driving force for FA production. In this scheme, a cytosolic **malate dehydrogenase (MDH)** reduces oxaloacetate to malate (oxidizing NADH), and a cytosolic **malic enzyme (MAE)** subsequently transfers hydrogen to NADP, whereby malate is converted to pyruvate, which then re-enters mitochondria and undergoes a new cycle (Fig. 1). Additional overexpression of the endogenous citrate transporter Ctp1 appeared to have no beneficial effect on the FA production, suggesting that the export of citrate into the cytosol may not have been a limiting factor in this approach. In a follow-up study (Yu et al. 2018), several strategies to enhance the citrate-oxaloacetate shuttle were tested. Overexpression of the endogenous (mitochondrial) PDH components to enhance the mitochondrial synthesis of AcCoA did not lead to an increase in FA titers in this study. However, the overexpression of the **mitochondrial pyruvate carrier** subunits Mpc1 and Mpc3 and introduction of a heterologous CS in addition to the native one appeared to increase the flux through the citrate-oxaloacetate shuttle, measured as an increase in FA titers by about 20%.

An incremental improvement was achieved by introducing an additional ACL gene from *Aspergillus nidulans* (a murine ACL was already present in the engineered strain). Simultaneously the isocitrate dehydrogenase Idp2 and the citrate-α-ketoglutarate antiporter Yhm2, which also accepts oxaloacetate instead of α-ketoglutarate, were overexpressed. Idp2 was supposed to oxidize citrate to α-ketoglutarate and thereby provide additional NADPH for FA synthesis. This, however, can only occur at the expense of the cytosolic citrate and therefore AcCoA pool. Yhm2 overexpression was intended to increase the antiport of citrate with oxaloacetate. Unfortunately, since three genes were introduced at once as compared to the reference strain, their individual contribution to the improvement cannot be disentangled.

A further substantial improvement of FA synthesis in this strain background was achieved by **diverting the glucose flux from glycolysis to the oxidative PPP** to increase the NADPH pool. To this end, the expression of the phosphoglucose isomerase (*PGI1*) was downregulated, whereas the endogenous genes encoding PPP enzymes glucose-6-phosphate dehydrogenase (*ZWF1*), 6-phosphogluconate dehydrogenase (*GND1*), transketolase (*TKL1*), and transaldolase (*TAL1*) were concomitantly overexpressed. A downregulation of the mitochondrial isocitrate dehydrogenase *IDH2* to prevent the degradation of citrate in the TCA led to a further improvement but only in a strain with the upregulated PPP. In its precursor strain, lowering Idh2 activity had no beneficial effect, and this notion is consistent with previous results, obtained in a strain that contained the citrate-oxaloacetate shuttle, but was not engineered for an increased supply of NADPH (Tang et al. 2013). Collectively, these studies show that sufficient reducing power is essential for efficient FA production.

E. Strategies to Manipulate NADPH Level Independently of Precursor Supply Routes

Besides abovementioned possibilities of overexpressing enzymes that yield NADPH (e.g., Ald6, Zwf1) or creating transhydrogenase cycles (e.g., oxaloacetate-malate-pyruvate), the level of this cofactor can be manipulated through interventions into glycolysis or amino acid metabolism. One possibility to increase the pool of NADPH is bypassing the main source of cellular NADH—the glyceraldehyde-3-phosphate dehydrogenase (GAPDH) reaction—by a NADP-dependent enzyme. The expression of a heterologous non-phosphorylating **NADP-dependent GAPDH** in strains engineered for polyhydroxybutyrate (Kocharin et al. 2013) or FAEE (Shi et al. 2014a) production led to considerable product yield improvements, but not in a strain designed to produce fatty alcohols (d'Espaux et al. 2017). It has to be noted that—besides different product pathways—the genetic interventions into the carbon metabolism of these strains largely differed, suggesting that the applicability of this strategy is context-dependent. One important consideration is that the activity of a non-phosphorylating GAPDH will reduce the ATP yield of glycolysis, which, as outlined above, ultimately will lead to a reduction of the maximally attainable yield. The implementation of a phosphorylating NADP-dependent GAPDH does not suffer from this downfall, but its utility for production of oleochemicals in *S. cerevisiae* could not yet be demonstrated (d'Espaux et al. 2017). A complementary approach is **reducing the NADPH consumption** through competing pathways. Based on the observation that a large proportion of NADPH is oxidized by NADP-dependent glutamate dehydrogenase Gdh1, d'Espaux et al. (2017) demonstrated that a deletion of the *GDH1* gene is beneficial for fatty alcohol production.

F. Engineering Malonyl-CoA Supply

MalCoA, the extender unit for FA synthesis via FAS, is synthesized by carboxylation of AcCoA in an ATP-dependent manner by the endogenous enzyme **AcCoA carboxylase (Acc1)**. In some studies, the native form of Acc1 was overexpressed to boost the production of FA derivatives, which indeed led to higher product titers (Shin et al. 2012; Runguphan and Keasling 2014). As an alternative approach, Wang et al. (2014b) overexpressed a plant MalCoA synthetase, which ligates malonic acid and CoA to generate MalCoA directly. This improved the productivity of strains engineered for polyketide production and increased the total FA amount. Since the native Acc1 is known to be inactivated by phosphorylation by the kinase Snf1 upon glucose depletion, different strategies were developed to circumvent this regulatory mechanism, including relocalization of the mitochondrial ACC from *S. cerevisiae* (Hfa1) into the cytosol (d'Espaux et al. 2017) and site-directed **mutagenesis to prevent phosphorylation** (Choi and Da Silva 2014; Shi et al. 2014b). All of them led to remarkable improvements in yields of MalCoA-derived products.

The specific activity of the phosphorylation-resistant mutant Acc1^{S1157A} was higher compared to wild-type Acc1 even when measured in extract from glucose-grown cells (Choi and Da Silva 2014), suggesting that the protein might be partly phosphorylated also on this carbon source (not only on ethanol). In another study, a second (in silico predicted) phosphorylation site (S659) was mutated to alanine in addition to S1157, which led to an increase in specific activity and total FA yields compared to a control harboring the S1157A mutation alone (Shi et al. 2014b). Considering that Snf1 is activated by AMP (Sanz 2003), we hypothesize that phosphorylation resistance may become even more relevant on glucose-grown cells when the PDH bypass, which produces AMP in the ACS reaction (Fig. 1), is overexpressed.

Another important consideration for strain design is the **sensitivity of FA chain lengths to the AcCoA/MalCoA ratio.** Elevated concentration of MalCoA is known to favor the elongation of the FA carbon chains by FAS (Sumper et al. 1969). Indeed, the expression of Acc1^{S1157A} (Choi and Da Silva 2014; Hofbauer et al. 2014; Besada-Lombana et al. 2017) and Acc1$^{S1157A,\ S659A}$ (Zhou et al. 2016b) led to a higher proportion of longer FA chains. This demonstrates that, especially if tight chain length control is desired, the relative abundance of the precursor molecules must be subtly balanced. By affecting the lipid composition of the membranes, the ACC hyperactivity not only changes the product profiles but also appears to **negatively influence cell growth,** which was attributed to an imbalanced synthesis of long-chain FA or depletion of intermediates (Shi et al. 2014b). On the other hand, the expression of Acc1^{S1157A} was shown to improve the resistance of yeast to medium-chain FA (see Sect. IV.B). All these observations demonstrate that balancing the ACC activity is a non-trivial task and several factors such as the AcCoA abundance, product chain length, and toxicity must be taken into account. The development of product-specific biosensors (see Sect. IV.C) will accelerate the screening of strains producing the desired chain length in a high-throughput manner, which will greatly facilitate the balancing of AcCoA and MalCoA supply.

To summarize, the endogenous supply of cytosolic precursors (AcCoA and MalCoA), redox cofactors (NADH or NADPH), and energy equivalents (ATP) via the PDH bypass is not sufficient for production of FA and derivatives in *S. cerevisiae* beyond the proof-of-concept level. Although there are no systematic studies comparing the FA yields in dependence on different precursor supply routes, stoichiometric analyses (van Rossum et al. 2016b) and experimental data outlined above show that there is no "one best" precursor supply pathway valid for all AcCoA-derived products. Since the choice of the FA elongation system (FAS vs. reverse β-oxidation; see Sect. III) dictates the redox cofactor requirement, the appropriate precursor supply pathway must be chosen accordingly. Moreover, to minimize the amount of carbon that needs to be diverted for energy supply, the native PDH bypass must be fully replaced by a combinatorial precursor supply configuration, as predicted by stoichiometric analyses for different model compounds (van Rossum et al. 2016b) and experimentally shown for the isoprenoid farnesene (Meadows et al. 2016). Considering that very promising improvements in FA productivity, reaching total yields up to 30% of the theoretical maximum, have been recently achieved (Yu et al. 2018), it is conceivable that similar combinatorial approaches can lead to further advances in the production of oleochemicals.

III. Strategies for Chain Length Control

The majority of microbially synthesized FAs are 14–20 carbon atoms in length (referred to as long-chain FA). Mainly saturated and monounsaturated FA with 16 or 18 carbon atoms are synthesized in *S. cerevisiae* (Cottrell et al. 1986). Engineering microbes like *E. coli*, *S. cerevisiae*, or oleaginous yeasts (e.g., *Yarrowia lipolytica*) toward FA and FA derivatives with alternating chain length is desired due to their compelling properties as biofuels; for the cosmetic, healthcare, and pharmaceutical industries; or as platform compounds (Lennen and Pfleger 2013).

In microorganisms however, biosynthesis of **short- (≤C4), medium- (C6–C12), or very long- (≥C22) chain FA** is rare and low in quantity. For this reason strategies for chain length control through a focus on the main anabolic (FA biosynthesis) and catabolic (β-oxidation) pathways have been developed.

A. Manipulation of the Natural Fatty Acid Biosynthesis and Elongation Machinery

Fatty acid biosynthesis follows a conserved chemistry among different types of FAS enzymes which differ in their general organization. **Type I FAS** systems (mainly eukaryotic) carry all catalytic domains in one multifunctional complex (Schweizer and Hofmann 2004). However, fungal type I FAS greatly differ from animal type I FAS in their architecture (Leibundgut et al. 2008).

In type I systems, FA biosynthesis is initiated by the transfer of the starter unit AcCoA to the acyl-carrier protein (ACP) by the acetyltransferase (AT) domain of FAS. Subsequently, the ACP domain delivers the acetyl residue to the ketosynthase (KS) domain. Similarly, the extension unit MalCoA is loaded onto the free ACP by the malonyl-palmitoyl transferase (MPT) domain. Acetyl and malonyl moieties are condensed by the KS domain to a β-ketoacyl intermediate, which is then further processed. The β-ketoacyl intermediate is reduced by the ketoacyl reductase (KR) domain to form 3-hydroxyacyl-ACP, which is dehydrated by the dehydratase (DH) domain and eventually reduced by the enoyl reductase (ER) domain to a saturated acyl chain containing two additional carbon atoms. The reductive steps at the KR and ER domains are strictly NADPH-dependent. After completion of one cycle, the acyl chain is either condensed with another malonyl moiety for further elongation or released by the MPT domain as a CoA-bound thioester in fungal FAS (Lomakin et al. 2007). After release of acyl-CoA esters from the FAS complex, **thioesterases (TE)** cleave the thioester bond between the acyl chain and CoA to release free FA.

As a strategy for enrichment of short- and medium-chain FA, expression of short- and medium-chain-specific TE for early termination of FA biosynthesis has been established in multiple studies (Leber and Da Silva 2014; Fernandez-Moya et al. 2015; Xu et al. 2016; Zhu et al. 2017b). Combined with fungal FAS enzymes, soluble TE only have limited or no access to FAS-bound acyl-CoA or acyl-ACP esters, and protein engineers have therefore developed strategies to locate TE within the FAS complex. For instance, some type I FAS (e.g., from *Rhodosporidium toruloides* and *Aplanochytrium kerguelense*) harbor two redundant ACP domains, which are located inside of the FAS scaffold. It is therefore possible to replace one of them by a short-chain TE to enable a direct access to acyl intermediates inside the reaction chamber. The implementation of this strategy proved to be more efficient than the expression of free TE and increased short- and medium-chain FA production by 3- to 15-fold (Zhu et al. 2017b). In *Y. lipolytica* FAS, replacement of the MPT domain by a short- and medium-chain-specific TE resulted in an increase of up to 29% of C12–C14 FA of the total FA content (Xu et al. 2016).

As an alternative approach, the yeast FAS has been replaced by non-fungal enzymes. Human FAS (hFAS) carries its own TE domain, which releases the acyl moieties directly from the ACP (Leibundgut et al. 2008) and has a more flexible structure than yeast FAS (Brignole et al. 2009). Expression of hFAS, in which its own TE domain was deleted, in combination with the short-chain-specific TE CpFatB1 (*Cuphea palustris*) or TEII (*Rattus norvegicus*) as free proteins enabled the production of C6–C10 FA (Leber and Da Silva 2014). Substituting the TE domain of hFAS by TEII in the same polypeptide chain substantially increased the production of short- and medium-chain FA.

In contrast to type I FAS, the catalytic domains of the **bacterial type II FAS** system are expressed as single enzymes. This allows a simpler manipulation of the subunits, and acyl intermediates are freely accessible for TE. Expression of the *E. coli* system (acpS, acpP, fabB, fabD, fabG, fabH, fabI, fabZ) in combination with the TE fatB from *Ricinus communis* in a FAS-deficient *S. cerevisiae* strain significantly increased the total FA titer and shifted the FA profile toward C14 FA production (Fernandez-Moya et al. 2015).

A minimally invasive strategy to rewrite the chain length control relies on **rational engineering of yeast FAS** by site-directed mutagenesis. For instance, an early release of shorter acyl chains (C6–C10 or C14) was achieved by introducing bulky residues like tryptophan into the KS domain to shorten the acyl-ACP binding channel in *S. cerevisiae* (Gajewski et al. 2017; Zhu et al. 2017b) or *Y. lipolytica* (Rigouin et al. 2017) FAS. Furthermore, it was shown that decreasing the affinity of the MPT domain for malonyl moieties by introducing one point mutation disfavors the chain elongation and promotes the release of shorter-chain (C6–C10) FA (Gajewski et al. 2017). One additional mutation was introduced into the AT domain for a more frequent priming with AcCoA, as this was expected to shift the product profile toward shorter chains; however, this modification proved only effective in combination with mutations in the MPT and/or KS domain. Single or multiple mutations in the three domains caused different ratios of C6–C10 FA, with some of the mutants exhibiting a considerable specificity, e.g., for C8 FA (Gajewski et al. 2017). Thus, rational engineering of FAS holds a great promise for narrowing down the product profile to the desired chain length.

Very long-chain FAs (C22–C26) are present in *S. cerevisiae* at low abundance (Welch and Burlingame 1973) but are essential compounds, e.g., for sphingolipids (Oh et al. 1997). After their release from the FAS complex, FAs are elongated at the ER membrane from (long-chain) acyl-CoA precursors. **Elongases** with different product specificities (Elo1, Elo2, and Elo3) carry out the elongation of acyl-CoA intermediates with MalCoA. Elo1 elongates medium- and long-chain compounds (C14–C16) to C18 FA, Elo2 elongates compounds up to 22 C atoms, and Elo3 elongates compounds up to 26 C atoms (Toke and Martin 1996; Oh et al. 1997). Reminiscent of cytosolic FA biosynthesis, two reductions and one dehydration step are necessary to complete very long-chain FA synthesis. To selectively increase the pool of C22, overexpression of *ELO2* and deletion of *ELO3* are necessary (Yu et al. 2017), whereas overexpression of only *ELO3* is sufficient for enrichment of C26 FA (Wenning et al. 2017). Alternatively, FAS I from mycobacteria has been shown to naturally generate FA with 22–26 carbon atoms (Kaneda et al. 1995). Expression of FAS I from *Mycobacterium vaccae* in an Elo2-/Elo3-deficient strain increased the C22 FA pool by fourfold (Yu et al. 2017).

B. Reversal of β-Oxidation as an Orthogonal Pathway for Fatty Acid Biosynthesis

The **β-oxidation** cycle naturally is an FA degradation pathway (for review see Hiltunen et al. 2003). In each turn of the cycle, FAs are truncated by removing two carbon atoms from the FA chain, thereby generating AcCoA.

Reversing all reactions of the β-oxidation can consequently be used as an alternative synthetic pathway for FA production. The single reactions of the β-oxidation are **equilibrium-balanced,** and the functionality of the enzymes in the reverse direction has been demonstrated (Dellomonaco et al. 2011; Clomburg et al. 2012). The reverse β-oxidation pathway starts with the condensation of two AcCoA to acetoacetyl-CoA by a **thiolase**, followed by the reduction of the β-ketogroup to 3-hydroxyacyl-CoA by a **reductase/dehydrogenase**, dehydration to trans-enoyl-CoA by a **hydratase/dehydratase**, and a final reduction by another reductase/dehydrogenase, yielding butyryl-CoA. In this cycle, NADH or NADPH can serve as **electron donors** for the reductive steps (depending on the choice of heterologous enzymes). By iterating this reaction sequence, the acyl chain can be elongated by two carbon atoms per cycle. Depending on the desired chain length and functional groups of the products, the reverse β-oxidation can be **terminated through different enzymes**. For instance, TE can release free FA from the CoA-bound form, and aldehyde/alcohol dehydrogenases can reduce the acyl-CoA esters to n-alcohols (see Sect. V).

Elongation beyond a length of four carbon atoms requires multiple turns of reverse β-oxidation, which is hampered by a **competition between the thiolase and the termination enzymes**. Thus, although the synthesis of minor amounts of longer-chain FA via reverse β-oxidation could be demonstrated at the proof-of-concept level in *E. coli*, the product yields gradually decreased with each iteration of the cycle (Dellomonaco et al. 2011; Clomburg et al. 2012). Follow-up studies therefore focused on improving the selectivity of reverse β-oxidation for longer-chain products. To this end, the core pathway was expanded by thiolases that accept acyl-CoA intermediates with various chain lengths, such as the β-ketothiolase BktB, which proved suitable to produce C6–C10 compounds in multiple studies (Dekishima et al. 2011; Clomburg et al. 2015; Kim et al. 2015). The chain length specificity of termination enzymes (i.e., TE for free FA) was shown to be the second key determinant for selective production of longer acyl chains (Clomburg et al. 2015; Kim and Gonzalez 2018). To prevent a premature termination of the cycle, the deletion of endogenous TE genes was critical in *E. coli* strains. These observations are likely transferable to *S. cerevisiae*. To provide sufficient carbon and reducing equivalents, heterologous reverse β-oxidation pathways were expressed in the yeast cytosol. As a proof of concept, n-butanol production, for the synthesis of which only one functional turn of the cycle is needed, has been targeted. Using a variety of enzymes from different organisms and in combination with different precursor supply pathways (see above), titers of up to 1 g L^{-1} could be reached (Lian and Zhao 2015; Schadeweg and Boles 2016a, b). Less effort has been dedicated to the production of **medium-chain products** in *S. cerevisiae* so far, but the feasibility could be demonstrated by combining reverse β-oxidation with a medium-chain-specific TE, CpFatB1, thereby enabling the production of C6–C10 FA, albeit at a low yield (Lian and Zhao 2015). It is certain that by transferring the expanded enzyme toolbox developed in *E. coli* (Kim and Gonzalez 2018) and fine-tuning the expression of endogenous TE, many of which were recently characterized (e.g., Kruis et al. 2018), the production of short- and medium-chain FA via reverse β-oxidation will be further optimized in *S. cerevisiae* in the future.

Directly compared, canonical FA biosynthesis and reverse β-oxidation both bear certain advantages and drawbacks. The most obvious advantage of FA biosynthesis is a rather strict chain length control that can be easily manipulated by protein engineering of FAS, as outlined above, whereas specific production of longer carbon chains via reverse β-oxidation is still very challenging. While the FAS functional modules operate as a perfectly synchronized machine within one macromolecular complex, it is difficult to fine-tune the activity of individual reverse β-oxidation enzymes. At the current stage of technology, reverse β-oxidation can compete with FA biosynthesis in the product range of up to ten carbon atoms. On the other hand, FA biosynthesis is energetically more expensive than reverse β-oxidation, since each elongation cycle by FAS requires one MalCoA that is synthesized at the expense of one ATP molecule. In contrast, reverse β-oxidation has the advantage of using only AcCoA for elongation steps, which also makes it insensitive to the AcCoA/MalCoA ratio, a parameter that has a significant influence on the chain length control by FAS and overall cellular fitness. Whereas the reverse β-oxidation can be designed to use the easily available NADH, FAS is strictly dependent on the less abundant NADPH. Thus, the redox cofactor supply of reverse β-oxidation requires far less interventions into the central carbon metabolism compared to the FA biosynthesis route and offers a larger choice of potential precursor supply routes (see Sect. II, Fig. 1).

IV. Chassis Engineering

To achieve high FA production levels in *S. cerevisiae*, several obstacles have to be addressed within the chosen production pathway. Blocking the degradation of FA has been achieved through the disruption of different enzymes involved in β-oxidation. An issue that remains to be tackled is the toxicity of some FA. We will discuss different methods that have or could be employed to engineer more robust chassis. Furthermore, several biosensor systems have been developed, which can be used in high-throughput screenings to further speed up the search for best performing strains.

A. Elimination of By-product Formation and Fatty Acid Degradation

To increase FA pools, it is important not only to increase precursor pools (see Sect. II, Fig. 1) but also to downregulate competing pathways as well as to prevent degradation through specific gene deletions or overexpressions.

1. Prevention of β-Oxidation

The degradation, i.e., β-oxidation, of the already synthesized free FA or their activated forms, the acyl-CoAs, takes place in the peroxisomes (Hiltunen et al. 2003; van Roermund et al. 2003). In *S. cerevisiae*, β-oxidation of saturated FA is a cyclic mechanism mainly catalyzed by the three enzymes Pox1, Fox2, and Pot1 (Hiltunen et al. 2003). Pox1 is an acyl-CoA oxidase catalyzing the first step of the degradation of an acyl-CoA molecule. The dehydrogenated intermediate is then modified by the multifunctional enzyme Fox2, which acts as an enoyl-CoA hydratase as well as a 3-hydroxyacyl-CoA dehydrogenase, followed by thiolytic cleavage through Pot1, a 3-ketoacyl-CoA thiolase. These cyclic steps are repeated until the FA molecule is completely degraded (Hiltunen et al. 2003).

Prior to peroxisomal β-oxidation, free FAs are activated to their **acyl-CoA form**. In *S. cerevisiae*, this reaction can be catalyzed by five enzymes, Faa1, Faa2, Faa3, Faa4, and Fat1 (Black and DiRusso 2007). Long-chain FAs are primarily activated by Faa1 and Faa4 in the cytosol (Scharnewski et al. 2008) and transported into the peroxisomes by the heterodimeric ATP-binding cassette transporter Pxa1/Pxa2 (Hettema et al. 1996). Fat1 is a multifunctional enzyme, which imports long-chain FA into the peroxisomes and also has an acyl-CoA synthetase activity for very long-chain FA (Zou et al. 2002). Medium-chain FAs, on the other hand, are assumed to enter peroxisomes by passive diffusion or spontaneous flipping and are then activated by the peroxisomal Faa2 (Knoll et al. 1994; Hettema et al. 1996; Hettema and Tabak 2000). The role of Faa3 is not entirely clear yet. It showed some activity on very long-chain FA; however, its in vitro activity on long-chain FA was much lower than that of Faa1 and Faa2 (Johnson et al. 1994; Knoll et al. 1994).

The effects of disrupted FA activation, β-oxidation, or both, on FA titers have been studied extensively. In *S. cerevisiae* strains engineered for increased FA production, a positive effect on free FA titers was detected when deleting *FAA1* and *FAA4* separately, as well as together; however, only minimal increases were observed when deleting *POX1* (Li et al. 2014; Runguphan and Keasling 2014). Another group found that the two triple deletion mutants Δ*faa2*Δ*pxa1*Δ*pox1* and Δ*faa1*Δ*faa4*Δ*fat1* produced more free FA than the original strain (Leber et al. 2015). The combination of all six knockouts in one strain showed the highest titers with 1.3 g L^{-1} free FA, corresponding to 18% of the maximum theoretical yield (Leber et al. 2015).

Apart from these rather general β-oxidation disruptions, there have also been approaches for a **chain length-specific disruption of β-oxidation**, involving *FAA2*, *PEX11*, and *ANT1* (Leber et al. 2016). Pex11 is thought to transport short- and medium-chain FA into the peroxisomes, where they are activated by

Faa2 (Knoll et al. 1994; van Roermund et al. 2000). Ant1 is an adenine nucleotide transporter, which exchanges AMP by ATP across the peroxisomal membrane, thereby providing the energy for the acyl-CoA activation of the free FA (van Roermund et al. 2001). To decrease β-oxidation of specifically short-chain FA, a strain with three knockouts (*Δfaa2Δant1Δpex11*) was generated, which displayed a much stronger increase in hexanoic and octanoic acid levels compared to the "full" β-oxidation-deficient *Δfaa2Δpxa1Δpox1* mutant (Leber et al. 2016). This observation was attributed to the disruption of the AcCoA recycle mechanism in the latter strain (Leber et al. 2016). Overall, the ideal combination of fatty acyl-synthetase and β-oxidation gene knockouts must be evaluated carefully, depending on the desired chain length of the final FA. Nevertheless, most reported gene knockouts of this pathway have turned out to be beneficial for increasing FA titers (Buijs et al. 2015; Zhou et al. 2016b).

2. Disruption of Triacylglycerol and Steryl Ester Synthesis

A competing pathway that can decrease FA levels is the incorporation into storage lipids, primarily triacylglycerols (TAGs) or steryl esters (SEs). In *S. cerevisiae*, TAGs are synthesized from diacylglycerols and fatty acyl-CoAs by the acyltransferases Dga1 and Lro1 (Oelkers et al. 2000, 2002). Steryl ester synthesis of sterols and fatty acyl-CoAs is catalyzed by the acyl-CoA:sterol acyltransferases Are1 and Are2 (Yang et al. 1996). In engineered yeast strains, the knockout of one or several of these four TAG/SE genes has led to increased production of FA derivatives such as 1-hexadecanol (Tang and Chen 2015), fatty alcohols (d'Espaux et al. 2017; Tang et al. 2017), and FAEEs (Valle-Rodríguez et al. 2014). In another strain optimized for FA production (by disruption of β-oxidation and acyl-CoA-activating enzymes), TAG synthesis was increased by overexpressing Dga1. This was combined with an overexpression of the lipid recycle via the triacylglycerol lipase Tgl3, leading to 2.2 g L^{-1} extracellular free FA (Leber et al. 2015).

B. Engineering Yeast Product Tolerance and Excretion

Some FA as well as some of their derivatives are **toxic** to *S. cerevisiae*. In general, yeast is a robust cell factory that can sense stress and adapt its metabolism accordingly. However, to avoid loss of cell viability and ensure economically competitive yields, yeast strains are needed, which are tolerant to high product concentrations even at low pH (Deparis et al. 2017). Over the last years, much progress has been made to understand the underlying principles of the toxicity of these compounds, and random as well as rational approaches have been applied to improve yeast robustness and thereby product yields.

1. Toxicity Mechanisms and Transport of Fatty Acids and Derivatives

FAs that are toxic to *S. cerevisiae* include **hexanoic, octanoic, and decanoic acids** (Lafon-Lafourcade et al. 1984; Viegas et al. 1989; Liu et al. 2013). At acidic pH, which is the common fermentation condition, the undissociated FA can enter the cells by passive diffusion (Viegas 1997). In the neutral cytosol, they dissociate, thereby causing a decrease of the intracellular pH and an accumulation of the toxic anions (Viegas et al. 1989, 1998; Viegas 1997). Furthermore, they disturb the plasma membrane integrity, causing changes in membrane composition, permeability, and fluidity (Alexandre et al. 1996; Legras et al. 2010; Liu et al. 2013). To generate strains with higher tolerance, it is important to ensure a **rapid transport/secretion** of FA out of the cell, as well as to avoid their re-entrance into the cells. A transcriptome analysis of octanoic- and decanoic acid-stressed cells, respectively, revealed a partly overlapping response mechanism, which was similar to an oxidative stress response, but it also showed a compound-specific activation of genes encoding transcription factors and transporters (Legras et al. 2010). The role of the identified transporters Tpo1 and Pdr12 in octanoic and decanoic acid efflux was further analyzed in growth tests with knockout mutants (Legras et al. 2010). An overexpression of such efflux-involved transporters is one approach that could increase product yields—in the case, that secretion is a bottleneck. For example, overexpression of Pdr12 was shown to increase the secretion of short branched-chain FA at early time points of production (Yu et al. 2016). However, transporter overexpression can also have unwanted side effects, such as slower growth, depleting ATP from other cellular processes or altering the plasma membrane composition, as was hypothesized based on other Pdr12 overexpression studies (Nygård et al. 2014). Due to a lack of yeast efflux pumps known to specifically transport FA and their derivatives, another viable approach is the screening of heterologous transporters. In a recent study, 12 human or *Arabidopsis thaliana* (putative) transporters were screened, and one of them, human FATP1, improved overall cell fitness and fatty alcohol (Hu et al. 2018) as well as 1-alkene (Zhou et al. 2018) production and secretion in *S. cerevisiae*. For alkane resistance of *S. cerevisiae*, endogenous efflux pumps, namely, Snq2 and Pdr5, as well as heterologous transporters, namely, *Y. lipolytica* ABC2 and ABC3, were shown to have a positive effect (Chen et al. 2013a; Ling et al. 2013).

2. Methods for Increasing Yeast Robustness

To avoid the re-entrance of the products, the robustness of the **plasma membrane** needs to be improved. Octanoic acid, for instance, disrupts the plasma membrane composition, leading to membrane leakage and cell death (Legras et al. 2010; Liu et al. 2013). This effect was reduced by rationally engineering plasma membrane composition through increasing the oleic acid content, either by external supply (Liu et al. 2013) or by expression of a mutated AcCoA carboxylase, $Acc1^{S1157A}$ (Besada-Lombana et al. 2017). The increase in the average chain length of membrane FA, as well as higher *cis*-monounsaturated FA levels, was shown to provide higher tolerance to toxic FA (Liu et al. 2013; Besada-Lombana et al. 2017).

As tolerance is usually a complex phenotype, which cannot solely be improved by rational engineering of single genes, **adaptive laboratory evolution (ALE)** is an interesting alternative. In an ALE experiment, a strain is grown over many generations, with increasing concentrations of the toxic compound, for selection of cells with an enhanced tolerance phenotype. Causal mutations in the final strain are identified by whole genome sequencing (Dragosits and Mattanovich 2013; Mans et al. 2018). When performed with *E. coli*, an evolved strain not only showed higher tolerance to octanoic acid but also produced higher titers, which was attributed to changes in membrane composition and fluidity (Royce et al. 2015). A similar approach would be possible for *S. cerevisiae*, for which ALE has been performed successfully, for example, for increased resistance to high temperatures (Caspeta et al. 2014) and acetic acid (González-Ramos et al. 2016) or alcohol tolerance (González-Ramos et al. 2013; Davis López et al. 2018), but not for short- or medium-chain FA tolerance.

A method that can lead to similar results as ALE was termed **global transcription machinery engineering, gTME** (Alper et al. 2006). It can induce a remodeling of the transcriptome and therefore target polygenic traits, such as tolerance. gTME relies on the random mutagenesis of a transcription factor that regulates the transcription of several genes and was successfully applied to enhance *S. cerevisiae* tolerance to ethanol (Alper et al. 2006), as well as to adapt to growth on lignocellulosic hydrolysates through improved xylose utilization (Liu et al. 2011). When using this method in combination with an appropriate selection for mutants with increased growth in the presence of the toxic FA, it could be a valuable tool for future tolerance engineering but has so far not been applied to *S. cerevisiae* for FA tolerance.

C. Dynamic Pathway Control and Biosensors

The metabolic engineering of yeast to produce a desired compound can result in an imbalanced metabolism due to the disruption of a tightly regulated, complex biological system, which has evolved for growth and survival. However, inherent control components of this system, such as promoters, transcription factors, or gene copy number, can be utilized for a dynamic pathway control, thereby minimizing metabolic imbalances and increasing product titers (Michener et al. 2012; Schallmey et al. 2014; Shi et al. 2018).

1. Altering Gene Expression Level and Timing

The level and timing of pathway gene expression is crucial in metabolic engineering and can be adjusted, for instance, by using constitutive **promoters** of different strengths or promoters that are inducible/repressible (Da Silva and Srikrishnan 2012). This strategy was applied successfully, for example, to increase 1-alkene production and secretion in *S. cerevisiae*. By replacing the strong e*TDH3* promoter for expression of the membrane-bound enzyme PfUndB with the *GAL7* promoter, which is activated upon glucose depletion, 1-alkene **production was decoupled from growth** (Zhou et al. 2018). The same strategy, i.e., to separate growth from production by expressing pathway genes under the control of carbon source-dependent promoters, was also effective in increasing docosanol production (Yu et al. 2017). A similar approach was applied for *S. cerevisiae* short- and medium-chain FA production, again with a glucose-repressed promoter. Here, *FAS1* and *FAS2* were expressed on a low copy plasmid under control of the alcohol dehydrogenase II promoter of *S. cerevisiae*, p*ADH2*, leading to increased short- and medium-chain FA titers (Gajewski et al. 2017). In another approach, several promoters of different strengths were tested for downregulating the expression of *IDH2* (carbon flux redistribution into free FA) and *PGI1* (increasing NADPH supply), respectively, as the deletion of either gene led to growth defects. The expression of the genes was reduced by using weaker promoters, leading to increased long-chain FA production in *S. cerevisiae* (Yu et al. 2018). These examples underpin the importance of an appropriate control of gene expression and have much potential to be further exploited for increasing short- and medium-chain FA production—especially as promoters and terminators of different strengths have been described in great detail in recent years (Alper et al. 2005; Da Silva and Srikrishnan 2012; Curran et al. 2013; Lee et al. 2015).

2. Biosensors

Currently, the analysis of short- and medium-chain FA titers is time-consuming and laborious and is usually achieved through chromatography-based methods. A more rapid and convenient alternative are biosensors. They detect the concentration of a molecule—ideally over a wide range of concentrations—and transform it into an easily detectable, quantifiable output, such as growth rate or fluorescence, eventually enabling **high-throughput screenings**, as depicted in Fig. 2 (Michener et al. 2012; Schallmey et al. 2014; Shi et al. 2018). One example of in vivo biosensors is transcription factors, which, in nature, are inevitable for the dynamic control of gene expression. Once their inducing molecules and target promoters are known, transcription factors can be used in metabolic engineering to regulate production pathway expression or even live-monitor titers. Several bacterial **transcription factor-based systems** have been adapted to yeast (Teo et al. 2013; Li et al. 2015; Skjoedt et al. 2016; Wang et al. 2016), especially dynamic sensor-regulator systems of the FA intermediate MalCoA (Johnson et al. 2017). The prokaryotic transcription factor-based **FapR-*fapO*** system has been engineered in *S. cerevisiae* and enabled the sensing of intracellular MalCoA levels (Li et al. 2015; David et al. 2016). This system has been expanded to dynamically control production by coupling the expression of 3-hydroxypropionic acid path-

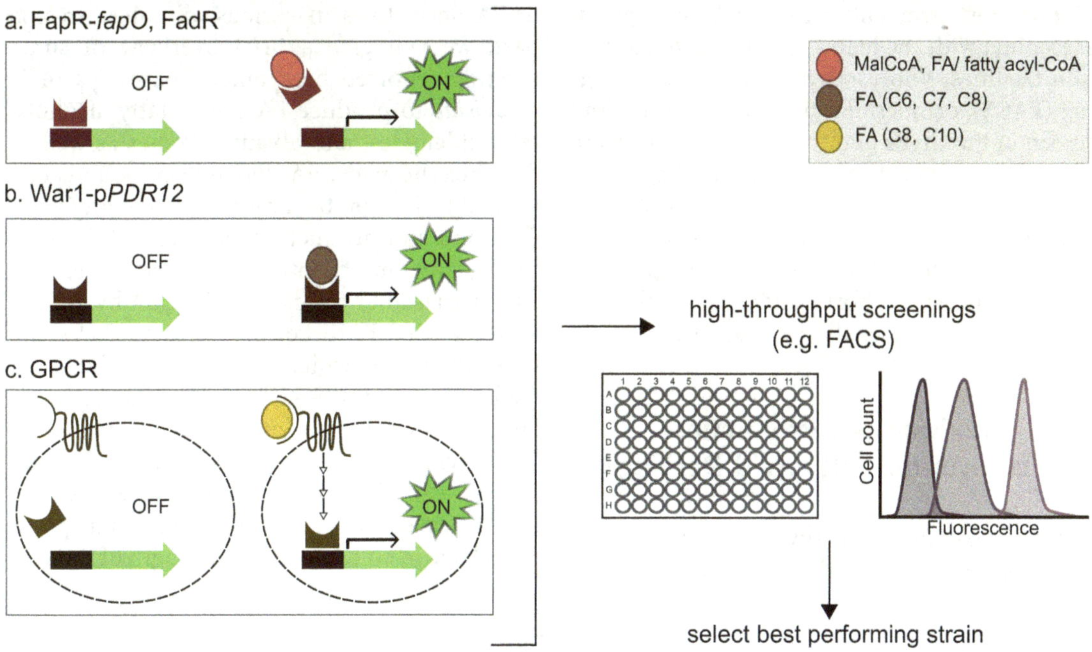

Fig. 2 Yeast biosensors developed for high-throughput screenings of fatty acids or pathway intermediates. (**a**) FapR and FadR are bacterial repressors (OFF state) which have been adapted to yeast for biosensing. Once they are bound by MalCoA (FapR) or FA/fatty acyl-CoA (FadR), expression is enabled, e.g., of a downstream-located fluorescent protein (ON state). (**b**) The yeast endogenous *PDR12* promoter (p*PDR12*) is regulated by the transcription factor War1 that is thought to constitutively bind to the promoter. War1p changes to an active form upon C6, C7, or C8 FA presence. (**c**) A G-protein-coupled receptor responsive to C8 and C10 FA was coupled to the yeast mating pathway for signaling and GFP expression as a quantifiable output. For high-throughput screenings with biosensors, high-throughput cultivation platforms and/or screening methods, such as fluorescence-activated cell sorting (FACS), are needed to select best performing strains

way genes to the intracellular MalCoA level (David et al. 2016). Another bacterial regulation system is based on the transcriptional repressor **FadR**, which is repressed in the presence of FA/ fatty acyl-CoAs and cannot bind to specific operator sites (Zhang et al. 2012). This property was exploited to monitor FA levels in *S. cerevisiae* by coupling it to synthetic yeast promoters and GFP expression (Teo et al. 2013). While these MalCoA sensor-regulator systems represent a promising method to fine-tune intracellular precursor supply, systems that directly sense the final, excreted FA, constitute another valuable tool for metabolic engineering. However, reports about such biosensors are scarce to date. The first short- and medium-chain FA yeast biosensor was shown to respond to octanoic and decanoic acid via heterologous G-protein-coupled receptors that were linked to the yeast mating pathway for signaling and GFP expression as a concentration-dependent output (Mukherjee et al. 2015). Besides the rather low linear and dynamic ranges of this biosensor, it was not proven to be functional in culture broth—an indispensable feature for an applicable biosensor. Another biosensor, which was developed for para-hydroxybenzoic acid, was based on the weak acid-inducible ***PDR12* promoter** which is regulated by the transcription factor War1 (Williams et al. 2017). Our group recently adapted this system for the sensing of hexanoic, heptanoic, and octanoic acid reaching high linear and dynamic ranges (Baumann et al. 2018). This biosensor can sense short- and medium-chain FA in *S. cerevisiae* culture broth, which facilitates the monitoring of end-product concentrations and opens the path to high-throughput screenings of producer strain libraries. High-throughput screening, with the help of appropriate cultivation facilities, **fluorescence-activated cell sorting (FACS)**, or microfluidics, permits the rapid screen of hundreds and thousands of strains to identify best performing cells (Becker et al. 2004; Dietrich et al. 2010; Wang et al. 2014a). High-throughput screening technologies have the potential to substantially speed up metabolic engineering efforts and strain selection in the next years (Schallmey et al. 2014).

V. *S. cerevisiae* as a Production Platform for Fatty Acid Derivatives

Several biosynthetic pathways have recently been engineered to produce a broad spectrum of FA derivatives in various microbes. In this part, we will give a brief overview of some recently reported biosynthetic pathways in *S. cerevisiae* to produce FA esters, fatty alcohols, fatty aldehydes, and alkanes/alkenes as well as dicarboxylic acids. FA-derived compounds can be synthesized on the one hand from fatty acyl-CoA, the end product of the yeast FA biosynthesis, and on the other hand from free FA, which arise after the cleavage of CoA by endogenous yeast TE. In contrast to acyl-CoA—the biosynthesis of which is tightly regulated by feedback inhibition—free FA can be accumulated to much higher levels in yeast (Foo et al. 2017; Teixeira et al. 2017). Both precursors can be processed by several heterologous enzymes as illustrated in Fig. 3. All downstream pathways have in common that the length of the acyl

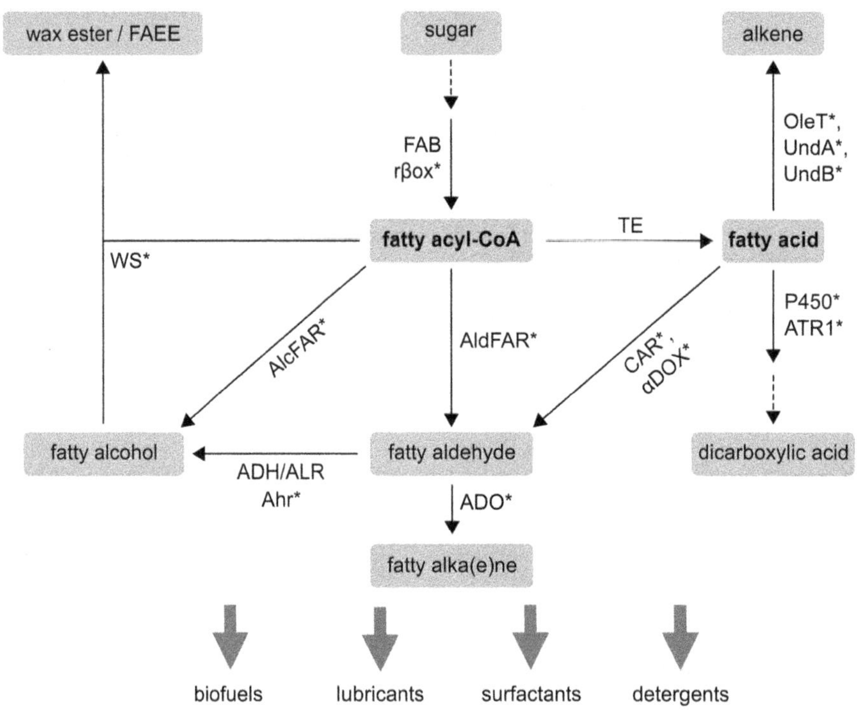

Fig. 3 Enzymatic routes for the production of FA-derived compounds in *S. cerevisiae*. FA-derived compounds can be synthesized from the end product of FA biosynthesis (FAB) or reverse β-oxidation (rβox), the fatty acyl-CoA, and from free FA by several heterologous enzymes in yeast. FAs result from the cleavage of CoA by endogenous thioesterases (TE). Heterologous enzymes are marked with an asterisk. *OleT* H_2O_2-dependent P450 FA decarboxylase, *UndA* medium-chain FA-preferring nonheme iron oxidase, *UndB* membrane-bound desaturase-like enzyme, *P450* cytochrome P450 CYP94C1, *ATR1* cytochrome reductase from *A. thaliana*, *CAR* carboxylic acid reductase from *M. marinum*, *αDOX* α-dioxygenase from *O. sativa* (rice), *FAR* fatty acyl-CoA reductase, *ADO* aldehyde deformylating oxygenase, *ADH* alcohol dehydrogenase, *ALR* aldehyde reductase, *Ahr* aldehyde reductase from *E. coli*, *WS* wax ester synthase

chain in the end product is determined by the acyl chain of the FA precursor and by the specificity of the downstream enzymes.

A. Alkenes

Free FA can be converted to terminal alkenes by a one-step decarboxylation pathway, catalyzed by an H_2O_2-dependent cytochrome P450 enzyme, OleT (Chen et al. 2015; Zhou et al. 2018); a medium-chain FA-preferring nonheme iron oxidase, UndA (Zhu et al. 2017a; Zhou et al. 2018); or a membrane-bound desaturase-like enzyme, UndB (Zhou et al. 2018), together with a reduction system. The major challenge when applying these iron-dependent enzymes is the inefficient electron transfer, resulting in low titers in yeast due to inefficient cofactor or reducing systems (Zhou et al. 2018).

B. Dicarboxylic Acids

Utilizing the **ω-oxidation pathway** in *S. cerevisiae* enables the production of ω-hydroxy- and α, ω-dicarboxylic acids (ω meaning "last" C atom when counted from the carboxyl group) from free FA. Such α, ω-dicarboxylic acids can serve as raw material for commodities and polymers (Han et al. 2017). Free FA can be terminally hydroxylated by a cytochrome P450 enzyme, followed by an oxidation to a carboxyl group by alcohol dehydrogenases and aldehyde dehydrogenases. Feeding of medium-chain FA to an *S. cerevisiae* strain expressing the cytochrome P450 enzyme CYP94C1, together with a cytochrome reductase, ATR1 from *A. thaliana*, allowed the production of ω-hydroxy- and α, ω-dicarboxylic acids with chain lengths ranging from C10 to C16 (Han et al. 2017).

C. Fatty Aldehydes, Alkanes, and Alcohols

For the production of alkanes and alcohols, a fatty aldehyde intermediate is used as a precursor. There are several possible routes for the synthesis of **fatty aldehydes**. For instance, FA can be reduced through action of a carboxylic acid reductase (CAR) from *Mycobacterium marinum* (Zhou et al. 2016b; Tang et al. 2017; Henritzi et al. 2018), or it can be oxidatively decarboxylated by an α-dioxygenase (α-DOX) from *Oryza sativa* (rice) (Jin et al. 2016; Foo et al. 2017). To be active, CAR requires a phosphopantetheinylation by a phosphopantetheinyl transferase (Akhtar et al. 2013). A third possibility is the reduction of CoA-bound FA to the corresponding aldehyde by fatty acyl-CoA reductases (AldFARs). Compared to CAR, it was shown that AldFAR-type enzymes are rather inefficient in yeast (Buijs et al. 2015; Zhou et al. 2016b). Aldehydes can be converted to **odd-chain fatty alkanes** (C_{n-1}) by cyanobacterial aldehyde deformylating oxygenases (ADO) or to **fatty alcohols** by endogenous alcohol dehydrogenases (ADHs) and aldehyde reductases (ALR) (Buijs et al. 2015; Zhou et al. 2016b; Kang et al. 2017; Zhu et al. 2017a). Zhou et al. (2016b) showed that the expression of a CAR together with its activating enzyme, phosphopantetheine transferase NpgA from *A. nidulans*, and an ADO, led to the synthesis of long-chain alkanes (Zhou et al. 2016b). To supply sufficient electrons, a reducing system was expressed additionally (Buijs et al. 2015). As previous studies showed that CAR also shows high activity toward medium-chain FA, Zhou et al. (2016a) produced medium-chain alkanes by screening different ADO orthologs and engineering their substrate binding sites (Zhu et al. 2017a). However, it is reported that ADO was not able to compete with the alcohol-forming enzymes (ADHs/ALRs) due to its low catalytic efficiency (Buijs et al. 2015; Zhou et al. 2016b; Foo et al. 2017; Kang et al. 2017; Zhu et al. 2017a). Another **two-step pathway to produce alkanes** was demonstrated by Foo et al. (2017). Expression of an α-DOX from rice led to the production of long odd-chain fatty aldehyde intermediates, which can subsequently be deformylated to even-chain alkanes (C_{n-2}) by ADO (Foo et al. 2017). The advantage of a dioxygenase is that it uses dioxygen instead of NADPH for the production of aldehydes (Foo et al. 2017). Odd-chain fatty aldehydes produced by α-DOX can also be oxidized to odd-chain fatty alcohols through endogenous yeast ADHs/ALRs (Jin et al. 2016). Zhou et al. (2016b) explored ADHs/ALRs of yeasts and could show that Adh5 plays an important role in the production of long-chain alcohols, as overexpressing *ADH5* together with CAR increased the amount of long-chain alcohols (Zhou et al. 2016b). So far, the expression of fatty alcohol pathways resulted in the production of a pool of long- or medium-chain fatty alcohols with different chain lengths. Recently, our group was able to selectively produce the diesel-like alcohol 1-octanol in *S. cerevisiae* by combining a C8 FA-producing FAS with a two-step reduction pathway composed of CAR together with the phosphopantetheinyl transferase Sfp from *Bacillus subtilis* and the aldehyde reductase Ahr from *E. coli* (Henritzi et al. 2018). It has been shown that Ahr accepts a broad range of aliphatic aldehydes with chain lengths from C4 to C16 (Akhtar et al. 2013).

Another way to produce fatty alcohols is a **one-step reduction** of the fatty acyl-CoA, catalyzed by fatty acyl-CoA reductases (AlcFARs) (Runguphan and Keasling 2014). Zhou et al. (2016b) expressed the CAR/Adh5 pathway together with a fatty acyl-CoA reductase to produce fatty alcohols simultaneously from free FA and acyl-CoAs (Zhou et al. 2016b). Other labs have also reported the functional expression of several heterologous AlcFARs (Feng et al. 2015; d'Espaux et al. 2017). The advantage of one-step reactions, like the direct conversion from fatty acyl-CoAs to alcohols by AlcFARs or the FA decarboxylation pathway for alkene production, is the reduced intermediate metabolite loss and the circumvention of toxic intermediates (Chen et al. 2015). Besides medium-chain and long-chain FA-derived chemicals, **very long-chain FA-derived products** also play an important role as ingredients for lubricants, detergents, polymers, photographic film-processing agents, coatings, cosmetics, and pharmaceuticals (Yu et al. 2017). Yu et al. (2017) were able to selectively produce docosanol (C22) from very long-chain FA by rewiring the native FA elongation system and overexpressing a heterologous mycobacteria FAS I system, which provides high levels of C22 FA as direct precursor, together with a specific AlcFAR (Yu et al. 2017).

D. Wax Esters and Fatty Acid Ethyl Esters

Further compounds of interest include wax esters, which are typically esters of long-chain FA and long-chain alcohols and are used in personal care products, lubricants, or coatings (Wenning et al. 2017). They can be synthesized by **wax ester synthases (WS)** from alcohols and fatty acyl-CoA thioesters (Shi et al. 2012; Runguphan and Keasling 2014). Esters of ethanol and FA with chain lengths ranging from C14 to C20 (**FAEE**) represent suitable diesel fuels (Shi et al. 2012). Shi et al. (2012) could functionally express different WSs from different species and characterize their substrate preference. Unfortunately, some of the WSs from bacteria are bifunctional enzymes, which function as WS and acyl-CoA:diacylglycerol acyltransferase (DGAT), resulting in TAG formation, leading to a depletion of the acyl-CoA precursor pool (Shi et al. 2012). Eriksen et al. (2015) investigated the heterologous expression of a FAS I from *Brevibacterium ammoniagenes* coupled with a WS/DGAT to produce FAEEs. This strategy has the advantage of providing additional FA for growth supplementation and to supply the FA needed for FAEE synthesis (Eriksen et al. 2015). The microbial synthesis of wax esters by the esterification of a (very) long-chain fatty acyl-CoA with a primary, very long-chain fatty alcohol was reported by Wenning et al. (2017). The group combined the expression of a heterologous FAR with the expression of a plant WS to synthesize different jojoba-like wax esters. The used FAR reduced long-chain fatty acyl-CoAs, which resulted from de novo FA biosynthesis and elongation, to a long-chain alcohol. The plant-derived WS enabled the esterification of this alcohol with a long-chain fatty acyl-CoA in *S. cerevisiae* (Wenning et al. 2017).

Overall, there are still many **obstacles** that have to be overcome when using these heterologous enzymes, such as the challenging expression in the heterologous yeast host, low catalytic efficiency, inefficient electron transfer, or loss of intermediates by competing pathways. These issues have to be tackled in order to increase yields and titers of FA-derived compounds. One promising strategy could be a compartmentalization into peroxisomes by a peroxisomal targeting of pathway enzymes (Sheng et al. 2016; Zhou et al. 2016a). Similarly, blocking competing pathways through deletion of specific genes is a common strategy. For instance, the deletion of *HFD1* was shown to be a crucial step for alkane and alcohol biosynthesis (Buijs et al. 2015; Zhou et al. 2016b). *HFD1* encodes an aldehyde dehydrogenase involved in sphingolipid degradation and coenzyme Q biosynthesis and catalyzes the oxidation of fatty aldehydes to FA (Zhu et al. 2017a). Additionally, many engineering strategies for increasing the supply of the precursor AcCoA and free FA also led to an increase in the production of FA-derived products (Zhou et al. 2016b; Teixeira et al. 2017).

VI. Optimization of Fermentation Conditions

Yeast FA titers keep rising through a plethora of metabolic engineering efforts; however, general production process optimization is just as important to unravel the full potential of producer strains. In oleaginous yeasts, like *Y. lipolytica*, lipid overproduction starts with the exhaustion of a primary nutrient, i.e., when entering stationary growth (Beopoulos et al. 2009). Consequently, **nitrogen and/or glucose limitation** has been utilized to increase lipid production in *S. cerevisiae* (Thompson and Trinh 2014; Yu et al. 2018). Such limiting conditions led to high lipid titers in a highly engineered *S. cerevisiae* strain, producing 33.4 g L^{-1} free FA in a fed-batch fermentation—the highest reported free FA titer by microbial fermentation to date (Yu et al. 2018). These results also emphasize the importance of the transition from flask cultivation to **fed-batch cultivation** in a fermenter. The latter permits a tight control of cultivation parameters, such as pH, aeration, and nutrient supply, thereby enabling higher titers of FA and derivatives (Thompson and Trinh 2014; Zhou et al. 2016b; Yu et al. 2018).

In a very recent study, the effect of "forced" FA synthesis on the metabolism and physiology of an engineered *S. cerevisiae* strain was analyzed (Gossing et al. 2018). Such a systematic characterization could help to determine the crucial parameters for improving yeast lipid production. Besides the expected increase in β-oxidation and storage lipids, the analyzed strain also showed higher levels of oxidative stress and decreased amino acid levels (Gossing et al. 2018). Therefore, a viable approach could be an adjustment of **media composition**, e.g., by providing higher amounts of **amino acids**, or overexpressing genes to improve the amino acid uptake or synthesis.

Several previous studies have demonstrated a positive effect of amino acid supplementation on lipid accumulation. The addition of methionine led to increased palmitoleic acid production (Kamisaka et al. 2015), whereas high leucine levels led to increased lipid accumulation in a Δ*snf2* strain (Kamisaka et al. 2007). The additional supplementation of several other amino acids has been demonstrated to contribute to ethanol tolerance in yeast (Hirasawa et al. 2007; Sekine et al. 2007; Pham and Wright 2008; Yoshikawa et al. 2009). Nevertheless, it is also important to consider the genetic background of the FA producing strain at hand, as working with auxotrophic strains can decrease growth (Baganz et al. 1997; Çakar et al. 1999; Basso et al. 2010). Additionally, for some strain series, such as the BY strains, the importance of sufficient amino acid supply for optimal growth has been emphasized (Hanscho et al. 2012).

Vitamins, such as biotin and pantothenate, also play pivotal roles for normal yeast growth and are essential in FA synthesis (Suomalainen and Keränen 1963; Tehlivets et al. 2007). The addition of pantothenate, for example, was used as a metabolic switch to regulate the synthesis of β-farnesene, an AcCoA-derived sesquiterpene (Sandoval et al. 2014). This method could possibly be transferred to FA production processes, as they, similarly to β-farnesene, depend on CoA intermediates.

The **fermentation temperature** has a major influence not only on yeast growth but also on FA tolerance and production (Piper 1995; Viegas and Sá-Correia 1995; Viegas 1997). An engineered *S. cerevisiae* produced more palmitoleic acid at low temperatures (20–25 °C) (Kamisaka et al. 2015), and the oleaginous yeast *Metschnikowia pulcherrima* was shown to produce high lipid levels at low temperatures and pH

(Santamauro et al. 2014). However, at lower temperatures, octanoic and decanoic acid are also more toxic (Viegas and Sá-Correia 1995; Viegas 1997). Furthermore, low temperatures might not be feasible in large-scale production, in which high gravity can lead to elevated temperatures, and cooling of big fermentation plants is cost-intensive (Gibson et al. 2007). Therefore, it might be more desirable to generate thermotolerant *S. cerevisiae* strains by introducing traits of thermotolerant yeasts, such as *Kluyveromyces marxianus* (Cernak et al. 2018), or even thermotolerant oleaginous yeasts, like *R. toruloides* (Wu et al. 2018). Another important cultivation parameter for FA production is the **pH** and media buffering. Short- and medium-chain FAs are more toxic to *S. cerevisiae* at lower pH values, causing decreased viability and yields (Viegas et al. 1989). Therefore, it was shown that the addition of potassium phosphate buffer to complex medium elevated short- and medium-chain FA production substantially (Gajewski et al. 2017).

A promising approach to avoid the re-uptake of the products by the cells and simultaneously decrease the effects of product toxicity is an **in situ extraction**. The addition of dodecane to the culture of a long-chain fatty alcohol-producing *S. cerevisiae* was successfully applied for the extraction of these compounds (Runguphan and Keasling 2014; d'Espaux et al. 2017). Upon addition of dodecane to a culture broth of a 1-octanol producing *S. cerevisiae* strain, however, production was reduced—likely due to the extraction of the precursor octanoic acid by dodecane (Henritzi et al. 2018). Therefore, when considering in situ extraction, an agent needs to be found, which is highly specific for each product, and does not remove important pathway intermediates from the culture.

Current efforts for yeast FA production were carried out in minimal (Zhou et al. 2016b; Yu et al. 2018) as well as complex medium (Leber et al. 2015; Gajewski et al. 2017). To be truly more sustainable than petroleum-based production, and to prevent a competition with food supply, starting materials, such as **lignocellulosic biomass** from agricultural waste, need to be utilized. To extract fermentable sugars from such biomass, it first must undergo pretreatment, in the process of which fermentation inhibitors arise. Therefore, the ideal FA producing yeast needs to be resistant to these inhibitors, as well as to be able to utilize glucose *and* five-carbon sugars, such as xylose, as carbon sources (Peralta-Yahya et al. 2012). D'Espaux et al. (2017) presented the first—and, so far, only—report about FA/alcohol production with *S. cerevisiae* solely from lignocellulosic feedstock. They fed the non-food crops sorghum and switchgrass, which were pretreated with ionic liquids, to an engineered *S. cerevisiae* strain and obtained up to 0.7 g L^{-1} fatty alcohols (d'Espaux et al. 2017). By combining the different engineering and cultivation strategies mentioned above, a further increase in FA titers produced from lignocellulosic biomass can be foreseen.

VII. Conclusions

Much progress has been made in recent years in the development of strategies for microbial production of FA and their derivatives from renewable feedstocks. The available literature shows that extensive interventions into the central carbon metabolism, relying on expression of heterologous pathways and manipulation of the activity of endogenous enzymes, are necessary to enable a high yield of oleochemicals in different chassis organisms. The highest yields of microbially produced oleochemicals reported to date were achieved in *S. cerevisiae*, owing to the great body of knowledge regarding the physiology of this yeast (0.1 g free FA/g glucose, i.e., approximately 30% of the theoretical yield). Moreover, FA production in *S. cerevisiae* has become more chain length-specific, and many FAs and a great variety of derivatives have already been produced successfully in lab-scale fermentations. The development of biosensors and high-throughput screening methods has become increasingly important and will considerably accelerate the development for well-performing strains. Once such challenges as metabolic imbalances and product toxicity have been tackled, yields, titers, and productivity could reach economically viable levels before long.

Acknowledgments L.B., F.W., and M.O. received funding from the European Union's Horizon 2020 research and innovation program under Grant Agreement No. 720824. S.B. received funding from the German Federal Ministry of Food and Agriculture following a decision of the German Bundestag under the Grant Number 22026315. The responsibility for the content of this publication lies with the authors. The authors thank Eckhard Boles for helpful comments on the manuscript.

References

Akhtar MK, Turner NJ, Jones PR (2013) Carboxylic acid reductase is a versatile enzyme for the conversion of fatty acids into fuels and chemical commodities. Proc Natl Acad Sci U S A 110:87–92

Alexandre H, Mathieu B, Charpentier C (1996) Alteration in membrane fluidity and lipid composition, and modulation of H+-ATPase activity in *Saccharomyces cerevisiae* caused by decanoic acid. Microbiology 142:469–475

Alper H, Fischer C, Nevoigt E, Stephanopoulos G (2005) Tuning genetic control through promoter engineering. Proc Natl Acad Sci U S A 102:12678–12683

Alper H, Moxley J, Nevoigt E, Fink GR, Stephanopoulos G (2006) Engineering yeast transcription machinery for improved ethanol tolerance and production. Science 314:1565–1568

Baganz F, Hayes A, Marren D, Gardner DCJ, Oliver SG (1997) Suitability of replacement markers for functional analysis studies in *Saccharomyces cerevisiae*. Yeast 13:1563–1573

Basso TO, Dario MG, Tonso A, Stambuk BU, Gombert AK (2010) Insufficient uracil supply in fully aerobic chemostat cultures of *Saccharomyces cerevisiae* leads to respiro-fermentative metabolism and double nutrient-limitation. Biotechnol Lett 32:973–977

Baumann L, Rajkumar AS, Morrissey JP, Boles E, Oreb M (2018) A yeast-based biosensor for screening of short- and medium-chain fatty acid production. ACS Synth Biol 7:2640–2646

Becker S, Schmoldt H-U, Adams TM, Wilhelm S, Kolmar H (2004) Ultra-high-throughput screening based on cell-surface display and fluorescence-activated cell sorting for the identification of novel biocatalysts. Curr Opin Biotechnol 15:323–329

Beopoulos A, Chardot T, Nicaud J-M (2009) *Yarrowia lipolytica*: a model and a tool to understand the mechanisms implicated in lipid accumulation. Biochimie 91:692–696

Besada-Lombana PB, Fernandez-Moya R, Fenster J, Da Silva NA (2017) Engineering *Saccharomyces cerevisiae* fatty acid composition for increased tolerance to octanoic acid. Biotechnol Bioeng 114:1531–1538

Black PN, DiRusso CC (2007) Yeast acyl-CoA synthetases at the crossroads of fatty acid metabolism and regulation. Biochim Biophys Acta 1771:286–298

Bocanegra JA, Scrutton NS, Perham RN (1993) Creation of an NADP-dependent pyruvate dehydrogenase multienzyme complex by protein engineering. Biochemistry 32:2737–2740

Brignole EJ, Smith S, Asturias FJ (2009) Conformational flexibility of metazoan fatty acid synthase enables catalysis. Nat Struct Mol Biol 16:190–197

Buijs NA, Zhou YJ, Siewers V, Nielsen J (2015) Long-chain alkane production by the yeast *Saccharomyces cerevisiae*. Biotechnol Bioeng 112:1275–1279

Çakar ZP, Sauer U, Bailey JE (1999) Metabolic engineering of yeast: the perils of auxotrophic hosts. Biotechnol Lett 21:611–616

Cardenas J, Da Silva NA (2016) Engineering cofactor and transport mechanisms in *Saccharomyces cerevisiae* for enhanced acetyl-CoA and polyketide biosynthesis. Metab Eng 36:80–89

Caspeta L, Chen Y, Ghiaci P, Feizi A, Buskov S, Hallström BM, Petranovic D, Nielsen J (2014) Altered sterol composition renders yeast thermotolerant. Science 346:75–78

Cernak P, Estrela R, Poddar S, Skerker JM, Cheng Y-F, Carlson AK, Chen B, Glynn VM, Furlan M, Ryan OW, Donnelly MK, Arkin AP, Taylor JW, Cate JHD (2018) Engineering *Kluyveromyces marxianus* as a robust synthetic biology platform host. MBio 9:1–16

Chen B, Ling H, Chang MW (2013a) Transporter engineering for improved tolerance against alkane biofuels in *Saccharomyces cerevisiae*. Biotechnol Biofuels 6:21

Chen Y, Daviet L, Schalk M, Siewers V, Nielsen J (2013b) Establishing a platform cell factory through engineering of yeast acetyl-CoA metabolism. Metab Eng 15:48–54

Chen B, Lee D-Y, Chang MW (2015) Combinatorial metabolic engineering of *Saccharomyces cerevisiae* for terminal alkene production. Metab Eng 31:53–61

Choi JW, Da Silva NA (2014) Improving polyketide and fatty acid synthesis by engineering of the yeast acetyl-CoA carboxylase. J Biotechnol 187:56–59

Clomburg JM, Vick JE, Blankschien MD, Rodríguez-Moyá M, Gonzalez R (2012) A synthetic biology approach to engineer a functional reversal of the β-oxidation cycle. ACS Synth Biol 1:541–554

Clomburg JM, Blankschien MD, Vick JE, Chou A, Kim S, Gonzalez R (2015) Integrated engineering of β-oxidation reversal and ω-oxidation pathways for the synthesis of medium chain ω-functionalized carboxylic acids. Metab Eng 28:202–212

Cottrell M, Viljoen BC, Kock JLF, Lategan PM (1986) The long-chain fatty acid compositions of species representing the genera *Saccharomyces*, *Schwanniomyces* and *Lipomyces*. Microbiology 132:2401–2403

Curran KA, Karim AS, Gupta A, Alper HS (2013) Use of expression-enhancing terminators in *Saccharomyces cerevisiae* to increase mRNA half-life and improve gene expression control for metabolic engineering applications. Metab Eng 19:88–97

d'Espaux L, Mendez-Perez D, Li R, Keasling JD (2015) Synthetic biology for microbial production of lipid-based biofuels. Curr Opin Chem Biol 29:58–65

d'Espaux L, Ghosh A, Runguphan W, Wehrs M, Xu F, Konzock O, Dev I, Nhan M, Gin J, Reider Apel A, Petzold CJ, Singh S, Simmons BA, Mukhopadhyay A, García Martín H, Keasling JD (2017) Engineering high-level production of fatty alcohols by *Saccharomyces cerevisiae* from lignocellulosic feedstocks. Metab Eng 42:115–125

Da Silva NA, Srikrishnan S (2012) Introduction and expression of genes for metabolic engineering applications in *Saccharomyces cerevisiae*. FEMS Yeast Res 12:197–214

David F, Nielsen J, Siewers V (2016) Flux control at the malonyl-CoA node through hierarchical dynamic pathway regulation in *Saccharomyces cerevisiae*. ACS Synth Biol 5:224–233

Davis López SA, Griffith DA, Choi B, Cate JHD, Tullman-Ercek D (2018) Evolutionary engineering improves tolerance for medium-chain alcohols in *Saccharomyces cerevisiae*. Biotechnol Biofuels 11:36

de Jong BW, Shi S, Siewers V, Nielsen J (2014) Improved production of fatty acid ethyl esters in *Saccharomyces cerevisiae* through up-regulation of the ethanol degradation pathway and expression of the heterologous phosphoketolase pathway. Microb Cell Factories 13:39

Dekishima Y, Lan EI, Shen CR, Cho KM, Liao JC (2011) Extending carbon chain length of 1-butanol pathway for 1-hexanol synthesis from glucose by engineered Escherichia coli. J Am Chem Soc 133:11399–11401

Dellomonaco C, Clomburg JM, Miller EN, Gonzalez R (2011) Engineered reversal of the β-oxidation cycle for the synthesis of fuels and chemicals. Nature 476:355–359

Deparis Q, Claes A, Foulquié-Moreno MR, Thevelein JM (2017) Engineering tolerance to industrially relevant stress factors in yeast cell factories. FEMS Yeast Res 17:861

Dietrich JA, McKee AE, Keasling JD (2010) High-throughput metabolic engineering: advances in small-molecule screening and selection. Annu Rev Biochem 79:563–590

Dragosits M, Mattanovich D (2013) Adaptive laboratory evolution—principles and applications for biotechnology. Microb Cell Factories 12:64

Eriksen DT, HamediRad M, Yuan Y, Zhao H (2015) Orthogonal fatty acid biosynthetic pathway improves fatty acid ethyl ester production in *Saccharomyces cerevisiae*. ACS Synth Biol 4:808–814

Feng X, Lian J, Zhao H (2015) Metabolic engineering of *Saccharomyces cerevisiae* to improve 1-hexadecanol production. Metab Eng 27:10–19

Fernandez-Moya R, Da Silva NA (2017) Engineering *Saccharomyces cerevisiae* for high-level synthesis of fatty acids and derived products. FEMS Yeast Res 17:87

Fernandez-Moya R, Leber C, Cardenas J, Da Silva NA (2015) Functional replacement of the *Saccharomyces cerevisiae* fatty acid synthase with a bacterial type II system allows flexible product profiles. Biotechnol Bioeng 112:2618–2623

Foo JL, Susanto AV, Keasling JD, Leong SSJ, Chang MW (2017) Whole-cell biocatalytic and de novo production of alkanes from free fatty acids in *Saccharomyces cerevisiae*. Biotechnol Bioeng 114:232–237

Gajewski J, Pavlovic R, Fischer M, Boles E, Grininger M (2017) Engineering fungal de novo fatty acid synthesis for short chain fatty acid production. Nat Commun 8:14650

Gibson BR, Lawrence SJ, Leclaire JPR, Powell CD, Smart KA (2007) Yeast responses to stresses associated with industrial brewery handling. FEMS Microbiol Rev 31:535–569

González-Ramos D, van den Broek M, van Maris AJA, Pronk JT, Daran J-MG (2013) Genome-scale analyses of butanol tolerance in *Saccharomyces cerevisiae* reveal an essential role of protein degradation. Biotechnol Biofuels 6:48

González-Ramos D, Gorter de Vries AR, Grijseels SS, van Berkum MC, Swinnen S, van den Broek M, Nevoigt E, Daran J-MG, Pronk JT, van Maris AJA (2016) A new laboratory evolution approach to select for constitutive acetic acid tolerance in *Saccharomyces cerevisiae* and identification of causal mutations. Biotechnol Biofuels 9:19

Gossing M, Smialowska A, Nielsen J (2018) Impact of forced fatty acid synthesis on metabolism and physiology of *Saccharomyces cerevisiae*. FEMS Yeast Res 18:153

Han L, Peng Y, Zhang Y, Chen W, Lin Y, Wang Q (2017) Designing and creating a synthetic omega oxidation pathway in *Saccharomyces cerevisiae* enables production of medium-chain α, ω-dicarboxylic acids. Front Microbiol 8:1123

Hanscho M, Ruckerbauer DE, Chauhan N, Hofbauer HF, Krahulec S, Nidetzky B, Kohlwein SD, Zanghellini J, Natter K (2012) Nutritional requirements of the BY series of *Saccharomyces*

cerevisiae strains for optimum growth. FEMS Yeast Res 12:796–808

Henritzi S, Fischer M, Grininger M, Oreb M, Boles E (2018) An engineered fatty acid synthase combined with a carboxylic acid reductase enables de novo production of 1-octanol in *Saccharomyces cerevisiae*. Biotechnol Biofuels 11:150

Hettema EH, Tabak HF (2000) Transport of fatty acids and metabolites across the peroxisomal membrane. Biochim Biophys Acta 1486:18–27

Hettema EH, van Roermund CW, Distel B, van den Berg M, Vilela C, Rodrigues-Pousada C, Wanders RJ, Tabak HF (1996) The ABC transporter proteins Pat1 and Pat2 are required for import of long-chain fatty acids into peroxisomes of *Saccharomyces cerevisiae*. EMBO J 15:3813–3822

Hiltunen JK, Mursula AM, Rottensteiner H, Wierenga RK, Kastaniotis AJ, Gurvitz A (2003) The biochemistry of peroxisomal β-oxidation in the yeast *Saccharomyces cerevisiae*. FEMS Microbiol Rev 27:35–64

Hirasawa T, Yoshikawa K, Nakakura Y, Nagahisa K, Furusawa C, Katakura Y, Shimizu H, Shioya S (2007) Identification of target genes conferring ethanol stress tolerance to *Saccharomyces cerevisiae* based on DNA microarray data analysis. J Biotechnol 131:34–44

Hofbauer HF, Schopf FH, Schleifer H, Knittelfelder OL, Pieber B, Rechberger GN, Wolinski H, Gaspar ML, Kappe CO, Stadlmann J, Mechtler K, Zenz A, Lohner K, Tehlivets O, Henry SA, Kohlwein SD (2014) Regulation of gene expression through a transcriptional repressor that senses acyl-chain length in membrane phospholipids. Dev Cell 29:729–739

Hu Y, Zhu Z, Nielsen J, Siewers V (2018) Heterologous transporter expression for improved fatty alcohol secretion in yeast. Metab Eng 45:51–58

Jin Z, Wong A, Foo JL, Ng J, Cao Y-X, Chang MW, Yuan Y-J (2016) Engineering *Saccharomyces cerevisiae* to produce odd chain-length fatty alcohols. Biotechnol Bioeng 113:842–851

Johnson DR, Knoll LJ, Rowleyo N, Gordon JI (1994) Genetic analysis of the role of *Saccharomyces cerevisiae* acyl-CoA synthetase genes in regulating protein N-myristoylation. J Biol Chem 269 (27):18037–18046. http://www.jbc.org/content/269/27/18037.full.pdf

Johnson AO, Gonzalez-Villanueva M, Wong L, Steinbüchel A, Tee KL, Xu P, Wong TS (2017) Design and application of genetically-encoded malonyl-CoA biosensors for metabolic engineering of microbial cell factories. Metab Eng 44:253–264

Kamisaka Y, Tomita N, Kimura K, Kainou K, Uemura H (2007) *DGA1* (diacylglycerol acyltransferase gene) overexpression and leucine biosynthesis significantly increase lipid accumulation in the Δ*snf2* disruptant of *Saccharomyces cerevisiae*. Biochem J 408:61–68

Kamisaka Y, Kimura K, Uemura H, Yamaoka M (2015) Addition of methionine and low cultivation temperatures increase palmitoleic acid production by engineered *Saccharomyces cerevisiae*. Appl Microbiol Biotechnol 99:201–210

Kaneda K, Imaizumi S, Yano I (1995) Distribution of C22-, C24- and C26-α-unit-containing mycolic acid homologues in mycobacteria. Microbiol Immunol 39:563–570

Kang M-K, Zhou YJ, Buijs NA, Nielsen J (2017) Functional screening of aldehyde decarbonylases for long-chain alkane production by *Saccharomyces cerevisiae*. Microb Cell Factories 16:320

Kim S, Gonzalez R (2018) Selective production of decanoic acid from iterative reversal of β-oxidation pathway. Biotechnol Bioeng 115:1311–1320

Kim S, Clomburg JM, Gonzalez R (2015) Synthesis of medium-chain length (C6-C10) fuels and chemicals via β-oxidation reversal in *Escherichia coli*. J Ind Microbiol Biotechnol 42:465–475

Knappe J, Schacht J, Möckel W, Höpner T, Vetter H, Edenharder R (1969) Pyruvate formate-lyase reaction in *Escherichia coli*. The enzymatic system converting an inactive form of the lyase into the catalytically active enzyme. Eur J Biochem 11:316–327

Knoll LJ, Johnson DR, Gordon JI (1994) Biochemical studies of three *Saccharomyces cerevisiae* acyl-CoA synthetases, Faalp, FaaZp, and Faa3p. J Biol Chem 269:16348–16356

Kocharin K, Chen Y, Siewers V, Nielsen J (2012) Engineering of acetyl-CoA metabolism for the improved production of polyhydroxybutyrate in *Saccharomyces cerevisiae*. AMB Express 2:52

Kocharin K, Siewers V, Nielsen J (2013) Improved polyhydroxybutyrate production by *Saccharomyces cerevisiae* through the use of the phosphoketolase pathway. Biotechnol Bioeng 110:2216–2224

Koopman F, Beekwilder J, Crimi B, van Houwelingen A, Hall RD, Bosch D, van Maris AJA, Pronk JT, Daran J-M (2012) De novo production of the flavonoid naringenin in engineered *Saccharomyces cerevisiae*. Microb Cell Factories 11:155

Kozak BU, van Rossum HM, Luttik MAH, Akeroyd M, Benjamin KR, Wu L, de Vries S, Daran J-M, Pronk JT, van Maris AJA (2014a) Engineering acetyl coenzyme A supply: functional expression of a bacterial pyruvate dehydrogenase complex in the cytosol of *Saccharomyces cerevisiae*. MBio 5: e01696–e01614

Kozak BU, van Rossum HM, Benjamin KR, Wu L, Daran J-MG, Pronk JT, van Maris AJA (2014b) Replacement of the *Saccharomyces cerevisiae* acetyl-CoA synthetases by alternative pathways for cytosolic acetyl-CoA synthesis. Metab Eng 21:46–59

Kozak BU, van Rossum HM, Niemeijer MS, van Dijk M, Benjamin K, Wu L, Daran J-MG, Pronk JT, van Maris AJA (2016) Replacement of the initial steps of ethanol metabolism in *Saccharomyces cerevisiae* by ATP-independent acetylating acetaldehyde dehydrogenase. FEMS Yeast Res 16:fow006

Krivoruchko A, Serrano-Amatriain C, Chen Y, Siewers V, Nielsen J (2013) Improving biobutanol production in engineered *Saccharomyces cerevisiae* by manipulation of acetyl-CoA metabolism. J Ind Microbiol Biotechnol 40:1051–1056

Krivoruchko A, Zhang Y, Siewers V, Chen Y, Nielsen J (2015) Microbial acetyl-CoA metabolism and metabolic engineering. Metab Eng 28:28–42

Kruis AJ, Gallone B, Jonker T, Mars AE, van Rijswijck IMH, JCM W-R, Smid EJ, Steensels J, Verstrepen KJ, SWM K, van der Oost J, Weusthuis RA (2018) Contribution of Eat1 and other alcohol acyltransferases to ester production in *Saccharomyces cerevisiae*. Front Microbiol 9:345

Lafon-Lafourcade S, Geneix C, Ribéreau-Gayon P (1984) Inhibition of alcoholic fermentation of grape must by fatty acids produced by yeasts and their elimination by yeast ghosts. Appl Environ Microbiol 47:1246–1249

Leber C, Da Silva NA (2014) Engineering of *Saccharomyces cerevisiae* for the synthesis of short chain fatty acids. Biotechnol Bioeng 111:347–358

Leber C, Polson B, Fernandez-Moya R, Da Silva NA (2015) Overproduction and secretion of free fatty acids through disrupted neutral lipid recycle in *Saccharomyces cerevisiae*. Metab Eng 28:54–62

Leber C, Choi JW, Polson B, Da Silva NA (2016) Disrupted short chain specific β-oxidation and improved synthase expression increase synthesis of short chain fatty acids in *Saccharomyces cerevisiae*. Biotechnol Bioeng 113:895–900

Lee ME, DeLoache WC, Cervantes B, Dueber JE (2015) A highly characterized yeast toolkit for modular, multipart assembly. ACS Synth Biol 4:975–986

Legras JL, Erny C, Le Jeune C, Lollier M, Adolphe Y, Demuyter C, Delobel P, Blondin B, Karst F (2010) Activation of two different resistance mechanisms in *Saccharomyces cerevisiae* upon exposure to octanoic and decanoic acids. Appl Environ Microbiol 76:7526–7535

Leibundgut M, Maier T, Jenni S, Ban N (2008) The multienzyme architecture of eukaryotic fatty acid synthases. Curr Opin Struct Biol 18:714–725

Lennen RM, Pfleger BF (2013) Microbial production of fatty acid-derived fuels and chemicals. Curr Opin Biotechnol 24:1044–1053

Li X, Guo D, Cheng Y, Zhu F, Deng Z, Liu T (2014) Overproduction of fatty acids in engineered *Saccharomyces cerevisiae*. Biotechnol Bioeng 111:1841–1852

Li S, Si T, Wang M, Zhao H (2015) Development of a synthetic malonyl-CoA sensor in *Saccharomyces cerevisiae* for intracellular metabolite monitoring and genetic screening. ACS Synth Biol 4:1308–1315

Lian J, Zhao H (2015) Reversal of the β-oxidation cycle in *Saccharomyces cerevisiae* for production of fuels and chemicals. ACS Synth Biol 4:332–341

Lian J, Si T, Nair NU, Zhao H (2014) Design and construction of acetyl-CoA overproducing *Saccharomyces cerevisiae* strains. Metab Eng 24:139–149

Ling H, Chen B, Kang A, Lee J-M, Chang M (2013) Transcriptome response to alkane biofuels in *Saccharomyces cerevisiae*: identification of efflux pumps involved in alkane tolerance. Biotechnol Biofuels 6:95

Liu H, Liu K, Yan M, Xu L, Ouyang P (2011) gTME for improved adaptation of *Saccharomyces cerevisiae* to corn cob acid hydrolysate. Appl Biochem Biotechnol 164:1150–1159

Liu P, Chernyshov A, Najdi T, Fu Y, Dickerson J, Sandmeyer S, Jarboe L (2013) Membrane stress caused by octanoic acid in *Saccharomyces cerevisiae*. Appl Microbiol Biotechnol 97:3239–3251

Liu W, Zhang B, Jiang R (2017) Improving acetyl-CoA biosynthesis in *Saccharomyces cerevisiae* via the overexpression of pantothenate kinase and PDH bypass. Biotechnol Biofuels 10:97

Lomakin IB, Xiong Y, Steitz TA (2007) The crystal structure of yeast fatty acid synthase, a cellular machine with eight active sites working together. Cell 129:319–332

Mans R, Daran J-MG, Pronk JT (2018) Under pressure: evolutionary engineering of yeast strains for improved performance in fuels and chemicals production. Curr Opin Biotechnol 50:47–56

Marella ER, Holkenbrink C, Siewers V, Borodina I (2018) Engineering microbial fatty acid metabolism for biofuels and biochemicals. Curr Opin Biotechnol 50:39–46

Meadows AL, Hawkins KM, Tsegaye Y, Antipov E, Kim Y, Raetz L, Dahl RH, Tai A, Mahatdejkul-Meadows T, Xu L, Zhao L, Dasika MS, Murarka A, Lenihan J, Eng D, Leng JS, Liu C-L, Wenger JW, Jiang H, Chao L, Westfall P, Lai J, Ganesan S, Jackson P, Mans R, Platt D, Reeves CD, Saija PR, Wichmann G, Holmes VF, Benjamin K, Hill PW, Gardner TS, Tsong AE (2016) Rewriting yeast central carbon metabolism for industrial isoprenoid production. Nature 537:694–697

Membrillo-Hernandez J, Echave P, Cabiscol E, Tamarit J, Ros J, Lin EC (2000) Evolution of the adhE gene product of *Escherichia coli* from a functional reductase to a dehydrogenase. Genetic and biochemical studies of the mutant proteins. J Biol Chem 275:33869–33875

Michener JK, Thodey K, Liang JC, Smolke CD (2012) Applications of genetically-encoded biosensors for the construction and control of biosynthetic pathways. Metab Eng 14:212–222

Mukherjee K, Bhattacharyya S, Peralta-Yahya P (2015) GPCR-based chemical biosensors for medium-chain fatty acids. ACS Synth Biol 4:1261–1269

Nielsen J, Larsson C, van Maris A, Pronk J (2013) Metabolic engineering of yeast for production of fuels and chemicals. Curr Opin Biotechnol 24:398–404

Nygård Y, Mojzita D, Toivari M, Penttilä M, Wiebe MG, Ruohonen L (2014) The diverse role of Pdr12 in resistance to weak organic acids. Yeast 31:219–232

Oelkers P, Tinkelenberg A, Erdeniz N, Cromley D, Billheimer JT, Sturley SL (2000) A lecithin cholesterol acyltransferase-like gene mediates diacylglycerol esterification in yeast. J Biol Chem 275:15609–15612

Oelkers P, Cromley D, Padamsee M, Billheimer JT, Sturley SL (2002) The *DGA1* gene determines a second triglyceride synthetic pathway in yeast. J Biol Chem 277:8877–8881

Oh CS, Toke DA, Mandala S, Martin CE (1997) *ELO2* and *ELO3*, homologues of the *Saccharomyces cerevisiae ELO1* gene, function in fatty acid elongation and are required for sphingolipid formation. J Biol Chem 272:17376–17384

Paddon CJ, Westfall PJ, Pitera DJ, Benjamin K, Fisher K, McPhee D, Leavell MD, Tai A, Main A, Eng D, Polichuk DR, Teoh KH, Reed DW, Treynor T, Lenihan J, Jiang H, Fleck M, Bajad S, Dang G, Dengrove D, Diola D, Dorin G, Ellens KW, Fickes S, Galazzo J, Gaucher SP, Geistlinger T, Henry R, Hepp M, Horning T, Iqbal T, Kizer L, Lieu B, Melis D, Moss N, Regentin R, Secrest S, Tsuruta H, Vazquez R, Westblade LF, Xu L, Yu M, Zhang Y, Zhao L, Lievense J, Covello PS, Keasling JD, Reiling KK, Renninger NS, Newman JD (2013) High-level semi-synthetic production of the potent antimalarial artemisinin. Nature 496:528–532

Peralta-Yahya PP, Zhang F, del Cardayre SB, Keasling JD (2012) Microbial engineering for the production of advanced biofuels. Nature 488:320–328

Pfleger BF, Gossing M, Nielsen J (2015) Metabolic engineering strategies for microbial synthesis of oleochemicals. Metab Eng 29:1–11

Pham TK, Wright PC (2008) The proteomic response of *Saccharomyces cerevisiae* in very high glucose conditions with amino acid supplementation. J Proteome Res 7:4766–4774

Piper PW (1995) The heat shock and ethanol stress responses of yeast exhibit extensive similarity and functional overlap. FEMS Microbiol Lett 134:121–127

Pronk JT, Yde Steensma H, van Dijken JP (1996) Pyruvate metabolism in *Saccharomyces cerevisiae*. Yeast 12:1607–1633

Rigouin C, Gueroult M, Croux C, Dubois G, Borsenberger V, Barbe S, Marty A, Daboussi F, André I, Bordes F (2017) Production of medium chain Fatty acids by *Yarrowia lipolytica*: combining molecular design and TALEN to engineer the fatty acid synthase. ACS Synth Biol 6:1870–1879

Rodriguez S, Denby CM, van Vu T, Baidoo EEK, Wang G, Keasling JD (2016) ATP citrate lyase mediated cytosolic acetyl-CoA biosynthesis increases mevalonate production in *Saccharomyces cerevisiae*. Microb Cell Factories 15:169

Royce LA, Yoon JM, Chen Y, Rickenbach E, Shanks JV, Jarboe LR (2015) Evolution for exogenous octanoic acid tolerance improves carboxylic acid production and membrane integrity. Metab Eng 29:180–188

Runguphan W, Keasling JD (2014) Metabolic engineering of *Saccharomyces cerevisiae* for production of fatty acid-derived biofuels and chemicals. Metab Eng 21:103–113

Rupilius W, Ahma S (2006) The changing world of oleochemicals. Palm Oil Dev 44:21–28

Sandoval CM, Ayson M, Moss N, Lieu B, Jackson P, Gaucher SP, Horning T, Dahl RH, Denery JR, Abbott DA, Meadows AL (2014) Use of pantothenate as a metabolic switch increases the genetic stability of farnesene producing *Saccharomyces cerevisiae*. Metab Eng 25:215–226

Santamauro F, Whiffin FM, Scott RJ, Chuck CJ (2014) Low-cost lipid production by an oleaginous yeast cultured in non-sterile conditions using model waste resources. Biotechnol Biofuels 7:34

Sanz P (2003) Snf1 protein kinase: a key player in the response to cellular stress in yeast. Biochem Soc Trans 31:178–181

Sarria S, Kruyer NS, Peralta-Yahya P (2017) Microbial synthesis of medium-chain chemicals from renewables. Nat Biotechnol 35:1158–1166

Schadeweg V, Boles E (2016a) Increasing n-butanol production with *Saccharomyces cerevisiae* by optimizing acetyl-CoA synthesis, NADH levels and trans-2-enoyl-CoA reductase expression. Biotechnol Biofuels 9:1

Schadeweg V, Boles E (2016b) n-Butanol production in *Saccharomyces cerevisiae* is limited by the availability of coenzyme A and cytosolic acetyl-CoA. Biotechnol Biofuels 9:1

Schallmey M, Frunzke J, Eggeling L (2014) Looking for the pick of the bunch: high-throughput screening of producing microorganisms with biosensors. Curr Opin Biotechnol 26:148–154

Scharnewski M, Pongdontri P, Mora G, Hoppert M, Fulda M (2008) Mutants of *Saccharomyces cerevisiae* deficient in acyl-CoA synthetases secrete fatty acids due to interrupted fatty acid recycling. FEBS J 275:2765–2778

Schmidt JH (2015) Life cycle assessment of five vegetable oils. J Clean Prod 87:130–138

Schweizer E, Hofmann J (2004) Microbial type I fatty acid synthases (FAS): major players in a network of cellular FAS systems. Microbiol Mol Biol Rev 68:501–517. Table of contents

Sekine T, Kawaguchi A, Hamano Y, Takagi H (2007) Desensitization of feedback inhibition of the *Saccharomyces cerevisiae* γ-glutamyl kinase enhances proline accumulation and freezing tolerance. Appl Environ Microbiol 73:4011–4019

Sheng J, Stevens J, Feng X (2016) Pathway compartmentalization in peroxisome of *Saccharomyces cerevisiae* to produce versatile medium chain fatty alcohols. Sci Rep 6:177

Shi S, Valle-Rodríguez JO, Khoomrung S, Siewers V, Nielsen J (2012) Functional expression and characterization of five wax ester synthases in *Saccharomyces cerevisiae* and their utility for biodiesel production. Biotechnol Biofuels 5:7

Shi S, Valle-Rodríguez JO, Siewers V, Nielsen J (2014a) Engineering of chromosomal wax ester synthase integrated *Saccharomyces cerevisiae* mutants for improved biosynthesis of fatty acid ethyl esters. Biotechnol Bioeng 111:1740–1747

Shi S, Chen Y, Siewers V, Nielsen J (2014b) Improving production of malonyl coenzyme A-derived metabolites by abolishing Snf1-dependent regulation of Acc1. MBio 5:321

Shi S, Ang EL, Zhao H (2018) In vivo biosensors: mechanisms, development, and applications. J Ind Microbiol Biotechnol 45:491–516

Shiba Y, Paradise EM, Kirby J, Ro D-K, Keasling JD (2007) Engineering of the pyruvate dehydrogenase bypass in *Saccharomyces cerevisiae* for high-level production of isoprenoids. Metab Eng 9:160–168

Shin G-H, Veen M, Stahl U, Lang C (2012) Overexpression of genes of the fatty acid biosynthetic pathway leads to accumulation of sterols in *Saccharomyces cerevisiae*. Yeast 29:371–383

Skjoedt ML, Snoek T, Kildegaard KR, Arsovska D, Eichenberger M, Goedecke TJ, Rajkumar AS, Zhang J, Kristensen M, Lehka BJ, Siedler S, Borodina I, Jensen MK, Keasling JD (2016) Engineering prokaryotic transcriptional activators as metabolite biosensors in yeast. Nat Chem Biol 12:951–958

Smidt O, Du Preez JC, Albertyn J (2012) Molecular and physiological aspects of alcohol dehydrogenases in the ethanol metabolism of *Saccharomyces cerevisiae*. FEMS Yeast Res 12:33–47

Stincone A, Prigione A, Cramer T, Wamelink MMC, Campbell K, Cheung E, Olin-Sandoval V, Grüning N-M, Krüger A, Tauqeer Alam M, Keller MA, Breitenbach M, Brindle KM, Rabinowitz JD, Ralser M (2015) The return of metabolism: biochemistry and physiology of the pentose phosphate pathway. Biol Rev Camb Philos Soc 90:927–963

Sumper M, Oesterhelt D, Riepertinger C, Lynen F (1969) Die Synthese verschiedener Carbonsäuren durch den Multienzymkomplex der Fettsäuresynthese aus Hefe und die Erklärung ihrer Bildung. Eur J Biochem 10:377–387

Suomalainen H, Keränen AJA (1963) The effect of biotin deficiency on the synthesis of fatty acids by yeast. Biochim Biophys Acta 70:493–503

Tang X, Chen WN (2015) Enhanced production of fatty alcohols by engineering the TAGs synthesis pathway in *Saccharomyces cerevisiae*. Biotechnol Bioeng 112:386-392

Tang X, Feng H, Chen WN (2013) Metabolic engineering for enhanced fatty acids synthesis in *Saccharomyces cerevisiae*. Metab Eng 16:95–102

Tang X, Feng L, Chen L, Chen WN (2017) Engineering *Saccharomyces cerevisiae* for efficient biosynthesis of fatty alcohols based on enhanced supply of free fatty acids. ACS Omega 2:3284–3290

Tehlivets O, Scheuringer K, Kohlwein SD (2007) Fatty acid synthesis and elongation in yeast. Biochim Biophys Acta 1771:255–270

Teixeira PG, Ferreira R, Zhou YJ, Siewers V, Nielsen J (2017) Dynamic regulation of fatty acid pools for improved production of fatty alcohols in *Saccharomyces cerevisiae*. Microb Cell Factories 16:1185

Teo WS, Hee KS, Chang MW (2013) Bacterial FadR and synthetic promoters function as modular fatty acid sensor- regulators in *Saccharomyces cerevisiae*. Eng Life Sci 13:456–463

Thompson RA, Trinh CT (2014) Enhancing fatty acid ethyl ester production in *Saccharomyces cerevisiae* through metabolic engineering and medium optimization. Biotechnol Bioeng 111:2200–2208

Toke DA, Martin CE (1996) Isolation and characterization of a gene affecting fatty acid elongation in *Saccharomyces cerevisiae*. J Biol Chem 271:18413–18422

Valle-Rodríguez JO, Shi S, Siewers V, Nielsen J (2014) Metabolic engineering of *Saccharomyces cerevisiae* for production of fatty acid ethyl esters, an advanced biofuel, by eliminating non-essential fatty acid utilization pathways. Appl Energy 115:226–232

van Roermund CW, Tabak HF, van den Berg M, Wanders RJ, Hettema EH (2000) Pex11p plays a primary role in medium-chain fatty acid oxidation, a process that affects peroxisome number and size in *Saccharomyces cerevisiae*. J Cell Biol 150:489–498

van Roermund CW, Drissen R, van den Berg M, Ijlst L, Hettema EH, Tabak HF, Waterham HR, Wanders RJ (2001) Identification of a peroxisomal ATP carrier required for medium-chain fatty acid β-oxidation and normal peroxisome proliferation in *Saccharomyces cerevisiae*. Mol Cell Biol 21:4321–4329

van Roermund CWT, Waterham HR, Ijlst L, Wanders RJA (2003) Fatty acid metabolism in *Saccharomyces cerevisiae*. Cell Mol Life Sci 60:1838–1851

van Rossum HM, Kozak BU, Niemeijer MS, Duine HJ, Luttik MAH, Boer VM, Kötter P, Daran J-MG, van Maris AJA, Pronk JT (2016a) Alternative reactions at the interface of glycolysis and citric acid cycle in *Saccharomyces cerevisiae*. FEMS Yeast Res 16: fow017

van Rossum HM, Kozak BU, Pronk JT, van Maris AJA (2016b) Engineering cytosolic acetyl-coenzyme A supply in *Saccharomyces cerevisiae*: pathway stoichiometry, free-energy conservation and redox-cofactor balancing. Metab Eng 36:99–115

van Rossum HM, Kozak BU, Niemeijer MS, Dykstra JC, Luttik MAH, Daran J-MG, van Maris AJA, Pronk JT (2016c) Requirements for carnitine shuttle-mediated translocation of mitochondrial acetyl moieties to the yeast cytosol. MBio 7:517

Viegas C (1997) Effects of low temperatures (9–33 °C) and pH (3.3–5.7) in the loss of *Saccharomyces cerevisiae* viability by combining lethal concentrations of ethanol with octanoic and decanoic acids. Int J Food Microbiol 34:267–277

Viegas CA, Sá-Correia I (1995) Toxicity of octanoic acid in *Saccharomyces cerevisiae* at temperatures between 8.5 and 30°C. Enzym Microb Technol 17:826–831

Viegas CA, Rosa MF, Sá-Correia I, Novais JM (1989) Inhibition of yeast growth by octanoic and decanoic acids produced during ethanolic fermentation. Appl Environ Microbiol 55:21–28

Viegas CA, Almeida PF, Cavaco M, Sá-Correia I (1998) The H(+)-ATPase in the plasma membrane of *Saccharomyces cerevisiae* is activated during growth latency in octanoic acid-supplemented medium accompanying the decrease in intracellular pH and cell viability. Appl Environ Microbiol 64:779–783

Vorapreeda T, Thammarongtham C, Cheevadhanarak S, Laoteng K (2012) Alternative routes of acetyl-CoA synthesis identified by comparative genomic analysis: involvement in the lipid production of oleaginous yeast and fungi. Microbiology (Reading, England) 158:217–228

Waks Z, Silver PA (2009) Engineering a synthetic dual-organism system for hydrogen production. Appl Environ Microbiol 75:1867–1875

Wang BL, Ghaderi A, Zhou H, Agresti J, Weitz DA, Fink GR, Stephanopoulos G (2014a) Microfluidic high-throughput culturing of single cells for selection based on extracellular metabolite production or consumption. Nat Biotechnol 32:473–478

Wang Y, Chen H, Yu O (2014b) A plant malonyl-CoA synthetase enhances lipid content and polyketide yield in yeast cells. Appl Microbiol Biotechnol 98:5435–5447

Wang M, Li S, Zhao H (2016) Design and engineering of intracellular-metabolite-sensing/regulation gene circuits in *Saccharomyces cerevisiae*. Biotechnol Bioeng 113:206–215

Wattanachaisaereekul S, Lantz AE, Nielsen ML, Nielsen J (2008) Production of the polyketide 6-MSA in yeast engineered for increased malonyl-CoA supply. Metab Eng 10:246–254

Welch JW, Burlingame AL (1973) Very long-chain fatty acids in yeast. J Bacteriol 115:464–466

Wenning L, Yu T, David F, Nielsen J, Siewers V (2017) Establishing very long-chain fatty alcohol and wax ester biosynthesis in *Saccharomyces cerevisiae*. Biotechnol Bioeng 114:1025–1035

Williams TC, Xu X, Ostrowski M, Pretorius IS, Paulsen IT (2017) Positive-feedback, ratiometric biosensor expression improves high-throughput metabolite-producer screening efficiency in yeast. Synth Biol 2:1–13

Wu C-C, Tsai Y-Y, Ohashi T, Misaki R, Limtong S, Fujiyama K (2018) Isolation of a thermotolerant *Rhodosporidium toruloides* DMKU3-TK16 mutant and its fatty acid profile at high temperature. FEMS Microbiol Lett 365(21). https://doi.org/10.1093/femsle/fny203

Xu P, Qiao K, Ahn WS, Stephanopoulos G (2016) Engineering *Yarrowia lipolytica* as a platform for synthesis of drop-in transportation fuels and oleochemicals. Proc Natl Acad Sci U S A 113:10848–10853

Xue S-J, Chi Z, Zhang Y, Li Y-F, Liu G-L, Jiang H, Hu Z, Chi Z-M (2018) Fatty acids from oleaginous yeasts and yeast-like fungi and their potential applications. Crit Rev Biotechnol 38:1049–1060

Yang H, Bard M, Bruner DA, Gleeson A, Deckelbaum RJ, Aljinovic G, Pohl TM, Rothstein R, Sturley SL (1996) Sterol esterification in yeast: a two-gene process. Science (New York, NY) 272:1353–1356

Yoshikawa K, Tanaka T, Furusawa C, Nagahisa K, Hirasawa T, Shimizu H (2009) Comprehensive phenotypic analysis for identification of genes affecting growth under ethanol stress in *Saccharomyces cerevisiae*. FEMS Yeast Res 9:32–44

Yu A-Q, Pratomo Juwono NK, Leong SSJ, Chang MW (2014) Production of fatty acid-derived valuable chemicals in synthetic microbes. Front Bioeng Biotechnol 2:1219

Yu A-Q, Pratomo Juwono NK, Foo JL, Leong SSJ, Chang MW (2016) Metabolic engineering of *Saccharomyces cerevisiae* for the overproduction of short branched-chain fatty acids. Metab Eng 34:36–43

Yu T, Zhou YJ, Wenning L, Liu Q, Krivoruchko A, Siewers V, Nielsen J, David F (2017) Metabolic engineering of *Saccharomyces cerevisiae* for production of very long chain fatty acid-derived chemicals. Nat Commun 8:15587

Yu T, Zhou YJ, Huang M, Liu Q, Pereira R, David F, Nielsen J (2018) Reprogramming yeast metabolism from alcoholic fermentation to lipogenesis. Cell 174:1549–1558.e14

Zhang F, Carothers JM, Keasling JD (2012) Design of a dynamic sensor-regulator system for production of chemicals and fuels derived from fatty acids. Nat Biotechnol 30:354–359

Zhang Y, Nielsen J, Liu Z (2018) Metabolic engineering of *Saccharomyces cerevisiae* for production of fatty acid-derived hydrocarbons. Biotechnol Bioeng 110:87

Zhou YJ, Buijs NA, Zhu Z, Gómez DO, Boonsombuti A, Siewers V, Nielsen J (2016a) Harnessing yeast peroxisomes for biosynthesis of fatty-acid-derived biofuels and chemicals with relieved side-pathway competition. J Am Chem Soc 138:15368–15377

Zhou YJ, Buijs NA, Zhu Z, Qin J, Siewers V, Nielsen J (2016b) Production of fatty acid-derived oleochemicals and biofuels by synthetic yeast cell factories. Nat Commun 7:34

Zhou YJ, Hu Y, Zhu Z, Siewers V, Nielsen J (2018) Engineering 1-alkene biosynthesis and secretion by dynamic regulation in yeast. ACS Synth Biol 7:584–590

Zhu Z, Zhou YJ, Kang M-K, Krivoruchko A, Buijs NA, Nielsen J (2017a) Enabling the synthesis of medium chain alkanes and 1-alkenes in yeast. Metab Eng 44:81–88

Zhu Z, Zhou YJ, Krivoruchko A, Grininger M, Zhao ZK, Nielsen J (2017b) Expanding the product portfolio of fungal type I fatty acid synthases. Nat Chem Biol 13:360–362

Zou Z, DiRusso CC, Ctrnacta V, Black PN (2002) Fatty acid transport in *Saccharomyces cerevisiae.* Directed mutagenesis of FAT1 distinguishes the biochemical activities associated with Fat1p. J Biol Chem 277:31062–31071

15 Fungi Involved in the Biodeterioration and Bioconversion of Lignocellulose Substrates

Barry Goodell[1]

CONTENTS

I. Introduction

Fungi that cause deterioration of wood are primarily filamentous in form, and the body of the fungus consists of fine (2–10 μM), hair-like, elongated cells that branch and connect end to end to form an interconnected network (Schmidt 2006). Individual strands of the fungus are known as hyphae, whereas when hyphae grow together in layers on the surface of wood, they are called mycelia or a mycelial matt. Mycelia can also form unique structures such as sporophores (fruiting bodies) in sexually-capable fungi. These fruiting bodies may take the form of cups or discs in some Ascomycota species, while in the Basidiomycota, mushrooms or bracket structures on the side of wood substrates are common. Common/colloquial names for the sporophores of Basidiomycota growing from wood include conks, punks, shelf fungi, and others.

Mold fungi, stain fungi, soft rots, brown rots, and white rots are the groupings typically used for fungi that cause different types of wood degradation. Mold fungi are generally found in the Ascomycota. Traditionally many of the mold fungi were classified as fungi imperfecti or Deuteromycetes (not a formal taxa), and as time has progressed with new molecular techniques, many of these fungi have been able to be classified as either Ascomycota or Basidiomycota. As will be discussed later, mold fungi do not penetrate wood surfaces more than one or two cells, and they do not metabolize structural wood components. Because mold fungi do not degrade cellulose, lignin, or hemicellulose to an appreciable extent, they also do not cause appreciable mass loss as mold infection advances.

Stain fungi are able to penetrate into the sapwood of wood and sometimes living trees, and they feed on the sugars and starches primarily found in the parenchyma cells. They often initiate on the surface of wood, and in that regard in early staining stages, the boundaries between these two types of fungi are blurred, and stain fungi can be confused with mold fungi. Although not common, some fungal species classified generally as stain fungi have also been found to attack the structural components of wood (Encinas and Daniel 1996). In general, however, stain fungi cause only limited mass loss and cause only limited structural damage to wood (Zabel and Morrell 2020).

[1] University of Massachusetts, Amherst, MA, USA; e-mail: bgoodell@umass.edu

Genetics and Biotechnology, 3rd Edition
The Mycota II
J.P. Benz, K. Schipper (Eds.)

Decay and Depolymerization of Lignocellulose Cell Walls

Wood decay fungi including brown rots, white rots, and soft rot fungi generally have the capacity to attack both holocellulose and lignin components. The type of attack will vary as outlined for the specific decay types below, but as decay progresses, the fungi not only depolymerize all or portions of the lignocellulose, but they metabolize particularly the oligosaccharide and sugar fractions of the substrate once adequate depolymerization has occurred to allow depolymerized fractions to be small enough to allow passage through the fungal cell wall and membrane. As depolymerization of holocellulose and lignin occurs, mechanical property loss occurs in the wood (Highley 1988). In some cases, depolymerization is paced to match metabolism by the fungus, and in such cases, mass loss of the wood will parallel mechanical property loss (Goodell 2014). More often however, as in the initial stages of decay by brown rot fungi, rapid depolymerization of the wood cell wall occurs with little metabolism of the cell wall components, so mechanical property loss can advance significantly before large mass loss has occurred (Highley and Illman 1991). Concurrent with mass loss as decay progresses, shrinkage of wood occurs. In some decay types, the wood cell wall itself develops a porous structure so that bulk shrinkage may be less noticeable as mass loss increases. However, in brown rot of softwoods, it has been noted that the wood cell wall collapses upon itself (Daniel 2014) with rearrangement of modified lignin (Goodell et al. 2017). This ultimately results in the cubical, friable appearance of brown rotted wood as this type of decay advances. Shrinkage is also observed in other types of decay but is often masked because of micro-porosity development within the cell walls or when cells delaminate at the middle lamellae uniformly to allow a more uniform appearance of the wood.

Hyphal Progression During Decay

Fungal hyphae are small enough to penetrate into the cell lumen void space of wood cells (Fig. 1). In the decay fungi and stain fungi, the hyphae also travel through and penetrate pit

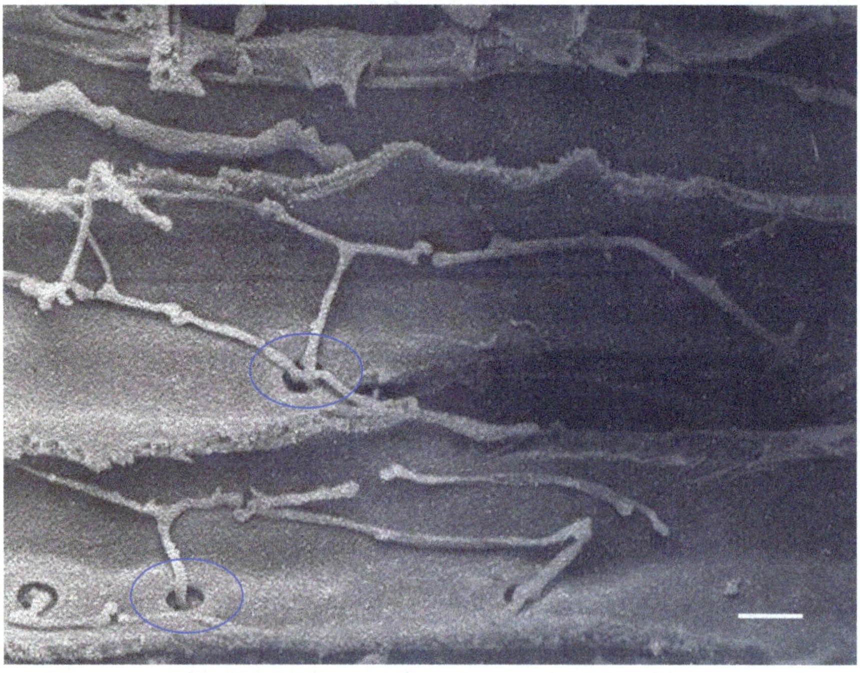

Fig. 1 SEM image of wood decay fungal hyphae ramifying through the tracheid cell walls of a softwood species. In early stages of wood degradation, decay fungi will preferentially penetrate through the bordered pits of wood (blue circles) to traverse from one wood cell to another. Scale bar = 10 μM

membranes such as bordered pits and ray parenchyma pits. In these fungi, bore holes are also typically apparent as the fungal hyphae have the capability of boring transversely into the wood cell walls to pass from one fiber/tracheid to another. The hyphae must ramify across the surface of the wood in the case of molds, or on the surface and throughout interior cells of wood in the case of stain and decay fungi, in order to obtain nutrients, which are required for fungal survival. Extracellular enzymes, and low molecular weight (LMW) fungal metabolites, are secreted by the fungal hyphae to solubilize compounds ranging from simple oligosaccharides to polymerized lignin. The goal of most fungi is to obtain sugars or short-chain oligosaccharides that can be absorbed or actively transported into the fungal thallus. Degraded and modified fragments of lignin can also be metabolized by some fungi, and the hyphae of these types of fungi again are responsible for secretion of metabolites that can depolymerize lignin. But lignin typically is either segregated from cellulose via a biochemical mechanism (brown rots) or is metabolized (white rots) primarily as a means to access the carbohydrate fractions of wood by the fungi.

Fungal Decay and Moisture Content

All fungi require moisture to grow in or on the surface of wood. Generally, the moisture content (MC) required for decay fungi to grow into wood and secrete metabolites into the wood cell wall must be above the fiber saturation point (Zabel and Morrell 2020). This is because enzymes and other degradative components of the fungal LMW metabolites involved in fungal decay processes must be able to diffuse from the fungus through an extracellular fungal matrix (ECM) or biofilm layer which surrounds all fungal hyphae. The ECM is the gateway through which all extracellular metabolites must pass to then further diffuse into the wood cell wall. Liquid water must therefore be present in the lumen of the wood cells for this type of diffusion to occur at the surface of the fungal hyphae. The wood cell wall must also be at, or near, fiber saturation to allow LMW fungal metabolites and ions to diffuse within the wall as a prerequisite for some types of decay initiation. Mold fungi only require moisture to be present on the surface of wood and other substrates, to grow. Many architectural manuals indicate that mold will grow on wood as low as 20% MC. Technically, mold growth is unlikely to occur at less than 30% MC (the fiber saturation point) because, as with all degradative fungi, some amount of liquid water is needed for extracellular enzyme secretion. Practically speaking however, the 20% MC figure is useful because, particularly when the wood MC is in equilibrium with moisture in the air in an enclosed or poorly ventilated space, then liquid water can condense out in an enclosed or poorly ventilated area. Depending on temperature, the relative humidity (RH) required to maintain 20% MC in wood is close to 90% RH. Over a broad range of temperature ranging from 3 to 32 °C a decrease in temperature of only ~ 2 °C will allow the dew point to be reached, and condensation of liquid water at the surface of wood under such conditions can then occur. Although small amounts of surface water on wood would not allow decay in any form to advance, some molds and even surface stains could establish growth, as staining and mold can develop in a matter of a few days at moderate temperatures above approximately 10 °C.

As a caveat to the discussion on MC limitations for fungal growth, very wet wood (from 150 to 200+ % MC) will also limit or stop fungal growth, and most decay fungi have a MC optimum for decay in the range between 50 and 100% MC (Fig. 2) (Zabel and Morrell 2020). MC optima exist for other fungal species as well. For many soft rot fungi, MC optima for decay have been reported to be much higher, "near saturation"; however, aeration of wood samples undergoing soft rot attack has also been assessed to be critical for optimal soft rot decay. This suggests that some soft rot fungi may be able to survive and thrive under conditions where they can pull dissolved oxygen from water. Relative to MC optima for wood decay fungi in different density wood species, it has been observed that lower density woods will continue to decay at high MC levels that stop decay in high-density species. This is because lower-density wood species have proportion-

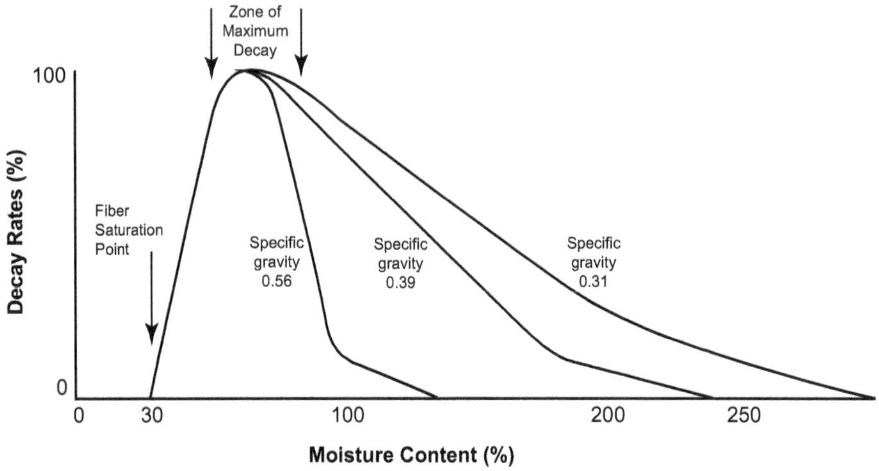

Fig. 2 Decay optima vary with moisture content (MC) and also with the density of wood species. Many other factors, including fungal species, are important in determining rates of decay, but it is important to recognize that aggressiveness of decay can vary simply by altering moisture content, particularly in higher density wood species. Figure used with permission from Academic Press. Zabel and Morrell 1991 (Zabel and Morrell 2020)

ally greater cell lumen volumes compared to denser wood species. This is important because, like most eukaryotes, wood decay fungi require oxygen for growth. Prior research has shown that some decay fungi are capable of decaying wood with oxygen levels even lower than 1% (Highley et al. 1983); however, optimal levels were found at atmospheric oxygen levels (21%) or higher (Kazemi et al. 2001).

II. Wood Decay Fungi: Basidiomycota Decay and Ascomycota Soft Rot Decay

A. Introduction to Decay Types

Basidiomycota fungal genera that decay wood are generally classed informally as either white rot or brown rot fungi although some genomes of genera within these classifications are known to be intermediate between classic model white rot and brown rot species (Riley et al. 2014). Soft rot fungi are also included in this section because they also deconstruct and depolymerize lignocellulose polymers. Wood decay fungi initially seek sites within or on the surface of wood where simple sugars and starches are available, such as in the parenchyma of the wood or other lignocellulosic materials. The use of the simple substrates in initial colonization of woody substrates allows these fungi to develop greater biomass and move through the substrate without the need for expenditure of greater amounts of energy which is needed for the more complex tasks of deconstructing lignocellulose. With most species, when conditions are right and particularly when the simple sugars, starches, and fatty acids are depleted, the cellular machinery for deconstruction of wood cell walls is activated. As depletion of simple compounds progresses, surface pectins and hemicellulose are typically the first polysaccharides to be attacked because these are the most accessible structural components of the wood cell wall and they are generally arranged in a more accessible amorphous structure at the molecular level. As decay further progresses, fungi use different mechanisms, depending on the type, for the depolymerization of cellulose, lignin, and the more tightly bound hemicellulose. The more minor constituents of pectins and extractives that are embedded in the cell wall also are depolymerized and metabolized as cell wall degradation continues. For deconstruction of cellulose and/or lignin, inducible extracellular enzymes are secreted to depolymerize or hydrolyze specific chemical linkages in substrates. Glycosyl hydrolase (GH) enzymes, for example, are enzymes that hydrolyze glycosidic bonds in either cellulose or hemicellulose. There are many different types of GHs and many families of this types of enzyme are now known (Williams 2017). Other enzymes that may be present in some fungal species will target specific linkages in lignin or hemicellulose, such as fungal esterases, that have high affinity for acetylated carbohydrates as found in some hemicelluloses. Some fungal enzymes are also specific for deconstruction of specific types of pectin in the bordered pits or middle lamella, while other enzymes in white rot fungi have broad specificity in degrading aromatic compounds.

White rot fungi primarily employ extracellular degradative enzymatic systems in the degradation of lignocellulose substrates, but in some cases LMW mediators or metal radical ions are known, or have been proposed, to work in conjunction with white rot enzymes to explain the types of degradation patterns observed at the nanoscale. Brown rot fungi decaying wood use a combination of a LMW catalytic system, that can penetrate into the wood cell wall in early stages of attack by the fungus, together with a more limited suite of cellulase enzymes. The pattern of attack by this LMW catalytic system is consistent with attack by hydroxyl radicals ($^{\bullet}OH$), and the most prominent theory on the LMW degradation mechanism involves a mechanism for hydroxyl radical attack (Goodell et al. 2017; Suzuki et al. 2006), which is known as the chelator-mediated Fenton (CMF) system (Kent et al. 2018; Goodell et al. 2017; Arantes and Milagres 2006; Nurika et al. 2019). Brown rot fungi discarded their machinery for production of most peroxidases (lignin degrading enzymes) and also many of their carbohydrate active enzymes (CAZymes) and oxidoreductases as they evolved the LMW non-enzymatic system to biocatalytically deconstruct cellulose and lignin. Although as noted previously, there is intergradation between brown rot and white rot fungi (Riley et al. 2014), this chapter will maintain traditional definitions of the two for convenience, to allow the reader to understand the differences at either end of the spectra.

Basidiomycota fungi evolved approximately 295 million years ago to produce Class II—peroxidase enzymes (PODs); enzymes which allowed these particular progenitors of current day white rot fungi to deconstruct lignin. Basidiomycota species prior to this possessed enzymes capable of depolymerizing and metabolizing cellulose, and therefore they were able to decompose grasses and sedges. However, as tracheophytes evolved as the first plants to incorporate lignin into to their cell walls about 375 million years ago, the fungal species at that time were unable to deconstruct this new lignified plant cellular material. Lignification allowed plants to grow larger and taller. Researchers have suggested that the gap of approximately 90 million years between lignin production in plants and the ability of fungi to decay those plants is in-part responsible for the buildup of vegetation which resulted in current-day coal seams in many locations on earth (Floudas et al. 2012; Eastwood 2014).

It is important to note the different roles of enzymes and LMW catalytic systems in the wood degradation process. Although extracellular enzymes are often casually discussed in the scientific literature as being capable of penetrating lignified plant cell walls, there is a large body of literature that demonstrates that, although enzymes can erode the surfaces of plant cell walls at the molecular level, they are too large to penetrate the intact structure of secondary plant cell walls for more than a nanometer at best (Tepfer and Taylor 1981; Flournoy et al. 1991). In plants such as corn and *Arabidopsis*, where secondary walls do not exist, and in particular when delignification procedures have been used (Ding et al. 2012), enzymatic erosion and penetration of one to a few elementary microfibril layers have been observed. However, in intact, heavily lignified, secondary cell walls of wood, enzymes do not readily penetrate. This has been demonstrated in white rot fungi and brown rot fungi, via the use of many different probes of sizes simulating those of enzymes (Stone and Scallan 1967; Tepfer and Taylor 1981; Kleman-Leyer et al. 1992; Arantes et al. 2012). In the soft rot Ascomycota, it has been suggested that LMW mediator compounds are involved with some enzymatic attack, but this has not been well studied in this type of decay. Brown rot fungi, and some types of white rot fungi, employ LMW catalytic systems that allow small metabolites to penetrate relatively deep into cell walls, and even into the middle lamella, to catalyze reactions that depolymerize and solubilize cell wall components and overcome the limitations imposed by the bulkiness and size of enzymes relative to the unmodified pore structure of the wood secondary cell wall (Highley and Illman 1991; Arantes et al. 2012). The importance of non-enzymatic radical-generating mechanisms, and the lack of enzyme penetration into intact lignified cell walls, is sometimes ignored by pathologists and microbial physiologists studying plant cell walls in the scientific literature, leading to erroneous conclusions about how cell wall degradation and even microbial pathogenesis involving cell walls occurs.

B. Brown Rot Fungi

Brown rot fungi comprise only about 6% of all known Basidiomycota species, yet they degrade approximately 80% of wood, by mass, in the northern hemisphere and other regions of the world (Eastwood 2014). Although brown rot fungi are present globally, in nature they tend to attack softwoods preferentially or to be "generalists" (Krah et al. 2018), and therefore, they have greater impact in the northern hemisphere (primarily boreal forests) of the world and in other regions where softwoods predominate. Brown rot wood decay fungi are perhaps the most destructive organisms of wood on earth, and for this reason, considerable space in this chapter will be devoted to their action.

Brown rot fungi are classified as such because of the brown color of the wood residue which remains after fungal degradation has progressed to advanced stages. The wood in this stage is also classically described as being friable, crumbly, checked across the grain and sometimes having a cross-hatched appearance (Fig. 3). Early decay or incipient decay stages of brown rot, defined as less than 10% mass loss, often will appear visually unchanged from undecayed wood other than appearing to be wet in some areas. Because brown rot fungi have a unique LMW catalytic mechanism for initiating decay in wood, the LMW components can diffuse through the wood cell wall rapidly to depolymerize both lignin and holocellulose components. It is this chelator-mediated Fenton (CMF) depolymerization, particularly of the crystalline cellulose backbone of the elementary fibrils of wood, that dramatically reduces the mechanical properties of wood undergoing brown rot attack. In early work, Wilcox (1978) reviewed literature demonstrating that 70% of both modulus of elasticity and modulus of rupture could be lost with degradation by brown rot fungi at a level of only 10% mass loss. Wood at this stage would not appear brown and crumbly, and this is the key reason why early brown rot decay stages are considered dangerous from a structural perspective, because it can be impossible visually to tell how much strength remains in a wood sample and the extent that the wood has been attacked by brown rot fungi is unknown. For this reason, it is critically important to keep untreated wood protected from moisture when wood is used in either interior or exterior structural applications.

As reviewed above, even in initial stages of brown rot degradation of wood, both holocellulose and lignin are depolymerized. The fact that lignin is extensively depolymerized by brown rot fungi has been recognized for some time. In the 1990s, it was also recognized that hydroxyl radicals would rapidly depolymerize and then repolymerize lignin in modified form (Barr and Aust 1994; Goodell et al. 1997). Researchers also confirmed that lignin depolymerization occurred by the action of hydroxyl radicals in brown rot fungi (Yelle et al. 2011) and postulated further that repolymerization of the modified lignin occurred: "via radical coupling of the phenolic units that become enriched during attack on the aromatic rings by $^{\bullet}OH$." Because lignin is entangled with hemicellulose and encrusts the cellulose elementary fibrils in wood, depolymerization of lignin is an essential byproduct of hydroxyl radical attack

Fig. 3 Brown rotted wood decay from the collection of Holzforschung München, Technical University of Munich. The interior wood is in an advanced stage of brown rot degradation. At this stage, the wood is brown in color and often has a cross-hatched or cubical appearance. From visual appearance alone is not possible to determine if the exterior wood of this piece has undergone early stages of brown rot degradation and has lost any mechanical properties. (Image courtesy of Ralf Rosin, Holzforschung München, TUM)

on lignocellulose (see below) in CMF brown rot degradation. Modification of the chemistry of the lignin phenolic units would also not be possible without depolymerization. Further, the modified lignin residue that is described in older literature as "slightly modified lignin" is actually quite extensively modified in that aliphatic side-chain cleavage and demethoxylation have occurred, and then the majority of the modified aromatic subunits have repolymerized to produce the brown, crumbly polymerized mass that is observed in heavily brown-rotted wood. These wood residues should therefore be described as containing *extensively* modified lignin, even though—chemically—there are similarities between the original lignin and brown-rotted lignin.

Genome sequencing of many brown rot decay fungi has occurred only within the past decade, and prior to that, brown rot fungi were generally considered to be primitive species which had not evolved a complete suite of enzymes to deconstruct holocellulose or lignin. The reason why brown rot fungi evolved from the white rot progenitors, discarding many of their extracellular enzyme systems in the process and evolving the LMW catalytic mechanism, is unknown. However, several researchers have postulated that this shift from reliance solely on extracellular enzymes to reliance primarily on a system like the CMF system is more favorable from the perspective of obtaining energy from a recalcitrant substrate like wood (Martinez et al. 2009; Eastwood 2014; Arantes and Goodell 2014). This more recent evolution of brown rot genera is the reason that these fungi currently represent only a small percentage of all Basidiomycota wood decay species compared to the predominant white rot species. However, despite evolving later, the brown rot fungi have exploited an environmental niche relative rapidly, and the vast majority of softwoods in the world are degraded by brown rots. This suggests that shifting to a LMW degradative system, with a smaller complement of CAZymes, provides an evolutionary advantage to the brown rots. As discussed by Eastwood (Eastwood 2014): "*The brown rot mechanism [. . .] has evolved from a white rot ancestry at least 5 times. This suggests that there is a strong selection pressure for ability to cast off the energetically expensive need to depolymerize lignin in conifer-dominated habitats.*"

1. Non-enzymatic Mechanisms of Brown Rot Fungal Wood Decay

As noted previously, brown rot fungi lack the full complement of CAZymes and oxidoreductases necessary to deconstruct cellulose. Early researchers studying fungal attack of wood recognized the dilemma of having a class of fungi that aggressively decomposed wood, but did not have the biochemical machinery known to be required at the time for this degradation. In the 1960s, Cowling (Cowling 1964; Pettersson et al. 1963; Cowling 1961) noted that "readily diffusible enzyme systems" must be responsible for the rapid depolymerization of cellulose in brown rots, but 40 years later Cowling expressed regret for describing the active component as a very small cellulase rather than a LMW catalytic compound of a more general type (personal communication). Highley, and also Nicholas and their coworkers in the 1980s and 1990s, noted that brown rot decay resembled wood that was treated with Fenton reagent ($Fe^{2+} + H_2O_2 \rightarrow Fe^{3+} + {}^{\bullet}OH + OH^-$); however, results using Fenton treatment of wood alone, or Fenton treatment with cellulolytic enzymes, were not able to fully mimic the action of brown rot (Highley and Illman 1991; Schmidt et al. 1981; Jin et al. 1990a; Parra et al. 1998). The excellent research of these early researchers demonstrated that Fenton chemistry alone was not responsible for brown rot degradation and that the non-enzymatic mechanism in brown rot was more complex. Some research on brown rot has ignored this early research, with non-enzymatic brown rot mechanisms being described as simple "Fenton" chemistry, which does not reflect the complexity of the chelator-mediated mechanism.

Research in the 1990s explored the role of cellobiose dehydrogenase, with the finding that this enzyme may play a role in iron reduction and subsequent generation of hydroxyl radicals; however, this work has been discounted because of the limited number (one) of brown rot species found to produce cellobiose dehydrogenase. Instead, there are currently two viable working theories that describe how a LMW catalytic system functions in brown rot fungi to solubilize wood cell wall components in advance of fungal enzymatic action occurring in the wood cell lumen:

(i) The laboratory of Tanaka and Enoki has posited that pyridine coenzymes provide electrons to LMW glycopeptides which can then penetrate the wood cell wall to generate hydroxyl radicals (Kido et al. 2015). It remains unclear whether the reductants required for the glycopeptide systems persist in extracellular environments, and it must also be explored whether it is feasible for glycopeptides to diffuse into the wood cell wall to reduce iron and then diffuse back out to the coenzyme surrounding the fungal hyphae to complete the redox cycle. Because enzymes are too large to penetrate intact secondary wood cell walls, pyridine coenzymes and other enzymes would also not be able to penetrate, at least until advanced stages of decay when cell wall porosity is increased. Further, as noted by Hatakka and Hammell (Hatakka and Hammel 2011) in discussing research with glycopeptides isolated from a white rot fungus (Tanaka et al. 2007): "*the inferred molecular masses of the encoded peptides are around 14 kDa, much larger than reported for the substances that were first isolated from colonized wood.*" Still, genomic analysis has found comparatively high levels of genes encoding for Fe^{3+}-reducing glycopeptides in some brown rot species (Riley et al. 2014), and continued research on this theory will be important to demonstrate what role expressed glycopeptides may play in brown rot degradation of wood.

(ii) Work in the early 1990s through the present by Jellison and Goodell, Hammel and Houtman, and other groups has demonstrated that brown rot fungi produce LMW compounds that reduce iron in a catalytic (repeated) manner, and these compounds participate in oxidative reactions that depolymerize both cellulose and lignin. This hypothesis has been increasingly supported by the fungal degradation research community over the last 25 years. Initial reports described these compounds as siderophores because they were isolated using procedures adapted from those used to isolate siderophores from bacteria (Jellison et al. 1991). Because siderophore receptor sites were not identified on the fungal cell membranes and the catalytic function of siderophores that was observed in the compounds isolated from brown rot fungi has not been widely reported, the terminology of LMW "chelators" has been used rather than "siderophore." The term "chelator-mediated Fenton" (CMF) system is now used to describe the system (Goodell et al. 2017; Cragg et al. 2015; Arantes et al. 2012; Kent et al. 2018) with the seminal work on the CMF mechanism in fundamental form described in the mid-1990s (Goodell et al. 1997). Two types of chelators, oxalate and a phenolate/hydroquinone, are involved with some brown rot fungi including perhaps the most widely studied brown rot fungus *Gloeophyllum trabeum*. In this mechanism, oxalate first sequesters oxidized iron within the low pH environment of the fungal extracellular matrix (ECM) produced within the wood cell lumen. The fungus regulates oxalate concentrations carefully in the cell lumen, and it has been proposed that oxalate-bound iron diffuses into the wood cell wall at a rate regulated by both intracellular and extracellular fungal mechanisms. Oxalate will release iron to phenolate/hydroquinone chelators in the environment of the wood cell wall where pH is higher and oxalate concentration is lower than in the immediate environment surrounding the fungal hyphae. Numerous groups have demonstrated the production of 2,5-dimethoxyquinone (2,5-DMHQ) and other chelators such as phenolate/hydroquinone-type chelators. These chelators, including variegatic acid in some species (Zhu et al. 2016), are produced by brown rot fungi and are capable of multiple iron reduction (MIR) similar to that described which occurs with methoxylated phenolic compounds derived from lignin (Tamaru et al. 2019). These compounds are proposed to diffuse from the fungal hyphae into the wood cell wall where the higher pH of the cell wall (5.5–6.0) has then been shown to permit a transfer of iron from oxalate to the redox-cycling chelators. Iron reduction then spontaneously occurs in the higher pH environment. Once this chemistry is initiated, a sustained reduction reaction will occur to generate a stream of hydroxyl radicals (Tamaru et al. 2019). The production of hydrogen peroxide occurs naturally in the presence of ferrous iron (Barb et al. 1951; Varela and Tien 2003). The generation of hydrogen peroxide (30 μM concentration) directly from water in microdroplet form has also been reported to occur (Lee et al. 2019), and similar reactions could also potentially generate hydrogen peroxide in the microvoids of the wood cell wall, particularly when transition metals are present. The presence of peroxide assumes adequate oxygenation of the cell wall although the later systems demonstrates peroxide generation directly from water microdroplets in oxygen-limited environments. In any case, because oxygen is a prerequisite for fungal growth, an oxygenated system would be present in most active fungal environments.

Research suggests that early upregulation of a large group of LMW compounds and enzymes that may generate LMW catalysis occurs prior to upregulation of glycoside hydrolase enzymes (Zhang et al. 2016). The upregulation of these metabolites is timed so that subsequent upregulation of CAZymes follows in a staggered manner (Zhang et al. 2016). These findings are consistent with both glycopeptide and CMF working theories, and the staggered upregulation hypothesis also does not preclude mechanisms that allow spatial diffusion of LMW metabolites as has been described for the CMF mechanism. Polyketide synthases (PKS) have also been found to be upregulated in the brown rot fungi (Riley et al. 2014), and this may be important due to the role these synthases play in the production of LMW metabolites ranging from siderophores to fungal antibiotics. Some of these LMW metabolites derived from polyketide pathways are structurally related to the LMW compounds isolated from brown rot fungi that have been proposed for redox chemistries involved in non-enzymatic degradation schemes (Paszczynski et al. 1999; Kerem et al. 1999; Goodell et al. 1997).

The generation of hydroxyl radicals is a consistent feature that is recognized in the LMW catalytic system in brown rots and that is also proposed to occur in the glycopeptide hypothesis in these fungi. The hydroxyl radical is the most potent oxidant known in biological systems, and it is therefore important that it be generated in a location where it cannot damage fungal hyphae. Relative to the CMF mechanism, hydroquinones, catecholates, and similar redox-cycling compounds are greatly affected by pH, and under low pH conditions with free iron, they can be expected to reduce multiple moles of iron which is then able to generate hydroxyl radicals (Tamaru et al. 2019; Goodell et al. 1997) as has been demonstrated in the CMF mechanism. Further exploration which advances our understanding of how hydroxyl radicals are generated by CMF or other mechanisms must take into account either how the fungus is protected from hydroxyl radical action (Zhu et al. 2016) or how hydroxyl radical generation is spatially controlled to react within the wood cell wall (Goodell et al. 2017). More research is needed in this area to explore non-enzymatic catalytic mechanisms in the brown rot fungi and in microorganisms in general. This type of mechanism for the non-enzymatic generation of hydroxyl radicals at a distance from the organism to solubilize a substrate while also preventing oxidative damage to the microorganism itself is so far unknown outside of the brown rot fungi.

2. Enzymatic Degradation in the Brown Rot Fungi

Riley et al. (Riley et al. 2014) published a paradigm-shifting paper discussing a new approach to the classification of brown rot and white rot fungi. As might be expected for fungi that do not metabolize lignin, lignin-degrading peroxidase enzymes have been lost in the brown rots, although the genomes of some brown rot species have been shown to encode for laccases. Brown rot fungi have a reduced number of enzymes acting on crystalline cellulose, with cellobiohydrolase enzymes absent or lacking a critical cellulose-binding domain (Riley et al. 2014). Lytic polysaccharide monooxygenase (LPMO) enzymes are generally also reduced in the brown rot fungi compared to white rot species. The Gloeophyllales and Boletales families have one and two cellobiose dehydrogenase enzymes, respectively, but other brown rot families and genera analyzed have lost these enzymes completely. Endoglucanases are considered to be one of the most important enzymes in brown rot fungi, yet they are reduced in number compared to the white rots. A clean division between the brown rots and white rots cannot be drawn relative to the number and types of hemicellulase and pectinase enzymes that are produced by these species (Riley et al. 2014). Additional research must still confirm particular aspects of the research demonstrating the interaction between non-enzymatic and enzymatic brown rot mechanisms; however, with the reduction in CAZyme complement including LPMO and endoglucanases in brown rots, the non-enzymatic systems clearly take on a priority role in the brown rot fungi.

More study is needed to understand whether interactions between enzymes and metabolomic compounds are important in brown rot degradation, for example, if some of the LMW compounds produced by the fungi are important as electron donors for some enzymes like LPMOs. It has been reported that lignin derivatives produced during brown rot degradation are active in multiple iron reduction, and it is proposed that these compounds also function as electron donors for enzymes such as LPMOs (Tamaru et al. 2019; Westereng et al. 2015; Frommhagen et al. 2017). Understanding whether temporal staging in expression of LMW metabolites and extracellular enzymes is important requires further study. Currently, it is unclear if staging of non-enzymatic and enzymatic systems is required to protect extracellular enzymes from non-enzymatic radical generating systems or if generation of the non-enzymatic radicals within the wood cell wall where enzymes have no accessibility allows adequate protection of extracellular enzymes. Brown rot fungi may have additional mechanisms that have yet to be discovered, but certainly, the use of dual mechanisms (both spatial separation and temporal expression staging) to protect critical enzymes is possible.

3. Dry Rot

"Dry rot" is a colloquial term that has been used to describe a particular brown rot decay type caused by the genus *Serpula* with some validity, as this genus has the ability to channel water from several meters away to the site of active decay. Therefore, although the wood still requires water to decay the wood, the requirement that the wood be wet before decay initiation does not hold with this type of brown rot fungus. *Serpula lacrymans* is common in Europe and Asia, whereas *Serpula* (*Meruliporia*) *incrassata* is the most common of the dry rot species in North America (Boddy 2016). *Serpula* dry rots are true brown rot fungi, but their unique ability to channel water from sources at a distance from the wood being infected is permitted by an elongated hyphal mass known as a rhizomorph structure. Multiple rhizomorphs are often present and have been reported to extend as much as 10 m from the site of decay to reach sources of water. Interior decay of secondary wood structures with fungal rhizomorphs extending to wet soil sites exterior to the structure have been personally observed by the author.

The term "dry rot" is also sometimes used by laypersons as a misnomer, particularly in North America, to describe brown-rotted wood caused by genera other than *Serpula*. In these cases, the wood has been decayed by a brown rot fungus in a wetted state, and it has the appearance of brown crumbly wood in the dry state. The term may be incorrectly used in this way by laypersons observing wood that has been wetted at some point in its history, and decay had initiated and progressed to an advanced state before water was removed and the wood dried out, but it is not an accepted use by professionals in the field. As discussed generally for fungi that attack wood, the presence of an adequate amount of water, in amounts greater than the fiber saturation point (FSP) of the wood, is necessary for all types of fungal degradation to occur. Brown rot decay cannot proceed in the absence of water.

C. White Rot Fungi

Additional background information on white rot degradation of wood and the properties of white-rotted wood can also be found in Sect. IV of this chapter on biotechnological applications of wood degrading fungi. White rots are classified in different ways, but the most common is to physically classify the fungus by the appearance of the decayed wood. Two distinct types of white rot from this macroscopic perspective are (**a**) decayed wood that appears whitish or bleached in a uniform manner and is variously known as "spongy white rot," "stringy white rot," or because it is the most common type, just "white rot" (Fig. 4a) and (**b**) decayed wood that is variously known as "pocket white rot" or

Fig. 4 (**a**) The classic stringy and bleached appearance of white-rotted wood in a hardwood. This type of decay is often colloquially referred to as "spongy white rot" or "stringy white rot." (**b**) Pocket white rot in a hardwood on the slopes of Mount Kilimanjaro with the author's hand for scale. Note the presence of white-rotted wood and fungal mycelium within the pockets

"pecky white rot" where the wood undergoes advanced decay in discrete locations to produce isolated zones of severely degraded and bleached wood, with these zones or pockets typically filled with white mycelium or bleached wood residue (Fig. 4b). White rot fungi can also be categorized based on the chemistry of components that have been removed and specifically whether holocellulose or lignin is preferentially attacked within the wood cell wall. In this type of classification, (a) where cellulose, hemicellulose, and lignin wood components are all oxidized and metabolized at approximately the same rate, the decay is classified as "simultaneous white rot," whereas in (b) "selective white rot," the literature has suggested different scenarios, and it is likely that in different selective white rot species, slightly different rates in substrate removal preference occur. Typically, hemicellulose and lignin are attacked preferentially to cellulose in selective white rot, allowing the cellulose to remain relatively undegraded. However, in some reports, cellulose is also attacked, but it is removed at a slower rate than in simultaneous white rot. As noted in the introduction, in nature there is an intergradation of different decay types (Riley et al. 2014), and the same fungal species can produce different strains of both simultaneous and selective white rot decays (Otjen and Blanchette 1986; Otjen et al. 1987).

1. White Rot Polysaccharide Degradation Mechanisms

White rot fungi are characterized by their ability to produce a complete enzymatic system capable of degrading hemicellulose and cellulose and to directly or indirectly oxidize and mineralize lignin (Aust 1995; Abdel-Hamid et al. 2013; Cragg et al. 2015). Both endoglucanases and exoglucanases that can act synergistically on crystalline cellulose are produced. Enzymatic systems for breakdown of holocellulose include endo-1,4-β-glucanase, β-glucosidases, cellobiohydrolases, cellobiose dehydrogenases, and LPMO enzymes (Baldrian and Valaskova 2008) as well as xylosidase, xylanase, acetyl xylan esterase, glucuronidase, and arabinofuranosidase, with these later enzymes being necessary for complete depolymerization and oxidation of hemicellulose (Rytioja et al. 2014).

The non-enzymatic processes known in white rot fungi, particularly for hemicellulose depolymerization and selective white rot attack, are less well understood, but LMW compounds moving into the wood cell wall in advance of enzyme action have been well documented by Daniel's laboratory (Kim et al. 2015; Daniel 2014) and are reviewed below for specific enzymes: Relative to holocellulase-active enzymes, white rot fungi have a complete complement of CAZymes including the endo-acting enzymes that also are possessed by brown rot fungi. The white rot fungi also possess processive exo-acting cellulases, also known as cellobiohydrolases (CBH). The cellulase enzymes in white rot fungi are primarily in the glycoside hydrolase (GH) families which attack cellulose hydrolytically, but also include the LPMO enzymes (auxiliary activities (AA) enzyme family 9—formerly GH61) which oxidatively cleave cellulose. White rot fungi have many more enzymes that are active on crystalline cellulose, in particular enzymes in the GH6 and GH7 (CBH enzyme) families. In addition, they have many more cellulose-binding module (CBM1) family enzymes. For most GH enzymes to be optimally functional, inclusion of a CBM as part of the enzyme is important. Seven known families of hemicellulase enzymes and eleven pectinase enzymes are encoded in the Basidiomycota, but as discussed in the section on brown rot fungi, there are no notable differences between white rot and brown rot species relative to the enzyme families for hemicellulases and pectinases in the two types of decay fungi (Riley et al. 2014).

2. White Rot Lignin Degradation Mechanisms

Because of the unique nature of the lignin-degrading enzymes in white rot fungi, this section provides greater detail on these systems. Enzymes of importance include lignin peroxidases, versatile peroxidases, manganese peroxidases, dye-decolorizing peroxidases, and laccase. Individual white rot fungal species will possess one or more of these enzymes.

The degradation of lignin in the white rot fungi is biochemically complex. Lignin-degrading enzymes and the biochemical mechanisms employed by these enzymes to oxidize lignin with LMW compounds have been well described in a number of reviews (Tuor et al. 1995; Leonowicz et al. 1999). A recent review of *Trametes* showed that the genome encodes for nine putative lignin peroxidases, seven putative short manganese peroxidases, and two putative versatile peroxidases; however only one manganese peroxidase and one versatile peroxidase isozyme were consistently expressed under different growth conditions (Vasina et al. 2017). This suggests the importance of these particular two isozymes but also indicates that the fungus does not rely on these enzymes alone, and it possesses several "back-up strategies." Some enzymes have the capability of reacting directly with lignin as a substrate, while others use mediators or LMW compounds involved in electron transfer. Some enzymes are also metal-dependent, requiring chelated or a free transition metal to be in close proximity to the enzyme and substrate in ways not yet fully elucidated for all enzymes. Non-enzymatic systems in white rot fungi generally are unable to penetrate as deeply into the wood cell wall compared to the non-enzymatic extracellular metabolites associated with brown rot fungi. Instead, the LMW compounds in these systems appear to be more intimately associated with the enzymes and may be mediator radicals, oxygen radicals, electron-transfer agents, or chelate-radicals that are generated by action of the enzymes and/or transition metal cofactors. These LMW compounds penetrate the wood cell wall in advance of the enzymes (Daniel 2003; Arantes et al. 2011). In some selective white rot fungal degradation, altered cell wall structure associated with these LMW compounds has been observed using transmission electron microscopy (Fig. 5). The altered lignocellulose structure suggests that these LMW compounds penetrate completely through wood cell walls including the middle lamella regions in advance of enzymes (Messner et al. 2003; Daniel 2003, 2014). It has been reported as far back as the 1960s that the middle lamellae is often attacked early in some types of white rot degradation of wood, suggesting that some LMW mediators or long-lived radicals promoted by the white rot fungal attack are capable of penetrating deep into cell walls in early white rot attack to preferentially attack lignin-rich zones (Cowling 1961; Daniel 2014). However, despite these

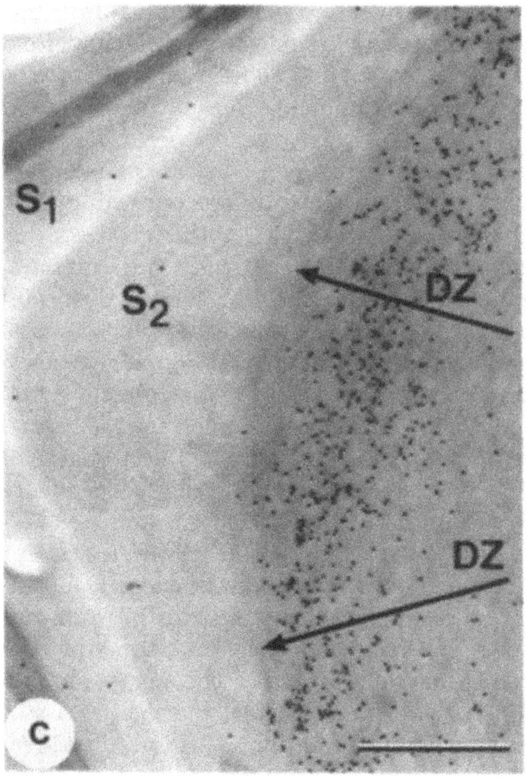

Fig. 5 Selective white rot degradation in wood structure showing S_1 and S_2 wood cell wall layers and the penetration of fungal lignin peroxidase enzymes (immunolabelled with gold and appearing as numerous small black spheres in the image). Darker zones of the wood cell wall indicate that cell wall modification is occurring, and this modification (DZ arrows) appears in advance of lignin peroxidase enzymatic penetration into the cell wall. This indicates the action of fungal LMW compounds in these zones. This type of low molecular weight radical in the degradation of wood is also associated with manganese peroxidase (MnP), versatile peroxidase, and laccase enzymes (Fackler et al. 2007) in selective white rot degradation. (Image courtesy of Dr. Geoffrey Daniel)

research findings on LMW compound penetration of wood cell walls in advance of cellulose- and lignin-degrading enzymes in white rot fungi, research on any deep-diffusing LMW oxidizing compounds in the white rot fungi is still quite limited, as reviewed below for specific lignin-degrading enzymatic systems. A better understanding of the nature of LMW compounds, and which particular enzyme systems they are associated with, would enhance understanding of fungal mechanisms and would also aid future research on utilization of fungal systems in biorefineries and in other bioprocessing efforts.

Lignin Peroxidases Lignin peroxidase (LiP) is able to oxidize both phenolic and non-phenolic lignin groups, with the latter composing 70–90% of the wood lignin fraction. Lignin peroxidase will oxidize only the exposed components of the wood cell wall at the lumen surface, and therefore it is proposed to oxidize smaller intermediates (mediators) such as veratryl alcohol. Given the size of the veratryl alcohol radical, it would have the capacity to penetrate into the wood cell wall from the lumen surface and may be involved in the gradual erosion of the wood cell from the S_3 layer outward. It follows that the affinity of the radical for non-polar substrates such as lignin could potentially provide a role for such a radical (assuming it is capable of diffusing away from the enzyme) for the selective attack on lignin in selective white rot degradation. Lignin peroxidase's ability to oxidize non-phenolic lignin has made it an attractive candidate for use in biopulping, dye decolorization, and biorefinery applications, but it has been more limited in these applications than desired. Several research groups (Mäkelä et al. 2015; Pérez-Boada et al. 2005; Fernandez-Fueyo et al. 2012) discussing the use of LiP have proposed that veratryl alcohol may function as a cation radical that can oxidize lignin substrates indirectly, whereas versatile peroxidase (below) is able to oxidize Mn^{2+} directly allowing direct enzymatic action on substrates. Although these researchers have speculated that the cation radical was closely associated with the enzyme, it is possible that lignin monomers functioning as cation radicals could function as LMW agents diffusing ahead of LiP.

Manganese Peroxidases Manganese peroxidases (MnP) are Class II fungal peroxidases capable of oxidizing Mn^{2+} to Mn^{3+}. Low redox potential phenolic compounds are able to be oxidized directly through this action. However, mediators such as unsaturated fatty acids or glutathione are required for the degradation of high redox potential non-phenolic compounds.

It has been reported that carboxylic acids can also function as a mediator in the degradation of high redox potential substrates (Qin et al. 2017). Some aliphatic acids such as malonate, lactate, and oxalate produced by white rot fungi can function as metal-chelating agents and increase the oxidation rate of Mn^{2+} to allow substrate attack (Hofrichter 2002). The chelated Mn^{3+} functions as a LMW diffusible redox mediator that can attack phenolic lignin to produce unstable radical species. The chelated Mn^{3+} is capable of diffusing into the wood cell wall to oxidize phenolic lignin compounds, and this is one of the potential LMW diffusible species observed by Daniel as cited above (Daniel 2014). MnP has a lower redox potential than LiP and under physiological conditions can only oxidize phenolic lignin structures.

Versatile Peroxidases Versatile peroxidases (VP) are considered to be hybrids between LiP and MnP (Knop et al. 2016), and they are enzymes with broad substrate activity. VP can directly oxidize non-phenolic substrates such as veratryl alcohol as well as phenolic substrates (Fernandez-Fueyo et al. 2012; Mäkelä et al. 2015). VPs are more widely produced in white rot fungi than originally considered when they were first cloned and sequenced in the year 2000 (Giardina et al. 2000). They are now known to be produced in many fungi, such as *Phanerochaete chrysosporium*, *Ceriporiopsis subvermispora*, *Pleurotus ostreatus*, *P. eryngii*, and *Bjerkandera* sp. (Floudas et al. 2012). VPs have been reported to oxidize phenolic and non-phenolic substrates directly without the aid of a mediator, and they have the capacity to oxidize aromatic substrates directly.

Dye-Decolorizing Peroxidases The dye-decolorizing peroxidases (DyP) were first reported in 1995 in a *Geotrichum* (Ascomycota) species but have recently become more widely reported in white rot fungi. The DyP-type peroxidases have no homology to other known peroxidases. They are unable to oxidize Mn^{2+}, and they are unique enough that they are classified in their own superfamily of heme peroxidases (EC 1.11.1.19). They are similar to VPs in that they can oxidize Reactive Black 5, phenols, and veratryl alcohol, but unlike VPs they also have the ability to oxidize recalcitrant anthraquinone dyes (ExPASy 2018; Lauber et al. 2017).

Laccase Unlike the peroxidases, the laccase enzyme employs a mechanism involving a four-electron reduction of oxygen to degrade lignin. Laccase is produced by many microbial organisms, both degradative and non-degradative, and may play multiple metabolic roles. Laccase alone is not capable of oxidizing non-phenolic lignin compounds but can oxidize "mediator" compounds to their radical forms, and in close proximity to laccase, these compounds then catalyze the oxidation of non-phenolic lignin. Natural mediators have been reported, including ortho-diphenols, para-diphenols, methoxysubstituted phenols, diamines, and benzenethiols. Since laccase is a robust enzyme, its potential for industrial use is high, and many uses have been studied ranging from the grafting of polymers to lignin to the production of fiberboard products (Mate Diana and Alcalde 2017). Several lignin-derived compounds (Camarero et al. 2005), and also hydroxamate siderophores from fungi, have been reported as useful in mediating laccase-aided delignification processes (Niku-Paavola et al. 2003). Fungi are generally reported to produce hydroxamic acid siderophores in the broader literature; however, research on wood-degrading fungi has isolated many catecholate compounds that are "siderophore-like" in their structure. Given that the structure of lignin has more catecholate properties, it would be useful also to assess the catecholate/phenolate compounds similar to siderophores that are produced by wood-decaying Basidiomycota species as laccase mediator compounds. Several LMW lignin-derivative compounds have now been reported to have metal reduction capabilities and have been postulated to also function in electron-transfer capacity for extracellular enzymes (Tamaru et al. 2019). Laccase-derived lignin fragments have also been found to promote the activity of lytic polysaccharide monooxygenases (LPMOs) for potential enhancement of biorefinery activities (Brenelli et al. 2018).

D. Soft Rot/Ascomycota Fungi

Soft rot fungal degradation is traditionally considered to be a type of decay where the outer few millimeters of wood surface wood are attacked in relatively wet environments by Ascomycota genera. In some cases, the fungal species causing soft rot attack of wood may still informally be classified as fungi imperfecti, when these fungi have not yet been examined using molecular systematics. The surface attack by soft rot often occurs when the wood is in a waterlogged or partially submerged condition such as in driftwood or industrial cooling tower slats, resulting in a loss of surface density (hardness). The traditional perspective (since the 1950s) of how soft rot fungi attack wood is changing however, as it is now known that soft rot fungal attack can occur in both wet and dry environments. Further, it has also been observed that these fungi can penetrate several centimeters deep into the wood, particularly when the wood has been buried, as observed in field tests and in-service poles (Greaves 1977).

Researchers have pointed out that in wet environments, the exclusion of oxygen deep in wood probably prevents growth of soft rot fungi other than at the surface of wood (Worrall et al. 1991). Soft rot fungi seem to have a more selective requirement for oxygen than Basidiomycota species and some bacteria, and therefore decay fungi or bacteria that attack wood would outcompete soft rot fungi in many environments.

All wood undergoing decay ultimately has reduced density, so to some degree the term soft rot is misnomer. Further, previous reports (Daniel and Nilsson 1998) have also noted that soft rotted wood, particularly in chromated copper arsenate (CCA wood preservative)-treated poles, is quite hard at the surface. Upon drying, soft rotted wood also becomes brown and develops surface checks across the grain as the wood shrinks. Although the decay may be superficial, the surface appearance may be similar to brown rot decay. The wood may assume a weathered appearance, similar to unpainted "barn board" (Zabel et al. 1985). Several common genera of soft rot include *Chaetomium*, *Humicola*, and *Lecythophora*. Soft rot in North America is less commonly reported as a structurally degrading microorganism compared to brown rot, and it is more widely reported in other environments, ranging from tropical to temperate and even polar environments (Blanchette et al. 2004).

Two different types of attack on the wood cell wall can be produced by soft rot fungi. **Type I soft rot** is characterized by cavity formation, which typically occurs in the S_2 layer of the wood cell wall (the thickest layer of the secondary cell wall of lignified woody tracheids or fibers—Fig. 6). These cavities may also be formed in some cases in the adjacent S_1 layers, often in the same cell. **Type II soft rot** is a general erosion of the wood cell wall layers similar to that seen in white rot, and similarly, starting from the S_3-lumen interface and working outward. Often, particularly in hardwoods, both Type I and Type II attack can be produced by the same fungus in the same sample. Soft rot fungi are sensitive to high levels of lignin and preferentially grow in low-lignin hardwoods such as *Populus* or *Betula*. Lignin levels in the wood also impact the pattern of attack which can occur within the wood cell wall (Daniel and Nilsson 1998).

Type I Soft Rot:

Soft rot fungi typically initiate Type I attack using fine microhyphae, to penetrate from the lumen and bore perpendicularly into the wood cell wall (Unligil and Chafe 1974; Daniel 2014; Hale and Eaton 1985). The fungal microhyphae reorient and realign in the direction of the S_2 cellulose microfibrils when they reach the S_2 layer. T-branching of the hyphae also occurs once the hyphae are within the S_2 layer, with the two hyphal tips of the branches then growing in opposite directions, spiraling in the long direction of the fiber. At the start of cavity formation, the microhyphae extend longitudinally into the S_2 cell wall via development of what

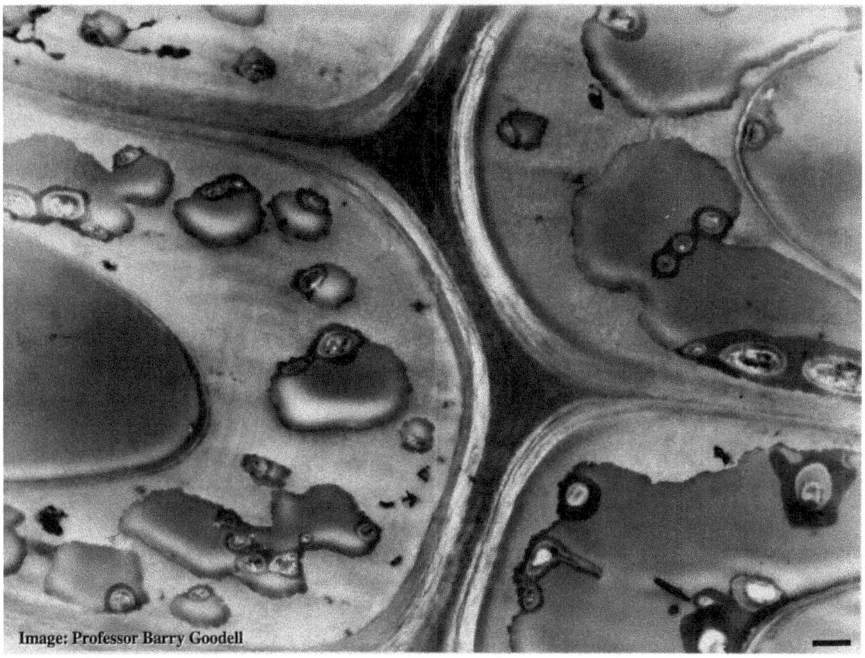

Fig. 6 Transmission electron micrograph of southern pine wood (xs) that was used as a cooling tower slat which was then decayed by Type I soft rot. Cavity formation shows the fungal hyphae growing within the S_2 layer of the wood cell walls and producing cavities which surround individual hyphae. The cavities often coalesce in this heavily degraded sample

is known as a proboscis hypha. The process of enzymatic secretion and cavity formation then is used to produce a series of longitudinally oriented diamond-shaped or elongated conical cavities in what is known as Type I attack (Fig. 6). Multiple hyphae often invade a wood cell wall and can completely riddle the wall in advanced stages of degradation. The conically pointed ends of the cavities are formed presumable because of the way that the extracellularly secreted enzymes interact with the crystalline structure of cellulose in the cell wall.

Soft rot fungi are capable of producing a complete complement of CAZymes, including exo-1,4-β-glucanase, endo-1,4-β-glucanase, and 1,4-β-glucosidase (Tanaka et al. 1992; Goodell et al. 2008). Research using a variety of agar tests shows a number of cellulase, hemicellulase, pectinases, and laccase enzymes to be produced in many of 27 freshwater soft rot ascomycetes tested (Simonis et al. 2008), with several of these species not having been commonly reported in the soft rot literature previously. Earlier work also indicated that laccase was produced by certain soft rot fungi, suggesting that there is at least a limited ability to enzymatically (Tanaka et al. 2000) degrade wood cell wall lignin. Other research indicates that laccase is the only lignin-degrading enzyme produced by at least one soft rot species (Liers et al. 2011). Researchers (Daniel and Nilsson 1998) have hypothesized that a radical-generating system may also function in conjunction with a system for enzymatically oxidizing lignin, which would suggest laccase mediator involvement. They have also reported that cavity formation may be "flattened" on one side where the fungus encounters more heavily-lignified wood and that lignin-rich wood is not as readily degraded as wood that is cellulose-rich. However, Type I soft rot fungi have been observed to lose cavity-formation capability when moving into delignified wood cell walls. This suggests that the orientation of cellulose elementary fibrils, and also the surrounding encapsulating lignin matrix, may play a role in cavity formation.

Type II Soft Rot:

Type II soft rot attack is less common, with only a few species reportedly capable of performing this type of soft rot alone. Type II soft rot appears similar to white rot at the microscale, as the wood cell wall is thinned from the lumen surface (S_3), outward, until only the middle lamella remains (Daniel and Nilsson 1998). This type of soft rot decay rarely occurs in softwoods, with hardwoods being attacked preferentially—with those species that possess a combination of syringyl- and guaiacyl-type lignins and generally having lesser amounts of total lignin in the cell walls, more readily attacked. Most species of soft rot fungi isolated from softwood poles are capable of producing both Type I and Type II soft rot attack with one, *Phialophora fastigiata*, producing Type II attack followed sporadically by Type I attack with longer exposure in laboratory testing (Morrell and Zabel 1985).

III. Staining and Mold Fungal Degradation of Wood

This section reviews fungi that inhabit the surface of wood or that penetrate primarily into the parenchyma of sapwood and cause only limited amounts of degradation to the non-structural wood components. It is important to recognize that other causes of discoloration and surface coloration of wood and wood products occur, and casual observation by laypersons is often not enough to distinguish fungal stain and mold from these other causes.

A. Mold Fungi

Mold fungi typically are classified in the Ascomycota, but where sequencing has not been done and the sexual stage of the mold fungus is unknown, many are still classified as fungi imperfecti. Mold fungi inhabit the surface of wood and feed on simple surface sugars and starches. Typically, the mycelial growth is hyaline (translucent), and the spores are pigmented, which discolors the surface of wood resulting in a degraded appearance in finished lumber or other appearance-graded wood products. Most mold fungi grow on a variety of surfaces, and they require only surface moisture and a nutrient source for growth. The same species can often grow on inert objects like glass or plastic when those materials are coated or spotted with a thin layer of plant or animal residues that contain enough nutrient to support growth of the mold. Mold fungi such as *Trichoderma/Hypocrea* are commonly used in bioprocessing and have been engineered to produce special enzymes for biorefinery and other industrial applications.

When growing on the surface of wood, sugars from parenchyma will substitute for the sugar residues from other plant or animal residues to enable fungal mold growth. The presence of mold on structural lumber does not cause any structural damage, and in many areas of the world, structural lumber with mold growth is marketed and classified with no defect; the same as un-molded wood. However, for esthetic reasons and also because of growing concern about mold spores from a human and animal health perspective (World-Health-Organization 2009), it may be desirable to prevent mold from occurring on wood or remediating any mold growth. As reviewed earlier in the introduction of this chapter, some architectural manuals indicate that fungi—including mold—will grow on wood at only 20% MC. Technically, this is incorrect as the mold fungi require liquid moisture to survive and sporulate; however, at 20% MC with the cycling of temperature, water can condense on the surface of wood to permit mold growth. In these localized zones of the wood surface, liquid water in the surface capillaries would occur, permitting mold growth.

B. Staining of Wood by Fungi

As noted in the section on soft rot fungi, some stain fungi can intergrade with, and cause, soft rot under certain conditions. Stain fungi can also intergrade with mold fungi, and in early stages of stain fungal growth, when these fungi may grow just on the surface of wood and produce spores, they may be considered to be mold fungi. However, true stain fungi have the ability to penetrate into wood to seek out sugars, starches, but also wood resins and waxes (Mäkelä et al. 2015) in the parenchyma cells. Some stain fungi also have the ability to remove the more amorphous polysaccharide components such as pectins and hemicellulose. Staining fungi typically produce fine bore holes in wood (Fig. 7a) although these are smaller diameter bore holes than seen with wood

a

b

47

Fig. 7 (**a**) Fine bore holes produced by blue stain fungi in pine sapwood tracheids. The bore holes (circled in yellow) with size-constricted hyphae are typically much smaller compared to those produced by Basidiomycota decay fungi. (**b**) Blue stain in early stages of development on pine wood. Coloration of the hyphae is developing in the sapwood, but not the heartwood. Similarly, perithecial fruiting body development also is present on the surface of the sapwood and the hyaline/clear spore mass droplets are being exuded at the tip of the perithecia. At the right, the spore droplets have coalesced to form a larger pool of spore exudate

decay Basidiomycota fungi, as the hyphae become constricted in size as they penetrate the wood cell wall. The penetration of the hyphae may be in part via mechanical action as opposed to enzymatic/catalytic degradation of the wall to produce the bore hole. The formation of appressoria to initiate the bore hole production is evident in blue staining hyphae. Daniel points out that wood species or types that are less lignified, such as tension wood in hardwoods, can be more readily attacked by stain fungi, and hemicellulose is more readily removed by stain fungi in this type of wood (Encinas and Daniel 1995; Daniel 2016).

Fungal stain genera commonly include *Ceratocystiopsis*, *Ophiostoma*, and *Grosmannia*, with genomic information increasingly separating and reclassifying some stain fungi into new genera (De Beer et al. 2013a, b). In species that produce a sexual stage, these fungi produce dark colored, flask-shaped, spore-bearing structures known as perithecia that are produced on the surface of wood. Spore-containing asci are produced in the base of the perithecia, and as the perithecia matures, the asci and spores are exuded through the neck of the perithecia for dissemination. Typically, the spore mass is discharged in a moist viscous fluid droplet which rests at the top of the opening in the perithecia neck (Fig. 7b). Although spores can be dispersed by wind and rain water, many species of fungal stain that produce perithecia have adapted to be disseminated by beetles. The height of the perithecia neck for particular fungal species has evolved to match the height and anatomy of particular beetles or other insects and arachnids (particularly mites). The viscous fluid mass of spores is often described as a sticky slime, which adheres to the body or legs of the insects/arachnids as they feed upon the fungal mycelium and spores. The spores are then disseminated as the vector travels to other sites. An excellent review of wood staining by fungi and other agents (Uzunovic et al. 2008) also reports that log harvesting equipment can be a major factor in fungal stain dissemination. Spores will germinate on suitable wood surfaces, but the wood must be at or above the fiber saturation point, at least in localized surface areas, for stain fungi to grow and penetrate.

Because these fungi must first gain a foothold in wood by accessing simple monomeric or oligosaccharide substrates in the sapwood, they typically do not penetrate deeply into heartwood, as simple sugars and starches have been depleted in the heartwood, and fungistatic or fungitoxic phenolic extractives are present instead. For this reason, these fungi are also sometimes known as "sapstain" fungi. As the fungal hyphae mature in wood, they produce melanin in their cell walls, which is thought to be useful in the protection of fungal hyphae from sunlight UV exposure and resulting oxygen radical damage. This melanization also occurs in some sporophores and is presumed also to protect against sunlight. These fungi may cause the wood to take on the appearance of different colors, and in some cases the wood can appear blue, green, yellow, red, or other shades of the spectrum. Blue staining is perhaps most noticeable, and it is caused when fungal hyphae grow in the parenchyma, especially the ray parenchyma, just under the surface of the wood. As the fungal hyphae melanize, they take on a dark brown or black coloration. This pigmentation is reflected through the crystalline layers of cellulose, and as the light is scattered by the crystalline cellulose, different colors may be apparent. Most often a blue or gray color is apparent to the eye, and for this reason this type of wood is often described as "blue-stained." The appearance of different colors is an optical illusion, much in the way that dark soot particles in smoke can appear blue, or other colors, as they scatter and reflect light.

Stain discoloration can occur under favorable conditions in as little as 3 days in lumber. Some softwood genera with wide sapwood bands and large resin canals such as pines (*Pinus*) will stain more readily than wood species that have more heartwood and smaller/absent resin canals such as spruce (*Picea*) or fir (*Abies*). Staining typically occurs at temperatures above 15 °C, and moving wood to a cold environment to arrest stain development is a potential remediation method in some cases. However staining at lower temperatures has been reported in some areas where the fungi have apparently adapted to colder climates (Miller and Goodell 1981). Prevention of stain development can be achieved by keeping logs wet to exclude oxygen needed by the fungi for growth. In mill yards, water sprinklers above log decks are sometimes used for this purpose. Biocidal treatments to protect against wood staining can also be used, although rapid processing to get logs sawn and the lumber then dried to 19% MC or less is preferred from both an environmental and expense standpoint whenever possible.

Fungal staining in structural lumber and wood products is allowed in many countries as no change in modulus of rupture or modulus of elasticity occurs in this wood. However, some wood products specifically exclude fungal stained wood. Also, international shipments of wood with fungal stain (and mold) have been limited in some cases (NeLMA 2016; FAO 2016). High-temperature kiln drying of wood can kill the fungi, and for some countries, this is a permissible solution if international shipment of wood with fungal stain is a concern. Fungal stain is also a concern in certain wood products that may undergo shock loading as is the case in ladder stock where a misplaced foot, when ascending or descending, may result in a rapid loading of the step below. The removal of hemicellulose in heavily stained wood causes loss of toughness or shock resistance by as much as 15–30% (Forest-Products-Laboratory 2010) and hence, this wood is not allowed by code in structural applications that may experience shock loading or that require toughness.

IV. Biotechnological Applications of Decay Fungi

A. White Rot Fungi

As noted earlier in this chapter, the term "white rot fungi" refers to a group of organisms which attack all of the major wood constituents, primarily via enzymatic action. Often white-rotted wood, especially in late stages of decay, is characterized by residual wood material which appears bleached. In some common types of white rot decay, the wood will become soft and develop a "stringy" character where the softened wood fibers can be easily separated allowing the wood to be peeled apart. In other types of white rot, pockets of softened, deteriorated zones of wood may appear. Like all wood-degrading fungi, white rot fungi require oxygen, water, and a suitable wood substrate for growth. In general, the preferred substrates for white rot fungi are hardwood species or bamboo. Compared to brown rot attack, enhanced growth with corresponding greater mass loss will typically occur when hardwoods are inoculated with white rot fungi because these fungi have the ability to metabolize lignin rather than just depolymerizing and modifying lignin as occurs in the brown rots.

White-rotted wood has historically been used for a variety of applications, ranging from insulated paneling in Russian refrigerator trucks/vehicles to cattle feed. In the later application, the selective white rot fungi free the cellulose from the lignin fraction of the wood or fiber, improving the digestibility in ruminant animals. White-rotted wood that has been naturally decayed by *Ganoderma* spp. in southern Chile has been used as food by indigenous peoples and as cattle feed for centuries (Dill and Kraepelin 1986). Biotechnological and biorefinery applications of white rot fungi have been explored for more than 30 years (Blanchette et al. 1988; Wolfaardt et al. 2004; Eriksson and Kirk 1985) with the initial focus being to free cellulose from lignin to enhance "biopulping" prospects. Composite wood products have also been bonded together by using lignin, and the lignin on the surface of wood fibers, after activation by white rot fungal enzymes including laccase and peroxidase enzymes (Widsten and Kandelbauer 2008), with laccase enhancement through the use of mediators. White rot fungi also have a long history of testing for use in applications in the remediation of pollutants and xenobiotics (Barr and Aust 1994; Magan et al. 2010), at least in aerobic environments with adequate moisture for fungal growth.

More recently, interest in biorefineries and the production of cellulose-derived sugars for fermentation and direct conversion to biofuels and platform chemicals has been a targeted use for white rot fungi. Proposed uses for selective white rot fungi include using their potential for selective delignification of biomass in biotechnological applications including biorefineries (Dashtban et al. 2010; Baba et al. 2011). In this regard, white rot species including *Ceriporiopsis subvermispora*, *Phlebia subserialis*, *Dichomitus squalens*, *Pycnoporus cinnabarinus*, and many others now have been tested for potential use in biorefinery applications (Capolupo and

Faraco 2016). Pretreatment by these organisms has been shown to reduce the energy required in refining wood chips in mechanical pulp production and to improve certain strength and brightness properties of paper. White rot fungi such as *Phanerochaete chrysosporium*, *Phanerochaete crassa*, and *Pleurotus pulmonarius* have been studied as potential bio-bleaching agents for both Kraft pulp and sulfite pulp. The CAZymes and lignin-degrading enzymes from these fungi, and others discussed earlier in this chapter, have been studied and cloned into yeast and bacterial vectors for use in biorefinery applications. Further advances in this area with the application of enzyme suites, including the use of lytic polysaccharide monooxygenase (LPMO) enzymes from decay fungi to enhance sugar and phenolic yield from biomass is expected in the future (Bissaro et al. 2018).

Given the need for sustainable production of a wide range of products beyond pulp and paper however, interest in the past 5 years has also been on the conversion of biomass, including lignin, to produce platform chemicals for biopolymers. The use of lignin as an "economic pull-through" product to provide additional high-value products over and above the holocellulose-derived products is of increasing interest in the field. This has opened up potential for the use of white rot fungi, as well as brown rot fungi, in these applications.

B. Brown Rot Fungi

Less information is available relative to the use of brown rot fungi in biotechnological applications. In part this is because the removal of lignin from wood pulp was the primary focus of microbial/enzymatic conversion of lignocellulose in the 1970s and 1980s, and white rot fungi with peroxidases and laccases were the most appropriate biotechnological tools for this task. It was thought that brown rot fungi had no enzymes useful in this regard; however, more recent studies on delignification have showed that some types of LMPOs may interact with lignin degrading enzymes to promote the degradation and release of lignin (Li et al. 2019), suggesting that cellulase enzymes from either white rot or brown rot fungi may have a broader role in lignin deconstruction than was previously thought. Because of historical carry-over, white rot fungi have continued to be better studied, and their use as the primary agents in biotechnology processing has continued even as the field has shifted to the use of lignocellulose biomass in the production of platform chemicals (from both holocellulose and lignin) to produce downstream bio-based chemicals.

Brown rot fungal activity was previously demonstrated to be effective for use in activating lignin for production of laminated wood (Jin et al. 1991) and composite panels (Jin et al. 1990b). Additional research also demonstrated the feasibility of using brown rotted lignin as a high-quality, formaldehyde-free adhesive (Li and Geng 2005). The chelator-mediated Fenton (CMF) mechanism, derived from the brown rot fungi, has been considered for application in the activation of lignin on fiber surfaces to provide enhanced bonding of fiber-based products and has been patented for use in those applications (Qian et al. 2002, 2004; Goodell and Jellison 2008; Yelle et al. 2004).

Relative to pollutant and xenobiotic remediation, *Serpula* spp. brown rot was found to be effective in the removal of copper from preservative-treated wood as a remediation treatment (Kartal et al. 2015). Similarly CMF chemistry was demonstrated to be significantly more effective than Fenton chemistry alone in the degradation of complex pollutants like pentachlorophenol (Goodell et al. 1997) and DDT (Purnomo et al. 2008) and also in the decolorization of recalcitrant dyes (Goodell et al. 2002, 2004).

Brown rot CAZymes have been little explored for biorefinery applications, but a powerful LMPO has been isolated from the brown rot *Gloeophyllum* spp. and cloned into yeast for expression (Kojima et al. 2016). The combination of brown rot enzymatic and non-enzymatic mechanisms may lead to future biorefinery applications, particularly where non-fiber products are of primary interest, and downstream platform chemicals produced from oligosaccharides and modified lignin can be produced (Goodell et al. 2017).

In addition to using isolated enzymes or chemistries derived from brown rots, the brown rot fungi alone or in consortia have been used successfully to colonize biomass to pretreat lignocellulose biomass (bagasse and wood) to promote the release of sugars from these materials in downstream processing (Ray et al. 2010). The Mycologix LTD company (CompaniesHouse.Gov.UK 2013) was technically successful in the pretreatment of biomass in a commercial application, but rapid geopolitical changes resulting in decreases in the price of fossil hydrocarbons caused the company to declare insolvency after 4 years of operation. Given the success of this process however, as the economics for bioconversion changes in the future, similar processes using brown rot fungi may be developed for use in bio-based platform chemical production. Although the need for liquid fuels may lessen as other forms of energy to power vehicles increases in popularity, the need for bioconversion processes in the production of high-energy density liquid fuels from lignocellulose in certain types of transit (long-range aerospace and oceanic transport) will continue, and as the use of fossil fuels inevitably dwindles, required pretreatments for biomass to produce this type of commercial biofuel will likely increase. The use of brown rot systems for this type of biomass preparation in biorefineries may therefore find application in the future.

V. Conclusions

This chapter has focused on fungi and fungal systems that decay and degrade lignocellulose materials. These fungi include the predominant aerobic microorganisms on earth involved in the depolymerization of cellulose and lignin. Fungi reviewed include those that cause brown rot, white rot, soft rot, stain and mold of wood, and woody biomass. Most wood-degrading fungi possess extracellular enzymatic systems for the degradation of lignocellulose, but some of the more aggressive degraders of lignocellulose, the brown rot fungi, use a non-enzymatic CMF mechanism, and perhaps other non-enzymatic mechanisms, to depolymerize the components of the woody cell walls. A more limited suite of CAZymes is then subsequently secreted by the brown rot fungi to permit further wood cell wall deconstruction. The white rot fungi deconstruct lignin primary via the use of lignin-degrading enzymes, which in some cases require LMW mediator compounds. Soft rot fungi are unique in the manner in which they are capable of tunneling into the wood cell wall and, in Type I attack, produce a unique string of diamond-shaped cavities that follows the microfibril angle of the dominant wood cell wall layer.

Staining fungi have the ability to penetrate, primarily into sapwood tissue, where they colonize parenchyma cells containing starches and other readily digested substrates. The hyphae of wood-staining fungi have the ability to pigment, resulting in the coloration of woody substrates.

Mold fungi generally do not penetrate into wood cells deeply, and these fungi remain on the surface of wood where they metabolize surface residues that may be deposited from external sources (wind and water deposits) or that have diffused from deeper in the wood. Mold fungi generally do not have colored hyphae, but the mold spores can be darkly pigmented.

All the fungi discussed in this chapter have the capacity to be used in biotechnological applications. Only the Basidiomycota white rot and brown rot biotechnology applications and future potential are overviewed in this chapter, as the Ascomycota applications are addressed in other chapters of this book. Historical applications including the use of lignin-oxidizing enzymes from white rot fungi, and their use in biorefining and bio-pulping applications are briefly reviewed. The potential of both classes of fungi as well as their enzymes and the non-enzymatic CMF mechanism in brown rots is discussed for biotech applications, including their use in biorefineries, and pollutant/xenobiotic remediation.

References

Abdel-Hamid AM, Solbiati JO, Cann IKO (2013) Chapter One—Insights into lignin degradation and its potential industrial applications. In: Sariaslani S, Gadd GM (eds) Advances in applied microbiology, vol 82. Academic, San Diego, CA, pp 1–28. https://doi.org/10.1016/B978-0-12-407679-2.00001-6

Arantes V, Goodell B (2014) Current understanding of brown-rot fungal biodegradation mechanisms: a review. In: Deterioration and protection of sustainable biomaterials, vol 1158. ACS symposium series, vol 1158. American Chemical Society, Washington, DC, pp 3–21. https://doi.org/10.1021/bk-2014-1158.ch001

Arantes V, Milagres AMF (2006) Degradation of cellulosic and hemicellulosic substrates using a chelator-mediated Fenton reaction. J Chem Technol Biotechnol 81(3):413–419. https://doi.org/10.1002/jctb.1417

Arantes V, Milagres AMF, Filley T, Goodell B (2011) Lignocellulosic polysaccharides and lignin degradation by wood decay fungi: the relevance of nonenzymatic Fenton-based reactions. J Ind Microbiol Biotechnol 38(4):541–555. https://doi.org/10.1007/s10295-10010-10798-10292

Arantes V, Jellison J, Goodell B (2012) Peculiarities of brown-rot fungi and biochemical Fenton reaction with regard to their potential as a model for bioprocessing biomass. Appl Microbiol Biotechnol 94(2):323–338. https://doi.org/10.1007/s00253-012-3954-y

Aust SD (1995) Mechanisms of degradation by white rot fungi. Environ Health Perspect 103(Suppl 5):59–61

Baba Y, Tanabe T, Shirai N, Watanabe T, Honda Y, Watanabe T (2011) Pretreatment of Japanese cedar wood by white rot fungi and ethanolysis for bioethanol production. Biomass Bioenergy 35:320–332. https://doi.org/10.1016/j.biombioe.2010.08.040

Baldrian P, Valaskova V (2008) Degradation of cellulose by basidiomycetous fungi. FEMS Microbiol Rev 32(3):501–521. https://doi.org/10.1111/j.1574-6976.2008.00106.x

Barb WG, Baxendale JH, George P, Hargrave KR (1951) Reactions of ferrous and ferric ions with hydrogen peroxide. Part I. The ferrous ion reaction. Faraday Soc Trans 47:462–500

Barr DP, Aust SD (1994) Mechanisms white rot fungi use to degrade pollutants. Environ Sci Technol 28(2):79A–87A

Bissaro B, Várnai A, Røhr ÅK, Eijsink VGH (2018) Oxidoreductases and reactive oxygen species in conversion of lignocellulosic biomass. Microbiol Mol Biol Rev 82(4):e00029–e00018. https://doi.org/10.1128/mmbr.00029-18

Blanchette RA, Burnes TA, Leatham GF, Effland MJ (1988) Selection of white-rot fungi for biopulping. Biomass 15(2):93–101. https://doi.org/10.1016/0144-4565(88)90099-6

Blanchette RA, Held BW, Jurgens JA, McNew DL, Harrington TC, Duncan SM, Farrell RL (2004) Wood-destroying soft rot fungi in the historic expedition huts of Antarctica. Appl Environ Microbiol 70(3):1328–1335

Boddy L (2016) Chapter 9—Interactions with humans and other animals. In: The fungi, 3rd edn. Academic Press, Boston, pp 293–336. https://doi.org/10.1016/B978-0-12-382034-1.00009-8

Brenelli L, Squina FM, Felby C, Cannella D (2018) Laccase-derived lignin compounds boost cellulose oxidative enzymes AA9. Biotechnol Biofuels 11(1):10. https://doi.org/10.1186/s13068-017-0985-8

Camarero S, Ibarra D, Martínez MJ, Martínez ÁT (2005) Lignin-derived compounds as efficient laccase mediators for decolorization of different types of recalcitrant dyes. Appl Environ Microbiol 71(4):1775–1784. https://doi.org/10.1128/AEM.71.4.1775-1784.2005

Capolupo L, Faraco V (2016) Green methods of lignocellulose pretreatment for biorefinery development. Appl Microbiol Biotechnol 100(22):9451–9467. https://doi.org/10.1007/s00253-016-7884-y

Cowling EB (1961) Comparative biochemistry of the decay of sweetgum sapwood by white-rot and brown-rot fungi. USDA, Washington, DC

Cowling EB (1964) Microorganisms and microbial enzyme systems as selective tools in wood anatomy. In: Cote Jr WA (ed) Proceedings, Upper Saranac Lake, NY, pp 341–368

Cragg SM, Beckham GT, Bruce NC, Bugg TDH, Distel DL, Dupree P, Etxabe AG, Goodell BS, Jellison J, McGeehan JE, McQueen-Mason SJ, Schnorr K, Walton PH, Watts JEM, Zimmer M (2015) Lignocellulose degradation mechanisms across the Tree of Life. Curr Opin Chem Biol 29:108–119. https://doi.org/10.1016/j.cbpa.2015.10.018

Daniel G (2003) Microview of wood under degradation by bacteria and fungi. In: Wood deterioration and preservation, vol 845. ACS symposium series, vol 845. American Chemical Society, Washington, DC, pp 34–72. https://doi.org/10.1021/bk-2003-0845.ch004

Daniel G (2014) Fungal and bacterial biodegradation: white rots, brown rots, soft rots, and bacteria. In: Nicholas Darrel D, Goodell B, Schultz Tor P (eds) Deterioration and protection of sustainable biomaterials. ACS symposium series, vol 1158. American Chemical Society, Washington, DC, pp 23–58. https://doi.org/10.1021/bk-2014-1158.ch002

Daniel G (2016) Chapter 8—Fungal degradation of wood cell walls: origins, functions, and applications. In: Kim YS, Funada R, Singh AP (eds) Secondary xylem biology. Academic Press, Boston, pp

131–167. https://doi.org/10.1016/B978-0-12-802185-9.00008-5

Daniel G, Nilsson T (1998) Chapter 3. Development in the study of soft rot and bacterial decay. In: Bruce A, Palfreyman J (eds) Forest products biotechnology. Taylor and Francis, London, pp 235–250

Dashtban M, Schraft H, Syed TA, Qin W (2010) Fungal biodegradation and enzymatic modification of lignin. Int J Biochem Mol Biol 1(1):36–50

De Beer Z, Seifert K, Wingfield M (2013a) The ophiostomatoid fungi: their dual position in the Sordariomycetes. The ophiostomatoid fungi: expanding frontiers. CBS Biodiversity Ser 12:1–19

De Beer ZW, Seifert K, Wingfield M (2013b) A nomenclator for ophiostomatoid genera and species in the ophiostomatales and microascales. Ophiostomatoid fungi: expanding frontiers. CBS Biodiversity Ser 12:261–268

Dill I, Kraepelin G (1986) Palo Podrido: model for extensive delignification of wood by *Ganoderma applanatum*. Appl Environ Microbiol 52(6):1305–1312

Ding S-Y, Liu Y-S, Zeng Y, Himmel ME, Baker JO, Bayer EA (2012) How does plant cell wall nanoscale architecture correlate with enzymatic digestibility? Science 338(6110):1055–1060. https://doi.org/10.1126/science.1227491

Eastwood DC (2014) Evolution of fungal wood decay. In: Deterioration and protection of sustainable biomaterials. ACS symposium series, vol 1158. American Chemical Society, Washington, DC, pp 93–112. https://doi.org/10.1021/bk-2014-1158.ch005

Encinas O, Daniel G (1995) Wood Cell Wall Biodegradation by the Blue Stain Fungus *Botryodiplodia theobromae* Pat. Material und Organismen 29 (4):255–272

Encinas O, Daniel G (1996) Decay capacity of different strains of the blue stain fungus *Lasiodiplodia theobromae* on various wood species. Material und Organismen 30(4):239–258

Eriksson K-E, Kirk TK (1985) Biopulping, biobleaching and treatment of kraft bleaching effluents with white-rot fungi. In: Moo-Young M (ed) Comprehensive biotechnology, vol 4. Pergamon Press, Toronto, pp 271–294

ExPASy (2018) Enzyme entry EC 1.11.1.19. Dye decolorizing peroxidase. https://enzyme.expasy.org/EC/1.11.1.19

Fackler K, Gradinger C, Schmutzer M, Tavzes C, Burgert I, Schwanninger M, Hinterstoisser B, Watanabe T, Messner K (2007) Biotechnological wood modification with selective white-rot fungi and its molecular mechanism. Food Technol Biotechnol 45:269–276

FAO (2016) Regulation of wood packaging material in international trade. International Standard for Phytosanitation Measures. ISPM 15. Food and Agriculture Organization of the United Nations: International Plant Protection Convention, Viale delle Terme di Caracalla, 00153 Rome, Italy

Fernandez-Fueyo E, Ruiz-Duenas FJ, Miki Y, Martinez MJ, Hammel KE, Martinez AT (2012) Lignin-degrading peroxidases from genome of selective ligninolytic fungus *Ceriporiopsis subvermispora*. J Biol Chem 287(20):16903–16916. https://doi.org/10.1074/jbc.M112.356378

Floudas D, Binder M, Riley R, Barry K, Blanchette RA, Henrissat B, Martínez AT, Otillar R, Spatafora JW, Yadav JS, Aerts A, Benoit I, Boyd A, Carlson A, Copeland A, Coutinho PM, de Vries RP, Ferreira P, Findley K, Foster B, Gaskell J, Glotzer D, Górecki P, Heitman J, Hesse C, Hori C, Igarashi K, Jurgens JA, Kallen N, Kersten P, Kohler A, Kües U, Kumar TKA, Kuo A, LaButti K, Larrondo LF, Lindquist E, Ling A, Lombard V, Lucas S, Lundell T, Martin R, McLaughlin DJ, Morgenstern I, Morin E, Murat C, Nagy LG, Nolan M, Ohm RA, Patyshakuliyeva A, Rokas A, Ruiz-Dueñas FJ, Sabat G, Salamov A, Samejima M, Schmutz J, Slot JC, St. John F, Stenlid J, Sun H, Sun S, Syed K, Tsang A, Wiebenga A, Young D, Pisabarro A, Eastwood DC, Martin F, Cullen D, Grigoriev IV, Hibbett DS (2012) The paleozoic origin of enzymatic lignin decomposition reconstructed from 31 fungal genomes. Science 336(6089):1715–1719. https://doi.org/10.1126/science.1221748

Flournoy DS, Kirk TK, Highley TL (1991) Wood decay by brown-rot fungi: changes in pore structure and cell wall volume. Holzforschung 45(5):383–388

Forest-Products-Laboratory (2010) Wood handbook—wood as an engineering material. General Technical Report FPL-GTR-190. U.S. Department of Agriculture, Forest Service, Forest Products Laboratory, Madison, WI, p 508

Frommhagen M, Mutte SK, Westphal AH, Koetsier MJ, Hinz SWA, Visser J, Vincken J-P, Weijers D, van Berkel WJH, Gruppen H, Kabel MA (2017) Boosting LPMO-driven lignocellulose degradation by polyphenol oxidase-activated lignin building blocks. Biotechnol Biofuels 10:121–121. https://doi.org/10.1186/s13068-017-0810-4

Giardina P, Palmieri G, Fontanella B, Rivieccio V, Sannia G (2000) Manganese peroxidase isoenzymes produced by *Pleurotus ostreatus* grown on wood sawdust. Arch Biochem Biophys 376(1):171–179. https://doi.org/10.1006/abbi.1999.1691

Goodell B (2014) Current understanding of brown-rot fungal biodegradation mechanisms: a review. In: Deterioration and protection of sustainable biomaterials. ACS symposium series, vol 1158. American Chemical Society, Washington, DC, pp 3–21. https://doi.org/10.1021/bk-2014-1158.ch001

Goodell B, Jellison J (2008) Oxidation using a non-enzymatic free radical system mediated by redox cycling chelators. US Patent 7,396,974B2

Goodell B, Jellison J, Liu J, Daniel G, Paszczynski A, Fekete F, Krishnamurthy S, Jun L, Xu G (1997)

Low molecular weight chelators and phenolic compounds isolated from wood decay fungi and their role in the fungal biodegradation of wood. J Biotechnol 53:133–162

Goodell B, Qian Y, Jellison J, Richard M, Qi W (2002) Lignocellulose oxidation by low molecular weight metal-binding compounds isolated from wood degrading fungi: a comparison of brown rot and white rot systems and the potential application of chelator-mediated Fenton reactions. Prog Biotechnol 21:37–47. Biotechnology in the Pulp and Paper Industry. Elsevier Press

Goodell B, Qian Y, Jellison J, Richard M (2004) Decolorization and degradation of dyes with mediated Fenton reaction. Water Environ Res 76(6):2703–2707

Goodell B, Qian Y, Jellison J (2008) Fungal decay of wood: soft rot—brown rot—white rot. In: Development of commercial wood preservatives. ACS symposium series, vol 982. American Chemical Society, Washington, DC, pp 9–31. https://doi.org/10.1021/bk-2008-0982.ch002

Goodell B, Zhu Y, Kim S, Kafle K, Eastwood D, Daniel G, Jellison J, Yoshida M, Groom L, Pingali SV, O'Neill H (2017) Modification of the nanostructure of lignocellulose cell walls via a non-enzymatic lignocellulose deconstruction system in brown rot wood-decay fungi. Biotechnol Biofuels 10:179. https://doi.org/10.1186/s13068-017-0865-2

Greaves H (1977) An illustrated comment on the soft rot problem in Australia and Papua New Guinea. Holzforschung 31(3):71–79. https://doi.org/10.1515/hfsg.1977.31.3.71

Hale MD, Eaton RA (1985) The ultrastructure of soft rot fungi. I. Fine hyphae in wood cell walls. Mycologia 77(3):447–463. https://doi.org/10.2307/3793202

Hatakka A, Hammel KE (2011) Fungal biodegradation of lignocelluloses. In: Hofrichter M (ed) Industrial applications. The mycota (A comprehensive treatise on fungi as experimental systems for basic and applied research), vol 10. Springer, Berlin, pp 319–340. https://doi.org/10.1007/978-3-642-11458-8_15

Highley TL (1988) Cellulolytic activity of brown-rot and white-rot fungi on solid media. Holzforschung 42 (4):211–216

Highley TL, Illman BL (1991) Progress in understanding how brown-rot fungi degrade cellulose. Biodeterior Abstr 5(3):231–244

Highley T, Bar-Lev S, Kirk T, Larsen M (1983) Influence of O2 and CO2 on wood decay by heartrot and saprot fungi. Phytopathology 73:630–633

Hofrichter M (2002) Review: lignin conversion by manganese peroxidase (MnP). Enzyme Microb Technol 30(4):454–466. https://doi.org/10.1016/S0141-0229(01)00528-2

Jellison J, Chandhoke V, Goodell B, Fekete FA (1991) The isolation and immunolocalization of iron-binding compounds produced by *Gloeophyllum trabeum*. Appl Microbiol Biotechnol 35:805–809

Jin L, Schultz TP, Nicholas DD (1990a) Structural characterization of brown-rotted lignin. Holzforschung 44(2):133–138

Jin L, Sellers T, Schultz TP, Nicholas DD (1990b) Utilization of lignin modified by brown-rot fungi. I. Properties of flakeboard produced with a brown-rotted lignin modified phenolic adhesive. Holzforschung Int J Biol Chem Phys Technol Wood 44. https://doi.org/10.1515/hfsg.1990.44.3.207

Jin L, Nicholas DD, Schultz TP (1991) Wood laminates glued by enzymatic oxidation of brown-rotted lignin. Holzforschung 45(6):467–468

Kartal SN, Terzi E, Yılmaz H, Goodell B (2015) Bioremediation and decay of wood treated with ACQ, micronized ACQ, nano-CuO and CCA wood preservatives. Int Biodeterior Biodegrad 99:95–101. https://doi.org/10.1016/j.ibiod.2015.01.004

Kazemi SM, Dickinson DJ, Murphy RJ (2001) Effects of initial moisture content on wood decay at different levels of gaseous oxygen concentrations. J Agric Sci Technol 3:293–304

Kent MS, Zeng J, Rader N, Avina IC, Simoes CT, Brenden CK, Busse ML, Watt J, Giron NH, Alam TM, Allendorf MD, Simmons BA, Bell NS, Sale KL (2018) Efficient conversion of lignin into a water-soluble polymer by a chelator-mediated Fenton reaction: optimization of H2O2 use and performance as a dispersant. Green Chem 20(13):3024–3037. https://doi.org/10.1039/C7GC03459H

Kerem Z, Jensen KA, Hammel KE (1999) Biodegradative mechanism of the brown rot basidiomycete *Gloeophyllum trabeum*: evidence for an extracellular hydroquinone-driven Fenton reaction. FEBS Lett 446(1):49–54

Kido R, Takeeda M, Manabe M, Miyagawa Y, Itakura S, Tanaka H (2015) Presence of extracellular NAD+ and NADH in cultures of wood-degrading fungi. Biocontrol Sci 20. https://doi.org/10.4265/bio.20.105

Kim JS, Gao J, Daniel G (2015) Cytochemical and immunocytochemical characterization of wood decayed by the white rot fungus *Pycnoporus sanguineus* I. preferential lignin degradation prior to hemicelluloses in Norway spruce wood. Int Biodeterior Biodegrad 105:30–40. https://doi.org/10.1016/j.ibiod.2015.08.008

Kleman-Leyer K, Agosin E, Conner AH, Kirk TK (1992) Changes in molecular size distribution of cellulose during attack by white rot and brown rot fungi. Appl Environ Microbiol 58(4):1266–1270

Knop D, Levinson D, Makovitzki A, Agami A, Lerer E, Mimran A, Yarden O, Hadar Y (2016) Limits of versatility of versatile peroxidase. Appl Environ Microbiol 82(14):4070–4080. https://doi.org/10.1128/AEM.00743-16

Kojima Y, Varnai A, Ishida T, Sunagawa N, Petrovic DM, Igarashi K, Jellison J, Goodell B, Alfredsen G, Westereng B, Eijsink VG, Yoshida M (2016) A lytic polysaccharide monooxygenase with broad xyloglucan specificity from the brown-rot fungus *Gloeophyllum trabeum* and its action on cellulose-xyloglucan complexes. Appl Environ Microbiol 82(22):6557–6572. https://doi.org/10.1128/aem.01768-16

Krah F-S, Bässler C, Heibl C, Soghigian J, Schaefer H, Hibbett DS (2018) Evolutionary dynamics of host specialization in wood-decay fungi. BMC Evol Biol 18(1):119. https://doi.org/10.1186/s12862-018-1229-7

Lauber C, Schwarz T, Nguyen QK, Lorenz P, Lochnit G, Zorn H (2017) Identification, heterologous expression and characterization of a dye-decolorizing peroxidase of *Pleurotus sapidus*. AMB Express 7:164. https://doi.org/10.1186/s13568-017-0463-5

Lee JK, Walker KL, Han HS, Kang J, Prinz FB, Waymouth RM, Nam HG, Zare RN (2019) Spontaneous generation of hydrogen peroxide from aqueous microdroplets. Proc Natl Acad Sci USA. https://doi.org/10.1073/pnas.1911883116

Leonowicz A, Matuszewska A, Luterek J, Ziegenhagen D, Wojtaś-Wasilewska M, Cho N-S, Hofrichter M, Rogalski J (1999) Biodegradation of lignin by white rot fungi. Fungal Genet Biol 27(2–3):175–185. https://doi.org/10.1006/fgbi.1999.1150

Li K, Geng X (2005) Formaldehyde-free wood adhesives from decayed wood. Macromol Rapid Commun 26 (7):529–532. https://doi.org/10.1002/marc.200400594

Li F, Ma F, Zhao H, Zhang S, Wang L, Zhang X, Yu H (2019) A lytic polysaccharide monooxygenase from a white-rot fungus drives the degradation of lignin by a versatile peroxidase. Appl Environ Microbiol 85(9):e02803–e02818. https://doi.org/10.1128/aem.02803-18

Liers C, Arnstadt T, Ullrich R, Hofrichter M (2011) Patterns of lignin degradation and oxidative enzyme secretion by different wood- and litter-colonizing basidiomycetes and ascomycetes grown on beech-wood. FEMS Microbiol Ecol 78 (1):91–102. https://doi.org/10.1111/j.1574-6941.2011.01144.x

Magan N, Fragoeiro S, Bastos C (2010) Environmental factors and bioremediation of xenobiotics using white rot fungi. Mycobiology 38(4):238–248. https://doi.org/10.4489/MYCO.2010.38.4.238

Mäkelä MR, Marinović M, Nousiainen P, Liwanag AJM, Benoit I, Sipilä J, Hatakka A, de Vries RP, Hildén KS (2015) Chapter Two—Aromatic metabolism of filamentous fungi in relation to the presence of aromatic compounds in plant biomass. In: Sariaslani S, Gadd GM (eds) Advances in applied microbiology, vol 91. Academic Press, New York, pp 63–137. https://doi.org/10.1016/bs.aambs.2014.12.001

Martinez D, Challacombe J, Morgenstern I, Hibbett D, Schmoll M, Kubicek CP, Ferreira P, Ruiz-Duenas FJ, Martinez AT, Kersten P, Hammel KE, Wymelenberg AV, Gaskell J, Lindquist E, Sabat G, Bon-Durant SS, Larrondo LF, Canessa P, Vicuna R, Yadav J, Doddapaneni H, Subramanian V, Pisabarro AG, Lavín JL, Oguiza JA, Master E, Henrissat B, Coutinho PM, Harris P, Magnuson JK, Baker SE, Bruno K, Kenealy W, Hoegger PJ, Kües U, Ramaiya P, Lucas S, Salamov A, Shapiro H, Tu H, Chee CL, Misra M, Xie G, Teter S, Yaver D, James T, Mokrejs M, Pospisek M, Grigoriev IV, Brettin T, Rokhsar D, Berka R, Cullen D (2009) Genome, transcriptome, and secretome analysis of wood decay fungus *Postia placenta* supports unique mechanisms of lignocellulose conversion. Proc Natl Acad Sci USA. https://doi.org/10.1073/pnas.0809575106

Mate Diana M, Alcalde M (2017) Laccase: a multipurpose biocatalyst at the forefront of biotechnology. Microb Biotechnol 10(6):1457–1467. https://doi.org/10.1111/1751-7915.12422

Messner K, Fackler K, Lamaipis P, Gindl W, Srebotnik E, Watanabe T (2003) Overview of white-rot research: where we are today. ACS Symp Ser 845:73–96. https://doi.org/10.1021/bk-2003-0845.ch005

Miller DJ, Goodell B (1981) Blue staining in ponderosa pine sapwood at moderate and low temperatures. Forest Prod J 31(2):54–55

Morrell J, Zabel RA (1985) Wood strength and weight losses caused by soft rot fungi isolated from treated southern pine utility poles. Wood Fiber Sci 17 (1):132–143

Mycologix LTD (2013) CompaniesHouse.Gov.UK. Accessed 13 Sept 2019

NeLMA (2016) Beware of using blue-stained wood in wood packaging materials. http://www.nelma.org/beware-of-using-blue-stained-wood-in-wood-packaging-materials/

Niku-Paavola ML, Anke H, Poppius-Levlin K, Viikari L (2003) Siderophores as natural mediators in laccase-aided degradation of lignin. In: Applications of enzymes to lignocellulosics. ACS Symposium Series, vol 855. American Chemical Society, Washington, DC, pp 176–190. https://doi.org/10.1021/bk-2003-0855.ch011

Nurika I, Eastwood D, Bugg T, Barker G (2019) Biochemical characterization of *Serpula lacrymans* iron-reductase enzymes in lignocellulose breakdown. J Ind Microbiol Biotechnol. https://doi.org/10.1007/s10295-019-02238-7

Otjen L, Blanchette RA (1986) A discussion of microstructural changes in wood during decomposition by white rot basidiomycetes. Can J Bot 64:905–911

Otjen L, Blanchette R, Effland M, Leatham G (1987) Assessment of 30 white rot basidiomycetes for selective lignin degradation. Holzforschung Int J Biol Chem Phys Technol Wood 41(6):343. https://doi.org/10.1515/hfsg.1987.41.6.343

Parra C, Rodriguez J, Baeza J, Freer J, Duran N (1998) Iron-binding catechols oxidating lignin and chlorolignin. Biochem Biophys Res Commun 251 (2):399–402

Paszczynski A, Crawford R, Funk D, Goodell B (1999) De novo synthesis of 4,5-dimethoxycatechol and 2,5-dimethoxyhydroquinone by the Brown Rot Fungus *Gloeophyllum trabeum*. Appl Environ Microbiol 65(2):674–679

Pérez-Boada M, Ruiz-Dueñas FJ, Pogni R, Basosi R, Choinowski T, Martinez MJ, Piontek K, Martínez AT (2005) Versatile peroxidase oxidation of high redox potential aromatic compounds. Site-directed mutagenesis, spectroscopic and crystallographic investigation of three long-range electron transfer pathways. J Mol Biol 354:402. https://doi.org/10.1016/j.jmb.2005.09.047

Pettersson G, Cowling EB, Porath J (1963) Studies on cellulolytic enzymes. Biochim Biophys Acta 67:1–8

Purnomo AS, Kamei I, Kondo R (2008) Degradation of 1,1,1-trichloro-2,2-bis (4-chlorophenyl) ethane (DDT) by brown-rot fungi. J Biosci Bioeng 105 (6):614–621. https://doi.org/10.1263/jbb.105.614

Qian Y, Goodell B, Genco JM (2002) The effect of a chelator mediated fenton system on the fiber and paper properties of hardwood kraft pulp. J Wood Chem Technol 22(4):267–284. https://doi.org/10.1081/WCT-120016262

Qian Y, Goodell B, Jellison J, Felix CC (2004) The effect of hydroxyl radical generation on free-radical activation of TMP fibers. J Polym Environ 12(3):147–155. https://doi.org/10.1023/b:jooe.0000038546.65047.35

Qin X, Sun X, Huang H, Bai Y, Wang Y, Luo H, Yao B, Zhang X, Su X (2017) Oxidation of a non-phenolic lignin model compound by two Irpex lacteus manganese peroxidases: evidence for implication of carboxylate and radicals. Biotechnol Biofuels 10 (1):103. https://doi.org/10.1186/s13068-017-0787-z

Ray MJ, Leak DJ, Spanu PD, Murphy RJ (2010) Brown rot fungal early stage decay mechanism as a biological pretreatment for softwood biomass in biofuel production. Biomass Bioenergy 34 (8):1257–1262. https://doi.org/10.1016/j.biombioe.2010.03.015

Riley R, Salamov AA, Brown DW, Nagy LG, Floudas D, Held BW, Levasseur A, Lombard V, Morin E, Otillar R, Lindquist EA, Sun H, LaButti KM, Schmutz J, Jabbour D, Luo H, Baker SE, Pisabarro AG, Walton JD, Blanchette RA, Henrissat B, Martin F, Cullen D, Hibbett DS, Grigoriev IV (2014) Extensive sampling of basidiomycete genomes demonstrates inadequacy of the white-rot/brown-rot paradigm for wood decay fungi. Proc Natl Acad Sci USA 111 (27):9923–9928. https://doi.org/10.1073/pnas.1400592111

Rytioja J, Hildén K, Yuzon J, Hatakka A, de Vries RP, Mäkelä MR (2014) Plant-polysaccharide-degrading enzymes from basidiomycetes. Microbiol Mol Biol Rev MMBR 78(4):614–649. https://doi.org/10.1128/MMBR.00035-14

Schmidt I (2006) Wood and tree fungi: biology, damage, protection, and use. Springer, Berlin, p 346. ISBN-13: 978-3642068751

Schmidt CJ, Whitten BK, Nicholas DD (1981) A proposed role for oxalic acid in nonenzymatic wood decay by brown-rot fungi. Proc Am Wood Preserv Assoc 77:157–164

Simonis JL, Raja HA, Shearer CA (2008) Extracellular enzymes and soft rot decay: are ascomycetes important degraders in fresh water? Fungal Divers 31:135–146

Stone JE, Scallan AM (1967) The effect of component removal upon the porous structure of the cell wall of wood. II. Swelling in water and the fiber saturation point. Tappi 50(10):496–501

Suzuki MR, Hunt CG, Houtman CJ, Dalebroux ZD, Hammel KE (2006) Fungal hydroquinones contribute to brown rot of wood. Environ Microbiol 8(12):2214–2223. https://doi.org/10.1111/j.1462-2920.2006.01160.x

Tamaru Y, Yoshida M, Eltis LD, Goodell B (2019) Multiple iron reduction by methoxylated phenolic lignin structures and the generation of reactive oxygen species by lignocellulose surfaces. Int J Biol Macromol 128:340–346. https://doi.org/10.1016/j.ijbiomac.2019.01.149

Tanaka H, Fuse G, Enoki A (1992) Soft rot of wood caused by three microfungi grown on four media. Material und Organismen 27(3):157–170

Tanaka H, Itakura S, Enoki A (2000) Phenol oxidase activity and one-electron oxidation activity in wood degradation by soft-rot deuteromycetes. Holzforschung 54(5):463–468. https://doi.org/10.1515/HF.2000.078

Tanaka H, Yoshida G, Baba Y, Matsumura K, Wasada H, Murata J, Agawa M, Itakura S, Enoki A (2007) Characterization of a hydroxyl-radical-producing glycoprotein and its presumptive genes from the white-rot basidiomycete *Phanerochaete chrysosporium*. J Biotechnol 128(3):500–511. https://doi.org/10.1016/j.jbiotec.2006.12.010

Tepfer M, Taylor IEP (1981) The permeability of plant cell walls as measured by gel filtration chromatography. Science 213(August 14):761–763

Tuor U, Winterhalter K, Fiechter A (1995) Enzymes of white-rot fungi involved in lignin degradation and ecological determinants for wood decay. J Biotechnol 41(1):1–17. https://doi.org/10.1016/0168-1656(95)00042-O

Unligil HH, Chafe SC (1974) Perforation hyphae of soft rot fungi in the wood of white spruce [*Picea glauca* (Moench.) Voss.]. Wood Sci Technol 8:27–32

Uzunovic A, Byrne T, Gignac M, Yang D-Q (2008) Wood discolourations & their prevention: with an emphasis on bluestain. Special publication SP-50. FPInnovations, Canada

Varela E, Tien M (2003) Effect of pH and oxalate on hydroquinone-derived hydroxyl radical formation during brown rot wood degradation. Appl Environ Microbiol 69(10):6025–6031. https://doi.org/10.1128/AEM.69.10.6025-6031.2003

Vasina DV, Moiseenko KV, Fedorova TV, Tyazhelova TV (2017) Lignin-degrading peroxidases in white-rot fungus *Trametes hirsuta* 072. Absolute expression quantification of full multigene family. PLoS One 12(3):e0173813. https://doi.org/10.1371/journal.pone.0173813

Westereng B, Cannella D, Wittrup Agger J, Jørgensen H, Larsen Andersen M, Eijsink VGH, Felby C (2015) Enzymatic cellulose oxidation is linked to lignin by long-range electron transfer. Sci Rep 5:18561. https://doi.org/10.1038/srep18561. https://www.nature.com/articles/srep18561#supplementary-information

Widsten P, Kandelbauer A (2008) Adhesion improvement of lignocellulosic products by enzymatic pretreatment. Biotechnol Adv 26(4):379–386. https://doi.org/10.1016/j.biotechadv.2008.04.003

Wilcox WW (1978) Review of literature on the effects of early stages of decay on wood strength. Wood Fiber 9(4):252–257

Williams S (2017) CAZypedia carbohydrate-active enzymes. https://www.cazypedia.org/index.php/Glycoside_Hydrolase_Families. Accessed 30 Dec 2017

Wolfaardt F, Taljaard JL, Jacobs A, Male JR, Rabie CJ (2004) Assessment of wood-inhabiting Basidiomycetes for biokraft pulping of softwood chips. Bioresour Technol 95(1):25–30. https://doi.org/10.1016/j.biortech.2004.01.007

World-Health-Organization (2009) Dampness and mould: WHO guidelines for indoor air quality. http://www.euro.who.int/__data/assets/pdf_file/0017/43325/E92645.pdf

Worrall JJ, Anagnost SE, Wang CJK (1991) Conditions for soft rot of wood. Can J Microbiol 37:869–874

Yelle D, Goodell B, Gardner D, Amirbahman A, Winistofer P, Shaler S (2004) Bonding of wood fiber composites using a synthetic chelator-lignin activation system. Forest Prod J 54(4):73–78

Yelle DJ, Wei D, Ralph J, Hammel KE (2011) Multidimensional NMR analysis reveals truncated lignin structures in wood decayed by the brown rot basidiomycete *Postia placenta*. Environ Microbiol 13(4):1091–1100. https://doi.org/10.1111/j.1462-2920.2010.02417.x

Zabel RA, Morrell JJ (2020) Wood microbiology: decay and its prevention, 2nd edn. Academic Press, San Diego, CA

Zabel RA, Lombard FF, Wang CJK, Terracina F (1985) Fungi associated with decay in treated southern pine in the eastern United States. Wood Fiber Sci 17:75–91

Zhang J, Presley GN, Hammel KE, Ryu J-S, Menke JR, Figueroa M, Hu D, Orr G, Schilling JS (2016) Localizing gene regulation reveals a staggered wood decay mechanism for the brown rot fungus *Postia placenta*. Proc Natl Acad Sci USA 113(39):10968–10973. https://doi.org/10.1073/pnas.1608454113

Zhu Y, Mahaney J, Jellison J, Cao J, Gressler J, Hoffmeister D, Goodell B (2016) Fungal variegatic acid and extracellular polysaccharides promote the site-specific generation of reactive oxygen species. J Ind Microbiol Biotechnol:1–10. https://doi.org/10.1007/s10295-016-1889-5

16 Biotechnology of Marine Fungi: New Workhorses and Applications

Annemarie Kramer[1], Antje Labes[1]

CONTENTS

I. Introduction

The diversity in the fungal tree of live and their manifold ecological functionalities are not yet reflected in the applied variety of taxonomic groups and the diversity of products on the market. Recently, new ecological niches and habitats including the oceans came into focus to increase the diversity of fungi available for biotechnology (Manohar and Raghukumar 2013). Marine fungi have been neglected for a long time and therefore the knowledge of them is still quite rare. But early screening campaigns revealed that marine fungi are an excellent source for new natural products, indicating their **potential for biotechnological application**. Especially fungi from unique and extreme marine environments, such as the deep sea, are potentially a promising source for rare and unusual compounds (Fenical and Jensen 1993; Vignesh et al. 2011).

The latest large review on marine fungi and their biotechnological potential was published in 2015 (Bonugli-Santos et al. 2015), reflecting on the findings from basic research. This knowledge is rarely transferred into the scientific and bioengineering community working with terrestrial fungi and their biotechnological application.

Here, we aim to illustrate the marine fungal diversity and its potential for biotechnology, describe the current stage of technological realization, and highlight special challenges in working with marine fungi. We hope to provide access to this amazing source to the fungal biotech and scientific community.

II. Marine Fungi: A Disputed Group with High Diversity

A. Taxonomic and Phylogenetic Considerations

Within the marine microorganisms, fungi are still a poorly investigated group. For many years, the presence of fungi in marine habitats was negated, mainly as a consequence of their

[1] Flensburg University of Applied Sciences, Flensburg, Germany; e-mail: antje.labes@hs-flensburg.de

Genetics and Biotechnology, 3rd Edition
The Mycota II
J.P. Benz, K. Schipper (Eds.)

cell biology and feeding strategies and partly because of the **bias of research** (Richards et al. 2012). However, the number of isolated fungal strains from the marine environment illustrates the presence of marine fungi. They have been found in association with algae (Raghukumar 2006), sponges (Wiese et al. 2011), and corals (Golubic et al. 2005), as well as in deep-sea mud (Takami et al. 1997, Nagahama 2003) and sediments (Damare et al. 2006), even in anoxic (Stoeck et al. 2003) and methane hydrate-bearing deep-sea sediments (Lai et al. 2007). Living fungi were isolated from marine sediments with an estimated age from 0.18 to 0.43 million years (Gupta 2002, Raghukumar et al. 2004b). In 1979, Kohlmeyer and Kohlmeyer distinguished between two groups of marine fungi, i.e., the obligate and the facultative marine fungi (Kohlmeyer and Kohlmeyer 1979). Due to the fact that many taxa isolated from marine habitats are already known from terrestrial habitats, Hyde et al. stated marine fungi to be rather a **physiological and ecological** group, but certainly not a systematical and taxonomical one (Hyde et al. 2000).

Recent molecular studies on marine mycology revealed the formation of **environmental clusters** within the major fungal phyla and showed that marine fungi occupy a central position in a large number of marine habitats (Manohar and Raghukumar 2013). The rising acceptance of the significance of marine fungi is reflected by the increasing number of publications over the last 60 years. To enhance the currently incomplete understanding of fungal evolutionary diversity (Hawksworth 2001; Jones 2011; Jones et al. 2011), the investigation of the fungal evolution in marine habitats, especially in the deep sea, is essential. These habitats are considered to be hot spots of early evolution (Manohar and Raghukumar 2013).

In the past, the identification of novel strains from marine habitats was impeded by fungal taxonomy based on the morphological description of cultivable strains from terrestrial origin (Richards et al. 2012). Nowadays, the use of additional molecular identification techniques has led to a renewal within the fungal tree of life. Holomorphic relationships were disclosed, as for the genus *Aspergillus* and eight related teleomorph genera (Pitt and Samson 2007), illustrating the obscurity of the nomenclature of fungi.

In 2011, the eighteenth International Botanical Congress in Melbourne, Australia, released the International Code of Nomenclature for algae, fungi and plants, the so-called Melbourne Code (Hawksworth 2011). The code favors the **one-fungus-one-name-rule**, with respect to the teleomorph forms. Besides the deficient nomenclature within the group of fungi, the reconstruction of the tree of life refuted the formerly assumed close relationship of fungi and plants. Fungi are an independent group equal to plants and animals, to the latter of which they are more closely related (Hedges et al. 2004) as both groups belonging to the *Opisthokonta* (Adl et al. 2012). The last common ancestor was a unicellular organism living in the ocean, equipped with a flagellum (Buckley 2008). Molecular data revealed that fungi colonized the land up to 1 billion years ago (Parfrey et al. 2011).

Metagenomics approaches identified a diverse collection of marine fungi, including sequences branching close to chytrids (flagellated protozoa related to fungi), filamentous hypha-forming fungi, and multicellular fungi. The majority of the sequences branched with ascomycete and basidiomycete yeasts (Richards et al. 2012).

With respect to the cultivated diversity, the availability of novel marine genera is still limited. From a total of approximately 30 studies on secondary metabolites from fungi of the deep sea, one-third of the production strains were identified as belonging to *Penicillium* and one-third to *Aspergillus*, and the remainder was distributed to different other genera (Wang et al. 2015). The high proportion of members of the genera *Penicillium* and *Aspergillus* may reflect the high abundance of representatives of these two genera in the samples. Alternatively, the commonly used isolation media favor the growth of these fast-growing groups, reducing the probability to isolate fungi present in low abundance and with other nutrient requirements.

B. Fungal Role in the Marine Environment

Fungi are widely distributed in marine environments from the deep sea to polar ice covers. They occur in sediments and are found in all kinds of living and dead organic matter (Hyde et al. 2000; Jones 2011). Their numbers in the water column of the oceans are low compared to bacteria, and most of the studies on marine fungi have been made with those associated with marine sediments, specific substrates like driftwood or with living macroorganisms such as algae, corals, and sponges (Imhoff 2016).

Cultivation-based data and molecular functional studies demonstrated that fungi in the ocean cover a variety of ecological roles. As known for the terrestrial habitats, many marine fungi act as **saprophytes**, especially degrading macromolecular material as the starter for the microbial food chain (Scheu 2002). Fungi, besides directly serving as nutrient source, also help in removing hardly digestible compounds present in plant parts (Raghukumar et al. 2004b). The fibrous lignocellulose tissues may be softened by the lignocellulose-degrading enzymes of the saprophytic fungi. Fungi might also supply essential nutrients to detritivores. Thereby, fungi play a crucial role in the chemical cycling of carbon and nitrogen in the sea sediments (Karthikeyan et al. 2014). Especially deep-sea fungi were postulated to significantly contribute to aggregate formation and carbon contribution (Raghukumar et al. 2004b).

Despite the rise of metagenome studies especially focusing on the marine environment, metagenome data specifically targeting fungi are missing. The use of NGS techniques opened a new era for marine fungal biodiversity research, tapping an unexpected fungal diversity and richness in the marine realm (e.g., Amend et al. 2012; Redou et al. 2014; Richards et al. 2012; Hassett and Gradinger 2016; Rämä et al. 2014). Generally, the fungal abundance in marine systems can vary greatly. The available studies suggest fungi to be represented in sediments and associated to macroorganisms with up to 30% of the obtained sequences but also show high abundances in the open water column. The fungal sequences found do illustrate the gap between molecular and cultural approaches; especially the deep lineages of fungi are not well represented in the cultural collections, and the overlap of the described diversity is still limited. However, the molecular studies underline the untapped potential of fungal diversity in the oceans.

Marine fungi can be repeatedly isolated from living organisms. Both plants and animals harbor a number of fungi, often with representatives of *Acremonium*, *Aspergillus*, *Fusarium*, *Penicillium*, *Phoma*, and *Trichoderma* (Wang 2006). Due to their accumulation within the animal or plant, usually a large number of fungal species can be isolated, which increases the probability to find representatives of less common taxa. For example, fungi belonging to the less common genera *Beauveria*, *Botryosphaeria*, *Epicoccum*, *Tritirachium*, and *Paraphaeosphaeria* have been obtained from marine sponges (Zhang et al. 2009). The fungi do occur on surfaces but also in specialized parts of macroorganisms. For instance, it was shown that the sponge *Tethia aurantia* harbors different fungal communities in its different tissue layers (Wiese et al. 2011; Thiel et al. 2007).

It is assumed that associations to macroorganisms are based on different ecological roles including mutual or even symbiotic beneficial interaction, saprophytic activity on injured parts of the host and pathogenicity (Richards et al. 2012; Yarden 2014). Aspergillosis in marine animals is a well-studied example of fungal pathogenicity. *Aspergillus sydowii* is a pathogen, e.g., of the Caribbean sea fan (*Gorgonia* spp.) as well as other gorgonian corals. The impact of disease ranges from local mass impact to partial tissue loss and eventual recovery. *A. sydowii* is known as a common terrestrial soil fungus but was shown to prosper in both terrestrial and marine environments. Although *A. sydowii* is common and cosmopolitan, it had not previously been recognized causing disease in terrestrial plants or animals. It was concluded that isolates taken from diseased corals have acquired specific pathogenic potential not seen in isolates from other sources, because non-marine strains of *A. sydowii* did not cause disease in sea fans (Kim and Rypien 2015).

A wide number of fungi were described in association with marine plants, including mangroves. On the one hand, endophytic lifestyle seems to be quite common (Rashmi et al. 2019). On the other hand, pathogenic, e.g., oomycetes and chytrids frequently occurred and induced prevalence of disease in algal aquacultures. These infections can destroy the populations of host plants to a great extent (Li et al. 2010), mainly known from Asian countries cultivating a number of algal species since long time. Frequent re-isolation of fungi and specific occurrence of fungal species in association with specific macroorganisms point toward a number of host-microbe interactions beyond pathogenicity and opportunistic commensalism. Cultured fungi isolated from sessile marine animals and algae have been demonstrated to be capable of producing novel chemicals, which have the potential to affect the marine hosts and their microbiome (as reviewed in Imhoff et al. 2011; Rateb and Ebel 2011; Raghukumar 2008). Also, the determination of abundant bioactive compounds of fungal origin in the **holobiont** indicates their roles in host-fungal and fungal-microbe interactions.

In any case, a thorough understanding of the roles marine fungi do have in the ecosystem will help to prospect for novel genes and products and facilitate realization of biotechnological application of marine fungi, their enzymes, and metabolites.

III. Biotechnology of Marine Fungi: Challenges and Opportunities

A. Cultivation of Marine Fungi: Same But Different?

Any biotechnological production depends on the fungal physiology, as well as on culture medium composition. **Carbon and nitrogen sources** play an important role in biotechnological production. This is well established for numerous fungi from terrestrial environments and was confirmed for marine fungi. Complex substrates, such as starch, casein, pectin, malt extract, wheat bran, olive oil, xylan, and sugarcane bagasse, were used for the cultivation of marine fungi from a variety of marine sources. The optimum temperature and pH of most of these fungi (and subsequently their enzymes) ranges from 20 to 70 °C and from 3 to 9, respectively (Kjer et al. 2010; Golubic et al. 2005; Imhoff 2016).

Marine fungal strains of genera also described from terrestrial habitats have approximately the same optimum conditions for growth and use the same primary pathways for energy and biomass production. However, the **demand of salt** during cultivation can be crucial as a consequence of adaptation to ocean salinity (e.g., D'Souza-Ticlo et al. 2009; Chen et al. 2011 for enzymes from marine fungi). Salt demand means not only presence of sodium chloride in the appropriate concentration. **Sulfate**, one of the major constituents of sea water but also minor elements including heavy metals, is equally important for proper growth media composition of many, especially the obligate marine fungi. Accordingly, many researchers use "**artificial sea water**" for cultivation of these strains (Kjer et al. 2010; Jones and Jennings 1964).

The diversity of marine fungi is not adequately represented in the cultured diversity. Many studies have focused on just a few genera, mainly *Penicillium*, *Aspergillus*, *Fusarium*, and *Cladosporium*. This might be due to the relatively easiness of isolation of these groups, which also dominate isolation campaigns from terrestrial habitats. But still, even when comparing marine strains of these groups to terrestrial relatives, different secondary metabolite biosynthesis pathways can be found (Imhoff 2016). The high diversity seen at the genus level extends further to the subgenus level. For example, representatives of *Penicillium* are among the most studied fungi and represent important drug producers. Nevertheless, many new secondary metabolites are continuously found within strains of marine origin of this genus as shown in reviews by Rateb and Ebel (2011) and Blunt et al. (2015). However, many of the novel and unknown environmental fungal taxa identified in molecular studies are likely to be difficult to propagate in culture, either because they are outcompeted by spores of (terrestrial) fungi also contained in the environmental samples, or alternatively their life cycle is dependent on a symbiotic interaction (Del Campo et al. 2014).

B. Availability of Strain Collections

For the optimal and sustainable use of the marine fungal sources, it is essential to store the microbial cultures. Culture collections keep strains available for further investigations and production processes. Conservation of living cells while reducing metabolism, e.g., by **cryoconservation,** is crucial in order to prevent genetic and phenotypic changes induced by repeated passaging of the cultures. Culture collections thus play a vital role in the conservation and sustainable use of microbial resources including fungi (Daniel and Prasad 2010). They also provide the authentic biological material needed for research in the form of **reference strains** (Fig. 1). Until now, only a limited number of marine fungi are deposited in **culture collections with public access.** As a consequence, only a few voucher sequences are available although they are a prerequisite for an accurate sequence-based identification of marine fungi. Bearing in mind the huge number of so far undescribed taxa and strains, there is a strong need for further extension of culture collections with marine fungal specimen and adopted software to store the different kinds of data related to a cultured taxon (Pena and Malm 2012). Nevertheless, some small specific collections of marine fungi exist (Table 1).

C. Genetic Background as a Prerequisite for Biotechnological Application

The overall potential of marine fungi for biotechnological application can only be estimated. A better description of the **phylogenetic diversity** of marine fungi, the **biosynthetic potential** of strains, and the **phylogeny of natural products biosynthesis** are needed. Therefore, much emphasis has to be given to determine the phylogenetic position of the fungal strains in order to enable correlation of the phylogenetic relationship to the desired product or, alternatively, to the genetic potential. Accordingly, an increasing number of genome sequences of fungi are currently completed demonstrating an overall tremendous biosynthetic capacity of fungi. High numbers of biosynthetic gene clusters are coding for secondary metabolites in a single genome (e.g., Kumar et al. 2018). For the majority of these

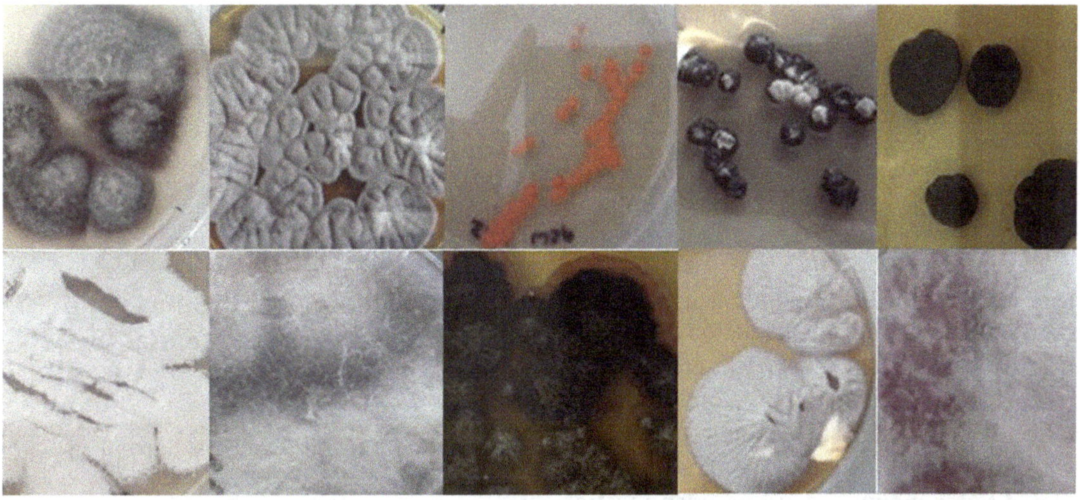

Fig. 1 Examples of unclassified strains from a marine fungal strain collection at Flensburg University of Applied Sciences, Flensburg, Germany, illustrating the morphological diversity of marine fungi

Table 1 Strain collections comprising a significant number of marine fungal strains

Collection	Focus organisms	Web contact[a]	Description
American type culture collection (ATCC)	Filamentous fungi and yeasts, representing over 7600 fungal species	www.lgcstandards-atcc.org/en/Products/Cells_and_Microorganisms/Fungi_and_Yeast.aspx	Offers storage and management of biomaterials
Biological Resource Center, NITE (NBRC)	Filamentous fungi and yeasts	www.nite.go.jp/en/nbrc/	General deposit possible
Canadian Collection of Fungal Cultures (CCFC)	Over 350,000 fungal and fungal plant disease specimens	https://www5.agr.gc.ca/eng/scientific-collaboration-and-research-in-agriculture/agriculture-and-agri-food-research-centres-and-collections/canadian-collection-of-fungal-cultures-daomc/?id=1503505359059	Small subset of national culture collection comprises marine isolates
mFSC Flensburg strain collection of marine fungi	Unclassified marine fungal isolates	https://hs-flensburg.de/forschung/fue/forschungsinstitute/zait	Collection from different marine habitats, in preparation for storage in public collection
Kiel collection of marine microbes	Marine fungi	www.geomar.de	Collection of marine fungi at Helmholtz Centre for Ocean Research, GEOMAR, not public
Marbank	Marine organisms from Norwegian waters	https://en.uit.no/forskning/forskningsgrupper/gruppe?p_document_id=405720	Offers storage of samples, marine products and customized sampling
National Center for Genetic Engineering and Biotechnology (BIOTEC)	2486 fungal species[b]	www.thai2bio.net	Small subset of marine isolates
United Kingdom National Culture Collection	Central access point for a variety of collections: fungi, yeast, pathogenic fungi	www.ukncc.co.uk	Small subset of national culture collection comprising also marine isolates
Westerdijk Fungal Biodiversity Institute	More than 50,000 fungal strains	www.westerdijkinstitute.nl/	Small subset of national culture collection comprising also marine isolates

Adapted from Reich and Labes (2017)
[a]Accessed 12th February 2020
[b]Number obtained 12th February 2020

gene clusters, the corresponding natural products remain to be identified, but the ever-improving analytical tools will certainly help to elucidate new compounds. Combination of molecular methods and genomic approaches will also contribute to solve the problems related to taxonomy and species identification of marine fungi (Reich and Labes 2017; Kramer et al. 2016).

IV. Biotechnological Products and Processes: From Potential to Application

A. Bioactive Natural Products: Secondary Metabolites

Communication with the host and within associated microbial communities was demonstrated to be mediated by secondary metabolites. Therefore, associations of fungi and macroorganisms are of special interest with respect to the discovery of bioactive natural products (Rashmi et al. 2019). Accordingly, many screening campaigns led to the discovery and subsequent identification of new compounds with interesting bioactivities, including **antibiotic, antiviral, antitumoral**, etc. Marine fungi produce **structurally unique bioactive natural products** that are not found in terrestrial counterparts. Kong et al. showed that large portions of marine scaffolds are novel (Kong et al. 2010). Auranomides A and B, quinazolin-4-ones substituted with a pyrrolidin-2-iminium moiety from the marine-derived fungus *Penicillium aurantiogriseum*, or aspergillols A and B with a C–C fusion of an anthraquinone and orcinol unit from a deep-sea *Aspergillus versicolor* represent new scaffolds from marine habitats (Song et al. 2012; Wu et al. 2016). Probably, these compounds constitute only a small portion of the possible repertoire, as drug discovery efforts often involve a specific focus, either on disease target, taxonomic groups or on habitats, but do not aim to describe the full **genetic biosynthetic potential.**

Today, only a few compounds made their way to the pharmaceutical market. Marine fungal compounds do face the same challenges in drug discovery and drug development as any other fungal compounds. However, in the 1950s, **cephalosporin C**, a **β-lactam type natural antibiotic**, was discovered from a *Cephalosporium* (later reclassified as *Acremonium*) strain obtained at the Sardinian coast (Newton and Abraham 1955; Abraham et al. 1953). Gliotoxin was identified as a new type of the **antibiotic diketopiperazine** produced by a marine deep-sea sediment *Aspergillus* sp. strain (Okutani 1977).

Marine fungi did also prove to be effective producers of natural products with antitumoral activity (Pejin and Karaman 2017). However, halamid, an anticancer compound of marine fungal origin (trade name plinabulin) is the only one already being in clinical testing phase 2 and 3 for different indications (for review see Pereira et al. 2019). Nevertheless, the current chemotherapeutic clinical pipeline will be fed with marine-derived fungal agents, which are in preclinical stage at the moment.

A steadily increasing number of new, active fungal natural products have been identified from the marine environment, proving them to be a prolific source for bioactive compounds: Blunt et al. listed over 200 natural products from marine-sourced fungi (excluding those from mangroves) in the year 2013 (Blunt et al. 2015). **Cosmetic and cosmeceutical applications** might also be taken into account for marine fungal secondary metabolites (Agrawal et al. 2018). Here, biotechnological approaches can help to close the developmental gap between discovery and production in order to solve compound availability issues in drug development (as reviewed for antibiotics in Silber et al. 2016).

B. Enzymes

Marine fungi can be used for the biotechnological production of **salt- and osmostable enzyme variants** but do also provide new enzymatic activities. Fungi foster, e.g., the detrital processes by production of extracellular degradative enzymes (Sikes et al. 2009). These fungal enzymes from marine origin can be used in biotechnological applications. Extracellular degradative enzymes such as cellulases, xylanases, and ligninases of several terrestrial fungi already have found biotechnological applications in paper and pulp industries, food production, and clothing industry. Fungi isolated from living and decaying marine plant material are capable of degrading "classical" plant macromolecules, like cellulose, cellobiose, lipids, pectin, starch, xylan, and tannic acid (Gessner 1980). In addition, marine fungi can degrade the special carbohydrates found in marine plants, such as alginates, sulfated polysaccharides, laminarin, fucoidan, etc. (as early described as in Schaumann and Weide 1990). These enzymes can be applied in transforming these sugars for energetic use of algal remaining's (e.g., for biogas production) especially from aquacultures. Also, fungal polygalacturonases are useful enzymes for clarification of fruit juices in the food industry (Sandri et al. 2011).

The removal of color residues is a major challenge in paper and textile industries. Effluents from such plants can possess alkaline pH and high salt content. Marine fungi may be ideally suited for the bioremediation of such effluents (Verma et al. 2010): White-rot fungi containing laccase and other enzymes capable of decolorization were isolated from mangroves. Laccase production was shown in several marine fungi in the presence of sea water (D'Souza-Ticlo et al. 2009). However, salt-tolerant lignin-degrading enzymes have not been sufficiently explored for their biotechnological applications.

Marine fungi have also been discussed as source for enzymes or for whole cell application in the **degradation of polycyclic aromatic hydrocarbons** (PAHs). PAHs are widely distributed in the environment and may persist for extended periods of time with toxic, mutagenic, and carcinogenic effects (for review see Bonugli-Santos et al. 2015). The ligninolytic and the monooxygenase system of cytochrome P-450 may be involved in PAH degradation by filamentous fungi (Haritash and Kaushik 2009). The use of marine-derived fungi for the **bioremediation of polluted saline environments** will be facilitated by their tolerance to saline conditions.

Additionally, deep-sea fungi growing under extreme conditions are a good source for industrially useful enzymes with novel properties (Synnes 2007). Proteases constitute one group of the several extracellular enzymes being produced by these fungi; elevated pressure resistance, high pH stability and activity in the presence of several commercial detergents and high salt content were shown for fungal proteases from deep-sea sediments. These features are desirable for commercial application, e.g., as detergents additive for cold wash (Raghukumar 2008).

C. Compatible Solutes

All organisms in marine environments have to adapt to the high salinity of seawater causing significant osmotic pressure. Seawater organisms therefore need to maintain water potentials lower than that of their environment in their cells to enable water uptake. Cells developed a number of strategies to deal with salts: effective efflux machineries, antiporter proteins, and the production of compatible solutes (Forsyth et al. 1971; Ahmadi et al. 2016). The latter strategy is frequently observed in marine fungi. They accumulate osmolytes such as **glycerol, mannitol, polyol, and trehalose** (Blomberg and Adler 1992). These compounds are of technological interest due to their **stabilizing effects, stress-protective and therapeutic activities** in cosmeceuticals and pharmaceuticals but also as **cryoprotectants** (Jadhav et al. 2018).

D. "Detox" Activities

Deep-sea hot vent systems are often characterized by high heavy metal concentration. Fungi thriving in these habitats need to cope with these conditions by producing detoxification compounds, often extracellular sugars such as chitosan-like molecules. These can be applied in cosmetic industry (Brown et al. 2016). Also, enzymes from these fungi show high resistance against heavy-metal intoxication. For instance, deep sea *Cryptococcus* strain was tolerant to $CuSO_4$ up to a concentration of 50 mM and showed high activity of superoxide dismutase, an enzyme responsible for scavenging superoxide radicals (Miura et al. 2002).

V. How to Design Biotechnological Processes for True Marine Fungi

Bonugli-Santos et al. (2015) reviewed the biotechnological production of marine enzymes including bioprocess strategies adopted for the cultivation of marine organisms for enzyme production. Large-scale production (e.g., in bioreactors) of glucoamylase, superoxide dismutase, lignin peroxidase, chitinase, protease, and glutaminase by marine fungal strains including, e.g., *Aureobasidium pullulans*, *Penicillium janthinellum*, and others, has been reported in the literature (Sarkar et al. 2010). These enzymes are produced in bioreactors largely through submerged-state fermentation, and the conditions related to bioreactor production are quite comparable to processes using terrestrial strains. Specific studies for establishing strategies of marine enzyme purification are scarce; the same holds true for processes of secondary metabolite production.

The approach to bring marine fungi into biotechnological application is similar to other microbial processes: After the optimum culture conditions are defined on a small scale (in general, Erlenmeyer flasks in shaking conditions are used), the cultivation has to be transferred to bioreactor-based systems and subsequent scale-up must be performed. Substrate consumption, product formation, and cellular biomass are important factors, which should be considered, and quantified, for appropriate scale-up studies. Special attention in fungal production processes has to be given to **pellet formation** (Krull et al. 2013).

Generally, the unique features of marine environment do have relevance to marine fungal biotechnology. A consideration of the unique properties of the marine environment is important for marine biotechnology: Biotechnological production processes are influenced by the special adaptations of organisms to their environment. The physical factors relevant for cultivation that influence the growth of marine fungi most are **salinity** (including low water potential, high concentrations of sodium ions and high osmotic pressure), **slightly alkaline pH,** partly **oligotrophic nutrient conditions** (especially in the water column), and **high hydrostatic pressure** in combination with **low temperature**, especially in the deep-sea environments (Richards et al. 2012; Bonugli-Santos et al. 2015). Sea water in average has a salinity of 33–35 ppt. Freshwater in comparison contains less than 0.5 ppt salts. Sea waters in estuaries and brackish areas, like in the Baltic Sea or in river deltas, can show wide salinity gradients, requiring special adaptation of the inhabiting organisms. Also, hypersaline waters occur, e.g., in the Dead Sea with salinities of 50–100 ppt. The presence of high levels of salt in seawater are challenging with respect to classical performance of biotechnological processes in stainless steel vessels, as corrosion becomes a severe issue. Even more, early experiments demonstrated that salinity optima for growth in some marine fungi show upward shifts with increasing incubation temperature (Damare and Raghukumar 2008). This interaction of salinity and temperature is termed ***Phoma* pattern** since it was first described in the marine species of *Phoma* (Lorenz and Molitoris 1990). Interestingly, only few reports have investigated the impact of varying salt concentrations on production of marine fungal metabolites. Initial findings show that some marine fungal species exhibit increased growth with increasing seawater concentration in the medium (Masuma et al. 2001). In addition to the osmotic effects on the cell, secondary metabolite production might be sensitive to the seawater composition, i.e., concentration of specific salts, which could have implications for tank reactor cultivation of marine fungi.

In contrast to terrestrial fungi, which generally grow best at pH 4.5–6.0, marine fungi were demonstrated to grow and produce various extracellular enzymes at pH 7–8 (Damare et al. 2006; Raghukumar et al. 2004a). Accordingly, the pH-control in biotechnological processes will need adaptations, when marine strains of a well-known terrestrial species are cultivated. Pressurized conditions are of importance for the cultivation of deep-sea fungi. Low temperature optima in combination with high pressures are of interest, as the catalytic properties of enzymes may be altered (Eisenmenger and Reyes-De-Corcuera 2009).

Bioprocess engineering in marine biotechnology follows the path from **discovery to commercialization** with a variety of possible starting points, but a full value chain remains mainly theoretical, as quantitative biotechnological engineering studies on marine fungi are virtually non-existing in the literature. Silber et al. (2016) reviewed the biotechnological realization of marine fungal antibiotics and demonstrated that biotechnology has a vast potential for sustainable production of antibiotics. Biotechnology, as studied and developed with terrestrial representatives, may help to expand and understand the chemical space in a targeted manner, provide classical full fermentative and semi-synthetic processes, and may include metabolic engineering manipulating the genetic background as a basis for generation of "biological" derivatives (Silber et al. 2016). For terrestrial fungi, the technical and economic feasibility of large-scale processes has been proven many times. Successful transfer into stirred-tank-reactor-system for antibiotic production from marine fungi has been demonstrated, e.g., for the biosynthesis of the tetramic acid compounds ascosetin and lindgomycin by an Arctic marine fungus of the Lindgomycetaceae family (Wu et al. 2015) exhibiting a novel chemical composition.

Sarkar et al. (2010) provided bioprocess data for fermentations of marine fungi for enzyme production. Based on their analysis, there is a demand for further adaptations of scale and inclusion of knowledge on the biology of the fungi into the design of the production processes (e.g., salinity adaptations or pressure for deep-sea organisms). Development of full fermentative processes is required (Sarkar et al. 2010). The increasing knowledge gained on all regulatory levels by means of "omics" techniques will provide the necessary insights. Knowledge on the regulatory genes can be applied in genetic approaches to activate natural product production. Due to the lack of model organisms upon marine fungi, genetic tools for heterologous expression and for homologous manipulation of marine fungi are still in their infancy. The recently successful overexpression of genes leading to synthesis of enniatins, cyclodepsipeptides originally isolated from *Halosarpheia* sp., in *Aspergillus niger* demonstrated reprogramming and transplanting of biosynthetic pathways into established models as a further tool for such approaches (Zobel et al. 2016).

Inclusion of new methods into process development will significantly contribute to further improvements. The use of immobilized cells in a repeated batch tower reactor led to a significant increase of yield in the production of cephalosporin (Ozcengiz and Demain 2013). Niche-mimic bioreactors might be one opportunity to enhance productivity as well as the development of specialized reactors in industrial scale.

VI. Conclusions

A majority of culture-dependent and culture-independent fungi recovered from marine resources show homology to terrestrial species, which obfuscated the ecological role of marine fungi. The presence of novel fungal sequences with less than 97% similarity to previously identified fungal sequences in public databases and culture studies demonstrating obligate marine culture conditions has proven the multiphyletic ecological group of marine fungi.

These fungi show unique cultivation properties and features. In nature, they fulfill a number of ecological functions, comparable to terrestrial fungi. Their unique features may be applied in biotechnology for new products but also for redesign of existing biotechnological processes with more robust culture conditions, as marine fungi did adapt to harsh conditions including high pressure, high osmolarity, cold, toxic heavy metal concentrations, etc.

Only few examples of industrial processes using marine fungi as production organisms are available, but the number of studies highlighting the potential is impressive for enzymes, secondary metabolites, and other possible products. Industrial-scale biotechnological processes should be feasible from scientific, technical, and economic perspectives. This viability needs to be demonstrated through the development of process concepts based on fermentation including metabolic engineering, purification, molecular design, and synthesis data. Some obstacles remain, which need to be solved. Especially, the reduction of the salinity in culture media is a technological challenge.

To realize the potential, we suggest to:

1. Apply products of marine fungi but also establish new and robust workhorses in biotechnology. The biodiversity of filamentous fungi from marine sources is mostly untapped for these applications.
2. Bring together the different scientific communities working on marine and terrestrial fungi and learn from existing processes. Interdisciplinarity between technologists, ecologists, and marine scientist will help

to close gaps in the knowledge of marine fungi relevant for biotechnological productions.

In summary, marine fungal biotechnology can deliver new production strains, enzymes, and products for a sustainable use of marine resources and deeper insights into technological and ecological questions.

Conflicts of Interest The author declares no conflict of interest.

References

Abraham EP, Newton GG, Crawford K, Burton HS, Hale CW (1953) Cephalosporin N: a new type of penicillin. Nature 171:343

Adl SM, Simpson AG, Lane CE, Lukes J, Bass D, Bowser SS, Brown MW, Burki F, Dunthorn M, Hampl V, Heiss A, Hoppenrath M, Lara E, Le Gall L, Lynn DH, McManus H, Mitchell EA, Mozley-Stanridge SE, Parfrey LW, Pawlowski J, Rueckert S, Shadwick RS, Schoch CL, Smirnov A, Spiegel FW (2012) The revised classification of eukaryotes. J Eukaryot Microbiol 59:429–493. https://doi.org/10.1111/j.1550-7408.2012.00644.x

Agrawal S, Adholeya A, Barrow CJ, Deshmukh SK (2018) Marine fungi: an untapped bioresource for future cosmeceuticals. Phytochem Lett 23:15–20. https://doi.org/10.1016/j.phytol.2017.11.003

Ahmadi N, Baroiller J, D'Cotta Carreras H, Morillon R (2016) Adaptation to salinity. In: Torquebiau E (ed) Climate change and agriculture worldwide. Springer, Dordrecht, pp 45–58

Amend AS, Barshis DJ, Oliver TA (2012) Coral-associated marine fungi form novel lineages and heterogeneous assemblages. ISME J 6:1291–1301

Blomberg A, Adler L (1992) Physiology of osmotolerance in fungi. In: Rose AH (ed) Advances in microbial physiology. Academic Press, London, pp 145–212

Blunt JW, Copp BR, Keyzers RA, Munro MH, Prinsep MR (2015) Marine natural products. Nat Prod Rep 32:116–211. https://doi.org/10.1039/c4np00144c

Bonugli-Santos RC, dos Santos Vasconcelos MR, Passarini MRZ, Vieira GAL, Lopes VCP, Mainardi PH, dos Santos JA, de Azevedo DL, Otero IVR, da Silva Yoshida AM, Feitosa VA, Pessoa A, Sette LD (2015) Marine-derived fungi: diversity of enzymes and biotechnological applications. Front Microbiol 6:269. https://doi.org/10.3389/fmicb.2015.00269

Brown D, Brunt K, Rehmann N (2016) Chitosan biopolymer from fungal fermentation for delivery of chemotherapeutic agents. Mater Matter 11:3

Buckley M (2008) The Fungal Kingdom - diverse and essential roles in earth's ecosystem. American Academy of Microbiology, Tucson, AZ

Chen HY, Xue DS, Feng XY, Yao SJ (2011) Screening and production of ligninolytic enzyme by a marine-derived fungal *Pestalotiopsis* sp. J63. Appl Biochem Biotechnol 165:1754–1769. https://doi.org/10.1007/s12010-011-9392-y

Damare S, Raghukumar C (2008) Fungi and macroaggregation in deep-sea sediments. Microb Ecol 56:168–177. https://doi.org/10.1007/s00248-007-9334-y

Damare S, Raghukumar C, Raghukumar S (2006) Fungi in deep-sea sediments of the Central Indian Basin. Deep Sea Res Pt I 53:14–27. https://doi.org/10.1016/j.dsr.2005.09.005

Daniel HM, Prasad GS (2010) The role of culture collections as an interface between providers and users: the example of yeasts. Res Microbiol 161:488–496

Del Campo J, Sieracki ME, Molestina R, Keeling P, Massana R, Ruiz-Trillo I (2014) The others: our biased perspective of eukaryotic genomes. Trends Ecol Evol 29:252–259. https://doi.org/10.1016/j.tree.2014.03.006

D'Souza-Ticlo D, Sharma D, Raghukumar C (2009) A thermostable metal-tolerant laccase with bioremediation potential from a marine-derived fungus. Mar Biotechnol 11:725–737. https://doi.org/10.1007/s10126-009-9187-0

Eisenmenger JM, Reyes-De-Corcuera J (2009) High pressure enhancement of enzymes: a review. Enzyme Microb Technol 45:331–347

Fenical W, Jensen PR (1993) Marine microorganisms: a new biomedical resource. In: Attaway DH, Zaborsky OR (eds) Pharmaceutical and bioactive natural products. Springer, New York, pp 419–457

Forsyth PM, Shindler BD, Gochnauer BM, Kushner JD (1971) Salt tolerance of intertidal marine bacteria. Can J Microbiol 17:825–828

Gessner RV (1980) Degradative enzyme production by salt-marsh fungi. Botanica Mar 23:133–139

Golubic S, Radke G, Le-Campion-Alsumard T (2005) Endolithic fungi in marine ecosystem. Trends Microbiol 13:229–235

Gupta SM (2002) Pyloniid stratigraphy – a new tool to date tropical radiolarian ooze from the central tropical Indian Ocean. Mar Geol 184:85–93. https://doi.org/10.1016/s0025-3227(01)00276-6

Haritash A, Kaushik C (2009) Biodegradation aspects of polycyclic aromatic hydrocarbons (PAHs): a review. J Hazard Mater 169:1–15

Hassett BT, Gradinger R (2016) Chytrids dominate arctic marine fungal communities. Environ Microbiol 18:2001–2009

Hawksworth DL (2001) The magnitude of fungal diversity: the 1.5 million species estimate revisited. Mycol Res 105:1422–1432

Hawksworth DL (2011) A new dawn for the naming of fungi: impacts of decisions made in Melbourne in

July 2011 on the future publication and regulation of fungal names. IMA Fungus 2:155–162. https://doi.org/10.5598/imafungus.2011.02.02.0610.3897/mycokeys.1.2062

Hedges SB, Blair JE, Venturi ML, Shoe JL (2004) A molecular timescale of eukaryote evolution and the rise of complex multicellular life. BMC Evol Biol 4:2. https://doi.org/10.1186/1471-2148-4-2

Hyde KD, Sarma VV, Jones EBG (2000) Morphology and taxonomy of higher marine fungi. In: Hyde KD, Pointing SB (eds) Marine mycology - a practical approach, 1st edn. Fungal Diversity Press, Hong Kong, pp 172–204

Imhoff JF (2016) Natural products from marine fungi - still an underrepresented resource. Mar Drugs 14:1–19. https://doi.org/10.3390/md14010019

Imhoff JF, Labes A, Wiese J (2011) Bio-mining the microbial treasures of the ocean: new natural products. Biotechnol Adv 29:468–482. https://doi.org/10.1016/j.biotechadv.2011.03.001

Jadhav K, Kushwah B, Jadhav I (2018) Insight into compatible solutes from halophiles: exploring significant applications in biotechnology. In: Singh J, Sharma D, Kumar G, Sharma NR (eds) Microbial bioprospecting for sustainable development. Springer, Singapore, pp 291–307

Jones EBG (2011) Are there more marine fungi to be described? Botanica Mar 54:343–354. https://doi.org/10.1515/bot.2011.043

Jones GEB, Jennings DH (1964) The effect of salinity on the growth of marine fungi in comparison with non-marine species. Trans Br Mycol Soc 47:619–625. https://doi.org/10.1016/S0007-1536(64)80041-3

Jones MD, Forn I, Gadelha C, Egan MJ, Bass D, Massana R, Richards TA (2011) Discovery of novel intermediate forms redefines the fungal tree of life. Nature 474:200–203. https://doi.org/10.1038/nature09984

Karthikeyan P, Kanimozhi K, Senthilkumar G, Panneerselvam A, Asho G (2014) Optimization of enzyme production in *Trichoderma viride* using carbon and nitrogen source. Int J Curr Microbiol App Sci 3:88–95

Kim K, Rypien K (2015) Aspergillosis of Caribbean sea fan corals, *Gorgonia* spp. In: Woodley CM, Downs CA, Bruckner AW, Porter JW, Galloway SB (eds) Diseases of coral. Wiley, Hoboken, NJ, pp 236–241

Kjer J, Debbab A, Aly AH, Proksch P (2010) Methods for isolation of marine-derived endophytic fungi and their bioactive secondary products. Nat Protoc 5:479–490. https://doi.org/10.1038/nprot.2009.233

Kohlmeyer J, Kohlmeyer E (1979) Marine mycology: the higher fungi. Academic Press, New York

Kong DX, Jiang Y-Y, Zhang H-Y (2010) Marine natural products as sources of novel scaffolds: achievement and concern. Drug Discov 15:884–886

Kramer A, Labes A, Imhoff JF (2016) Phylogenetic relationship and secondary metabolite production of marine fungi producing the cyclodepsipeptides scopularide A and B. Mar Biotechnol 18:466–474. https://doi.org/10.1007/s10126-016-9707-7

Krull R, Wucherpfennig T, Esfandabadi ME, Walisko R, Melzer G, Hempel DC, Kampen I, Kwade A, Wittmann C (2013) Characterization and control of fungal morphology for improved production performance in biotechnology. J Biotechnol 163:112–123. https://doi.org/10.1016/j.jbiotec.2012.06.024

Kumar A, Sorensen JL, Hansen FT, Arvas M, Syed MF, Hassan L, Benz JP, Record E, Henrissat B, Poggeler S, Kempken F (2018) Genome sequencing and analyses of two marine fungi from the North Sea unraveled a plethora of novel biosynthetic gene clusters. Sci Rep 8:10187. https://doi.org/10.1038/s41598-018-28473-z

Lai X, Cao L, Tan H, Fang S, Huang Y, Zhou S (2007) Fungal communities from methane hydrate-bearing deep-sea marine sediments in South China Sea. ISME J 1:756–762. https://doi.org/10.1038/ismej.2007.51

Li W, Zhang T, Tang X, Wang B (2010) Oomycetes and fungi: important parasites on marine algae. Acta Oceanol Sin 29:74–81

Lorenz R, Molitoris H-P (1990) Published. Phoma-pattern. Combined influence of salinity and temperature on growth of marine fungi. In: Reisinger A, Bresinsky A (eds) Fourth international mycological congress. IMC 4, Regensburg. 136/3

Manohar CS, Raghukumar C (2013) Fungal diversity from various marine habitats deduced through culture-independent studies. FEMS Microbiol Lett 341:69–78. https://doi.org/10.1111/1574-6968.12087

Masuma R, Yamaguchi Y, Noumi M, Omura S, Namikoshi M (2001) Effect of sea water concentration on hyphal growth and antimicrobial metabolite production in marine fungi. Mycoscience 42:455–459

Miura T, Abe F, Inoue A, Usami R, Horikoshi K (2002) Superoxide dismutase is involved in high tolerance to copper in the deep-sea yeast, *Cryptococcus* sp. N6. Biotechnol Lett 24:1069–1074. https://doi.org/10.1023/A:1016082530108

Nagahama T (2003) *Cryptococcus surugaensis* sp. nov., a novel yeast species from sediment collected on the deep-sea floor of Suruga Bay. Int J Syst Evol Microbiol 53:2095–2098. https://doi.org/10.1099/ijs.0.02712-0

Newton GGF, Abraham EP (1955) Cephalosporin C, a new antibiotic containing sulphur and D-aminoadipic acid. Nature 175:548–548

Okutani K (1977) Gliotoxin produced by a strain of *Aspergillus* isolated from marine mud. Bull Jap Soc Sci Fish 43:995–1000

Ozcengiz G, Demain AL (2013) Recent advances in the biosynthesis of penicillins, cephalosporins and clavams and its regulation. Biotechnol Adv 31:287–311. https://doi.org/10.1016/j.biotechadv.2012.12.001

Parfrey LW, Lahr DJ, Knoll AH, Katz LA (2011) Estimating the timing of early eukaryotic diversification with multigene molecular clocks. Proc Natl Acad Sci USA 108:13624–13629. https://doi.org/10.1073/pnas.1110633108

Pejin B, Karaman M (2017) Antitumor natural products of marine-derived fungi. In: Mérillon JM, Ramawat K (eds) Fungal metabolites. Reference series in phytochemistry. Springer, Cham, pp 1–28

Pena C, Malm T (2012) VoSeq: a voucher and DNA sequence web application. PLoS One 7:e39071. https://doi.org/10.1371/journal.pone.0039071

Pereira RB, Evdokimov NM, Lefranc F, Valentao P, Kornienko A, Pereira DM, Andrade PB, Gomes A (2019) Marine-derived anticancer agents: clinical benefits, innovative mechanisms, and new targets. Mar Drugs 17:1–21. https://doi.org/10.3390/md17060329

Pitt JI, Samson RA (2007) Nomenclatural considerations in naming species of *Aspergillus* and its teleomorphs. Stud Mycol 59:67–70. https://doi.org/10.3114/sim.2007.59.08

Raghukumar C (2006) Algal-fungal interactions in the marine ecosystem: symbiosis to parasitism. In: Tewari A (ed) Recent advances on applied aspects of Indian marine algae with reference to global scenario. CSMCRI, Bhavnagar, pp 366–385

Raghukumar C (2008) Marine fungal biotechnology: an ecological perspective. Fung Div 31:19–35

Raghukumar C, Muraleedharan U, Gaud VR, Mishra R (2004a) Xylanases of marine fungi of potential use for biobleaching of paper pulp. J Ind Microbiol Biotechnol 31:433–441. https://doi.org/10.1007/s10295-004-0165-2

Raghukumar C, Raghukumar S, Sheelu G, Gupta SM, Nath BN, Rao BR (2004b) Buried in time: culturable fungi in a deep-sea sediment core from the Chagos Trench, Indian Ocean. Deep Sea Res Part 1 Oceanogr Res Pap 51:1759–1768. https://doi.org/10.1016/j.dsr.2004.08.002

Rämä T, Norden J, Davey ML, Mathiassen GH, Spatafora JW, Kauserud H (2014) Fungi ahoy! Diversity on marine wooden substrata in the high North. Fungal Ecol 8:46–58

Rashmi M, Kushveer JS, Sarma VV (2019) Secondary metabolites produced by endophytic fungi from marine environments. In: Jha S (ed) Endophytes and secondary metabolites. Reference series in phytochemistry. Springer, Cham, pp 491–526

Rateb ME, Ebel R (2011) Secondary metabolites of fungi from marine habitats. Nat Prod Rep 28:290–344. https://doi.org/10.1039/c0np00061b

Redou V, Ciobanu MC, Pachiadaki MG, Edgcomb V, Alain K, Barbier G, Burgaud G (2014) In-depth analyses of deep subsurface sediments using 454-pyrosequencing reveals a reservoir of buried fungal communities at record-breaking depths. FEMS Microbiol Ecol 90:908–921

Reich M, Labes A (2017) How to boost marine fungal research: a first step towards a multidisciplinary approach by combining molecular fungal ecology and natural products chemistry. Mar Genomics 36:57–75. https://doi.org/10.1016/j.margen.2017.09.007

Richards TA, Jones MD, Leonard G, Bass D (2012) Marine fungi: their ecology and molecular diversity. Annu Rev Mar Sci 4:495–522. https://doi.org/10.1146/annurev-marine-120710-100802

Sandri IG, Fontana RC, Barfknecht DM, da Silveira MM (2011) Clarification of fruit juices by fungal pectinases. LWT Food Sci Technol 44:2217–2222. https://doi.org/10.1016/j.lwt.2011.02.008

Sarkar S, Pramanik A, Mitra A, Mukherjee J (2010) Bioprocessing data for the production of marine enzymes. Mar Drugs 8:1323–1372. https://doi.org/10.3390/md8041323

Schaumann K, Weide G (1990) Enzymatic degradation of alginate by marine fungi. Hydrobiologia 204 (205):589–596

Scheu S (2002) The soil food web: structure and perspectives. Eur J Soil Biol 38:11–20. https://doi.org/10.1016/S1164-5563(01)01117-7

Sikes EL, Uhle ME, Nodder SD, Howard ME (2009) Sources of organic matter in a coastal marine environment: evidence from n-alkanes and their δ13C distributions in the Hauraki Gulf, New Zealand. Mar Chem 113:149–163. https://doi.org/10.1016/j.marchem.2008.12.003

Silber J, Kramer A, Labes A, Tasdemir D (2016) From discovery to production: biotechnology of marine fungi for the production of new antibiotics. Mar Drugs 14:1–20. https://doi.org/10.3390/md14070137

Song F, Ren B, Yu K, Chen C, Guo H, Yang N, Gao H, Liu X, Liu M, Tong Y, Dai H, Bai H, Wang J, Zhang L (2012) Quinazolin-4-one coupled with pyrrolidin-2-iminium alkaloids from marine-derived fungus *Penicillium aurantiogriseum*. Mar Drugs 10:1297–1306. https://doi.org/10.3390/md10061297

Stoeck T, Taylor GT, Epstein S (2003) Novel eukaryotes from the permanently anoxic Cariaco Basin (Caribbean Sea). Appl Environ Microbiol 69:5656–5663

Synnes M (2007) Bioprospecting of organisms from the deep sea: scientific and environmental aspects. Clean Technol Environ Policy 9:53–59

Takami H, Inoue A, Fuji F, Horikoshi K (1997) Microbial flora in the deepest sea mud of the Mariana Trench. FEMS Microbiol Lett 152:279–285

Thiel V, Neulinger SC, Staufenberger T, Schmaljohann R, Imhoff JF (2007) Spatial distribution of sponge-associated bacteria in the Mediterranean sponge *Tethya aurantium*. FEMS Microbiol Ecol 59:47–63. https://doi.org/10.1111/j.1574-6941.2006.00217.x

Verma AK, Raghukumar C, Verma P, Shouche YS, Naik CG (2010) Four marine-derived fungi for bioremediation of raw textile mill effluents. Biodegradation 21:217–233. https://doi.org/10.1007/s10532-009-9295-6

Vignesh S, Raja A, James RA (2011) Marine drugs: implication and future studies. Int J Pharmacol 7:22–30. https://doi.org/10.3923/ijp.2011.22.30

Wang G (2006) Diversity and biotechnological potential of the sponge-associated microbial consortia. J Ind Microbiol Biotechnol 33:545–551. https://doi.org/10.1007/s10295-006-0123-2

Wang YT, Xue YR, Liu CH (2015) A brief review of bioactive metabolites derived from deep-sea fungi. Mar Drugs 13:4594–4616. https://doi.org/10.3390/md13084594

Wiese J, Ohlendorf B, Blumel M, Schmaljohann R, Imhoff JF (2011) Phylogenetic identification of fungi isolated from the marine sponge *Tethya aurantium* and identification of their secondary metabolites. Mar Drugs 9:561–585. https://doi.org/10.3390/md9040561

Wu B, Wiese J, Labes A, Kramer A, Schmaljohann R, Imhoff JF (2015) Lindgomycin, an unusual antibiotic polyketide from a marine fungus of the Lindgomycetaceae. Mar Drugs 13:4617–4632. https://doi.org/10.3390/md13084617

Wu Z, Wang Y, Liu D, Proksch P, Yu S, Lin WT (2016) Antioxidative phenolic compounds from a marine-derived fungus *Aspergillus versicolor*. Tetrahedron 72:50–57

Yarden O (2014) Fungal association with sessile marine invertebrates. Front Microbiol 5:228. https://doi.org/10.3389/fmicb.2014.00228

Zhang Y, Mu J, Feng Y, Kang Y, Zhang J, Gu PJ, Wang Y, Ma LF, Zhu YH (2009) Broad-spectrum antimicrobial epiphytic and endophytic fungi from marine organisms: isolation, bioassay and taxonomy. Mar Drugs 7:97–112. https://doi.org/10.3390/md7020097

Zobel S, Boecker S, Kulke D, Heimbach D, Meyer V, Süssmuth RD (2016) Reprogramming the biosynthesis of cyclodepsipeptide synthetases to obtain new enniatins and beauvericins. ChemBioChem 17:283–287. https://doi.org/10.1002/cbic.201500649

17 The Biotechnological Potential of Anaerobic Gut Fungi

Veronika Flad[1], Diana Young[1], Susanna Seppälä[2], Casey Hooker[3], Noha Youssef[4], Sabine M. Podmirseg[5], Magdalena Nagler[5], Matthew Reilly[6,7], Yuanfei Li[8], Kateřina Fliegerová[9], Yanfen Cheng[8], Gareth W. Griffith[10], Mostafa Elshahed[4], Kevin V. Solomon[3], Michelle A. O'Malley[2], Michael K. Theodorou[7]

CONTENTS

[1] Bavarian State Research Center for Agriculture (LfL), Central Department for Quality Assurance and Analytics, Freising, Germany; e-mail: veronika.flad@lfl.bayern.de; diana.young@lfl.bayern.de
[2] Department of Chemical Engineering, University of California, Santa Barbara, Santa Barbara, CA, USA; e-mail: sseppala@ucsb.edu; momalley@ucsb.edu
[3] Agriculture and Biological Engineering, Purdue University, West Lafayette, IN, USA; e-mail: hookerc@purdue.edu; kvs@purdue.edu
[4] Department of Microbiology and Molecular Genetics, Oklahoma State University, Stillwater, OK, USA; e-mail: noha@okstate.edu; mostafa@okstate.edu
[5] Institute of Microbiology, Universität Innsbruck, Innsbruck, Austria; e-mail: Sabine.Podmirseg@uibk.ac.at; Magdalena.Nagler@uibk.ac.at
[6] Department of Biology, University of York, York, UK; e-mail: matthew.reilly@york.ac.uk
[7] Agriculture Centre for Sustainable Energy Systems, Harper Adams University, Newport, Shropshire, UK; e-mail: mtheodorou@harper-adams.ac.uk
[8] Laboratory of Gastrointestinal Microbiology, Nanjing Agricultural University, Nanjing, China; e-mail: yanfencheng@njau.edu.cn
[9] Institute of Animal Physiology and Genetics, v.v.i, Academy of Sciences of the Czech Republic, Prague 4, Czech Republic; e-mail: fliegerova@iapg.cas.cz
[10] Institute of Biological, Environmental and Rural Sciences, Aberystwyth University, Penglais, Aberystwyth, Ceredigion, Wales, UK; e-mail: gwg@aber.ac.uk

I. Introduction

Anaerobic gut fungi belong to the early diverging phylum ***Neocallimastigomycota*** and are commonly found in the digestive tract of ruminants and hindgut fermenters. Animal hosts include a wide range of important livestock and wild animals such as cattle, sheep, horses, and elephants (Liggenstoffer et al. 2010). Instead of producing biomass-degrading enzymes of their own, these large herbivores rely on the synergistic action of an intestinal microbial community, composed of fungi, cellulolytic bacteria, protozoa, and archaea (Peng et al. 2016). The *Neocallimastigomycota* fungi are key members of these microbial communities, as they function as initial colonizers of the ingested plant fibers (lignocellulose) and contribute to the biomass breakdown process both physically and enzymatically (Podolsky et al. 2019). Digestion of such recalcitrant mixtures requires the synergistic action of a large

Genetics and Biotechnology, 3rd Edition
The Mycota II
J.P. Benz, K. Schipper (Eds.)

number of diverse enzymes, and anaerobic fungi meet this challenge via the secretion of a vast array of intricately organized and environmentally regulated enzymes (Haitjema et al. 2017; Solomon et al. 2016a, 2018; Youssef et al. 2013).

In this chapter, we briefly describe the phylogeny, unusual life style, isolation procedures, and growth requirements of these uniquely anaerobic fungi. We further discuss how "omics" technologies facilitate the identification of useful proteins and metabolic pathways and give examples of successful efforts to transfer gut fungal proteins to model microbes. Further, we provide an overview of current and future biotechnological uses of anaerobic fungi, including a survey of their intellectual property landscape. We conclude that the understudied *Neocallimastigomycota* are a rich and largely untapped source of bioactivities that can be harnessed for diverse biotechnological applications, from animal nutrition to bioprocessing technologies that seek to convert renewable biomass into valuable fuels and chemicals.

A. Background and Taxonomy

The *Neocallimastigomycota* were first proposed as fungi following Orpin's seminal work in the 1970s analyzing the morphology of rumen "flagellates" that proliferated on fiber-rich diets in the guts of ruminant animals (Orpin 1975). Subsequent studies conclusively demonstrated that these were true fungi with a chitinous cell wall, which challenged the previously held dogma that all fungi required oxygen to grow (Orpin 1977). It is estimated that the anaerobic fungi from the phylum *Neocallimastigomycota* diverged from other fungi in the late Cretaceous period (66 $\pm$ 10 million years ago), around the time when grasses and mammalian grazers emerged (Emerling et al. 2018; Prasad et al. 2005; Wang et al. 2019). Similar to their relatives in the basal fungal phyla *Chytridiomycota* and *Blastocladiomycota*, *Neocallimastigomycota* alternate between **motile zoospores** and **vegetative thalli** (James et al. 2006) as part of their **life cycle** (Fig. 1). In the animal gut, fungal zoospores (Fig. 1a) are released from mature sporangia and swim toward and attach to ingested plant material, likely attracted by soluble sugars and phenolic acids (Wubah and Kim 1996). Following attachment, the zoospores encyst and germinate, penetrating the plant surface and developing either to rhizoidal or bulbous thalli (Fig. 1d). Anaerobic fungi thrive in herbivores via enzymatic degradation of plant carbohydrates, and they directly utilize sugar-rich breakdown products from degradation—this leads to expansion of fungal rhizoids/holdfasts, and zoosporangia develop and enlarge. Finally a new generation of zoospores is produced and matured in the zoosporangium (Fig. 1f), with the timing of zoospore release likely triggered by ingestion of fresh forage, though the mechanistic underpinnings remain unclear (Orpin 1977). The full life cycle is reported to take 8–32 h (Mountfort and Orpin 1994; Theodorou et al. 1996a).

Anaerobic fungal taxa have historically been grouped according to thallus morphology and the number of flagella per zoospore, as exemplified in Fig. 2 (Wang et al. 2017). For example, fungi with **monocentric** thalli have one center of growth, while **polycentric fungi** have multiple centers of growth—they are readily distinguishable via light microscopy. Based on these morphological features, six genera were originally identified: *Neocallimastix* (rhizoidal, monocentric, polyflagellate), *Piromyces* (rhizoidal, monocentric, monoflagellate), *Anaeromyces* (rhizoidal, polycentric, monoflagellate), *Orpinomyces* (rhizoidal, polycentric, polyflagellate), *Caecomyces* (bulbous, monocentric, monoflagellate), and *Cyllamyces* (bulbous, polycentric, monoflagellate) (Griffith et al. 2010).

Recently multiple additional genera have been isolated and described: *Buwchfawromyces* (Callaghan et al. 2015), *Feramyces* (Hanafy et al. 2018), *Liebetanzomyces* (Joshi et al. 2018), *Oontomyces* (Dagar et al. 2015), *Pecoramyces* (Hanafy et al. 2017), *Agriosomyces*, *Aklioshbomyces*, *Capellomyces*, *Ghazallomyces*, *Joblinomyces*, *Khoyollomyces* and *Tahromyces* (Hanafy et al. 2020). All are monocentric and,

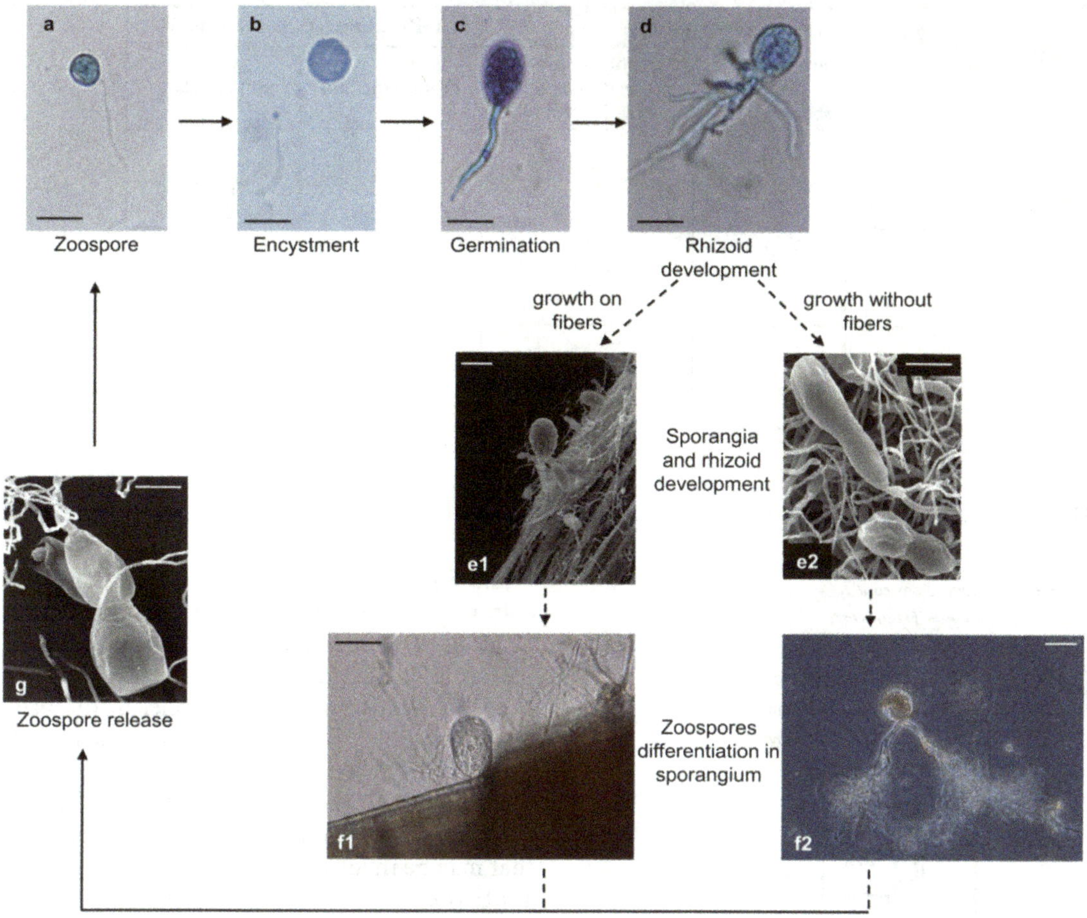

Fig. 1 Anaerobic fungal life cycle, as illustrated via the growth and maturation of *Pecoramyces* sp. (isolate C1A) grown in liquid medium enriched with corn stover or wheat straw (e1, f1) and liquid medium with cellobiose and xylan as soluble carbon sources (e2, f2). Light microscopy after staining with lactophenol cotton blue using an Olympus BX51 microscope, scale bar 20 μm (**a**, **b**, **c** and **d**). Scanning electron microscopy pictures examined using FEI Quanta 600 scanning electron microscope, scale bar 50 μm (**e1**, **e2**, **g**). Light microscopy without staining using a Leica DM 2000 LED, scale bar 50 μm (**f1**, **f2**)

with the exception of *Feramyces* and *Ghazallomyces*, form monoflagellate zoospores. Culture-independent diversity surveys based on the DNA of the internal transcribed spacer region (ITS) did suggest that additional clades were present, and the first DNA metabarcoding study of anaerobic fungi (Liggenstoffer et al. 2010) identified the existence of several novel clades, some of which have subsequently been isolated into pure cultures (Fig. 3). Due to the high level of intragenomic variability in the ITS region, the adjacent large ribosomal subunit (LSU; D1/D2 variable regions; Fig. 3) is now currently favored for accurate phylogenetic reconstructions (Dagar et al. 2011).

Although the niche of anaerobic fungi is the digestive tract of large mammalian herbivores, most laboratory isolates have been obtained from herbivore feces. Moreover, fungi in feces and post-rumen digesta remain viable when dried in air at ambient temperature (Davies et al. 1993) and even in sun-baked feces (Milne et al. 1989). Although populations tend to decline slowly after drying, the size of the anaerobic fungal community in fresh feces, quantified as thallus forming units, was esti-

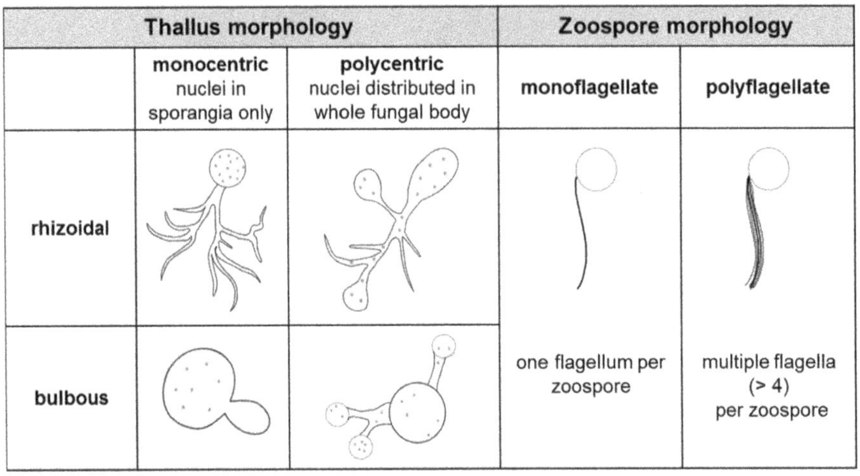

Fig. 2 Morphological criteria for classification of anaerobic fungi; Grey dots represent nuclei

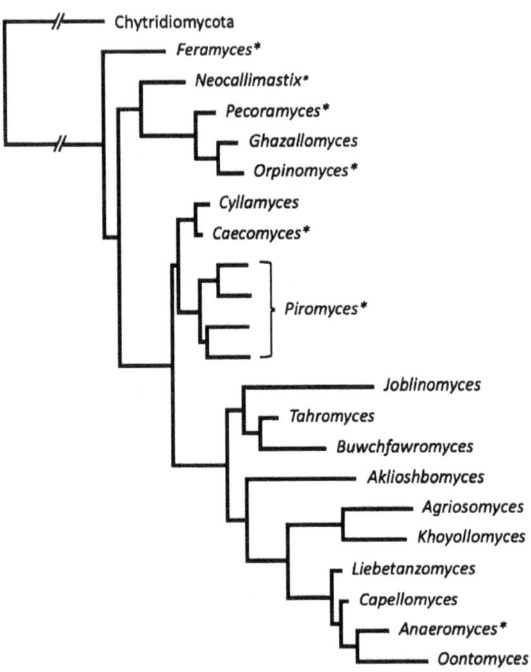

Fig. 3 Phylogenetic tree of phylum *Neocallimastigomycota* based on the large (28S) ribosomal subunit, with only the currently named genera included. Genera with published genome and/or transcriptome data are indicated by *asterisk*

mated to be equivalent to that in the rumen (Davies et al. 1993; Theodorou et al. 1990). However, despite many repeated attempts, it has not been possible to isolate anaerobic fungi from dried rumen digesta nor were they found in the dried residues from laboratory cultures. To account for these findings, Davies et al. (1993) amended the generally accepted life-cycle of the gut fungi to include an additional aero-tolerant survival cycle. In the proposed scheme, the survival cycle is activated when conditions for vegetative growth become less favorable. There have been several independent observations of a **resistant structure** that may be involved in aerobic survival (Joblin 1981; Nielsen et al. 1995; Wubah et al. 1991). These studies provide the most convincing evidence that zoosporangia in feces (rather than zoospores) are aero-tolerant and responsible for seeding the viable fungal cultures in the laboratory. The fact that viable cultures of anaerobic fungi cannot be isolated from dried rumen contents, whereas they are readily obtained from dried feces, suggests that these structures in both environments, although similar in appearance, are physiologically different.

B. Culture Methodology

Traditional culture collections commonly do not house anaerobic fungi, and most laboratories that work with them isolate their own strains directly from source material (Haitjema et al. 2014). Anaerobic fungi have been successfully enriched and isolated from herbivore feces or gut content samples using lignocellulosic substrates (Fig. 4a) or plant sugars (Fig. 4b;

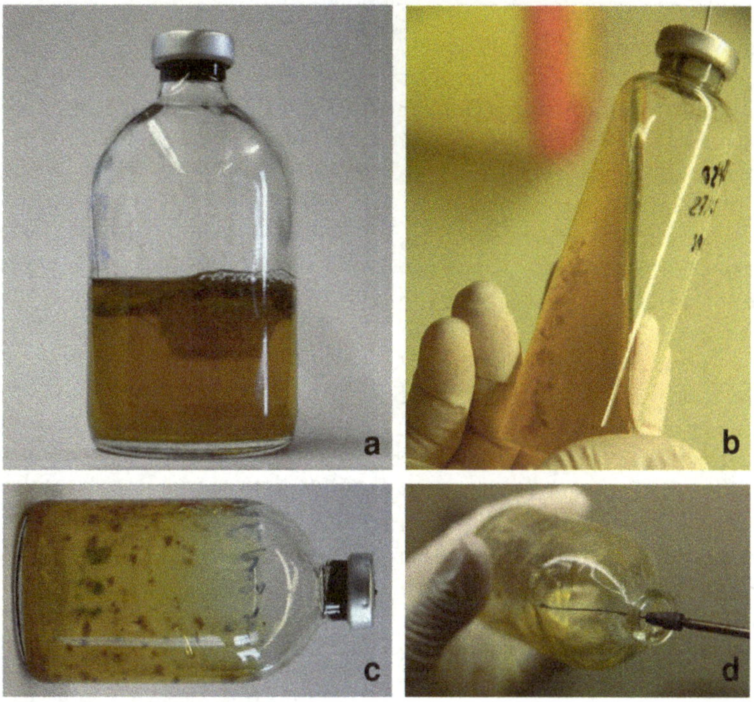

Fig. 4 Cultivation of anaerobic fungi (**a**) growing on wheat straw forming a floating mat; (**b**) aggregation of rhizoid in medium without fibers; (**c**) agar roll-tube with fungal colonies; (**d**) picking a single colony from a roll-tube for purification

such as cellobiose, xylan) as a carbon source. A step-by-step description of the procedure can be found in Peng et al. (2018).

In brief, the fecal sample is mixed to a slurry (ca. 10% w/v) in liquid medium that contains antibiotics to prevent growth of prokaryotes. Commonly used antibiotics include a mix of penicillin G (or Kanamycin), ampicillin,and streptomycin C (effective against bacteria), and/or chloramphenicol for suppressing both bacteria and methanogens. The slurry is serially diluted (1:10, 1:100, and 1:1000), and 1 ml aliquots are used to inoculate 9 ml of liquid culture medium containing plant fiber. Fungal growth is monitored daily and is identified by the occurrence of floating mats of fiber, which occur due to gas production (Fig. 4a). For bulbous species that are less effective in producing floating mats, positive growth is detected using a microscope. Anaerobic fungal enrichments are further purified by several passages through agar containing roll-tubes (Fig. 4c) from which single colonies are picked (Fig. 4d) to inoculate fresh liquid culture medium (Haitjema et al. 2014).

Subsequent routine sub-culturing is performed in either complex liquid medium containing up to 15% of rumen fluid or chemically defined medium (Caldwell and Bryant 1966; Davies et al. 1993; Lowe et al. 1985). Cultivation is typically done in small (50–100 mL) gas-tight serum bottles or Hungate tubes (10–20 mL), under 100% CO_2 or a N_2/CO_2 (70%/30%) atmosphere and in the dark at 39°C (Haitjema et al. 2014; Lowe et al. 1985).

The most common substrates that are used to cultivate anaerobic fungi in laboratory conditions vary highly in their lignin, cellulose, and hemicellulose contents (Table 1). Nonetheless,

Table 1 Composition of lignocellulosic substrates used for anaerobic fungal growth

Substrate	Cellulose (% DM)	Hemicellulose (%DM)	Lignin (%DM)
Corn stover	44	29	22
Corncob	36–45	35–41	14–15
Grass silage	25–40	26–50	9–30
Maize silage	24	19	3
Rice straw	32–47	18–28	5.5–24
Ryegrass	43	28	7
Wheat straw	33–45	20–32	8–22

Filya (2004); Hassan et al. (2018); Jin et al. (2011)

anaerobic fungi are able to degrade many of these materials with comparable efficiency, even when composed of high levels of phenolic compounds inhibitory to other fungi (Hooker et al. 2018; Skyba et al. 2013). To date, milled wheat straw has been the preferred substrate to grow anaerobic fungi in laboratory conditions, but other types of agricultural waste (such as, maize silage, grass silage, grasses, rice straw, hay, reed canary grass, ryegrass at earing stage, maize stems, soybean hulls, and palm press fiber) also promote anaerobic fungal growth (Griffith et al. 2009; Mountfort and Orpin 1994; Trinci et al. 1994; Zhu et al. 1996). Young forages, rich in leaf and poor in stems (such as, ryegrass at the leafy stage) that contain low amounts of crude fibers and high amounts of pectin and starch, were found to be unfavorable for anaerobic fungal growth (Mountfort and Orpin 1994).

Long-term viability of gut fungal cultures requires routine sub-culturing every 3–14 days (dependent on the species) (Haitjema et al. 2014; Peng et al. 2018). For long-term storage and to avoid genetic drift by repeated sub-culturing, anaerobic fungi can be cryo-preserved (Solomon et al. 2016b). However, the cryo-preservation methodology has not been standardized and is not always successful (Callaghan et al. 2015; Haitjema et al. 2014; Peng et al. 2018; Solomon et al. 2016b). The culture methodology may be improved by establishing techniques that allow continuous cultivation that better resembles the conditions in the rumen ecosystem. **Semi-continuous cultivation** of anaerobic fungi has been performed in continuous-flow cultures containing up to 8% (w/v) fibrous substrate (Zhu et al. 1996; Zhu et al. 1997). Teunissen et al. (1992) also used a semi-continuous culture for harvesting anaerobic fungal enzymes. In these systems, effluent is usually removed discretely or continuously from the culture vessel and replaced by an equivalent amount of fresh culture medium. Anaerobic fungi showed higher enzyme production and increased degradation of plant-fibers in continuous cultivation systems, possibly due to the removal of inhibitory fermentation end products (see Sect. II.A) in culture effluents (Zhu et al. 1996).

C. Physiology and Metabolism

Anaerobic fungi have evolved strategies to thrive in oxygen-free habitats where they degrade a wide range of substrates: (1) They possess a large arsenal of lignocellulolytic enzymes that are secreted either freely or organized in multi-protein complexes called fungal **cellulosomes.** (2) Energy production occurs in **hydrogenosomes** instead of mitochondria. Anaerobic fungi rely on substrate-level phosphorylation (mixed-acid fermentation), rather than oxidative phosphorylation (respiration) (Mountfort and Orpin 1994). (3) Their cell membranes contain tetrahymanol rather than sterols (ergosterol in most fungi), since synthesis of these requires oxygen in the oxidation of the squalene precursor.

Tetrahymanol, a pentacyclic sterol-like compound, is formed directly by the cyclisation of squalene via squalene hydro-lyase, a process similar to the synthesis of the pentacyclic hopanoids in bacteria (Youssef et al. 2013).

1. Some Biomass-Degrading Enzymes Are Organized in Cellulosomes

The hydrolytic strategy of anaerobic fungi is based on the extensive production of **carbohydrate-active enzymes (CAZymes)** that they secrete to the surrounding environment. CAZymes include all enzymes involved in carbohydrate assembly (glycosyltransferases) and breakdown (glycoside hydrolases, polysaccharide lyases, carbohydrate esterases) (Lombard et al. 2014). Each carbohydrate polymer requires a certain set of CAZymes (Fig. 5). Whether anaerobic fungi have evolved strategies to enzymatically depolymerize and/or catabolize lignin remains unclear but their ability to solubilize linkages in the lignin-polysaccharide complex has been reported (Joblin and Naylor 1989), rendering cellulose and hemicellulose accessible for degradation. For information on the individual enzymatic reactions, see Theodorou et al. (1996b).

Notably, whereas aerobic fungi typically secrete enzymes into the environment, anaerobic fungi are unique among the fungal kingdom

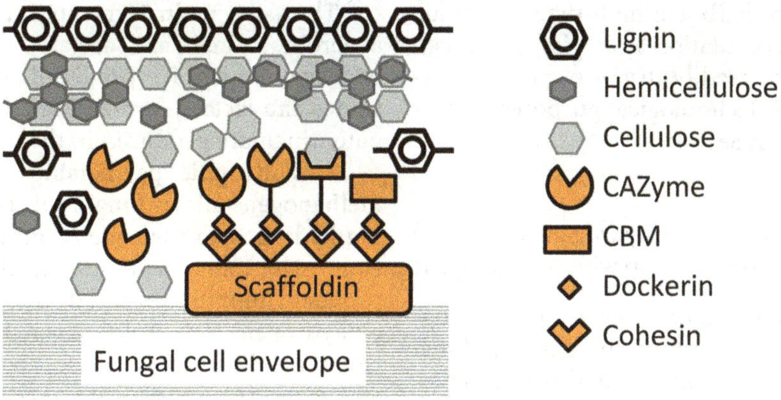

Fig. 5 Anaerobic fungi produce both freely diffusive and cellulosome-complexed CAZymes to depolymerize plant biomass. Fungal cellulosomes are assembled onto non-catalytic scaffoldin domains and dockerin-fused enzymes are tethered to the scaffold through interactions with complementary cohesin domains. *CBM* carbohydrate-binding modules

as they organize up to 2/3 of their secreted enzymes in fungal **cellulosomes** (Haitjema et al. 2017; Henske et al. 2017). Cellulosomes (Fig. 5) are multi-protein superstructures in which a large variety of CAZymes are bound together, allowing substrate channeling and concerted, efficient enzyme activities, even with reduced enzyme loads (Bayer et al. 2004). Such complexes were first described in hydrolytic bacteria from the genus *Clostridium* (Lamed et al. 1983) and evidence for cellulosomes in anaerobic fungi (*Neocallimastix frontalis*) was reported as early as 1992 (Wilson and Wood 1992). More recently, the intricate interactions of the fungal cellulosomal components were uncovered through **multi-"omics"** analyses, which showed that fungal cellulosomes are highly divergent from their bacterial counterparts—both in sequence/structure and in enzymatic function (Haitjema et al. 2017). In cellulosomes, the enzymes are stitched together by a "plug-and-socket" mechanism: a subset of the secreted CAZymes and carbohydrate binding modules (CBMs) contain so called **dockerin domains** interacting with **cohesin domains** that are part of a very large non-catalytic scaffoldin protein (Fig. 5; Gilmore et al. 2015). The dockerin-tagged CBMs are believed to bring the entire cellulosome into close proximity to the substrate, adding to the advantage of organizing the enzymes into these synergistic structures (Gilmore et al. 2015; Seppälä et al. 2017). While in bacteria the dockerin and cohesin domains are specific to each bacterial species, fungi appear to leverage conserved dockerins, cohesins, and scaffoldins across the anaerobic fungi (Haitjema et al. 2017). Most cellulosome-associated fungal proteins feature one or more N- or C-terminal dockerin domain, often organized in tandem repeats whose function remains unknown (Haitjema et al. 2017).

Anaerobic fungal cellulosomes are evolutionarily chimeric as it appears that fungi have adopted useful strategies from their bacterial counterparts through **horizontal gene transfer** of the enzymatic components (Haitjema et al. 2017). In this way, the anaerobic fungal enzymatic machinery shows several advantages compared to their aerobic fungal and bacterial counterparts. Anaerobic fungal enzymes exhibit high catalytic activity and substrate affinities (Ljungdahl 2008; Morrison et al. 2016a; Wang et al. 2011), and many of the enzymes are multi-functional and have dual endoglucanase-cellobiohydrolase (Liu et al. 1997), endoglucanase- lichenase (Chen et al. 1998; Eberhardt et al. 2000; Qiu et al. 2000; Ye et al. 2001), cellulase-xylanase (Liu et al. 1997; Morrison et al. 2016b), or β-glucosidase-β-xylosidase activities (Morrison et al. 2016a). Overall, anaerobic fungi harbor a multitude of unique enzymes and degradation strategies, which can be harnessed to readily degrade specific feedstocks or for biotechnological purposes, as further discussed in Sect. II.

2. Anaerobic Fungi Have Hydrogenosomes and Live in Close Relationship with Methanogens

Physiological and genomic investigations suggest that energy generation in anaerobic fungi is mediated by fungal **hydrogenosomes**, although this hypothesis is yet to be confirmed by molecular studies. Hydrogenosomes are ATP-producing organelles found across a range of unrelated anaerobic eukaryotes that use protons rather than oxygen as the terminal electron acceptor. They are believed to have evolved from degenerate mitochondria (Boxma et al. 2004; Hackstein et al. 2019; Müller et al. 2012), and recent genomic studies have confirmed the fungal mitochondrial origin of some hydrogenosomal import proteins (Haitjema et al. 2017; Youssef et al. 2013). Hydrogenosomes perform three major metabolic tasks:

(i) **Pyruvate metabolism**: pyruvate (from cytosolic plant sugar metabolism) is converted to formate and acetyl-CoA using pyruvate/formate lyase (PFL) (Boxma et al. 2004). Marvin-Sikkema et al. (1993) demonstrated the presence of a hydrogenosome membrane-associated pyruvate/ferredoxin oxidoreductase (PFO), similar to that found in hydrogenosomes of *Trichomonas* spp., and suggested that acetyl-CoA from cytosolic PFL activity was transported to the hydrogenosome.
(ii) **ATP production**: Acetyl-CoA is converted to acetate via succinyl-CoA synthetase, initiating ATP production exclusively via substrate level phosphorylation.
(iii) **Regeneration of reduced electron carriers**: NAD(P)H is re-oxidized by an NAD(P)H dehydrogenase, which transfers electrons to an iron-only hydrogenase producing hydrogen (Haitjema et al. 2017; Kameshwar and Qin 2018; Youssef et al. 2013).

Thus, the **main fermentation products** of anaerobic fungi are acetate, formate, and hydrogen which are produced in hydrogenosomes and lactate, formate, ethanol, and succinate which are produced in the cytosol.

A **syntrophic partnership** with archaeal methanogens allows fungi to thrive by siphoning hydrogen to methane (Conrad 1999; Peng et al. 2016; Ushida et al. 1997), alleviating end-product inhibition caused by accumulation of hydrogen. It is well-known that there is a close relationship between anaerobic fungi and methanogens, with **interspecies hydrogen transfer** resulting in methane production and more efficient re-generation of oxidized nucleotides (NAD^+, $NADP^+$) (Bauchop and Mountfort 1981; Cheng et al. 2009). Acetate and formate are also substrates for methanogenesis, and Li et al. (2019) have recently shown that, when *Methanobrevibacter thaueri* and *Pecoramyces* sp. F1 are grown together in batch culture, the majority of formate produced by the fungus is used for methanogenesis. More details on such microbial interactions and their biotechnological potential are reported in Sect. III.C.

II. Identification of Useful Enzymes

Anaerobic fungi offer a wealth of novel proteins and enzymes for diverse biotechnology applications. Chief among these are potentially superior lignocellulose-degrading enzymes, which offer opportunities to overcome a significant bottleneck during the production of second generation biofuels and chemicals (Seppälä et al. 2017). Anaerobic fungal genomes acquired by next-generation sequencing techniques continue to refine our understanding of the biosynthetic potential of anaerobic fungi and suggest new enzymes for platform biochemical production pathways such as isoprenoid and secondary metabolism, and fatty acid production (Grigoriev et al. 2011; Grigoriev et al. 2014; Hillman et al. 2017). Here, we briefly discuss recent progress on the adoption of emerging "omics" technologies to characterize the enzyme systems of anaerobic fungi, translation of these enzymes to heterologous production, and progress towards direct genetic manipulation of the fungi.

A. "Omics" Analyses Identify Enzymes and Their Regulation Patterns

While extensive work has shown that anaerobic fungi possess significant hydrolytic potential, the detailed function and regulation of many of the individual genes at the transcriptional level has only recently been understood. Advances in next-generation sequencing (Wilken et al. 2019) enabled the first genomes and transcriptomes for a small collection of isolated anaerobic fungi. The **first draft genome** of the *Neocallimastigomycota* was published in 2013. The genome of *Orpinomyces* sp. C1A (later re-classified as *Pecaromyces* sp. C1A) revealed a large abundance of CAZymes (>350 glycoside hydrolase enzymes, >90 carbohydrate esterases and an array of dockerin domain-containing proteins associated with cellulosome assembly) (Youssef et al. 2013). Later, **high-resolution genomes** of *Anaeromyces robustus*, *Neocallimastix californiae*, and *Piromyces finnis* revealed that these organisms have the largest array and diversity of biomass-degrading enzymes in the fungal kingdom (Haitjema et al. 2017; Solomon et al. 2016a). Indeed, horizontal gene transfer has recently been shown to play a crucial role in expanding the substrate utilization capabilities and electron disposal pathways in the *Neocallimastigomycota* (Murphy et al. 2019). Anaerobic fungi readily adapt gene expression in response to substrate complexity and may provide a source of enzymes tailored to degrade feedstocks based upon their composition (Couger et al. 2015; Henske et al. 2018a; Solomon et al. 2016a; Youssef et al. 2013). **Carbon catabolite repression** (Montenecourt and Eveleigh 1977; Strauss et al. 1995) is responsible for regulation of diverse CAZymes in anaerobic fungi (Henske et al. 2018a; Solomon et al. 2016a). Low concentrations (0.5 mg/ml) of various sugars (sucrose, galactose, xylose, fructose and arabinose) were sufficient to hinder enzyme production (Henske et al. 2018a). It has however also been shown that several of the secreted enzymes are highly transcribed even when cultured on glucose or other soluble sugars (Couger et al. 2015; Wang et al. 2011). The presence of complementary **natural antisense transcripts** for several CAZymes even in the presence of glucose suggests that some enzymes are also post-transcriptionally regulated (Solomon et al. 2018). Moreover, **epigenetic regulation** may also play a role as adenine methylation appears especially prevalent near transcriptional start sites in anaerobic fungi. However, no clear link between DNA methylation and gene expression has been identified (Mondo et al. 2017). Nonetheless, determining and manipulating many of the regulatory strategies of anaerobic fungi will be essential to develop these organisms to overexpress CAZymes for bioenergy purposes.

B. Systems for Heterologous Enzyme Expression

In spite of the small number of sequenced anaerobic fungal genomes, numerous **CAZymes,** and proteins have been successfully cloned and produced in other organisms, including *Escherichia coli*, *Saccharomyces cerevisiae*, and *Pichia pastoris* (Table 2). Due to the limited number of physical isolates, only five genera in the *Neocallimastigomycota* have been sourced for enzymes: *Anaeromyces*, *Neocallimastix*, *Orpinomyces*, *Piromyces*, and *Pecoramyces*. The majority of these studies relied on functional screening of cDNA libraries to source enzymes from anaerobic fungi but the recent availability of anaerobic fungal genomes and transcriptomes has enabled "omics"-guided approaches for the identification of highly expressed key enzymes.

The identified genes sometimes undergo codon optimization prior to being cloned into an appropriate vector (Morrison et al. 2016b; Wang et al. 2011). Protein products are examined for validation of predicted activities (Liu et al. 1997), and their biochemical and kinetic properties are characterized (Morrison et al. 2016a; Morrison et al. 2016b; Tsai et al. 2003). Determining these properties is important for efficient consolidated

Table 2 Production of selected anaerobic fungal proteins in heterologous hosts

Activity	Species/strain	Expression host	References
Endoglucanase (endo-1,4-β-glucanase)	*Neocallimastix patriciarum*	*Clostridium beijerinckii* NCIMB 8052	López-Contreras et al. (2001)
Cellobiohydrolase	*Orpinomyces* sp.Y102	*Escherichia coli*	Chen et al. (2014)
β-D-glucosidase	*Orpinomyces* sp. PC-2	*Saccharomyces cerevisiae*	Li et al. (2004)
	Piromyces sp. strain E2	*Pichia pastoris*	Harhangi et al. (2002)
		Saccharomyces cerevisiae	O'Malley et al. (2012)
Xylanases (endo-β 1 ,4-xylanase)	*Neocallimastix frontalis*	*Kluyveromyces lactis* and *Penicillium roqueforti*	Durand et al. (1999)
	Orpinomyces sp. PC-2	*Hypocrea jecorina*	Li et al. (2007)
B-xylosidase	*Pecoramyces ruminantium* strain C1A	*Escherichia coli*	Morrison et al. (2016a)
Xylose isomerase	*Piromyces* sp. strain E2	*Saccharomyces cerevisiae*	Kuyper et al. (2003); O'Malley et al. (2012)
Xylulokinase	*Piromyces* sp. strain E2	*Saccharomyces cerevisiae*	O'Malley et al. (2012)
Glycosyl hydrolase 48	*Piromyces* sp. strain E2	*Saccharomyces cerevisiae*	O'Malley et al. (2012)
Fungal Scaffoldin (ScaA)	*Piromyces finnis* *Anaeromyces robustus* *Neocallimastix californiae*	*Escherichia coli*	Haitjema et al. (2017)
Acetyl xylan esterase	*Neocallimastix patriciarum*	*Escherichia coli*	Dalrymple et al. (1997)
	Orpinomyces sp. PC-2	*Escherichia coli*	Blum et al. (1999)
Feruloyl esterase	*Anaeromyces mucronatus*	*Escherichia coli*	Gruninger et al. (2016)
	Piromyces equi	*Trichoderma reesei*	Poidevin et al. (2009)
Mannanase	*Piromyces* sp.	*Escherichia coli*	Ali et al. (1995)
Galacto-(α-1,2)-arabinosidase	*Neocallimastix frontalis*	*Escherichia coli*	Jones et al. (2017)
β-1,2-arabinobiosidase	*Neocallimastix frontalis*	*Escherichia coli*	Jones et al. (2017)
Lichenase	*Orpinomyces* sp. PC-2	*Escherichia coli*	Chen et al. (1997)
Swollenin	*Pecoramyces ruminantium* strain C1A	*Escherichia coli*	(Morrison et al. 2016b)
Celpin	*Piromyces* sp. strain E2	*Saccharomyces cerevisiae*	O'Malley et al. (2012)
Cylophillin	*Orpinomyces* sp. PC-2	*Escherichia coli*	Chen et al. (1995); Chen et al. (2006)
CBM29	*Piromyces equi*[a]	*Nicotiana tabacum*	Obembe et al. (2007)

[a]Proper taxonomic description is missing for this isolate

bioprocessing in industrial organisms (such as *Saccharomyces cerevisiae*) as these hosts grow under conditions that are different from those where the enzymes evolved (den Haan et al. 2015; la Grange et al. 2010).

Apart from gut fungal CAZymes, other proteins have been successfully cloned. Xylose isomerase from *Piromyces* sp. E2 was successfully produced in *Saccharomyces cerevisiae*, allowing it to grow on C5 as well as C6 sugars (Kuyper et al. 2003). Swollenin from *Pecoramyces ruminantium* strain C1A was produced in *Escherichia coli* and its addition to a lignocellulolytic enzyme cocktail enhanced overall glucan and xylan hydrolysis yields from lignocellulosic biomass (Morrison et al. 2016b). Finally, Haitjema (2017) produced fragments of cellulosomal protein (ScaA) in *Escherichia coli* and used ELISA assays to demonstrate its strong binding activity to dockerin domains, firmly suggesting that ScaA acts as a scaffolding in these multi-enzyme complexes.

C. Progress Toward Direct Genetic Manipulation of Anaerobic Fungi

As many anaerobic fungal enzymes are post-translationally modified (Haitjema et al. 2017), biotechnology would benefit from enhanced native production enabled by a robust **genome engineering toolkit** (Podolsky et al. 2019). Such a toolkit should include robust transformation methods for genetic DNA manipulation, including the introduction of foreign DNA and plasmids with appropriate promoters, terminators, and other biological parts needed for controlled gene expression. However, genetic engineering of anaerobic gut fungi is still in its infancy—very few genetic regulatory elements have been identified, versatile selection markers, and reporter genes are lacking, and chromosomal integration methods are nonexistent. An early transformation attempt in *Neocallimastix frontalis* used biolistic bombardment of mature fungal biomass, which transiently expressed the bacterial beta-glucuronidase gene from an *Escherichia coli* plasmid modified with the putative *Neocallimastix frontalis* enolase promoter (Durand et al. 1997). Recently, a method for the collection of viable, competent, and developmentally synchronized anaerobic fungal zoospores under strict anaerobic conditions was developed (Calkins et al. 2016). The collected spores were shown to be naturally competent for exogenous nucleic acids supplied to the culture medium. Subsequently, an **RNAi approach** leveraged this natural competency. Chemically synthesized siRNAs targeting D-lactate dehydrogenase (ldhD) gene were added to germinating spores of *Pecoramyces ruminantium* strain C1A, resulting in marked gene silencing and indicating the possibility of gene knockdown approaches in anaerobic fungi (Calkins et al. 2018).

Finally, advances in gene editing technologies promise new opportunities to manipulate the **extremely AT-rich genomes** of anaerobic fungi (Zetsche et al. 2015). While many fungal species incorporate exogenous DNA with high efficiency through non-homologous end-joining pathways, site-specific knockout or gene-integration powered by nucleases will be necessary to avoid pleiotropic effects. The efficiency and accuracy of these gene-editing events is enhanced in multiple fungi by removing endogeneous non-homologous end-joining pathways (Catalano et al. 2011). Coupled with emerging gut fungal transformation methods, these techniques are poised to revolutionize stable engineering of anaerobic fungi for a range of biotechnological applications.

III. Anaerobic Fungi and Their Applications in Agriculture and Biotechnology

Anaerobic fungal activities may also be directly leveraged for biotechnology to improve the economic viability of farming operations. Anaerobic fungi are integral to the digestion of fiber-rich animal feed in important livestock species including cattle, goats, and sheep (Gruninger et al. 2014). Thus, anaerobic fungi may be manipulated in the animal gut to enhance nutrition of forage (Gordon and Phillips 1998). Similarly, bioprocesses may be designed to exploit the ability of anaerobic fungi to ferment ethanol from agricultural residues and improve methane (biogas) formation via hydrogen syntrophy with methanogens. Below we discuss deployment of anaerobic fungi and their enzymes into animal feed, bioenergy production from agricultural wastes by bioethanol production, and anaerobic digestion.

A. Anaerobic Fungi for Animal Nutrition

Beyond bio-based chemical production, anaerobic fungal CAZymes have the potential to improve the utilization of lignocellulosic feedstocks in herbivores. Two conceivable strategies have been proposed: using fungal strains to **pretreat feedstock** and including fungal strains in animal feed as a **probiotic supplement**. The aim of the pretreatment process is to promote colonization of anaerobic fungi on the feed and thus enhance its availability to the rumen microbial flora. Such pretreatment could theoretically improve feedstock quality not just for cattle, but for other animals that may not possess anaerobic fungi such as swine and poultry (Gruninger et al. 2014). To use the treated substrate, pressing, drying, and sterilization procedures would be necessary. Palatability and energy/cost studies would be needed to determine if such an approach is practically feasible.

The application of anaerobic fungi as probiotics for ruminants has been studied in buffalo (Paul et al. 2004; Saxena et al. 2010), buffalo calves (Paul et al. 2011; Tripathi et al. 2007), and Holstein-Friesian calves (Dey et al. 2004). In the latter study, the control group was not fed with fungus-free medium or inactivated fungal culture; thus the results of this study are inconclusive. In all other studies, adequate control diets were applied. Diets directly supplemented with anaerobic fungal cultures improved the rumen fermentation characteristics (pH, volatile fatty acids, and nitrogen content) and enhanced the ruminal microbial population and CAZyme activities. As a consequence, increased animal growth rate, digestibility of nutrients, and nitrogen retention was observed. In lactating buffalos, administration of anaerobic fungi also increased milk fat by up to 6% (Saxena et al. 2010).

The effect of administration of anaerobic fungal cultures (*Orpinomyces* sp.) and enzymes on ruminal digestion was studied also in sheep (Lee et al. 2000). Fungal enzymes (hydrolases produced by *Orpinomyces* sp.) did not influence fermentation parameters. In contrast, a significant positive response was obtained with administration of a live fungal culture. Collectively, these studies indicate that anaerobic fungi can be used as a probiotic to enhance diet digestibility leading to higher production. Nevertheless, this approach must take account of economic considerations and the need to repeatedly administrate oral-dosages of anaerobic fungi to maintain the desired response (Ribeiro et al. 2017).

B. Bioethanol Production with Anaerobic Fungi

Bioethanol production has increased dramatically over the last two decades (Balat and Balat 2009), and there is a critical need to transition from "first-generation" bioethanol derived from sugar cane and maize to **"second-generation" fuels** derived from lignocellulosic residues (Mood et al. 2013). The bioethanol process consists of four stages: pretreatment, enzymatic hydrolysis, fermentation, and distillation (Aditiya et al. 2016). Most bioethanol is commercially produced by fermentations involving adapted or genetically modified strains of *Saccharomyces cerevisiae*, *Zymomonas mobilis*, and *Escherichia coli* (Yang et al. 2007) that cannot natively degrade lignocellulose. The limiting factor in the production of bioethanol from second generation crops is the initial release of fermentable sugars from the recalcitrant lignocellulosic biomass (Dashtban et al. 2009). This is commonly overcome by the application of costly mechanical and/or chemical pretreatments such as milling, alkaline or acid hydrolysis, ionic liquid pretreatment or steam explosion (Sharma et al. 2017). The unique ability of anaerobic fungi to infiltrate and enzymatically degrade untreated lignocellulosic substrates makes them attractive candidates for alternative microbial conversion of plant biomass prior to bioethanol production. For example, *Anaeromyces robustus* and *Neocallimastix californiae* have been employed in a batch fermentation to drive growth of *Saccharomyces cerevisiae* by the release of sugars generated by fungal hydrolysis of lignocellulose (Henske et al. 2018b). Additionally, *Pecoramyces ruminantium* has been used in a saccharification-fermentation approach for hydrolysis of different alkaline pretreated lignocellulosic substrates prior to ethanol production by *Escherichia coli* (Ranganathan et al. 2017). Yields of ethanol were (approximately one third) lower than that achieved after using commercial enzyme cocktails but the authors conclude that their approach is potentially more cost-effective than enzyme technology, and yields are potentially optimizable through the use of selected/modified strains and enhanced process design (Ranganathan et al. 2017).

Another limitation to the production of bioethanol from lignocellulosic biomass is the inability of many ethanol-producing yeasts to ferment 5-carbon sugars such as xylose and arabinose, which are abundant in hemicellulose. Acquisition of the ability to ferment pentose sugars by yeasts is a major priority for bioethanol production from second generation energy crops (Yang et al. 2007). It has long been known that anaerobic fungi deconstruct hemicellulose and ferment both pentose and hexose sugars (Theodorou et al. 1996b). In an attempt to overcome the inability of yeast to ferment pentose sugars, the xylose isomerase gene from *Piromyces sp.* E2 was successfully expressed in *Saccharomyces cerevisiae* (Yang et al. 2007). However, ethanol production was still hampered due to the clear preferences of the isomerization equilibrium towards xylose formation. Kuyper et al. (2005) overcame this limitation by supplementing genes for all of the enzymes involved in the conversion of xylulose to glycolytic intermediates. In order to minimize xylitol production, the endogenous gene encoding aldose reductase was deleted. Under anaerobic conditions, the engineered yeast strain entirely consumed mixtures of glucose and xylose sequentially, with glucose as the preferred substrate.

C. Biogas Production with Anaerobic Fungi

Biogas and digestate production by anaerobic digestion (AD) of plant biomass is a commercially viable strategy to generate sustainable renewable energy and fertilizer/soil improver (Strzalka et al. 2017), but common strategies omit anaerobic fungi. During AD a mixed microbial consortium, comprising of bacteria (such as hydrolytic and syntrophic acetate oxidizing *Clostridia*) and methanogenic archaea converts substrates such as lignocellulosic residues, animal manure/slurry, and food processing waste to biogas (Schnürer 2018). Biogas, the main energy product from AD, is a mixture of methane (50–75%) and carbon dioxide (25–50%) with traces of other gases including hydrogen, hydrogen sulfide, and ammonia. Biogas can be upgraded to bio-methane, for injection to the gas network for electricity production and use in domestic, commercial, and industrial heating or use as a fuel for road transport (Schnürer 2016).

To achieve an adequate amount and quality of biogas, and more importantly a high methane yield, the process conditions in biogas reactors should try to resemble those found in the natural model for biogas plants, the rumen. In the rumen the conditions are constant temperatures (such as 39°C), low redox potential (−0.4 V), nearly neutral pH values, an anoxic environment and constant influx of fresh or undigested feed (Weimer et al. 2009). All these parameters are very similar to those that occur in mesophilic anaerobic digesters (Weimer et al. 2009). However, it should be noted that longer hydraulic retention times are necessary in biogas reactors to achieve sufficient digestion of feedstocks. Compared to the rumen anaerobic fungi are usually missing in the biogas producing microbial consortium. While anaerobic fungi do not produce bio-methane as a fermentation end product, they do produce biochemical substrates which can be used by methanogenic archaea (Cheng et al. 2009; see below).

Therefore, **bioaugmentation** of biogas reactors with anaerobic fungi or microbial consortia derived from rumen fluid (Nagler et al. 2019) could help to enhance and accelerate the conversion of lignocellulose to biogas. To date, most bioaugmentation research has involved the addition of anaerobic fungi to laboratory-scale AD bioreactors. These bioaugmentation trials are summarized in Table 3.

While the results in Table 2 appear encouraging, many studies are inconclusive due to the inclusion of spent culture medium alongside anaerobic fungal cultures in augmented reactors. This confounded bias was avoided in recent studies employing the semi-continuous fermentation of maize silage and anaerobic sludge treated with a 1:1 mixture of *Anaeromyces* KF8 and *Piromyces* KF9 (Procházka et al. 2012) and the two-stage batch fermentation of hay with two individual *Neocallimastix frontalis* isolates (Dollhofer et al. 2018). In both experiments, a control reaction with the respective heat-inactivated culture was performed. In these augmented trials, increased biogas production (Procházka et al. 2012) or accelerated biomass degradation and higher initial biogas yields (Dollhofer et al. 2018) were achieved.

Table 3 Summary of bioaugmentation publications with anaerobic fungi

Inocula used	Reactor setup	Substrate	Result	Reference
Orpinomyces sp. *Anaeromyces* sp. *Piromyces* sp.	Batch, fed-batch; semicontinuous fermentation	Microcrystalline cellulose, maize silage, grass silage	Biogas production	Procházka et al. (2012)
Two individual *Neocallimastix frontalis* strains	Two-phase batch fermentation	Hay	LCB degradation, VFA and initial biogas production	Dollhofer et al. (2018)
Mix of *N. frontalis*, *Orpinomyces* sp., *Piromyces* sp.	Batch reactors	Cow manure, granular sludge	CH_4 production and substrate degradation	Yildirim et al. (2017)
Anaeromyces sp.	Batch reactors	Microalgae biomass		Aydin et al. (2017)
Mix of *Neocallimastix* sp., *Orpinomyces* sp., and mixed bacterial culture (F210)	Single-phase, two-phase batch fermentation	Wheat straw, mushroom spent straw	CH_4 production	Ferraro et al. (2018)
Piromyces rhizinflata YM600	Two-phase system	Corn silage, cattail	LCB degradation, VFA and H_2 production, enhanced initial CH_4 production	Nkemka et al. (2015)

LCB lignocellulosic biomass feedstocks; *VFA* volatile fatty acids

Another important concern regarding the utilization of anaerobic fungi for biogas production is their short survival time in digesters compared to rapidly growing prokaryotic competitors. Several studies (Table 3) have monitored the survival of anaerobic fungi in augmented reactors by molecular means (Dollhofer et al. 2018; Procházka et al. 2012; Yildirim et al. 2017). Signals for the presence of anaerobic fungi were found to decrease in augmented anaerobic sludge and were not detectable after 7–10 days. Molecular screening has also been used to verify the presence of anaerobic fungi via transcriptional activity in full-scale agricultural biogas plants operated with at least ~30% cattle-manure and/or cattle-slurry (Dollhofer et al. 2017; Young et al. 2018). In these commercial biogas plants, anaerobic fungal DNA belonging to the genera *Anaeromyces*, *Neocallimastix*, *Orpinomyces*, *Caecomyces*, *Cyllamyces*, and *Piromyces* was detected along with four unclassified anaerobic fungal clades. Furthermore, limited transcription of anaerobic fungal GH5 endoglucanases was determined on an mRNA level. These results indicate that potentially active anaerobic fungi are introduced into the reactor via ruminant-feces, but that the fungi were not likely able to survive given the conditions in the biogas plants.

An alternative to the bioaugmentation strategy reviewed above is to utilize stable **co-cultures** of anaerobic fungi and methanogenic archaea for biogas production. In these co-cultures, methanogens have been found attached to the surface of fungal rhizoids and sporangia (Bauchop and Mountfort 1981; Jin et al. 2011; Leis et al. 2014), which is likely to improve interspecies hydrogen transfer from fungal hydrogenosomes to methanogens and preserve fungal viability. Co-cultures of *Methanobrevibacter* and *Methanobacterium* combined with representatives of the genera *Piromyces*, *Neocallimastix*, *Orpinomyces*, *Caecomyces*, and *Anaeromyces* have been reported (Jin et al. 2011; Swift et al. 2019). Co-cultures produce more hydrogen, formate, acetate, methane, and often bolster enzyme activity compared to fungal monoculture (Bauchop and Mountfort 1981; Jin et al. 2011; Li et al. 2016). Thus, co-cultured methanogens may alter the metabolism and lignocellulose degrading activity of the anaerobic fungus. The fiber-degrading and methane-producing ability of the co-cultures are consistently stable and can last for several years in in vitro transfers (Bauchop and Mountfort 1981; Li et al. 2016).

Microbial consortia dominated by species of *Clostridia* in syntrophic relationship with methanogenic archaea are relatively ineffective at degrading recalcitrant lignocellulosic substrates (Bauchop and Mountfort 1981; Irvine and Stewart 1991; Joblin et al. 1989). In contrast, stable co-cultures of anaerobic fungi and methanogens may offer faster rates of biogas production and reduced hydraulic retention times; both significant advantages for commercial biogas production. However, our understanding of the role and survival of anaerobic fungi in conventional biogas processes is limited, and scale-up techniques have not yet been realized. Nonetheless, fungus-methanogen co-culture offers the possibility of consolidated bioprocessing, using a single platform for the complete deconstruction of lignocellulose and its conversion to biogas. Hence this area of research is likely to receive increasing attention in the future.

IV. Current and Future Opportunities for Exploitation

Beyond basic science research, the anaerobic fungi have been attracting attention for their translational potential to a wealth of industries from bio-based processing to animal health and welfare and beyond. Novel technologies based on engineered anaerobic fungal strains and/or processing strategies have seen a number of intellectual property filings worldwide, as discussed below. We further provide a brief overview for exploitation of unknown secondary metabolites and membrane proteins sourced from anaerobic fungi, which are useful to bio-based chemical production and strain engineering.

A. Potential Source of Commodity via Secondary Metabolites

Anaerobic fungi perform mixed acid fermentation for energy generation and thus produce formate, acetate, lactate, ethanol, CO_2, and hydrogen during the conversion of plant biomass (Mountfort and Orpin 1994). They are able to effectively degrade lignocellulosic biomass (Seppälä et al. 2017) and thus could help to exploit such abundant and cheap substrates for **organic acid production**. Several studies indicate the potential of anaerobic fungi to produce organic acids (Dollhofer et al. 2018; Nkemka et al. 2015), but the yield, the concentration, and the productivity need to be determined to decide if their utilization could be an economical approach to produce platform chemicals (Sauer et al. 2008). Furthermore, anaerobic fungi survive in a highly complex ecosystem alongside numerous bacteria, methanogenic archaea, and protozoa. To thrive in this competitive environment, it is likely that they produce secondary metabolites to boost their chances for survival. Progress in integrated-omics (Hillman et al. 2017) combined with the screening of a diverse set of anaerobic fungal isolates could thus potentially deliver a plethora of novel natural products for biotechnological use.

B. Characterization and Metabolic Engineering with Membrane Proteins

Current biotechnology uses easily engineered microorganisms, such as *Saccharomyces cerevisiae*, to convert lignocellulosic biomass into a wide range of fuels and chemicals. A major limitation is the poor ability of yeast to efficiently co-utilize glucose and xylose, the second most abundant sugar in plant biomass (Yang et al. 2007). As discussed in Sect. II.B, exogenous enzymes and pathways have been engineered into yeast that permit the conversion of xylose into ethanol, but poor xylose uptake into the cells remains a severe bottleneck (Young et al. 2011). Therefore, **membrane proteins** such as transporters for underutilized sugars that can be used to bolster transport in yeast are highly sought after (recently reviewed in Hara et al. (2017)). Primary sequence analyses of a dataset comprising >60,000 transcripts sourced from three strains of anaerobic fungi suggest that besides biomass-degrading enzymes, anaerobic fungi harbor a wide variety of membrane-embedded receptors and transporters that the fungi likely use to sense and sequester carbohydrates from the environment (Seppälä et al. 2016). For example, the study found that anaerobic fungi appear to have G-protein coupled receptors with putative carbohydrate-binding domains. Furthermore, the study found that all major eukaryotic carbohydrate transporter classes are represented in anaerobic fungi: Major Facilitator Superfamily (MFS), the Sodium Solute Symporter family, and the Sugars Will Eventually be Exported Transporter family (SWEET). The occurrence of SWEETs is frequent in plant and bacterial systems but is rare in fungi and so far exclusive to lower lineages, such as *Chytridiomycota*. Unexpectedly, the vast majority of identified transporter components were found to be homologous to substrate-binding proteins (SBPs) that are typically associated with prokaryotic ATP-binding cassette (ABC) transporters (Ter Beek et al. 2014). In prokaryotes, the extracellular SBPs deliver small molecules to ATP-driven uptake systems, and whereas it is likely that the fungi acquired these proteins by horizontal gene transfer, their function in the eukaryotic host remains unknown (Haitjema et al. 2017). Further experimental characterization is necessary, as very few membrane proteins from anaerobic fungi have been heterologously produced to date (Haferkamp et al. 2002; van der Giezen et al. 1998). Recently, it was shown that fluoride transporters from anaerobic fungi bolster solvent tolerance of yeast (Seppälä et al. 2019). In conclusion, anaerobic fungi are a promising source of receptors and transporters to expand metabolite uptake modulate tolerance in model microbes, and these proteins should be assessed for their use in bioprocessing applications.

C. Opportunities Identified from Published Patents

The Espacenet tool from the European Patent Office was used to search the worldwide collection of published applications for (a) the number of anaerobic fungal patents filed, (b) their date of publication, (c) the location of the inventors, and (d) the exploitation area of the published patent. The search terms used were directed at patent titles and abstract only. All meaningful combinations of the following search terms were used to search the database: anaerobic, gut, fungi, fungus, rumen, and all named fungal genera. With these search terms, 384 published patents were identified. On removal of duplicates and erroneously listed applications, 51 patents were found that directly involved the anaerobic fungi or where anaerobic fungi were used as a "catch all" term alongside other microbial groupings. To place this number of patents into context, replacing the search terms with *Trichoderma* or *Aspergillus* resulted in patent lists of 3968 and > 10,000, respectively. The majority of anaerobic fungal patents were filed by China (28%), the USA (23%), Taiwan (14%), the UK (9%), Australia (7%), Canada (5%), the Netherlands (4%), and Turkey (4%). Japan, South Africa, South Korea, and Russia accounted for the remaining patents. Chinese and Taiwanese inventors were responsible for the majority of recent (2001–2018) patents.

The first peer review article on the anaerobic fungi was published in 1975 (Orpin 1975). The first patent to mention anaerobic fungi appeared 14 years later in 1989 (JPH01289488 (A)). In this patent, a Japanese inventor listed anaerobic fungi alongside other fungi in a "catch-all" application for water cleanup. The UK inventors Hazelwood and Gilbert (1993) published the first patent directly attributable to the anaerobic fungi (WO9325693 (A1)). This patent was concerned with the expression of truncated recombinant xylanases from *Neocallimastix patriciarum* and their potential use in the pulp and paper industries.

Approximately 70% of **patents** are concerned with **molecular biology-based applications**, broadly focused on gene identification, gene characterization, cloning, and expression of recombinant enzymes and their derivatives in heterologous hosts. Xylanase, esterase, and cellulase genes from *Neocallimastix* and *Piromyces* are described in many patents alongside the enhancement of transformed enzyme activities, for example, to produce xylanases with improved thermal stability and pH tolerance. Expression hosts are mostly *Escherichia coli*, *Saccharomyces cerevisiae*, and *Pichia pastoris*. The latter species is the methylotrophic yeast widely employed for protein/enzyme production using recombinant DNA techniques. In one patent (US5948667 (A)), plant cells of Canola were transformed with a xylanase gene from *Neocallimastix patriciarum* with the intention of improving the digestion of crop residues by ruminant livestock. Another patent

(AU2016366227 (A1)) is concerned with consolidated bioprocessing, using a single platform for the complete deconstruction of lignocellulose. Where mentioned, the agricultural and industrial outlets for these enzymes included enhancing the nutritional value of livestock feeds and the need for enzymes as alternatives to caustic chemicals in the paper and pulping industry.

The remaining 30% of patents do not use molecular methodologies and are concerned with water cleanup or supplements and additives to enhance anaerobic fungal activity in the rumen or improve the nutritional value of livestock feeds. Six of the more recent patents from China describe the use of anaerobic fungal cultures with or without co-cultured methanogens in green chemistry applications for sustainable chemical production, the production of acetic acid, formic acid, methane, ethanol, and enzymes. The two bioaugmentation patents from Turkey are concerned with the addition of anaerobic fungi for enhanced methane production in laboratory-scale bioreactors.

It is not possible to judge the extent of exploitation of anaerobic fungi in industrial processes from a listing of their published patents. However, for the 30% of patents concerned with non-molecular inventions, due to limitations in large-scale anaerobic fungal culture technology, exploitation of these inventions appears remote at the present time. For the 70% majority of inventions, those concerned with gene characterization and the transformation and activity of novel processes in heterologous hosts, some exploitation is already apparent, and progress will be dependent upon improvements over what is achievable in conventional systems.

V. Conclusions

During the last 50 years, substantial progress in anaerobic gut fungal research has been achieved. This work has improved our understanding of their life cycle, evolutionary history, and ecological roles. Anaerobic gut fungi possess multiple unique properties that render them exceptional for biotechnological applications involving plant biomass deconstruction, which also bridge to direct applications in agriculture, biofuel, and bio-based chemical production. Insights into high-resolution genomes of anaerobic fungi provided by the recent application of "omics" techniques offer new opportunities to identify biomass-degrading enzymes, metabolic pathways, and other biotech-relevant proteins and small molecules sourced from the *Neocallimastigomycota*. Nevertheless, anaerobic fungi have not yet been widely employed in large-scale industrial processes owing to their relatively slow and strict growth requirements, along with difficulties in their genomic manipulation. Future adoption of anaerobic fungi (and their proteins and pathways) in biotechnology relies on continued investment in their understanding at all levels—from the need to isolate and characterize additional strains, to the development of universal genomic editing strategies for strain engineering. Closing the existing gaps of knowledge on anaerobic gut fungal ecology, physiology, and genomics will remove current obstacles hampering their utilization and allow their biotechnological application in the future.

References

Aditiya HB, Mahlia TMI, Chong WT, Nur H, Sebayang AH (2016) Second generation bioethanol production: a critical review. Renew Sustain Energy Rev 66:631–653. https://doi.org/10.1016/j.rser.2016.07.015

Ali BRS, Zhou L, Graves FM, Freedman RB, Black GW, Gilbert HJ, Hazlewood GP (1995) Cellulases and hemicellulases of the anaerobic fungus Piromyces constitute a multiprotein cellulose-binding complex and are encoded by multigene families. FEMS Microbiol Lett 125(1):15–22. https://doi.org/10.1111/j.1574-6968.1995.tb07329.x

Aydin S, Yildirim E, Ince O, Ince B (2017) Rumen anaerobic fungi create new opportunities for enhanced methane production from microalgae biomass. Algal Res 23:150–160. https://doi.org/10.1016/j.algal.2016.12.016

Balat M, Balat H (2009) Recent trends in global production and utilization of bio-ethanol fuel. Appl Energy 86:2273–2282. https://doi.org/10.1016/j.apenergy.2009.03.015

Bauchop T, Mountfort DO (1981) Cellulose fermentation by a rumen anaerobic fungus in both the absence and the presence of rumen methanogens. Appl Environ Microbiol 42(6):1103–1110. https://www.ncbi.nlm.nih.gov/pubmed/16345902

Bayer EA, Belaich JP, Shoham Y, Lamed R (2004) The cellulosomes: multienzyme machines for degradation of plant cell wall polysaccharides. Annu Rev Microbiol 58:521–554. https://doi.org/10.1146/annurev.micro.57.030502.091022

Blum DL, Li X-L, Chen H, Ljungdahl LG (1999) Characterization of an acetyl xylan esterase from the anaerobic fungus *Orpinomyces* sp. strain PC-2. Appl Environ Microbiol 65(9):3990–3995. https://aem.asm.org/content/aem/65/9/3990.full.pdf

Boxma B, Voncken F, Jannink S, Van Alen T, Akhmanova A, Van Weelden SWH, Van Hellemond JJ, Ricard G, Huynen M, Tielens AGM, Hackstein JHP (2004) The anaerobic chytridiomycete fungus Piromyces sp. E2 produces ethanol via pyruvate:formate lyase and an alcohol dehydrogenase E. Mol Microbiol 51(5):1389–1399. https://doi.org/10.1046/j.1365-2958.2003.03912.x

Caldwell DR, Bryant MP (1966) Medium without rumen fluid for nonselective enumeration and isolation of rumen bacteria. Appl Microbiol 14(5):794–801. https://www.ncbi.nlm.nih.gov/pubmed/5970467

Calkins S, Elledge NC, Hanafy RA, Elshahed MS, Youssef N (2016) A fast and reliable procedure for spore collection from anaerobic fungi: application for RNA uptake and long-term storage of isolates. J Microbiol Methods 127:206–213. https://doi.org/10.1016/j.mimet.2016.05.019

Calkins SS, Elledge NC, Mueller KE, Marek SM, Couger MB, Elshahed MS, Youssef NH (2018) Development of an RNA interference (RNAi) gene knockdown protocol in the anaerobic gut fungus *Pecoramyces ruminantium* strain C1A. PeerJ 6: e4276. https://doi.org/10.7717/peerj.4276

Callaghan TM, Podmirseg SM, Hohlweck D, Edwards JE, Puniya AK, Dagar SS, Griffith GW (2015) *Buwchfawromyces eastonii* gen. nov., sp. nov.: a new anaerobic fungus (Neocallimastigomycota) isolated from buffalo faeces. MycoKeys 9:11–28. https://doi.org/10.3897/mycokeys.9.9032

Catalano V, Vergara M, Hauzenberger JR, Seiboth B, Sarrocco S, Vannacci G, Kubicek CP, Seidl-Seiboth V (2011) Use of a non-homologous end-joining-deficient strain (delta-ku70) of the biocontrol fungus *Trichoderma virens* to investigate the function of the laccase gene lcc1 in sclerotia degradation. Curr Genet 57(1):13–23. https://doi.org/10.1007/s00294-010-0322-2

Chen H, Li XL, Ljungdahl LG (1995) A cyclophilin from the polycentric anaerobic rumen fungus *Orpinomyces* sp. strain PC-2 is highly homologous to vertebrate cyclophilin B. Proc Natl Acad Sci USA 92(7):2587–2591. https://doi.org/10.1073/pnas.92.7.2587

Chen H, Li XL, Ljungdahl LG (1997) Sequencing of a 1,3-1,4-beta-D-glucanase (lichenase) from the anaerobic fungus *Orpinomyces* strain PC-2: properties of the enzyme expressed in Escherichia coli and evidence that the gene has a bacterial origin. J Bacteriol 179(19):6028–6034. https://doi.org/10.1128/jb.179.19.6028-6034.1997

Chen H, Li X-L, Blum DL, Ljungdahl LG (1998) Two genes of the anaerobic fungus *Orpinomyces* sp. strain PC-2 encoding cellulases with endoglucanase activities may have arisen by gene duplication. FEMS Microbiol Lett 159(1):63–68. https://doi.org/10.1111/j.1574-6968.1998.tb12842.x

Chen H, Li XL, Xu H, Ljungdahl LG, Cerniglia CE (2006) High level expression and characterization of the cyclophilin B gene from the anaerobic fungus *Orpinomyces* sp. strain PC-2. Protein Pept Lett 13(7):727–732. https://doi.org/10.2174/092986606777790511

Chen YC, Chen WT, Liu JC, Tsai LC, Cheng HL (2014) A highly active beta-glucanase from a new strain of rumen fungus *Orpinomyces* sp.Y102 exhibits cellobiohydrolase and cellotriohydrolase activities. Bioresour Technol 170:513–521. https://doi.org/10.1016/j.biortech.2014.08.016

Cheng YF, Edwards JE, Allison GG, Zhu WY, Theodorou MK (2009) Diversity and activity of enriched ruminal cultures of anaerobic fungi and methanogens grown together on lignocellulose in consecutive batch culture. Bioresour Technol 100 (20):4821–4828. https://doi.org/10.1016/j.biortech.2009.04.031

Conrad R (1999) Contribution of hydrogen to methane production and control of hydrogen concentrations in methanogenic soils and sediments. FEMS Microbiol Ecol 28(3):193–202. https://doi.org/10.1111/j.1574-6941.1999.tb00575.x

Couger MB, Youssef NH, Struchtemeyer CG, Liggenstoffer AS, Elshahed MS (2015) Transcriptomic analysis of lignocellulosic biomass degradation by the anaerobic fungal isolate Orpinomyces sp. strain C1A. Biotechnol Biofuels 8(1):208. https://doi.org/10.1186/s13068-015-0390-0

Dagar SS, Kumar S, Mudgil P, Singh R, Puniya AK (2011) D1/D2 domain of large-subunit ribosomal DNA for differentiation of *Orpinomyces* spp. Appl Environ Microbiol 77(18):6722–6725. https://doi.org/10.1128/AEM.05441-11

Dagar SS, Kumar S, Griffith GW, Edwards JE, Callaghan TM, Singh R, Nagpal AK, Puniya AK (2015) A new anaerobic fungus (*Oontomyces anksri* gen. nov., sp. nov.) from the digestive tract of the Indian camel (*Camelus dromedarius*). Fungal Biol 119 (8):731–737. https://doi.org/10.1016/j.funbio.2015.04.005

Dalrymple BP, Cybinski DH, Layton I, McSweeney CS, Xue G-P, Swadling YJ, Lowry JB (1997) Three *Neocallimastix patriciarum* esterases associated with the degradation of complex polysaccharides are members of a new family of hydrolases. Microbiology 143(8):2605–2614. https://doi.org/10.1099/00221287-143-8-2605

Dashtban M, Schraft H, Qin W (2009) Fungal bioconversion of lignocellulosic residues; opportunities & perspectives. Int J Biol Sci 5(6):578–595. https://doi.org/10.7150/ijbs.5.578

Davies DR, Theodorou MK, Lawrence MI, Trinci AP (1993) Distribution of anaerobic fungi in the digestive tract of cattle and their survival in faeces. J General Microbiol 139:1395–1400. https://doi.org/10.1099/00221287-139-6-1395

den Haan R, van Rensburg E, Rose SH, Görgens JF, van Zyl WH (2015) Progress and challenges in the engineering of non-cellulolytic microorganisms for consolidated bioprocessing. Curr Opin Biotechnol 33:32–38.. https://doi.org/10.1016/j.copbio.2014.10.003

Dey A, Sehgal JP, Puniya AK, Singh K (2004) Influence of an anaerobic fungal culture (*Orpinomyces* sp.) administration on growth rate, ruminal fermentation and nutrient digestion in calves. Asian Australas J Anim Sci 17(6):820–824. https://doi.org/10.5713/ajas.2004.820

Dollhofer V, Callaghan TM, Griffith GW, Lebuhn M, Bauer J (2017) Presence and transcriptional activity of anaerobic fungi in agricultural biogas plants. Bioresour Technol 235:131–139. https://doi.org/10.1016/j.biortech.2017.03.116

Dollhofer V, Dandikas V, Dorn-In S, Bauer C, Lebuhn M, Bauer J (2018) Accelerated biogas production from lignocellulosic biomass after pre-treatment with *Neocallimastix frontalis*. Bioresour Technol 264:219–227. https://doi.org/10.1016/j.biortech.2018.05.068

Durand R, Rascle C, Fischer M, Fèvre M (1997) Transient expression of the β-glucuronidase gene after biolistic transformation of the anaerobic fungus *Neocallimastix frontalis*. Curr Genet 31(2):158–161. https://doi.org/10.1007/s002940050190

Durand R, Rascle C, Fèvre M (1999) Expression of a catalytic domain of a *Neocallimastix frontalis* endoxylanase gene (xyn3) in *Kluyveromyces lactis* and *Penicillium roqueforti*. Appl Microbiol Biotechnol 52(2):208–214. https://doi.org/10.1007/s002530051510

Eberhardt RY, Gilbert HJ, Hazlewood GP (2000) Primary sequence and enzymic properties of two modular endoglucanases, Cel5A and Cel45A, from the anaerobic fungus *Piromyces equi*. Microbiology 146(8):1999–2008. https://doi.org/10.1099/00221287-146-8-1999

Emerling CA, Delsuc F, Nachman MW (2018) Chitinase genes (*CHIA*s) provide genomic footprints of a post-Cretaceous dietary radiation in placental mammals. Sci Adv 4(5):eaar6478. https://doi.org/10.1126/sciadv.aar6478

Ferraro A, Dottorini G, Massini G, Mazzurco Miritana V, Signorini A, Lembo G, Fabbricino M (2018) Combined bioaugmentation with anaerobic ruminal fungi and fermentative bacteria to enhance biogas production from wheat straw and mushroom spent straw. Bioresour Technol 260:364–373.. https://doi.org/10.1016/j.biortech.2018.03.128

Filya I (2004) Nutritive value and aerobic stability of whole crop maize silage harvested at four stages of maturity. Anim Feed Sci Technol 116(1/2):141–150. https://doi.org/10.1016/j.anifeedsci.2004.06.003

Gilmore SP, Henske JK, O'Malley MA (2015) Driving biomass breakdown through engineered cellulosomes. Bioengineered 6(4):204–208. https://doi.org/10.1080/21655979.2015.1060379

Gordon GL, Phillips MW (1998) The role of anaerobic gut fungi in ruminants. Nutr Res Rev 11(1):133–168. https://doi.org/10.1079/NRR19980009

Griffith GW, Ozkose E, Theodorou MK, Davies DR (2009) Diversity of anaerobic fungal populations in cattle revealed by selective enrichment culture using different carbon sources. Fungal Ecol 2 (2):87–97. https://doi.org/10.1016/j.funeco.2009.01.005

Griffith GW, Baker S, Fliegerová K, Liggenstoffer A, van der Giezen M, Voigt K, Beakes G (2010) Anaerobic fungi: Neocallimastigomycota. IMA Fungus 1 (2):181–185. http://www.ncbi.nlm.nih.gov/pubmed/22679578

Grigoriev IV, Cullen D, Goodwin SB, Hibbett D, Jeffries TW, Kubicek CP, Kuske C, Magnuson JK, Martin F, Spatafora JW, Tsang A, Baker SE (2011) Fueling the future with fungal genomics. Mycology 2 (3):192–209. https://doi.org/10.1080/21501203.2011.584577

Grigoriev IV, Nikitin R, Haridas S, Kuo A, Ohm R, Otillar R, Riley R, Salamov A, Zhao X, Korzeniewski F, Smirnova T, Nordberg H, Dubchak I, Shabalov I (2014) MycoCosm portal: gearing up for 1000 fungal genomes. Nucleic Acids Res 42 (D1):D699–D704. https://doi.org/10.1093/nar/gkt1183

Gruninger RJ, Puniya AK, Callaghan TM, Edwards JE, Youssef N, Dagar SS, Fliegerova K, Griffith GW, Forster R, Tsang A, McAllister T, Elshahed MS (2014) Anaerobic fungi (phylum Neocallimastigomycota): advances in understanding their taxonomy, life cycle, ecology, role and biotechnological potential. FEMS Microbiol Ecol 90(1):1–17. https://doi.org/10.1111/1574-6941.12383

Gruninger RJ, Cote C, McAllister TA, Abbott DW (2016) Contributions of a unique β-clamp to substrate recognition illuminates the molecular basis of exolysis in ferulic acid esterases. Biochem J 473 (7):839–849. https://doi.org/10.1042/bj20151153

Hackstein JHP, Baker SE, van Hellemond JJ, Tielens AGM (2019) Hydrogenosomes of anaerobic fungi: an alternative way to adapt to anaerobic environments. In: Tachezy J (ed) Hydrogenosomes and mitosomes: mitochondria of anaerobic eukaryotes. Springer, Cham, pp 159–175. https://doi.org/10.1007/978-3-030-17941-0_7

Haferkamp I, Hackstein JH, Voncken FG, Schmit G, Tjaden J (2002) Functional integration of mitochondrial and hydrogenosomal ADP/ATP carriers in the *Escherichia coli* membrane reveals different biochemical characteristics for plants, mammals and anaerobic chytrids. Eur J Biochem 269(13):3172–3181

Haitjema CH, Solomon KV, Henske JK, Theodorou MK, O'Malley MA (2014) Anaerobic gut fungi: advances in isolation, culture, and cellulolytic enzyme discovery for biofuel production. Biotechnol Bioeng 111(8):1471–1482. https://doi.org/10.1002/bit.25264

Haitjema CH, Gilmore SP, Henske JK, Solomon KV, de Groot R, Kuo A, Mondo SJ, Salamov AA, LaButti K, Zhao Z, Chiniquy J, Barry K, Brewer HM, Purvine SO, Wright AT, Hainaut M, Boxma B, van Alen T, Hackstein JHP, Henrissat B, Baker SE, Grigoriev IV, O'Malley MA (2017) A parts list for fungal cellulosomes revealed by comparative genomics. Nat Microbiol 2:17087. https://doi.org/10.1038/nmicrobiol.2017.87

Hanafy RA, Elshahed MS, Liggenstoffer AS, Griffith GW, Youssef NH (2017) *Pecoramyces ruminantium*, gen. nov., sp. nov., an anaerobic gut fungus from the feces of cattle and sheep. Mycologia 109(2):231–243. https://doi.org/10.1080/00275514.2017.1317190

Hanafy RA, Elshahed MS, Youssef NH (2018) *Feramyces austinii*, gen. nov., sp. nov., an anaerobic gut fungus from rumen and fecal samples of wild Barbary sheep and fallow deer. Mycologia 110(3):513–525. https://doi.org/10.1080/00275514.2018.1466610

Hanafy RA, Lanjekar VB, Dhakephalkar PK, Callaghan TM, Dagar SS, Griffith GW, Elshahed MS, Youssef NH (2020) Seven new Neocallimastigomycota genera from wild, zoo-housed, and domesticated herbivores greatly expand the taxonomic diversity of the phylum. Mycologia. https://doi.org/10.1080/00275514.2019.1696619

Hara KY, Kobayashi J, Yamada R, Sasaki D, Kuriya Y, Hirono-Hara Y, Ishii J, Araki M, Kondo A (2017) Transporter engineering in biomass utilization by yeast. FEMS Yeast Res 17(7). https://doi.org/10.1093/femsyr/fox061

Harhangi HR, Steenbakkers PJM, Akhmanova A, Jetten MSM, van der Drift C, Op den Camp HJM (2002) A highly expressed family 1 β-glucosidase with transglycosylation capacity from the anaerobic fungus *Piromyces* sp. E2. Biochimica et Biophysica Acta 1574(3):293–303. https://doi.org/10.1016/S0167-4781(01)00380-3

Hassan SS, Williams GA, Jaiswal AK (2018) Emerging technologies for the pretreatment of lignocellulosic biomass. Bioresour Technol 262:310–318. https://doi.org/10.1016/j.biortech.2018.04.099

Hazlewood GP, Gilbert HJ (1993) Recombinant xylanases. International patent number: WO9325693. https://patents.google.com/patent/WO1993025693A1/en

Henske JK, Gilmore SP, Knop D, Cunningham FJ, Sexton JA, Smallwood CR, Shutthanandan V, Evans JE, Theodorou MK, O'Malley MA (2017) Transcriptomic characterization of *Caecomyces churrovis*: a novel, non-rhizoid-forming lignocellulolytic anaerobic fungus. Biotechnol Biofuels 10(1):305. https://doi.org/10.1186/s13068-017-0997-4

Henske JK, Gilmore SP, Haitjema CH, Solomon KV, O'Malley MA (2018a) Biomass-degrading enzymes are catabolite repressed in anaerobic gut fungi. AIChE J. https://doi.org/10.1002/aic.16395

Henske JK, Wilken SE, Solomon KV, Smallwood CR, Shutthanandan V, Evans JE, Theodorou MK, O'Malley MA (2018b) Metabolic characterization of anaerobic fungi provides a path forward for bioprocessing of crude lignocellulose. Biotechnol Bioeng 115(4):874–884. https://doi.org/10.1002/bit.26515

Hillman ET, Readnour LR, Solomon KV (2017) Exploiting the natural product potential of fungi with integrated -omics and synthetic biology approaches. Curr Opin Syst Biol 5:50–56. https://doi.org/10.1016/j.coisb.2017.07.010

Hooker CA, Hillman ET, Overton JC, Ortiz-Velez A, Schacht M, Hunnicutt A, Mosier NS, Solomon KV (2018) Hydrolysis of untreated lignocellulosic feedstock is independent of S-lignin composition in newly classified anaerobic fungal isolate, *Piromyces* sp. UH3-1. Biotechnol Biofuels 11:293. https://doi.org/10.1186/s13068-018-1292-8

Irvine HL, Stewart CS (1991) Interactions between anaerobic cellulolytic bacteria and fungi in the presence of *Methanobrevibacter smithii*. Lett Appl Microbiol 12(2):62–64. https://doi.org/10.1111/j.1472-765X.1991.tb00504.x

James TY, Kauff F, Schoch CL, Matheny PB, Hofstetter V, Cox CJ, Celio G, Gueidan C, Fraker E, Miadlikowska J, Lumbsch HT, Rauhut A, Reeb V, Arnold AE, Amtoft A, Stajich JE, Hosaka K, Sung G-H, Johnson D, O'Rourke B, Crockett M, Binder M, Curtis JM, Slot JC, Wang Z, Wilson AW, Schüßler A, Longcore JE, O'Donnell K, Mozley-Standridge S, Porter D, Letcher PM, Powell MJ, Taylor JW, White MM, Griffith GW, Davies DR, Humber RA, Morton JB, Sugiyama J, Rossman AY, Rogers JD, Pfister DH, Hewitt D, Hansen K, Hambleton S, Shoemaker RA, Kohlmeyer J, Volkmann-Kohlmeyer B, Spotts RA, Serdani M, Crous PW, Hughes KW, Matsuura K, Langer E, Langer G, Untereiner WA, Lücking R, Büdel B, Geiser DM, Aptroot A, Diederich P, Schmitt I, Schultz M, Yahr R, Hibbett DS, Lutzoni F, McLaughlin DJ, Spatafora JW, Vilgalys R (2006) Reconstructing the early evolution of fungi using a six-gene phylogeny. Nature 443(7113):818–822. https://doi.org/10.1038/nature05110

Jin W, Cheng YF, Mao SY, Zhu WY (2011) Isolation of natural cultures of anaerobic fungi and indigenously associated methanogens from herbivores and their bioconversion of lignocellulosic materials to methane. Bioresour Technol 102 (17):7925–7931. https://doi.org/10.1016/j.biortech.2011.06.026

Joblin K (1981) Isolation, enumeration, and maintenance of rumen anaerobic fungi in roll tubes. Appl Environ Microbiol 42(6):1119-1122. http://aem.asm.org/content/42/6/1119.full.pdf

Joblin KN, Naylor GE (1989) Fermentation of woods by rumen anaerobic fungi. FEMS Microbiol Lett 65(1–2):119–122. https://doi.org/10.1111/j.1574-6968.1989.tb03608.x

Joblin KN, Campbell GP, Richardson AJ, Stewart CS (1989) Fermentation of barley straw by anaerobic rumen bacteria and fungi in axenic culture and in co-culture with methanogens. Lett Appl Microbiol 9(5):195–197. https://doi.org/10.1111/j.1472-765X.1989.tb00323.x

Jones DR, Uddin MS, Gruninger RJ, Pham TTM, Thomas D, Boraston AB, Briggs J, Pluvinage B, McAllister TA, Forster RJ, Tsang A, Selinger LB, Abbott DW (2017) Discovery and characterization of family 39 glycoside hydrolases from rumen anaerobic fungi with polyspecific activity on rare arabinosyl substrates. J Biol Chem 292(30):12606–12620. https://doi.org/10.1074/jbc.M117.789008

Joshi A, Lanjekar VB, Dhakephalkar PK, Callaghan TM, Griffith GW, Dagar SS (2018) *Liebetanzomyces polymorphus* gen. et sp. nov., a new anaerobic fungus (Neocallimastigomycota) isolated from the rumen of a goat. MycoKeys 40(40):89–110. https://doi.org/10.3897/mycokeys.40.28337

Kameshwar AKS, Qin W (2018) Genome wide analysis reveals the extrinsic cellulolytic and biohydrogen generating abilities of Neocallimastigomycota fungi. J Genomics 6:74–87. https://doi.org/10.7150/jgen.25648

Kuyper M, Harhangi HR, Stave AK, Winkler AA, Jetten MSM, Laat WTAM, Ridder JJJ, Op den Camp HJM, Dijken JP, Pronk JT (2003) High-level functional expression of a fungal xylose isomerase: the key to efficient ethanolic fermentation of xylose by *Saccharomyces cerevisiae*? FEMS Yeast Res 4(1):69–78. https://doi.org/10.1016/S1567-1356(03)00141-7

Kuyper M, Hartog MM, Toirkens MJ, Almering MJ, Winkler AA, Dijken JP, Pronk JT (2005) Metabolic engineering of a xylose-isomerase-expressing *Saccharomyces cerevisiae* strain for rapid anaerobic xylose fermentation. FEMS Yeast Res 5(4-5):399–409. https://doi.org/10.1016/j.femsyr.2004.09.010

la Grange DC, den Haan R, van Zyl WH (2010) Engineering cellulolytic ability into bioprocessing organisms. Appl Microbiol Biotechnol 87 (4):1195–1208. https://doi.org/10.1007/s00253-010-2660-x

Lamed R, Setter E, Kenig R, Edward B (1983) The cellulosome: a discrete cell surface organelle of *Clostridium thermocellum* which exhibits separate antigenic, cellulose-binding and various cellulolytic activities. In: Biotechnology bioengineering symposium, vol 13. https://www.researchgate.net/profile/Bayer_Edward/publication/236400605_The_cellulosome_A_discrete_cell_surface_organelle_of_Clostridium_thermocellum_-which_exhibits_separate_antigenic_cellulose-binding_and_various_cellulolytic_activities/links/545524ca0cf2cf51647dd28c.pdf

Lee SS, Ha JK, Cheng K-J (2000) Relative contributions of bacteria, protozoa, and fungi to in vitro degradation of orchard grass cell walls and their interactions. Appl Environ Microbiol 66(9):3807–3813. https://doi.org/10.1128/AEM.66.9.3807-3813.2000

Leis S, Dresch P, Peintner U, Fliegerová K, Sandbichler AM, Insam H, Podmirseg SM (2014) Finding a robust strain for biomethanation: anaerobic fungi (Neocallimastigomycota) from the Alpine ibex (*Capra ibex*) and their associated methanogens. Anaerobe 29:34–43. https://doi.org/10.1016/j.anaerobe.2013.12.002

Li X-L, Ljungdahl LG, Ximenes EA, Chen H, Felix CR, Cotta MA, Dien BS (2004) Properties of a recombinant β-glucosidase from polycentric anaerobic fungus *Orpinomyces* PC-2 and its application for cellulose hydrolysis. Appl Biochem Biotechnol 113–116:233–250. http://www.ncbi.nlm.nih.gov/pmc/articles/PMC5890932/

Li XL, Skory CD, Ximenes EA, Jordan DB, Dien BS, Hughes SR, Cotta MA (2007) Expression of an AT-rich xylanase gene from the anaerobic fungus *Orpinomyces* sp. strain PC-2 in and secretion of the heterologous enzyme by *Hypocrea jecorina*. Appl Microbiol Biotechnol 74(6):1264–1275. https://doi.org/10.1007/s00253-006-0787-6

Li Y, Jin W, Cheng Y, Zhu W (2016) Effect of the associated methanogen methanobrevibacter thaueri on the dynamic profile of end and intermediate metabolites of anaerobic fungus *Piromyces* sp. F1. Curr Microbiol 73(3):434–441. https://doi.org/10.1007/s00284-016-1078-9

Li Y, Li Y, Jin W, Sharpton TJ, Mackie RI, Cann I, Cheng Y, Zhu W (2019) Combined genomic, transcriptomic, proteomic, and physiological characterization of the growth of *Pecoramyces* sp. F1 in monoculture and co-culture with a syntrophic methanogen. Front Microbiol 10(435):435. https://doi.org/10.3389/fmicb.2019.00435

Liggenstoffer AS, Youssef NH, Couger MB, Elshahed MS (2010) Phylogenetic diversity and community structure of anaerobic gut fungi (phylum Neocallimastigomycota) in ruminant and non-ruminant herbivores. ISME J 4(10):1225–1235. https://doi.org/10.1038/ismej.2010.49

Liu JH, Selinger LB, Hu YJ, Moloney MM, Cheng KJ, Beauchemin KA (1997) An endoglucanase from the anaerobic fungus *Orpinomyces joyonii*: characterization of the gene and its product. Can J Microbiol 43(5):477–485. https://www.ncbi.nlm.nih.gov/pubmed/9165703

Ljungdahl LG (2008) The cellulase/hemicellulase system of the anaerobic fungus Orpinomyces PC-2 and aspects of its applied use. Ann N Y Acad Sci 1125(1):308–321. https://doi.org/10.1196/annals.1419.030

Lombard V, Ramulu HG, Drula E, Coutinho PM, Henrissat B (2014) The carbohydrate-active enzymes database (CAZy) in 2013. Nucleic Acids Res 42 (Database issue):D490–D495. https://doi.org/10.1093/nar/gkt1178

López-Contreras AM, Smidt H, van der Oost J, Claassen PAM, Mooibroek H, de Vos WM (2001) *Clostridium beijerinckii* cells expressing *Neocallimastix patriciarum* glycoside hydrolases show enhanced lichenan utilization and solvent production. Appl Environ Microbiol 67(11):5127–5133. https://doi.org/10.1128/aem.67.11.5127-5133.2001

Lowe SE, Theodorou MK, Trinci AP, Hespell RB (1985) Growth of anaerobic rumen fungi on defined and semi-defined media lacking rumen fluid. J Gen Microbiol 131(9):2225–2229. http://mic.sgmjournals.org/content/131/9/2225.full.pdf

Marvin-Sikkema FD, Pedro Gomes TM, Gottschal JC, Prins RA, Marvin-Sikkema FD (1993) Characterization of hydrogenosomes and their role in glucose metabolism of *Neocallimastix* sp. L2. Arch Microbiol 160(5). https://doi.org/10.1007/bf00252226

Milne A, Theodorou MK, Jordan MGC, King-Spooner C, Trinci APJ (1989) Survival of anaerobic fungi in feces, in saliva, and in pure culture. Exp Mycol 13 (1):27–37. https://doi.org/10.1016/0147-5975(89)90005-4

Mondo SJ, Dannebaum RO, Kuo RC, Louie KB, Bewick AJ, LaButti K, Haridas S, Kuo A, Salamov A, Ahrendt SR, Lau R, Bowen BP, Lipzen A, Sullivan W, Andreopoulos BB, Clum A, Lindquist E, Daum C, Northen TR, Kunde-Ramamoorthy G, Schmitz RJ, Gryganskyi A, Culley D, Magnuson J, James TY, O'Malley MA, Stajich JE, Spatafora JW, Visel A, Grigoriev IV (2017) Widespread adenine N6-methylation of active genes in fungi. Nat Genet 49 (6):964–968. https://doi.org/10.1038/ng.3859. http://www.nature.com/ng/journal/v49/n6/abs/ng.3859.html#supplementary-information

Montenecourt BS, Eveleigh DE (1977) Preparation of mutants of *Trichoderma reesei* with enhanced cellulase production. Appl Environ Microbiol 34 (6):777–782. http://www.ncbi.nlm.nih.gov/pmc/articles/PMC242747/

Mood SH, Golfeshan AH, Tabatabaei M, Jouzani GS, Najafi GH, Gholami M, Ardjmand M (2013) Lignocellulosic biomass to bioethanol, a comprehensive review with a focus on pretreatment. Renew Sustain Energy Rev 27:77–93. https://doi.org/10.1016/j.rser.2013.06.033

Morrison JM, Elshahed MS, Youssef N (2016a) A multifunctional GH39 glycoside hydrolase from the anaerobic gut fungus *Orpinomyces* sp. strain C1A. PeerJ 4:e2289. https://doi.org/10.7717/peerj.2289

Morrison JM, Elshahed MS, Youssef NH (2016b) Defined enzyme cocktail from the anaerobic fungus *Orpinomyces* sp. strain C1A effectively releases sugars from pretreated corn stover and switchgrass. Sci Rep 6:29217. https://doi.org/10.1038/srep29217

Mountfort D, Orpin C (1994) Anaerobic fungi: biology, ecology, and function. CRC, Boca Raton, FL

Müller M, Mentel M, van Hellemond JJ, Henze K, Woehle C, Gould SB, Yu RY, van der Giezen M, Tielens AG, Martin WF (2012) Biochemistry and evolution of anaerobic energy metabolism in eukaryotes. Microbiol Mol Biol Rev 76(2):444–495. https://doi.org/10.1128/mmbr.05024-11

Murphy CL, Youssef NH, Hanafy RA, Couger MB, Stajich JE, Wang Y, Baker K, Dagar SS, Griffith GW, Farag IF, Callaghan TM, Elshahed MS (2019) Horizontal gene transfer as an indispensable driver for Neocallimastigomycota evolution into a distinct gut-dwelling fungal lineage. Appl Environ Microbiol. https://doi.org/10.1128/aem.00988-19

Nagler M, Kozjek K, Etemadi M, Insam H, Podmirseg SM (2019) Simple yet effective: microbial and biotechnological benefits of rumen liquid addition to lignocellulose-degrading biogas plants. J Biotechnol 300:1–10. https://doi.org/10.1016/j.jbiotec.2019.05.004

Nielsen BB, Zhu W-Y, Trinci APJ, Theodorou MK (1995) Demonstration of zoosporangia of anaerobic fungi on plant residues recovered from faeces of cattle. Mycol Res 99(4):471–474. https://doi.org/10.1016/S0953-7562(09)80648-5

Nkemka VN, Gilroyed B, Yanke J, Gruninger R, Vedres D, McAllister T, Hao X (2015) Bioaugmentation with an anaerobic fungus in a two-stage process for biohydrogen and biogas production using corn silage and cattail. Bioresour Technol 185:79–88. https://doi.org/10.1016/j.biortech.2015.02.100

Obembe OO, Jacobsen E, Timmers J, Gilbert H, Blake AW, Knox JP, Visser RGF, Vincken J-P (2007) Promiscuous, non-catalytic, tandem carbohydrate-binding modules modulate the cell-wall structure and development of transgenic tobacco (*Nicotiana tabacum*) plants. J Plant Res 120(5):605–617. https://doi.org/10.1007/s10265-007-0099-7

O'Malley MA, Theodorou MK, Kaiser CA (2012) Evaluating expression and catalytic activity of anaerobic fungal fibrolytic enzymes native topiromyces sp E2 in *Saccharomyces cerevisiae*. Environ Progr Sustain Energy 31(1):37–46. http://onlinelibrary.wiley.com/doi/10.1002/ep.10614/abstract

Orpin CG (1975) Studies on the rumen flagellate *Neocallimastix frontalis*. J Gen Microbiol 91(2):249–262. https://doi.org/10.1099/00221287-91-2-249

Orpin CG (1977) On the induction of zoosporogenesis in the rumen phycomycetes *Neocallimastix frontalis, Piromonas communis* and *Sphaeromonas communis*. J Gen Microbiol 101(2):181–189. https://doi.org/10.1099/00221287-101-2-181

Paul SS, Kamra DN, Sastry VR, Sahu NP, Agarwal N (2004) Effect of anaerobic fungi on in vitro feed digestion by mixed rumen microflora of buffalo. Reprod Nutr Dev 44 (4):313–319.. http://www.ncbi.nlm.nih.gov/pubmed/15535463

Paul SS, Deb SM, Punia BS, Das KS, Singh G, Ashar MN, Kumar R (2011) Effect of feeding isolates of anaerobic fungus *Neocallimastix* sp. CF 17 on growth rate and fibre digestion in buffalo calves. Arch Anim Nutr 65(3):215–228. https://doi.org/10.1080/1745039X.2011.559722

Peng XN, Gilmore SP, O'Malley MA (2016) Microbial communities for bioprocessing: lessons learned from nature. Curr Opin Chem Eng 14:103–109. https://doi.org/10.1016/j.coche.2016.09.003

Peng X, Swift CL, Theodorou MK, O'Malley MA (2018) Methods for genomic characterization and maintenance of anaerobic fungi. In: de Vries RP, Tsang A, Grigoriev I (eds) Fungal genomics. Methods in molecular biology, vol 1775. Humana, New York

Podolsky IA, Seppälä S, Lankiewicz TS, Brown JL, Swift CL, O'Malley MA (2019) Harnessing nature's anaerobes for biotechnology and bioprocessing. 10 (1). https://doi.org/10.1146/annurev-chembioeng-060718-030340

Poidevin L, Levasseur A, Paës G, Navarro D, Heiss-Blanquet S, Asther M, Record E (2009) Heterologous production of the *Piromyces equi* cinnamoyl esterase in *Trichoderma reesei* for biotechnological applications. Lett Appl Microbiol 49(6):673–678. https://doi.org/10.1111/j.1472-765X.2009.02734.x

Prasad V, Strömberg CAE, Alimohammadian H, Sahni A (2005) Dinosaur coprolites and the early evolution of grasses and grazers. Science 310 (5751):1177–1180. https://doi.org/10.1126/science.1118806

Procházka J, Mrázek J, Štrosová L, Fliegerová K, Zábranská J, Dohányos M (2012) Enhanced biogas yield from energy crops with rumen anaerobic fungi. Eng Life Sci 12(3):343–351. https://doi.org/10.1002/elsc.201100076

Qiu X, Selinger B, Yanke LJ, Cheng KJ (2000) Isolation and analysis of two cellulase cDNAs from *Orpinomyces joyonii*. Gene 245 (1):119–126. https://doi.org/10.1016/S0378-1119(00)00028-7

Ranganathan A, Smith OP, Youssef NH, Struchtemeyer CG, Atiyeh HK, Elshahed MS (2017) Utilizing anaerobic fungi for two-stage sugar extraction and biofuel production from lignocellulosic biomass. Front Microbiol 8(635):635. https://doi.org/10.3389/fmicb.2017.00635

Ribeiro GO, Oss DB, He Z, Gruninger RJ, Elekwachi C, Forster RJ, Yang W, Beauchemin KA, McAllister TA (2017) Repeated inoculation of cattle rumen with bison rumen contents alters the rumen microbiome and improves nitrogen digestibility in cattle. Sci Rep 7(1):1276. https://doi.org/10.1038/s41598-017-01269-3

Sauer M, Porro D, Mattanovich D, Branduardi P (2008) Microbial production of organic acids: expanding the markets. Trends Biotechnol 26(2):100–108. https://doi.org/10.1016/j.tibtech.2007.11.006

Saxena S, Sehgal JP, Puniya AK, Singh K (2010) Effect of administration of rumen fungi on production performance of lactating buffaloes. Benef Microbes 1 (2):183–188. https://doi.org/10.3920/bm2009.0018

Schnürer A (2016) Biogas production: microbiology and technology. In: Hatti-Kaul R, Mamo G, Mattiasson B (eds) Anaerobes in biotechnology. Advances in biochemical engineering/biotechnology, vol 156. Springer, Cham. https://doi.org/10.1007/10_2016_5

Schnürer A (2018) Microbiology of the biogas process. Swedish University of Agricultural Sciences

Seppälä S, Solomon KS, Gilmore SP, Henske JK, Rieth MD, O'Malley MA (2016) Mapping the membrane proteome of anaerobic gut fungi using RNA-Seq. Biophys J 110 (3, Suppl 1):58a–59a. https://doi.org/10.1016/j.bpj.2015.11.382

Seppälä S, Wilken SE, Knop D, Solomon KV, O'Malley MA (2017) The importance of sourcing enzymes from non-conventional fungi for metabolic engineering and biomass breakdown. Metab Eng 44:45–59. https://doi.org/10.1016/j.ymben.2017.09.008

Seppälä S, Yoo JI, Yur D, O'Malley MA (2019) Heterologous transporters from anaerobic fungi bolster fluoride tolerance in *Saccharomyces cerevisiae*. Metab Eng Commun 9:e00091. https://doi.org/10.1016/j.mec.2019.e00091

Sharma HK, Xu C, Qin W (2017) Biological pretreatment of lignocellulosic biomass for biofuels and bioproducts: an overview. Waste Biomass Valorization 10(2):235–251. https://doi.org/10.1007/s12649-017-0059-y

Skyba O, Douglas CJ, Mansfield SD (2013) Syringyl-rich lignin renders poplars more resistant to degradation by wood decay fungi. Appl Environ Microbiol 79(8):2560–2571. https://doi.org/10.1128/aem.03182-12

Solomon KV, Haitjema CH, Henske JK, Gilmore SP, Borges-Rivera D, Lipzen A, Brewer HM, Purvine SO, Wright AT, Theodorou MK, Grigoriev IV, Regev A, Thompson DA, O'Malley MA (2016a) Early-branching gut fungi possess a large, comprehensive array of biomass-degrading enzymes. Science 351(6278):1192–1195. https://doi.org/10.1126/science.aad1431

Solomon KV, Henske JK, Theodorou MK, O'Malley MA (2016b) Robust and effective methodologies for cryopreservation and DNA extraction from anaerobic gut fungi. Anaerobe 38:39–46. https://doi.org/10.1016/j.anaerobe.2015.11.008

Solomon KV, Henske JK, Gilmore SP, Lipzen A, Grigoriev IV, Thompson D, O'Malley MA (2018) Catabolic repression in early-diverging anaerobic fungi is partially mediated by natural antisense transcripts. Fungal Genet Biol 121:1–9. https://doi.org/10.1016/j.fgb.2018.09.004

Strauss J, Mach RL, Zeilinger S, Hartler G, Stöffler G, Wolschek M, Kubicek CP (1995) Crel, the carbon catabolite repressor protein from *Trichoderma reesei*. FEBS Lett 376 (1):103–107. https://doi.org/10.1016/0014-5793(95)01255-5

Strzalka R, Schneider D, Eicker U (2017) Current status of bioenergy technologies in Germany. Renew Sustain Energy Rev 72:801–820. https://doi.org/10.1016/j.rser.2017.01.091

Swift CL, Brown JL, Seppälä S, O'Malley MA (2019) Co-cultivation of the anaerobic fungus *Anaeromyces robustus* with *Methanobacterium bryantii* enhances transcription of carbohydrate active enzymes. J Ind Microbiol Biotechnol. https://doi.org/10.1007/s10295-019-02188-0

Ter Beek J, Guskov A, Slotboom DJ (2014) Structural diversity of ABC transporters. J Gen Physiol 143 (4):419–435. https://doi.org/10.1085/jgp.201411164

Teunissen MJ, Baerends RJS, Knelissen RAG, Op den Camp HJM, Vogels GD (1992) A semi-continuous culture system for production of cellulolytic and xylanolytic enzymes by the anaerobic fungus *Piromyces* sp. strain E2. Appl Microbiol Biotechnol 38 (1):28–33. https://doi.org/10.1007/bf00169414

Theodorou MK, Gill M, King-Spooner C, Beever DE (1990) Enumeration of anaerobic chytridiomycetes as thallus-forming units: novel method for quantification of fibrolytic fungal populations from the digestive tract ecosystem. Appl Environ Microbiol 56(4):1073–1078. http://aem.asm.org/content/56/4/1073.full.pdf

Theodorou MK, Mennim G, Davies DR, Zhu WY, Trinci AP, Brookman JL (1996a) Anaerobic fungi in the digestive tract of mammalian herbivores and their potential for exploitation. Proc Nutr Soc 55 (3):913–926. https://doi.org/10.1079/pns19960088

Theodorou MK, Zhu W-Y, Rickers A, Nielsen BB, Gull K, Trinci APJ (1996b) Biochemistry and ecology of anaerobic fungi. In: Howard DH, Miller JD (eds) Human and animal relationships. Springer, Berlin, pp 265–295. https://doi.org/10.1007/978-3-662-10373-9_14

Trinci APG, Davies DR, Gull K, Lawrence MI, Nielsen BB, Rickers A (1994) Anaerobic fungi in herbivorous animals. Mycol Res 98. https://doi.org/10.1016/s0953-7562(09)80178-0

Tripathi VK, Sehgal JP, Puniya AK, Singh K (2007) Effect of administration of anaerobic fungi isolated from cattle and wild blue bull (*Boselaphus tragocamelus*) on growth rate and fibre utilization in buffalo calves AU - Tripathi, Vimal K. Arch Anim Nutr 61(5):416–423. https://doi.org/10.1080/17450390701556759

Tsai C-F, Qiu X, Liu J-H (2003) A comparative analysis of two cDNA clones of the cellulase gene family from anaerobic fungus *Piromyces rhizinflata*. Anaerobe 9(3):131–140. https://doi.org/10.1016/S1075-9964(03)00087-8

Ushida K, Newbold CJ, Jouany JP (1997) Interspecies hydrogen transfer between the rumen ciliate *Polyplastron multivesiculatum* and *Methanosarcina barkeri*. J Gen Appl Microbiol 43(2):129–131. https://www.ncbi.nlm.nih.gov/pubmed/12501346

van der Giezen M, Kiel JAKW, Sjollema KA, Prins RA (1998) The hydrogenosomal malic enzyme from the anaerobic fungus *Neocallimastix frontalis* is targeted to mitochondria of the methylotrophic yeast *Hansenula polymorpha*. Curr Genet 33 (2):131–135. https://doi.org/10.1007/s002940050318

Wang T-Y, Chen H-L, Lu M-Y, Chen Y-C, Sung H-M, Mao C-T, Cho H-Y, Ke H-M, Hwa T-Y, Ruan S-K, Hung K-Y, Chen C-K, Li J-Y, Wu Y-C, Chen Y-H, Chou S-P, Tsai Y-W, Chu T-C, Shih C-C, Li W-H, Shih M-C (2011) Functional characterization of cellulases identified from the cow rumen fungus *Neocallimastix patriciarum* W5 by transcriptomic and secretomic analyses. Biotechnol Biofuels 4 (1):24. https://doi.org/10.1186/1754-6834-4-24

Wang X, Liu X, Groenewald JZ (2017) Phylogeny of anaerobic fungi (phylum Neocallimastigomycota), with contributions from yak in China. Antonie Van Leeuwenhoek 110(1):87–103. https://doi.org/10.1007/s10482-016-0779-1

Wang Y, Youssef N, Couger M, Hanafy R, Elshahed M, Stajich JE (2019) Comparative genomics and divergence time estimation of the anaerobic fungi in herbivorous mammals. bioRxiv 401869 [Preprint]. 31 Aug 2018 [cited 30 Mar 2019]. Available from https://doi.org/10.1101/401869

Weimer PJ, Russell JB, Muck RE (2009) Lessons from the cow: what the ruminant animal can teach us about consolidated bioprocessing of cellulosic biomass. Bioresour Technol 100(21):5323–5331. https://doi.org/10.1016/j.biortech.2009.04.075

Wilken SE, Swift CL, Podolsky IA, Lankiewicz TS, Seppälä S, O'Malley MA (2019) Linking 'omics' to function unlocks the biotech potential of non-model fungi. Curr Opin Syst Biol 14:9–17.. https://doi.org/10.1016/j.coisb.2019.02.001

Wilson CA, Wood TM (1992) The anaerobic fungus Neocallimastix frontalis: isolation and properties of a cellulosome-type enzyme fraction with the capacity to solubilize hydrogen-bond-ordered cellulose. Appl Microbiol Biotechnol 37(1):125–129.

https://link.springer.com/article/10.1007/BF00174216

Wubah DA, Kim DS (1996) Chemoattraction of anaerobic ruminal fungi zoospores to selected phenolic acids. Microbiol Res 151(3):257–262. https://doi.org/10.1016/S0944-5013(96)80022-X

Wubah DA, Fuller MS, Akin DE (1991) Resistant body formation in *Neocallimastix* Sp., an anaerobic fungus from the rumen of a cow. Mycologia 83(1):40–47. https://doi.org/10.1080/00275514.1991.12025977

Yang S-T, Liu X, Zhang Y (2007) Chapter 4—Metabolic engineering—applications, methods, and challenges. In: Yang S-T (ed) Bioprocessing for value-added products from renewable resources. Elsevier, Amsterdam, pp 73–118. https://doi.org/10.1016/B978-044452114-9/50005-0

Ye XY, Ng TB, Cheng KJ (2001) Purification and characterization of a cellulase from the ruminal fungus *Orpinomyces joyonii* cloned in *Escherichia coli*. Int J Biochem Cell Biol 33(1):87–94. https://doi.org/10.1016/S1357-2725(00)00068-6

Yildirim E, Ince O, Aydin S, Ince B (2017) Improvement of biogas potential of anaerobic digesters using rumen fungi. Renew Energy 109:346–353. https://doi.org/10.1016/j.renene.2017.03.021

Young E, Poucher A, Comer A, Bailey A, Alper H (2011) Functional survey for heterologous sugar transport proteins, using *Saccharomyces cerevisiae* as a host. Appl Environ Microbiol 77(10):3311–3319. https://doi.org/10.1128/AEM.02651-10

Young D, Dollhofer V, Callaghan TM, Reitberger S, Lebuhn M, Benz JP (2018) Isolation, identification and characterization of lignocellulolytic aerobic and anaerobic fungi in one- and two-phase biogas plants. Bioresour Technol 268:470–479. https://doi.org/10.1016/j.biortech.2018.07.103

Youssef NH, Couger MB, Struchtemeyer CG, Liggenstoffer AS, Prade RA, Najar FZ, Atiyeh HK, Wilkins MR, Elshahed MS (2013) The genome of the anaerobic fungus *Orpinomyces* sp. strain C1A reveals the unique evolutionary history of a remarkable plant biomass degrader. Appl Environ Microbiol 79(15):4620–4634. https://doi.org/10.1128/AEM.00821-13

Zetsche B, Gootenberg JS, Abudayyeh OO, Slaymaker IM, Makarova KS, Essletzbichler P, Volz SE, Joung J, van der Oost J, Regev A, Koonin EV, Zhang F (2015) Cpf1 is a single RNA-guided endonuclease of a class 2 CRISPR-Cas system. Cell 163(3):759–771. https://doi.org/10.1016/j.cell.2015.09.038

Zhu W-Y, Theodorou MK, Longland AC, Nielsen BB, Dijkstra J, Trinci APJ (1996) Growth and survival of anaerobic fungi in batch and continuous-flow cultures: ecology. Anaerobe 2(1):29–37. https://doi.org/10.1006/anae.1996.0004

Zhu W-Y, Theodorou MK, Nielsen BB, Trinci APJ (1997) Dilution rate increases production of plant cell-wall degrading enzymes by anaerobic fungi in continuous-flow culture. Anaerobe 3(1):49–59. https://doi.org/10.1006/anae.1997.0070

Biosystematic Index

Genetics and Biotechnology, 3rd Edition
The Mycota II
J.P. Benz, K. Schipper (Eds.)

Subject Index

Genetics and Biotechnology, 3rd Edition
The Mycota II
J.P. Benz, K. Schipper (Eds.)

CPSIA information can be obtained
at www.ICGtesting.com
Printed in the USA
LVHW022043161120
671833LV00005B/174

9 783030 499235